Chemistry and Physics of Fracture

NATO ASI Series

Advanced Science Institutes Series

A series presenting the results of activities sponsored by the NATO Science Committee, which aims at the dissemination of advanced scientific and technological knowledge, with a view to strengthening links between scientific communities.

The series is published by an international board of publishers in conjunction with the NATO Scientific Affairs Division

A Life Sciences	Plenum Publishing Corporation
B Physics	London and New York
C Mathematical	D. Reidel Publishing Company
and Physical Sciences	Dordrecht, Boston, Lancaster and Tokyo
D Behavioural and Social Sciences	Martinus Nijhoff Publishers
E Applied Sciences	Dordrecht, Boston, Lancaster
F Computer and Systems Sciences	Springer Verlag
G Ecological Sciences	Berlin, Heidelberg, New York, London,
H Cell Biology	Paris, and Tokyo

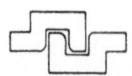

Series E: Applied Sciences – No. 130

Chemistry and Physics of Fracture

edited by

R.M. Latanision
Massachusetts Institute of Technology
Cambridge, MA, USA

and

R.H. Jones
Battelle Northwest Laboratories
Richland, WA, USA

1987 **Martinus Nijhoff Publishers**
Dordrecht / Boston / Lancaster
Published in cooperation with NATO Scientific Affairs Division

Proceedings of the NATO Advanced Research Workshop on
"Chemistry and Physics of Fracture",
Bad Reichenhall, FRG,
June 23–July 1, 1986

Library of Congress Cataloging-in-Publication Data

NATO Advanced Research Workshop on "Chemistry and
    Physics of Fracture" (1986 : Bad Reichenhall,
    Germany)
    Chemistry and physics of fracture.

    (NATO ASI series. Series E, Applied sciences ;
no. 130)
    "Proceedings of the NATO Advanced Research Workshop
on 'Chemistry and Physics of Fracture,' Bad Reichenhall,
FRG, June 23–July 1, 1986"--T.p. verso.
    Includes bibliographies and index.
    1. Fracture mechanics--Congresses. 2. Corrosion and
anti-corrosives--Congresses. 3. Stress corrosion--
Congresses. I. Latanision, R. M. II. Jones, R. H.
III. North Atlantic Treaty Organization. Scientific
Affairs Division. IV. Title. V. Series.
TA409.N4 1986   620.1'122   87-15415

ISBN-13: 978-94-010-8140-5   e-ISBN-13: 978-94-009-3665-2
DOI: 10.1007/978-94-009-3665-2

Distributors for the United States and Canada: Kluwer Academic Publishers,
P.O. Box 358, Accord Station, Hingham, MA 02018-0358, USA

Distributors for the UK and Ireland: Kluwer Academic Publishers, MTP Press Ltd,
Falcon House, Queen Square, Lancaster LA1 1RN, UK

Distributors for all other countries: Kluwer Academic Publishers Group, Distribution
Center, P.O. Box 322, 3300 AH Dordrecht, The Netherlands

Preface

For many years it has been recognized that engineering materials that are tough and ductile can be rendered susceptible to premature fracture through their reaction with the environment. Over 100 years ago, Reynolds associated hydrogen with detrimental effects on the ductility of iron. The "season cracking" of brass has been a known problem for decades, but the mechanisms for this stress-corrosion process are only today being elucidated. In more recent times, the mechanical properties of most engineering materials have been shown to be adversely affected by hydrogen embrittlement or stress-corrosion cracking. Early studies of environmental effects on crack growth attempted to identify a unified theory to explain the crack growth behavior of groups of materials in a variety of environments. It is currently understood that there are numerous stress-corrosion processes some of which may be common to several materials, but that the crack growth behavior of a given material is dependent on microstructure, microchemistry, mechanics, surface chemistry, and solution chemistry.

Although the mechanism by which various chemical species in the environment may cause cracks to propagate in some materials but not in others is very complex, the net result of all environmentally induced fracture is the reduction in the force and energy associated with the tensile or shear separation of atoms at the crack tip. Herein lies both the simplicity and complexity of the atomistics of crack tip processes: while it is possible to analyze the energetics of the separation of atom pairs or small groups of atoms, experimental crack growth measurements cannot presently be made at the atomic level and therefore involve the tensile and/or shear separation of millions of atoms. Furthermore, the interaction of the crack with complex microstructural features such as dislocations, dispersed phases, solute atoms, interfaces and free surfaces contributes to the nature of crack growth as a collective phenomenon and can be a major factor controlling crack propagation in a given material-environment combination. The way in which chemical species in the environment interact can also complicate the atomistics of crack growth processes. For example, whether a species such as hydrogen influences crack growth by its adsorption and concentration on the crack tip surface or by its presence within the material ahead of the crack remains an open issue today.

Nonetheless, the atomistics of fracture can be divided into unit processes that deal with the chemistry, mechanics, and microstructural aspects of fracture. For instance, identifying the differences in the chemical, physical, or mechanical states between propagating and non-propagating cracks can yield considerable information about the critical conditions for crack growth. Likewise, the kinetics of fracture processes can be studied by rate theory, with the rate-determining step describing the critical mechanism in crack propagation.

To provide a forum for international and interdisciplinary discussion on this complex and controversial subject, the Advanced Research Workshop on Chemistry and Physics of Fracture was held from 23 June through 1 July 1986 at the Hotel Axelmannstein in Bad Reichenhall, FRG. Sixty delegates from fifteen countries were in attendance, representing a wide range of disciplines and research interests: metallurgy, solid-state and solution chemistry, physics, fracture mechanics, surface science and electrochemistry. The conference format and organization followed that used successfully in two previous NATO workshops on similar topics, the Advanced Study Institute on Surface Effects in Crystal Plasticity, 1975, and the Advanced Research Institute on Atomistics of Fracture, 1981. The first four days were devoted to introductory and overview lectures on theory of fracture, solid-state chemistry and physics of fracture, solution chemistry, and structure and properties of interfaces. After a one-day break, three days of workshop sessions and a summary lecture were held. Workshop topics included novel aspects of fracture, intergranular embrittlement, hydrogen embrittlement, and stress corrosion and corrosion fatigue. Each workshop session began with a survey lecture, which was followed by short contributed papers. Afterwards, the delegates were divided into four independent working groups to address (1) key issues within topics that were selected from a list developed by the workshop organizers, or (2) issues raised by the group. Edited summaries of the deliberations of these groups were prepared by the session chairmen and are published in the proceedings as workshop summaries.

There was ample time for discussion following each lecture, and the discussions became livelier with each session. It was often necessary for the chairman to cut off discussion, as the number of questions far exceeded the time available to deal with them. Discussion continued during breaks and meals, a genuine proof of the interest and enthusiasm of the delegates. Much more than at past gatherings, individuals from various disciplines were noticeably willing to listen and talk to one another -- this alone would have made the workshop a pleasure and success.

While the sessions were occasionally long and arduous, there was time to relax and enjoy the scenery in the Bavarian Alps around Bad Reichenhall. Breaks in the sessions were alternated between morning, afternoon and evening. Activities included sightseeing in Salzburg, Austria; a unique visit to the salt mine in Bad Reichenhall; an opportunity to hear the Bartok-Quartett from Budapest; a day-long tour to the Konigsee and Berchtesgaden and other interesting places. The photographs included in this volume are certain to remind the delegates of both the technical sessions and the beautiful surroundings of the workshop.

Many thanks are due to members of the organizing committee for their help in developing and conducting this workshop and to the delegates for their significant contributions to the success of the workshop. This workshop was organized under the aegis of the Double Jump Programme of the NATO Scientific Affairs Division with the aim of promoting closer international cooperation between universities and industries on the projects of industrial interest. We are, therefore, grateful to NATO and to our Double Jump partners and to the U.S. Army Research Office for providing the financial resources without which this workshop could not have been held. We believe that industry will find that the new, first-principle understanding that is emerging may well lead to improved performance in the case of contemporary materials and, ultimately, to the evolution of new engineering materials designed from the molecular stage to resist embrittlement. Advanced technologies of all kinds demand industrial materials capable of performing in increasingly more hostile circumstances. This workshop and the spirit of international cooperation that it has stimulated will, we believe, serve as a guide in meeting technology's challenges.

Finally, we acknowledge with great pleasure the superb assistance of Lisa Kaminski and Christa Parsons, who handled logistical problems and large volumes of typing during the course of the workshop, and Pam Whiting, Marji Cochran, and Connie Beal, who assisted in the preparation of the Proceedings.

December 1986.   R.M. Latanision
                                            R.H. Jones

ORGANIZATION
Chairmen:
  R. M. Latanision
      Massachusetts Institute of Technology, U.S.A.

  R. H. Jones
      Battelle Northwest Laboratories, U.S.A.

Organizing Committee:
  M. Daw, Sandia National Laboratory, Livermore, CA, U.S.A.
  D. J. Duquette, Rensselaer Polytechnic Institute, Troy, NY, U.S.A.
  M. E. Eberhart, Los Alamos National Laboratory, Los Alamos, NM, U.S.A.
  T. E. Fischer, Exxon Research & Engineering Co., Annandale, NJ, U.S.A.
  G. S. Frankel, IBM Watson Research Center, Yorktown Heights, NY, U.S.A.
  J. P. Hirth, Ohio State University, Columbus, OH, U.S.A.
  R. H. Jones, Battelle Northwest Laboratories, Richland, WA, U.S.A.
  J. F. Knott, Cambridge University, Cambridge, England, U.K.
  R. M. Latanision, Massachusetts Institute of Technology, Cambridge, MA,
      U.S.A.
  P. Neumann, Max-Planck-Institut für Eisenforschung, Dusseldorf, FRG
  J. R. Rice, Harvard University, Cambridge, MA, U.S.A.
  A. W. Thompson, Carnegie-Mellon University, Pittsburg, PA, U.S.A.
  R. Thomson, National Bureau of Standards, Gaithersburg, MD, U.S.A.
  G. Whitesides, Harvard University, Cambridge, MA, U.S.A.

Sponsors:
  North Atlantic Treaty Organization
  U.S. Army Research Office
  Battelle Northwest Laboratories
  Cabot Corporation
  Martin-Marietta Corporation
  Rockwell International
  Shell Oil Company

Contents

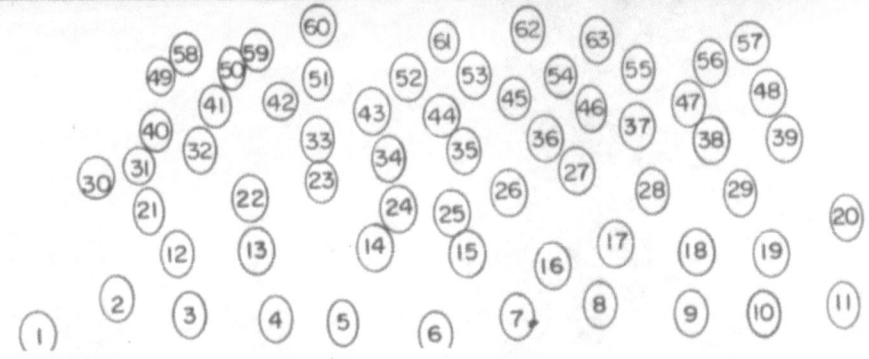

1. L.M. Brown	17. H. Hanninen	33. W.W. Gerberich	49. D.J. Duguette
2. J.R. Smith	18. A.S. Argon	34. K. Nisancioglu	50. J.P. Hirth
3. J.R. Rice	19. T. Watanabe	35. B. Lichter	51. M. Hashimoto
4. H.-J. Grabke	20. H. Bompard	36. S. Altintas	52. I.M. Bernstein
5. D.R. Baer	21. J.P. Fidelle	37. R. Thomson	53. A.W. Thompson
6. C. Parsons	22. D.G. Pettifor	38. M. Daw	54. M. Ferreira
7. L. Kaminski	23. A.T. Cole	39. J.F. Knott	55. D.D. Macdonald
8. R.H. Jones	24. X.G. Zhang	40. F.R.N. Nabarro	56. R.M. McMeeking
9. R.M. Latanision	25. T.E. Fischer	41. W. Zheng	57. A. Turnbull
10. M.O. Spiedel	26. S.M. Ohr	42. M. Finnis	58. M.E. Eberhart
11. W. Losch	27. J. Strehlau	43. S. Tahtinen	59. T. Reuschle
12. J. Oudar	28. H.-J. Engell	44. J. Lumsden	60. D.D. Vvedensky
13. P. Marcus	29. K. Sieradzki	45. D. Alexander	61. S.J. Thorpe
14. H.K. Birnbaum	30. P. Haasen	46. M. Pourbaix	62. T. Harris
15. C.J. Altstetter	31. J. Chene	47. S. Kayali	63. R. Newman
16. H. Mughrabi	32. J.Th.M. DeHosson	48. J.W. Hancock	

Grandhotel
AXELMANNSTEIN

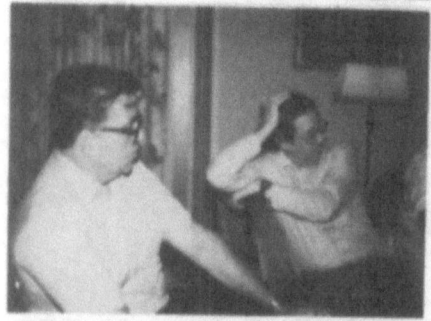

Introductory Lectures

Session Chairman: R. M. Latanision

CHEMISTRY AND PHYSICS OF FRACTURE
(An Overview)

A.S. ARGON

Massachusetts Institute of Technology
Department of Mechanical Engineering
Cambridge, MA 02139
USA

1. INTRODUCTION

Fracture, the more or less abrupt separation of a solid under stress, is of unquestionable technological importance. In load bearing structures, it is to be avoided, while in some other circumstances, such as in machining, rock drilling, demolition, etc., it is to be promoted under control. In either case, for its control, the fracture phenomenon is in need of as precise a quantitative understanding as is possible. This understanding requires different but usually complementary approaches. Here, in this overview, we will only give an overall perspective of the phenomenon, which will be discussed in much greater detail and with more precision in the specific contributions by others that follow.

Although in some instances fracture in materials is preceeded by extensive quasi-uniform degradation or damage, in most cases, the damage is highly localized at crack tips which progress through the material under stress by affecting only a thin layer around the fracture surfaces. The quasi-uniform degradation is characteristic of very heterogeneous materials with a large variability in local properties. In either the localized or the dispersed case, but particularly in the former, an important aspect of the fracture process is how action is brought to bear on the crack tip elements of volume by the surrounding field of deforming solid. The crack, for whatever reason for its presence, produces a field concentration which strongly depends on the elastic or inelastic behavior of the solid. Whenever the separation process in the crack plane has no easy alternatives, but proceeds, e.g., by plastic cavitation, initiated from widely separated hard particles, fracture may become so difficult that a major fraction of the cracked part is forced to undergo large plastic distortions. This disperses the action away from the crack tip and makes the separation there even more difficult. This results in high energy absorbing structures. In such relatively ideal behavior, the complexity of the material separation process lies in finding large strain solutions for the local crack tip damage process, nested in the deformation field of the background. This behavior will be of interest here only to the extent that it furnishes the operational approaches to how the action from the distant field is focused in the crack plane in the most difficult cases of tough materials.

The subjects of more central concern here are those where fracture in the crack plane is comparatively easy and can develop to the local critical conditions with little deformation in the background field. Alternatively, this lack of toughness may derive from intrinsic properties of the material such as an ability to undergo cleavage fracture, or may result from a variety of extrinsic environmental modifiers that tend to embrittle the material. It is in these latter aspects where the understanding of fracture requires the best combined considerations of the mechanics of distant field deformation, together with the chemistry and physics of the mechanism of separation. These problems of brittle behavior have two aspects that are best considered as complementary: the actual mechanism of separation transforming a continuous solid into a fractured one and the transport problem of bringing to the crack tip the extrinsic agency, producing the degraded behavior. The rate dependence of these two complementary aspects is usually quite different. Thus, in cleavage fracture, while an increased deformation rate does not permit time for the inelastic crack tip stress relaxation processes to develop and results in brittle response, in environmentally-controlled embrittlement an increased deformation rate does not permit time for the embrittling agency to travel to the crack tip and results in crack tip blunting and toughness. These mechanistic and transport considerations of embrittled behavior have rarely been treated in the desired complementary manner. While considerable control can be established over the embrittling behavior purely by understanding the kinetics of the transport of the degrading agency, full appreciation of the local mechanism itself may provide the missing clues for its elimination altogether, by modification of the chemistry or the controlling microstructural features.

The problems of concern that result in unexpected brittle behavior of structures remain, as before, the abrupt ductile to brittle transition in fracture in cleavable solids, intergranular embrittlement resulting from segregation of cohesion impairing species, hydrogen embrittlement, stress corrosion cracking, and liquid metal embrittlement. In the previous NATO workshop on the Atomistics of Fracture [1], very important progress was made in the elucidation of both the mechanisms and the rate controlling aspects of these unexpected brittle behaviors. Since that time, much additional progress has been made that focuses on the key aspects of these phenomena. Here, in this overview, we will very briefly summarize the understanding in these phenomena and make some projections into areas where new developments will be necessary.

2. INTRINSIC BRITTLE BEHAVIOR

In a fundamental study, Rice and Thomson [2] have attributed the intrinsic cleavability of a solid to the existence of very large energy barriers to the initiation of dislocations from atomically sharp cracks under stress, favoring planar separation without any accompanying inelastic deformation. This permitted them to classify only the face centered cubic metals of Ag, Aℓ, Au, Cu, Ni, and Pb as non-cleavable, intrinsically ductile solids, while most other materials, including the body centered cubic metals of Fe, W, the compounds of LiF, MgO, the semi-conducting solids, such as Si, and even the close packed metal Zn, as being intrinsically brittle with crack tip energy barriers to dislocation emission being in the hundreds of electron volts. Although this classification largely conforms to familiar practice with these solids, the calculated magnitudes of the energy barriers to dislocation emission in the cleavable solids would preclude any intrinsic plastic response below their respective melting points. The experimental evidence in all these cleavable solids is against this, and indicates that in each case, dislocations can be emitted from the perfect crystal at the crack tip, impeded by radically lower energy barriers. Thus, the cleavage response results only when brittle cracks traverse through material elements at velocities too high for dislocation emission, or for emitted dislocations to leave the vicinity of the crack tip to permit further emissions. Recent considerations and modeling [3] indicate that when the dislocation emission process is considered as a two-parameter activation process in which both the radius of the loop and the Burgers displacement of the dislocation are variables, the saddle point configuration is governed primarily by the latter, and the energy barrier is radically lowered. Experiments with Si have actually shown that in this case, the brittle to ductile transition is governed not even by emission of dislocations anymore, but by the kinetics of their glide, away from the crack tip into the background [4]. This demonstrates that even in this rather simple intrinsic behavior, much better information is necessary on the ideal shear response of solids that must come from fundamental considerations of the physics of cohesion and shear collapse. Such information is now attainable through atomistic modeling that is currently in progress in several groups [5-7]. These developments on intrinsic behavior are of fundamental importance, since they guide the developments in extrinsic behavior, where cohesion impairing embrittling species are present at the crack tip. In the solution of these crack tip problems, it is essential that the basic physics of cohesion and the details of the mechanistic aspects be combined with the methodology of non-linear mechanics. Ultimately, these problems where the local non-linearity of the deformation field and the presence of the crack surface strongly influence the atomistic response of the solid must be solved by computer simulation. While many fine examples of such simulation have been published recently (e.g. see [8,9]), none of these are complete enough to be considered fully satisfactory. In view of this and because of the comlexities of such simulations, semi-quantitative approaches examining trends in the elastic constants of elements in the periodic table and how they are affected are useful to rationalize observed behavior [10].

In problems of brittle crack propagation, the crack tip initiated process of dislocation emission and blunting are complemented by crack tip shielding that results from the visco-plastic polarization of background dislocations in response to the passage of the travelling crack. Such visco-plastic shielding of travelling cleavage cracks has been con-sidered in some detail recently by Freund and Hutchinson [1] as a continuum response. It has indicated that in previously dislocated material where the dislocations are not locked, some background shielding is always possible, even in the case of cracks moving with near relativistic velocities.

3. INTERGRANULAR EMBRITTLEMENT

With the exception perhaps of some intermetallic compounds such as Ni$_3$Al, where the structure of grain boundaries in the pure substance result in brittle separation and boron doped boundaries alleviate this behavior, [12], in nearly every other case, embrittlement at grain broundaries involves the segrega-tion of a cohesion impairing species to the grain boundary. Because of this, the thermodynamics and kinetics of segregation of trace impurities to grain boundaries, interfaces and free surfaces have been studied in considerable detail [13]. Segre-gation of atomic species to internal interfaces can occur in a large variety of combinations. Thus, it is not sufficient to study the segregation of only a single species, but it is neces-sary to consider also interaction of separate species among themselves in co-segregation problems. While some co-segrega-tion has been studied in considerable detail, a systematic study of the mechanical effects of the segregation has not yet been carried out under controlled conditions.

The brittle behavior of interfaces that have been "pre-contaminated" to equilibrium levels require an approach quite similar to that taken by Rice and Thomson [2] for intrinsic brittleness, where the impaired fracture toughness of the interface prescribes levels of crack tip stress intensity that probes the ideal shear strength by dislocation emission in possible planes in neighboring grains. Although detailed con-siderations for specific grain boundaries with known reduced fracture toughness, attending to the proper kinematics of alternative slip initiation in the neighboring grains have not been carried out, some such work is now in progress on bi-crystals, both at the theoretical and experimental levels [14].

The segregation or adsorption on separating interfaces concurrent with the propagation of a crack along the interface is clearly a rate phenomenon where the propagation of the crack at low levels of toughness has to await the arrival of the co-hesion impairing species to the crack tip. While this phenome-non is of a non-equilibrium nature requiring an approach based on kinetics, considerable progress has been made in the deter-mination of the non-equilibrium cohesive interface properties by a thermodynamics approach [15]. The combination of the quasi-static separation process of the contaminated interface with the details of the competing dislocation emission from the crack tip into the adjoining grains still awaits solution.

The propagation of an intergranular crack in the presence of concurrent adsorption or segregation will ultimately require computer simulation for the proper accounting of the important crack tip non-linear behavior with the adsorption and its effects at the crack tip. Such simulations have been attempted [16], and improved versions are now in progress [17].

The thermodynamics and kinetics of grain boundary embrittlement has received much more attention, for clear operational reasons, than the understanding of the actual atomic or molecular mechanisms of the cohesion impairment that is produced by the segregation. Nevertheless, some progress has been made here too by computations of alterations produced by impurity species in the binding orbitals of atoms situated in polyhedral sites characteristic of the grain boundary structure [18]. At the present, such computations have only provided clues to the embrittlement, but have not given the strain dependence of the total free energy of the contaminated boundary required for quantitative predictions.

It is useful to recall here that in brittle separation across a cleavage plane or a partially embrittled interface, which may be accompanied by crack tip shielding due to viscoplastic relaxations around the crack, the crack tip energy release rate appears as a multiplicative factor in the total energy release rate [19]. Thus, any reduction in the crack tip energy release rate due to the segregation or adsorption has a substantial adverse effect on the total fracture toughness.

4. HYDROGEN EMBRITTLEMENT

The embrittlement produced by hydrogen in normally ductile metals has received concentrated attention from both the mechanistic point of view of the effect of hydrogen in producing loss of cohesion and the point of view of its transport through and storage in the lattice.

The clearest effect of hydrogen has been in the systems such as Nb and V, where it produces brittle hydrides under stress. Much recent research (for a summary see e.g. [20]) has established that in these systems, an applied tensile stress increases the solvus temperature and results in the precipitation of hydrides. The process is reversible, so that the removal of the stress results in the re-dissolution of the hydrides. There has been little evidence that hydrides have lower cohesion than the surrounding metal. Instead, the evidence indicates that the dislocation mobility in the hydrides is quite restricted and that much of the embrittlement is a result of this high plastic resistance of the hydride, coupled with a cleavage mode of separation in it. Thus, in this type of embrittlement, both the stress and the plastic strains play a role. The former affects the precipitation of a hydride phase, while the latter, through a plastic drag, stresses the hydride further to produce cleavage in it. The rate of crack advance is governed by the rate of formation of cleavable hydrides at the crack tip, which in turn, is controlled by the transport rate of hydrogen to the crack tip. This transport exhibits unusual complexities as the hydrogen diffuses both through the lattice and along dislocations and can be trapped at dislocations and incoherent interfaces [21]. This complex behavior often results in complex transients.

Impurities are known to modulate the effect of hydrogen in several ways. Interstitials such as C, N, O, produce hardening of the host metal lattice and promote a more rapid buildup of plastic drag around hydrides and thus hasten their cleavage. On the other hand, substitutional impurities tend to trap more hydrogen, increase the solubility of the hydrogen in the metal, and thereby retard the stress induced rise of the solvus.

Thus, as a whole, the hydride-produced hydrogen embrittlement is comparatively well understood, requiring only better information on the properties of hydrides.

In metals such as Fe and Ni, where stable hydrides have not been observed in stressed or unstressed samples, the role of hydrogen in producing embrittlement is not very clear. The research of Birnbaum [20] has established a number of clues. Thus, e.g., embrittlement occurs often preferentially on prior austenite boundaries where temper embrittlement producing species such as S, P, As, or Sb are present, and that critical local concentrations of H appear to be necessary for the effect. There is also considerable evidence of slip plane decohesion on the {110} planes in Fe and the {111} planes in Ni. Not much is presently known on the relation to this of the dislocation content of these planes and possibly the altered core structure of dislocations. There is little evidence for hydrogen induced loss of cohesion in the metal lattice, although this has often been postulated as the principle mechanism of hydrogen embrittlement.

In view of this generally unsatisfactory level of understanding of the role of hydrogen in the embrittlement of Fe and Ni, there has been considerable emphasis given to finding operational cures of the effect through rather detailed studies of the diffusion of hydrogen through the metal lattice, its trapping at dislocations, and transport by them to sinks [21]. Such experiments and theoretical models have indicated, however, that there is not much buildup of hydrogen pressure at sinks, making it not too plausible that the embrittlement is prominently related to cleavage cracking by pressurized blisters. Clearly, this is an area where further research is necessary, particularly in the more promising directions of providing explanations for the enhanced cracking of prior austenite boundaries and slip planes.

5. STRESS CORROSION CRACKING

The most complex of the embrittling phenomena is stress corrosion cracking, which is likely to combine some aspects of intergranular embrittlement, and hydrogen embrittlement with electrochemical aspects of local anodic dissolution, passivation by films, and the like. While the electrochemistry of corrosion and passivation of unstressed planar surfaces is well understood, the preferential occurrence of these phenomena at crack tips under stress and undergoing local plastic deformation is not too clear.

Of the important driving forces of the phenomenon, the nature of stress intensification at the crack tip is the best understood. The understanding of the electrochemical and associated transport phenomena inside the crack channel are comparatively little understood even though their importance in governing the focused action of the electrochemical processes at the crack tip must be abundantly clear. Understanding of the rate mechanism of the stress corrosion cracking problem demands a full level of understanding of both the mechanical and electrochemical concentration of action at the crack tip.

Since in many SCC stituations fracture is intergranular, it is necessary to suspect either a concurrent intergranular embrittlement or enhanced electrochemical processes occurring preferentially along such boundaries with somewhat different chemistry. Some work in this direction of apportioning the effect to two separate mechanisms has been done [22], but more is necessary. In transgranular stress corrosion cracking, the fracture surfaces are remarkably smooth and cleavage like. This suggests a direct analogue to the cleavage problem, where the electrolyte, in the least, produces loss of cohesion across the "cleavage" plane. Additional mechanisms such as preferential removal of certain alloying species at the tip of the crack by the electrolytic action, concentrated anodic dissolution have also been proposed, but not definitively established.

In view of the complexity of the phenomenon in commercial alloys, it is most desirable to undertake research in more model systems that lend themselves to better control.

6. LIQUID METAL EMBRITTLEMENT

Of all the embrittlement phenomena, the one that produces the most dramatic effects, almost instantly, is liquid metal embrittlement [24].

The embrittlement process is most powerful in certain specific couples, such as Al and Ga, or Hg. It is much more effective along grain boundaries of polycrystalline metals, but occurs also in somewhat less dramatic fashion along cleavage planes in single crystals. The mechanisms that have been proposed for it range from variants of the Rice-Thomson problem, where the liquid metal produces cohesion impairment at the crack tip along the grain boundary or along a cleavage plane to propagate the crack before the crack can undergo plastic blunting [25]. Other suggestions have included a form of microductile dimple fracture initiated by a fine distribution of planar slip processes emanating from the crack tip and in some way, producing cavities just in front of the crack [26]. While these suggestions have some support in fracture surface features, they are often kinematically deficient and do not explain the rapidity of the process. Moreover, they are not compatible with the observations of an almost complete absence of accompanying plastic flow.

In view of the above, an interesting suggestion of Cahn [27], comparing separation by liquid metal attack to the phenomenon of thickening of an interface by a disjoining pressure is worth some serious consideration. In this phenomenon, e.g., a soap film is separated into two surfaces by injecting water into the body of the film. Alternatively a column of two fluids A, that are separated by an emulsion film of

immissible fluid B, are separated further by the thickening of
this film, by the injection of fluid B into the film. The pre-
sure required to inject the fluid into the soap film or into
the emulsion B is the "disjoining pressure", and the work of
separation is the product of this disjoining pressure with a
critical displacement of atomic dimensions that produces two
distinct non-interacting surfaces or interfaces. In a solid,
the separation is likely to involve some additional work necessary for overcoming the elastic resistance of the matrix.

7.   CONCLUSIONS

The very brief introductions to the embrittling phenomena
that we gave above were intended only as the most rudimentary
reminders of places where new advances will be necessary. Such
advances can come only through interdisciplinary efforts, combining the basic physics of cohesion and decohesion with the
chemistry of possible crack tip reactions, and finally, with
the mechanics of the accompanying background deformation. New
levels of sophistication in these approaches, providing many of
the sought answers will be unfolded in the following chapters
of this book.

REFERENCES

1. Latanision, R.M. and Pickens, J.R. (editors), "Atomistics
    of Fracture", Plenum Press, New York (1983).
2. Rice, J.R. and Thomson, R., Phil. Mag. 29, 73 (1974).
3. Argon, A.S., Acta Met., in the press.
4. Brede, M. and Haasen, P., in this volume.
5. Rose, J.H., Smith, J.R., and Ferrante, J., Phys. Rev.,
    B28, 1835 (1983).
6. Daw, M.S. and Baskes, M.I., Phys. Rev., B29, 6443 (1984).
7. Pettifor, D.G., in "Atomistics of Fracture", edited by
    Latanision, R.M., and Pickens, J.R., Plenum Press, New
    York, p. 281 (1983).
8. de Celis, B., Argon, A.S., and Yip S., J. Appl. Phys.,
    54, 4864 (1983).
9. Daw, M.S. and Baskes, M.I., Phys. Rev., Lett. 50, 1285
    (1983).
10. Eberhart, M.E., Latanision, R.M., and Johnson, K.H., Acta
    Met., 33, 1769 (1985).
11. Freund, L.B. and Hutchinson, J.W., J. Mech. Phys. Solids,
    33, 169 (1985).
12. Lin, C.T., White, C.L., and Horton, J.A., Acta Met., 33,
    213 (1985).
13. Guttmann, M., in "Atomistics of Fracture, edited by
    Latanision, R.M. and Pickens, J.R., Plenum Press, New
    York, p. 465 (1983).
14. Rice, J.R., private communication, to be published.
15. Hirth, J.P. and Rice, J.R., Met. Trans., A11, 1501 (1980).
16. Daw, M.S., Baskes, M.I., Bisson, C.L., and Wolfer, W.G.,
    in this volume.
17. Daw, M.S. and Baskes, M.I., private communication, to be
    published.

18. Eberhart, M.E., Johnson, K.H., Messner, R.P., and Briant, C.L., in "Atomistics of Fracture", edited by Latanision, R.M. and Pickens, J.R., Plenum Press, New York, p. 255 (1983).

19. Rice, J.R., in "Proceedings of the First International Conference on Fracture", edited by Yokobori, T., et al, Japanese Society for Strength and Fracture of Materials: Sendai, Japan, vol. 1, p. 309 (1966).

20. Birnbaum, H.K., in "Atomistics of Fracture", edited by Latanision, R.M. and Pickens, J.R., Plenum Press, New York, p. 733 (1983).

21. Hirth, J.P. and Johnson, H.H., in "Atomistics of Fracture", edited by Latanision, R.M. and Pickens, J.R., Plenum Press, New York, p. 771 (1983).

22. Parkins, R.N., in Atomistics of Fracture", edited by Latanision, R.M. and Pickens, J.R., Plenum Press, New York, p. 969 (1983).

23. Pugh, E.N., in "Atomistics of Fracture", edited by Latanision, R.M. and Pickens, J.R., Plenum Press, New York, p. 997 (1983).

24. Shchukin, E.D., in "Surface Effects in Crystal Plasticity", edited by Latanision, R.M. and Fourie, Noordhoff, Leyden, p. 701 (1977).

25. Westwood, A.R.C., Preece, C.M., and Kamdar, M.H., Trans. ASM, 60, 763 (1967).

26. Lynch, S.P., in "Atomistics of Fracture", edited by Latanision, R.M. and Pickens, J.R., Plenum Press, New York, p. 955 (1983).

27. Cahn, J.W., in "Atomistics of Fracture", edited by Latanision, R.M. and Pickens, J.R., "Plenum Press, New York, p. 427 (1983).

DISCUSSION

Comment by M. Daw:

You discussed the free energy required for dislocation emission from a crack tip, and also the catalytic effect of H on the fracture process. Is it possible also that H chemisorbed on the crack surface could directly influence the dislocation emission free energy?

Reply:

Yes indeed. There is now ample evidence that hydrogen trapping at dislocation reduces at least the core energy and thereby the dislocation line energy. Therefore, all line energy controlled processes become easier to accomplish.

WHAT, IF ANYTHING, CAN CHEMISTRY OFFER TO FRACTURE MECHANICS?

George M. Whitesides* and Thomas X. Neenan

Department of Chemistry
Harvard University
Cambridge, MA   02138

Chemistry is primarily concerned with the structures and properties of bonds between atoms and with relating these bonds and their properties to the structure, reactivity, and other properties of molecules. The principle concerns of chemistry have thus been individual molecules and molecular reactivity. Attention is now slowly turning to the structure, reactivity, and properties of solids: that is, to collections of molecules and to very large molecules. Why slowly? Certainly not because the problems in the solid state, or more broadly in materials science, are unimportant. The objectives of rationalizing properties such as tensile modulus, fracture strength, corrosion resistance, electrical conductivity using atomic-level structural information, dielectric constant, and thermal conductivity are clearly simultaneously immensely interesting scientifically and important technologically.[1,2] They are also, unfortunately, very difficult to attain. Moreover, the difficulty often is of an annoying sort--that is, it is associated with defects. The properties of large, regular ensembles of atoms and bonds can often be calculated with a high degree of precision using standard techniques of statistical mechanics.[3] What cannot be readily calculated is the often critical influence of unknown impurities or defects of unknown composition and structure on the properties of interest. In general, only average or highly perfect structure is available. Even in cases where impurities are intentionally introduced (as in the doping of semiconductors), the measured properties of the final material represent an average of the constituent structures. Detailed structural information concerning defects is only now becoming available.[4]

Fracture represents a particular problem for chemists. The calculation of tensile modulus of a single crystal is a tractable problem, given sufficient information about structure and single-bond properties. The calculation of the stress at which failure occurs in a real brittle solid is not. Fracture is fundamentally a kinetic problem and is intimately associated with rates of crack initiation, crack stopping, loading, and energy dissipation. The thermodynamics of fracture--that is, the energy required to create the new fracture surfaces--is one limiting useful quantity, but not, in general, of high predictive value. Understanding the relation of defects in real solids to the ultimate materials properties of these solids at the level of individual bond properties is a challenge of large magnitude.

By way of example, Scheme I shows a sketch of a chemical process that is very much simpler than fracture, but at the outer limit of what is presently considered a practical level of complexity by individuals concerned with detailed rationalization of reactivity in terms of bond making and bond breaking processes. This scheme outlines a homogeneous catalytic cycle: that of the hydrogenation of ethylene by hydrogen using a rhodium-based catalyst.[5,6,7] In some distant way, this catalytic process and fracture of a rhodium single crystal share a number of common features. Both involve the creation of vacant coordination sites on rhodium (in the

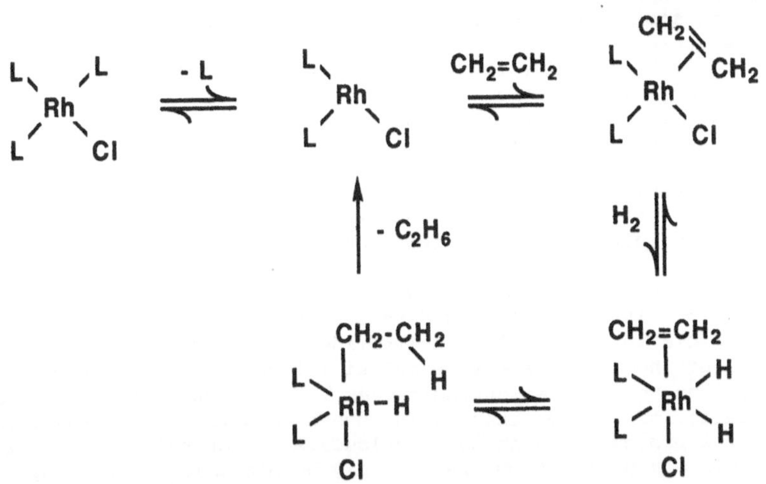

Scheme I. Schematic mechanism for hydrogenation of ethylene by
dihydrogen catalyzed by soluble rhodium(I) complex:
L = $(C_6H_5)_3P$.

case of fracture, by cleaving rhodium-rhodium bonds; in the case of the
catalytic cycle, by dissociation of a phosphine ligand L from the rhodium
center). Both involve "reconstruction" around the rhodium following the
bond breaking: that is, the immediate environment of the atoms remaining
bonded to the rhodium shifts to minimize the free energy of the system.
Both involve adsorption of new molecular species onto the vacant site on
rhodium (in the case of fracture, dioxygen and water from air; in the case
of catalytic hydrogenation, ethylene and hydrogen). In both various
transformations subsequently occur that involve these adsorbed species.
For the hydrogen reaction, all of these transformations are critical to the
overall functioning of the catalytic system. For fracture, it is unclear
how many of these reactions (if any) are critical to the question of the
stress under which the crystal fractures or yields. In the case of
fracture, we acknowledge that we have very little idea of the relative
rates, reversibility, or energetics of individual steps in the process
leading to the breaking of individual rhodium-rhodium bonds, and less idea
whether this process--as opposed, say, to failure in adhesion between
rhodium and an embedded particle or breakage of a surface film--is the step
that ultimately determines whether fracture occurs under given
circumstances. We also stand little chance of modelling effectively
catastrophic fracture in a single crystal using a homogeneous
organometallic system since no organometallic compounds are known that
undergo multiple (i.e. $10-10^3$) sequential metal-metal bond scissions.

In contrast, in the catalytic reaction, we believe that we have a
very *good* idea of what happens and know in fair detail which steps
determine the overall catalytic throughput in the system. The details of
our understanding of the catalytic system are unimportant for this
discussion, but from these details, and related details drawn from a large
number of other studies,[8] it is possible to draw several cautionary
inferences about conclusions from atomic-level investigations of catalysis.
These cautionary inferences can certainly be extended to the more complex
process of fracture as well.

1) *Any intermediate that can be observed is irrelevant.* This inference is semi-facetious, but it has a hidden grain of truth. That is, any species that can be detected in this type of catalytic cycle is probably present in sufficiently high concentration and is sufficiently stable and long-lived that it does not represent the highly reactive species that determine the course of catalysis. The same generalization may hold for many materials systems having properties dominated by defects. Often only the small, hard-to-detect anomalies in structures are the ones that truly determine the systems properties; the more easily accessible average or theoretical structure represents only a limiting value and is not directly helpful in determining real properties.

2) *Entropy may be as important as enthalpy.* The correctness of this statement for a system involving dissociation of ligands (as in the catalytic cycle in Scheme I) is self-evident, but the fact remains that a great deal of theory in catalysis is still focused on enthalpic considerations and ignores entropic ones.[9] The reason for this disregard of entropy is, of course, that it is much more difficult to calculate the entropy of a process than it is to calculate the enthalpy. In certain systems, disregard of entropy is probably permissible; in others, particularly where solvents or solvent effects are involved or where association or dissociation is important, entropy may dominate. Particularly for considerations of fracture in organic solids, entropy may prove to be as important as it is in catalysis.

3) *New catalysts are developed empirically.* Despite a long and highly successful (from an academic point of view) effort to understand and rationalize catalysis,[10] new catalysts are usually developed by trial and error: that is, by preparing new, thermodynamically stable compositions of matter and passing reactive mixtures of substances over these catalysts to detect reactions that depend upon the catalytic substances.[11] Theoretical considerations and mechanistic understanding has proved highly valuable in catalysis in *incremental* improvement of existing systems; it has not proved useful in inventing new ones. In an analogous sense, mechanisms and theory may be highly useful in rationalizing and incrementally improving the properties of some materials systems, but probably will not lead to new ones in the immediate future.

4) *Mechanisms are based on or inferred from studies of kinetics.* In complex systems, it is difficult enough to rationalize the energetics of starting materials and products based on their structures, much less to infer structures and energies of intermediates lying between starting materials and products, and even less the structures of transition states connecting these intermediates.[12] Although obtaining information concerning a complete reaction coordinate for a process is always difficult, it is also necessary to have a real understanding of the process. The word "process" implies a kinetic phenomenon: that is, the transformation of the system from one state to another during a time that is significant to the observer. It is not possible to rationalize a process without understanding the rate of the process, and thus the details of the structures and energetics of the intermediates lying between starting materials and products. Fracture is, of course, a process, and must ultimately be approached in the same way, if it is to be explained in full atomic detail.

A problem that occurs repeatedly in trying to apply chemical considerations based on bonding to the consideration of processes as complicated as the fracture of organic solids is that of the proper size of the domain that must be considered in discussing the critical phenomena underlying the process of interest. A certain amount of success has been achieved in attempting to model the surface of heterogeneous catalysis using soluble organometallic clusters.[13] The fraction of an organic solid involves considerably more complexity. A covalent organic solid is a collection of a very large number of individual, directed, covalent bonds (for example, for diamond, ~2 x 10^{23} bonds/cm^3); and fracture involves a small fraction of these bonds, but still a large number of them (at least 10^{14}/cm^2). A molecule involves a much smaller number of bonds (10–10^4) of which only 1-5 might break or form in a typical reaction. Simply on the basis of the number of bonds involved, fracture represents a much more difficult problem. In trying to understand fracture, chemists must first begin to address the problem of the quantity of the solid--that is, the number of bonds and atoms--that must be considered in atomic detail to prove or disprove any hypothesis concerning correlations between atomic and macroscopic structure. Is fracture initiated by the failure of a single bond? By concerted failure of 100 bonds? By failure of all of the bonds in a region of 10^3 nm^2? *How large is the problem?* And what are the characteristics of the region to be examined (whatever its size)? And how are these characteristics to be identified? The interior of a perfect single crystal? The region surrounding a crack tip? A representative region in an oxide-covered surface? The region including a critical void or microscopic embedded particle? Since fracture is often dominated by defects, the anomalies in structure are often the most important features of the system. These anomalies are also, of course, the most difficult features to detect and characterize, and thus to discuss in chemical detail.

Given the severe problems in making plausible connections between atomic level information in solids (at any level of theory or experiment) and macroscopic materials properties of that solid, what can chemistry realistically be expected to offer to fracture and materials science? In fact, should chemistry be expected to be an active participant in this area at all? In broad terms, the answer to the latter question is certainly "yes." Organic solids (particularly polymers) are the most rapidly evolving of the important classes of structural materials. Composites often contain organic components, and critical auxiliary components in materials (adhesion promoters, barrier film forming agents, corrosion inhibitors) are often molecular entities. Moreover, chemistry has and will continue to make important contributions to a number of areas of materials science entirely apart from organic solids: organometallic precursors for sol-gel-derived ceramics[14] and pre-ceramic precursors[15] for ceramics provide examples.

In more specific terms, we suggest a number of areas in which current events in chemistry offer analytical techniques, synthetic procedures, or conceptual insights of real relevance to fracture:

1) *Control of interfaces and thin films.* Fracture is a process that creates new interfaces; initiation of fracture often occurs at interfaces. Understanding the influence of environmental effects on the energetics of interfaces, and the kinetics of their creation, is a subject of real opportunity for chemistry.[16] The characterization and control of surface films as passivating systems, or as weak or strong components[17] in a system are also, in principle, under synthetic control. Adhesion promoters and compatibilizing agents for many heterogeneous systems can be designed using chemical techniques.[18,19]

2) *Analytical methods.* One of the areas of most rapid advance in chemistry and physics has been the development of new analytical techniques applicable to the study of materials at the microscopic scale. The range

of vacuum physics spectroscopies[21]--X-ray photoelectron spectroscopy, Auger spectroscopy, secondary ion mass spectroscopy, Rutherford backscattering spectroscopy,[22] and many others--make it possible for the first time to characterize in useful detail many materials systems. Transmission electron microscopy can now image single atoms in many circumstances.[23] New techniques, especially scanning tunneling microscopy[24] and force balance methods,[25,26] offer an unprecedented degree of detail concerning structure and potential functions for surfaces.

The large majority of the new instrumental techniques are most easily applicable to surfaces or thin films. Many important materials properties originate, of course, in some part of the system other than the surface. There are slowly emerging new techniques for examining structure in a microscopic scale deeper in a solid. Most of these techniques are based on the availability of high intensity X-ray sources. EXAFS,[27] low-angle X-ray reflectography,[28,29] and X-ray tomography[30] all provide examples. In certain types of systems, techniques such as nuclear magnetic resonance spectroscopy can also be applied to examining the interior structure of solids,[31] as can certain of the newer acoustical methods.

An important problem in the area of analytical methodology remains, however, the imaging of the interior structure of solids. Techniques such as single-fiber pullout[32] (based on optical dichroism of poorly understood origin) are primitive but provide one of the few available windows into the changes accompanying strain in the interior of solids.

3) *New materials and structures.* Chemistry continues to provide techniques of unparalleled utility and flexibility for the preparation of many new types of materials, especially those based on organic or organometallic precursors. Polymer alloys and phase-separated systems provide one outstanding example;[33] high-strength polymer fibers[34,35] based on liquid crystals provide a second. The expertise is available in chemistry to make almost any conceivable structure; the guidance needed to use this synthetic expertise efficiently is not.

4) *Control of impurity and defects; new processing techniques.* It goes without saying that chemistry is an essential element in materials processing. Detailed studies increasingly suggest that the strength of many high-performance fibers (for example, PAN-based carbon fibers or poly"dimethylsilane"-silicon carbide fibers) is dominated by impurities in the polymer[36]: gel or dust particles being particularly important. Thus, the preparation of highly uniform systems, and the processing of these systems to give defect-free structures is an area in which chemistry and chemical engineering can play an important role.

5) *Conceptual models.* The value of even simple chemical theory in understanding local properties of materials is clear.[37] Organometallic chemistry offers an enormous range of possible structures for metal-containing systems, and thoughtful examination of these illustrations provides at worst a highly developed intuition concerning structural types that appear to be energetically favored. More complex theory offers greater detail, but typically at the cost of examining a smaller system.

One of the major problems in theory is, in fact, to find the best strategies for integrating the many different types that are presently available, ranging from *ab initio* theory for very small isolated systems at the one extreme through semi-empirical and molecular mechanics calculations, thorugh statistical methods, to finite element calculations. How does one transfer results from each level of detail to the next highest level of systems integration? What features are essential to be transferred correctly, and which can be neglected?

Given the problems faced in connecting (or even recognizing!) relations between microscopic structure and macroscopic properties, it is perhaps no wonder that the most important advances in chemistry as applied to materials have come on the one hand from strongly physical methods for analysis or theory that provide static structural detail concerning solids, and on the other hand from largely empirical synthetic or preparative activity that has provided new solids to be examined for useful properties. The connection between these analytical/theoretical methods and the synthetic methods has been extremely limited.

We suggest that an alternative, hybrid approach--the so-called "physical-organic" approach--is just beginning to be applied usefully in materials science. This approach integrates synthesis and physical measurements: hypotheses concerning the relation between structure and properties are tested by synthesizing new structures, rather than by taking advantage of higher precision and greater detail in physical measurement. This method is particularly useful when applied to organic or organometallic materials, where it is possible to achieve very high variability in structure with relatively modest synthetic effort. The physical-organic approach is fundamentally science by analogy: it does not generate quantities of numbers in a form satisfying to those who are trained in the physics paradigm. It has, however, proved highly successful in solving complex problems in chemistry and biology, especially concerning the kinetics of organic reactions, catalytic mechanisms, and rational drug design.[38] We illustrate this approach with a project in progress in our own laboratory.

The objective of this work is to prepare new organic solids having particularly high thermal stability, resistance to chemical corrosion, hardness, and thermal conductivity. The basis for this project is a qualitative consideration of the extraordinary breadth of properties displayed by organic solids containing only carbon and hydrogen (Figure 1).

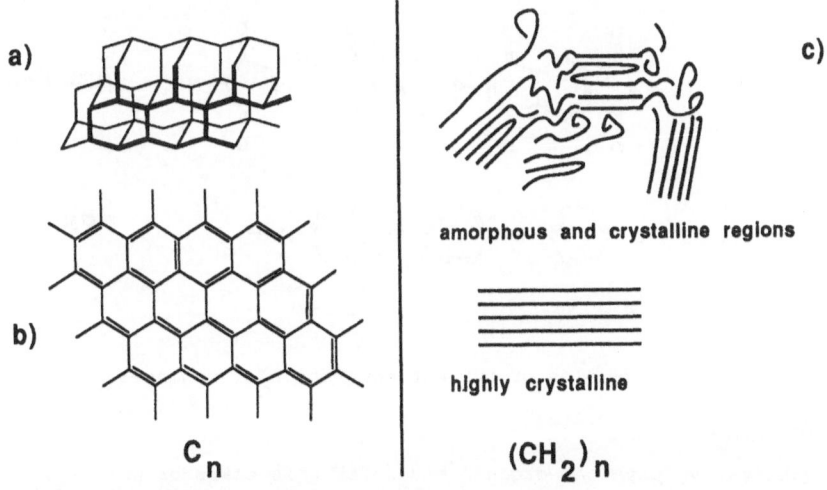

Figure 1. Organic solids containing only carbon and hydrogen:
(a) diamond; (b) graphite; and (c) schematic
representation of an organic polymer such as
polyethylene showing amorphous and crystalline
regions.

18

At one extreme is diamond. Diamond is both the hardest known solid and the best thermal conductor at room temperature. It is an electrical insulator and can be optically transparent. It contains a very high density of directed, strong, sp^3 carbon-carbon bonds. It is brittle in fracture. Graphite represents a second extreme. It has very high tensile modulus in the graphitic plane, and very low modulus perpendicular to it. It is sufficiently slippery to be used as a lubricant. It is an excellent electrical conductor, and a black solid. Polyethylene and related compounds represent a third example. This class of polymers are typically waxy solids with a density of ~1.5 g/cm³ (as compared with 2.3 g/cm³ for graphite and 3.5 g/cm³ for diamond). They are soft, insulating (both thermally and electrically) materials that can be processed readily.[39] They are typically semi-transparent. Their structures consist of a mixture of amorphous and crystalline regions, and physical properties depend strongly on the relative proportions of these regions and on the orientation of the crystalline regions. Typical polyethylene film has low tensile strength. Ultra-drawn polyethylene has very high tensile strength.

The concept that aligned, covalent bonds form the basis for structurally useful solids has, of course, been heavily exploited in making the current generation of engineering plastics.[40] Figure 2 gives examples.

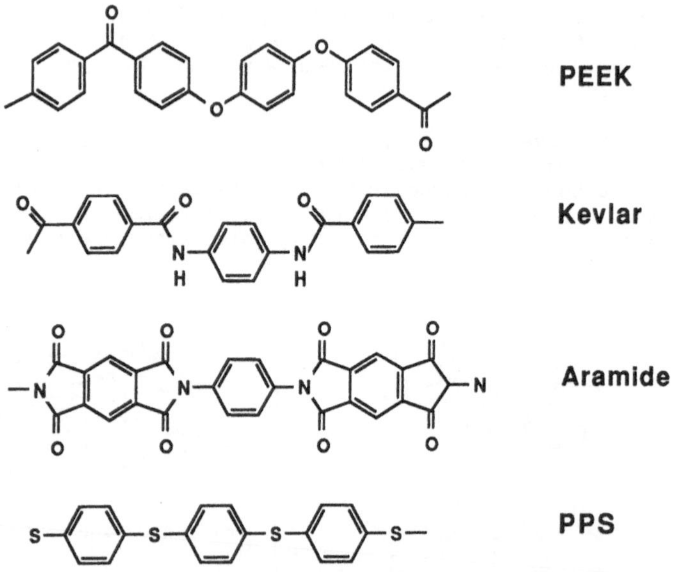

Figure 2. Examples of current generation of engineering plastics.

The qualitative physical-organic basis for this class of polymers is straightforward. All consist of highly stable, aromatic units joined in a relatively rigid form by strong covalent bonds. The examples shown represent a small number of those currently being developed; the major distinctions between them have less to do with ultimate physical properties than with economics and processability. This class of polymers has, as a group, three disadvantages. The first is that almost all are very difficult to process. The second is that most are formed in the liquid crystalline state, in order to obtain maximum tensile strength along the axis of the aligned, extended polymeric chains. Since interchain interactions are weak, they are typically weak perpendicular to the direction of preferred orientation, and fibers prepared from these materials tend to fail in compression. This perpendicular weakness is also reflected by limited resistance to damage by organic solvents. Finally, the ceiling operating temperatures of these materials (~350 °C) will probably not be sufficient for many future needs.

We and others[41] have set out to explore the idea that introducing heavy cross-linking between chains may provide a way of preparing three-dimensional, highly cross-linked solids having high strength and good solvent resistance. Our approach (Figure 3) has been to prepare polymers

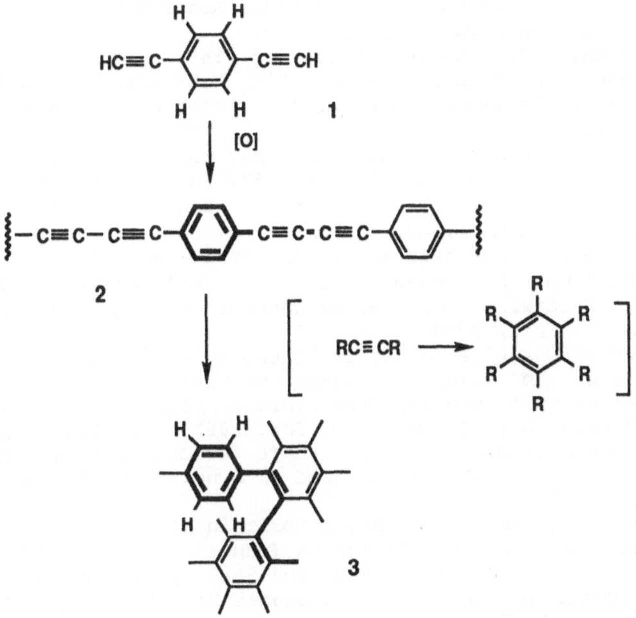

Figure 3. Synthesis of dense carbon solids from acetylenic precursors. Under appropriate processing conditions, the linking diacetylene units in 2 may be cyclotrimerized to yield highly cross-linked structures such as 3.

having a high density of diacetylene (C≡C-C≡C) units. Properly formulated, these polymers can be processed, molded, and then cross-linked either catalytically or thermally. The resulting solids have, in fact, a high density of covalent bonds, and can be considered to be a type of glassy carbon. Preliminary results indicate that these solids are very stable thermally, very resistant to damage by solvents, and (for organic materials) very hard.

The detailed results of this project are less important than the approach. The project is based on reasoning by analogy, and has very little quantitative calculation involved. The process involves an examination of a group of interesting materials--diamond, graphite, and polyethylene--and the inference from these structures of structural features that may underly the desirable physical properties. These hypotheses are tested by synthesizing new solids incorporating as many of the structural features identified by examination of existing systems as possible.

ACKNOWLEDGMENTS

This work was supported by the National Science Foundation, Grants CHE 85-08702 and DMR 83-16979.

REFERENCES

1. Cowan DO; Wiygul FM: C&E News, 28, July 21, 1986.
2. Opportunities in Chemisry, National Academy Press: Washington, 1985.
3. Cohen ML: Science 1986, 234, 549.
4. Bowen HK: Sci. American 1986, 255, 169.
5. Halpern J; Okamoto T: Inorg. Chim. Acta 1984, 89, L53.
6. Tolman CA; Meakin PZ; Lindner DL; Jesson J-P: J. Am. Chem. Soc. 1974, 96, 2762.
7. Halpern J; Okamoto T.; Zakhariev J.: J. Mol. Catal. 1977, 2, 65.
8. Rooney JJ: J. Mol. Catal. 1985, 31, 147. Muetterties EL; Krause MJ: Angew. Chem. Intern. Ed. Engl. 1983, 22, 135.
9. Low JJ; Goddard III WA: J. Am. Chem. Soc. 1986, 108, 6115. Saillard J-Y; Hoffmann R: . J. Am. Chem. Soc. 1984, 106, 2006.
10. Parshall GW: Homogeneous Catalysis. Wiley-Interscience, New York, 1980. Satterfield CN: Heterogeneous Catalysis in Practice. McGraw-Hill, New York, 1980.
11. Shilov AE: Activation of Saturated Hydrocarbons by Transition Metal Complexes. D. Reidel: Hingham, MA, 1984.
12. For an example involving the hydroformylation of alkenes see Heck RF; Breslow DS: J. Am. Chem. Soc. 1961, 83, 4023. Alemdarogly NH; Pennsger JLM; Oltay E: Monatsheft für Chemie 1976, 107, 1153.
13. Muetterties EL; Rhodin TN; Band E; Brucker CF; Pretzer WR: Chem. Rev. 1979, 79, 91.
14. Woodhead JL; Segal DL; Chem in Britain 1984, 310.
15. Brinker CJ; Clark DE; Ulrich DR (eds): Better Ceramics through Chemistry II. Materials Research Society Symposia Proceedings, Vol. 73: Pittsburgh, PA, 1986. Trefonas P; West R; Miller RD: J. Am. Chem. Soc. 1985, 107, 2737. Seyferth D; Wiseman GH: J. Am. Ceram. Soc. 1984, 67, C132-133.
16. Kardos JL: J. Adhesion 1973, 5, 119.
17. Kardos JL: Chemtech 1984, 430.
18. Angus JC; Stultz JE; Shiller PJ; MacDonald JR; Mirtech MJ; Domitz S: Thin Solid Solids 1984, 118, 311-20.
19. Hergenrother PM: Chemtech 1984, 496.
20. Pluddemann EP: Silane Coupling Agents. Plenum Press: New York, 1982.
21. Feldman LC; Mayer JW: Fundamentals of Surface and Thin Film Analysis. North-Holland, 1986.
22. Evans CA; Strathman MD: Ind. Res. Dev. 1983, 25, 99.
23. Willims DB: Practical Electron Microscopy in Materials Science. Verlag Chemie International: Weinheim, 1984.
24. Quate CF: Physics Today 1986, August, 26.
25. Marra J; Israellachvili J; Biochemistry 1985, 24, 4608.
26. Pashley RM; McGuiggan PM; Ninham BW; Evans DF: Science 1985, 229, 1088.

27. Lei PA; Citrin PH; Eisenberger P; Kincaid PM: Rev. Mod. Phys. 1981, 53, 769.
28. Pomerantz M; Segmüller A; Netzer L; Sagiv J: Thin Solid Films, 1985, 132, 153.
29. Pomerantz M; Segmüller A: Thin Solid Films 1980, 68, 33.
30. Baumann KJ; Kennedy WH; Herbert DL: J. Compos. Mater. 1984, 18, 536.
31. Duijvestijn MJ; Van Der Lugt C; Snidt J; Wind RA; Ziln KW; Staplin DC: Chem. Phys. Letters 1983, 102, 25.
32. Hadjis N; Piggott MR; J. Mater. Sci. 1977, 12, 358.
33. Shiomi T; Karasz FE; MacKnight WJ: Macromolecules 1986, 19, 2274. Ten Brinke G; Karasz FE; MacKnight WJ: Macromolecules 1983, 16, 1827. Sperling LH; Manson JA; Yenwo GM; Devia N; Pulido JE; Conde A, In: Polymer Alloys; Klemper D; Frusch KC (eds). Plenum: New York, 1977.
34. Stille JK: Contemp. Top. Polym. Sci. 1984, 5, 209.
35. Baer E: Sci. American 1986, 255, 179.
36. Hughes JDH: Carbon 1986, 24, 551.
37. Burdett JK; McLarnan TJ; Hughbanks T: J. Am. Chem. Soc. 1986, 106, 3101. Burdett JK; McLarnan TJ: J. Solid State Chem. 1984, 53, 382. Kertesz J; Hoffmann R: J. Am. Chem. Soc. 1984, 106, 3453.
38. Lowry TH; Richardson KS: Mechanism and Theory in Organic Chemisry, 2nd Edition. Harper & Row: New York, 1981. March J: Advanced Organic Chemistry. Wiley-Interscience: New York, 1985.
39. Bassett DC: Principles of Polymer Morphology. Cambridge University Press: London 1981. See also Allcock HR; Lampe FW: Contemporary Polymer Chemistry. Prentice-Hall: Englewood Cliffs, NJ, 1981.
40. Sivaram S: J. Sci. Ind. Res. 1982, 41, 599.
41. Dawson DJ; Fleming WW; Lyerla JR; Economy J, In: Reactive Oligomers, ACS Symposium Series 282, 63.

DISCUSSION

Comment by M. Pourbaix:

Should I understand from the beginning of your lecture that you are pessimistic about the possibility of a useful link between thermodynamics and kinetics? Do you think that the concept of affinity is of no practical interest?

Reply:

I am less optimistic that kinetics of reactions correlate with thermodynamics (that is, that exergonic reactions are necessarily fast, and that the reaction path followed is the most exergonic) with organic fracture processes than with those involving metals or ceramics, organic reaction mechanisms seldom are determined entirely by the thermodynamics of the processes involved, and often appear almost independent.

Comment by A. S. Argon:

Your statement that the very large differences in energies of different bonds between, say metals, and polymers with van der Waals interactions must give rise to very different behavior is true often only on the basis of overall energies. On a relative basis when properties are normalized with proper elastic properties of the material they become surely more comparable.

Reply:

The point of difference is less the magnitudes of the bond strengths, and more the relative magnitudes in the same system. For example, in metals, all bonds between an atom and its nearest neighbors have roughly the same energy. In polyethlylene, a hydrogen atom interacts with its directly bonded carbon with a high energy (~5 eV) and with hydrogens of a neighboring chain very weakly (~0.1 eV). This anisotropy in the organic system makes certain kinds of motions possible in the solid that do not occur in metals (chain repetition); on the other hand, cooperative motions such as dislocation propagation occur in certain metals but not in polyethylene.

MECHANICS OF BRITTLE CRACKING OF CRYSTAL LATTICES AND INTERFACES

JAMES R. RICE
Division of Applied Science, Harvard University, Cambridge, MA 02138, USA

1.INTRODUCTION

This is a review of some concepts in the mechanics of fracture appropriate
to interpreting atomistically brittle modes of cracking of crystal lattices and
of interfaces between crystalline phases, especially in solids which are
normally, or potentially, ductile. Crack tip modelling is first discussed
based on elastic-brittle concepts, assuming either that there is no near-tip
dislocation activity or, for the time being, neglecting it. That discussion
brings out the importance of the Griffith condition and its generalization to a
cracking interface (e.g., an embrittled grain boundary), and includes an
account of the thermodynamics of interfacial decohesion in presence of
(possibly embrittling) segregants. It is suggested that the phenomenology of
grain boundary embrittlement by P and Sn in Fe, and of "cohesion
enhancement" by C, can be at least partly understood as an effect of the
various segregants on altering the work necessary to separate an interface
against atomic cohesive forces.

Dislocation interactions with the crack tip typically control whether a
potential atomically-brittle decohesion mode of fracture actually occurs.
Also, if such does not occur (or occurs only along the interfaces of
void-nucleating inclusions), then dislocation plasticity processes provide the
means by which fracture can ultimately occur through void growth to
coalescence, or localized shear, or some combination of the two.

The next focus of the discussion here is on crack tip interaction with dislocations. This is addressed in the context of emission of a single
dislocation from the crack tip. Whether such emission or, instead, atomic
decohesion first occurs is thought to be fundamental to whether a crystal (or
interface) is intrinsically cleavable. Finally, some concepts in the modelling
of multidislocation processes at a crack tip are reviewed with emphasis on
using continuum plasticity to model processes at the single crystal scale,
based on constitutive relations that build-in the notions of a limited set of
allowable slip systems and a resolved shear stress to control flow on each.
Viscoplastic effects in maintaining a crack tip stress concentration suitable
for continued cleavage cracking in normally ductile solids are also discussed.

There is an extensive literature, not discussed here, on the analysis of
crack tip plastic phenomena over a scale that is large compared to microstructure, as addressed by classical macroscopic plasticity theory as intended
for the (homogenized) polycrystal scale. Such is the primary focus of
inelastic fracture mechanics as developed for prediction of crack growth in
engineering structures. That branch of the subject has also served as a basis
for analyses of stress states prevailing at cleavage nucleation from carbides
in the complex steel microstructure ahead of a macrocrack or notch, and for

ductile hole growth failures in the crack tip strain field. It is a somewhat more mature area than those mentioned above and an excellent review is given in the 1985 book by Kanninen and Popelar[1]. We do draw on some recent results on the viscoplastic dynamics of crack tip stressing[2].

2. ELASTIC-BRITTLE CRACK THEORY

We recall that when a cracked solid is analyzed within linear elastic theory, with the crack represented as a traction-free surface, there results a near-tip stress field of form (see fig. 1)

$$\sigma_{\alpha\beta} = K_M \, r^{-1/2} \, f_{\alpha\beta}{}^M(\theta) + \ldots \tag{2.1}$$

where there is summation on the repeated M over modes I, II, III (associated with cartesian directions 2,1,3) and where the dots represent terms which are bounded at the crack tip ($r=0$). The f's are universal functions of polar angle θ and, on the plane $x_2 = 0$ ($\theta = 0$) prolonging the crack surface the non-zero components of $f_{2\beta}{}^M(0)$, corresponding to traction components $\sigma_{2\beta}$ on that plane, are $f_{22}{}^I(0) = f_{21}{}^{II}(0) = f_{23}{}^{III}(0) = 1/\sqrt{2\pi}$ with the usual normalization. Here x_1 is in the direction of crack growth. The three K's are called stress intensity factors and, for simple circular or tunnel cracks, are proportional to certain remotely applied stress components and to the square root of crack size.

The previous result applies for any homogeneous linear solid, isotropic or not, but when the crack lies along an interface between elastically dissimilar materials a more complicated form results[3] with the r and θ dependence above replaced by the real part of complex expressions of the form $r^{-1/2+i\epsilon} \, F_{\alpha\beta}{}^M(\theta)$. This form implies an oscillatory variation (with r) of the $r^{-1/2}$ amplitude of terms giving the stress field, as $r \rightarrow 0$, but for short cracks as arise in microscale applications to grain boundaries or phase interfaces, the region predicted to be dominated by those oscillations is often smaller than a lattice spacing[4].

Another useful concept of elastic-brittle theory, not limited to linear solids, is that of the energy release rate G. It is simplest to introduce in the 2D context of a flat crack under plane strain and/or anti-plane strain. Let Φ be the strain energy (Helmholtz free energy) associated with the elastic deformation of the crack-containing body at fixed temperature T, per unit thickness of the body, and suppose the body is loaded with tractions proportional to generalized force Q, with work-conjugate displacement q on a unit thickness basis. Then G is defined as the "configurational force" conjugate to crack length a, by

$$Q \, dq - G \, da = d\Phi . \tag{2.2}$$

This defines G, e.g., $G = -\partial\Phi/\partial a$ at fixed q, on the basis of continuum elastic solutions for the cracked solid. The energy release rate can also be calculated by setting $G = J$ where, in the present 2D context, the integral J is path-independent (for a homogeneous material, or crack on an interface between homogeneous materials) and given by

$$J = \int_\Gamma [\phi n_1 - n_\alpha \sigma_{\alpha\beta} \partial u_\beta / \partial x_1] \, ds \tag{2.3}$$

Here, in an interpretation which holds good at finite elastic deformation, the coordinates x_1, x_2, path Γ (fig. 1) and length s along it, and outer unit normal n are measured off in the unstressed reference configuration, ϕ is the strain energy per unit volume of reference configuration, u is displacement and σ is the nominal stress tensor such that $n_\alpha \sigma_{\alpha\beta}$ is force per unit area of reference configuration.

For the linear homogeneous solid there results in general the expression

$$G = \Lambda_{MN} K_M K_N \tag{2.4}$$

reducing to

$$G = [(1-\nu)/2\mu](K_I^2 + K_{II}^2) + (1/2\mu)K_{III}^2 \tag{2.5}$$

in the isotropic case (ν = Poisson ratio, μ = shear modulus). For the general anisotropic material the matrix Λ is $(1/8\pi)$ times the inverse of a matrix α that appears later in the pre-logarithmic energy factor for a line dislocation having the same direction as the crack tip[5].

The criterion for crack growth in the absence of plasticity would seem to be, essentially, that of Griffith,

$$G = 2\gamma_{int} \tag{2.6}$$

where $2\gamma_{int}$ is the reversible work of separating, against atomic cohesive forces, the interface (or, for a homogeneous material the lattice plane) along which the crack spreads. This is most readily seen by writing the total free energy F of the crack-containing body as $\Phi + 2\gamma_{int}a$ and requiring, as an equilibrium condition, that $dF = Q\,dq$ under the presumed isothermal conditions. Then eq. (2.2) leads directly to (2.6). From an irreversible thermodynamic viewpoint it may be asserted that $\dot{F} \leq Q\dot{q}$ (superposed dots mean time rates) during any isothermal process. This is equivalent to saying the entropy of the cracked solid and its thermally equilibrating heat reservoir, taken together as an isolated system, is nondecreasing, and leads to the inequality

$$(G - 2\gamma_{int})\,\dot{a} \geq 0 \ . \tag{2.7}$$

That is, $G \geq 2\gamma_{int}$ for crack growth and the difference $G - 2\gamma_{int}$ accounts (in the present idealized situation without plasticity) for dissipation into phonons during the unstable atomic scale jumps of the crack associated with lattice trapping[6,7]. The same thermodynamic inequality applies when a crack surface forms in a fluid environment from which a species can adsorb onto the crack walls, but in that case it has been shown[7] that $2\gamma_{int}$ is to be replaced by

$$(2\gamma_{int})_o - 2 \int_{-\infty}^{\mu} \Gamma_s(\mu')\,d\mu' \ , \tag{2.8}$$

as could be anticipated by the Gibbs adsorption equation. Here the first term refers to the clean interface or lattice plane, presumed to be initially without adsorbate, and $2\Gamma_s(\mu)$ is the total coverage of the two crack walls when they are at equilibrium with adsorbate at chemical potential μ in the fluid environment. This describes, macroscopically, effects which may also be treated at the molecular level in terms of the adsorbing species weakening cohesive bonds at the crack tip[6], although the kinetics of the process are not addressed.

As another perspective on the elastic-brittle case, let us reject the singular crack model in favor of one for which the crack walls gradually separate under the high stresses at the crack tip until they are pulled out of range and the stress $\sigma (= \sigma_{22})$ falls to zero. Fig. 2a shows schematic plots of σ and the opening gap $\delta [= u_2(x_1, 0^+) - u_2(x_1, 0^-)]$ near the tip of a crack at limiting conditions; σ has fallen to zero for $x_1 < x_1'$ and δ is essentially zero for $x_1 > x_1''$. Then, by a well known application of the J integral[8], it may be shown that for paths Γ which do not cut across the interval of x_1 axis between x_1' and x_1'',

$$J = - \int_{x_1'}^{x_1''} \sigma \, \partial \delta / \partial x_1 \, dx_1 \, , \qquad (2.9)$$

and also, if the distance $x_1'' - x_1'$ is much smaller than crack length or other overall geometric dimensions for the cracked solid, we have $J = G$ where G is the crack tip energy release rate as calculated for the singular crack of the same length under identical external loading. Thus, if $\sigma = \sigma(\delta)$ as sketched in fig. 2b, the condition of crack advance reads

$$G = - \int_{x_1'}^{x_1''} \sigma(\delta) \, \partial \delta / \partial x_1 \, dx_1 = \int_0^{\delta'} \sigma(\delta) \, d\delta = 2\gamma_{int} \qquad (2.10)$$

where δ' is the opening at which the crack walls are pulled out of stress range of one another and we recognize that the area under the σ versus δ curve merely defines $2\gamma_{int}$ as introduced earlier.

Thus this model confirms the Griffith criterion. We shall shortly examine the effect of interfacial chemical composition on the σ versus δ relation which enters the model but first let us note what additional insights come from discrete lattice models for crack tips. It has, for example, been claimed on occasion in the past literature that calculated results for lattice models violate the requirement $G \geq 2\gamma_{int}$ for crack growth or $G \leq 2\gamma_{int}$ for crack healing. However, closer examination[6,7] reveals that such claims are not sustained when there are no mechanical inconsistencies in treatment of non-linearities and when G is properly calculated for the generally anisotropic elastic solid which provides the correct long wavelength or continuum limit for the lattice model adopted. The significant difference is that in the lattice models, employed for force laws which are consistent with the creation of no dislocations at the crack tip, the value of G at crack growth exceeds $2\gamma_{int}$ by a fractionally small amount (in three dimensional modelling) associated with the lattice trapping effect, and G at onset of crack healing is smaller than $2\gamma_{int}$ by a similarly small amount.

The existing understanding of such plastic processes as dislocation emission from crack tips and near tip sources, and the viscoplastic dynamics of crack tip relaxation, does not denigrate the significance of $2\gamma_{int}$ as a measure of the resistance to atomic decohesion. To be sure, the macroscale G (versus the local value at the tip of any cleavage-like separation that may occur) will in those circumstances usually be greater by an order of magnitude or more than $2\gamma_{int}$, but the presumption is that so long as the fracture mechanism is cleavage-like, $2\gamma_{int}$ controls the intensity of the local G required at the decohering crack tip bonds. Hence $2\gamma_{int}$ has what is sometimes called a "valve" effect on the much larger macroscale G, which includes the energy flow to near tip plastic processes. Stated alternatively, the crack tip is

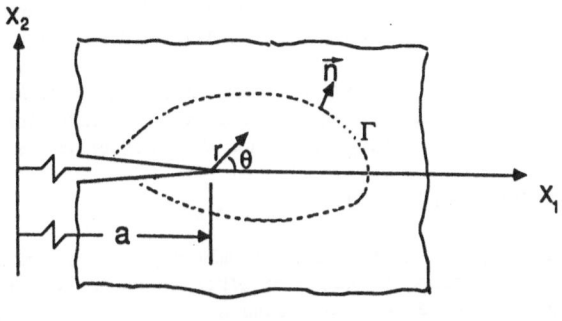

Figure 1

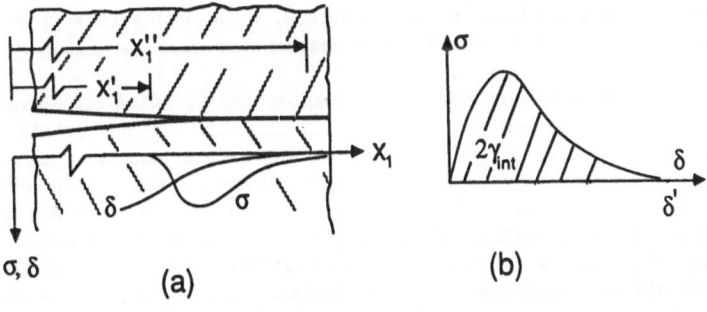

(a)   (b)

Figure 2

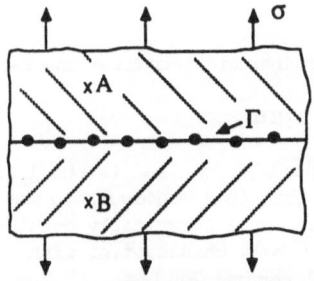

Figure 3

then said to be "screened" by the dislocation plasticity in its vicinity. Thus, irrespective of the fact that $2\gamma_{int}$ is usually minute by comparison to the measurable macroscale G, an appropriate route to understanding some of the solute and environmental chemistry effects on fracture is via their effects on $2\gamma_{int}$. This is taken up next but provides obviously an incomplete descriptor of such effects since transport and reaction kinetics are neglected and since chemical factors can also influence dislocation emission and motion.

3. INTERFACIAL SEGREGATION AND COHESION

Consider a uniformly stressed, and uniformly separating, interface as a thermodynamic system described in terms of Gibbsian excess quantities. For example, in fig. 3, define δ as the excess of the total displacement of point A relative to point B over that accountable by homogeneous straining of the two phases in which points A and B reside. Then $\sigma d\delta$ is the excess of work over that accountable as work in straining the adjoining phases, and this work expression will obey the laws of thermodynamics in terms of analogously defined excess internal energy, heat adsorption, free energy ($\bar{f}$ per unit area of undeformed interface), entropy ($\bar{s}$), etc. There is another work mode involving strains in the plane of the interface, with conjugate work by the surface stress tensor, but we shall not consider such strains here.

In fracture as affected by solute segregation one is often concerned with interfaces that are out of composition equilibrium with the bulk, both before and after separation. For simplicity we consider here a single segregant, which is present at concentration Γ per unit reference area of interface and which is present only at extremely small concentrations in the adjoining bulk phases (so that Γ is unambiguously defined). We deal with non-equilibrium separation by adopting the constitutive approximation[9] that $\sigma = \sigma(\delta,\Gamma,T)$ and $\bar{f} = \bar{f}(\delta,\Gamma,T)$. Within this approximation one regards, e.g., $\bar{f}(\infty,\Gamma,T)$ as the same function of Γ irrespective of whether it refers to the energy of a pair of surfaces that have been freshly created by a fracture and contain total solute Γ or to a pair of free surfaces which have reached a state of composition and reconstruction equilibrium at which solute Γ is present (the temperature T at the interface is understood to be the same in both cases). The model described by this approximation is thus one for which all states of the interface are at local equilibrium, but not necessarily at equilibrium with the bulk phases.

Then, by the standard thermodynamic formalism one has

$$d\bar{f} = \sigma d\delta - \bar{s} dT + \mu d\Gamma \qquad (3.1)$$

where the μ is defined by $\partial \bar{f}(\delta,\Gamma,T)/\partial \Gamma$ and is the local chemical potential of the interfacial solute. Within this framework we may identify two limiting cases of isothermal separation[9]. The classically considered one is separation at composition equilibrium with bulk phases, i.e. with μ = constant, for which the work $\int \sigma d\delta$ of interfacial separation is

$$(2\gamma_{int})_{\mu=const} = \bar{\gamma}(\infty,\mu,T) - \bar{\gamma}(\delta_b,\mu,T) \qquad (3.2)$$

$$= (2\gamma_{int})_o - \int_{-\infty}^{\mu} [2\Gamma_s(\mu') - \Gamma_b(\mu')]d\mu' ,$$

where $\gamma(\delta,\mu,T) = \bar{f} - \mu\Gamma$, $\delta = \delta_b(\mu,T)$ on the unstressed boundary, and (at the fixed T considered) $\Gamma = 2\Gamma_s(\mu)$ and $\Gamma = \Gamma_b(\mu)$ describe the relations between Γ

and μ on the two free surfaces $(\delta = \infty)$ created by separation and on the unstressed boundary or interface $(\delta \doteq \delta_b)$. Similarly, for separation at constant composition Γ, the work $\int \sigma d\delta$ is

$$(2\gamma_{int})_{\Gamma = const} = \bar{f}(\infty, \Gamma, T) - \bar{f}(\delta_b, \Gamma, T) \tag{3.3}$$

$$= (2\gamma_{int})_o - \int_0^\Gamma [\mu_b(\Gamma') - \mu_s(\Gamma'/2)] d\Gamma'$$

Here $\delta = \delta_b(\Gamma, T)$ on the unstressed boundary and $\mu = \mu_s(\Gamma/2)$ and $\mu = \mu_b(\Gamma)$ describe the two relations between Γ and μ mentioned above. Often $(2\gamma_{int})_o$, which refers to the interface with $\Gamma = 0$ (hence $\mu = -\infty$), is written as $2\gamma_s - \gamma_b$.

The Langmuir-McLean model gives for the form for either of the relations between Γ and μ

$$\mu = \Delta h - T \Delta s_{vibr} + RT \ln [\Gamma/(\Gamma_1 - \Gamma)] \tag{3.4}$$

where Δh, Δs_{vibr} and Γ_1 (denoting full coverage) will be different according to whether the relation refers to the pair of separated surfaces or to the unstressed boundary. The reference state from which the enthalpy (indistinguishable from energy here) and vibrational entropy changes, Δh and Δs_{vibr}, are reckoned is immaterial but is conveniently regarded as the bulk phase at the same T, so that $(\Delta h)_s$ and $(\Delta h)_b$ are negative. Further, Δh can, in principle, be determined from high-temperature adsorption isotherms for which the surfaces or boundary are at composition equilibrium with the bulk, so that then $\mu = RT \ln x$ also (x = bulk concentration). The procedure is that the high temperature data over a limited range of T at which equilibrium can be attained, and at low to moderate Γ, are fitted to

$$\Gamma/(\Gamma_1 - \Gamma) = x \exp[\Delta g^\circ(T)/RT] \tag{3.5}$$

to define Δg° $(= \Delta h - T \Delta s_{vibr})$, and Δh and Δs_{vibr} are determined separately using $\Delta s_{vibr} = d\Delta g^\circ/dT$. Grabke[10] describes this procedure for some solutes in Fe, and also explains the elaborations upon it when there is site competition between multi-component segregants. One assumes, compatibly with the Langmuir-McLean modelling, that the Δh and Δs_{vibr} values so determined are independent of T (except at very low T for the latter) and hence applicable at the temperature of fracture, which is usually below the range of T for which composition equilibrium between the interface and bulk can be attained.

Detailed forms for the two limiting works of separation, $2\gamma_{int}$, defined by (3.2) and (3.3) have been given in the literature[9] based on the Langmuir-McLean form (3.4). For separation at constant composition one finds, for low temperature fracture processes, that the entropy terms make only small contributions, and thus that

$$(2\gamma_{int})_{\Gamma = const} \approx (2\gamma_{int})_o - \Gamma[(\Delta h)_b - (\Delta h)_s] \tag{3.6}$$

This is a convenient expression and shows that the difference between adsorption energies on the coherent boundary and on the separated surfaces,

$(\Delta h)_b - (\Delta h)_s$, is a critical quantity measuring the embrittlement potential of a given solute. The quantities can in a few cases be estimated from higher temperature adsorption data (Table 1). In the future it is to be hoped that quantum-based electronic calculations, as discussed by other contributors to this conference, will give sufficiently reasonable estimates of $(\Delta h)_b$ and $(\Delta h)_s$ so as to be useful in alloy design.

The discussion thus far is based on the approximation of local equilibrium at the separating interface and has neglected differences between surfaces freshly created by fracture and surfaces which are in a possibly reconstructed long-time equilibrium state. For example, it is sometimes argued[11] that an embrittling segregant such as S in Ni acts to redistribute charge and reduce the strength of near-interface metal-metal bonds, thus encouraging an atomic-scale picture of brittle intergranular fracture in which the crack avoids separating the segregant-metal bonds along the interface but, rather, runs nearby and roughly parallel to the interface by separating the adjacent weakened metal-metal bonds. (If this picture applies at all, actual fractures must involve frequent shifting of the crack from one to the other side of the interface, since matched grain boundary facets usually give comparable Auger spectra[10,12], suggesting a not very different distribution of solute on the two fracture surfaces.)

Without passing judgement on the validity or not of that atomic-scale picture of interface-avoidance, evidently involving large departures from local equilibrium at the interface during separation, we can prove that it requires more work to separate the interface than does a hypothetical reversible separation for which there is local equilibrium at the interface throughout the separation. Thus, while motivated as an explanation of embrittlement, the atomic-scale interface-avoidance scenario does not in fact give the most brittle separation imaginable in presence of a given amount of segregant on the interface. Of course, there is no reason that nature should choose the most brittle process of separation, so such is not necessarily an argument against the interface- avoidance concept.

The proof is simply stated with reference to fig. 4 where there is depicted an isolated system consisting in its initial state of a boundary of unit area with segregant Γ and a heat reservoir at temperature T. By some generaly irreversible process involving the doing of work W, the boundary is separated. Ultimately the segregated solute atoms reposition themselves along the separated interface, and the interface may also reconstruct, so that a final equilibrium state is reached. Both the initial and final equilibrium states are local, in that they refer to equilibrium of the solute within the interface region; the temperature considered is assumed to be such that composition equilibrium with the bulk cannot be attained. The second law requires that W $\geq W_{rev}$, the work of a hypothetical reversible separation at local equilibrium within the interface region, and that work of reversible separation is defined by the free energy change between the initial and final local equilibrium state:

$$W \geq W_{rev} = \bar{f}(\infty,\Gamma,T) - \bar{f}(\delta_b,\Gamma,T). \tag{3.7}$$

Here the notation used when modelling with the local equilibrium approximation is used to emphasize, through comparison to (3.3), that the lower bound to the work of reversible separation is precisely what is given by (3.3) and, in the low T limit, by (3.6).

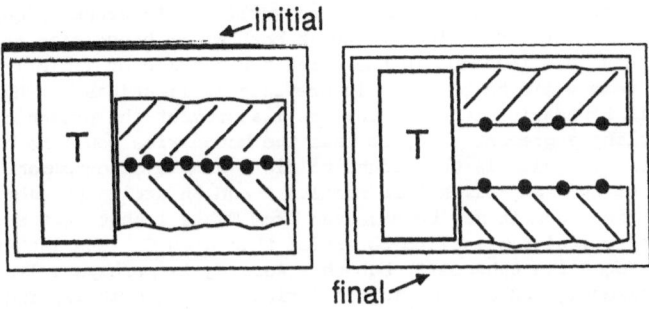

Figure 4

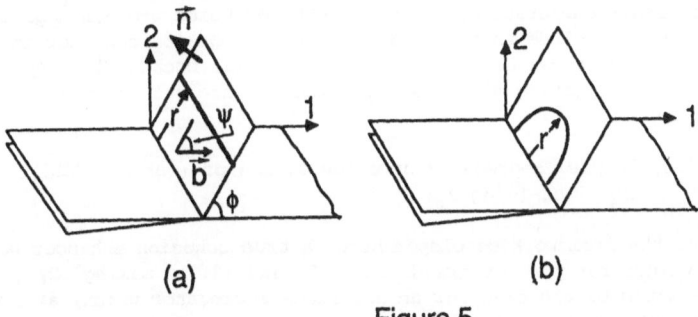

(a)   (b)

Figure 5

H. Grabke presented at this meeting values of segregation enthalpies at surfaces or grain boundaries for various elements in Fe, based on high temperature equilibrium adsorption as discussed earlier. Apparently, only in the cases of C,P and Sn are the values known at both grain boundaries and free surfaces, and these are recorded in the first two columns of Table 1. Some but not all of the values are to be found in previous publications[10]. The result for C on grain boundaries required for its measurement the presence of another embrittling segregant (Si), so that the boundaries could be broken and studied by Auger; the data is then found to fit a simple site-competition generalization of the Langmuir-McLean model. P and Sn are known embrittlers of Fe grain boundaries, and it may be observed from Table 1 that they show a very much higher theoretical embrittling potential than does C. The latter, C, is often called a "cohesion enhancer", but this wording is perhaps a little imprecise. Apparently, pure Fe grain boundaries are not brittle, and neither are they embrittled by the presence of C. Thus, there is no evidence in that sense that C increases cohesion, and the theoretical estimates here suggest that it actually slightly degrades cohesion. In fact, the observed beneficial effects[10] of C are in cases for which the grain boundaries are already brittle due to other segregants, whose deleterious effects are reduced or eliminated by addition of C. The cause may be explained from the second column of Table 1: C has an unusually high binding energy, $-(\Delta h)_b$, to the grain boundary and thus successfully competes for sites with other segregants. (The competition is more accurately described in terms of $(\Delta g^\circ)_b$ rather than $(\Delta h)_b$, since it takes place at higher temperatures for which entropy terms are non-negligible). Thus we tentatively conclude that C is not a true cohesion enhancer in Fe, but rather slightly degrades cohesion; it is beneficial because it displaces from the boundary segregating elements which would cause far greater degradation of cohesion.

To generalize, helpful segregants have low values of $[(\Delta h)_b - (\Delta h)_s]$ and high values of $[-(\Delta h)_b]$ or $[-(\Delta g^\circ)_b]$.

According to the framework developed here, a true cohesion enhancer would show negative values for the integrands in (3.2) and (3.3), making $2\gamma_{int} > (2\gamma_{int})_o$. This would be the case for an anomalous segregator which, at a given potential, segregates more abudently to an unstressed grain boundary to a pair of free surfaces. It has been suggested recently[13] that B in slightly off-stoichiometric compositions of the ordered alloy Ni_3Al shows such anomalous segregation and it does appear to be a cohesion enhancer in that case since the alloy without B shows brittle grain boundary failure, but is ductilized by B. It would be interesting to have reliable quantum-electronic based results for $(\Delta h)_b$ and $(\Delta h)_s$ in this case; their difference in the form as in the last column of Table 1 is inferred to be negative.

The discussion here has focused on separation at constant segregant amount, Γ. However, when segregants are mobile, as may be the case for some H embrittlement processes at room temperature, and for S in Fe in the high temperature situation of stress-relief cracking, separation at conditions approaching constant segregant potential, μ, may occur. This is always the more embrittling situation, for it may be shown[9] that so long as $d\Gamma/d\mu > 0$ on the pair of free surfaces, $2\gamma_{int}$ for slow separation at constant μ is less than that for fast separation at constant Γ.

TABLE 1.   Segregation enthalpies (kJ/mol), after H.J. Grabke[10]:

system	$-(\Delta h)_s$	$-(\Delta h)_b$	$(\Delta h)_b - (\Delta h)_s$
C in Fe	85	57	28
P in Fe	180	34	146
Sn in Fe	>200	22	>178

4. DISLOCATION EMISSION FROM A CRACK TIP

There is now a large literature on this topic[14]. In the context of under-standing whether a given solid or interface is intrinsically cleavable, we may follow the Kelly-Tyson-Cottrell and Rice-Thomson approach, as reinterpreted by Mason, and ask the question: Given an atomistically sharp crack in a crystal lattice, or along an interface, are conditions first reached to emit a dis-location from the crack tip or, rather, to grow the crack by cleavage-like bond decohesion? Thus, two values of the local crack tip G are identified, G_{disl} for dislocation emission and G_{cleave} $(=2\gamma_{int})$ for cleavage-like decohesion; configurations with $G_{cleave} < G_{disl}$ are regarded as intrinsically clevable whereas those with $G_{disl} < G_{cleave}$ are assumed to fail by a more plastic mechanism.

The procedure of estimating G_{disl} is outlined subsequently. First some relevant comments and reservations are reviewed:

1. In working out the crack tip competition thus far, it has generally been assumed that the mechanisms are non-interacting. That is, the possible effects of incipient shear rearrangements near the tip on the cleavage criterion, and of large bond stretching on the dislocation emission criterion, have been neglected.

2. While the cleavage criterion (sections 2,3) is on relatively solid ground, the emission criterion has, thus far, been worked out only in the context of linear elasticity and, further, predictions of emission conditions depend on poorly characterized parameters describing the core energies and configurations of very-near-tip dislocations and, in some treatments, the ledge energy associated with the dislocated step left at the crack tip. An incipient loop is generally predicted to become unstable at a radius of one to two Burgers vectors, and thus is well outside the range of confidence in continuum modelling of dislocation phenomena. This is a fertile area for clarification by atomistic calculations, such as by the embedded atom method[15] or other techniques[6]. The continuum model may be used with greater confidence to set a lower bound to G_{disl}, as discussed later.

3. The emission calculations as thus far formulated inevitably assume that the crack tip lies along the intersection of a slip plane with the fracture plane. It is possible but not obvious that an otherwise oriented crack would tend to have its tip rotate locally to such an orientation. Calculations[15] for an embedded atom model intended to represent Ni confirm that a crack tip con-strained against rotation can undergo brittle cleavage whereas the same model leads to dislocation emission for a crack whose tip lies in an appropriately oriented slip plane.

4. The crack tips at which the cleavage versus emission competition is critical are, in many cases, cracks which have nucleated at a brittle site and are propagating rapidly along a brittle pathway, such as through or on the interface of a second phase particle or through a region affected by corrosion, before testing the potentially ductile host material. Thus there will often be an essential dynamic element to the process and cracks judged as dislocation emitting when stationary may not be so when propagating. There are attempts at analysis of these effects,[16,2] including the possibilities that dislocations are not nucleated at all from the rapidly moving tip or that they are freely nucleated but cannot be moved out from the near tip region in great enough speed and numbers to relax the cleavage-producing stress.

5. A final but extremely important reservation is that for lattices or interfaces that are judged to be intrinsically cleavable in the sense of the competition so far discussed, the issue of whether cleavage actually occurs may be dominated by plastic flow associated with pre-existing, rather than tip nucleated, dislocations. For example, in the absence of readily cracked brittle phases, crack nucleation will require the generation of large internal stresses by dislocation pile-ups or other entanglements, or sometimes by local shape transformations, and hence will have a strong dependence on slip character and parameters such as dislocation density, temperature and strain rate which affect hardness. For similar reasons, once a micro-crack is nucleated, it will be able affect hardness. For similar reasons, once a micro-crack is nucleated, it will be able to propagate by a bond-decohesion mode only if dislocations from external sources neither move so as to intersect and blunt-out the crack tip nor act to reduce general stress levels near an unblunted tip to values smaller than what must be maintained to continue bond-decohesion. That is, to the extent that a local energy release G to an atomistically sharp crack can be defined[2,17] in presence of extensive plasticity from nearby pre-existing sources, the plastic flow must not so relax stress that a local G equal to G_{cleave} cannot be maintained.

With these various reservations in mind, let us now examine the calculation of G_{disl} for an atomistically sharp stationary crack which lies along the interface between two homogeneous but arbitrarily anisotropic linear elastic phases (e.g., two grains, or two parts of the same crystal if the "interface" is merely a lattice plane) and has tip lying in a slip plane in one of the phases. The slip plane is shown in the upper phase, at angle ϕ with the interface and with unit normal n in figure 5; b is the Burgers vector for a dislocation on that plane and it makes an angle ψ with the normal to the crack tip.

The simplest approach is to consider a straight line dislocation lying parallel to the crack tip at distance r from it (fig. 5a). In this case the remarkably general result is that the radial component of self force (i.e., image-like configurational force) attracting the dislocation to the tip in an unloaded solid is[18]

$$f_r^{self} = -\alpha_{\beta\gamma} b_\beta b_\gamma / r .$$

(4.1)

where $\alpha_{\beta\gamma} b_\beta b_\gamma$ is the pre-logarithmic factor in the expression

$$\Phi = \alpha_{\beta\gamma} b_\beta b_\gamma \ln(r_\infty/r_o)$$

(4.2)

(r_∞ and r_0 are outer and inner cut-off radii) for the strain energy per unit length of an identically oriented straight dislocation line in an uncracked homogeneous solid of elastic properties identical to those for the phase in which the dislocation resides. The only non-zero components of the matrix α when that material is isotropic are

$$(4.3)$$

When the solid is loaded there is also the radial force $f_r^{load} = n_\beta \sigma_{\beta\gamma} b_\gamma$ where here $\sigma_{\alpha\beta}$ is the elastic crack tip stress field of (2.1). The result can be put in the form

$$f_r^{load} = K_M S_M \, b/\sqrt{r} \qquad (4.4)$$

where here b is the magnitude of the Burgers vector and the S's are geometric factors such that each term in the sum $K_M S_M/\sqrt{r}$ is the resolved shear stress on the slip plane due to a particular mode. Also, for the general interface crack, we have here neglected the oscillatory singularity[3,4]. When the adjoining solids in fig. 5 can be regarded as elastically identical and isotropic, the S factor for mode I loading is

$$S_I = \sqrt{(1/8\pi)} \, \cos(\phi/2) \, \sin\phi \, \cos\psi \qquad (4.5)$$

and attains its greatest possible value of 0.15 when $\phi = 70.5^\circ$ and $\psi = 0$.

We can bound from below the critical value of $K_M S_M$ for nucleation in the following way: Choose for r any value which is large enough for continuum dislocation theory, incorporating a line dislocation model, to apply near the tip. Then by identifying the value of $K_M S_M$ such that $f_r^{load} + f_r^{self} = 0$ at that r, and by noting that the positive term decays more slowly with r than does the negative ($1/\sqrt{r}$ versus $-1/r$), we see that $(K_M S_M)_{disl}$ at nucleation must exceed the value identified as corresponding to $f_r = 0$. This is because f_r will be negative for arbitrarily smaller r and the dislocation will be drawn back to the tip rather than emitted from it. Thus

$$(K_M S_M)_{disl} \geq \alpha_{\beta\gamma} b_\beta b_\gamma /b\sqrt{r} \qquad (4.6)$$

or, for the isotropic solid under mode I loading,

$$(K_I)_{disl} \geq \frac{\mu b}{(1-\nu)\,\sqrt{2\pi r}} \frac{(1-\nu\sin^2\psi)}{\cos(\phi/2)\sin\phi\cos\psi} \qquad (4.7)$$

Clearly, the smaller that r is chosen the less likely is the underlying hypothesis that continuum dislocation concepts are valid at that r, and hence the less certainly valid is the lower bound. A valid r could not be smaller than a core cut-off size r_0 and the right sides of (4.6) or (4.7) with r replaced by r_0 is sometimes taken as an approximation of uncertain accuracy to the nucleation condition.

The set-up for a more elaborate model of nucleation is shown in fig. 5b, where an emergent dislocation loop is shown as a semi-circle of radius r. This geometry was studied in the original Rice-Thomson work and has recently been brought to a high level of development[4,19], taking account of elastic anisotropy for crystals and of the possibility of nucleation by emitting one partial dislocation at a time for fcc solids. Recent developments[4,19c] also include a more exact account of the three dimensional elastic interaction between an emergent dislocation loop and a crack. Thus, for an emergent full dislocation of Burgers vector b in a loaded solid one has energy

$$U = \pi r [E_{core} + \alpha b^2 \ln(8mr/e^2 b)] + 2r \ E_{ledge} - 3.5 \ b \ r^{3/2} \ K_M S_M \qquad (4.8)$$

Here the last term contains the effect of the applied loading, the term before it gives the energy (E_{ledge} per unit length) of the ledge created at the tip by the emergent dislocation, and the first two terms, apart from the quantity m, represent one half of the total energy $2\pi r \ [E_{core} + \alpha b^2 \ \ln(8r/e^2 b)]$ of a full circular loop of radius r in an uncracked solid. In the latter expression αb^2 is the average around the circular loop of $\alpha_{\beta\gamma} b_\beta b_\gamma$ for straight dislocations having the direction of the local loop tangent [$\alpha = (2-\nu)\mu/8\pi(1-\nu)$ $\approx \mu/10$ for an isotropic solid] and E_{core} is the average core energy per unit length, alternatively expressible in terms of a cut-off radius as $E_{core} = \alpha b^2 \ln(b/r_o)$. The term m which appears in (4.8), but not in the expression fo: a full circular loop in an uncracked solid, is necessary to describe the exact elastic interaction between a dislocation loop and a crack; its value depends on ϕ and ψ and is estimated[4,19c] to lie between approximately 1.2 and 1.9 for representative cases of partial and full dislocation nucleation in fcc crystals and along symmetric tilt interfaces under mode I loading. Thus m corrects the approximation in previous work[14,19a] of writing the energy of an emergent loop at a crack tip as half the energy of a full loop (the emergent loop and its geometrical image relative to the crack tip) in an uncracked solid. The loop nucleation condition given by simultaneous solution of dU/dr = 0 and $d^2 U/dr^2 = 0$ is

$$(K_M S_M)_{disl} = 0.76 \ \alpha\sqrt{mb} \ \exp \ [(\pi E_{core} + 2E_{ledge})/2\pi\alpha b^2] \qquad (4.9)$$

$$= 0.76 \ \alpha b\sqrt{m/r_o} \ \exp \ (E_{ledge}/\pi\alpha b^2)$$

and the corresponding loop radius is

$$r = (2.51 \ r_o/m) \ \exp(-2E_{ledge}/\pi\alpha b^2) \qquad (4.10)$$

(These expressions appear with m replaced by unity in ref. 19a. Thus the predicted values of G_{disl} in fig. 2 of that work, for nucleation from crack tips on [110] symmetric tilt interfaces in Cu, should be increased by the factor m and the inferred values of r_o in Table 1 of that work, to rationalize experimental values of K_{III} for screw dislocation nucleation in various crystals, should be reinterpreted as values of r_o/m.)

It is interesting to note that when the ledge term is small compared to $\pi\alpha b^2$, as is generally expected, the estimate of $(K_M S_M)_{disl}$ just made is not very different from the result of evaluating the right side of (4.6) with $r=r_o$. It should be remarked, however, that both the result just given and that discussed next for partials may overestimate $(K_M S_M)_{disl}$ by an as yet unquantified amount due to the assumption of a definite (semicircular) loop shape.

The case of nucleating dislocations one partial at a time in fcc solids is addressed in a similar manner[4,19b]. Then the energy U of the first partial (whichever turns out to be easier to nucleate) is taken as above, based on the S's, b, E_{core} and E_{ledge} for that partial, but with the energy $\pi r^2 \gamma_{sf}/2$ of the stacking fault added to (4.8). Once the first partial is emitted and has enlarged to a stable (because of the γ_{sf} term) loop size, the energy of a new emergent loop, consisting of the second partial, is estimated with a similar set of terms as in U above, except that now a term $-\pi r^2 \gamma_{sf}/2$ for elimination of the stacking fault and an interaction energy between the first and second loops also appear[4,19b]. In general, with use of consistent core energy estimates[20] for partial and associated full dislocations it is found[4] that for tensile mode I loading the nucleation loads for the first and second partials bracket that estimated from the full dislocation analysis,

$$(K_I)_{disl}^{1st\ partial} < (K_I)_{disl}^{full} < (K_I)_{disl}^{2nd\ partial} . \qquad (4.11)$$

This suggests that even though the calculation predicts a lower load to nucleate a full dislocation than a pair of partials, the prediction for the pair of partials is to be regarded as the more plausible since, once the first is nucleated, the possibility of having the full dislocation emerge all at once has disappeared.

We may note the following: First, there is no unique value of G_{disl} for general mixed mode loading, since the emission criterion is phrased in terms of a critical value for the combination $K_M S_M$ (or pair of such combinations based on the sets of S's for a pair of partials); G is not uniquely determined by such combinations. Second, the emission criterion, like the cleavage criterion, is susceptible to alteration by segregated solutes on an interface and by environmental species. The effects enter through the values chosen for E_{ledge} and E_{core} (or r_o), although it is interesting that they do not affect the bound in (4.6). For example, it must be presumed that adsorption onto the crack tip ledge left by the emerging dislocation will lower E_{ledge}. In addition, as emphasized by Seah and Hondros[14f], E_{ledge} for an interface with solute segregation may be lower than that for the pure interface due to the different character of bonds disrupted in forming the step. Also, if a species available environmentally or in solid solution tends to segregate along the dislocation core, and has enough mobility to keep up with the emerging dislocation, then E_{core} is lowered. Those effects all serve to ease the dislocation nucleation threshold, although they must be expected to be much less significant than solute and environment effects on reducing the cleavage threshold by affecting $2\gamma_{int}$; for example, the lower bound (4.6) for dislocation nucleation is independent of the solute and environment effects.

Finally, the impression that crack tips always respond in either a distinctly brittle or ductile manner is dispelled by the work of Vehoff, Neumann and coworkers[21]. They show for Fe-3%Si crystals that the relative amounts of what appears to be simultaneously occuring cleavage decohesion and ductile crack opening (whether due primarily to tip nucleated dislocations or to internal sources is not clear) can be varied over a wide range. They quantify the contributions by the sharpness of the crack tip opening angle in cyclic loading, and show that the angle can be varied significantly by variation of temperature, loading rate, or environmental H.

5. MULTI-DISLOCATION AND CRYSTAL PLASTICITY EFFECTS AT CRACK TIPS

The discussions on brittle cracking have been qualified by reference to the control exerted by plastic response over whether such cracking does in fact occur. Unfortunately there is no comprehensive analysis as yet of a crack initiating and propagating by bond decohesion in presence of extensive nearby plastic flow, although important insights have been gained from several attempts at analysis of such situations[2,16,17]. Here we discuss briefly two recent types of work, based on continuum plasticity theory, which may be useful

38

to increased understanding.

The first involves elastic-plastic analysis of crack growth processes at the single crystal scale, based on constitutive relations that build-in the notions of a limited set of allowable crystal slip systems and a critical resolved shear stress to activate each. Unfortunately, the studies as so far published[22] in this area are limited to ideally plastic crystals deforming by anti-plane strain (i.e., mode III). The extent to which similar concepts will apply for tensile cracks is not yet fully clear, and neither is it known to what extent moderate amounts of hardening, whether due to deformation or to viscoplastic effects, will modify the ideally plastic results. Nevertheless, it may be useful to review the results for anti-plane shear. It is found[22] that for a stationary crack subjected to increasing load, the crack tip plastic zone collapses into discrete planes of discontinuity emanating from the crack tip, with elastic response elsewhere. The displacement is discontinuous across those planes (i.e., they correspond to localized plastic shear zones) and so is the stress state, which is now bounded at the crack tip. Also, near-tip plastic flow during quasistatic crack growth is found to be confined to similarly oriented planes of discontinuity emanating from the moving tip. In that case the near-tip stress distribution readjusts so that stresses are fully continuous across those planes, but the particle velocity, or displacement rate, is discontinuous there, and hence a finite plastic strain accumulates as a material point is traversed by one of the discontinuity planes as it moves through the material along with the tip.

For appropriately oriented cracks, such as a crack on the (010) plane growing in the [101] direction in a fcc crystal with standard <110> {111} type slip systems, the predicted orientations of the discontinuity planes coincide with the orientations of {111} slip planes which are active in the plastic flow. What is remarkable, however, is that the predicted orientations of the discontinuity planes do not always coincide with those of slip planes, even when there are slip planes which intersect the crack tip. As an example, for the same (010) crack plane and [101] growth direction in a bcc crystal with <111>{110},{211} or {321} type slip, the predicted planes of discontinuity emanating from the crack tip are found to lie perpendicular to the sip planes which are active in producing the plastic flow (which are $(10\bar{1})$ planes in that case). Also, in other cases such as a crack on (010) growing in the [100] direction in an fcc crystal, the predicted orientation of the discontinuity planes bears no such simple relation to what is now a pair of slip systems ($[1\bar{1}0]$ slip on $(11\bar{1})$ and (111) planes) which are simultaneously active in producing the predicted flow.

Some items of possible significance relating to the different modes of plastic relaxation just discussed involve the necessity or not of there being activation of internal sources of dislocations to produce the plastic flow pattern predicted for relaxing the crack, and also the effect of dislocations emanating from internal sources on displacing material points at the crack tip (or, to pursue an analogy to the tensile case, on blunting-out the crack tip). For example, in the first fcc case, the planar plastic zones coincide with slip planes. The dislocations necessary to accomplish the predicted plastic deformation could be generated from internal sources along those slip planes. If such sources are not abundant , the necessary dislocations could instead be nucleated from the crack tip or from highly stressed sources very near to it, and could glide out from the tip along slip planes (much like in an anti-plane shear version of fig. 5b) to accomplish the predicted plastic relaxation pattern. By contrast, for the bcc case and the latter fcc case, the predicted plastic flow pattern can be accomplished only by the activation of internal sources; in those cases the crack tip does not lie in slip planes of the active systems and hence the necessary dislocations cannot be nucleated at the tip and be swept out from it along slip planes. Thus, if the temperature or loading rate is such that those internal sources have no time to act, the crack tip stress concentration cannot be relaxed as predicted at the tip and, instead, some pattern of plastic flow will result involving a less relaxed stress state. For similar reasons, dislocation loops nucleated from internal sources can, in the first case, expand towards the tip and accomplish the anti-plane shear analog of blunting-out the tip, whereas such is not feasible in the other cases discussed since expanding loops on the active slip planes are not then carried in towards the tip.

Thus, if making the analogy to mode I cracks is not too extreme, we should expect that some crack and crystal geometries will, like the first mode III fcc case discussed, be very effective at relieving near tip stresses by appropriate dislocation nucleation, and at tip blunting by impingement of internally generated dislocations into the tip region. Other crack and crystal geometries will be less successful in relaxing the crack, especially when rate and temperature make difficult the operation of the pervasive internal sources required, and will not involve dislocation motion along planes which impinge on and result in blunting of the tip. Presumably, the latter situations would be judged more conducive to brittle cracking.

In fact, preliminary and as yet unpublished work by the author on the near-tip asymptotic analysis of stationary and growing cracks in ideally plastic crystals in mode I plane strain, using methods based on previous work[23], lends some support to expectations based on an analogy to the mode III cases. As examples, a crack growing on (010) in the [101] direction in a fcc crystal is predicted to have localized shearing along the {111} slip planes which intersect along the crack tip. However, the same crack tip location when at the head of a crack on a (101) plane, growing in the [010] direction, in fcc is predicted to have localized shearing along planes emanating from the crack tip that run perpendicular to the families of {111} slip planes which are active in producing the flow. That is, the predicted flow in the second case must rely on pervasive sources and cannot, like in the first case, be supplied by dislocations that are swept out form the tip along slip planes. Also, dislocation loops nucleated from internal sources will in the first case, but not the second, tend to impinge upon and blunt the tip. The bcc case of a crack on (010), growing along the [101] direction, is similar to the second fcc case just discussed in that localized shear is predicted along planes emanating from the crack tip that run perpendicular to, rather than along, the {211} planes which are then active in producing the flow. By contrast, a second bcc case, with the crack on (101) and growing along [010], is analogous to the first fcc case discussed above, in that now the predicted localized shear occurs on planes coincident with the active {211} slip planes which intersect at the tip.

It is of interest but is possibly only coincidental that of the mode I cases just discussed, the first bcc case and second fcc case, identified as conducive to brittle cracking according to the present considerations, coincide with what appear to be observed cleavage systems. For example (010) is the normal cleavage plane in bcc Fe and, in the case of slow internal cracking of H charged Fe-3%Si, for which cracks are reported[24] to grow along [101], dislocation etching near crack tips suggests substantial activity on families of {211} planes which do not feed dislocations in towards the crack tip. Also, the facets identified as microscopic cleavage bursts[16d] in the transcrystalline scc of fcc Cu have the (101) orientations, although the preferred growth direction, if any, along them is not known.

As a final topic we may note that once a rapidly propagating brittle crack is started, perhaps through failure of a brittle phase, its possibility for subsequent continuation through a normally ductile cyrstalline lattice may be strongly dependent on the high strain rate viscoplastic properties of that lattice. This situation has been analyzed recently by Freund and Hutchinson[2] ir an elaboration of concepts introduced previously by Hart[17]. Briefly, if a material is described by elastic-viscoplastic constitutive relations, for which the plastic strain rate ε^p increases less rapidly with stress σ than as proportionality to σ^3 at high stress, then the continuum solution retains a stress field singularity of the type (2.1) at the crack tip and hence results in a finite energy flow G_{tip} to the tip. The local G_{tip} should, presumably, be equated to $2\gamma_{int}$ for crack growth to occur, at least in an intrinsically cleavable material.

By considering steady state mode I plane strain crack growth with only a small scale plastic yield zone at the tip, it may be shown that[2]

$$G_{tip} = G_{far} - \dot{a}^{-1} \int_A \sigma_{\alpha\beta} \dot{\varepsilon}_{\alpha\beta}{}^p \, dA - \int_{-h}^{+h} \phi^e \, dx_2 \qquad (5.1)$$

where coordinates are attached as in fig. 1. Here $\dot{a}$ is crack speed, A is the plastically active area of the $x_1 x_2$ plane near the moving tip, in which stresses $\sigma_{\alpha\beta}$ and plastic strain rates $\dot{\varepsilon}_{\alpha\beta}{}^p$ occur, and ϕ^e is the elastic strain energy density in the residually stressed plastic wakes of height h bordering the crack surfaces. Also, G_{far} is the far-field or macroscopic value of G that characterizes the total energy flow to the crack tip region; G_{far} is what would be inferred from a fracture test.

A complete solution of the problem would represent the stress field by (2.1) based on K^{far}, from which there is to be subtracted the residual-like stress field associated with the (initially unknown) distribution of plastic strains $\varepsilon_{\alpha\beta}{}^p$. This total stress field generates, by the asumed viscoplastic law, strain rates $\varepsilon_{\alpha\beta}{}^p$ which must be made consistent with the distribution of $\varepsilon_{\alpha\beta}{}^p$ on which the stress field is based. A simpler approximation, which may be improved upon[25], is to note that the total stress field reduces, near the tip, to (2.1) based on K^{tip}. This approximation to stress may be used to calculate $\dot{\varepsilon}_{\alpha\beta}{}^p$ everywhere in the plastically active region, and hence to convert (5.1) to an equation relating G_{tip} to G_{far} and to crack speed. Details are given in the references cited[2,25] where, in fact, elastodynamic generalizations of (2.1) and (2.5) for a propagating crack are used, and where the relation between $\dot{\varepsilon}^p$ and σ is assumed to become linear at high strain rates, consistently with a phonon drag mechanism in that range, but to involve obstacle or lattice-resistance controlled thermally activated flow at lower strain rates, with strongly non-linear stress dependence. These studies suggest the possibility of crack propagation with G_{far} greatly larger than G_{tip}, with the ratio of the two depending on crack speed as well as temperature and parameters of the assumed viscoplastic constitutive relation. It is not yet clear that this approach can explain the phenomenology of cleavage fracturing in Fe[16b], where G_{far} for crystals is inferred to be 7 to 9 times $2\gamma_{int}$ and to be of order 500 times greater for polycrystals. Nevertheless, it does seem to incorporate the ingredients necessary to ascertain whether plastic flow will allow maintenance of sufficient near-tip stress to continue to meet conditions for crack growth by bond-decohesion in cleavable materials.

6. ACKNOWLEDGEMENT

This study was supported by the NSF Materials Research Laboratory at Harvard University.

7. REFERENCES

1. M.F. Kanninen and C.H. Popelar, Advanced Fracture Mechanics (Oxford, 1985).
2. L.B. Freund and J.W. Hutchinson, J. Mech. Phys. Solids, 33, 169 (1985); L.B. Freund, J.W. Hutchinson and P.S. Lam, Eng. Fracture Mech, 23, 119 (1986).
3. J.W. Willis, J. Mech. Phys. Solids, 19, 353 (1971); D.L. Clements, Int. J. Eng. Sci., 9, 256 (1971).
4. P.M. Anderson, Ductile and Brittle Crack Tip Response (Ph.D. Dissertation, Harvard, 1986).
5. A.N. Stroh, Phil Mag., 3, 625 (1958); D.M. Barnett and R.J. Asaro, J. Mech. Phys. Solids, 20, 353 (1972).
6. R. Thomson, Atomistics of Fracture, 167 (Plenum, ed. Latanision and Pickens, 1983) and Solid State Physics, 39, 1 (1986); B.R. Lawn and T.R. Wilshaw, Fracture of Brittle Solids (Cambridge, 1975).
7. J.R. Rice, J. Mech. Phys. Solids, 26, 61 (1978) and Proc. 8th U.S. Nat. Congr. Appl. Mech., 191 (Western Periodicals, ed. Kelly, 1979).
8. G.P. Cherepanov, Appl. Math. Mech. (transl. PMM), 31, 476 (1967); J.R. Rice, Fracture: An Advanced Treatise, 2, 191 (Academic, ed. Liebowitz, 1968) and J. Appl. Mech., 35, 379 (1968); J.D. Eshelby, Inelastic Behavior of Solids, 77 (McGraw Hill, ed. Kanninen et al, 1970).
9. J.R. Rice, Effect of Hydrogen on Behavior of Materials, 455 (TMS-AIME, ed. Thompson and Bernstein, 1976); R.J. Asaro, Phil. Trans. Roy. Soc. London, A295, 150 (1980); J.P. Hirth and J.R. Rice, Met. Trans., 11A, 1502 (1980).
10. G. Tauber and H.J. Grabke, Ber. Bunsenges Phys. Chem., 82, 298 (1978); H. Erhart and H.J. Grabke, Metal Sci., 15, 401 (1981); H. Hänsel, L. Stratmann, H. Keller and H.J. Grabke, Acta Met., 33, 659 (1985); H.J. Grabke, presentation at 1986 NATO Conf. on Phys. and Chem. of Fracture.
11. M.E. Eberhart, K.H. Johnson, R.P. Messner and C.L. Briant, Atomistics of Fracture, 255 (Plenum, ed. Latanision and Pickens, 1983).
12. K. Tatsumi, N. Okumura and S. Funaki, Grain Boundary Structure and Related Phenomena, Supplement to Trans. Japan Inst. Metals, 427 (1986).
13. C.L. White, R.A. Padgett, C.T. Liu and S.M. Yalisov, Scripta Met., 18, 1417 (1984); C.T. Liu, C.L. White and J.A. Horton, Acta Met., 33, 213 (1985); C.L. White, J. Vac. Sci. Technol., A4, 1633 (1986).
14. (a) A. Kelly, W. Tyson and A. Cottrell, Phil. Mag., 15, 567 (1967); (b) J.R. Rice and R. Thomson, Phil. Mag. 29, 73 (1974); (c) D.D. Mason, Phil. Mag., 39, 455 (1979); (d) T. Yokobori, A.T. Yokobori Jr. and A. Kamei, Int. J. Fracture, 11, 781 (1975) and 12, 519 (1976); (e) S.M. Ohr, Mat. Sci. Eng., 72, 1 (1985); (f) M.P. Seah and D. Hondros, Atomistics of Fracture, 855 (Plenum, ed. Latanision and Pickens, 1983); (g) J.E. Sinclair and M.W. Finnis, ibid, 1047; (h) I.H. Lin and R. Thomson, Acta Met., 34, 187 (1986).
15. M.S. Daw and M.I. Baskes, Phys. Rev. Lett., 50, 1285 (1983) and Phys. Rev. B, 29, 6443 (1984); M.I. Baskes, Mat. Res. Soc. Bull, 9, no. 4, 14 (1986).
16. (a) M.L. Jokl, V. Vitek and C.J. McMahon Jr., Acta Met., 28, 1479 (1980); (b) J.F. Knott, Atomistics of Fracture, 209 (Plenum, ed. Latanision and Pickens, 1983); (c) P. Haasen, ibid, 707; (d) K. Sieradzki and R.C. Newman, Phil Mag., A51, 95 (1985), and R.C. Newman and K. Sieradzki, Scripta Met., 17, 621 (1983); (e) I.H. Lin and R.M. Thomson, J. Mat. Res., 1, 73 (1986).
17. E.W. Hart, Int. J. Solids Structures, 16, 807 (1980).

18. J.R. Rice, Fundamentals of Deformation and Fracture, 33 (Cambridge, ed. Bilby et al, 1985).

19. (a) P.M. Anderson and J.R. Rice, Scripta Met., 20, 1467 (1986); (b) J.S. Wang, P.M. Anderson and J.R. Rice, in press, Proc. 5th Int. Conf. on Mechanical Behavior of Materials, Beijing, 1987 (Pergamon); (c) P.M. Anderson and J.R. Rice, manuscript, 1986.

20. F. Prinz, H.O.K. Kirchner and G. Schoek, Phil. Mag., A38, 321 (1978); F. Prinz, A. Korner and H.O.K. Kirchner, Phil. Mag. A47, 441 (1983).

21. H. Vehoff and P. Neumann, Acta Met., 27, 915 (1979) and 28, 265 (1980); H. Vehoff, W. Rothe and P. Neumann, Proc. 5th Int. Cong. Fracture, 1, 265 (1981); H. Vehoff and W.Rothe, Acta Met., 11, 1781 (1983).

22. J.R. Rice and R. Nikolic, J. Mech. Phys. Solids, 33, 595 (1985).

23. J.R. Rice, Mechanics of Solids, 539 (Pergamon, ed. Hopkins and Sewell, 1982) and J. Mech. Phys. Solids, 21, 63 (1973); W.J. Drugan, J.R. Rice and T.L. Sham, J. Mech. Phys. Solids, 30, 447 (1982) and 31, 191 (1983).

24. A.S. Tetelman and W.D. Robertson, Acta Met., 11, 415 (1963).

25. P.A. Mataga, High Strain Rate Crack Growth (Ph.D. Dissertation, Harvard, 1986).

DISCUSSION

Comment by M. O. Speidel:

You have indicated that an "anomalous segregator" could toughen a material, and that boron was such an anomalous absorber in an intermetallic phase.
  Questions:
      1) Are there other such anomalous segregators?
      2) What makes an element a anomalous segregator?

Reply:

Thermodynamics does not preclude anomalous segregation, i.e., that a segregant at a given equilibrating potential locates more abundantly along a coherent grain boundary than along the pair of surfaces created after separation, or that for a fixed amount of segregant, the potential on the separated surfaces exceeds that on the grain boundary. However, it did not seem to be taken seriously that such segregation might actually occur until the recent report of Liu, White and co-workers at Oak Ridge that B seems to segregate anomalously in Ni_3Al. I do not know of another case, but presumably some workers are now looking for the effect. The phenomenon may be widespread but undetected because such segregations would make grain boundaries more resistant to separation and hence less susceptible to discovery by the Auger technique. I do not understand, in electronic terms, what makes a segregator anomalous.

While only anomalous segregators can increase the theoretical work of separation, regular segregators may in some circumstances also act to reduce intergranular embrittlement. For example, C seems to be a regular segregator in Fe, but one which (according to H. Grabke's data on segregation enthalpies) reduces the work of segregation far less than does an equal segregation of P or Sn. Further, C binds more strongly to the grain boundary than do either of the latter. Thus, when several segregators are present, increase of C may reduce intergranular brittleness not because C itself increases the theoretical work of separation, but rather because due to its strong binding energy to the grain boundary, it displaces other elements such as P or Sn which would cause a far more deleterious reduction in the theoretical work of separation. Of course, segregation might also affect the nucleation of dislocations from intergranular crack tips through as yet not well quantified mechanisms; and this as well as consideration of effects on the work of separation may be important to understanding segregant effects on grain boundary embrittlement.

Comment by R. H. Jones:

When considering the effects of the environment on fracture, is it possible to quantify the effect on E_{core}, E_{ledge}, and γ_{int}?

Reply:

Yes. We have two routes. The one which, given the progress revealed at this conference, will perhaps be the method of choice in a few years (if not already) is the use of direct electronic/atomic calculations. These can be applied to determine $2\gamma_{int}$, for example, in terms of appropriately defined differences in the thermodynamic functions for a coherent grain interface with, say, environmentally derived segregants along it and for the pair of free surfaces created by separation. Similarly, calculations revealing the core energy of a dislocation and the ledge energy of a step at a crack tip should be possible at least within the framework of simpler techniques like the Daw-Baskes embedded atom method.

The other route is experimental. Data on solute segregation isotherms for coherent grain interfaces and for free surfaces allow us to calculate segregation effects on $2\gamma_{int}$, at least to within the approximation that the surface resulting after fracture is not very different energetically from that obtained in equilibrium segregation studies. Otherwise we obtain only a lower bound on the work of separation. Similarly, if we know the segregation isotherm for a dislocation core, we can estimate segregation effects on the core energy. The ledge energy is much harder to estimate although in some cases we can use terrace-ledge-kink explanations of the orientation variation of surface thermodynamic properties to infer an energy value for ledges on a free surface. These, however, are not the same as ledges at a crack tip. One can also estimate the ledge energy by a simple model like that of Seah and Hondros at the previous conference, based on bond energies and the altered coordination of atoms near the ledge.

All this discussion has been phrased as if the environmentally induced local chemical compositions were known, although this will often require detailed modelling of the kinetics of transport and segregation near the crack tip, in a way which cannot be decoupled from the mechanics of stressing and deformation.

MATERIALS SCIENCE OF FRACTURE PROCESSES

J.F. Knott
Department of Metallurgy and Materials Science, University of Cambridge

INTRODUCTION

Experimental values of the fracture strengths of materials are commonly some two orders of magnitude smaller than the theoretical limit of approximately one-tenth of Young's modulus (0.1E). The low experimental values are attributed to the presence in materials of cracks or crack-like defects which are able to provide high stress-concentrations. The cracks may be present ab initio as a consequence of manufacturing processes; they may be nucleated by plastic deformation; or they may be introduced deliberately, e.g. by fatigue, to provide standard precracks in testpieces.

Materials may be classified as amorphous (glassy) or crystalline, and the paper describes the processes of crack extension in both amorphous and crystalline single-phase materials. Crack extension in materials which are macroscopic composites is not treated, but particular attention is paid to metallic alloys in general, and steels in particular. These possess a crystalline matrix and may contain small particles of ceramics (carbides, oxides), intermetallic compounds, or perhaps glassy phases (in some inclusions).

AMORPHOUS MATERIALS

Griffith[1] carried out his classic fracture experiments on an amorphous material: a soda-lime silicate glass. Silica occurs in a number of crystalline forms (quartz, tridymite, cristobalite), which involve networks of corner-sharing SiO_4 tetrahedra, and the role of Na^+, Ca^{++} or other Group I or Group II cations is to modify this network and to remove long-range order. The glassy structure is equivalent to that of a frozen liquid. Rapid cooling prevents the nucleation and growth of crystalline phases. Common inorganic glasses are based on silicate, alumino-silicate or boro-silicate systems, but glassy phases also occur in non-metallic inclusions in metals and as intergranular binder in sintered polycrystalline ceramics (e.g. SiC, SiN). Metallic glasses and amorphous inorganic semi-conductors are used for magnetic and electronic applications, but the other main type of amorphous material of interest from the mechanical point of view comprises long-chain organic polymers, such as PMMA or polystyrene, in which is often difficult to align the chains in crystalline form. Even under moderate rates of cooling, the chains remain tangled, like spaghetti heaped in a bowl, but also having side-groups oriented at fixed angles to the chain axes.

The lack of long-range order and crystalline structure means that amorphous materials cannot contain slip-planes and so crystal slip does not occur. In inorganic glasses, flow is of a viscous nature at moderately high temperatures, as exhibited during the drawing of a glass rod or tube. At low temperature, glass appears to behave as a perfectly brittle solid, but Marsh[2] has argued that, even here, fracture is preceded by a type of flow. Two main pieces of evidence are cited:

a) for smooth specimens, the measured values of fracture stress are coincident with values of flow stress derived from indentation experiments (using a theory of indentation based on the expansion of a cavity), suggesting that fracture is preceded by flow;

b) for precracked specimens in inert environments, the work of fracture is greater than the energy of the two surfaces created; similarly, the terminal velocities of cracks are less than would be expected for purely elastic behaviour.

These observations suggest that a measure of flow is associated with crack propagation in glass. They also imply that Griffith's experimental observations on the relationship between work of fracture and surface energy were influenced by effects of water-vapour on the conditions leading to the onset of crack instability.

Amorphous polymers exhibit two features of deformation which are absent in metals: these are pronounced non-linear elastic behaviour and crazing. In response to low stresses, the tangled chains straighten to give large, non-linear elastic strains: this implies that the side-groups can move relative to one another, i.e. that there is minimal cross-linking. As the stress is increased, the chains become aligned parallel to the tensile axis. The final stages of this process are relatively easy, as the chains finally untwist, and a "yield drop" may be observed in a displacement-controlled test. This is followed by chain straightening and inter-chain slipping (cold drawing) and then hardening.

Alternatively, small amounts of chain straightening on a microscopic scale may produce crazes. These are regions of material which have been "opened up" (having about 30% of the density of the rest of the polymer) and which are traversed by intertwined chains[3] whose ends are still ravelled in the tangles on either side of the craze. The spaghetti analogy could be pursued by supposing that a fork was twisted in the bowl and that several strands were drawn out in a group. The interlocking of individual strands in a polymer is, of course, enhanced by the side groups. Crazes are often observed to run ahead of cracks and must therefore be regarded as precursors to fracture. The incidence of crazing is strongly affected by environment: PMMA does not usually craze in air, but crazes readily when ethanol is fed into a crack tip under stress. Deformation in amorphous polymers may also occur by shear yielding, a process which is strongly pressure-dependent, presumably because a compression normal to the shear band makes it more difficult for coiled chains to untwist and for side-groups to slide past each other.

CRYSTALLINE SOLIDS

Most structural materials have a crystalline form, with a regular repeat of atoms or ionic groups throughout space. The classification of the symmetry of structure and properties is made in terms of seven crystal systems, defined by rotation axes or combinations of rotation axes and mirror planes. Metals are particularly simple crystals, and generally possess either cubic or hexagonal symmetry. The cubic metals may be either body-centred cubic b.c.c. (Na, K, α-Fe, Cr, Mo, Nb, V, Ta) or what is usually referred to as face-centred cubic, f.c.c. but is better termed cubic close-packed, c.c.p., to emphasise the metallic bonding (Cu, Ag, Au, Al, Ni, Pt, Ir). Common hexagonal metals are Zn, Cd, Mg, α-Ti, α-Zr, and Be. A number of metals exhibit allotropic forms: Co is hexagonal at low temperature and c.c.p at high temperature; Ti and Zr change from hexagonal to b.c.c. at high temperature, and Fe undergoes two transformations α(b.c.c) -> (c.c.p) at $910^{\circ}C$; -> δ (b.c.c) at $1390^{\circ}C$. By suitable alloying it is possible to stabilise the high-temperature form at room temperature, e.g. an austenitic steel such as 316 (18Cr 12Ni 2 Mo bal Fe) is c.c.p. at room temperature. Some common metals not possessing the simple cubic or hexagonal structure are Sn, U and Mn.

Slip deformation in metals is predicted to occur in close-packed directions on close-packed planes, if these exist.[4,5,6] The deformation is accomplished by the movement of dislocations and the Burgers vector, $\underline{b}$ (defined as the closure failure in perfect lattice after performing a circuit normal to the direction of the dislocation line vector, $\underline{l}$) is simply a lattice translation vector (LTV) for a unit, perfect, dislocation. The Burgers vectors of the unit dislocation are a/2<110>, slipping on {1$\bar{1}$1} planes in c.c.p.; a/3<11$\bar{2}$0>, slipping on <0001> planes in hexagonal metals of high axial ratio; and a/2<111>, slipping on {110} planes in b.c.c. The {1$\bar{1}$0} planes in b.c.c. are not close-packed, although they contain close-packed <111> directions. Slip on planes such as {10$\bar{1}$1} may be observed in hexagonal metals of low axial ratio. Experimentally, slip in metals is independent of the hydrostatic component of stress and this is achieved in close-packed cubic or hexagonal structures by splitting the unit dislocation into partial dislocations e.g.

on {1$\bar{1}$1} in c.c.p. a/2<110> -> a/6<21$\bar{1}$> + a/6<121>   1)
on (0001) in h.c.p. a/3<11$\bar{2}$0> -> a/3<10$\bar{1}$0> + a/3<01$\bar{1}$0>   2)

A number of dislocations have been suggested for b.c.c. metals: one of particular interest being the asymmetrical splitting of a unit a/2[111] screw dislocation into three a/6[111] partials on {1$\bar{1}$2} planes. Hexagonal and b.c.c. metals also deform fairly easily by mechanical twinning, promoted in b.c.c. metals by low temperatures and high strain rates. The twinning shears are <$\bar{1}$011> on {10$\bar{1}$2} planes in h.c.p. metals and a/6<111> on {1$\bar{1}$2} planes in b.c.c. metals. Deformation twins in c.c.p. metals are observed only under explosive rates of loading, but can be generated by the movement of a/6<112> partials on successive {111} planes. The conditions leading to deformation twinning are not well formulated: one possibility is that when dislocations are squeezed together (e.g. in a pile-up) under the action of shear stress and there is insufficient thermal activation for the cross-slip of unit dislocations, cross-slip of a partial can occur, in a manner that generates a twin.

The distinction between <u>symmetry</u> of properties, as defined by the lattice, and the nature of these properties, as a consequence of the type of bonding, is well drawn if the behaviour of ionic or covalent cubic crystals is contrasted with that of c.c.p. metals. Consider the NaCl structure (which is shared <u>inter alia</u> by LiF, MgO and Ti(C,N)), the diamond structure (Si, Ge) and the zinc blende structure (ZnS, GaAs and other III-V semi-conductors). These are all referred to face-centred-cubic lattices, as are Au, Cu and many other metals, but the lattice type simply characterises the symmetry of a number of physical properties. Strictly speaking, the symmetry is characterised by the crystallographic point group and structural units can degrade the lattice symmetry: NaCl and diamond possess full (holosymmetric) cubic symmetry (m3m) but zinc blende lacks a centre of symmetry ($\bar{4}$3m). The properties of these non-metallic structures are, however, critically dependent on bond type. Compounds with the NaCl structure are ionic, usually visualised in terms of one ionic species (e.g. Cl^-) located on f.c.c. lattice points and the other (Na^+) located at positions $00\frac{1}{2}$ with respect to each Cl^- ion. Each ion is then surrounded by six of the opposite species (octahedral coordination). In diamond, the bonding is strongly covalent: if one atom is located at each f.c.c. lattice point, a second atom is located at a position $\frac{1}{4}\frac{1}{4}\frac{1}{4}$ with respect to each lattice point. Each atom is then surrounded by four equi-distant neighbours (tetrahedral coordination, characteristic of carbon bonding.) In zinc blende, one species (e.g. S^{--}) is located at each lattice point and the other (Zn^{++}) occupies the $\frac{1}{4}\frac{1}{4}\frac{1}{4}$ positions.

In each structure, the Burgers vector of the unit dislocation is equal to a LTV, i.e. a/2<110>. The plastic response is, however, strictly dependent on bonding. In close-packed metals (with delocalised metallic bonding), slip occurs on close-packed {$\bar{1}$11} planes and dislocations move easily; i.e. the lattice friction, or Peierls-Nabarro stress, is low. In NaCl, there are no close-packed planes and if slip on {$\bar{1}$11} planes in <110> directions is examined, it is found that like charges are brought closer together. Strong electrostatic repulsions prevent this and the slip systems in the NaCl structure become <110> {$\bar{1}$10}, which are much fewer in number than the <110> {$\bar{1}$11} systems in c.c.p. This has consequences particularly with respect to the deformation of polycrystals. To transmit an arbitrary shape change from grain to grain, it is necessary for a material to possess at least five independent slip systems. These are readily available in c.c.p. and b.c.c. materials, which explains why most metals can be deformed to large plastic strains and still retain continuity. In the NaCl structure, however, there are insufficient systems and cracks can form at grain boundaries as a result of in-compatible deformation (facilitating the mining of the rock-salt for which the Salzburg region is famous!) Similar effects would occur in hexagonal metals were it not for the occurrence of twinning, or slip on non-basal planes. In diamond or zinc blende, the presence of a dislocation leaves a (charged) "dangling bond". Movement of the dislocation through the lattice involves the rupture of successive, directional, strong covalent bonds so that the Peierls stress is extremely high. The difference between the stress required to produce slip and that which produces fracture is then small. It should be noted also that the Peierls stress of b.c.c. transition metals increases markedly with decrease in temperature and that this is associated with a change in behaviour from ductile to brittle.

Cleavage fracture is common to many mineral systems, a number of which are now becoming of interest in device materials applications. It is usually the case that a cleavage facet is a plane of high reticular density, which presumably implies long (and not very strong) bonds normal to the cleavage planes. Good examples of cleavage rhombs, {10$\bar{1}$2} planes, are to be found in quartz or calcite, which are both trigonal. Some structures, such as mica, are layer silicates: sheets of strongly bonded silicate networks, linked by weak alkali metal/oxygen bonds. It then becomes easy to split the sheets apart, e.g. in the preparation of thin mica sheets for electrical insulation purposes. A more macroscopic version of the mica type of structure can be found in slates, shales and mudstones. Stratification during the sedimentary rock-forming processes leads to the formation of rather strong sheets with weak inter-lamellar bonding.

CRACK PROPAGATION IN "IDEAL" SOLIDS

The classical model for crack propagation, following Griffith[1], treats a central, through-thickness crack of length 2a lying normal to an applied stress, σ_{app}, in an infinite body. Writing E' for E, Young's modulus, in plane stress or $E/(1-\upsilon^2)$ in plane strain (where is Poisson's ratio), the elastic strain energy per unit thickness, U, associated with a half-crack, length a, is given, for a linear elastic material, (i.e. one in which stress, σ, is related to strain, ε, through the expression, $\sigma = E'\varepsilon$) by:

$$U = - \sigma_{app}^2 \pi a^2 / 2E' \qquad\qquad 3)$$

The surface energy per unit thickness, S, for the upper and lower crack surfaces, each of area a x unity, is given by:

$$S = +2\gamma a \qquad\qquad 4)$$

where γ is the surface energy per unit area. Griffith wrote the total energy per unit thickness, W, as:

$$W = U + S \qquad\qquad 5)$$

When plotted as a function of crack length, this exhibits a maximum when dW/da = 0, i.e. when (dU/da + dS/da) = 0. We have

$$- \sigma_{app}^2 \pi a / E' + 2\gamma = 0 \qquad\qquad 6)$$

or

$$\sigma_{app(F)} = \sigma_F = (2E'\gamma/\pi a)^{1/2} \qquad\qquad 7)$$

which is Griffith's equation for the fracture stress.

The great advantage of the Griffith approach is that it defocuses attention from the detail of atomic separation processes at the crack tip. It assumes that crack tip stresses (which are infinite for a sharp crack in an elastic solid) are always sufficient to sever bonds and simply considers the energy balance for propagation. The two terms in this balance comprise the decrease in stored elastic energy and the increase in energy associated with two free surfaces. It is of value to consider the differential (or "rate") terms in equation 6). The modulus of the first term dU/da is defined by G, and the condition for fracture can be thought of simply as achieving a critical value of G, G_{crit}, such that:

$$G_{crit} = 2\gamma \qquad \qquad 8)$$

The engineering application of the Griffith equation is effected by use of a virtual work argument, due to Irwin[7], which enables values of G to be calculated using stress analysis. The stress, $\sigma_{(r)}$, at a position r, θ ahead of a crack of length a in a body of width W, is given by:

$$\sigma_{(r)} = K(2\pi r)^{-1/2} f(\theta) \qquad \qquad 9)$$

where K is the stress intensity factor. For standard test-piece geometries, K is calculated from the applied stress and values of compliance function $Y(a/W)$: for the infinite body:

$$K = \sigma_{app}(\pi a)^{1/2} \qquad \qquad 10)$$

A virtual work argument, which calculates the change in strain energy for a crack extension $\delta a (\delta a \rightarrow 0)$ by equating it to the work done by the stress $\sigma_{(r)}$ decreasing to zero as it moves through crack-tip displacements, gives the identity:

$$G = K^2/E' \qquad \qquad 11)$$

A critical value of G is then equivalent to a critical value, K_{crit}, which is referred to as a material's <u>fracture toughness</u>.

For non-linear elastic material, the stress-strain relationship assumes the form:

$$(\sigma/\sigma_o) = (\varepsilon/\varepsilon_o)^N \qquad \qquad 12)$$

where σ_o, ε_o and N are constants. The stress ahead of a crack is given by:

$$\sigma_{(r)} = \sigma_o(EJ/\sigma_o)r^{-N/(1+N)}f(\theta) \qquad \qquad 13)$$

where J is the potential energy release rate, reducing to G for linear elastic material[8,9]. Fracture in non-linear elastic material is then conveniently characterised by a critical value of J.

The shape of the stress-strain curve for a material which exhibits elastic/strongly strain-hardening plastic behaviour can also be described by an expression of the form of equation 12). This is referred to as "power-law hardening" material. It is possible to calculate a value of

"J" in a manner similar to that employed for non-linear elastic material, and critical values, J_{Ic}, are commonly used to <u>characterise crack extension</u> in tough materials, but it must be emphasised that such values do <u>not</u> incorporate the property of a critical energy release rate.

SEPARATION PROCESSES AT A CRACK TIP

Models of brittle fracture in a precracked body usually assume that the crack is atomically sharp and examine local conditions at the tip to decide whether the crack can extend by simple separation of the atoms spanning the tip or whether the crack becomes blunted by the generation of local plastic flow, visco-elastic flow or crazing. Note that the assumption of an atomically sharp initial crack does <u>not</u> apply in the macroscopic sense to fracture toughness tests[10] carried out on metallic alloys. Here, the fatigue precracking procedure permits maximum stress intensity values, K_f, up to 0.7 K_{Ic}. Even for an embrittled high-strength steel, with K_{Ic} = 50 MPam$^{1/2}$ and yield strength 1.5 GPa, a clearly acceptable value of K_f = 30 MPam$^{1/2}$ implies that the initial crack-tip opening-displacement, δ, <u>before</u> the K_{Ic} test is performed, is 3μm, compared with an atomic separation of order 0.3 nm, i.e. the initial macroscopic crack is approx. 10^4 atomic spacings in width. As will be described in the following section, however, local crack-tip plasticity can nucleate sharp microcracks in brittle second-phase particles, ahead of the macrocrack, and it is these which eventually propagate to produce catastrophic failure.

A sharp crack, subjected to an applied tensile stress, develops both high tensile stresses and high shear stresses at its tip, and, for a crystalline solid, it is possible to contemplate a competition between the tensile fracture of the bond at the crack tip and the generation of slip dislocations from the tip[11,12]. In a real crystal, there will also be a network structure of grown-in dislocations, which gives rise to the possibility of the production of plastic flow by the operation of dislocation sources relatively remote from the crack tip. An annealed metal may typically contain 10^{12} dislocation lines m^{-2} (an average spacing of 1μm), whereas silicon of semi-conductor purity may contain only 10^8 lines m^{-2} (spacing 100μm). An additional factor in iron or steel is that the network dislocations may be pinned by carbon or nitrogen atoms.

Before treating the competition between fracture and blunting, it is of value to make two points concerning the Griffith energy balance. The first relates to the assumption that the work of fracture for a brittle solid may be equated simply to the surface energies of two free surfaces. A brittle material such as silicon has rigidly directed, tetrahedral, covalent bonds. When these are fractured, restructuring of the surfaces takes place to try to accommodate unfilled bonding orbitals, so that the configuration of bonds across the fracture plane prior to separation is different from that on either of the two separated surfaces. It is also important to note that the Griffith energy balance, in its critical "defocussing" of attention from events at the crack tip, considers global "before" and "after" states (from strained bonds to two surfaces), without concerning itself with any <u>mechanism</u> which might require an "activation energy" to enable the separation processes to occur[13]. This point will be taken up later.

The second point concerns crack extension in the presence of a surface-active species. Rice[14] distinguishes clearly between two kinds of separation: separation at <u>constant surface excess</u> and separation at <u>constant chemical potential</u>. The former relates to a situation where the species is "frozen in" along the fracture path, prior to separation, which occurs rapidly: the latter is associated with slow separation at a rate determined by the arrival of species at the crack tip, so that the chemical potential is constant. Rice shows that the value of "2γ" for separation at constant chemical potential is always less than the value of "2γ" for separation at constant surface excess, so that, if the species is embrittling, the work of fracture is reduced. An example of this effect is described in the final section of the paper.

The competition between bond fracture and crack blunting was analysed originally by Kelly, Tyson and Cottrell[11]. The ideal fracture strength of the crack-tip bond was taken as approx. $0.1E$ (where E is the value of modulus normal to the fracture plane) and it was assumed that dislocations were created at the crack tip at a stress of order 0.1μ, where μ is the shear modulus. The local crystallography of slip systems was also included. In essence, the analysis is able to distinguish between inherently brittle crystals, such as mica or diamond, in which it is easier to sever the crack-tip bond than to generate dislocations, and inherently ductile crystals, such as c.c.p. metals, in which it is easy to generate dislocations and move them away from the crack tip. Following this argument, glass and glassy polymers would be classed as inherently brittle materials, because they do not contain slip planes. Rice and Thompson[12] later examined the dislocation generation process in detail, taking account of the energy of formation of a dislocation loop in the presence of image forces. This produced essentially the same classification of brittle and ductile materials. An intriguing point is that both analyses conclude that iron is "borderline" with respect to ductile or brittle behaviour and this seems to relate well to the observed transitional behaviour of iron. It should, however, be noted that other b.c.c. metals exhibit transitions from brittle to ductile behaviour as a function of temperature, yet this is not predicted by the analyses. The feature of dislocation generation at stresses of order 0.1μ is that thermal activation even at room temperature has minimal influence on the generation process (kT at 300K is only approx. 0.025eV).

As will be described in the following section, some of the more reliable measurements[15,16] of the stresses required to propagate atomically sharp (micro-)cracks in iron(steel) imply values of the work-of-fracture lying in the range 9-14 Jm^{-2}, contrasting with a value for 2 of $2Jm^{-2}$. Additionally, however, the implied work-of-fracture values are independent of testing temperature (in the range 100-200K, at least). These observations are consistent with Smith's theory of micro-crack propagation[17], which shows that, if cleavage fracture is to be dependent on tensile stress (as demonstrated, for example, by the effects of notches on transition behaviour), it is essential for the work-of-fracture to increase as the crack nucleus increases in size. If the nucleus forms in a brittle particle, for which 2γ is, say, approx. $2Jm^{-2}$, it follows that the work of fracture in the ferrite grain must be greater than $2Jm^{-2}$.

These observations are similar to those made by Marsh[2] for glass, in which the work-of-fracture was also greater than 2γ, and it is tempting to seek a similar explanation, in terms of the mechanism which must be operated to effect the separation of the atoms which span the crack tip. Recall that the Griffith expression treats "before" and "after" states, but does not consider the mechanism of separation. If the "activation energy" associated with an intermediate (mechanistically necessary) configuration is higher than 2γ, the experimental value of the work-of-fracture will be greater than that given by the Griffith equation. Knott[13] has argued that for iron - a "borderline" material - the necessary change in crack tip configuration could be equated to the extremely localised movement of a few crack-tip dislocations. This would account for the extra work observed experimentally and would be compatible with the lack of temperature dependence, because stresses are still of the order of 0.1μ. Similar concepts are invoked in models which involve "lattice-trapped" dislocations. The need to achieve high displacements is demonstrated clearly in polymers which produce crazes ahead of the crack tip. Here, separation is effected only by the drawing-out of long strands of polymeric material in the craze.

All these arguments ignore the fact that real materials are not initially dislocation free, but contain a network of grown-in dis-locations, which could possibly move under the influence of the crack tip stress field. A typical dislocation density in annealed mild steel would be 10^{12} lines m^{-2}, which gives an average spacing of sources of $1\mu m$. In iron or steel the dislocations may be pinned by carbon or nitrogen atoms. It is not easy to decide whether sources around a sharp crack tip will operate before the crack propagates, but Knott[13] has suggested a method for calculating a lower bound for fully-pinned material by analogy with Cottrell's[18] interpretation of the Petch model[19] for the spread of yield from grain to grain at the lower yield stress. Here, the stress intensity (in shear) ahead of the slip-band is equated to that ahead of a mode II crack subjected to a shear stress ($\tau_{app} - \tau_i$), where τ_{app} is the applied shear stress and τ_i is the friction stress in shear. Then, the stress $\tau_{(r)}$ at a distance r ahead of the slip band, is given by:

$$\tau_{(r)} = (\tau_{app} - \tau_i)(d/4r)^{1/2} \tag{14}$$

where d is the grain diameter. It is assumed that yielding is triggered off in the unyielded grain when $\tau_{(r)}$ attains a critical value τ^* at a critical distance, r^*. Rearrangement of equation 14) then gives:

$$\tau_{app(Y)} = \tau_i + (4r^*)^{1/2}\tau^* d^{-1/2} \tag{15}$$

Multiplication by the Taylor factor (2.75 for b.c.c. iron) then gives the Petch equation. We see that the value of k_y in shear, k_y^s, may be equated to $(4r^*)\tau^*$. For fully-pinned dislocations, experimental observations give a value for k_y of $0.71 MNm^{-3/2}$. Although this value probably relates to spread of yield by dislocation creation at the common grain boundary, it also serves as a means of calculating a lower bound to τ^* for unpinning from a source. (Values of k_γ for lightly pinned material are less than 0.71 $MNm^{-3/2}$ (down to 0.05 $MNm^{-3/2}$) and are temperature dependent).

Write $\tau^* = k_\gamma^s/(4r^*)^{1/2} = 0.71/5.5(r^*)^{1/2} = 0.126/(r^*)^{1/2}$ and consider a crack of length 2a subjected to tensile loading. The shear stress at a distance r ahead of the crack is given by:

$$\tau = \sigma_{r\theta} = \{K/(8\pi r)^{1/2}\} \sin\theta \cos(\theta/2) \qquad 16)$$

which has a maximum value, τ_{max}, when $\theta = 70°32'$, of:

$$\tau_{max} = 0.385 \, K/(2\pi r)^{1/2} \qquad 17)$$

At fracture, the applied value of K is $(2E'\gamma)^{1/2}$ for elastic failure, or $(E'G_{crit})^{1/2}$ for quasi-elastic failure. If it is assumed that the nearest source is at a distance r* ahead of the crack tip (equal to r* ahead of a slip band), we may write:

$$\tau_{max} = 0.385(2E'\gamma)^{1/2} / (2\pi r^*)^{1/2} \qquad 18a)$$

or

$$\tau_{max} = 0.385 (E'G_{crit})^{1/2} / (2\pi r^*)^{1/2} \qquad 18b)$$

With = 1 Jm^{-2}, G_{crit} = 10 Jm^{-2}, E = 200 GNm^{-2}, ν = 0.3, the values of τ_{max} become $0.1/(r^*)^{1/2}$, equn 18a) and $0.23/(r^*)^{1/2}$, equn 18b) respectively. By comparison with the values of τ^* above, it is deduced that fully pinned sources will not be operated at the Griffith stress. At experimentally observed values of (microcrack) propagation stress, it appears that some nearby sources could be operated, but it must be appreciated that τ^* is a lower bound to the stress required to operate sources and that crystallographic constraints on slip have not been included. The balance between operation or non-operation of sources at the higher propagation stress could be fine.

The calculation pertains to fully pinned dislocations and the possibility exists that sources could be operated if the pinning were less heavy. The effect of this has been investigated by Bowen[28] by taking a mild steel in which the carbide size was first established by holding for a period of time at 700°C. Slow cooling from 700°C allowed dislocations to become fully pinned, whereas rapid quenching from 700°C effectively suppressed pinning. The local fracture stress over the range -140 to -100°C in the latter case was substantially increased, suggesting that operation of sources was indeed occurring, but implying that such operation does not occur in the usual annealed condition.

THE FRACTURE TOUGHNESS OF STRUCTURAL STEELS

The values of 2γ and work-of-fracture considered in the preceding section lay in the range 2-14 Jm^{-2}. In contrast, even a low value of plane strain fracture toughness, for example, K_{Ic} = 30 $MNm^{-3/2}$ for brittle cleavage in mild steel at 77K, equates to a very high value of critical strain energy release rate: for the assumed K_{Ic} value, G_{crit} = 4kJm^{-2}. This high value should **not** be interpreted as an energy absorbed in each increment of crack growth. It is best regarded as **precursor work**, assoc-

iated with the plastic deformation which occurs ahead of the fatigue crack
before fracture. This must necessarily occur, first to initiate sharp
cleavage microcrack nuclei in brittle second-phase particles; secondly, to
develop sufficient tensile stress by plastic constraint to propagate a
nucleus, probably, immediately after nucleation. It is likely that the
critical propagation event is associated with a microcrack nucleus running
at speed through the brittle particle and that the event itself has a
"go/no-go" character: either the local tensile stress is sufficient to
cause the microcrack to propagate at speed through the ferrite matrix or
it is not, and the microcrack arrests at the particle/matrix interface and
blunts plastically.

The fact that the high values of G_{crit} are associated with precursor
work may be appreciated in terms of the relationship between plastic zone
size and stress intensity. In plane strain, the maximum extent of plasticity, R_{IY}, is given by:

$$R_{IY} = 0.16(K/\sigma_Y)^2 \qquad\qquad 19)$$

where σ_Y is the yield stress. Suppose now that a fracture toughness
test-piece is loaded to a stress intensity of $0.999\ K_{Ic}$. The plastic zone
size is then approx. 99.8% of that at failure. If the specimen is unloaded from $0.999\ K_{Ic}$, there is some reverse plasticity near the crack
tip: on reloading to $0.999\ K_{Ic}$, the initial situation is recovered. All
the work expended in moving dislocations in the plastic zone up to 0.999
K_{Ic} (and, by extension, up to the onset of fracture at K_{Ic}) has occurred
prior to the catastrophic extension of the fatigue precrack. Earlier, it
was demonstrated that the precrack can in no sense be regarded as atomically sharp, being some 10^3-10^4 atomic spacings in width after fatiguing.
Clearly, the fatigue pre-cracking procedure employed in any standard
fracture toughness test tends not to produce atomically sharp cracks: the
interpretation is that these are produced by dislocations in the (precursor) plastic zone, usually in brittle second-phase particles.

In wrought steels, these particles are generally carbides. Detailed
studies have been made for low-carbon steels, in which microcracks have
been nucleated in grain-boundary carbides[21] or spheroidal carbides[22]. Use
has been made particularly of blunt-notched specimens, for which finite
element stress analyses are available to enable values of tensile stress
in the yielded zone below the notch to be calculated. Cleavage fracture
occurs when the maximum tensile stress attains a critical value, σ_F,
which, for spheroidal carbides, has been shown to be inversely
proportional to the square root of carbide radius, C_o[15]. This corresponds
to a local Griffith criterion, with an experimental work-of-fracture of
14 Jm^{-2}, which is a value similar to that observed for grain-boundary[16]
carbides. Recent research on quenched-and-tempered structural steels
again demonstrates a dependence of σ_F on $C_o^{-1/2}$, with a work-of-fracture
of approx. 9 Jm^{-2}.

The σ_F fracture stress criterion can be used in conjunction with an
elastic/plastic analysis of the stress distribution ahead of a sharp crack
to predict the critical size of plastic zone required to produce fracture

and, hence, K_{Ic}. The principle of this prediction may be understood simply by reference to an elastic stress distribution. From equation 9) by rearrangement, we have:

$$K = \sigma_{(r)}(2\pi r)^{1/2} \qquad\qquad 20)$$

If the critical value of K, K_{Ic}, is associated with a critical value of σ, σ_F, it is necessary also to specify a critical distance, r_{crit}, corresponding to the position of the active microcrack nucleus. In this elastic form, the model is equivalent to that used to treat the Petch equation, eqn 14) et seq: σ_F is equivalent to τ^* and r_{crit} is equivalent to r^*. In the original analysis of K_{Ic} (the so-called RKR model[21]) r_{crit} was found experimentally to be close to two grain diameters, but further work has shown that the critical distance results from a statistical averaging of potential microcrack nuclei. From dimensional considerations, the relationship between K_{Ic} and σ_F must always involve a parameter having dimensions of the square-root of length.

Although the basic RKR model and its development has been rather successful in its predictions, it has proved difficult to provide unambiguous evidence for the model, in terms of the propagation of a carbide microcrack nucleus at σ_F, because carbide initiation sites cannot be identified on fracture surfaces. Studies of cleavage fracture in ferritic weld metal[23,24,25,26] have, however, been able to demonstrate that cleavage propagates from hard, oxide or silicate inclusions which are present as deoxidation products. (The technique requires the careful tracing-back of "river-lines" on cleavage facets to the point of origin.) The critical nucleus can then be located both in notched bars and in sharply precracked fracture toughness testpieces. The results show a number of interesting features:

a) Fracture initiates at an inclusion whose size is towards the top of the distribution,
b) the location of the main initiation site in a blunt notched bar is in the plastic zone, at a position corresponding to a high tensile stress (>95% of the maximum value)
c) the actual value of critical tensile stress <u>at the initiation site</u> varies with the inverse square root of inclusion diameter (work-of-fracture approx. 9 Jm^{-2}) and is independent of test temperature
d) the RKR prediction of values of critical distance agrees with experimentally observed values only at relatively high temperatures (ca 170K) where the plastic zone size at failure is large and samples the inclusion distribution in a representative manner. At low temperatures, it appears that the high stresses in the small plastic zone can propagate microcracks from smaller inclusions than those sampled in the notched bar during the determination of σ_F.

In summary, the observations on weld metals broadly confirm the models which have been used to treat cleavage crack propagation from brittle second-phase particles and provide systems which can be used to test detailed predictions. Support is given for a local (microcrack propagation) work-of-fracture of order 10 Jm^{-2} and it is plausible that this is associated primarily with crack-tip atomic processes rather than the large-scale operation of nearby sources. There is need for further work in this area, to treat the dynamic aspects of microcrack nucleation and to explain observed effects of ferrite grain size. It appears that an alternative model system may also be available in titanium-treated ultra-low carbon bainitic steels: here, cleavage initiation sites can be traced to cuboid Ti(C,N) particles[27].

FRACTURE AT "CONSTANT CHEMICAL POTENTIAL"

This final section briefly treats a form of separation which appears to correspond to Rice's definition[14] of "separation at constant chemical potential". The experiments involve the segregation of free sulphur to a crack tip at temperatures around 500°C. Detailed results have been obtained mainly on a quenched-and-tempered 2.25 Cr 1Mo steel, although similar observations have been made on A533B and are presumably generic to structural steels of this type, if heat-treated to enable sulphur to be available and mobile at the test temperature[28,29,30,31].

"Conditioning" of the material is achieved initially by the austenit-ising and cooling procedures. Typically, an austenite grain size is first established by holding for a number of hours at a temperature of, say, 1300°C. If the material is quenched rapidly from this temperature, (condition 1300Q) virtually all the sulphur is held in solid solution. If it is cooled from 1300°C to 900°C and held at 900°C for a period of time before quenching (1300/900Q) the sulphur precipitates at the prior austenite grain boundaries as manganese sulphides and so is effectively removed from solution. The grain size is the same for both conditions and, apart from a slight change in solid solution hardening as a result of MnS formation, the yield stress also remains constant. The two conditions are then subjected to steady loads at temperatures in the range 450-550°C.

Observations have been made on notched bars of the Griffiths and Owen geometry and on fatigue precracked testpieces. In the notched bars, material given the 1300Q treatment begins to crack in an intergranular manner after a relatively short period of time, whereas that in the 1300/900Q condition does not. The initial cracking is observed <u>below</u> the notch root, where the triaxiality is high, and can be attributed to sulphur segregation, by use of Auger spectroscopy, AES, and laser ion mass analysis, LIMA. The precracked tests show that the crack growth rate as a function of stress intensity is strongly influenced by the prior treat-ment: in a vacuum of $\leq 1\mu$bar at $K=50MNm^{-3/2}$ at 500°C, the 1300Q treatment has a growth rate of 0.1 μms^{-1}; the 1300/900Q treatment has a growth rate of 40 μms^{-1}. Even in a hard vacuum (≤ 0.2 nbar) the fracture surface at 500°C is completely intergranular, yet, if a testpiece is unloaded and fractured at 77K, the low temperature fracture is completely transgranular cleavage. Sulphur is detected on the high-temperature intergranular facets by AES, but is absent on transgranular facets. It is concluded that the high temperature intergranular separation is controlled by the diffusion of sulphur from within the grain to a position of high tri-axiality just ahead of the crack tip, perhaps into expanded grain-boundary "deltahedra" sites. It is possible to calculate an activation energy for cracking, by measuring crack growth-rates at the same K value at different temperatures, and the value obtained experimentally is reasonably close to that for the diffusion of sulphur in iron (figures are not available for diffusion of sulphur in a dislocated martensite).

Experimental work in this area is continuing. The system not only seems to provide a good example of separation at constant chemical potential, but can be used to model static/cyclic interactions to predict growth-rates as a function of frequency in high-temperature fatigue tests. Use of the activation energy then enables combined effects of temperature and frequency to be explored. A major unexplained feature at present is that the growth rates are higher in air than in vacuum, even though sulphur is segregating in both cases. In A533B at 290°C, vacuum growth is transgranular, whereas growth in air is intergranular. These observations demonstrate effects of environment over and above the basic segregation phenomenon and therefore provide further examples of the crucial chemical /mechanical interactions which control crack growth rates and which are the raison d'être of this Workshop.

ACKNOWLEDGEMENTS

The author wishes to thank Professor D. Hull for provision of research facilities and Dr P. Bowen, M.B.D. Ellis, Dr C.A. Hippsley, Dr J.J. Lewandowski, Dr D.E. McRobie, Dr J.H. Tweed and M.G. Vassilaros for valuable discussions and for permission to use previously unpublished results.

REFERENCES

1   Griffith, A.A. Phil. Trans. Roy. Soc. 1920, A 221, p.163

2   Marsh, D. Proc. Roy. Soc. 1964, A 282, p.33

3   Kramer, J. and other authors in Adv. in Polymer Sci. 52/53 ed. H.H. Kausch, Springer-Verlag (Berlin) 1983

4   Hirth, J. and Lothe, J. "Theory of Dislocations" McGraw-Hill (New York) 1968

5   Nabarro, F.R.N. "Theory of Crystal Dislocations" Oxford (Clarendon Press) 1967

6   Cottrell, A.H. "Theory of Crystal Dislocations" Blackie and Son Ltd., 1964

7   Irwin, G.R. Trans ASME Jnl.Appl.Mech. 1957, 24, p.361

8   Rice, J.R. "Fracture: An Advanced Treatise" ed. H. Liebowitz, Academic Press 1968 Vol 2 p.191

58

9   Hutchinson, J. Jnl.Mech.Phys.Solids 1968, $\underline{16}$, pp.13 and 337

10   Knott, J.F. "Fundamentals of Fracture Mechanics" Butterworths (London) 1973

11   Kelly, A., Tyson, W.R. and Cottrell, A.H. Phil.Mag. 1967 $\underline{15}$ p.567

12   Rice, J.R. and Thomson, R. Phil.Mag 1974, $\underline{29}$, p.73

13   Knott, J.F. "Atomistics of Fracture" Ed. R.M. Latanision and J.R. Pickens (Plenum Publishing Corporation) p.209

14   Rice, J.R. "Effect of Hydrogen on Behaviour of Materials" ed. A.W. Thompson and I.M. Bernstein Met.Soc. AIME, 1976, p.455

15   Curry, D.A. and Knott, J.F. Met.Soc. 1978, $\underline{12}$, p.511

16   Bowen, P., Druce, S.G. and Knott, J.F. Acta Met, 1976, $\underline{34}$, p.1121

17   Smith, E. Proc. Conf. on Physical Basics of Yield and Fracture, Inst. Physics and Physical Society Oxford 1966, p.36

18   Cottrell, A.H. Symposium on the Relationship between the Structural and Mechanical Properties of Metals, Nat. Phys. Lab HMSO (London) 1963, p.456

19   Petch, N.J. Jnl. Iron and Steel Inst. 1953, $\underline{174}$, p.25

20   Bowen, P. private communication

21   Ritchie, R.O., Knott, J.F. and Rice, J.R. Jnl. Mech. Phys. Solids 1973, $\underline{21}$, p.395

22   Curry, D.A. and Knott, J.F. Met. Sci. 1979, $\underline{13}$, p.341

23   Tweed, J.H. and Knott, J.F. Met. Sci, 1983, $\underline{17}$, p.45

24   McRobie, D.E. and Knott, J.F. Mat. Sci and Tech. 1985, $\underline{1}$, p.357

25   Bowen, P., Ellis, M.B.D., Strangwood, M. and Knott, J.F., Proc. 6th Europ. Conf. on Fracture, Amsterdam 16-20 June 1986 (EMAS)

26   McRobie, D.E. "Cleavage Fracture in C/Mn Weld Metals" Ph.D thesis, Cambridge University, May 1985

27   Vassilaros, M.G. "Fracture Toughness of Ultra-Low Carbon Steel Plate and Heat Affected Zone" CPGS dissertation Cambridge University, March 1986

28   Bowen, P., Hippsley, C.A. and Knott, J.F. Acta Met, 1984, $\underline{32}$, p.637

29   Bowen, P., Hippsley, C.A. and Knott, J.F. Proc 6th Europ Conf. on Fracture Amsterdam 16-20 June 1986 (EMAS)

59

30 Lewandowski, J.J., Hippsley, C. A. and Knott, J.F., Acta Met. 1986 to
    be published.

31 Ellis, M.B.D., Lewandowski, J.J. and Knott, J.F. Proc. 7th Intl. Conf.
    Strength of Metals and Alloys, Pergamon 1985, p. 1087.

DISCUSSION

Comment by W. Gerberich:

You indicated that in many cases the first particle fracture event provides unstable fracture near the ductile-brittle transition. For the case
where you might have a more uniform distribution of carbides, assume one
carbide nucleates a few facets near the crack tip and then another set
form and arrest in another region, etc. It would seem that this and its
analogy in intergranular facets forming during hydrogen embrittlement
would produce a discontinuous fracture/zone much like the process zone you
described for a crazed polymer. Would not the local instability process
be modified by such a process zone?

Reply:

Perhaps the main reason that the fracture of the first (large) particle
is associated with total unstable fracture is that fracture toughness
tests are carried out under increasing stress intensity (constant loading
rate) so that energy is continually being fed into the system. In a
displacement-controlled system, any separation process which involves substantial local displacements ahead of the crack tip can cause the load to
drop (e.g. a "pop-in"). The discontinuous hydrogen fracture that you describe could be attributed either to a "driving effect" of hydrogen pressure or to an embrittling effect (due to a reduction of work-of-fracture).
Either of these effects could affect only limited "patches" of material,
due to limited ingress of hydrogen along favored boundaries or to
decreases in "driving force" (pressure) as the micro-cracks start to grow.
I can imagine a hydrogen crack growing in "patches" of the sort that you
describe at constant K. But I would have thought that if you started with
a material containing intergranular "patches" and increased the applied K
until failure occurred, the critical event would be able to be calculated
simply from K_{1C} and a stress intensity representative of initial stress
concentrator-plus "patches". A recent paper by Paul Bowen and myself
(Int. Jnl. Fracture 1985) treats the analogous problem of quench-cracks
formed at notches. At very low temperatures, the "notch & crack" stress
intensity, calculated from a weight function, characterizes fast fracture,
but, at high temperatures, the final fracture forms independently, some
distance ahead of the notch, and the quench cracks blunt out.

Comment by K. Sieradzki:

Is the microcrack density below a notch high enough so that microcrack
interactions can become important in determining crack nucleation?

Reply:

For the cases described, no.

Theory of Fracture

Session Chairmen: H. Mughrabi and L. M. Brown

PLASTIC PROCESSES AT CRACK TIPS

P. NEUMANN

Max-Planck-Institut für Eisenforschung GMBH
4000 Düsseldorf 1
Federal Republic of Germany

1. INTRODUCTION

Elasticity is a well developed theory and therefore cracks in elastic media are well understood. Only at an atomic scale, where the elasticity description breaks down, details of the atomic positions near sharp crack tips are a matter of discussions, still. Correspondingly, a sound physical model for crack advance is available within the scope of continuum elasticity: The elastic energy release rate due to crack growth must be equal to the energy required for forming the new surfaces.

The complexity of the picture changes drastically if plastic processes become significant at the crack tip. A continuum description of plasticity breaks down at a much coarser scale than elasticity since the elementary carrier of plasticity, the dislocations, are line defects which cannot end within a crystal. In addition they have a long range stress field which falls off with distance, r, only as 1/r. This long range dislocation interaction results in inhomogenous dislocation distributions (cell formation) on a scale of micrometers.

Because of these complications fracture processes with plasticity occurring at the crack tip are currently a very active field of research. In the following the effects of plasticity at crack tips will be discussed in a brief and tutorial way. First, crack tips with a few dislocations in their neighbourhood will be discussed, proceeding in further paragraphs to situations in which gross plasticity at crack tips is encountered.

2. CRACKS WITH ISOLATED DISLOCATIONS

When cracks or dislocations are present in a stressed solid, the stresses exert driving forces which try to move these stress singularities. These forces can be expressed according to Eshelby [1] by a contour integral along a contour which has to encircle only the singularity of interest. The Eshelby integral was re-formulated for sharp cracks by Rice [2]. In this case the contour has to surround the crack tip starting from one crack surface and ending on the other.

Fig. 1 describes a typical situation of a sharp crack surrounded by four dislocations. The contour shown does indeed surround the crack tip in the required way excluding - by the appropriate detours - all the screw dislocations which are lying nearby . This integral is independent of the contour used as long as the contour contains just the singularity of interest. The force on every dislocation can be obtained by integration along a circle containing just this one dislocation.

Any contour lying in the far field of the configuration will contain the crack tip as well as the plastic zone around the crack tip with all the dislocations (e.g. the contour in fig. 1 but without the dislocation detours). Therefore, if the applied loads and the resulting far field is used to obtain the driving force, it is the sum of the driving forces on all stress singularities within the contour, since the contributions of those parts of the contour which are passed twice in opposite directions do cancel.

$$F_{appl} = F_{crack} + \sum_i F_{disl\ i} \qquad [1]$$

Thus we obtain the important result that the driving force felt by the crack tip due to applied forces is modified by the presence of the dislocations in the surrounding. This effect is called shielding [3]. Depending on their Burgers vectors and their position, the dislocations can reduce (shield) or increase (anti-shield) the driving force felt by the crack tip. Quantitative results of the amount of shielding can be obtained only by detailed calculations and examples will by given in the following.

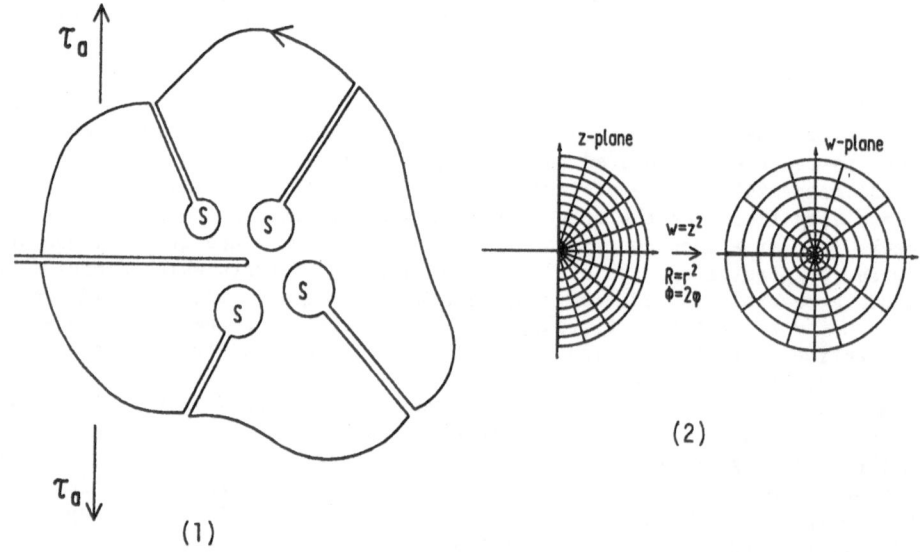

FIGURE 1: Contour for the Eshelby-Rice integral to determine the driving forces on the crack tip and neigbouring dislocations.

FIGURE 2: Conformal mapping from the half Z-plane onto the cracked w-plane by the function $w = z^2$.

One general result can be deduced, however, very simply: If it is assumed that dislocation motion is opposed at every place by a position independent friction stress, τ_f, and if all the dislocations are moving, then the driving force on the i-th dislocation is $b_i \tau_f$ and eq. [1] becomes

$$F_{appl} = F_{crack} + \tau_f \sum_i b_i \qquad [2]$$

If the sum of the Burgers vectors of the dislocations is equal to zero – which is quite likely if the dislocations have been produced by sources – then eq. [2] shows that $F_{appl} = F_{crack}$, i.e. there is no shielding. This result is quite useful to obtain some reference configuration with zero shielding.

In order to obtain quantitative expressions for the shielding effects of dislocations, the stress fields of dislocations close to the crack tip must be known. They can be given in closed form for sharp cracks. The method involved is most easily explained in the special case of anti-plane shear or mode III loading. In this mode only one component of the displacements, u_3, and four components of the stress tensor, $\sigma_{32} = \sigma_{23}$ and $\sigma_{31} = \sigma_{13}$ are non-vanishing. The latter ones are put together to form a complex function using complex coordinates $z = x + y$

$$\sigma(z) = \sigma_{32} + i\sigma_{31} \qquad [3]$$

The boundary conditions can be given by a distribution of external forces, $\underline{F}$, or by precribed displacements, $\underline{u}$. Only the z components, F_3, u_3 are non-vanishing. It turns out that the boundary condition can be written very simply as

$$\sigma dz = -dF_3 + i\mu du_3 \qquad [4]$$

where μ is the shear modulus and dz is the line element lying in the surface of the material with the convention that the material is to the left of dz. If the equilibrium equations as well as the equations of compatibility are written down in this notation it is easy to verify that these equations are just the Cauchy equations which guarantee that $\sigma(z)$ is complex differentiable at all z where there are no elastic singularities. Thus, the only requirement to obtain a solution in mode III is to find appropriate complex differentiable functions $\sigma(z)$ which fulfil the boundary condition [4]. This can be done most easily by using the Laurent expansion and determining the coefficients such that the boundary condition [4] is satisfied.

In this way it is easy to obtain the stress field of a screw dislocation at ζ:

$$\sigma(z) = \frac{B}{z-\zeta} + f(z) \qquad [5]$$

where $B = \mu b/2\pi$ and $f(z)$ is an arbitrary function which is complex differentiable at $z = \zeta$. The stress field of an arbitrary number of screw dislocations near an infinite plane surface is easily obtained by adding the stress fields plus that of the usual image dislocations.

$$\sigma(z) = K_a \sqrt{\frac{2}{\pi}} + \sum_i \frac{b_i\mu}{2\pi} \left(\frac{1}{z-\sqrt{\zeta_i}} - \frac{1}{z+\sqrt{\zeta_i}}\right) \qquad [6]$$

$\overline{\zeta_i}$ is the complex conjugate of ζ_i. In [6] the arbitrary function $f(z)$ of equ. [5] was selected as $f(z)=K_a\sqrt{2/\pi}$ in order to obtain a solution for an external loading, which produces an applied stress intensity K_a.

By a simple conformal mapping with the function $w = z^2$ (fig. 2) the half plane with the dislocations can be mapped onto the cracked w-plane. In the w-plane we do not use $\sigma(w)$ but $\sigma^*(w) = \sigma(z(w))\cdot z'(w)$ as a solution since the boundary conditions for $\sigma^*(w)$ are more easily related to those of $\sigma(z)$:

$$\sigma^*(w)\ dw = \sigma(z(w))\ dz \qquad [7]$$

Due to the conformal mapping $\sigma^*(w)$ is a complex differentiable function again and therefore a solution of the mapped problem. We thus obtain from [6] and [7]

$$\sigma^*(w) = \frac{K}{\sqrt{2\pi w}} + \sum_i \frac{\mu b_i}{4\pi}\left(\frac{2}{w-\zeta_i} + \frac{\sqrt{w}+\sqrt{\zeta_i}}{(w-\zeta_i)\sqrt{\zeta_i}} + \frac{\sqrt{w}-\sqrt{\zeta_i}}{(w-\zeta_i)\sqrt{\zeta_i}}\right) \tag{8}$$

with

$$K = K_a - \sum_i \frac{\mu b_i}{2\sqrt{2\pi}}\left(\frac{1}{\sqrt{\zeta_i}} + \frac{1}{\sqrt{\zeta_i}}\right) = K_a - \sum_i \frac{\mu b_i}{\sqrt{2\pi R_i}} \cos\frac{\Phi_i}{2} \tag{9}$$

and
$$\zeta_i = R_i \exp(2\pi i \Phi_i)$$

In [8] the first term represents the characteristic crack tip singularity, the second term is due to the screw dislocation singularities. The third and fourth terms are the transformed image dislocation terms which are, however, not singular any more. This deserves some discussion: In the untransformed configuration the image screw dislocations are outside the material. In the transformed configuration the image dislocations are still outside the material but only if the w-plane is correctly treated as a $\sqrt{z}$-Riemann plane. If the correct sheet of the Riemann plane is used only, to represent the inside of the material, the stress field of the image dislocations is non-singular there.

Equation [9] shows that the acting local stress intensity at the crack tip, with the local stress intensity K, deviates from the applied stress intensity K_a. The modification of K_a to K stems from the surrounding dislocations at ζ_i. Their contributions can be such that they reduce the local K below K_a. Then they are said to "shield" the crack tip. If they enhance K above the value of K_a they are said to be in an anti-shielding configuration.

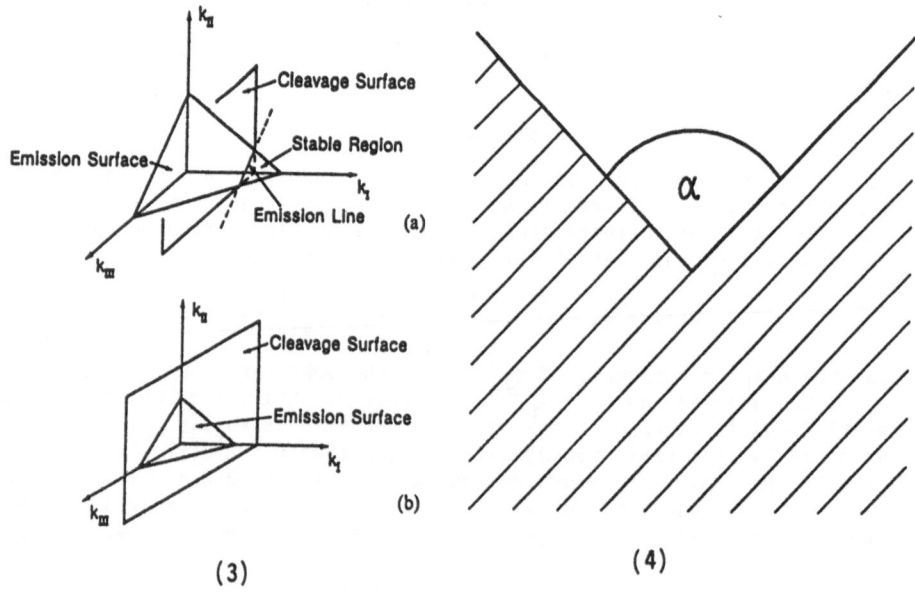

(3)

(4)

FIGURE 3: Crack stability diagram according to [4].

FIGURE 4: Wedge crack with a crack tip angle α.

More general treatments of this kind can be carried out in closed form for sharp cracks with arbitrary loading, characterized by K_I, K_{II}, K_{III}, and with straight dislocations parallel to the crack front but with arbitrary Burgers vectors [4]. With the local stress intensities K_I, K_{II}, K_{III} at hand, the considerations of Rice and Thomson [5] about the balance of decohesion fracture and dislocation emission can be extended. In the most simple model [4] the condition for emitting dislocations is assumed to depend linearly on K_I, K_{II}, and K_{III}. This condition is schematically shown in fig. 3. The Griffith criterion for crack extension due to decohesion is assumed to depend on K_I only. It is represented by a plane perpendicular to the K_I axis. Depending on the relative position of these two planes various situations for the competition between dislocation emission and crack extension can be discussed [4].

For the understanding of crack tips in ductile materials it would be very interesting to determine the stress singularities at wedge shaped cracks with a non-zero crack tip opening angle α (fig. 4). In mode III the stresses around such a wedge crack can be determined very simply by a conformal mapping with the function $z=w^{m/n}$. The resulting crack in the w-plane has a crack tip opening angle $\alpha=\pi(2-n/m)$ and the stress singularity is now $\sigma(w) = cw^{m/n-1}$ [6]. Similar configurations with mode I or mode II loading and with edge dislocations near the crack tip are very important configurations but they represent complicated problems and they are not solved yet.

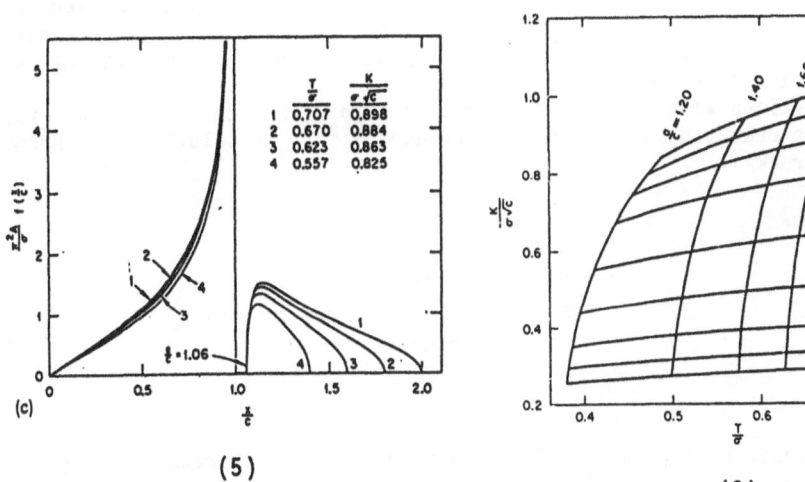

(5)

(6)

FIGURE 5: Dislocation distribution within a crack (crack dislocations) and ahead of the crack with a dislocation free zone from $x/c = 1$ to $e/c = 1.06$. T = applied stress σ = friction stress for the dislocations. The crack is treated as a region of the slip plane with zero friction stress. From [7].

FIGURE 6: Variation of the extend of the dislocation free zone with various parameters according to [7]. σ, c have the same meaning as in fig. 5 ,a, is the position of the outer end of the plastic zone, K is the local stress intensity at the crack tip.

The calculations of local stress intensities as discussed above are always obtained under the assumption that the dislocations are at arbitrary, but pre-determined positions. For describing realistic crack configurations it is of interest to calculate self-consistently equilibrium configurations in which the forces on every dislocation are smaller than the friction stress. Such calculations have been made only for dislocations on one or two slip planes. The best known configuration is that of the Dugdale Bilby Cottrell Swinden (DBCS) model. It comprises a single slip plane which is the extension of the crack plane. This configuration results in a one dimensional problem which is identical for mode III and mode II. The stress singularity is logarithmic and has no $1/\sqrt{x}$ term. Therefore it follows immediately from the Eshelby theorem that the driving force for crack growth is zero in the DBCS-model. This is obviously due to the shielding effect of the dislocations on the slip plane. They have a singular density at the crack tip and it is not surprising that they shield the crack tip completely yielding a vanishing local K. Chang and Ohr [7,8] have pointed out, however, that there exist other dislocation distributions in the DBCS configuration which are in equilibrium as well. They do not extend to the crack tip and are therefore characterized by a dislocation free zone ahead of the crack tip. Due to the existence of this dislocation free zone the shielding is not complete and there is a non-vanishing local K at the crack tip. The resulting dislocation distributions are shown in fig. 5. The crack extends from 0 to x = c. The density of the crack dislocations in this range is shown. From x = e to x = a the density of the real dislocations is non-zero. The dislocation free zone extends from x = c to x = e. The effects of the various parameters on the local stress intensity K is shown in fig. 6. σ is the friction stress exerted on the dislocations, T is the applied stress, and c, a, e have the meaning defined in fig. 5. It is obvious that K depends strongly on the extent of the dislocation free zone e-c.

Configurations with slip on two inclined slip planes emanating from the tip of sharp cracks have also been calculated [9,10]. No solutions are known for wedge cracks.

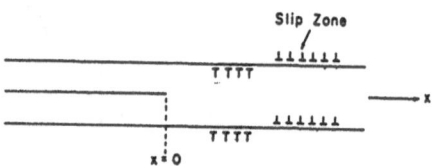

FIGURE 7: Double slip plane model of Weertman, Lin and Thomson [11].

Recently Weertman, Lin, and Thomson [11] have considered in some detail the double slip plane model as shown in fig. 7, where dislocations are present on two planes parallel to the crack plane. The sum of the Burgers vectors of all dislocations on both slip planes are assumed to be zero. As discussed earlier, the shielding effect of the dislocations is zero if the local stress at all dislocations is equal to the friction stress. The authors were successful to demonstrate that, starting from such a configuration, crack extension beyond x = 0 increases shielding, since the shielding dislocations are pushed ahead with the moving crack front, whereas the anti-shielding dislocations stay behind. Therefore, the double slip plane model is the most simple configuration in which - besides other interesting features - stability against crack extension due to dislocation shielding can be demonstrated.

From the above discussions it is evident that dislocations close to the crack can drastically change the crack tip behavior via shielding effects. There are at least three processes which must be considered:

1. The crack advances by a Griffith type decohesion process.

2. The crack tip emits dislocations without further decohesion advance. Due to the emitted dislocations the crack tip shape is altered, however. This shape change may be a considerable contribution to the total crack advance.

3. Dislocations from the immediate neighbourhood of the crack tip are attracted due to the crack tip field and escape out of the material at the crack tip. The crack tip shape is altered by these dislocation activities, too. Again this is genuine ductile crack advance.

In a heavily dislocated ductile material the third process will be overwhelming and lead to a blunt crack tip. The nearby dislocations shield the blunted crack tip from external stresses. Decohesion as well as dislocation emission are difficult to envision under such circumstances. If the crack progresses into regions of reduced dislocation density the likelyhood of capturing dislocations from the neighbourhood for further ductile crack advance is reduced. Simultaneously the shielding is reduced and due to the larger local stress intensity values dislocation emission from the crack tip or decohesion crack growth become more likely. Therefore the crack behaviour will be strongly influenced by the density and mobility of the dislocations close to the crack tip. A semi-quantitative treatment of such a situation was given by Ashby and Embury [12] and used to model the ductile brittle transition in steels. Lin and Thomson [13] have pointed out that slip activity on two inclined slip planes could lead to an automatic alternation of slip activity on the two planes as was described by Neumann [14,15] since dislocations on one plane hinder emission of more dislocations on the same plane much more than emission of dislocations on the other plane.

3. CRACK TIPS IN MATERIALS WITH HIGH DISLOCATION DENSITIES

The shape of crack tips in ductile materials is entirely determined by the plastic flow in the surrounding. Continuum mechanical solutions which assume isotropic plastic flow may be used to describe the macroscopic crack tip shape at a scale above that of the microstructure. The change in the crack tip shape is usually taken into account in ideal plasticity solutions [16] and in numerical solutions of more general constitutive equations [17]. In most of these treatments it is assumed that under certain conditions the crack is growing due to a decohesion prozess. Plastic strain enters the picture only via a formal fracture criterion. I.e. it is often postulated that decohesion crack growth is initiated whenever a critical strain is reached at a critical distance ahead of the crack tip. It is not simple to justify such a fracture criterion in physical terms.

For the sake of clarity it is most interesting to consider the limiting case in which the crack advances exclusively by ductile shape change without any decohesion type crack growth. The most simple case is the shear-off by highly localized shear along a single shear band. Obviously, in the absence of slip band decohesion, infinitely large strains are

70

required in the shear bands for complete shear-off. If two non-planar slip planes are involved, the situation becomes more complicated geometrically, but then purely ductile crack growth can be achieved with average shear strains of the order of unity only. Therefore the requirements for such ductile crack growth processes using two slip planes will be discussed more extensively in the following:

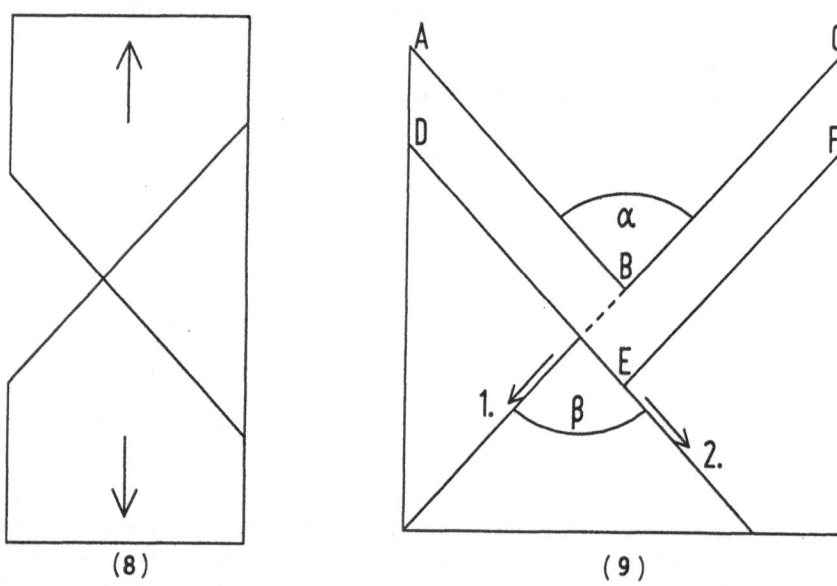

(8)   (9)

FIGURE 8: Complete slip line solution for a 90° notch after Onat and Prager [18].

FIGURE 9: Alternating slip model of ductile crack growth: Alternating slip on slip planes 1 and 2 produce crack front, b, out of crack front, a. The moving crack has a constant crack tip angle α which is equal to the angle β between the two activated slip planes. After [14].

Onat [18] has given an ideal plasticity solution which requires slip on two 45° slip planes emanating from a 90° wedge crack (fig.8). As a result of the shear activity the crack grows in a self-similar manner. Since most materials exhibit a certain coarsenes of slip, on a fine scale the continuum description should be abandoned in favour of a discontinuous one like that in fig. 9. This purely ductile crack advance due to the slip on alternating slip planes was observed in situ in a SEM by Vehoff and Neumann [19] (fig. 10). In both the configurations of fig. 8 and 9 the shear rates in the active slip planes are infinite; however, the average shear left behind the crack tip is exactly equal to one (for the case of the 90° wedge crack). Other more complicated ideal plasticity solutions involving fans at the crack tip have been proposed in order to describe various ductile crack growth configurations [20].

In the following a very simple kinematical solution will be discussed in some detail in order to illustrate the non-linear geometrical effects which are essential for the stationary crack tip shape of cracks growing only by slip and without any decohesion. This solution provides a self similar, purely ductile growth of the crack due to simultaneous slip on two

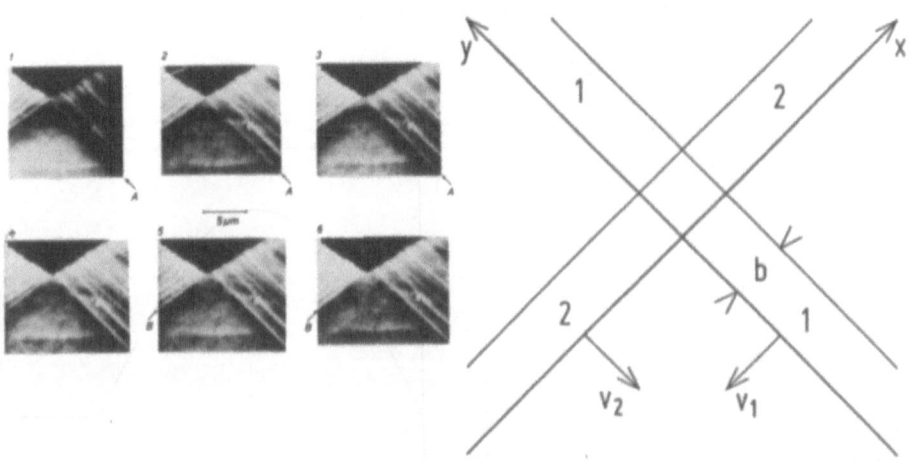

(10)   (11)

FIGURE 10: In situ observation of the alternating slip model of ductile crack growth. Crack tip in copper at successive increments of the crack tip opening displacement ($\Delta\delta$ = 0.4μm), showing the successive activation of slip plane A (fig. 23.1 to 23.3) and slip plane B (fig. 23.4 to 23.6) according to the alternating slip model (see fig. 9). Note the changing relative position of the crack tip and the slip lines due to the changing displacements. From Vehoff and Neumann [19]).

FIGURE 11: Two intersecting slip bands 1 and 2. Active slip occurs only within the two bands. Both bands are moving with respect to the material with a constant velocity v_1, v_2 respectively. The front of a moving Lüders band is an example of such moving shear bands.

intersecting moving slip bands of finite strain rates. The configuration is shown in fig. 11. There are two slip bands 1,2 which move through the material with velocities $\underline{v}_1$ and $\underline{v}_2$ with $\underline{v}_1 = v_b = \underline{v}_2$. Both shear bands have a finite width b and have a velocity profile given by some arbitrary function f(x) as indicated in fig. 12a. The resulting shear strain rate profile is given by $\dot{\gamma} = f'(x)$ and is shown schematically in fig. 12b. The strain left behind after the shear band has passed is $\gamma_{final} = v_t/v_b$. The velocity fields from the two shear bands as a function of x,y,t are given by

$$\underline{\dot{u}}^{(1)} = (0,-f(x-v_bt)); \qquad \underline{\dot{u}}^{(2)} = (-f(y-v_bt),0) \qquad [10]$$

It is useful to changed to a moving coordinate system according to

$$x \longrightarrow x-v_bt \qquad\qquad y \longrightarrow y-v_bt \qquad [11]$$

In the moving coordinate system the shear bands are stationary but material points below the shear bands are moving upward with the constant velocity $\sqrt{2} \cdot v_b$. The resulting velocity field due to the combined action of both shear bands in the moving coordinate system is

$$\underline{v} = (v_b - f(y), \; v_b - f(x)) \qquad\qquad [12]$$

From this the strain rate can be directly evaluated giving

$$\dot{\varepsilon}_{xy} = -\frac{1}{2}(f'(y) + f'(x)); \qquad \dot{\varepsilon}_{xx} = \dot{\varepsilon}_{yy} = 0 \qquad\qquad [13]$$

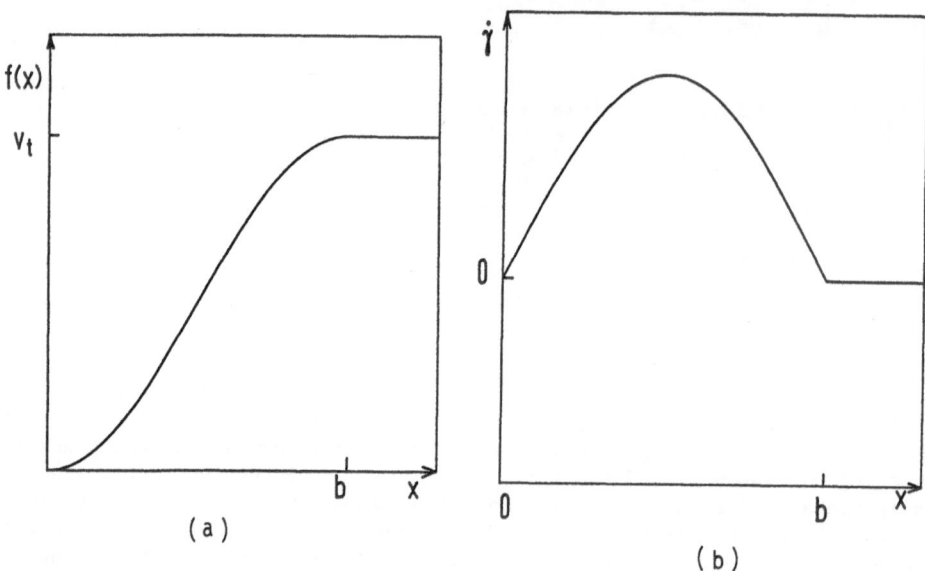

FIGURE 12: a) Shear velocity distribution within a shear band of thickness, b) Shear strain rate in the same shear band.

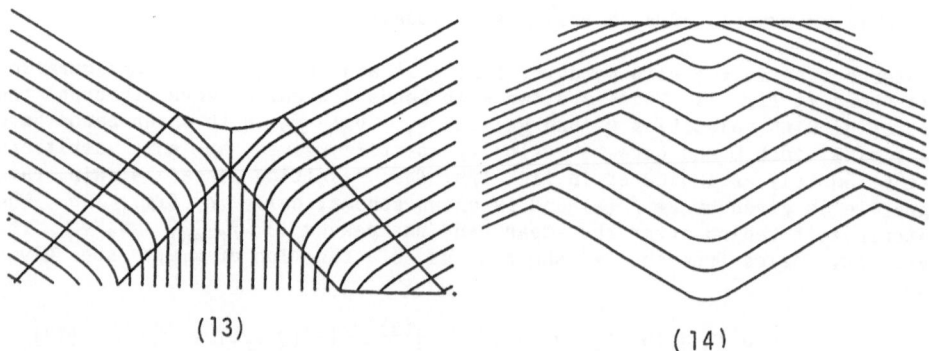

FIGURE 13: Stream lines of material points in that coordinate system in which the shear bands are stationary. γ final has the value 1.27.

FIGURE 14: Similar to fig. 13 but now successive positions and shapes of a straight line of material points are plotted in the original specimen coordinate system in which the material points below the shear bands are at rest. It is obvious how a wedge is formed due to the shear activity.

Such a velocity field usually does not lead to a growing crack. If, however, the critical value of $\gamma_{final} = v_t/v_b = 1$ is reached or exceeded, the shear bands produce a point of zero velocity and a growing blunt wedge crack is obtained. The streamlines according to such a velocity field are plotted in fig. 13 for a value of $\gamma_{final} = 1.27$ and for a piecewise linear $f(x)$. It is important to note that the rounded wedge crack grows simply by the appropriate shear movements and no decohesion fracture is involved at all. The only requirement is that the maximum shear strain γ_{final} produced by the shear bands is larger or equal to unity.

The existence of a point of zero velocity, however, implies a strain singularity since the material points which move through the shear bands and approach this point of zero velocity, will stay an infinite length of time in the shear band region. The form of this singularity can easily be calculated on the line $x = y$. For simplicity we introduce a new coordiante z along this line by defining $z = x\sqrt{2} = y\sqrt{2}$. Using [12] and [13] we obtain

$$v(z) = \sqrt{2}(v_b - f(\frac{z}{\sqrt{2}})); \qquad \dot{\varepsilon}_{xy}(z) = -f'(\frac{z}{\sqrt{2}}) = v'(z) \qquad [14]$$

Finally we may integrate the shear strain rate over time to obtain the shear strain:

$$\varepsilon_{xy}(z) = \int \dot{\varepsilon}_{xy}(z)dt = \int v'(z) \frac{dz}{v(z)} = \int_{\sqrt{2}v_b}^{v} \frac{dv}{v}$$

$$= \ln \frac{v}{\sqrt{2}v_b} = \ln (1 - \frac{f(\frac{z}{\sqrt{2}})}{v_b}) \qquad [15]$$

Obviously the integration can be carried out very simply even for arbitrary $f(x)$ resulting in a logarithmic strain singularity. This demonstrates that the logarithmic strain singularity obtained from more complicated complete solutions of moving cracks [16] may be interpreted as the result of the non-linear geometric superposition of intersecting shear bands.

The stream lines passing through the point of zero velocity obviously follow the stationary crack front shape. Fig. 14 shows how a straight line of material points is deformed by the action of the shear bands after successive increments in time, $\Delta t = i \cdot b/v_b$. The relative positions of these lines is shown as they are observed in that coordinate system in which the material points below the shear bands are at rest. The lines of fig. 14 may be interpreted as the successive shapes of the original specimen surface from before the slip bands started to intersect until after they intersected and formed the wedge.

Since these shape changes are a consequence of the shear activity on two inclined slip systems, any continuum solution which favours such anisotropic shear should show similar shape changes. Fig. 15 shows an example from Needleman and Tvergaard [21]. An initially round crack tip is deformed according to the J_2 corner theory and four stages of the resulting crack tip shapes are shown. It is obvious that a wedge type crack tip develops from the previously round crack tip. In crystalline solids the preferred direction of shear bands is naturally given by the crystallographic slip planes. The observation of ductile crack growth in single crystals of pure metals gives abundant evidence for the development of wedge type crack tips [19,22-25].

If there is just enough deformation in the shear bands to produce a point of zero velocity ($\gamma_{final} = v_t/v_b = 1$) the crack tip angle coincides with the angle between the active slip bands. Larger values of γ_{final} produce larger α. The width d of the active shear bands determines the radius of curvature at the crack tip. Experimental evidence shows that this radius of curvature is below the resolution of the SEM (see fig. 10).

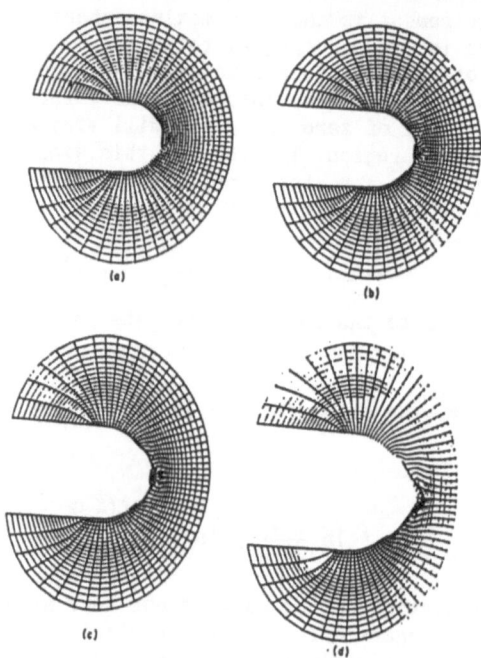

(a)   (b)

(c)   (d)

FIGURE 15: Sharpening of a round crack tip due to plastic deformation according to the J_2 corner theory. From [21].

From fig. 13 it is obvious that the crack tip angle, α, is always larger than the angle between the shear bands. Reduction of the crack tip angle below this value can be obtained only by reverseslip, i.e. if f(x) in fig. 12 is modified to that of fig. 16. The corresponding stream lines in fig. 17 are identical to those of fig. 13 up to their inflection points, but then the reverseslip bends them back and thus reduces the crack tip angle. If the reverse slip fully reverses the forward slip, a crack results with parallel flanks and round tip. Under the following conditions reverse slip can be expected:

1. In fatigue experiments reverse slip is induced during the compression phase of each cycle. This leads to complete reversal of the strain i.e. the crack tip angle becomes zero at the end of the compression.

2. The same configuration can be produced if a cutting knife is pressed into the material. Then there is no external loading but the knife exerts tractions on the crack tip area. They provide the driving force for the forward shear. However, the enclosing

elastic material exerts back stresses which drive the reverse slip
of the kind shown in fig. 17.

Under sustained tensile external loading which introduces the forward slip
under fully plastic conditions, it is difficult to imagine how the required
reverse slip could occur. This is in agreement with the known solutions of
ideal plasticity theory. Therefore the crack tip angle under these
conditions is equal to the angle between slip bands or smaller than this
value (due to decohesion events, see below). This is confirmed by
experimental findings of ductile crack growth in f.c.c. and b.c.c. single
crystals [19,22-27].

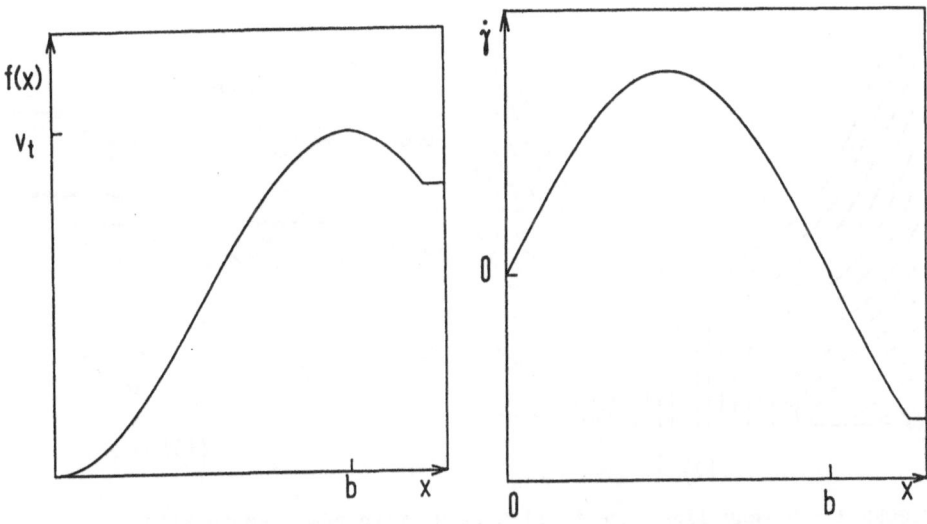

FIGURE 16: Similar to fig. 12 but now showing the shear velocity
distribution and the shear strain rate distribution in a shear band which
contains reverse slip.

Single crystals are ideally suited for ductile crack growth experiments
since they allow shear on slip planes which may extend all across the
ligament. Under these fully plastic conditions the final strain of 100% in
the shear bands can be easily accommodated for arbitrarily large crack
growth increments, i.e. wide shear bands. In a contained plasticity
situation such large strains can be accommodated only in the form of thin
shear bands. This is the reason why the ductile crack growth occurs so
readily under fatigue conditions. For the crack growth increment in the
tensile part of each cycle, Δa, shear bands with 100% shear strain but only
with a thickness, Δa, are required. In the compression part of each cycle
these shear bands are annihilated such that at any moment there is only one
pair of shear bands of thickness Δa or less present which can be easily
contained in an elastic surrounding. This is the mechanism of the so-called
stage II crack growth [14,15,28].
Under fully ductile conditions the crack growth rate per cycle is –
because of the large crack tip angle – equal to the crack tip opening,
which on the other hand – due to a general relation of linear fracture

mechanics is proportional to the square of the stress intensity range. Thus the well known Paris equation

$$\Delta a/\Delta N = C \ \Delta K^n \qquad \qquad [16]$$

is a direct consequence of this mechanism with an exponent n = 2. This value of n is frequently observed in the typical stage II regime. More detailed considerations and deductions of a crack growth law along these lines are due to Weertmann [29]. Furthermore the stress situation occuring due to the dislocation emission on alternating slip planes, the developing back stresses and the change of the crack tip geometry was simulated in great detail by Sinclair [30].

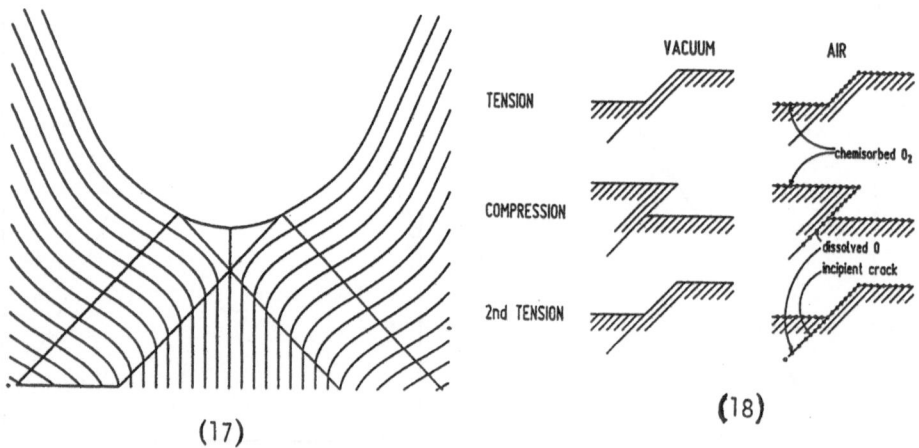

(17)

(18)

FIGURE 17: Stream lines due to slip bands with 50% reverse slip.

FIGURE 18: Crack formation along slip planes due to the combined action of cyclic strain and oxidation. After [32].

If the stress intensity range, ΔK, is too small for producing slip bands with a shear strain of $\gamma = 1$, then only one slip plane with $\gamma < 1$ can be activated. Crack growth along one slip plane is indeed observed experimentally (stage I crack growth). It cannot be readily understood in terms of plasticity since slip steps which form during one half cycle are annihilated statistically due to reverse slip in the other half cycle. Only irreversible surface production due to slip contributes to crack growth. But single slip is highly reversible. Such slip irreversibilities were sought in the process of cross slip [31] but could never be substantiated in a quantitative way. On the other hand it is well documented in most materials that the fatigue life time is increased by a factor of 10 if the tests are performed in vacuum. Therefore, there is strong experimental evidence [35] that slip on single slip systems leads to crack growth only in co-operation with an oxidizing or otherwise corrosively active environment. Thompson's model [32] which is shown in fig. 18 still seems to be the most appropriate one although a quantitative treatment is lacking. In essence it may be concluded that sustained local straining for a large number of cycles plus the presence of an active environment leads to crack initiation in fatigue.

In order to understand under which conditions sustained local straining may occur, the mechanism of fatigue hardening must be discussed first [25]: During cyclic loading dislocations of opposite signs are mixed thoroughly. Pile-ups of dislocations of one sign are not observed. Therefore, in pure metals fatigue hardening is due to the mutual trapping of dislocation of opposite sign (dipole formation). This typically leads to plastic strain amplitudes of a few per cent of the elastic strain amplitude at stress levels of about a quarter of the UTS. For the crack initiation process it is therefore important under which conditions this low level of plasticity cannot be maintained. There are several well known reasons for a such a break-down of the fatigue hardening.

1. If the stress level is high enough, the required slip plane distance of dislocations which can form stable dipoles may be as low as 50 nm. Such narrow dipoles tend to convert into vacancy rows. This is generally believed to be the mechanism for the persistent slip band formation in pure metals [33]. The resulting topography changes due to the enhanced slip and crack initiation were studied in detail [34,35].

2. In precipitation hardened alloys particles may be cut and even be temporarily dissolved or re-distributed. This is a powerful work-softening mechanism. Like any work-softening this will amplify fluctuations in strain to extensive strain localizations along slip planes. In aluminium alloys almost precipitate-free channels have been observed together with extensive stage I crack growth [36].

3. In older papers [37,38] as well in a very recent study [39] it was observed that in polycrystalline copper crack initiation happens at low amplitudes almost exclusivly at twin boundaries. The cracks form parallel to the twin boundaries and propagate in the twin boundaries. This may be explained in terms of localized strain again. Because under stress, twin boundaries can move laterally back and forth over a distance which is determined by the content of twinning dislocations in the boundary itself. As a consequence of the oscillating twin boundary, the region over which the twin boundary passes undergoes a cyclic strain range of 70 per cent (the twinning strain in f.c.c. metals).

The persistent slip bands in pure metals have been studied extensively because they are the locations of fatigue crack initiation in most metals. Crack initiation in these bands is studied best by a sectioning technique which cuts through the persistent slip bands. A sectioning technique was developed which preserves the edges with an accuracy beyond that of the SEM of 20 nm. Thus the profile of the persistent slip band topography can be studied as well as the crack initiation starting deep inside the persistent band. Since the sectioning is done at 90° with respect to the the original specimen surface there is no taper magnification of the profile and artifacts due to a large taper magnification are avoided. Furthermore the original undisturbed specimen surfaces can be observed in addition to the sectioning plane. Details of this technique are described in [34,35]. Fig. 19 shows a typical cut through a persistent slip band after 50,000 cycles at 0.2% plastic strain amplitude. The persistent slip band consists of triangular extrusions. From the valleys between neighbouring extrusions crack nuclei start to grow. The preferred locations for crack initiation are the interfaces between the persistent slip band and the matrix. By careful statistical studies it was well established that in the early stages of fatigue crack nucleation occurs only at one of the two interfaces between persistent slip band and matrix [34]. It is always that side of the persistent slip band where the slip planes form an obtuse angle with the specimen surface. An example is shown in fig. 19 where a typical stage I crack is visible at this location. Later in life, however, the predominant stage II cracks always start from the other side of the persistent slip band as indicated in fig. 20. A detailed theory explaining these preferred locations is still lacking.

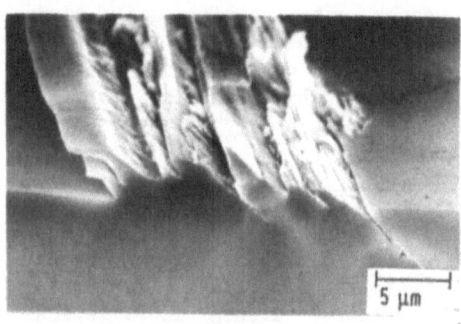
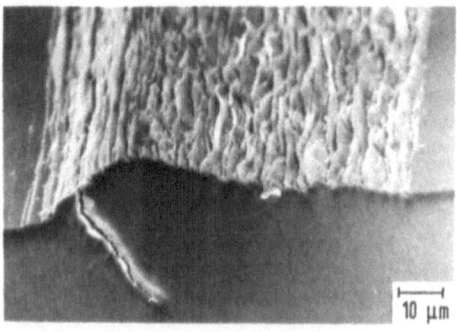

(19)   (20)

FIGURE 19: Persistent slip band sectioned at 90° with respect to the surface. A crack is visible at that location in the persistent slip band (obtuse angle site) which is a preferred crack initiation site early in life. From [34].

FIGURE 20: Persistent slip band with stage II crack at the other cracking location which is preferred later in life. From [34].

4. MIXTURES BETWEEN DECOHESION FRACTURE AND RUPTURE

In the previous paragraphs the pure cases of, decohesion fracture and ductile rupture have been discussed in detail. However, in most applications mixtures of these two pure cases will be observed. There is multiple experimental evidence that decohesion fracture occurs simultaneously with plasticity. This plasticity contributes to the crack advance via the ductile crack tip shape changes, as discussed above. In the ductile to brittle transition range of steels but also under environmentally enhanced embrittling situations (hydrogen embrittlement, stress corrosion cracking) the contributions of plasticity and decohesion processes to the crack advance can be of roughly the same magnitude. Under such circumstances it is desirable to measure in a quantitative way the contributions of ductility as well as that of decohesion to the fracture process. In the following various methods will be dicussed to assess quantitatively the amount of plasticity accompanying fracture.

The ductile to brittle transition which is commonly observed in steels can be observed in its purest form in silicon crystals at around 600°C. The transition occurs within a temperature range of a few degrees. The exact value depends on the loading rate. The experiments show convincingly that the transition occurs whenever the dislocation mobility is high enough to allow enough dislocations to be emitted from the moving crack and transported into the otherwise dislocation free material [40-42].

Dimple fracture is usually considered as a purely ductile fracture mode. This is definitely true for the ductile necking of the ligaments between voids which have formed ahead of the crack tip. However, the process of void formation is mainly a decohesion process which occurs at the interface between inclusions and the matrix. The interfaces quite commonly have a small cohesive strength. Without such inhomogeneities with reduced cohesive strength, voids would not form ahead of the crack tip. Instead, the crack would grow in a purely ductile manner with a crack tip angle of the order of 90° as demonstrated in single crystals of various pure metals and alloys. In this sense the decohesion processes leading to void formation are an essential part of dimple fracture which therefore must be considered to be a mixture of ductile and decohesive fracture.

Another type of failure is the shear band fracture which occurs after shear localization along the characteristics of the slip line field. The details of this fracture type is not known at all. Obviously, due to the high shear strains, the cohesive strength of the material are lowered in a yet unknown way. Therefore the ratio between decohesion and ductile failure is not known at all in shear band fracture.

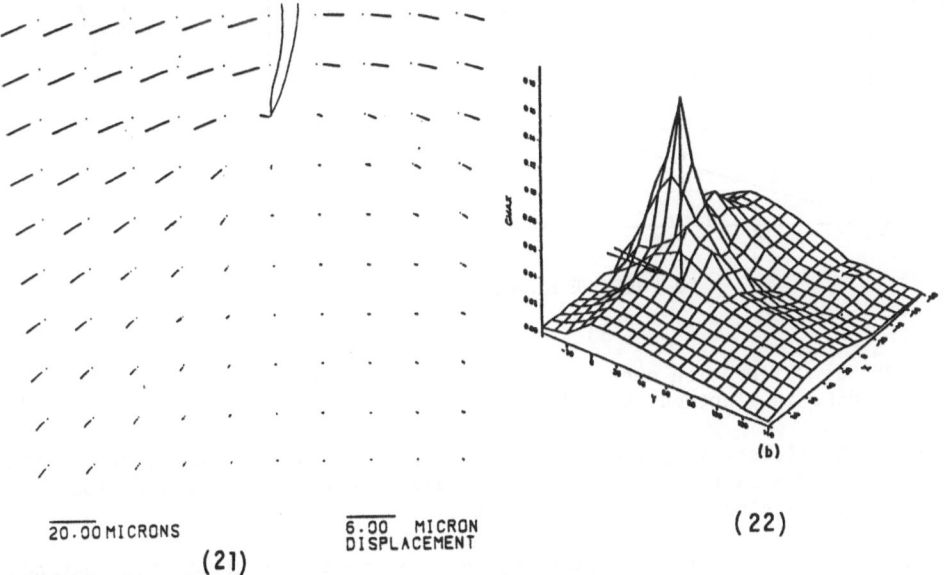

20.00 MICRONS   6.00 MICRON DISPLACEMENT   (22)

(21)

FIGURE 21: Displacements near a fatigue crack tip measured by the stereo evaluation method. From [43].

FIGURE 22: Strain distribution calculated from fig. 21 around a fatigue crack. From [43].

The most direct way to measure plasticity accompanying cracks is to measure the displacements on the side surfaces of the specimen. Davidson et al. [43] have developed a technique to evaluate small displacements from in situ SEM pictures taken at two different load levels. By using these two pictures as input for a stereographic image evaluation system, the lateral displacements are evaluated as height differences and from these the lateral displacements can be re-calculated. Fig. 21 shows an example of displacements measured in this way. Fig. 22 shows the strains evaluated from these displacements. A well defined peak of the strain distribution is visible at crack tip. Since most of the strains are plastic strains, the defined. Comparisons of these measured strain distributions with the solutions of continuum plasticity are being elaborated.

Plasticity which accompanies fracture can also be measured on the fracture surfaces by a quantitative evaluation of the width of electron channeling lines (fig. 23) [44-45]. The line broadening is, however, more directly related to the dislocation density than to the plastic strain. However, there are empirical relations between these two quantities such that the method can also be used to get an estimate of the plastic strain immediately below the fracture surfaces.

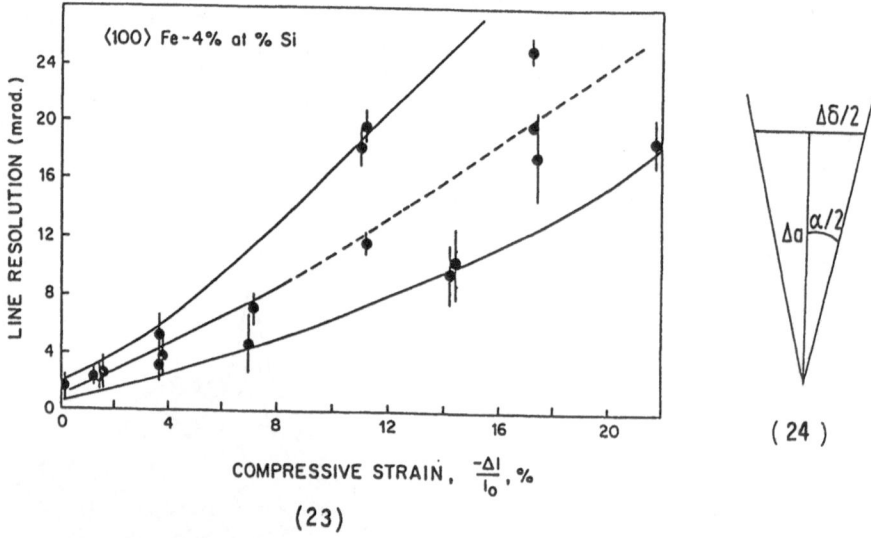

(23)

FIGURE 23: Calibration curve for determining compressive strains from the line widths in electron channeling patterns. From [45].

FIGURE 24: Illustrating the definition of the normalized crack growth rate $a_n = \operatorname{ctg} \alpha/2 = \Delta a/(\Delta\delta/2)$. It is the ratio between the decohesion and ductility contributions to crack growth.

Under fully plastic conditions the plasticity at the crack tip extends all across the ligament. Therefore, the plastic crack tip opening, $\Delta\delta$, can be measured conveniently at the specimen's ends via the plastic elongation of the specimen. The latter can be obtained with high precision from the total strain by subtracting the elastic strain which in turn is calculated from the applied stress. The crack advance can be measured independently by some potential drop technique or by a compliance technique. Then from the crack growth increments and from the plastic crack tip opening, $\Delta\delta$, the crack tip angle can be calculated according to fig. 24 as

$$\operatorname{ctg}(\frac{\alpha}{2}) = \frac{\Delta a}{\Delta\delta/2} \qquad [17]$$

The crack growth increment per crack opening increment $a_n = \Delta a/(\Delta\delta/2)$ is called the normalized crack growth rate and is inversely related to the crack tip angle $a_n = \operatorname{ctg} \alpha/2$ [23]. Both are quantitative measures of the ratio between ductile and decohesion fracture processes. If the crack advances by decohesion, the crack tip angle is obviously zero. In the ductile case, however, it is usually of the order of 90° (more precisely equal to the angle between active slip planes). If both process occur in an intimate mixture, as indicated in fig. 25, all intermediate values of the crack tip angle α can be realized. Careful observation of fracture surfaces in Fe2.6%Si [46] indicate that the mixing between decohesion and ductile cracking occurs on a scale less than the resolution of the SEM, i.e. pure decohesion without plasticity does not occur over distances exceeding 20 nm.

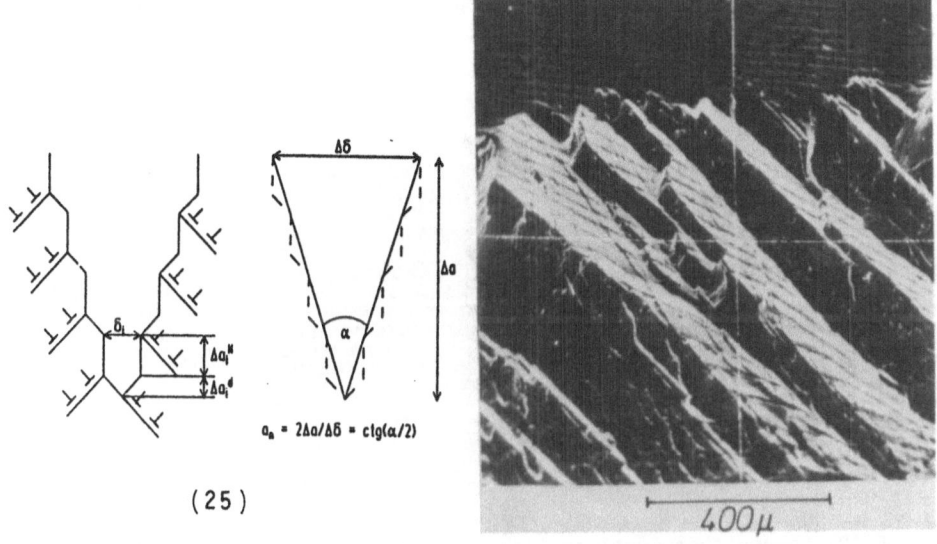

(25)

(26)

FIGURE 25: Mixture between decohesion type fracture ($\alpha = 0$) and ductile fracture ($\alpha = 90^O$) can produce effective crack tip angles α of any value between 0 and 90°. From [46].

FIGURE 26: Transition from purely ductile crack growth along a (110)-plane in Fe3%Si to quasi-cleavage crack growth along (100)-planes due to a change in the crack tip opening rate by a factor of two. The (100) facets show the usual cleavage appearance. From [47].

With this technique of measuring the crack tip angle, various embrittling effects were studied in a quantitative way. In Fe2.6%Si single crystals crack growth behavior was observed in that regime of loading rate and temperature in which the transition from ductile to brittle behavior occurs [24]. Ideal ductile crack growth is observed in a specimen with [110] tensile axis. Ideal ductile striations are formed on the (110) fracture plane in cyclic loading with large growth increments per cycle (Δa per cycle = 50 μm). If the crack opening rate is increased by a factor of 2 the fracture occurs in an apparently cleavage like manner on {100}-planes. They are inclined to the original (110) fracture surface and therefore crack growth along (100) leads to a strongly fragmented fracture surface, as shown in fig. 26.

If the specimen axis is chosen parallel to a [100] direction, flat fracture surfaces can be produced in the decohesion case as well as in the ductile case (fig. 27). In the latter case the slip planes are (2,-1,1) and (2,1,-1). They intersect along the crack front which is parallel to [011]. With this orientation the variation of the crack tip angle in the quasi-cleavage case could be measured as a function of the crack tip opening rate and temperature [24]. An activation energy can be determined by plotting versus 1/T the logarithm of that $\dot{\delta}$, which is required to produce a given crack tip angle α. The slope of this Arrhenius plot yields 0.68 eV, which is in good agreement with the activation energy of dislocation motion in Fe2.6%Si. This is strong evidence that the ductile brittle transition in Fe2.6%Si is determined by the possibility to move dislocations in the near tip region.

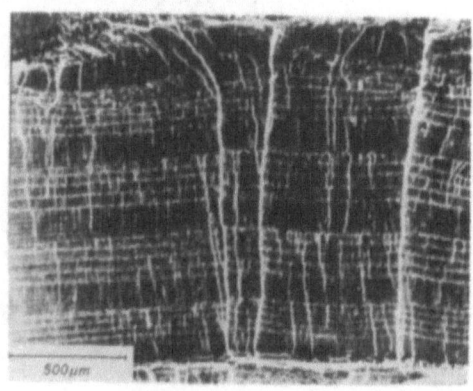

FIGURE 27: Change from ideal ductile crack growth along a (100)-plane in Fe3%Si to quasi-cleavage crack growth on the same plane. From [24].

In a similar way the variation of the crack tip angle in Fe2.6%Si above the ductile to brittle transition was measured as a function of hydrogen pressure, crack growth rate, and temperature [26]. Again a wide range of crack tip angles was measured reaching from 90 to 7°. Similar experiments were performed in nickel single crystals [27]. The results of these measurements can be accounted for quantitatively if the ratio between decohesion and ductile crack advance, $a_n = \text{ctg } \alpha/2$ is assumed to be proportional to the coverage of the crack tip surface with hydrogen or with the concentration of hydrogen in near-tip traps. All experiments were carried out at such low crack opening rates (from 0,1 $\mu m/s$ to 100 $\mu m/s$) that the hydrogen was in equilibrium at the crack tip with the external gas phase, i.e. the transport from the external gas phase to the crack tip was not rate limiting. From the quantitative evaluation of these measurements it could be concluded that there is a special site for the hydrogen at the crack tip which is responsible for the embrittling effect. This special site is postulated, since the values of the pre-exponential factor as well as the binding energy are not in quantitative agreement neither with those of adsorbed hydrogen at free surfaces nor with those of hydrogen in volume traps.

Also the change of the crack tip angle due to stress corrosion cracking was determined during intergranular stress corrosion fracture of Fe2.6%Si bi-crystals in carbonate solution. Crack tip angles from 90° to 2° were observed on a variety of high angle grain boundaries. The results can be described in a quantitative way by a simple slip step dissolution model. The influence of the various parameters of the grain boundary under consideration is being studied. The results strongly support the view that the fracture process is controlled by the accompanying plasticity. More details of these studies are reported in one of the workshops of this conference [48].

5. CONCLUDING REMARKS

It was tried to highlight in this paper the various aspect of crack tip plasticity as far as it is relevant to the cracking process. Since commercial alloys usually represent a compromise between strength and toughness, the fractures which are a mixture between ductile and decohesive failure are very relevant indeed. In order to understand these processes better, it seems very desirable to measure quantitatively both the ductile and the decohesive contributions to the cracking process in well characterized materials. Measurements of the crack tip angle seem to be one possible experimental approach.

REFERENCES

1. Eshelby JD: in: Prog. Solid State Phys. 3, 79, 1956.
2. Rice JR: in: "Fracture", ed. H.Liebowitz, Academic Press, 2, 191, 1968.
3. Thomson R: J. Mat. Sci. 13, 128, 1978.
4. Lin IH and Thomson R: Acta met., 34, 187, 1986.
5. Rice JR and Thomson R: Phil. Mag., 29, 73, 1974.
6. Ohr SM, Chang SJ: to be published.
7. Chang, SJ, Ohr SM: J. Appl. Phys., 52, 7174, 1981.
8. Ohr SM, Chang SJ: J. Appl. Phys., 53, 5645, 1982.
9. Riedel H: J. Mech. Phys. Sol., 24, 277, 1976.
10. Vitek V: J. Mech. Phys. Sol., 24, 263, 1976.
11. Weertman J, Lin IH, Thomson R: Acta metall., 31, 473, 1983.
12. Ashby MF, Embury JD: Scripta Met., 19, 557, 1985.
13. Lin IH, Thomson R: Scripta Met, to be published.
14. Neumann P: Z.f.Metallkunde, 58, 780, 1967.
15. Neumann P: Acta Met., 22, 1155 and 1167, 1974.
16. Rice JR, Sorensen EP: J. Mech. Phys. Solids, 26, 163, 1978.
17. Rice, JR, McMeeking RM, Parks DM and Sorensen EP: Com. Meths. Appl. Mech. Eng., 17/18, 411, 1979.
18. Onat ET, Prager W: J. Appl. Phys., 25, 491, 1954.
19. Vehoff H and Neumann P: Acta Met., 27, 915, 1979.
20. McClintock FA, in: Liebowitz (ed), Fracture, Vol. 3, 48, 1971.
21. Needleman A and Tvergaard V: ASTM STP 803, I-80, 1983.
22. Neumann P: Acta Met., 22, 1155, 1974.
23. Neumann P, Fuhlrott H and Vehoff H: ASTM STP, 675, 371, 1979.
24. Vehoff H and Neumann P: Acta Met., 28, 265, 1980.
25. Neumann P: Fatigue, in "Physical Metallurgy", Hrsg. R.W. Cahn, P. Haasen, Chapter 24, North Holland, Amsterdam, 1554, 1983.
26. Vehoff H, Rothe W: Acta Met., 31, 1781, 1983.
27. Vehoff H and Klameth KH: Acta Met., 33, 955, 1985.
28. Laird, C., in: Fatigue Crack Propagation, ASTM STP 415, 131, 1967.
29. Weertman J: IUTAM Sumposion on "Three Dimensional Constitutive Relations and Ductile Fracture", ed. S.Nemat-Nasser, North-Holland Publishing, Co., p.111, 1981.
30. Sinclair JE: 2. Int. Conf. on Fundamentals of Fracture, Gatlinburg, Tennesse, U.S.A. Nov. 4-7, 88, 1985.

31. Mott, NF: Acta metall., 6, 195, 1958.
32. Thompson N, Wadsworth NJ, and Louat N, Phil.Mag., 1, 113, 1956.
33. Essmann U, Gösele U, and Mughrabi H, Phil.Mag.A, 44, 405, 1981.
34. Hunsche A, Neumann P: Acta Met., 34, 207, 1986.
35. Hunsche A, Neumann P: Int. ASTM-Symp. on Fundamental Questions and Critical Experiments on Fatigue, Dallas-Fort Worth, Texas USA, ASTM STP, 1984.
36. Gerold V, Lerch BA and Steiner D: Z. f. Metallkde.,75, 547, 1984.
37. Stubbington CA, Forsyth PJE: J. Inst. Metals, 86, 90, 1957.
38. Boettner, RC, MCEvily AJ and Liu YC: Phil. Mag. 10, 95, 1964.
39. Tönnessen A, Neumann P: to be published.
40. St John C: Phil. Mag. 32, 1193, 1975.
41. Haasen P: in: Atomistics of Fracture, NATO Adv. Res. Inst., Plenum, New-York, 707, 1981.
42. Michot G, George A: ICSMA 7, 12-16 August 1985.
43. Davidson DL, Hudak Jr. SJ, to be published by ASTM, Proc. 18. Nat. Symp. on Fracure Mechanics, 1985.
44. Davidson DL: in: Scanning Electron Microscopy, O. Johari, ed., IITRI, Chicago, 41, 1969.
45. Gerberich WW, Wright AG, Kurman E and Peterson KA: submitted to Met. Trans., 1985.
46. Vehoff H, Neumann P: in "Hydrogen Degradation of Ferrous Alloys" R.A.Oriani, J.P. Hirth, and M. Smialowski, Eds., Noyes Publications, Park Ridge, NJ, USA, 686, 1985.
47. Neumann P, Vehoff H, and Fuhlrott H, ICF4, Waterloo, Canada, 1313, 1977.
48. Stenzel H, Vehoff H and Neumann P: This conference.

Discussion: Remark Concerning the Evidence of Cleavage
in our Experiments in Single Crystals of Fe2.6%Si.

P.Neumann and H.Vehoff

In the course of this conference the evidence of cleavage in our
experiments has been questioned. Therefore we would like to recollect, for
easy reference, the evidence as it is published in the literature. In
connection with the earlier experiments on the effect of loading rate and
temperature on the ductile to brittle transition in Fe2.6%Si single
crystals we published pictures (fig. 6 from (1), reproduced here as fig. 1)
which show (100) crack facets with a cleavage like appearence. In those
experiments the orientation was selected to produce ductile crack advance
with the help of alternating slip on the (1,0,-1) and (0,1,-1) slip planes.
They intersect along the [1,1,1] direction on the (1,-1,0) fracture plane.
If the crack opening rate, δ, is increased by a factor of two only, the
plane (1,-1,0) fracture surface becomes heavily fragmented. The fragments
are all (1,0,0) facets which look very much cleavage like.

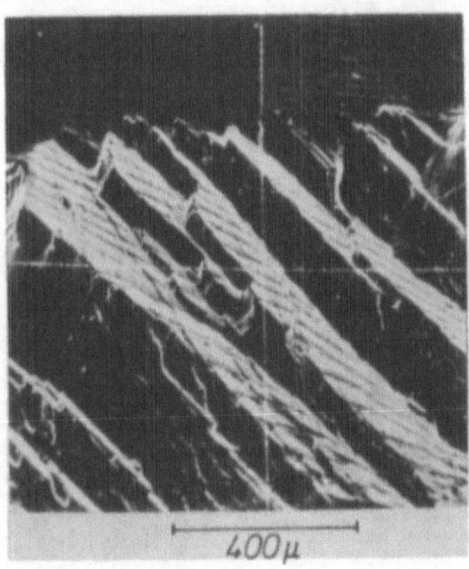

400μ

FIGURE 1: Ductile to brittle transition in FE2.6%Si due to a change in
crack opening rate by a factor of two. For the orientations see text. Crack
growth from top to bottom. From [1].

Corresponding experiments were done with the same orientations to study
the embrittling effect of gaseous hydrogen at rates much below the critical
rate for the ductile brittle transition in vacuum. The micrographs are
shown in fig. 5 from (3) and fig. 26-9 from (4) (reproduced here as fig.
2).

In subsequent experiments, with and without hydrogen, the (1,1,0)
ductile fracture surface was avoided in order to get ductile to brittle
transition without a change of the plane. Therefore the orientation was

chosen in such a way that slip occurs on the planes (2,1,-1) and (2,-1,1) which intersect along the [0,1,1] direction on the (1,0,0) fracture plane.

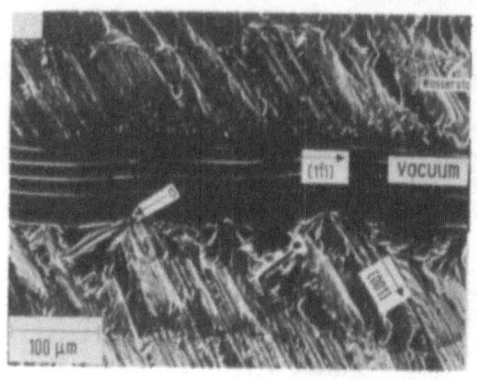

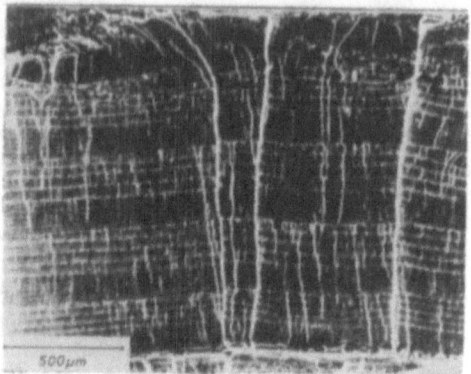

(2)

(3)

FIGURE 2: As fig. 1, now the transition is due to gaseous hydrogen. Crack growth from top to bottom. From [3,4].

FIGURE 3: As fig. 1, but now on a (100) plane. Crack growth from top to bottom. From [5,2].

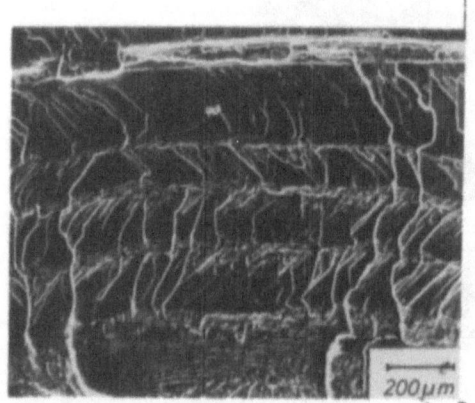

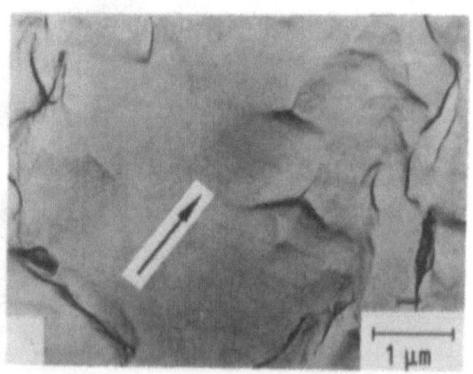

(4)

(5)

FIGURE 4: As fig. 3, but now showing fast pop-ins, which show up in the loading curve. Crack growth from top to bottom. From [5].

FIGURE 5: Secondary carbon replica (taken according to the procedure proposed by Lynch (5)) from a fracture surface in Fe2.6%Si in hydrogen. The arrow indicates the direction of crack growth. There is no evidence of voids. From [4].

Slip on these two planes produces ideal ductile crack growth on the (1,0,0) plane. An increase of the crack tip opening rate produces microscopically flat (1,0,0) quasi-cleavage planes on which the crack tip opening angle can be measured. An example of this kind was published (fig. 6 from (5) and fig. 11 from (2), reproduced here as fig. 3). The appearance of such fracture surfaces becomes more and more cleavage like with decreasing crack tip angle. At high loading rates occasionally so-called pop-ins occured. The resulting fracture surfaces are shown in fig. 5a from (5) (reproduced here as fig. 4). These pop-ins are fast crack jumps which show up even in the load elongation curve (fig. 5b from (5)) and are strong evidence for cleavage.

Because of the controversy whether micro-void formation could be responsible for the decreasing crack tip angle, secondary carbon replicas were taken from the quasi-cleavage facets in the manner which is described by Lynch (6). Following Lynch's prescription, the replicas are shadowed at a low angle with Germanium and the foils were also observed in the TEM under a low angle. In this way artifacts due to clustering of the deposited Germanium are amplified considerably. By taking these carbon replicas from flat tests surfaces, like fresh fracture surfaces of glass, we were able to determine the magnification regime at which the artifacts show up. We found that whenever ondulations were observed on the replicas taken from Fe2.6%Si we were either in the regime where artifacts did show up on glass surfaces or near ledges, where the shadowing conditions and the fracture mode was ill-defined. fig. 26-8 from (4) (partly reproduced here as fig. 5) shows examples of these secondary carbon replicas on which large featureless areas are visible. From careful examination of these carbon replica we had to concluded that there is no evidence of voids on the fracture surfaces in Fe2.6%Si. We also mailed Fe2.6%Si samples to Lynch, he took the replicas, and he agreed that there are large featureless areas on our fracture surfaces.

REFERENCES

1. P. Neumann, H. Vehoff and H. Fuhlrott, ICF4, Waterloo, Canada, 1977, 1313.
2. P. Neumann, H. Fuhlrott and H. Vehoff, ASTM STP 675, 1979, 371.
3. W. Rothe, H. Vehoff, P. Neumann, Third Int. Conf. on Hydrogen and Materials, Paris, 7-11.6.1982, Vol.2, 695.
4. H. Vehoff, P. Neumann, in "Hydrogen Degradation of Ferrous Alloys" R.A.Oriani, J.P. Hirth, and M. Smialowski, Eds., Noyes Publications, Park Ridge, NJ, USA, 1985, pp. 686-711.
5. H. Vehoff and P. Neumann, Acta Met., 28, 1980, 265.
6. S.P. Lynch, in: "Hydrogen Effects in Metals", (I.M.Bernstein and A.W.Thompson eds.), AIME, New York, 1981, p. 863.

DISCUSSION

Comment by R. C. Newman:

You state that the effect of hydrogen in embrittlement of Fe-Si is a very near-surface one. How does this length compare with the extent of an individual cleavage event?

Reply:

From the experimental evidence we can deduce that the embrittling effect of hydrogen does not reach further than 100 nm ahead of the crack tip.

The extent of the individual cleavage events, being defined as decohesion without dislocation activity, is not measurable directly but we can estimate that it is of the same order of magnitude.

Comment by S. M. Ohr:

During fatigue experiments, if dislocations generated during tensile part of the cycle survive unloading, these dislocations can anti-shield the crack during compresseive part of the cycle. This will create a large local k that can induce crack propagation. Have you seen any evidence of this?

Reply:

The loading geometry during compression is so complicated it is not clear that the anti-shielding phenomena of the kind you describe can happen readily. For fully plastic situations in single crystals and large crack advance rates per cycle the back slip in the compression phase is starting at the current crack tip and closing it. Thus the point where the crack is currently closing, is moving backwards from the crack tip. A sequence of these events is shown in Vehoff and Neumann (Acta Met. 1979, 915). I assume that the shielding events during the compression phase are compensating the shielding effects of the dislocations produce in the tension phase. Simultaneously crack closure develops. Therefore I do not think that exceedingly large local K's can develop in compression. At least we have seen no evidence of such K-values.

Comment by R. H. Jones:

Does your work on the effects of H_2 on the crack angle of Fe-Si support the concept of enhanced plasticity induced fracture?

Reply:

I think the answer to this question is clearly no, since enhancement of plasticity would simply lower the stress level at which ductile crack propagation is possible. Therefore, at a given stress level, when it competes with decohesion, the amount of plasticity occurring would be increasing and therefore the crack tip angle would be increasing. The effect of hydrogen is the other way around, however. The crack tip angle is decreased if hydrogen is added. Besides this argumentation there are experiments in which the plasticity is enhanced indeed, i.e., the experiments of Vehoff and Neumann (Acta Met. 1980, 265) in which the effect of crack tip opening rate and temperature without hydrogen was studied in Fe2.6%Si. Both a decrease in loading rate and an increase in temperature enhance the plasticity and in both cases the crack tip angle is increased.

Comment by T. Watanabe:

I am very much interested in Fe-Si bicrystal work you have mentioned. Why does the misorientation dependence of intergranular fracture differ between two types of bicrystals, depending on the site of notch?

Reply:

The slip planes with the largest Schmid factor are inclined at certain angles with respect to the grain boundary. Therefore, the angle between those slip planes and the crack plane is once α and in the other case when the crack comes from the other direction 180-α. Therefore in both cases the crystallographically most favoured slip planes fit differently into the stress field of the moving crack and thus different amounts of plasticity are accompanying the crack moving in one direction or the other. On the other hand the crack propagating on the interface in either direction encounters in both cases the same segregation on the grain boundary. Therefore this experiment is well suited to distinquish between effects of plasticity and of segregation on the crack growth.

Comment by K. Sieradzki:

You have described a cleavage criteria put forth by Lin and Thomson which relies primarily on the opening mode, K_I. The classical fracture criteria $G = 2\gamma$ is of course dependent on K_I, K_{II}, and K_{III}. Could you elaborate on this difference?

Reply:

If the total energy release rate is equated to twice the surface energy it is assumed that all the energy goes into the formation of new surfaces. We know, however, that energy may be dissipated during this process and transported away by phonons. In a similar vein only 10% of the total energy used for yielding is stored as self energy of dislocations. 90% of the energy is dissipated in heat. This may also be true, even if we consider at true cleavage crack in a disloation free material without the emission of dislocations (via phonon emission). Because of the unknown fraction of dissipated energy we must rely on our knowledge of the processes at the crack tip for deriving a fracture criterion. On an atomistic level I follow the arguments of Lin and Thomson that an opening action is necessary for decohesion crack advance, whereas shear components in the loading as expressed by K_{II} and K_{III} are not relevant.

Comment by H. Mughrabi:

You mentioned your observations of fatigue cracks at twin boundaries in copper polycrystals and your findings that, at reduced amplitudes, some disturbance (extrusions?, microcracks?) formed at the twins that you could not identify. I would like to point out that observations (SEM,TEM) exist which show that in copper persistent slip bands form preferentially on the [111] habit plane of twin boundaries, provided they are orientated suitably with respect to the applied stress.

Reply:

This is correct. N. Thompson (in Fracture, Wiley, 1959, p. 354) reported already this kind of evidence. However, in the modern literature, where crack initiation is related to grain boundary types, such an evidence is not reported.

Comment by R. Latanision:

You made the comment that, relative to experiments in vacuum, "chemi-sorbed" oxygen reduces the fatigue life of metals. While that is true phenomenologically, I am concerned about the word "chemisorbed." I don't think we really know very much at all about the chemistry of such environ-mental effects, and it is not clear that chemisorbed oxygen is involved. My point is simply that we need to be careful to separate phenomenology from mechanistics until the latter is better developed.

Reply:

I fully agree. I used the word "chemisorbed" in order to indicate that the oxygen is bound strongly enough to the copper such that it can be pulled into the material by slip and thus is able to weaken the material.

PLASTIC FLOW INSTABILITIES AT CRACK TIPS

Robert M. McMeeking
Materials Program, College of Engineering
University of California, Santa Barbara, CA 93106

INTRODUCTION

Instabilities of plastic flow can occur near crack tips for several reasons. In some cases they can be due to inherent features of the macroscopic phenomenological plasticity behavior and the unstable region spreads over distances large compared to the microstructure. For example, localization of flow in shear bands can occur in large scale yielding. In other cases, microscopic activity can set up geometric or material features which induce the instability on an initally small scale. An example of the latter circumstance is the nucleation of large voids from hard particles in the alloy near a blunt tip leaving an unstable highly strained ligament. These examples are but two of many possibilities that can operate at different size scales. In addition, the presence of an embrittling agent may play an important role in the nature of plasticity and so complicate the picture of crack tip instability. In this chapter, the continuum viewpoint of small scale failure processes involving plasticity near a monotonically loaded crack tip will be outlined. To be able to do this, we must first review recent developments in crack tip plasticity solutions. This will lead naturally to the work that has been done on modelling the ductile failure processes at crack tips. Most of these models involve a plastic instability of one kind or another. The result of the instability is crack advance and further complications arise due to the consequent nonproportional loading and reversal of stressing that material experiences near the moving crack tip. A further instability of the crack growth process itself can ensue causing the transition from slow to rapid propagation. Modelling of this behavior will be considered. Then towards the end of the chapter, some recent work on detailed modelling of environmental effects will be considered, specifically the influence of hydrogen in a model of near tip plasticity.

CRACK TIP PLASTICITY

The plastic flow at crack tips to be considered in this chapter is diffuse so that it will lead to blunting. Another possibility which we will not address is the idea that an atomistically sharp crack tip in a ductile material can be stable against the emission of dislocations. The mechanics of this situation have been studied by Rice and Thomson [1]. A sharp crack is retained if circumstances warrant and a brittle failure is possible. This model is thought to be relevant to environmental degradation in the sense that an embrittling species could tip the balance from dislocation emission and blunting to retention of sharpness and brittle behavior [2]. Instead of considering this problem, we will concentrate on the type of plastic flow that can be described by phenomenological plasticity and some instabilities that result from that kind of behavior. Furthermore, we will confine ourselves to situations of monotonically increasing loads and eliminate fatigue cycling from consideration. Although of substantial importance in crack growth, the detailed mechanics of fatigue crack propagation has not received the same attention as the nonfatigue case. In addition, the connection between the micromechanics and the micromechanisms is a much more difficult subject in the fatigue situation. Finally, we eliminate high temperature creep from consideration, although substantial progress has been made recently in the mechanics of crack growth when creep is the dominant plasticity mechanism. For a brief recent summary of this area see Hawk and Bassani [3].

As far as modelling crack tip plasticity is concerned, a distinction must be drawn' between cracks that are stationary and those that are growing quasistatically. The path dependent and irreversible nature of plastic flow means that crack propagation has a dominant effect on the crack tip stress and strain fields. The simplest situation where this is apparent in antiplane shearing (mode III) in an elastic-perfectly plastic material in small scale yielding. The stationary crack has a singularity for shear strain with a $1/r$ strength where r is the distance from the crack tip [4]. In contrast the steadily growing crack under quasistatic conditions has a log r singularity for shear strain and a plastic zone with a different shape [5]. In inplane deformation the situation is further complicated by blunting and distortion of the crack tip but similar differences between stationary and growing cracks prevail. However, crack growth can be intermittent even for monotonically loaded cracks, with steps of sudden growth interspersed with periods when the crack is stationary. This stationary pause can occur at constant applied load if an environmental degradation is occurring or, in the general case, as the loads increase. In this situation, the models for plastic flow at stationary cracks can be relevant. The relative importance of the stationary solution depends on whether the overall near tip distortions are dominated by those taking place when the crack is stopped or those occurring during propagation. Some cases of intermittent crack growth will be represented well at each step by models for stationary cracks while other cases will appear more like the models for continuous growth. Probably, there will also be cases in between which must be dealt with on an individual basis.

Another important consideration is the extent of the plasticity. If the applied loads are moderate, the plasticity will be confined to a region at the crack tip small compared to specimen size. The characteristics of the crack tip solutions will be determined entirely by Irwin's stress intensity factor K even though the elastic singularity solution is not relevant there [6]. This situation is known as small scale yielding. For larger loads, the plastic zone will spread over substantial portions of the body in a situation of large scale or general yield. Instabilities of a structural nature can ensue in general yield and a treatment of this must take into account geometric as well as material aspects of the problem. In this chapter we will generally consider only small scale yielding where the crack tip geometry is the only significant specimen attribute over and above the material behavior and magnitude of the applied loads.

The fundamental crack tip analysis for tensile opening (mode I) inplane strain for an isotropic material is that based on small geometry change and thus for a sharp crack tip. In perfect plasticity [6] in this model, the stresses very near the tip are independent of distance from the tip and in front of the crack involve a very high stress which is 3 times the yield stress in tension. The strains ahead of the crack are small, but singularly large (1/r) above and below the tip. For power law strain hardening, the HRR singularity prevails [7,8] in which the strains are less singular than in perfect plasticity but the stresses now become unbounded at the tip. Strain is still concentrated above and below the crack. In all these cases, the stress intensity factor K characterizes the crack tip fields, but, more directly, the J-integral [6] is path independent and can be used also as a characterizing parameter.

CRACK TIP BLUNTING

Rice and Johnson [8] studied crack tip blunting in mode I by using slip line solutions. The near tip behavior of elastic-perfectly plastic materials in plane strain approaches that of rigid-perfectly plastic materials and justifies the use of slip line theory. Their solution enforced smooth blunting of the crack tip and as a result a region of high strain develops ahead of the crack in contrast to the sharp crack case. The stress adjacent to the free surface is limited by yielding but builds up away from the tip due to triaxiality. The largest stress is still $3\sigma_o$ where σ_o is the yield stress in tension but instead of immediately ahead it is now located a distance approximately $2\delta_t$ ahead of the tip where δ_t is the crack tip opening displacement. The solution makes apparent the high strain and stress region near the tip in which plastic instabilities can occur. Indeed, in experiments Green and Knott [10] have observed failure in this region ahead of a blunt crack tip. The fracture developed along the theoretical principal shearing directions which form nearly logarithmic spirals from the crack.

McMeeking [11] used large deformation finite element analysis to model the blunting of a sharp crack under small scale yielding in plane strain conditions. He considered both nonhardening and hardening elastic-plastic materials. The constitutive law he used was von Mises yielding with isotropic hardening (J_2 flow theory) and accounts for rotation of the principal deformation directions. The small scale yielding solution was achieved by applying displacement boundary conditions remote from the crack tip to impose an asymptotic dependence on the mode I elastic crack tip singular stress field. Figure 1 shows the true stress $\sigma_{\theta\theta}$ (see inset of Fig. 1) normalized by the true tensile yield stress σ_o, vs the distance from the notch tip in the undeformed configuration for an elastic perfectly-plastic material. The distance is normalized by the current notch-opening, which allows results from the later increments of the finite element calculations to be plotted together. As the notch tip is approached at a given angle to the crack line in the undeformed configura-tion, the stress rises due to increasing strain. However, the hydrostatic stress cannot be maintained on the blunted notch surface and as a result there is a maximum for $\sigma_{\theta\theta}$, coinciding with a maximum for hydrostatic stress, some distance from the notch tip. Figure 1 also shows equivalent plastic strain plots from the later increments of the finite element solutions. The strains are clearly small except very close to the blunted tip. Outside the near-tip region, the large plastic strains are on lines at an angle to the crack plane and this is in agreement with the HRR solution [7,8]. For comparison, the stress and plastic strains ahead of the crack ($\theta = 0$) from the slip line solution of Rice and Johnson [9] have

94

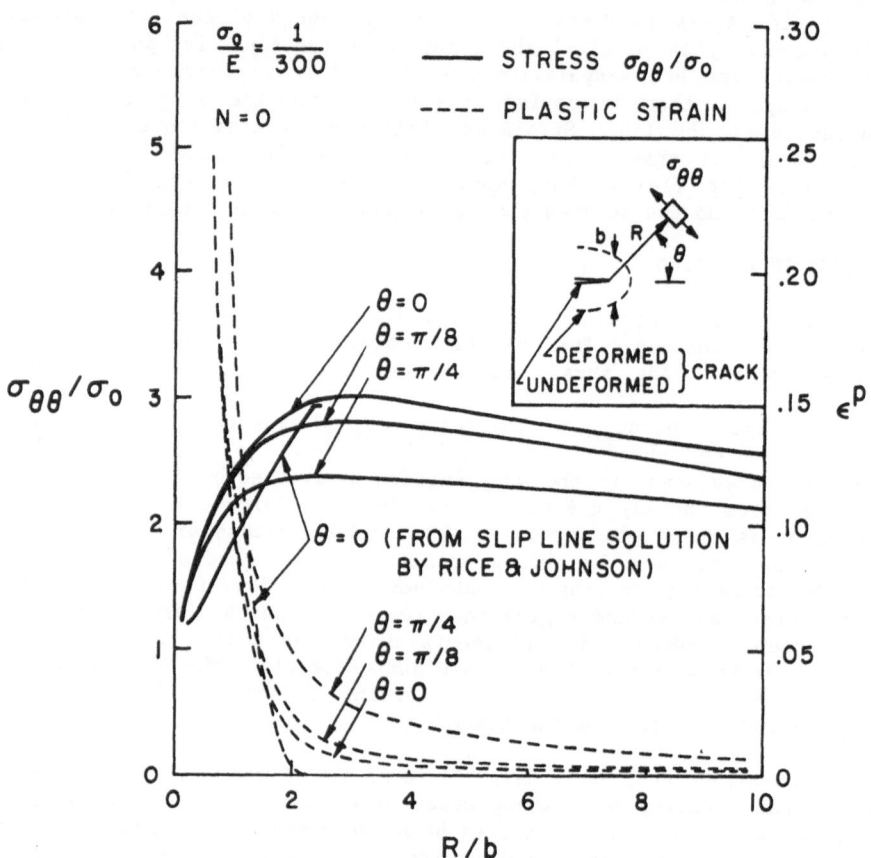

Figure 1. Stresses and tensile equivalent plastic strains near
a blunting crack tip in plane strain small scale
yielding in an elastic perfectly plastic material.

been plotted in Figure 1. The agreement between the finite element results and the slip line results are quite close as far as the position and magnitude of the stress maximum are concerned.

Similar finite element calculations for power-law hardening materials show that the magnitude of the stress at the maximum is higher and that the position of the stress maxima move closer to the notch tip as shown in Figure 2. Another hardening effect is an upturn in stress close to the notch surface, which arises from the elevation of flow stress by the large plastic strains in this area. In fact, when a sharp crack in a power-law hardening material is blunted to a finite width, infinitely large stress on the notch surface will arise, but only over a distance small compared to the blunted-crack width. This stress singularity arises because in the power-law hardening model infinite plastic strains produce infinite flow stresses, whereas a saturation to constant flow stress after large plastic strains is more physically realistic. In addition, few cracks are likely to be atomistically sharp and the finite radius at the tip of most real cracks will lead to only large but finite plastic strains at the tip.

It is important for instability considerations that the analysis of the crack tip strains for the hardening materials was carried out with stress strain curves that continuously rise. Although instabilities such as necks occur in strain hardening material, they are unlikely to develop in the highly constrained flow field in the near tip region in isotropically hardening materials. Localization in shear bands can be ruled out [12] for the intuitive reason that the continuous strain hardening will discourage accumulation of localized strain. The same reasoning shows that corners will not develop on the blunt tip shape because it would involve intense shearing at the vertex.

Thus it would seem that blunting crack tips in continuously strain hardening isotropic materials in small scale yielding which have not reached the propagation stage are rather stable. On the other hand, if the strain hardening is exhausted at finite strains then unstable possibilities exist. One is that shear bands can develop at 45° to the principal stress directions [12]. There have been no model calculations for blunting cracks in perfectly plastic undamaged isotropic materials in small scale yielding that have developed near tip shear bands, although Green and Knott [10] have observed them experimentally as mentioned before. It is possible that the algorithms and mesh arrangements used in calculations such as McMeeking's [11] are not well suited to picking up such shear banding. However, the near tip strain fields are highly inhomogeneous and it seems likely that an unstable mode of deformation would become apparent in the calculations. In fact, in the large scale yielding finite element analysis of McMeeking and Parks [13] for center crack panels, a shear band did develop and the strain field approached the pattern of 2 slip lines emanating from the notch that characterizes this geometry in large scale perfect plasticity [14], suggesting that algorithm and mesh arrangements that have been used are suitable for picking up instabilities. Thus a tentative hypothesis would be that absent a destabilizing feature of the material behavior, near tip shear band localizations will not occur in isotropically hardening materials with smooth yield surfaces in small scale yielding even if the hardening is exhausted. Some insight is to be gained from the slip line field of Rice and Johnson [9] shown in Figure 3. If localized shearing were to occur on a slip line such as the one from $(\alpha, \beta) = (-\pi/2, 0)$ to $(0,0)$, the velocity discontinuity

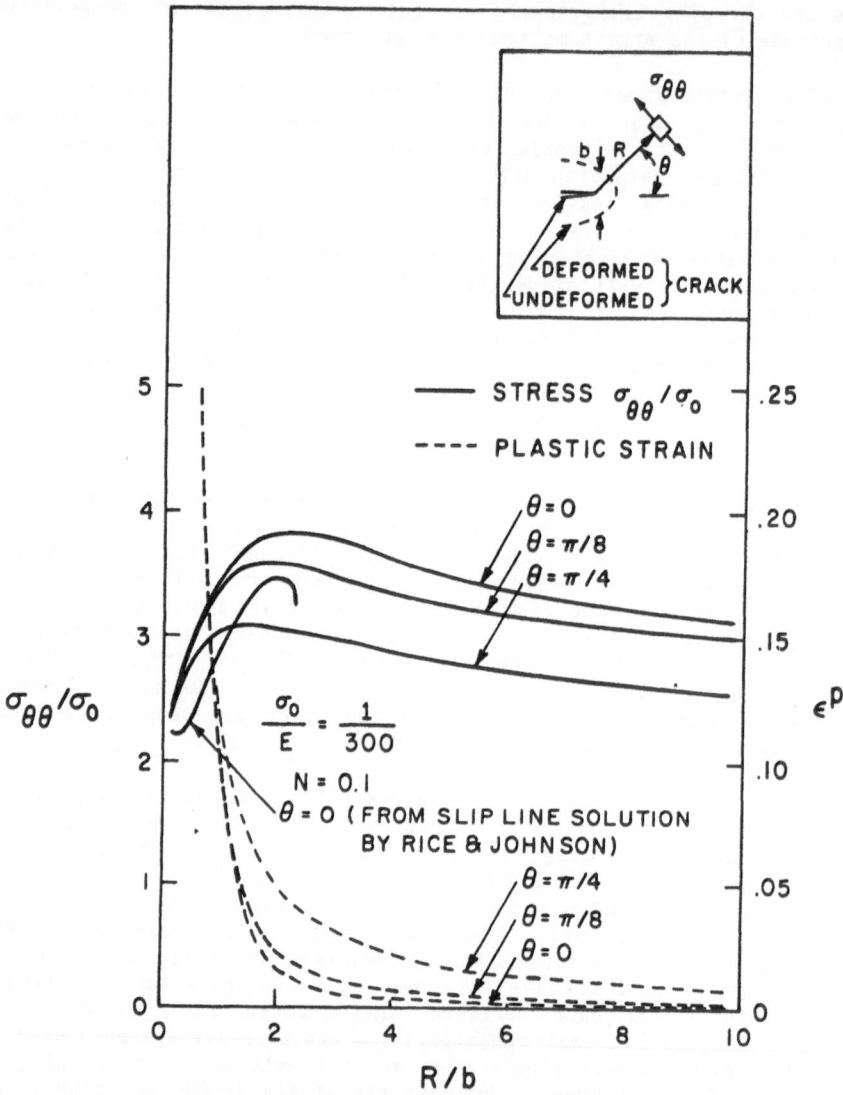

Figure 2. Stresses and tensile equivalent plastic strains near a blunting crack tip in plane strain small scale yielding in an elastic perfectly plastic material.

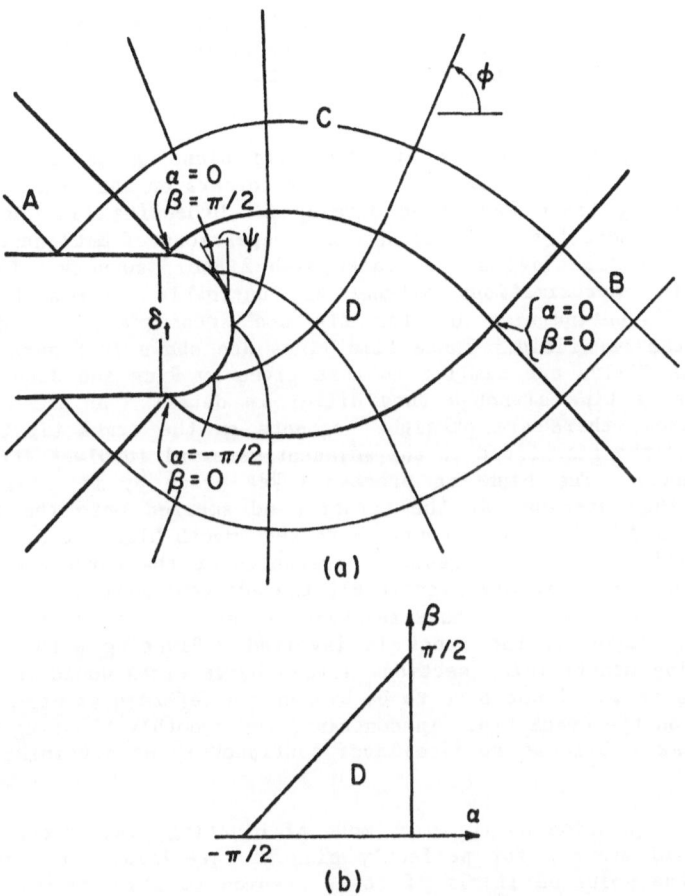

Figure 3. Slip line field for smooth blunting of a
crack in a rigid perfectly plastic material.

would have to be continued all the way along that slip line and into the region of small plastic strains further afield from the crack tip. Such intense shearing could not be accommodated further afield in the contained plastic zone and would seem to be ruled out in the analysis as strictly stated. Destabilizing features such as voids, particles or dislocation effects could promote localized flow, but generally they would introduce features that would alter the classical picture of crack tip continuum plasticity and blunting.

On the other hand, models of crack tip blunting in small scale yielding in perfect plasticity which involve corners on the crack tip and localized shearing there have been developed. McMeeking [15] used the methods of Rice and Johnson [9] to develop suggestions of McClintock [14] for near-tip slip line fields for cracks with 2 and 3 corners. They are valid near tip approximations without any untenable incompatibilities because the discontinuities on the tip are accommodated by diffuse straining in the far field. These slip lines are shown in Figures 4 and 5. The stress fields are similar to that given by Rice and Johnson for smoothly blunting tips although they differ in detail. As can be seen from the figures, there are straight segments on the crack tip between each corner and the flow field in the adjacent material involves straining by small amounts. The blunting process takes place by material being brought from the interior at the corners and smeared onto the tip to create new surface. This is in contrast to the smooth blunting case where the blunting takes place by a general stretching of the surface material and the strains are intensely high in all the adjacent material. As Rice [2] has pointed out, this may have relevance to stress corrosion cracking when a surface layer on the crack is involved. Blunting with a crack shape containing nonstraining sections joined by vertices would mean that the surface layer would not have to be broken and reformed at more than a few locations on the crack tip. In contrast, the smoothly blunting tip is likely to break a brittle surface layer continuously as straining takes place.

There is a question as to which mode of blunting, the smooth or the cornered, should prevail for perfectly plastic materials. It cannot be deduced from the solution itself if it is assumed to start from a mathematically sharp crack tip. The first instant of blunting would determine the shape which would form and it seems likely that shape would continue thereafter. However, if the starting configuration is blunt already and has cornered features then that shape might prevail thereafter. Thus the shape of a pre-fatigue crack, pit, or flaw is of some importance. Similarly, a crack tip with a brittle surface layer which cracks in a few locations might set off the corner blunting. In addition, one case could be destabilized into another perhaps by features of the microstructure that eventually favor one mode over the other. Observations do not settle the issue although the smooth case seems to be seen more often than one with corners. Clayton and Knott [16] have found localized shearing at many corners on a macroscopic notch tip while Rawal and Gurland [17] observed the smooth mechanism of blunting in the opening of a prefatigue crack with a shape very close to that shown in Figure 3. Further work is necessary to bring together the modelling of micromechanisms and micromechanics to elucidate the issues of smooth and nonsmooth blunting and possible unstable transitions between them.

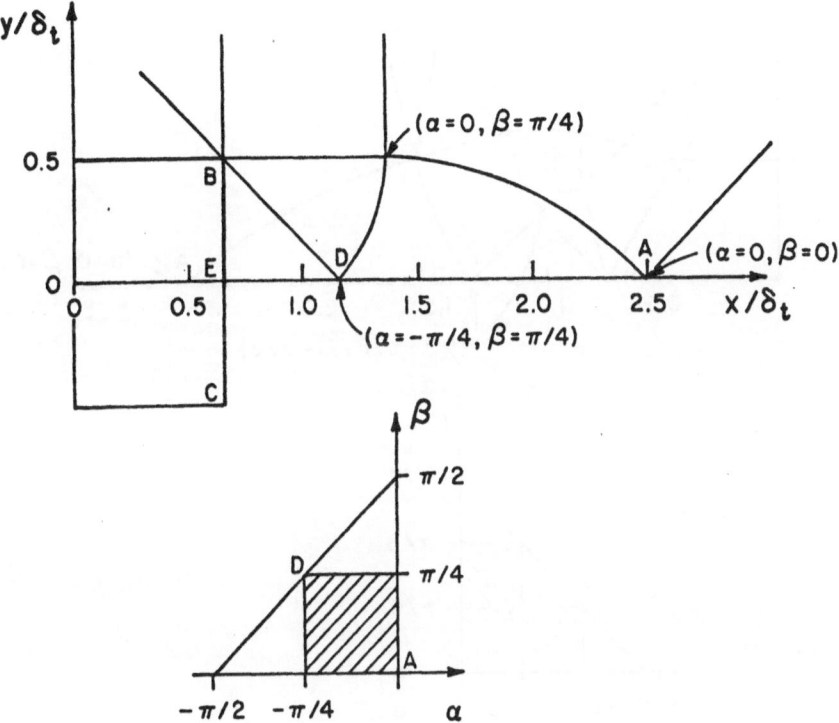

Figure 4. Slip line field for square bottomed blunting in a
rigid perfectly plastic material. Crack tip is BC.

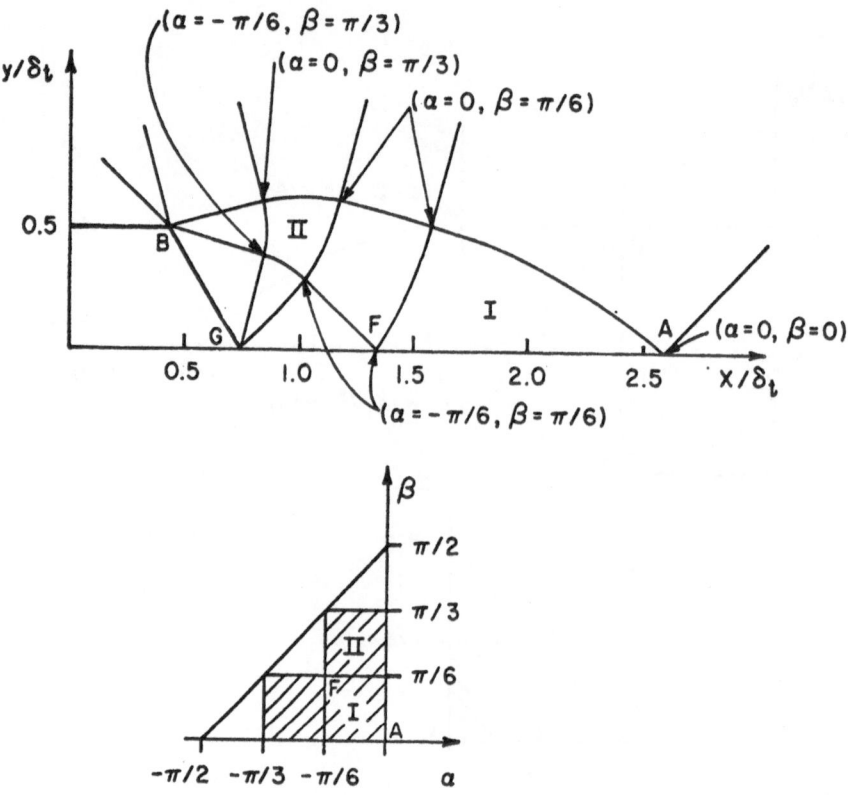

Figure 5.   Slip line field for blunting with 3 vertices.
Crack tip is BG and its mirror image.

DUCTILE RUPTURE AT CRACK TIPS

A major source of plastic instability at a crack tip can be the
generation of voids from inclusions and their subsequent growth and
coalescence in ductile fracture. McMeeking and Aravas [18,19] have
recently summarized the mechanics modelling in this area. Instabilities
of a variety of types can arise. One possibility is that void sheet
development gives rise to shear bands while another is that ligaments
between the crack and a void can fail in a number of unstable ways. One
important problem in the latter area is when a relatively weak inclusion
like MnS in steel gives rise to a large void at a low strain [20]. In
these circumstances a few voids only interact with the crack and modelling
can be limited to the blunting crack and a single or limited number of
voids. Although of some importance, the nucleation stage of the void is
not predominant and the growth and coalescence stages are of most
interest.

Aravas and McMeeking [21,22] used large deformation finite element
analysis to study the near crack tip growth of long cylindrical voids
aligned parallel to a mode I plane strain blunt crack under small scale
yielding conditions. The results of the calculations provide a reasonable
model for the behavior of holes generated by long stringers parallel to
the crack, like those in specimens cut in the long transverse direction of
a rolled plate. Two different configurations were analyzed: one with a
single hole ahead of the crack and one with two holes at 30° to the crack
line. Several values of the spacing to size ratio of the inclusions were
considered and the effects of this ratio on the conditions for fracture
initiation were examined. In a first set of calculations [21] the elastic-
plastic material was assumed to be fully dense and the possibility of
smaller-scale voids in the ligament between the large void and the crack
tip was not taken into account. The J_2 flow theory, suitably modified to
account for rotation of the principal deformation axes, was used to
describe the constitutive behavior of the material. Figure 6 shows the
deformed finite element mesh in the near tip region superposed on the
undeformed one (dashed lines) for the case of the void ahead of the crack.
The deformed configuration for the case of the two voids at 30° to the
crack plane is shown in Figure 7. In both cases the holes are pulled
towards the crack tip and change their shape to approximately elliptical
with the major axis radial to the crack. This shows that the effect on
the hole growth of the interaction of the neighboring free surfaces is
stronger than the effect of the mainly tensile stress field ahead of the
crack tip. The results of Aravas and McMeeking [21] show that the cylindrical holes ahead of the crack grow faster than those at 30° and this is
rather different from what has so far been inferred for the growth of
spherical voids [11].

In the figures, the behavior of the material between the void and the
crack as a necking unstable ligament is clear. It must be borne in mind
that the calculation is 2-dimensional and probably exaggerates the interaction of the void with the crack. However, the 3-dimensional behavior of
an initially spherical void with a blunting crack can be expected to have
a substantial element of void crack interaction. The constraint of the
material around the ligament will still force the ligament to stretch as
the crack blunts and incompressibility will force the void to expand
toward the crack. This will effectively cause the ligament to neck. In
addition, the hole will grow at a rate determined by the crack blunting

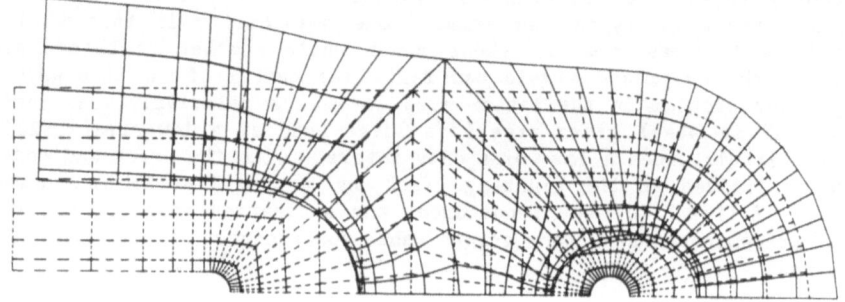

Figure 6. Near tip finite element mesh for a cylindrical hole
growing near a blunting crack tip. The original mesh
is in dashed lines, the deformed mesh is in full lines.

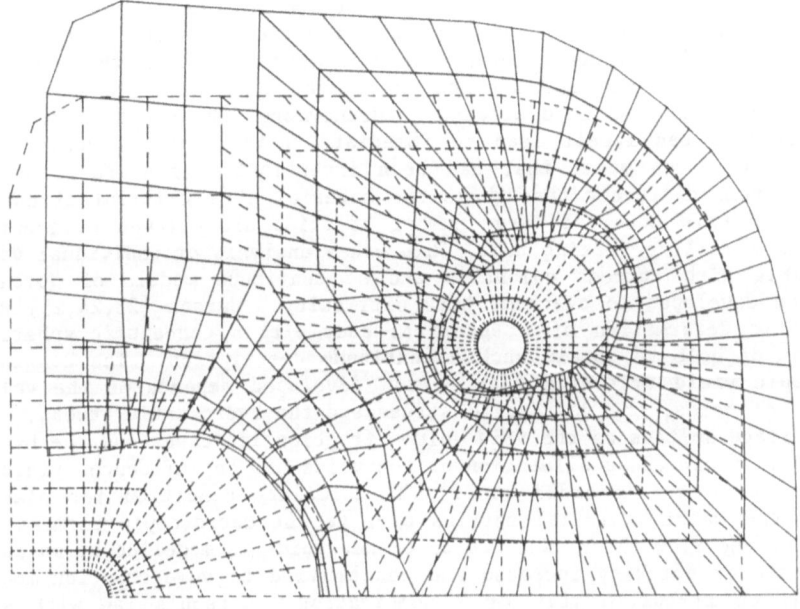

Figure 7. Near tip finite element mesh for a pair of
cylindrical holes symmetrically placed near a
blunting crack tip. The original mesh is dashed,
the deformed mesh is full.

and this can be thought of as a consequence of the unstable plastic flow of the ligament.

It can be inferred from Figures 6 and 7 that eventually the ligament would neck down to almost a chisel point but the amount of crack tip opening involved would then be substantial. Generally, another material instability will lead to failure of the ligament. This could be due to the nucleation of smaller voids from more tenacious smaller particles in the material. The nucleation event could be driven by some other unstable feature of the material and a proper model for that situation would have to take this behaviour into account. Furthermore, the creation of voids from particles can itself lead to immediate unstable shear localization [12, 23, 25]. One model that has been investigated is where voids are created ràther diffusely throughout the material at a critical stress or strain. In this model, a large number of voids are allowed to generate much like a failing necking bar. This makes analysis of individual voids intractable with current computers and so continuum models of porosity have been developed to overcome this difficulty. Gurson [23,24,25] used the plastic deformation of a spherical body with a concentric spherical inclusion or hole to obtain such a continuum description. The continuum macroscopic yield criterion that Gurson developed depends on the volume fraction of voids, the deviatoric stress and the hydrostatic stress. The yield surface softens as the void volume fraction increases and the hydrostatic stress can cause dilatation of the voids and so influences yielding in contrast to the unvoided case. The normality condition for plastic flow of the matrix material carries over to the macroscopic element and determines the flow rule. Nucleation of new voids is modelled by an additional rate of porosity increase and can be tied to void creation models depending on stress, strain or a combination. Gurson's law with some modifications have been used effectively to model the development of porosity and then a microcrack and failure in the neck of an initially unvoided tensile bar [26].

A difficulty of this model for void nucleation and porosity is that it is scale free. That is, the theory is based on a continuum of voids and in an inhomogeneous increasing stress field, void nucleation would commence at a single point and then spread. In any realistic situation, the particles from which the voids are nucleated are finitely spaced and while one by one void nucleation could occur, the associated plasticity should not be modelled on a continuum basis until many holes have been created. The resulting continuum model can apply only on a scale big enough to include many voids. Of course, the same difficulty applies in classical continuum plasticity where regions modelled must include many slip systems and many grains if polycrystals are involved, yet the models are used as if applying to infinitesimal elements of volume. It follows that the continuum porosity models should be used only if there are many voids per few grains or if a size scale for regions including many voids is inserted into the model. In all treatments of problems discussed in this chapter, the model is used scale free. Since the continuum plasticity model already involves the restriction that it is only valid if taken as the flow response of a volume element containing several grains, use of the continuum porosity model without a scale implies that there are many particles that nucleate voids within that collection of several grains. This restriction can be relaxed if response averaged over very large numbers of grains in materials with very fine microstructures are of interest, in which case nucleation of many voids on the scale of very many grains can be allowed for. When used in this manner, there is a danger that numerical calculations will be rather sensitive to mesh size and arrangement as far as the nucleation and growth of porosity is concerned. This problem is the standard error issue of solving partial differential equations on a grid, and the grid must be fine enough to provide results with a desirably low deviation from the exact solution. We assume that the numerical results discussed in this paper involve such low errors.

Aravas and McMeeking [22] took the nucleation of smaller-scale voids into account by using Gurson's [32,33,34] equations as modified by Tvergaard and Needleman [26], to model the constitutive behavior of the matrix material in a scale free manner. Using the modified Gurson equations and a method proposed by Tvergaard [27] to model material failure, they studied the formation and growth of the microcrack in the ligament between the larger hole and the crack tip. The developing microcrack is illustrated in Figure 8. In this way, the final stage of coalescence of the larger hole with the crack tip was analyzed in detail. The difference between the predicted COD at fracture initiation between the Gurson values and those based on results for the fully dense material is small. This shows that the results obtained using the fully dense elastic-plastic material together with some geometric criterion to predict localization are, numerically, fairly satisfactory.

It is usually assumed that a plane strain ligament such as the one modelled by Aravas and McMeeking [22] would fail in a process of localized shearing at 45° to the center line of the ligament [9]. There is no indication of such a failure mechanism in the results of Ref. [22]. This may be due to deficiencies of the mesh and a much finer mesh designed in an appropriate manner may produce the 45° shearing off. Indeed Tvergaard and Needleman [26] have obtained 45° shear band in plane strain tensile specimens. It should also be mentioned that the family of particles which nucleate voids and cause failure in the ligament may be those that harden the material by dislocation pinning and entanglement. When these particles come loose from the matrix, their hardening effect will be lost in addition to porosity being created. Even if only a few fail, some lowering of the macroscopic flow stress will result. Some of these effects may be accounted for by the adjustments that have been made to Gurson's laws [26, 27] but the effect may be more substantial than has been allowed for. The softening involved would be quite destabilizing and may be a key feature of the failure and shear banding of ductile ligaments near crack tips.

Calculations similar to those of Aravas and McMeeking [18,19] have been carried out by Aoki et al. [28]. They too used the Gurson laws to analyze the behavior of a failing ligament between the blunt crack and a large cylindrical hole. The hole grows into a shape different from that predicted in Ref. [18 & 19] and this discrepancy is unresolved. Some of the calculations in Ref. [28] concern strains due to microvoid growth around a blunt tip in the absence of a larger void. These results are probably relevant to very clean alloys or powder compacts with residual porosity. However, the comments previously made as to absence of size scale apply. In more typical cases, the role of microvoids is in the interaction between a larger hole and the blunt crack, as analyzed by Aravas and McMeeking [18,19]. Aoki et al. [29] have extended their calculations to mode II (in plane shearing) and mixed mode conditions. These calculations are of interest as they give finite deformation results for

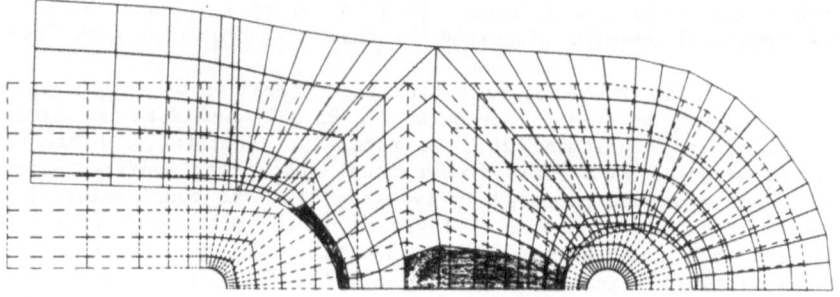

Figure 8.   Near tip finite element mesh for a hole growing near
a blunting crack tip. The material is modelled by
continuum dilatant plasticity representing the
nucleation, growth and coalescence of small voids.
Shaded regions have failed and the stress there is
essentially zero.

mixed mode blunt cracks, including the sharpening of one side of the blunt tip due to near tip shearing.

Needleman and Tvergaard [30] have used a blunt crack tip surrounded by the Gurson material in calculations of ductile rupture. In these calculations, however, there are islands of weak material distributed throughout the body. These islands nucleate damage at fairly low strains and should be thought of as colonies of inclusions which give rise to large voids in certain alloys such as those studied by Hancock and Mackenzie [31]. The remainder of the material nucleates voids at high strains and represents portions of the alloy containing small tenacious inclusions. The analysis is plane strain and so in the calculation the weak colonies form long cylinders through the material. As the crack is blunted, voids effectively form in the weak regions near the tip. Eventually, the ligaments between a nearby void and the blunt crack loses strength almost entirely and the crack has grown to absorb the first void. The calculations were continued so that coalescence with the crack of voids further afield was achieved as shown in Figure 9. In this way, continuing crack growth was induced. In some of the calculations, patterns of crack propagation were observed that were very similar to experimental shapes seen by Green and Knott [10]. Shear banding occurred also in the calculations, whereas such a behavior did not take place in the calculations of Aravas and McMeeking [19] or Aoki et al. [28]. However, the predictions of toughness phrased in terms of critical crack tip opening displacement for crack growth initiation are similar to those of Aravas and McMeeking [18,19] and all are in reasonable agreement with the data [18,30]. Thus the prediction of toughness is not very sensitive to the presence of localization in these particular calculations compared to the differences among the data. However, the morphology of crack growth is reproduced more effectively in the localized shearing of the calculations of Needleman and Tvergaard [30].

Figure 10 shows experimental results of several researchers together with the theoretical calculations of Rice and Johnson [9] and McMeeking [11], and the finite element results of Aravas and McMeeking [18,19] for the case of the single void ahead of the crack and for two different hardening exponents, namely N = 0 and 0.2. In Figure 10, b_f is the COD at fracture initiation, D is the average inclusion spacing and a_1^o is the average inclusion size. The results of Rice and Johnson [9] and McMeeking [11] perhaps provide an upper bound to the actual value of b_f/D because the interaction between the free surfaces of the blunting crack and the void was ignored in their calculations. On the other hand, the results of Aravas and McMeeking [18,19] provide perhaps a lower bound for b_f/D; the reason is that during the actual fracture process the nucleated holes at elongated inclusions have a length which is certainly smaller than the thickness of the specimen.

The estimates based on the finite element calculations for the perfectly-plastic material seem to agree well with the experimental results for prestrained En1A. The effect of prestraining is to reduce the hardening capacity of the material and to nucleate holes, if the conditions for nucleation are reached. For the En1A mild steel used by Green and Knott [10] the nucleation strain is of the order of the yield strain and the holes are nucleated relatively easily. Also Green and Knott [10] report that the extent of prestrain was sufficient to exhaust the hardening capacity of the material altogether. The specimens were also

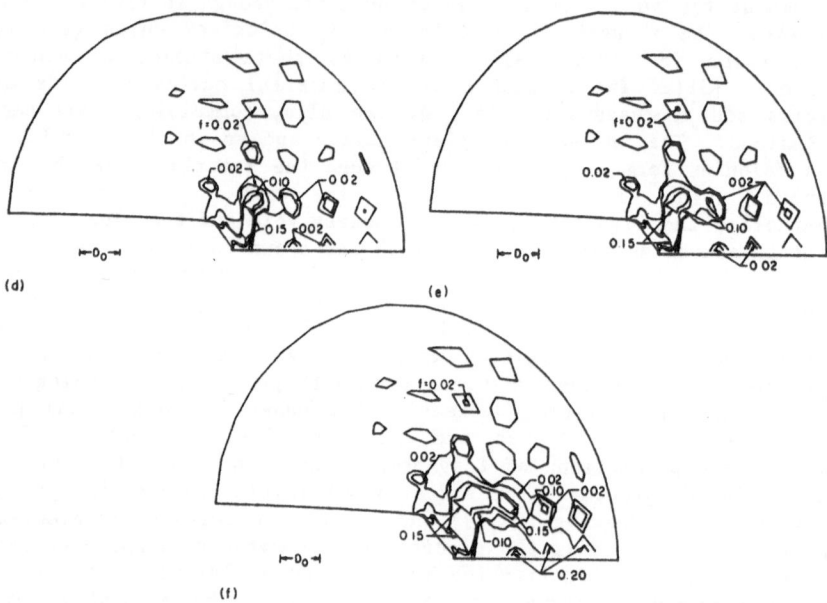

Figure 9. Plot of void volume fraction near a blunting crack
tip in a material with a continuum model of voids
nucleating, growing and coalescing. The crack is
propagating at first in a zig-zag manner but then
proceeds straight ahead. Plot is reproduced from
Needleman and Tvergaard [30] with permission.

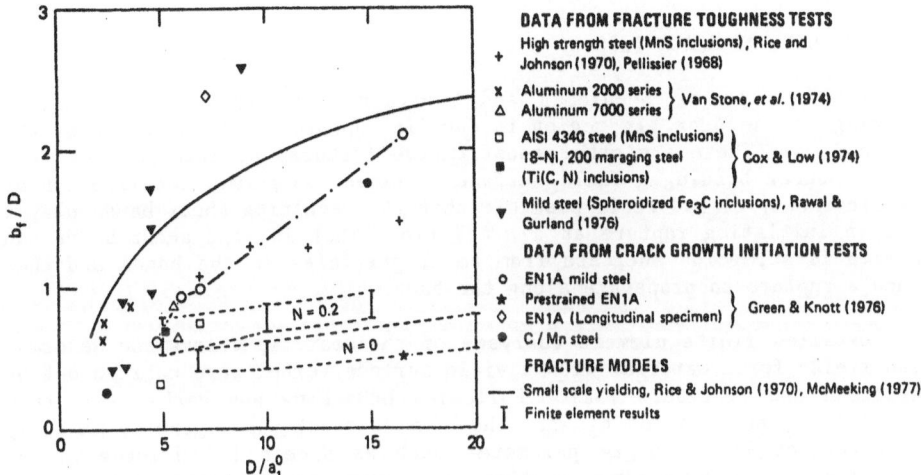

Figure 10. Data for COD at initiation of crack growth of fracture,
related to inclusion spacing D and particle size a.
Also, results for fracture from Rice and Johnson [9] ,
McMeeking [11] and Aravas and McMeeking [21].

cut from rolled plates in the long transverse direction so that the elongated inclusions were parallel to the crack front. This has the effect of making the material and inclusion conditions almost exactly the same as for the perfectly-plastic finite element 2-d calculations. A more detailed discussion of the experimental and finite element results can be found in [18]. Needleman and Tvergaard [30] have also plotted their results in a diagram like Figure 10. Their 3 points lie near 5 on the abscissa and 0.7 on the ordinate and so are comparable with the calculations of Aravas and McMeeking [18] as discussed previously.

SHEAR BANDING AT CRACK TIPS

The results of Needleman and Tvergaard [30] show that localized shearing in the near tip region in ductile rupture can be modelled by the Gurson laws. There are other constitutive features of material that can lead to shear banding. Yield surface vertices can permit localization of plastic flow [12] and it is possible that the resulting shear bands play a role in initiating rupture at a crack tip. That is, the shear bands can develop first, voids nucleate from small particles in the bands and then cause a rupture to propagate along the band.

Detailed finite element analyses of the near tip stress and deformation fields for a material with a yield surface vertex were carried out by Needleman and Tvergaard [32]. Following McMeeking and Parks [13], they focussed on the question of when the blunted crack tip region is uniquely characterized by a single parameter such as Rice's [6] J-integral, or equivalently, on the minimum specimen size requirements for valid J-tests. They carried out calculations for deeply cracked center-cracked panel (CCP) and single-edge crack bend (ECB) specimens. They based their calculations on the phenomenological corner theory of plasticity, termed J_2 corner theory, proposed by Christoffersen and Hutchinson [33]. In this model, a vertex is allowed to form at the current stress point on the yield surface as the material hardens. This can be thought of as introducing the effect of crystallographic multislip in the material. The results of Needleman and Tvergaard [32] show that the highly localized deformation accompanying shear banding leads also to near tip strain fields very different from that predicted by the classical smooth yield surface model. The initially smooth notch tip of the ECB specimen was flattened out and shear bands developed from the corners of this flattened region as shown in Figure 11. Similar results were found for the early deformation stages of the CCP specimen. In previous calculations by McMeeking and Parks [13], a tendency to form somewhat diffuse shear bands in CCP speciments of strain hardening materials was found at levels of load producing large scale yielding. The material used had a smooth yield surface and no such behavior was observed in small scale yielding in contrast to the findings of Needleman and Tvergaard [32] for the material with a yield surface vertex. Furthermore, McMeeking and Parks [13] observed no shear banding in the ECB case even at large scale yielding levels of load.

In view of the results of Needleman and Tvergaard [32], it does then seem likely that near tip shear bands can develop in small scale yielding due to effects associated with crystallographic multislip. These bands could well destabilize the material by nucleating voids and further modelling studies of this are called for, both for materials with and without large voids near the crack tip.

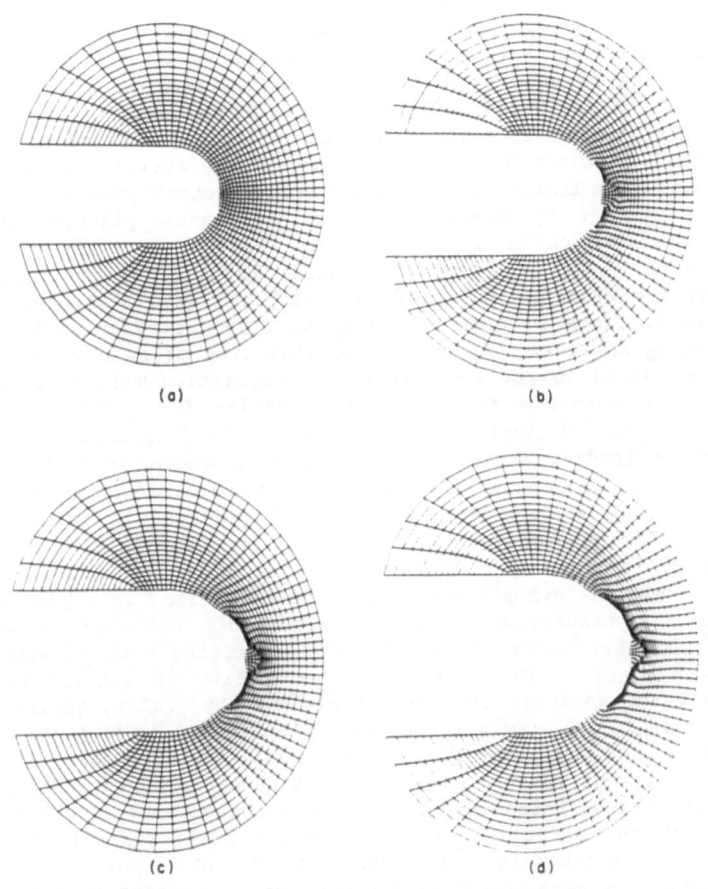

Figure 11. Deformed near tip mesh at four stages of deformation
for J_2 corner-theory for a crack in an edge cracked
beam in bending. Reproduced from Needleman and
Tvergaard [32] with permission.

ELASTIC PLASTIC CRACK GROWTH

Some authors have studied plastic instability at a crack tip from a somewhat more macroscopic point of view. In this treatment, the crack tip is considered to be sharp and so no details of crack tip structure and shape are involved. Furthermore, the microscopic processes leading to fracture and crack propagation are not considered, but a more phenomenological criterion is used to determine critical events. Typically the crack opening shape is used as a crack growth criterion on the basis that ductile processes of failure are largely strain controlled. Thus the near tip crack opening parameters measure the near tip strain levels and can be used as critical criteria for the operation of the ductile failure processes. Within this framework, the crack tip fracture problem in plane strain can be studied by consideration of an elastic-plastic material containing a crack which is loaded in mode I and small scale yielding. The loads are increased monotonically until the critical condition for propagation of the crack is met. At this stage, growth commences and it can be made to continue to grow according to some criterion such as that the crack opening angle be constant. The stability of such crack growth can now be considered in the sense that the propagation will be stable if the load to just maintain the critical condition for growth increases faster than the rate of increase caused by the loading system. In the case of a rigid loading system, the propagation would be stable under rising applied load as long as the load required to maintain the material crack growth continues to rise. If the load required to maintain the critical material condition for crack propagation goes through a maximum or approaches an asymptote, then as the applied load continues to rise through this level, the crack growth cannot be maintained in a quasistatic stable manner. A transition to dynamic, possibly unstable, processes would occur. In that sense an instability associated with plastic flow processes has occurred. This notion of instability is similar to that involved in R-curve analysis in plane stress used by Krafft, Sullivan and Boyle [34] and in plane strain large scale yielding by Paris et al. [35]. A distinction is that the plane strain small scale yielding case is highly constrained and free of influence of component geometry except through the magnitude of the stress intensity factor. As a consequence, one can think of the crack growth instability in plane strain small scale yielding as being a material instability. The question we shall pursue is which aspects of plasticity make the material more or less prone to this crack growth instability.

The most complete analysis for plane strain small scale yielding crack growth is that for perfect plasticity with a von Mises yield criterion and associated flow and was given by Drugan, Rice and Sham [36]. This follows on from work of Rice [37, 38] Cherepanov [39], Rice and Sorenson [40], Gao [41] and Rice, Drugan and Sham [42]. The near tip asymptotic stress field for such a growing crack is similar to the field for a stationary crack in the same circumstances. As shown in Figure 12, the stresses ahead of the crack are high, but fall in a nonproportional manner to yield stress levels behind the tip. As they do so, the state of stress changes from tension normal to the crack plane ahead of the crack, to pure shear adjacent to the tip and then to tension parallel to the crack plane behind the tip. This is the sequence of stress states that a material point close to the crack would experience as the crack tip grows by. This history of stress determines the elastoplastic strains. In turn, the strains determine the crack opening profile. The analysis of

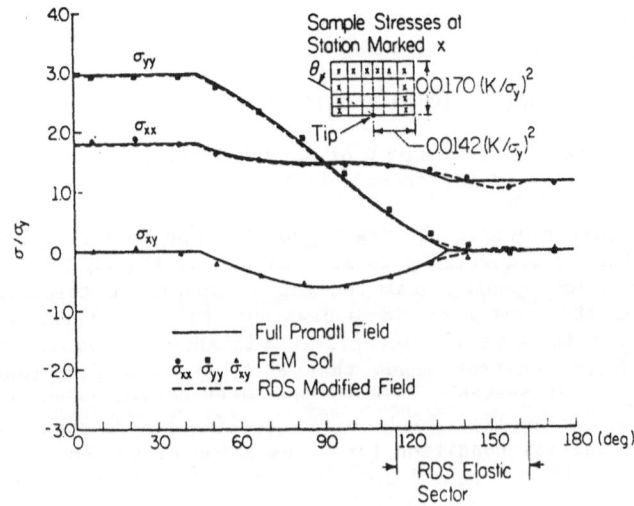

Figure 12. Stress field around a crack tip during steady state mode I plane strain crack growth in an elastic perfectly plastic material with Poisson's ratio 0.3. Finite element results from Ref [46] and an asymptotic solution (RDS) from Ref [42] are shown. Full Prandtl field applies to the stationary crack.

Drugan et al. [36] gives a near tip crack profile that can be characterized by

$$\delta = \beta(\sigma_o/E) \; r \; \ln(\rho/r) \tag{1}$$

where δ is the crack opening a distance r behind the current location of
the moving crack tip, β is a number approximately 5, σ_o is the yield
stress in tension, E is Young's modulus and

$$\rho = sEJ/\sigma_o^2 \; \exp[1 + (\alpha/\beta)(E/\sigma_o^2)(dJ/da)]. \tag{2}$$

In eq. (2) s $\sim$ 0.2, J is the J-integral of Rice [6] evaluated in the far
elastic field, $\alpha \sim$ 0.6 and a is the crack length.

The parameter ρ characterizes the crack geometry and so a propagation
criterion can be stated by requiring ρ to be constant during crack growth.
This means that the crack opening near the tip is invariant relative to
the growing tip and the criterion is equivalent to a critical crack
opening angle measured to some characteristic distance behind the tip.
The criterion that ρ is constant means that eq. (2) is a differential
equation for J during crack growth. This in turn will determine the value
of K associated with the applied loads since $J = (1-v^2)K^2/E$ where v is
Poisson's ratio. The initial condition for integration of the equation is
$J = J_{IC} = (1-v^2)K_{IC}^2/E$. If $\rho \leq$ seE J_{IC}/σ_o^2 (where e is the base of the
natural logarithm), dJ/da is zero or negative at the outset and under
rising load there would be an immediate instability of crack growth. If ρ
exceeds seE J_{IC}/σ_o^2, crack growth in a stable quasistatic manner is
possible. J will rise until dJ/da = 0 at which stage J will have the
value $J_{ss} = (1/se) \; \rho\sigma_o^2/E$. J_{ss} is the value of J at which crack growth can
occur in steady state. If the loads are increased above the level characterized by J_{ss}, unstable crack propagation will occur.

Hermann and Rice [43] have compared this model of crack growth with
data they obtained for a high-strength AISI-4140 steel with σ_o = 1,173
MPa. In this material they observed 5 to 10 mm of crack growth prior to
general yield in the specimens they tested. Integrals of eq. (2) with
s = 0.23, α = 0.65 and β = 5.08 were found to agree reasonably well with
plots of J versus a from sub-general yield experiments with J_{IC} = 35 kN/m
and ρ = 7.2 mm. The inferred value of J_{ss} was 80 kN/m which agrees
reasonably well with the plateau in the experimental curves.

Recall now, that the process of crack growth involves stresses at a
given material point which change substantially in a non-coaxial fashion.
In perfect plasticity, this means that the stress moves around the yield
surface which is unchanging in stress space. A feature of J_2 flow theory
as used by Drugan et al. [36] is that the material is elastically stiff to
changes of stress around the yield surface. Consequently, the strains
that accumulate and the profile of the crack that develops in that case
are a consequence of this stiff response. If for some reason the material
were less stiff to non-coaxial stress changes than J_2 flow theory, then
larger strains would accumulate and a greater crack opening would develop
for a given applied load. Thus if the crack opening profile is determined
by the material failure criterion as before, the instability level would
be reached at lower values of the applied load than in the case of J_2 flow
theory. Lam and McMeeking [44] have considered this issue following work
of Dean and Hutchinson [45] and Parks, Lam and McMeeking [46]. Lam and

McMeeking carried out finite element calculations for plane strain steady state crack growth in small scale yielding. They utilized a number of constitutive models including J_2 flow theory with isotropic strain hardening, a phenomenological corner plasticity theory [33] and Prager-Ziegler kinematic hardening plasticity [47, 48]. The purpose of using the corner theory and kinematic hardening material is that they are somewhat soft in response to non-coaxial stress changes of the type experienced in crack growth. As mentioned previously the corner theory material can be thought of as representing some of the effects of crystallographic slip in the polycrystalline material. The kinematic hardening has features similar to those of multisurface plasticity theories which can be used to represent the details of nonproportional inelastic response [49]. Pure kinematic hardening is an extreme form of this type of model, but can be studied for effects characteristic of that kind of response.

In the plastic response associated with the corner theory material [33], the initial yield surface is the von Mises form and thus is smooth. As plastic deformation takes place, strain hardening occurs and so the stress moves outwards from the original yield surface. A corner now develops on the yield surface at the current stress with conical surfaces subtended to an otherwise smooth yield surface which has expanded iso-tropically by latent hardening. If the stressing is proportional, the response is identical to the plasticity of a material with isotropic hardening of the same rate as the corner hardens. Thus for proportional loading, the material is plastically soft. This soft response also applies to all directions of stressing closer to proportional direction than the normals to the cone around the vertex. If severe reversals of stress are experienced, the material responds elastically and this would be the case if the direction of stressing lies within the cone subtended by the corner to the smooth part of the yield surface. This of course would be a stiff direction of straining. If the direction of stress change from the yield surface is outside and closer to the conical surfaces than the normals to that conical surface, then the response is intermediate between the soft plastic and the stiff elastic type. In plane strain small scale yielding crack growth, the excursions of stress involve directions in the intermediate range where the stiffness is somewhere between soft plastic and stiff elastic. This is in contrast to J_2 flow theory where in perfect plasticity all excursions are in the elastically stiff direction. When strain hardening is involved with J_2 flow theory, the response is slightly less stiff but considerably stiffer than for the corner theory material.

In the calculations of Lam and McMeeking [44], the crack opening profiles for steady state crack growth were computed for both isotropically strain hardening J_2 flow theory materials and corner theory materials with the same strain hardening in proportional loading. A crack opening angle can be defined as δ/r using the same notation as in eq. (1). The angle involved depends on the position r used to define it. However, a characteristic small material distance r_m can be introduced and the crack propagation criterion of steady near tip crack profile can be stated as $\delta/r_m = \delta_c/r_m$ where δ_c is a critical value. The same criterion can be used for the initiation of crack growth before which the crack will have a different profile in the near tip region than during crack growth. Thus the relationship between the crack opening at $r = r_m$ and K will change as was the case in the analysis of Drugan et al. [36] for perfectly plasticity. However, the calculations of Lam and McMeeking [44] give only the relationship between K in steady state (K_{ss}) when K has ceased increasing with crack propagation and K when crack propagation commences (K_c). The isotropic strain hardening results of Tracey [50] were used for the stationary crack. This is valid since the stationary crack solution involves only proportional stressing and so the behavior in the isotropic and corner theory cases is indistinguishable.

The results of Lam and McMeeking [44] for the ratio of $(K_{ss}/K_c)^2$ (equivalent to J_{ss}/J_{IC}) are shown in Figure 13 against the crack opening angle δ_c/r_m normalized by the yield strain $\varepsilon_y = \sigma_o/E$. Large values of K_{ss}/K_c imply that substantial amounts of stable growth will take place before K reaches K_{ss} and instability ensues. Thus, it can be seen that the nonhardening material, in this sense, is more stable than the strain-hardening cases. In turn the corner theory material is less stable than the equivalent isotropic strain hardening plasticity. This indicates that the effects caused by crystallographic slip as far as crack growth is concerned are destabilizing according to the model, as can be expected from the softness to noncoaxial response introduced by the crystallographic effects.

Similar consequences for crack growth stability arise from using Prager-Ziegler [47, 48] kinematic hardening. In the form used by Lam and McMeeking [44], the size of the yield surface is invariant, but a uniaxial stress history would induce linear strain hardening with a tangent modulus E_t. In nonproportional stress increments, the stress point tows the yield surface behind it but with the surface moving instantaneously in the direction of the surface normal at the current stress. Thus coaxial deviations at first induce a somewhat stiff response, but then softening occurs as the trailing yield surface moves more directly behind the stress and the surface normal aligns more with the stress increment direction. The results for K_{ss}/K_c obtained in the same way as before are shown in Figure 14 for two cases, one with a tangent modulus equal to 0.055E and the other in which the tangent modulus is 0.103E. The isotropic results were calculated by Lam and McMeeking [44] and involve a linearly strain hardening material, i.e. one for which the yield surface expands but its center remains fixed. As can be seen in the figure, the kinematic hardening does destabilize the crack growth somewhat. Thus, one might expect that a more precise model of the plastic response of polycrystalline metals with a multisurface yield criterion would lead to a reduced ability of those materials to sustain stable growth. The simple kinematic hardening law probably overestimates the destabilizing effect, so the real results are probably intermediate to the kinematic and isotropic estimates shown in Figure 14. Of course, the conclusions to be drawn from the work of Lam and McMeeking are dependent on the value of δ_c/r_m thought to be relevant to the failure processes operating in the material. Further work is necessary to shed light on this issue and to model in a more detailed fashion the phenomenon of stable crack growth in polycrystalline ductile metals.

HYDROGEN EMBRITTLEMENT

In this section, we will describe some initial results from incomplete work of Sofronis [51] on the coupled modelling of hydrogen diffusion and elastoplastic deformation in non-hydride formers such as iron. In fact the results obtained so far are all for iron. The model for diffusion of hydrogen through the iron takes into account the fact that

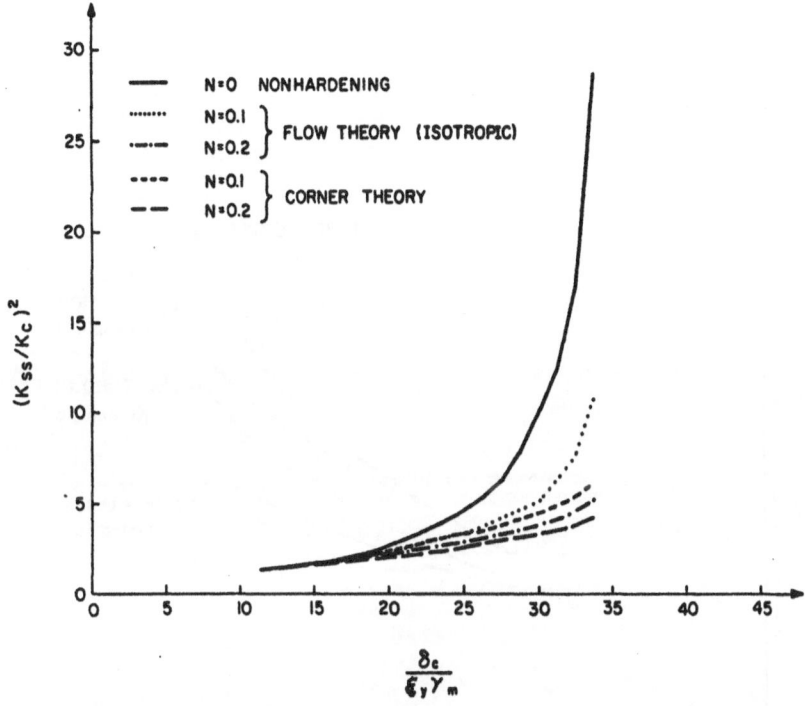

Figure 13. Ratio of stress intensity factor for steady state crack propagation to initiation value plotted against the critical crack opening angle δc/Ym. The results are for an isotropically strain hardening material and for one which develops a yield surface vertex. The number N is the exponent on strain in the hardening law.

118

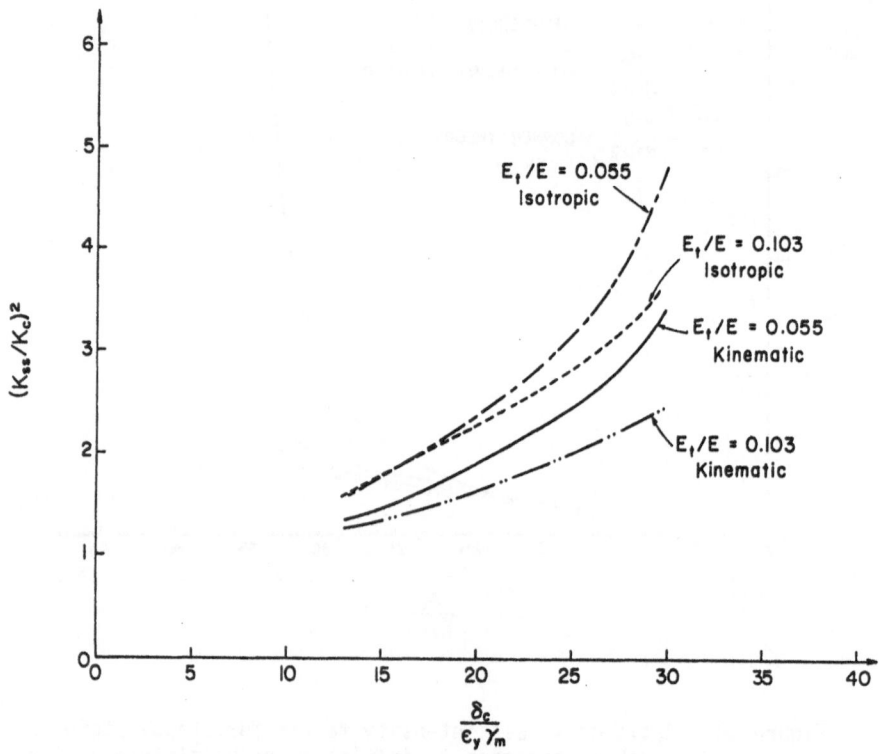

Figure 14. Ratio of stress intensity factor for propagation to initiation value versus critical crack opening angle. Results are for bilinear materials, one with isotropic hardening, the other with kinematic hardening.

dislocations and other microstructural features are strong traps for the hydrogen. The kinetics of trap filling is assumed to be rapid and Oriani's [52] model of trapped hydrogen populations in equilibrium with the local lattice population has been used. In fact, separate calculations of trap filling kinetics based on the work of McNabb and Foster [53] have shown that little difference results. The essential point of the Oriani equilibrium model of trap filling is that the fractional change of the trapped hydrogen concentration is proportional to the fraction of unoccupied traps and the fractional change of the lattice concentration. Since the dislocation density and other trap densities will depend on plastic deformation, the number of potential trap sites will increase with plastic strain. Experimental results of Kumnick and Johnson [54] on rolling have been used to determine the parameters of the model for the manner in which the number of trap sites increases with plastic deformation. At very large strains, the trap density saturates, giving an upper limit for hydrogen traps.

The diffusion equations used in the calculations are those summarized by Johnson and Lin [55] with a drift term to account for the change in chemical potential caused by dilatation of the lattice due to hydrostatic stress. When the equilibrium trap filling model is introduced, the diffusion equations can be written entirely in terms of lattice diffusion with an effective diffusion constant. When there are unfilled traps and the hydrogen concentration in the lattice is increasing, some of the hydrogen must enter the traps. This means that a smaller flux of hydrogen through the lattice results from a given chemical potential gradient. As a result the effective diffusion constant is lower than that for lattice diffusion alone as long as traps are filling. When the model of Oriani [52] for equilibrium between trap and lattice hydrogen populations is used, it is implicit that the traps fill very quickly. However, plastic deformation can create new traps very rapidly until the dislocation density saturates. Consequently, the effective diffusion constant can be kept to low levels by continuing plastic straining and steady state diffusion in the lattice cannot be achieved until plastic straining ceases or the dislocation density saturates.

Sofronis [51] has carried out calculations of hydrogen diffusion near a blunting crack tip in a strain-hardening elastoplastic material. These calculations go beyond, and provide much more detail, than the prior work of Kitigawa and Kojima [56]. Sofronis chose parameters for both the elastoplasticity and the hydrogen diffusion relevant to pure iron. Uniform equilibrium hydrogen concentrations in the unstrained lattice and the pre-existing traps were used as initial conditions. The crack surface was modelled as impermeable and the lattice concentration remote from the crack was held constant during straining. The crack was blunted to several times its original opening by a constant load rate and results for stress and strain similar to those of McMeeking [11] were obtained. Figure 15 shows the hydrogen concentration ahead of the crack after blunting from 10 microns open at the tip to 70 microns in 130 seconds. The concentration plotted is that prevailing while straining is going on. The significant aspect of the hydrogen distribution is that there is only a mild elevation of the lattice concentration at the peak hydrostatic stress location at a distance $R \sim 2b$ from the tip where b is the current crack opening displacement. In contrast, the trapped concentration is very high near the crack tip and falls away rapidly. This reflects the high dislocation density generated by the very large true strains of order

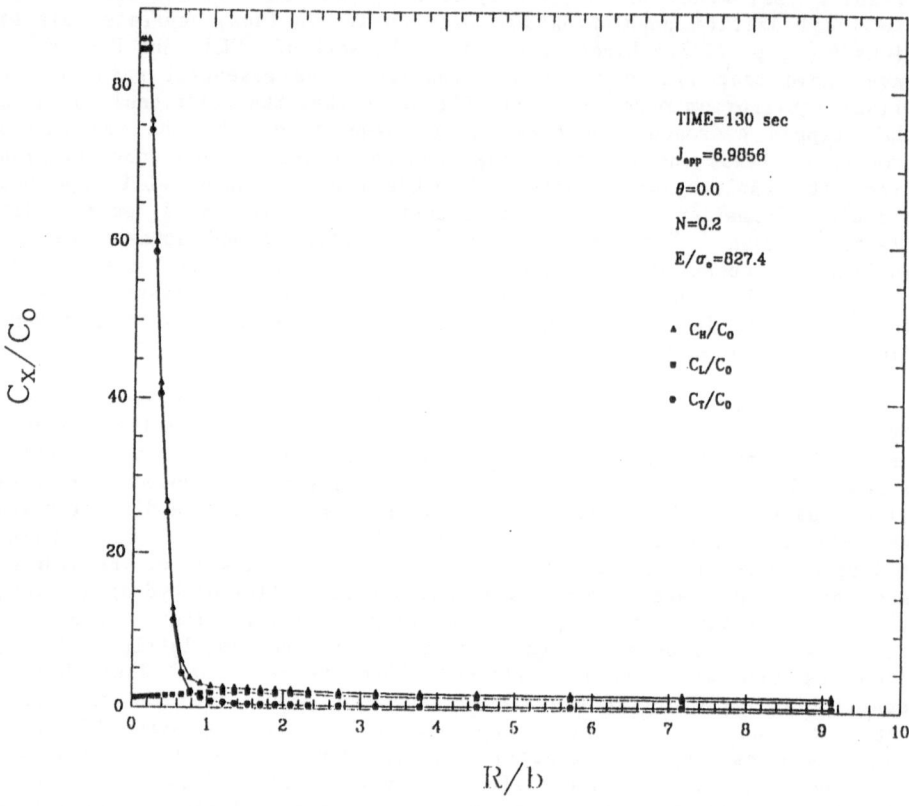

Figure 15. Hydrogen concentration ahead of a blunting crack tip in iron. Concentrations are plotted versus the distance from the tip normalized by the current crack tip opening displacement b. The concentrations are in the lattice (C_L) in the traps (C_T) and the total ($C_H = C_T + C_L$).

2 very near the crack tip. It should be noted that there is a region very close to the crack tip surface where the dislocation density has saturated and so the number of traps has reached its upper limit. The trapped hydrogen concentration there is saturated as indicated by the uniform value.

The mild elevation of hydrogen concentration near the hydrostatic stress maximum reflects the competition between the drift term and the continuing generation of traps. As new traps are created they pull hydrogen out of the lattice locally. The reduced lattice concentration that results in turn induces a flux of lattice hydrogen to that area and this competes with the pull of hydrogen to the hydrostatic stress region. As long as the load continues to increase and the crack blunts, the hydrogen is drawn to the crack tip from the hydrostatic stress region. When the load increase ceases, the plastic straining terminates and new traps are no longer created. Hydrogen flux to the hydrostatic stress region will eventually prevail and the steady state lattice distribution, exponentially dependent on hydrostatic stress, will be set up. It is clear that during loading, the strain rate will determine the extent to which the trapping mechanisms affects the lattice hydrogen distribution. At low strain rates, hydrogen can diffuse rapidly relative to the rate at which traps are created. Consequently, traps can be filled readily and the lattice concentration maintained. In that case, the hydrogen will be able to accumulate in the region of high hydrostatic stress. The concentration there will be much more pronounced than in Figure 15 and the loading-rate would be much lower than that used to obtain the results plotted in the Figure. At high loading rates, the lattice diffusion will be too slow to supply the traps and the high plastic strain region will exert an attraction for the hydrogen. This will have the effect of diminishing the relative hydrogen concentration in the high hydrostatic stress region. A further effect of strain rate, in which the dislocations move too fast for their trapped hydrogen atoms and leave them behind, has not been included in this model and so the comments concerning the effect of strain rate is not based on that consideration.

The fracture instability which leads to hydrogen influenced crack growth presumably depends on the hydrogen concentration either in the traps or in the lattice coupled with the stress or the strain or both. Birnbaum [57] has summarized the data and mechanisms for hydrogen related fracture. A number of possible theories can be modelled. However, there are insufficient results from Sofronis' work as yet to permit comprehensive modelling of the fracture process. That effort will be deferred to future work. It is clear that the type of analysis that Sofronis has carried out will provide a basis for detailed study of the interactions of hydrogen diffusion, stress and plastic strain during the fracture process. The results shown in Figure 15 indicate an effect of strain rate which could be quite significant during decohesion or enhanced dislocation motion leading to fracture. Consequently the influence of hydrogen on the crack tip elastoplastic fracture instability can be elucidated through work of this kind.

CLOSURE

In this paper, we have summarized some recent work on crack tip plastic instabilities associated with ductile fracture and crack propagation. We believe that the mechanics solutions involved have proved to be helpful to those seeking an understanding of the microscopic processes of those phenomena. We hope also that the as yet incomplete work on hydrogen diffusion coupled to crack tip blunting will also be useful in a similar way in hydrogen embrittlement studies once more results are available.

ACKNOWLEDGEMENT

This paper was prepared while the author was supported by a grant from the National Science Foundation MSM 85-04411.

REFERENCES

1. Rice, J.R. and Thomson, R., "Ductile Versus Brittle Behavior of Crystals," Phil. Mag., 29, 1974, p. 73.

2. Rice, J.R., "Mechanics Aspects of Stress Corrosion Cracking and Hydrogen Embrittlement," Proceedings of a Conference on Stress Corrosion Cracking and Hydrogen Embrittlement of Iron Base Alloys, Unieux-Firminy, France, June 12-16, 1973 (eds., R.W. Staehle et al.), National Association of Corrosion Engineers, Houston, Texas, 1977, p. 11 (see also other papers in same volume).

3. Hawk, D.E. and Bassani, J.L., "Transient Crack Growth Under Creep Conditions," J. Mech. Phys. Solids, 34, 1986, p. 191.

4. Hult, J.A.H. and McClintock, F.A., "Elastic-Plastic Stress and Strain Distribution Around Sharp Notches Under Repeated Shear," Proc. Ninth Int. Cong. Appl. Mech., 8, 1956, p. 51.

5. Chitaley, A.D. and McClintock, F.A., "Elastic-Plastic Mechanics of Steady Growth Under Anti-plane Shear," J. Mech. Phys. Solids, 19, 1971, p. 147.

6. Rice, J.R., "A Path Independent Integral and the Approximate Analysis of Strain Concentration by Notches and Cracks," J. Appl. Mech., 35, 1968, p. 379.

7. Rice, J.R. and Rosengren, G.F., "Plane Strain Deformation Near a Crack Tip in a Power Law Hardening Material," J. Mech. Phys. Solids, 16, 1968, p. 1.

8. Hutchinson, J.W., "Singular Behavior at the End of a Tensile Crack in a Hardening Material," J. Mech. Phys. Solids, 16, 1968, p. 13.

9. Rice, J.R. and Johnson, M.A., "The Role of Large Crack Tip Geometry Changes in Plane Strain Fracture," Inelastic Behavior of Solids, (eds. M.F. Kanninen et al.) McGraw-Hill, New York, 1970, p. 641.

10. Green, G. and Knott, J.F., "The Initiation and Propagation of Ductile Fracture in Low Strength Steels," Trans. ASME (J. Engg. Mater. Tech.) 98, 1976, p. 37.

11. McMeeking, R.M., "Finite Deformation of Crack-Tip Opening in Elastic-Plastic Materials and Implications for Fracture," J. Mech. Phys. Solids, 25, 1977, p. 357.

12. Rudnicki, J.W. and Rice, J.R., "Conditions for the Localization of Deformation in Pressure Sensitive Materials," J. Mech. Phys. Solids, <u>23</u>, 1975, p. 317.

13. McMeeking, R.M. and Parks, D.M., "On Criteria for J-Dominance of Crack Tip Fields in Large-Scale Yielding," Elastic Plastic Fracture, STP 668, Amer. Soc. Test. and Mater., Philadelphia, 1979, p. 175.

14. McClintock, F.A., "Plasticity Aspects of Fracture," Fracture: An Advanced Treatise, (ed. H. Liebowitz), Academic Press, New York, Vol. 3, 1971, p. 47.

15. McMeeking, R.M., "Blunting of a Plane Strain Crack Tip Into a Shape with Vertices," Trans. ASME (J. Engg. Mater. Tech.) <u>99</u>, 1977, p. 290.

16. Clayton, J.Q. and Knott, J.F., "Observations of Fibrous Fracture Modes in a Prestrained Low-Alloy Steel," Metal Sci., <u>10</u>, 1976, p. 63.

17. Rawal, S.P. and Gurland, J., "Observations on the Effect of Cementite Particles on the Fracture Toughness of Spheroidized Carbon Steels," Proc. 2nd Int. Conf. Mech Behav. Mater., Federation of Materials Societies, 1976, p. 1154.

18. McMeeking, R.M. and Aravas, N., "Models of Ductile Rupture," Proc. 10th Natl. Cong. Theo. Appl. Mech., 16-20 June, 1986, Austin, Texas, in press.

19. Aravas, N. and McMeeking, R.M. "On Crack Tip Blunting and Ductile Void Growth," Proc. 3rd Irish Durab. Frac. Conf. 26-27 Sep. 1985, Galway, in press.

20. Cox, T.B. and Low, J.R., "An Investigation of the Plastic Fracture of AISI 4340 and 18 Nickel-200 Grade Maraging Steels," Met. Trans. <u>5</u>, 1974, p. 1457.

21. Aravas, N. and McMeeking, R.M., "Finite Element Analysis of Void Growth Near a Blunting Crack Tip," J. Mech. Phys. Solids, <u>33</u>, 1985, p. 25.

22. Aravas, N. and McMeeking, R.M., "Microvoid growth and Failure in the Ligament Between a Hole and a Blunt Crack Tip," Int. J. Frac., <u>29</u>, 1985, p. 21.

23. Gurson, A.L., "Plastic Flow and Fracture Behavior of Ductile Materials Incorporating Void Nucleation, Growth and Interaction," Ph.D. Dissertation, Division of Engineering, Brown University, Providence, RI, 1975.

24. Gurson, A.L., "Continuum Theory of Ductile Rupture by Void Nucleation and Growth: Part I - Yield Criteria and Flow Rules for Porous Ductile Media," Trans. ASME (J. Engg. Mater. Tech.) <u>99</u>, 1977, p. 2.

25. Gurson, A.L., "Porous Rigid-Plastic Materials Containing Rigid Inclusions - Yield Function, Plastic Potential and Void Nucleation," Fracture 1977 (ed. D.M.R. Taplin) Pergamon, Oxford, 1977, Vol. 2, p. 357.

124

26. Tvergaard, V. and Needleman, A., "Analysis of the Cup-Cone Fracture in a Round Tensile Bar," Acta Met., 32, 1984, p. 157.

27. Tvergaard, V., "Material Failure by Void Coalescence in Localized Shear Bands," Int. J. Solids Struct., 18, 1982, p. 659.

28. Aoki, S., Kishimoto, K., Takeya, A. and Sakata, M., "Effects of Microvoids on Crack Tip Blunting and Initiation in Ductile Materials," Int. J. Fract., 24, 1984, p. 267.

29. Aoki, S., Kishimoto, K., Yoshida, T. and Sakata, M., "A Finite Element Study of the Near Crack Tip Deformation of a Ductile Material Under Mixed Mode Loading," Department of Physical Engineering Reprot, Tokyo Institute of Technology, 1985.

30. Needleman, A. and Tvergaard, V., "An Analysis of Ductile Rupture Modes at a Crack Tip," Division of Engineering Report, Brown University, Providence, RI, 1986.

31. Hancock, J.W. and Mackenzie, A.C., "On the Mechanisms of Ductile Failure in High-Strength Steels Subjected to Multi-axial Stress-States," J. Mech. Phys. Solids, 24, 1976, p. 147.

32. Needleman, A. and Tvergaard, V., "Crack-Tip Stress and Deformation Fields in a Solid with a Vertex on Its Yield Surface," Elastic Plastic Fracture Mechanics, STP 803, Vol. 1, Inelastic Crack Analysis, Amer. Soc. Test. and Mater., Philadelphia, 1983, p. 80.

33. Christoffersen, J. and Hutchinson, J.W., "A Class of Phenomenological Corner Theories of Plasticity," J. Mech. Phys. Solids, 27, 1979, p. 465.

34. Krafft, J.M., Sullivan, A.M. and Boyle, R.W., "Effect of Dimensions on Fast Fracture Instability of Notched Sheets," Proc. Crack Propag. Symp., Cranfield College, England, 1961.

35. Paris, P.C., Tada, H., Zahoor, A., and Ernst, H., "The Theory of Instability of the Tearing Mode of Elastic-Plastic Crack Growth," Elastic Plastic Fracture, STP 668, Amer. Soc. Test. Mater., Philadelphia, 1979, p. 5.

36. Drugan, W.J., Rice, J.R., and Sham, T-L., "Asymptotic Analysis of Growing Plane Strain Tensile Cracks in Elastic-Ideally Plastic Solids," J. Mech. Phys. Solids, 30, 1982, p. 447.

37. Rice, J.R., "Mathematical Analysis in the Mechanics of Fracture," Fracture; An Advanced Treatise (ed. H. Liebowitz) Academic Press, New York, 1968, Vol. 2, p. 191.

38. Rice, J.R., "Elastic-Plastic Models for Stable Crack Growth," Mechanics and Mechanisms of Crack Growth, Proc. Conf., Cambridge, England, April 1973, (ed. M.J. May) British Steel Corp. Physical Metallurgy Centre Publication, Sheffield, England, 1974, p. 14.

39. Cherepanov, G.P., Mechanics of Brittle Fracture, Nauka, Moscow, 1974 (Russian).

40. Rice, J.R., and Sorensen, E.P., "Continuing Crack-Tip Deformation and Fracture for Plane-Strain Crack Growth in Elastic-Plastic Solids," J. Mech. Phys. Solids, 26, 1978, p. 163.

41. Gao, Y-C., Acta mech. sin., 1, 1980, p. 48 (Chinese).

42. Rice, J.R., Drugan, W.J., and Sham T-L., "Elastic-Plastic Analysis of Growing Cracks," Fracture Mechanics: 12th Conf., STP 700, Amer. Soc. Test. Mater., Philadelphia, 1980, p. 189.

43. Hermann, L., and Rice, J.R., "Comparison of Experiment and Theory for Elastic-Plastic Plane Strain Crack Growth," Metal Science, 14, 1980, p. 285.

44. Lam, P.S., and McMeeking, R.M., "Analyses of Steady Quasistatic Crack Growth in Plane Strain Tension in Elastic-Plastic Materials with Non-isotropic Hardening," J. Mech. Phys. Solids, 32, 1984, p. 395.

45. Dean, R., and Hutchinson, J.W., "Quasi-static Steady Crack Growth in Small Scale Yielding," Fracture Mechanics: 12th Conf., STP 700, Amer. Soc. Test. Mater., Philadelphia, 1980, p. 383.

46. Parks, D.M., Lam, P.S., and McMeeking, R.M., "Some Effects of In-elastic Constitutive Models on Crack Tip Fields in Steady Quasistatic Growth," Advances in Fracture, Proc. 5th Int. Cong. on Fracture (Ed. D. Francois) Pergamon, Oxford, 1981, Vol. 5, p. 2607.

47. Prager, W. "The Theory of Plasticity: A Survey of Recent Achievements," Proc. Inst. Mech. Engineers, 169, 1955, p. 41.

48. Ziegler, H., "A Modification of Prager's Hardening Rule," Quarterly Appl. Math., 17, 1959, p. 55.

49. Phillips, A., "Combined Stress Experiments in Plasticity and Visco-plasticity: The Effects of Temperature and Time," Plasticity of Metals at Finite Strain: Theory, Computation and Experiment (Eds. E.H. Lee and R.L. Mallett) Division of Applied Mechanics, Stanford University and Department of Mechanical Engineering, Aeronautical Engineering and Mechanics, Rensselaer Polytechnic Institute, 1982, p. 230.

50. Tracey, D.M., "Finite Element Solutions for Crack-Tip Behavior in Small-Scale Yielding," Trans. ASME, J. Engg. Mater. Tech., 98, 1976, p. 146.

51. Sofronis, P., Ph.D. Dissertation, University of Illinois at Urbana-Champaign, 1986, in preparation.

52. Oriani, R.A., "The Diffusion and Trapping of Hydrogen in Steel," Acta Met., 18, 1970, p. 147.

53. McNabb, A., and Foster, P.K., "A New Analysis of the Diffusion of Hydrogen in Iron and Ferritic Steels," Trans. A.I.M.E., 227, 1963, p. 618.

126

54. Kumnick, A.J., and Johnson, H.H., "Deep Trapping States for Hydrogen in Deformed Iron," Acta Met., 28, 1980, p. 33.

55. Johnson, H.H., and Lin, R.W., "Hydrogen and Deuterium Trapping in Iron," Hydrogen Effects in Metals, (Eds. A.W. Thompson and J.M. Bernstein) TMS-AIME, New York, 1981, p. 3.

56. Kitigawa, H., and Kojima, Y., "Diffusion of Hydrogen Near an Elasto-Plastically Deformed Crack Tip," Atomistics of Fracture (Eds. R.M. Latinision and J.R. Pickens) Plenum Press, New York, 1983, p. 799.

57. Birnbaum, H.K., "Hydrogen Related Fracture of Metals," Atomistics of Fracture, (Eds. R.M. Latinision and J.R. Pickens) Plenum Press, New York, 1983, p. 733.

DISCUSSION

Comment by P. Neumann:

In spite of finding H at the blunted crack tip, you prefer the view that the crack advances by nucleation of voids ahead of the crack in the region of high hydrostatic stress. Did you consider and take into account the additional difficulty to nucleate a void if compared with the simple extension of the existing crack?

Reply:

In answer to your question, no. I was trying to say that the various hydrogen populations would affect different mechanisms. Thus, the hydrogen concentration alone is not necessarily the criterion to use to predict failure. The lattice concentration along with the hydrostatic stress may be the important combination, in which case the failure would occur in the interior of the material. If softening is important, that would tend to limit stress magnitudes; but I believe the high hydrostatic stresses will still occur in the interior. I do not know which mechanism, shear localization, decohesion or something else, is most important. I hope that modelling can be used to help sort out that question including the possibility of any interplay between mechanisms.

Comment by A. S. Argon:

In the discussion of crack tips interacting with large cavities with a porous material in between the crack and the cavities, you did not get much shear localization. Does this not imply that there may be shear localization by some other process such as non-normality conditions or other corner producing process which then probes particles in these bands and produces separation in shear zones? Furthermore, does this not translate directly into observations of Hirth on early decohesion of slip lines in the presence of hydrogen where the latter is first responsible for a strain softening process?

Reply:

Yes, I agree with you completely as far as the possibility of other mechanisms triggering localizations is concerned. That is what I meant by the "other features of the material" which would have ·to be introduced to improve the model. The continuum nucleation of voids and dilatancy model is one possibility for describing failure and succeeds in some cases and does less well in others. I think you are probably right about hydrogen induced softening triggering localization, and we hope to get some insight into this from our softening model.

Comment by W. Gerberich:

Could you address the figure which shows hydrogen trapping increasing rapidly toward the crack tip? If you have a plastically-blunted crack tip and the stresses are increasing away from the tip, since the pressure gradient would be increasing into the material, wouldn't the drift velocity be away from the crack tip rather than into it? Why then would there be more hydrogen in traps nearer the crack tip?

Reply:

The results shown were for a situation where the material has been previously charged with hydrogen. No transport was permitted in through the crack tip. Thus the hydrogen was being sucked into the crack tip region from the far field. During the calculation, the load was increased rather slowly and conditions close to equilibrium could be maintained. The lattice hydrogen reflected the pressure field with a maximum ahead of the crack. Dislocation traps were used and filled very quickly from the lattice hydrogen. The trapped hydrogen was distributed according to the plastic strain. If a higher rate of loading, diffusion through the crack tip, or other kinds of traps were introduced, I would expect very different results. We plan to model such situations in the near future.

Comment by H. Mughrabi:

You referred to a yield surface with a vertex that you attributed to crystallographic deformation effects. Are there also other reasons for such vertices, e.g., on continuum mechanics grounds?

Reply:

The vertex yield surface model is, of course, a continuum phenomenological plasticity device. The most obvious rationale is that it represents the effect which crystallographic multislip would have on the plastic flow and yielding. However, the phenomenological model is quite macroscopic and it can be argued that on a scale of several grains in a polycrystal the sharp vertex would not be apparent. A locally large but finite curvature of the yield surface is more likely to be relevant and such plasticity models are being used nowadays. Anelasticity may also help to keep yield surface curvatures high at least for highly non-proportional straining. Thus continuum mechanics grounds or a macroscopic viewpoint should he used to argue against a yield surface vertex. However, the vertex does model in an extreme form the effects of plasticity associated with nonproportional straining associated with growing a crack through the elastic-plastic material.

Comment by C. J. Altstetter:

Does a hydrogen atom in a trap site have the same effect on crack growth as a hydrogen atom in a normal interstitial site?

Reply:

Probably not. To answer the question properly, I would have to model the failure process well and determine what controls crack growth. From that I could think about the relative importance of the different hydrogen populations in the overall process. It is not going to be done by modelling alone, but I hope that the models can make some contributions to answering the question.

DISTRIBUTED DAMAGE PROCESSES IN FRACTURE

Anthony W. Thompson

Carnegie-Mellon University
Department of Metallurgical Engineering and Materials Science
Pittsburgh, PA 15213
U.S.A.

INTRODUCTION

Fracture events in materials occur in a wide variety of ways, ways which differ in both macroscopic and microscopic characteristics. One class of such characteristics has come to be called "distributed damage". The concept of distributed damage processes as part of a total fracture event implies multiple sites of cracking, void formation, or other damage phenomena, instead of a single crack which causes fracture by self-extension. An important consequence of the multiple damage concept has to do with fracture kinetics. The rates of damage formation at these multiple sites need not replicate the rate of crack motion, so that the familiar models of the crack tip as stationary in a field of moving material must become much more complex to represent varying damage rate phenomena. Finally, even the concept of fracture by crack motion can be less relevant, even irrelevant, in the presence of multiple damage. It is in principle possible for a specimen which is undergoing distributed damage processes to exhibit fracture, for example by stable growth and impingement of microvoids, without ever containing a distinct crack. Thus description of distributed damage fractures typically requires greater complexity (or sophistication), both in experimental viewpoint and also in modeling, compared to self-extending cracks. In addition, they often are markedly sensitive to microstructural parameters, and thus are of particular interest to the metallurgist and materials scientist.

The present paper first considers distributed damage from the perspective of a very familiar fracture example, ductile fracture or microvoid coalescence. This furnishes a useful background, because ductile fracture is fairly well understood, at least in broad outline, in both the analytical mechanics and physical behavior senses. The extensions of this understanding to more complex cases is then discussed, including recent interest in effects of distributed damage on crack tip driving forces, often called "crack tip shielding". The intent is not to provide a detailed review of any of these topics, but to identify the general features of what is known, as well as what appear to be the principal outstanding problems.

DUCTILE FRACTURE: ELEMENTARY PROCESSES

As has been described in a number of reviews [1-6], ductile fracture can be regarded as occurring with a sequence of three "elementary processes", which are the nucleation, growth and coalescence of microvoids. The latter stage gives its name to a common term for this fracture type, microvoid coalescence or MVC. A schematic view of this sequence, in its simplest form, is presented in Fig. 1. This division of the fracture sequence into elementary processes is convenient for discussion and for some analytical aspects, but it must be kept in mind, as is shown below, that actual fractures usually comprise multiple or overlapping processes over a wide range of strain, such as continued nucleation of new voids during growth of the previously-nucleated voids. Accordingly, most existing models for the elementary processes are necessarily considerable oversimplifications of reality. In particular, models for the elementary processes ordinarily cannot be used to describe the entire fracture sequence, even as an approximation.

130

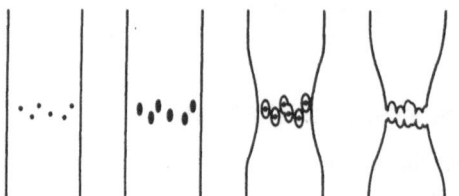

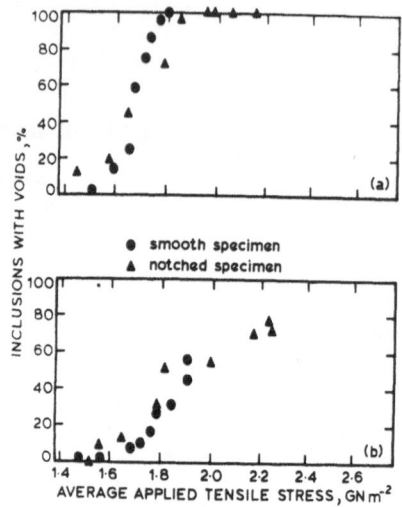

FIGURE 1. Schematic representation of ductile fracture process, left to right, from pre-existing particles or inclusions at left, to hole nucleation at particle poles with tensile stress, to transverse growth when necking (or other source of transverse stress) is present, to final fracture with half-voids or dimples on the two fracture surfaces (right).

FIGURE 2. Effect of stress on void nucleation at MnS inclusions in AISI 4340 steel of two purities, in smooth and notched specimens. (a) Commercial purity material, with large inclusions. (b) High-purity steel, with small inclusions. From Cox [16].

Microvoid Nucleation

It has been known since the pioneering work of Tipper [7] that microvoids in most real materials are largely nucleated at inclusions or other particles, and that these microvoids then grow and coalesce to comprise the fracture [7,8]. Subsequent work has elaborated this knowledge to a considerable extent [3,6,9]. Most evidence supports the view that interfacial strength is a dominant factor in nucleation. A number of earlier workers had suggested that the critical parameter was the energy of the new free surfaces formed by nucleation of a microvoid. It now seems clear that this is a necessary, not a sufficient, condition for nucleation, and is often satisfied at or near yield [3,6,11].

An extended Bridgman analysis, together with contributions from local stresses, as calculated by Argon and co-workers [10-12], lead to a simple relation, that states (in the original notation),

$$\sigma_{rr} = Y(\bar{\epsilon}^P) + \sigma_{T'}$$

where σ_{rr} is the radial stress on the particle, Y is the flow stress in the region of the particle, for the average plastic strain $\bar{\epsilon}^P$ of the region, and σ_T is the local hydrostatic stress. Nucleation occurs when the value of σ_{rr} reaches the interfacial strength. A method of using this relation for evaluation of the interfacial strength was devised by Argon and Im [12], and has been used in a number of subsequent investigations [13-15].

Among the strong indications of the role of stress (and indirectly, of the role of an approximately constant interfacial strength for a given microvoid nucleus) are the observations of Cox [16]. These are shown in Fig. 2, for both smooth and notched specimens of two purities of AISI 4340 steel. The notched specimens exhibited nucleation at much lower strains than the smooth specimens, but when the stresses were calculated the two sets of data were in good agreement for each steel (Fig. 2). It was also observed for the high-purity steel that the role of stress was different than for the steel with large inclusions. As has been reviewed [6], these data are consistent with a number of other observations.

There are, however, indications of a number of other factors which, though secondary and perhaps indirect compared to the interfacial strength, may still play a major role in special circumstances or when acting in concert [3-6,9,11,12,17]. These include the following. (i) Particle size. It has been observed by many workers that voids are nucleated first at the largest particles of a given type. Although largely an experimental phenomenon, albeit a persistent one [3,6], there may be analytical reasons to expect such behavior [11]. (ii) Particle shape. Both particle shape and volume fraction are known to be of importance to nucleation details; for example, an analysis for elongated particles has been presented [18]. Volume fraction is treated in more detail below, but increased volume fraction tends to increase nucleation density, for obvious reasons. (iii) Particle location and distribution. Location at grain boundaries or other potential fracture paths, or distributions which give rise to locally-increased volume fraction, can significantly affect the phenomenology of nucleation. (iv) State of stress, strain or both. The relation above indicates a direct role of stress state through σ_T, consistent with Bridgman's work [19], and there are also indications that the state of strain can be important [3,5,6]. (v) Interfacial structure and composition. These factors affect bond strength, but can be regarded as independent if heat treatment or other metallurgical changes affect them directly. For example, Hippsley and Druce presented a striking example of this kind by segregating phosporus to carbide particle interfaces in a steel [20]. (vi) Bond strength. With more detailed knowledge of bonding mechanisms, this factor could presumably be combined with the preceding one. Bonding strength would be expected to affect nucleation directly by increasing the required σ_{rr}, and this has been observed [21]. The same has been found in steels, for example, in comparing weakly-bonded MnS inclusions to the relatively strongly-bonded Fe_3C particles [3,6,15,22,23].

As the preceding list indicates, there remain a number of unresolved or incompletely understood issues regarding void nucleation. Both experimental and analytical work seems needed to address these more confidently.

Void Growth

Void growth has been of interest to mechanics analysts for a number of years, and the major work of McClintock [24,25], and the refinements of Rice and Tracey [26,27] and Rice and Johnson [28], make clear the major parameters in this phenomenon. Although McClintock's treatment is simplified in a number of ways, it appears to include the major parametric relations correctly. A widely-quoted equation from that work [25] is the following:

$$\epsilon_R = (1-n) \ln(L_i/2d_i) / \sinh[2\sigma_t(1-n)(2\bar{\sigma}/\sqrt{3})],$$

where ϵ_R is the predicted fracture strain, n is the work-hardening exponent, L_i and d_i are the initial spacing and size of holes (pre-existing holes were assumed), σ_t is the transverse stress, and $\bar{\sigma}$ is the equivalent stress. This relation indicates directly that a very large effect of transverse (or hydrostatic) stress on void growth is expected, as often observed. It seems often to be overlooked, however, that this relation substantially overstates the observed fracture ductility. McClintock, in a rarely-cited paper [25], made it clear that he was aware of this, and Gurland [29] has presented a graphical comparison between the McClintock calculation and the well-known data of Edelson and Baldwin [30], as shown in Fig. 3. The latter data [30] are addressed at greater length below. The reason for the overestimate (Fig. 3) may involve termination of stable void growth by strain localization [25], as discussed in the following section.

The foregoing relation identifies several of the essential parameters in void growth. However, there are several other factors as well. (i) The state of stress, strain or both. The McClintock relation, above, shows directly the role of stress state, but there are indications [6] that the state of strain can also be important, including the fact that growth cannot begin until plastic strains are present [3].   (ii) <u>Nucleation strain.</u> The McClintock relation assumes with pre-existing holes of non-zero size.   Most real materials will exhibit at least some period of strain before growth begins in earnest, particularly when nucleation occurs at a large value of strain. A number of authors have recently addressed this topic [17,31,32].   (iii) <u>Nucleation density.</u> As mentioned in discussing nucleation, the local density of nuclei can greatly affect local stress and strain, and thereby growth.   (iv) <u>Matrix flow relation.</u> This factor is discussed in more detail in the following section, but knowledge of the appropriate flow relation (or constitutive relation) for the matrix is essential to accurate description of growth, although it can be somewhat inaccessible experimentally.   (v) <u>Strain localization.</u> This factor is also discussed in the following section.   The tendency of matrix flow to localize strain into strong shear bands, thus shortening or terminating the void growth process, must be taken into account in calculating the growth strain.   Thus, as in the preceding section, there clearly are a number of issues here which deserve clarification. Experimental work on void growth in recent years has been sparse, perhaps because the mechanics appears complete, but additional work to address some of the issues listed above is especially needed.

Void Coalescence

This final stage in MVC is less well studied, with no clear mechanics basis for description. Moreover, it is a difficult experimental subject, as even a fairly stiff testing system may be compliant enough to make the specimen unstable near fracture, and it may not be possible to stop the fracture process merely by stopping the machine crosshead.   However, appropriately-stiff load trains for mechanical testing, or use of strain control (with extensometer strain measurement) in a closed-loop machine, can overcome these problems.   It is in any case already clear that void coalescence is a

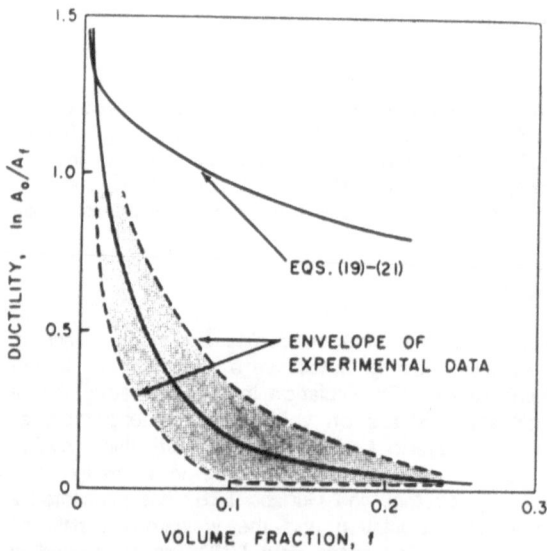

FIGURE 3. Comparison of McClintock's equations (19)–(21), which approach that given in the text for large relative hole-growth ratios, with the experimental data of Edelson and Baldwin [30]. Figure from Gurland [29].

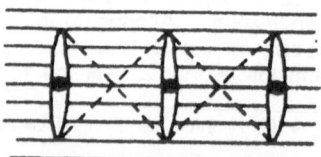

FIGURE 4. Criterion of Brown and Embury for termination of stable plastic void growth, as void length becomes equal to spacing, permitting a 45° shear line to be drawn between voids. From ref. 40.

physically complex process [3,6]. It can take place by direct impingement of voids, that is, by voids growing until they touch, as envisioned in Fig. 1 and elsewhere [24-27]; it can also occur by void sheet formation, as first recognized by Rogers [2,33]. Void sheets in turn can occur by profuse voiding in an existing slip band or strain localization. They can also occur if growing void density can trigger localization as the voided matrix becomes less resistant to localization. In either case, the coalescence process can arise in groups as well as pairs of voids, and either detailed observation or description of the attendant phenomena are challenging and accordingly rare.

From the mechanics perspective, there are two important issues. First, it is essential to identify the appropriate matrix flow relation. Whether or not localization is already proceeding, the void density will be high enough that the flow relation must be that for hole-containing material [34], often called "Gurson" material [35-37], though that mechanics is still emerging. The second issue relates to the onset of localization. Localization by shear intensification, or possibly by adiabatic shear, has been recognized for some time [38]. It appears experimentally that the profuse voiding usually precedes, rather than follows, strain localization (as concluded in Cox and Low's work [39], for example), but additional observations, as well as better mechanics criteria, would be welcome. As treated further below, fractography as well as sectioning of nearly-fractured specimens can be informative for the coalescence process.

One simple, microscale criterion for the onset of localization, proposed a number of years ago, is that of Brown and Embury [40], shown in Fig. 4. The concept of Fig. 4 is appealing, that when the void length is equal to the spacing, and 45° lines can be drawn between adjacent void tips, the material will be unstable against shear between the voids, i.e. to intervoid localization, thus terminating the fracture process. At least one subsequent work has offered "a degree of support" for this concept [41]. There are, however, a number of observations which contradict this criterion [42,43], by showing more closely spaced voids (or, equivalently, longer voids) than the proportions shown in Fig. 4. Additional work is clearly needed, both on localization criteria and on experimental observations of the coalescence process.

DUCTILE FRACTURE: ADDITIONAL CONSIDERATIONS

Having treated briefly the three processes proposed as components of the typical ductile fracture sequence, it is useful to consider some ways in which the fracture as a whole has been measured and analyzed. In so doing, the most important data are those of Edelson and Baldwin [30], on a variety of powder metallurgy materials with controlled volume fractions of various particles in copper matrices. A scatter band for those data is included in Fig. 3, but Fig. 5 shows the individual data, with the various materials identified. Although the original authors stated that their experimental materials exhibited a wide range of bonding strength between intermetallic or oxide particles and the copper matrix, in retrospect it is clear that all these particles were relatively weakly bonded. Indeed, all the data are consistent with the data for voids, prepared by incomplete densification of copper powder metallurgy compacts. It is also noteworthy that all the particles were least as large as 5 μm diameter, so no information about truly small particles is contained in these data. There are indications [44,45] that particles smaller than 1 μm diameter have a smaller effect on ductility than suggested by Fig. 5. In nickel containing 2 vol. pct. thoria, for example, Fig. 5 predicts a fracture strain of about unity, while published fracture strain data [46] range from 1.4 to over 2. Nevertheless, the striking and marked dependence of fracture strain on volume fraction of particles remains as a profound result of Edelson and Baldwin's work, and a continuing benchmark against which other work is measured.

It was stated under the heading of nucleation that particle shape and bonding could affect the nucleation process. Fig. 6 shows the data collected by Gladman, Holmes and McIvor [22] on steels, clearly indicating that carbide bonding, stronger than that of sulfides, and carbide or sulfide shape, can play a role in fracture strain. Unfortunately, no observations were reported in that work on nucleation strain or void growth, so that

134

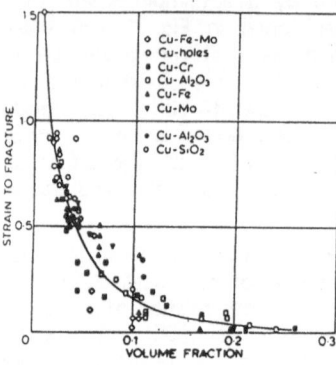

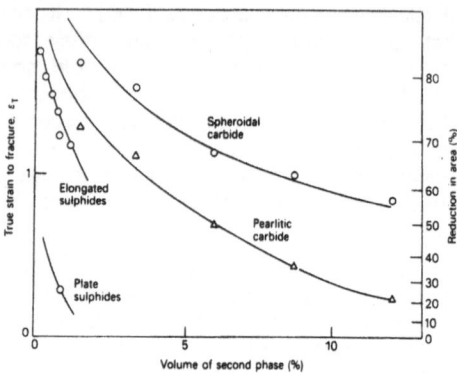

FIGURE 5. Dependence of fracture strain on volume fraction of particles, prepared by powder metallurgy methods of the alloys listed, as measured by Edelson and Baldwin. After [30].

FIGURE 6. Dependence of fracture strain on volume fraction of second-phase particles for sulfides and carbides in steels, as collected by Gladman, et al. [22].

it is not possible to separate the contributions of particle shape and bonding to the individual processes in ductile fracture.

The point was made above, and should be reiterated here, that there is no simple basis on which data like those of Figs. 5 and 6, comprising fracture strains, can be realistically compared to analytical expressions for the constituent processes of ductile fracture. Few if any instances of fracture occur solely as nucleation events; fractures which proceed from a virtually zero nucleation strain usually are not terminated by pure void impingement; and the coalescence process can only occur in voided material. Accordingly, it seems unrealistic to evaluate any expression for one of the constituent processes by comparing it to Figs. 5 or 6, unless very special experimental conditions have been devised (and verified). On the other hand, efforts have been made to combine criteria, or to separately evaluate them, for the constituent processes [17,31]. These efforts deserve expansion and further development.

The Crack Tip

Extension of the foregoing concepts to fracture behavior at the tip of a sharp crack can be regarded as a difficult exercise in states of stress and strain, as well as in questions of appropriate constitutive relations for material in the crack tip plastic zone, and in questions about the role of strain gradients and restricted plastic volumes in the plastic zone. However, given what is known about the role of stress and strain states in nucleation and growth, this is, at least qualitatively, a straightforward problem. It would be useful to have a careful, quantitative, and extensive experimental comparison of ductile fracture behavior for various specimen geometries, to demonstrate that these expectations are reasonably accurate, but there is no strong reason to doubt that they are. The questions of constitutive relations and of the volume and strain gradients of the plastic zone (and of the appropriate values of quantities like local fracture strain) are much more open in a fundamental sense and thus potentially much more troublesome. It is beginning to be clear, however, that certain opportunities exist to make improved measurements (or estimates) of these local quantities. Those opportunities are discussed below.

It is probably useful to emphasize at this point that actual experimental observations within plastic zones, although few in number, tend to reinforce the idea that local processes are variable in space and complex in pattern. Fig. 7, from Van Stone's work [6], illustrates the point. Without additional data of this kind, on materials which differ in mechanical behavior (including toughness), microstructure and fracture mode, it is difficult to generalize about the applicability of the many published approaches to plastic zone descriptions.

Process Zones

The process zone concept dates back at least to Krafft's 1964 paper [47], in which a physical picture like Fig. 8 was described and a rationale constructed for fracture toughness in terms of the zone dimension. The toughness equation was

$$K_{Ic} = En\sqrt{2\pi d_t},$$

where K_{Ic} is the plane strain fracture toughness, E is Young's modulus, n is the work-hardening exponent, and d_t is the process zone size. It was claimed that d_t was not a function of temperature or strain rate, but was a parameter characteristic of the material and the fracture mode. Later, Hahn and Rosenfield, in another well-known proposal, suggested a (dimensionally) somewhat similar toughness equation [48], in which K_{Ic} was proportional to $(\sigma_y E\lambda)^{1/2}$, where σ_y is the yield (or flow) strength and λ is the particle spacing. Subsequent work soon showed that this equation provides a poor description of toughness in at least some cases [49]. That there should be some parameter with the dimension of length in such an equation can be deduced from dimensional analysis, but just what that length should be is less clear. A recent review of this topic has been presented by Gerberich [50].

A more promising approach, for several reasons, is the "characteristic distance" concept of Ritchie, Knott and Rice or RKR [51]. Here the assumption is made that the fracture events on a local scale within the plastic zone must occur at some critical

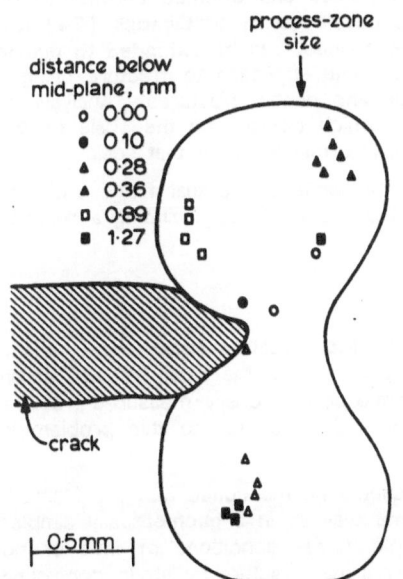

distance below
mid-plane, mm

o 0.00
• 0.10
▲ 0.28
▲ 0.36
□ 0.89
■ 1.27

process-zone size

crack

| 0.5mm |

FIGURE 7. Location of voids in plastic zone ahead of crack tip in ASTM A533B steel, from unpublished work of Van Stone presented in ref. 6.

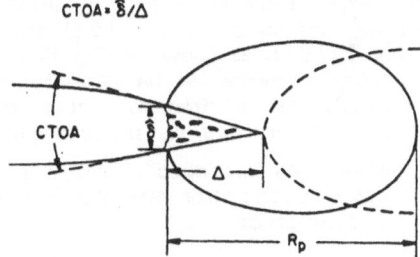

CTOA = δ/Δ

CTOA

Δ

R_p

FIGURE 8. Schematic of process zone, indicating "more intense" progress of fracture events within the plastic zone in region of width Δ, also called d_t. Also shown are crack tip opening angle (CTOA), crack tip opening δ, and plastic zone radius R_p. From Gerberich [50].

136

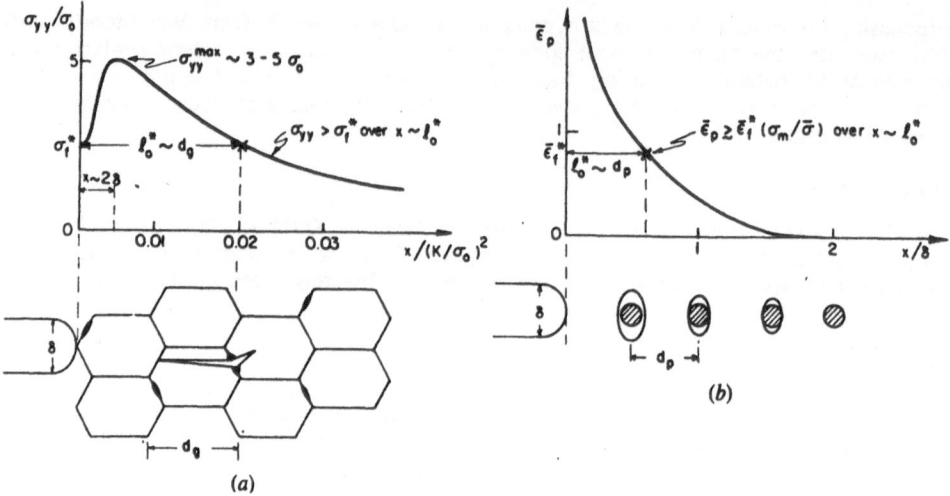

(a)

(b)

FIGURE 9. Schematic depiction of microscopic fracture criteria at crack tips. (a) Original RKR proposal for cleavage under stress control, as discussed in text. (b) Stress-modified strain-control model for microvoid coalescence. From Ritchie and Thompson [52].

stress which is present at a critical distance ahead of the crack tip. That distance is the characteristic distance. An illustration of the idea is shown in Fig. 9(a), which includes the notation of Ritchie and Thompson's review [52]. Fig. 9(a) is for cleavage fracture nucleated by cracking of grain boundary carbides, and the RKR result was that the characteristic distance was about two grain diameters. That distance was presumed to arise as a consequence of the need for a suitably sized and oriented carbide to be within range of the crack tip. By using concepts developed by McClintock [24] and McKenzie, Hancock and Brown [53], the same RKR concept can be extended to ductile fracture [52,54], as shown in Fig. 9(b). Thus the same approach to fracture using a characteristic distance can be taken for both brittle and ductile fractures. Analysis of "micromechanisms" of fracture, that is, mechanisms which operate on the scale of the materials' microstructure, can proceed with additional information about that scale.

Any equation developed in this way will have the same conceptual form, with the toughness expression (for generality in terms of the J integral [52]) including a distance and a fracture strain:

$$J_{Ic} \sim \sigma_0 \bar{\epsilon}_f \ell_0^*$$

where σ_0 is the local flow stress, $\bar{\epsilon}_f$ is the equivalent local fracture strain, and ℓ^* is the characteristic distance. The problem with this approach is that the local fracture strain need not be, and generally will not be, the same as the conventionally-measured fracture strain in smooth-bar tension or plane-strain tension. One solution to this problem is presented in the following section.

Before leaving the topic of process zones, it should be mentioned that the fracture events in a process zone may not ocur in the same way as in a geometrically simpler specimen such as a tensile specimen. For example, under conditions of little or no triaxiality and ample shear stresses, a particular material may fracture by strain-controlled MVC processes, while under the triaxial stresses, plane strain, and limited shear stress conditions at a crack tip, fracture may occur by a different (usually more brittle) mode, such as cleavage, which also is typically a stress-controlled fracture mode. Discussion of this point in detail is beyond the scope of this paper (it has been addressed elsewhere [55]). It is an important point to keep in mind, however, particularly when tensile−test information is used to model process zone events. It should be evident that such a procedure is risky unless detailed information is available about crack tip fracture processes.

Local Strains in Ductile Fracture

The local plasticity around void−nucleating particles should preserve a record of the post−strain initiation, as the microvoid lengthens in the loading direction. Thompson and Ashby [56] analyzed this problem with the use of the fracture surface "microroughness" parameter M suggested earlier [17]. Fig. 10(a) defines M as the ratio of depth to width of a microvoid on the fracture surface, h/w. Appropriate relationships from quantitative metallography (remembering that the fracture surface in general is neither planar nor necessarily a random section) permit the following relationship to be stated [56]: the ratio of h to the particle diameter d is given by the one−third power of $M^2/3f$, where f is the volume fraction of particles which are effective in nucleating voids. Since h/d is also a measure of the local strain, i.e. $\epsilon \cong \ln$ (h/d), the fracture strain should be given by

$$\epsilon_f \cong 1/3 \ln (M^2/3f).$$

Thus fracture surface measurement of M and knowledge of f gives a measure of fracture strain to insert in toughness relations. When toughness is experimentally measured, the characteristic distance is then obtained as the "free" parameter in the toughness expression. Additional work by Garrison and Thompson [32,57−59] has extended this idea, with the following results.

Void growth as a measure of local strain accumulation is based on the calculations of McMeeking [60,61], who used Rice and Tracey's results [26] to construct a relationship between the crack tip opening displacement δ and the distance of a void−nucleating particle from the crack tip, X_o. The dependence of void dimensions on the ratio of δ/X_o is shown in Fig. 10(b). This figure essentially is a prediction of observations on voids, whether in sectioned specimens or on the fracture surface, in connection with a measure of toughness, δ, and the particle spacing, which may be taken as X_o. Of particular interest is the linear relationships displayed, for both R_y and R_x. In general, R_x is more convenient to measure on fracture surfaces.

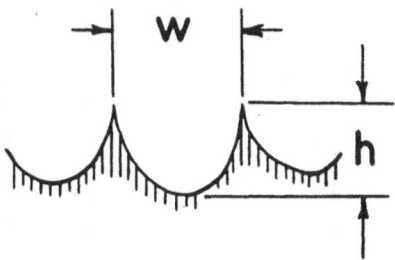

FIGURE 10(a). Definition of microroughness M, from depth h and width w of an individual microvoid. M = h/w.

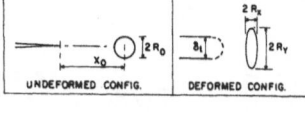

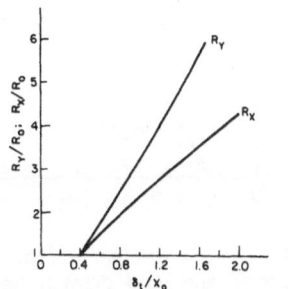

FIGURE 10(b). Relation between void growth and δ/X_o, from McMeeking's calculations. Figure after Garrison [59].

Garrison [59] then averaged R_x and R_y, the void dimensions in the projected plane of fracture, as R_v, the void radius. Then, following Rice and Johnson [28] and treating X_o as the nearest-neighbor distance among particles in three dimensions, one would expect the critical crack tip opening displacement δ_{lc} to scale as

$$\delta_{lc} \sim X_o (R_v/R_i),$$

where R_i is the initial void size, identified with the particle radius d/2. δ_{lc} in turn is related to J_{lc} as $\delta_{lc} = d_n (J_{lc}/\sigma_o)$, where d_n is a function of yield strain, work-hardening exponent n, and whether plane stress or plane strain conditions obtain, while σ_o is the flow stress, e.g. the average of yield and ultimate strengths. Measurements on a variety of tough steels, in various microstructural conditions, have been made and are shown in Fig. 11. Included are the McMeeking calculations from Fig. 10(b) for comparison. It is evident that these steels display amounts of void growth R_v/R_i and toughnesses δ_{lc} well in excess of the range of values calculated by McMeeking, but appear to retain a linear relation in this plot. More work is certainly of interest on this topic, and is continuing.

CRACK SHIELDING

This topic has been developed in detail by Ritchie and Cannon [62] and is summarized here for completeness. One perspective on crack behavior in materials is that the applied load(s) provide a driving force for crack motion, which in turn is resisted by the material surrounding the crack. In Ritchie's terminology [62,63], this resistance may be of two kinds, "intrinsic" (that is, the microstructure itself resists the fracture processes which make up crack motion) or "extrinsic", in which the crack tip is "shielded" from the far-field crack driving force, and motion is impeded by mechanical, microstructural and environmental factors which *locally* reduce the crack driving force. These are shown in Fig. 12. Item 1 in the figure shows crack deflection and meandering, which reduces crack driving force by lengthening the crack path, per unit

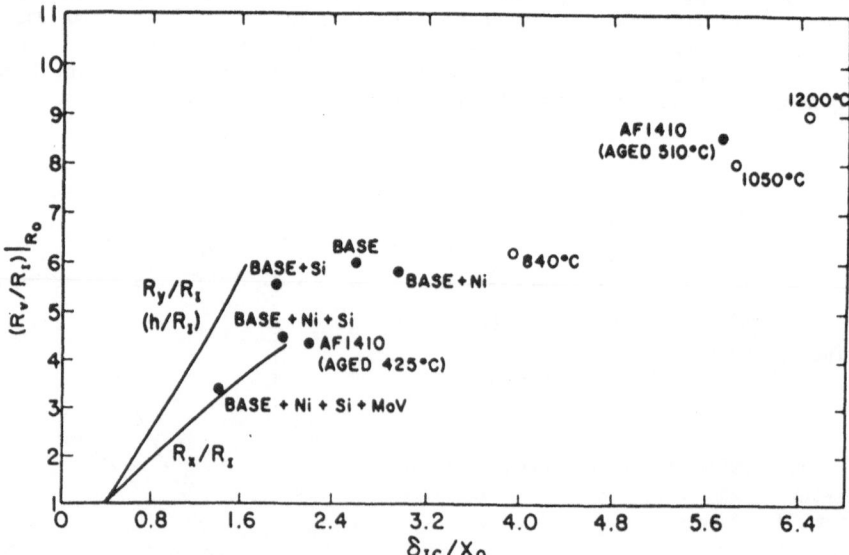

FIGURE 11. The calculated critical crack tip opening displacement, δ_{lc}, as a function of the spacing and void growth parameters, expressed as described in the text. Data are for 4340-type steels ("BASE") with alloying modifications, the 9 Ni-4 Co steel HP 9-4-20 (open circles), and the 14 Ni-10 Co steel AF1410 (unpublished data of Garrison). Also included are the McMeeking results from Fig. 10(b).

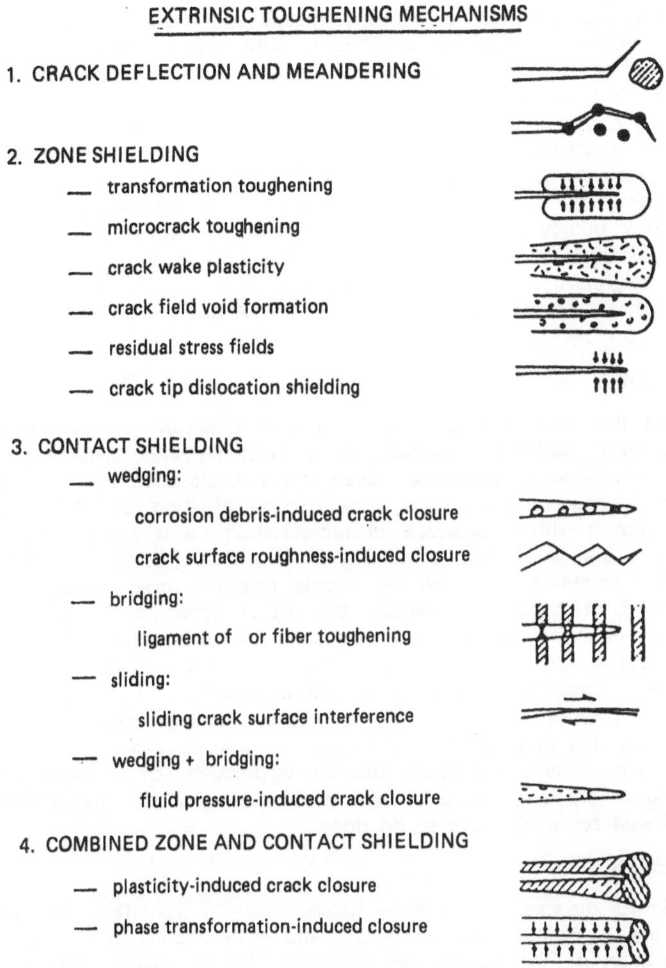

EXTRINSIC TOUGHENING MECHANISMS

1. CRACK DEFLECTION AND MEANDERING

2. ZONE SHIELDING
   ___ transformation toughening
   ___ microcrack toughening
   ___ crack wake plasticity
   ___ crack field void formation
   ___ residual stress fields
   ___ crack tip dislocation shielding

3. CONTACT SHIELDING
   ___ wedging:
       corrosion debris-induced crack closure
       crack surface roughness-induced closure
   ___ bridging:
       ligament of or fiber toughening
   ___ sliding:
       sliding crack surface interference
   ___ wedging + bridging:
       fluid pressure-induced crack closure

4. COMBINED ZONE AND CONTACT SHIELDING
   ___ plasticity-induced crack closure
   ___ phase transformation-induced closure

FIGURE 12. Schematic illustrations of several kinds of crack tip shielding, divided into four classes of extrinsic crack resistance mechanisms, as described in text. From Ritchie and Cannon [62].

advance. Item 2 shows several mechanisms of "zone shielding", in which more compliant or dilated zones surround the wake of the crack. Item 3 depicts "contact shielding", in which some of the load across the crack surfaces is transmitted by surface contact, bridging by unbroken material, or wedging. Item 4 shows combined zone and contact shielding processes which can increase crack closure. Similar concepts have been discussed in the context of process zones [50].

It should be emphasized that although the mechanical reasons for crack resistance are different between intrinsic and extrinsic mechanisms, either one *can* have its origin in microstructural features. Thus the importance of proper design, optimization, and control of microstructure is undiminished by the recognition of extrinsic crack resistance. However, the extrinsic–intrinsic distinction is a critical one if the correct decisions are to be made about desirable microstructures, whether in the context of material design, materials selection, or performance improvement.

In comparing the fracture behavior of most metals and alloys, typically characterized by relatively high values of fracture toughness, with the behavior of more brittle materials such as ceramics, intermetallic compounds, and many composite materials, it is essential to examine Fig. 12 carefully. For example, a weak interface between particles and matrix is usually damaging to fracture resistance in a high-toughness material, but in a brittle material the extrinsic effect of increased compliance in the process zone and along the crack wake can readily make weak particle-matrix interfaces quite desirable as a means of shielding the crack tip and thus increasing toughness. Similarly, the introduction of *either* ductile particles or fibers in a brittle matrix (e.g. glass filled with ductile aluminum particles, as in the work of Krstic [64,65]), *or* brittle fibers in a low- or medium-toughness matrix, can be effective in increasing effective toughness through crack shielding. There appear [50] to be a number of opportunities to model such phenomena accurately.

CONCLUDING REMARKS

The purpose of this paper has been to present an overview of distributed damage fracture concepts, with particular emphasis on a relatively-well understood example, ductile fracture or microvoid coalescence. Even in that topic area, it is clear that many issues, albeit probably second-order ones, remain to be clarified, and that a number of first-order issues remain without adequate characterization. It is encouraging that high-quality fundamental work is in fact continuing in this area [66,67]. Since there is extensive scope for additional work on the ductile fracture topic, usually regarded as fairly well developed, it should be evident that other types of distributed damage fracture can only be described as poorly understood. For the relatively brittle materials, crack tip shielding and extrinsic crack resistance mechanisms are of particular importance and will undoubtedly advance in degree of understanding. For higher-toughness, especially metallic, materials which exhibit distributed damage fractures other than microvoid coalescence, the propect for significant fundamental work in the foreseeable future is less sanguine. It is to be hoped that the opportunity of achieving significantly better understanding of cracking resistance and fracture toughness in those materials will provide a driving force for such work to be done.

ACKNOWLEDGEMENTS

I appreciate helpful discussions on these topics with I.M. Bernstein, J.R. Low, W.M. Garrison, R.O. Ritchie, N. Moody, and J.F. Knott, and experimental assistance from C.L. Thompson. Preparation of this paper was supported by the David Jackson Fund for Scholarly Endeavors and by the Association of American Railroads, through the Affiliated Laboratory at Carnegie Mellon University.

REFERENCES

1. Low J.R.: Progress in Materials Science, 1963, Vol. 12, 1-96.

2. Rogers H.C.: in Fundamentals of Deformation Processing (W.A. Backofen, et al., eds.), 199-255. Syracuse: Syracuse Univ. Press, 1964.

3. Rosenfield A.R.: Metall. Rev., 1968, Vol. 13, 29-40.

4. Knott J.F.: Fundamentals of Fracture Mechanics, Ch. 8. London: Butterworths, 1973.

5. Broek D.: Internat. Metals Reviews, 1974, Vol. 19, 135-180.

6. Van Stone R.H., T.B. Cox, J.R. Low and J.A. Psioda: Internat. Metals Reviews, 1985, Vol. 30, 157-179.

7. Tipper C.F.: Metallurgia, 1948-49, Vol. 39, 133-137.

8. Puttick K.E.: Phil. Mag., 1959, Vol. 4, 964-969.

9. Goods S.H., and L.M. Brown: Acta Met., 1979, Vol. 27, 1-15.

10. Argon A.S., J. Im, and A. Needleman: Metall. Trans. A, 1975, Vol. 6A, 815–824.

11. Argon A.S., J. Im and R. Safoglu: Metall. Trans. A, 1975, Vol. 6A, 825–837.

12. Argon A.S., and J. Im: Metall. Trans. A, 1975, Vol. 6A, 839–851.

13. Garber, R., I.M. Bernstein and A.W. Thompson: Scripta Met., 1976, Vol. 10, 341–45.

14. Cialone H., and R.J. Asaro: Metall. Trans. A, 1979, Vol. 10A, 367–375.

15. Garber R., I.M. Bernstein and A.W. Thompson: Metall. Trans. A, 1981, Vol. 12A, 225–234.

16. Cox T.B.: Ph.D. Thesis, Carnegie Mellon University, Pittsburgh, 1973.

17. Thompson A.W.: Acta Met., 1983, Vol. 31, 1517–1523.

18. Argon A.S.: J. Eng. Mater. Tech. (Trans. ASME, Series H), 1976, Vol. 98, 60–68.

19. Bridgman P.W.: Studies in Large Plastic Flow and Fracture. New York: McGraw-Hill, 1952.

20. Hippsley C.A. and S.G. Druce: Acta Met., 1983, Vol. 31, 1861–1872.

21. Fischmeister H.F., E. Navara and K.E. Easterling: Metal Sci. J., 1972, Vol. 6, 211–215.

22. Gladman T., B. Holmes and I.D. McIvor: in Effect of Second-phase Particles on the Mechanical Properties of Steel, 68–78. London: Iron and Steel Inst., 1971.

23. Joy G.D., and J. Nutting: ibid., 95–100.

24. McClintock F.A.: J. Appl. Mech. (Trans. ASME, Series E), 1968, Vol. 35, 363–371.

25. McClintock F.A.: in Ductility, 255–277. Metals Park: ASM, 1968.

26. Rice J.R., and D.M. Tracey: J. Mech. Phys. Solids, 1969, Vol. 17, 201–217.

27. Tracey D.M.: Eng. Fract. Mech., 1971, Vol. 3, 301–315.

28. Rice J.R., and M.A. Johnson: in Inelastic Behavior of Solids (M.F. Kanninen, W.F. Adler, A.R. Rosenfield and R.I. Jaffee, eds.), 641–672. New York: McGraw-Hill, 1970.

29. Gurland J.: in Composite Materials, Vol. 5: Fracture and Fatigue (L.J. Broutman, ed.), 45–91. New York: Academic Press, 1974.

30. Edelson B.I., and Baldwin, W.M.: Trans. ASM, 1962, Vol. 55, 230–250.

31. LeRoy G., J.D. Embury, G. Edward and M.F. Ashby: Acta Met., 1981, Vol. 29, 1509–1522.

32. Garrison W.M., and A.W. Thompson: Metall. Trans. A, in press.

33. Rogers H.C.: Trans. TMS-AIME, 1960, Vol. 218, 498–506.

34. Gurson A.L.: J. Eng. Mater. Tech. (Trans. ASME, Series H), 1977, Vol. 99, 2–15.

35. Berg C.A.: in Inelastic Behavior of Solids (M.F. Kanninen, W.F. Adler, A.R. Rosenfield and R.I. Jaffee, eds.), 171–209. New York: McGraw-Hill, 1970.

36. Tvergaard V., and A. Needleman: Acta Met., 1984, Vol. 32, 157–169.

37. Avaras G., and R.M. McMeeking: to be published (see McMeeking paper, this volume).

38. Backofen W.A.: in Fracture of Engineering Materials, 107-126. Metals Park: ASM, 1964.

39. Cox T.B., and J.R. Low: Metall. Trans., 1974, Vol. 5, 1457-1470.

40. Brown L.M., and J.D. Embury: in Proc. Third Int. Conf. on Strength of Materials and Alloys (ICSMA 3), 164-169. Cambridge: Inst. Metals, 1973.

41. Roberts W., B. Lehtinen and K.E. Easterling: Acta Met., 1976, Vol. 24, 745-758.

42. Bluhm J.I., and R.J. Morrissey: in Proc. First Int. Conf. on Fracture, Vol. 3, 1739-1780. Sendai: Japan Society for Strength and Fracture, 1966.

43. Park I.-G.: Ph.D. Thesis, Carnegie Mellon University, Pittsburgh, 1985.

44. Thompson A.W., and P.F. Weihrauch: Scripta Met., 1976, Vol. 10, 205-210.

45. Thompson A.W., and J.C. Williams: in Fracture 1977 (Proc. 4th Int. Conf. on Fracture), Vol. 2, 343-348. Waterloo: Univ. Waterloo Press, 1977.

46. Thompson A.W., and B.A. Wilcox: Scripta Met., 1972, Vol. 6, 689-696.

47. Krafft J.M.: Appl. Mater. Research, 1964, Vol. 4, 88-101.

48. Hahn G., and A. Rosenfield: in Applications-related Phenomena in Titanium Alloys (ASTM STP 432), 5-18. Philadelphia, Amer. Soc. Testing and Mater., 1968.

49. Brown W.F., and J.E. Srawley: in Review of Developments in Plane Strain Fracture Toughness Testing , W.F. Brown, ed. (ASTM STP 463), 216-248. Philadelphia: Amer Soc. Testing and Mater., 1970.

50. Gerberich W.W.: in Fracture: Interactions of Microstructure, Mechanisms, and Mechanics (J.M. Wells and J.D. Landes, eds.), 49-74. Warrendale: TMS-AIME, 1984.

51. Ritchie R.O., J.F. Knott and J.R. Rice: J. Mech. Phys. Solids, 1973, Vol. 21, 395-410.

52. Ritchie R.O., and A.W. Thompson: Metall. Trans. A, 1985, Vol. 16A, 233-248.

53. MacKenzie A.C., J.W. Hancock and D.K. Brown: Eng. Fract. Mech., 1977, Vol. 9, 167-188.

54. Ritchie R.O., W.L. Server and R.A. Wullaert: Metall. Trans. A, 1979, Vol. 10A, 1557-1570.

55. Thompson A.W.: Mater. Sci. and Tech., 1985, Vol. 1, 711-718.

56. Thompson A.W., and M.F. Ashby: Scripta Met., 1984, Vol. 18, 127-130.

57. Thompson A.W.: in Advances in Fracture Research (Proc. ICF 6), S.R. Valluri, D.M.R. Taplin, P.R. Rao, J.F. Knott and R. Dubey, eds., Vol. 2, 1393-1399. Oxford: Pergamon Press, 1984.

58. Garrison W.M.: Scripta Met., 1984, Vol. 18, 583-586.

59. Garrison W.M.: in Mechanical Properties and Phase Transformations in Engineering Materials (E.R. Parker Symposium), S.D. Antolovich, R.O. Ritchie and W.W. Gerberich, eds., 187-205. Warrendale: TMS-AIME, 1986.

60. McMeeking R.M.: J. Mech. Phys. Solids, 1977, Vol. 25, 357-81.

61. McMeeking R.M.: J. Eng. Mater. Tech. (Trans. ASME, Series H), 1977, Vol. 99, 290-97.

62. Ritchie R.O., and R.M. Cannon: Acta Met., submitted (also Report LBL-20656, Lawrence Berkeley Laboratory, University of California, Berkeley, Dec. 1985).

63. Ritchie R.O., and W. Yu: in Small Fatigue Cracks (R.O. Ritchie and J. Lankford, eds., 167–189. Warrendale: TMS-AIME, 1986.

64. Krstic V.D., and A.K. Khaund: in Advances in Fracture Research, Vol. 4 (D. Francois, ed.), 1577–1585. New York: Pergamon Press, 1982.

65. Krstic V.D., Phil. Mag., 1983, Vol. 48, 695–708.

66. Fisher J.R., and J. Gurland: Metal Sci. J., 1981, Vol. 15, 185–192.

67. Fisher J.R., and J. Gurland: Metal Sci. J., 1981, Vol. 15, 193–202.

DISCUSSION

Comment by T. Watanabe:

I have two comments concerning the control of intergranular fracture and crack deflection in fracture processes in polycrystals. We have found strong effects of grain boundary type or structure on intergranular fracture. (T. Watanabe: Res. Mechanica 11(1984):47-48, Met. Trans. 14A(1983): 531) Special types of boundaries such as low-angle boundaries and high-angle coincidence boundaries are very resistant to fracture, but high-angle general boundaries are not. So an increase of the frequency of special boundaries may control intergranular fracture. We also found the fracture mode can change, depending on the type of boundary in front of its propagating crack. This may be considered to be related to crack deflection in fracture processes in polycrystals.

Reply:

Although intergranular fracture was not considered in any detail in my paper, your point is a very interesting one. Details of interfacial structure (and, in some cases, the associated microchemistry) are usually not known, yet can be expected to play an important role in fracture processes. I am encouraged that you are addressing the structure directly, and I would suggest the inclusion of grain boundary composition measurements to complete the picture.

Comment by A. S. Argon:

When you referred to crack tip shielding, you created the impression of a notion of very general applicability. I think crack tip shielding when defined to imply a local crack tip K is only applicable to situations where the local process is of cleavage or brittle nature where crack tip singularity at crack advance has clear meaning. In ductile processes where the stress maximum moves into the interior, there is no crack tip region that can be characterized by a local K.

Reply:

I'm not sure I see how the local K is crucial to shielding. Changes in mechanical behavior near the crack can alter the net crack driving force, regardless of whether cracking is brittle or ductile. In relative terms, you may be right. In absolute terms, however, shielding contributions presumably depend on, say, process zone compliance (e.g. due to micro-cracks or microvoids), independent of fracture process. Thus a shielding contribution of, say, 2 MPa /m could arise from crack or void process zones. However, this contribution would look quite different to a brittle material with an intrinsic toughness of, say, 1 MPa /m, compared to a structural material with an intrinsic toughness of 50 MPa /m: 200% vs 4% improvement.

Comment by J. Hirth:

With regard to the size effect of particle decohesion, even with a local stress criterion, it would appear that a size effect would be expected. For small enough particles the release of strain energy would not suffice to supply the surface energy and decohesion would not occur. In stress terms, local stress relaxation would occur as void embryo formed so that the embryo would tend to close up rather than grow. Estimates of the critical events are only rough and would depend on interfacial cohe-sive energies, surface energies, etc. However, it would appear that theo-retically one would expect a trend of an increase in required stress for decohesion with a decrease in particle size.

Reply:

This suggestion is certainly reasonable, since as particle radius r decreases, the available strain energy goes down as r^3, while the required energy only goes down as r^2. There should thus be some lower limiting size for particles which can nucleate voids (Argon, et al., 1975, found r = 50Å), and particles near the lower limit could well show a size depen-dence. I think it remains unclear, though, why a 5μm particle would nucleate before one of, say, 2 μm size.

Comment by A. S. Argon:

On the subject of very small cavities on the scale of nanometer dimen-sions, caution is necessary. If such cavities can form by interface sepa-ration or by radiation damage, they will not have the local degrees of freedom for plastic expansion under stress as they may be far too small than sizes in which shape changes by means of crystal plasticity can be readily accomplished. Simply, the required local dislocation densities to provide the kinematics of shape changes will not be present.

Reply:

This is an interesting point. Certainly it appears that understanding of very small particle (or very small void) behavior is incomplete, with respect to this topic and others as well. It would be of value to know more about the limits to void size for plastic enlargement, which presum-ably would be a function of strain.

Comment by R. H. Jones:

If hydrogen can accelerate hole initiation and growth, doesn't this cause a restriction in the use of particles as hydrogen traps to increase resistance to hydrogen effects?

Reply:

In one sense, yes. However, it is important to consider the problem in a larger context. The particles may be trapping hydrogen which otherwise could cause brittle intergranular fracture, for example; this is the effect of ThO_2 particles in nickel. Presumably one would prefer ductile microvoid coalescence, albeit with somewhat reduced ductility, to brittle intergranular fracture. Moreover, particles chosen to have high nucleation strains could function as traps, reducing mobile hydrogen concentrations, while having a minimal effect on ductility.

Comment by J. F. Knott:

Two points:
    a) shear localization will be enhanced due to the presence of the free surfaces associated with the voids formed around non-metallic inclusions at low strains.
    b) note should be taken of the shapes of particles, such as carbides, on which fine-scale microvoids may be initiated. Spherical particles require high initiation strains, but plate- or lath-like carbides may initiate microvoids at much lower strains. In lightly tempered steels, these non-spherical carbides will additionally be stressed by transformation dislocations.

Reply:

I quite agree with both points. The first one I concurred with in the paper explicitly, the second one indirectly through the Gladman, et al. results. It is certainly important to keep in mind that lath-like carbides, or grain boundary carbide films, crack much more easily than spheres. It is to be hoped that comparable observations, preferably quantitative ones, will be generated for particle types in other alloy systems.

Comment by R. H. Jones:

Peter Haasen presented ductile-brittle transition behavior for silicon which was interpreted as being controlled by the mobility of dislocations and not their nucleation at crack tips. What do these results suggest relative to the Rice-Thomson model for the ductile-brittle transition?

Reply:

In his case, the dislocations were assumed to be emitted from the crack tip, and they would thus be in the shielding configuration. If for some reason, they were emitted asymmetrically from the tip, the shielding is such as to stabilize the cleavage at the tip. Hence, the crack can continue to cleave as the stress is raised to cancel the K_I shielding. Remember also, in the case of Si, the emission occurs by thermal fluctuations, so the crack is presumed to be mechanically stable against emission. Hence, it should always retain its cleavability.

There are some interesting questions regarding the nucleation of dislocations at the crack, when those dislocations cannot be moved, because of the large shielding which occurs when the dislocation remains very close to the crack. In this case, the crack may be nucleated past the shielding barrier, or the dislocation may simply jump the barrier back into the crack. In either case, the dislocation will disappear, but the crack will be slowed and show a nucleated slow crack growth. Three dimensional considerations would be important because of the width of the nucleated-dislocation loops, but I don't know what the observations are.

Comment by S. M. Ohr:

Would you clarify how cleavage can take place after the anti-shielding dislocations are neutralized by the shielding dislocations emitted from the crack tip?

Reply:

Neglecting the shielding dislocations, the stress conditions at the tip is the same as before. After the anti-shielding dislocations are neutralized, since we assumed the crack was initially a cleavage crack, it returns to this cleavage condition. Since the shielding dislocations do remain, the crack is partially shielded. When the external stress is then increased to overcome this shielding, the crack can again cleave. The shielding effect from these dislocations also induces K_{II}^D and K_{III}^D at the crack tip, but these terms stabilize the cleavage condition at the tip.

Comment by D. G. Pettifor:

Where does the bond strength between the particle and the matrix enter your model? The computer simulations with which you compared assumed that the nucleation of the void had already occurred.

Reply:

The bond strength is essentially reflected in the nucleation strain. The "growth" type of calculation does, as you mention, treat nucleation strain as zero, which is certainly not a general value, though it is appropriate for a number of important cases. A more general calculation would include nucleation strain explicitly.

Comment by R. M. Latanision:

Although Stan Lynch could not be with us at this meeting, his view of fracture and the issue of void formation are somewhat different than those which you've expressed. Could you speak on some of those matters of difference and give us your perspective on how the Lynch views fit into the whole perspective?

Reply:

There are two aspects of Lynch's work that may be relevant. One relates to local ductility. Lynch has very usefully insisted that many fractures which superficially appear brittle are in fact ductile on a local, fractographic scale. I entirely agree with those results and with that conclusion. The second aspect relates, as I understand Lynch, to control of plasticity in the crack tip region by conditions at the crack surface. This I think is less general, since I believe there are clear indications that events deep within the plastic zone, such as microvoid nucleation, can also be affected by, for example, hydrogen, and thus affect the fracture process. On balance, though, I applaud Lynch's emphasis on understanding plastic processes of local fracture.

Comment by P. Neumann:

Ali Argon's remark (1) is very important since it demonstrates that even a plastic growth mechanism will help voids to grow only if they have already critical size. Therefore, in the absence of weak inclusions, nucleation of voids is a problem and this should be considered carefully in any model of crack growth due to void formation and growth.

Reply:

I would agree that void nucleation can be an important part of the ductile fracture process (though in many technical materials there are ample numbers of weakly bonded particles). However, I wouldn't describe it as a problem except in the sense that nucleation occurs over a range of strains in most materials. The voids do not have to grow from zero size by plastic processes.

Comment by H. Birnhaum:

There are cases of shear localization in uniform structures, e.g. channeling of slip in radiation hardened material. This occurs because of softening of the material by the passage of dislocation. In a similar manner, the decrease in the resistance to dislocation motion by hydrogen can lead to strain localization at a crack tip with the deformation being localized to the regions where the hydrogen concentration is the highest. Since the dislocations also carry the hydrogen as an atmosphere, this leads to localization of the slip.

Reply:

I agree with your comment. In fact, localization can occur in material without any inhomogeneities (for mechanics reasons).

A Comparison of Void Growth and
Ductile Failure in Plane and
Axisymmetric States of Strain

J W Hancock
Department of Mechanical Engineering
University of Glasgow
Scotland

1. Abstract

Finite element studies of the plastic deformation of a model of a
periodic array of voids have shown that the deformation before strain
concentrates in the inter-void ligament leading to void coalescense,
is very much greater in axisymmetric than in plane states of strain.
In experimental tests of notched tensile specimens which produce plane
and axisymmetric states of strain in a range of stress states with
different hydrostatic components the strain to cause ductile failure
by void coalescense does not depend on the state of strain. The void
growth calculations and the experiments are reconciled by the view
that ductile failure initiates in the most favourable statistical
inhomogeneity in which the local strain state does not necessarily
correspond to that applied remotely.

2. Introduction

The growth of voids by plastic deformation has been widely studied
with the object of developing a quantitative understanding of the
ductile fracture process in structural metals.   Fundamental to this
problem is the behaviour of a single isolated void embedded in an
incompressible plastic matrix, for which the variational methods used
by Rice and Tracey[1] and Budiansky Slutsky and Hutchinson[2] have
emphasised the strong dependence of the void growth rate on the
triaxiality of the stress state and the non-linear behaviour of the
matrix.   This behaviour is reflected in the experimental observations
which show the strong dependence of the ductility of structural metals
on the state of stress [3,4].   Approximate calculations of the strain
to cause coalescense based on the behaviour of a single void naturally
show the correct stress state dependence but generally overestimate
the strains to cause failure.   To address this problem more detailed
calculations have been performed taking account of the interaction
between adjacent voids by Needleman[5] and Gurson[6].

A plane strain analysis of a doubly periodic array of cylindrical
voids, as illustrated in Fig.(1), has been given by Needleman[5] using
a large strain formulation which allowed the shape change of the void
to be examined.   In this case flow becomes localised into the

ligament between transversely adjacent voids at the load maximum. A somewhat different approach has been adopted by Gurson[6] who has smeared the effect of the individual voids out and represented the porous material as a dilating plastic continuum. On this basis Rudnicki and Rice[7] have examined the conditions for localisation with the idea that ductile failure can be regarded as bifurcation from a homogeneous flow field of a plastic continuum without the need to model the interaction between discrete voids. Two conditions for localisation arise, firstly, that of compatibility between the band and the bulk material: and a strain hardening requirement to ensure equilibrium between the localised band and the remote material. The compatability conditions are automatically satisfied for an incompressible material in plane strain, and in fact a general feature of the analyses of localisation even for dilating materials is that plane strain states are expected to be more prone to localisation than axi-symmetric states of strain.

In the present work the growth of void representative of a bi-periodic array examined following Needleman's work in plane strain conditions. The same void containing cell has been examined in axisymmetric conditions with the motive of examining the effect of strain state on the strain to cause void coalescense by localised flow. Finally, experiments following Clausing[8] and Hancock and Brown[9] have been performed on tensile specimens intended to give plane and axisymmetric conditions.

3. Numerical Method

The plane strain deformation of a bi-periodic array of cylindrical voids has been analysed by considering the unit cell shown in Fig (2). On the top boundary uniform vertical displacements are imposed while the lower and left hand boundaries have the usual symmetry conditions. The horizontal displacements on the right hand edge of the cell are constrained to be identical while still allowing the value of the displacement to be a free variable in the problem. This enables both compatability and equilibrium conditions to be maintained between similar cells adjacent to both the left hand and top edges and thus allows single cell to be representative of a bi-periodic array of cylindrical voids. The plane strain problem was analysed with a mesh comprising twelve 8 noded hybrid elastic-plastic elements provided by the finite element code ABAQUS[10].

The same mesh has been analysed in axisymmetric conditions and with similar boundary conditions in cylindrical coordinates. The cell is now cylindrical and contains a spherical void. The edges of the cell are again constrained to remain straight, but during the large geometry change with the value of the transverse displacement again left as a free variable deformation to simulate periodic boundary conditions. However such cylindrical cells do not tessellate properly to fill the available space while the axisymmetric nature of the problem is equivalent to smearing out the surrounding voids around the circumference of the cell. The axisymmetric problem, as posed is thus not a rigorous representation of a three dimensional array of voids under axisymmetric loading, but is considered to be sufficiently similar to show some of the appropriate effects while being a very much smaller computational problem.

In both plane and axisymmetric problem the matrix has been allowed to follow the power hardening law used by Needleman [5] is his original plane strain analysis of this problem. In uniaxial tension

$$e = \frac{\sigma}{E} \qquad \sigma < \sigma_o$$

$$e = \frac{\sigma^m}{E\sigma_o^{m-1}} \qquad \sigma > \sigma_o$$

Here σ_o is the initial yield stress and E is Young's modulus, and m is a constant which determines the strain hardening rate. Two strain hardening rates corresponding to m = 4, and m = 8 have been analysed to provide contact with Needleman's solutions [5].

4. Numerical Results

The plane strain analyses of periodic arrays of voids essentially confirm the results and conclusions of Needleman [5]. The circular cylindrical voids deform initially as ellipsoids in the manner expected of asingle isolated void. However at the axial load maximum, strain concentrates into the ligament between the transversely adjacent voids. The ligament then decreases markedly as the voids exhibit a high transverse growth rate and loose their elliptical shape as they approach coalescense. The deforming shape of a single quadrant of a voided cell is shown in Fig (2a) along with contours of equivalent plastic strain shown in Fig (2b). Denoting the edge length of the cell as l with an initial value of lo, flow localises into the intervoid ligament at an extension ratio (l/lo) of 1.11 for a strain hardening exponent m = 4 and at an extension ratio (l/lo) of 1.08 for a matrix strain hardening exponent of m = 8.

In contrast the developing shape of the spherical void in the axisymmetric cell is shown in Figs (3). Here flow only starts to localise in the transverse ligaments at extension ratios (l/lo) = 3.57 and 2.46 strain hardening exponents m = 4 and 8. The axisymmetric voided cell is thus very much resistant to localised flow than the plane strain voided cells.

5. Experimental Method and Materials

Axisymmetric and plane strain notched and un-notched tensile specimens with the dimensions shown in Figure (4) were tested under displacement control in a servo-hydraulic testing machine. The test method is described in detail by Hancock and Mackenzie (4) and Hancock and Brown (9). The tests were continued until local failure by void coalescense resulted in the formation of a distinct crack. The stress state and strain histories of material at the failure site was determined by reference to a finite element analysis of the notched specimens (9) using the displacements to relate the numerical and experimental results.

Two materials were examined. Swedish Iron is a commercially pure iron containing 1% volume fraction of iron oxide particles with a mean diameter of 5μm. The single generation of inclusions does not contribute greatly to the strength of the metal which has no structural use. For comparison tests were also carried out on a carbon manganese and steel denoted as grade 50D under British Standard 4360. This steel contains large non-metallic inclusions and has a pearlitic micro-structure, both of which contribute to the final fracture process. A detailed description of the behaviour of the Swedishĥhas been given by R Thomson (11) while Hancock and Brown(9) have given more details on the testing of the grade 50D steel.

6. Experimental Results

The experimental results from the tensile tests of notched and unnotched axisymmetric and plane strain specimens are shown in Fig (5) and Fig (6). These Figures show the deformation history of the site where failure intiated in terms of the development of the hydrostatic component of the stress state, denoted $\sigma_m = \frac{1}{3}\sigma_{KK}$, non dimensionalised with respect to current equivalent flow stress $\bar{\sigma}$, as a function of the equivalent plastic strain $\bar{e}^p$. The results show that as the triaxiality of the stress state ($\frac{\sigma_m}{\bar{\sigma}}$) increases the ductility decreases. The line through the end points of the deformation histories has been termed a failure locus and it is noteworthy that data from both plane and axisymmetric specimens lie on the same locus.

7. Discussion

A general feature of the analyses of flow localisation in materials with a wide range of constitutive relations is that localisation is inhibited by axisymmetric states of strain and favoured by plane states of strain. This basically arises because of the difficulty of maintaining compatability between a thin band in which large strain rates are possible and the bulk material which is only undergoing small strain rates in axisymmetric conditions. Although analyses

152

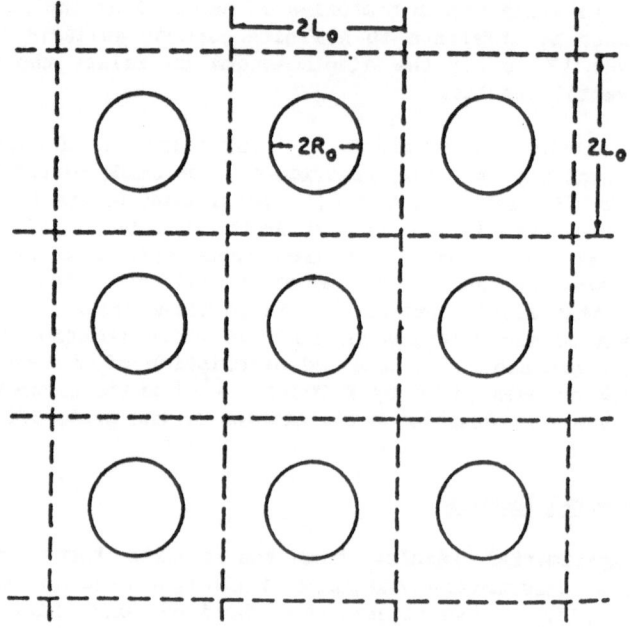

FIGURE 1. Schematic of double periodic array of cyclindrical voids as used by Needleman (5) for plane strain analysis.

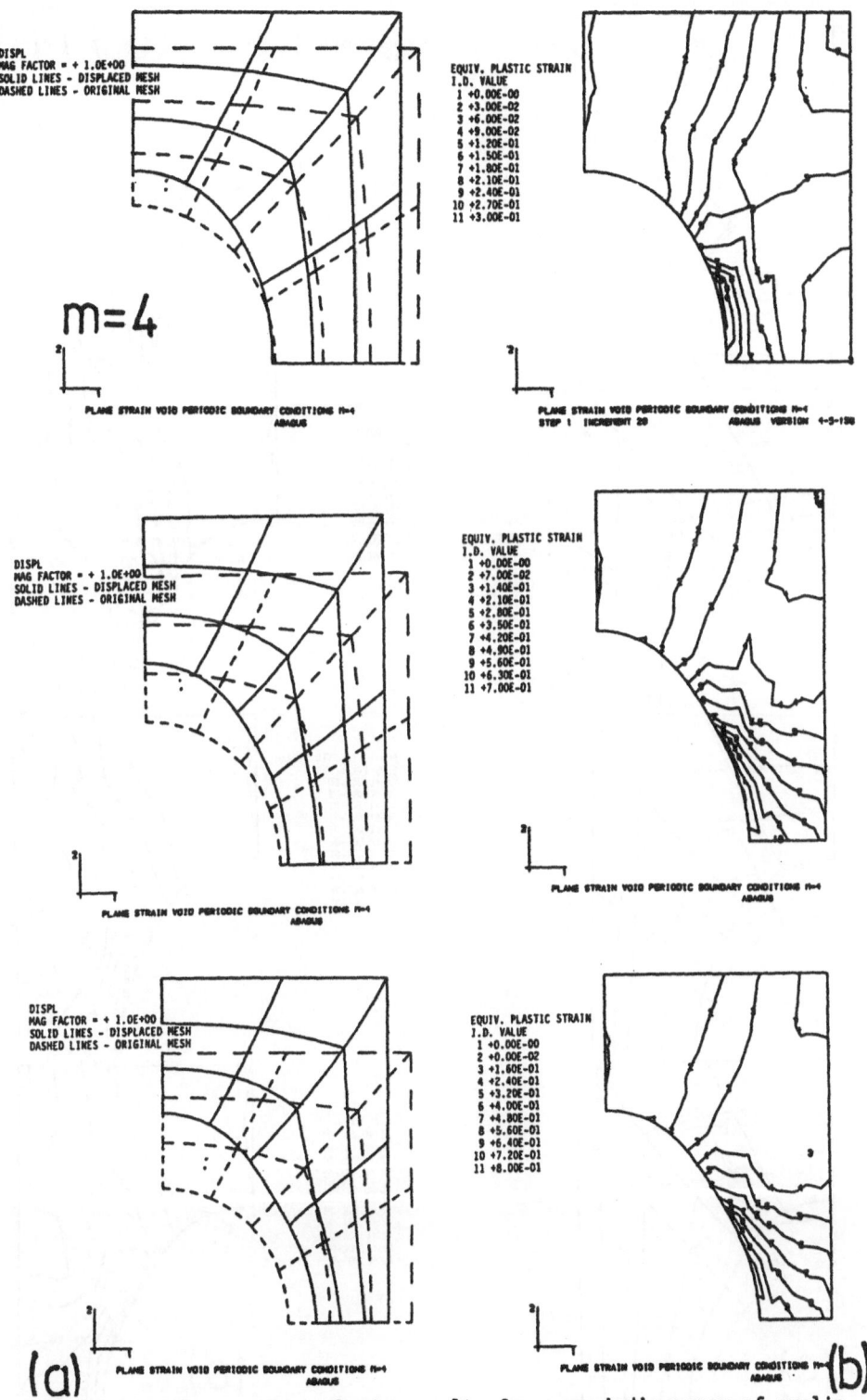

FIGURE 2. Plane strain analysis results for a periodic array of cyclindrical voids for strain hardening rates of a,b) m = 4 and c,d) m = 8.

154

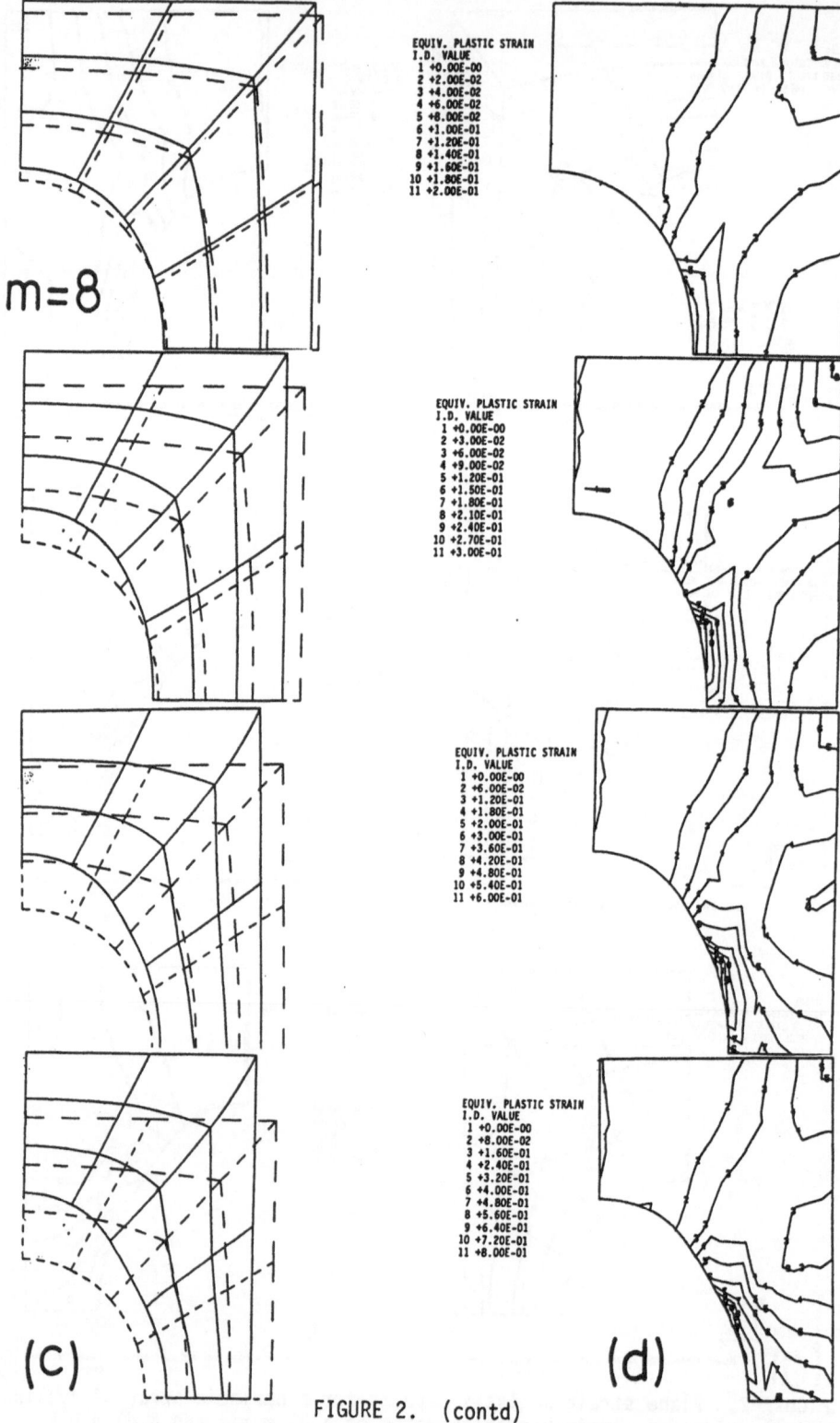

FIGURE 2. (contd)

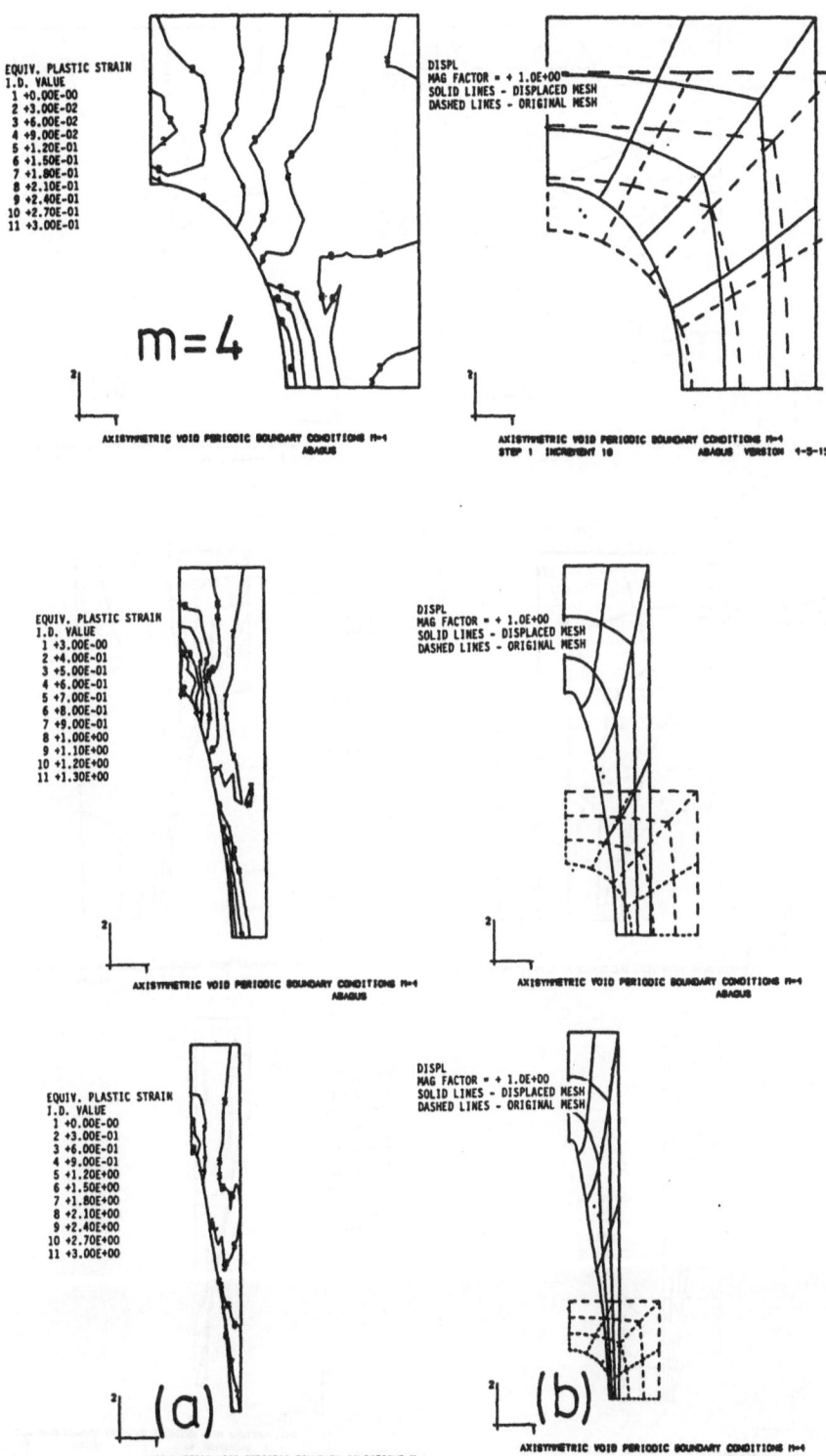

FIGURE 3. Axisymmetric strain analysis results for a periodic array of spherical voids for strain hardening rates a,b) m = 4 and c,d) m = 8.

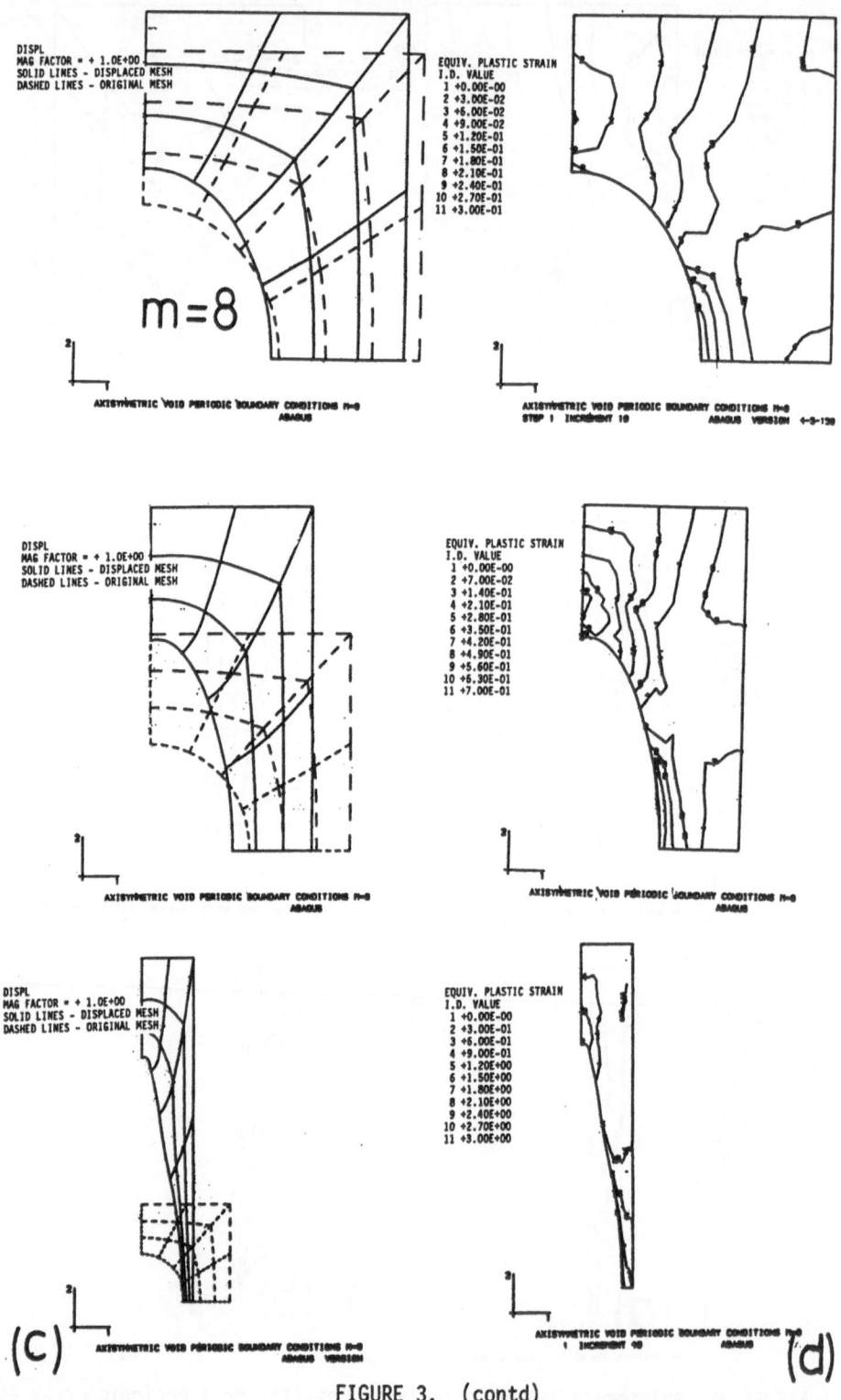

FIGURE 3. (contd)

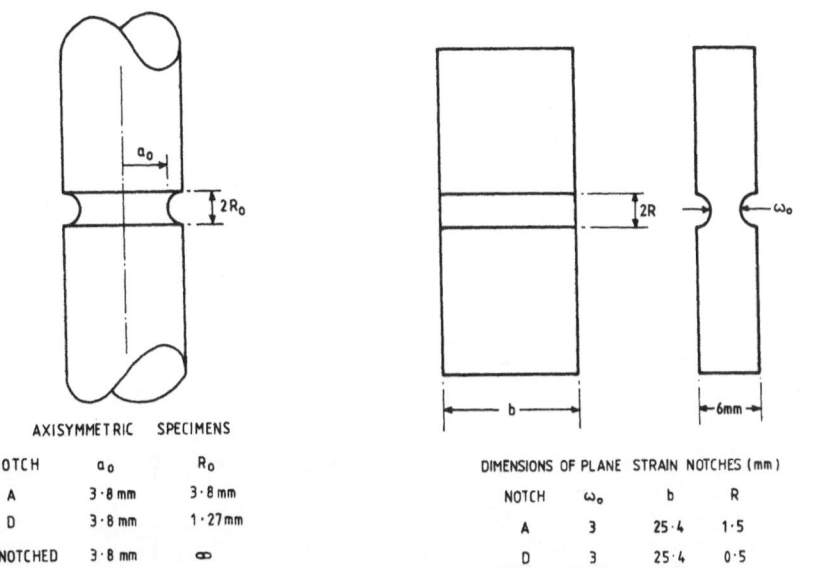

AXISYMMETRIC SPECIMENS

NOTCH	a_0	R_0
A	3·8 mm	3·8 mm
D	3·8 mm	1·27 mm
UNNOTCHED	3·8 mm	∞

DIMENSIONS OF PLANE STRAIN NOTCHES (mm)

NOTCH	ω_0	b	R
A	3	25·4	1·5
D	3	25·4	0·5

The dimensions of the notched axisymmetric specimens.   . The dimensions of the notched plane strain specimens.

FIGURE 4. Dimensions of a) notched axisymmetric specimens and b) notched plane strain specimens.

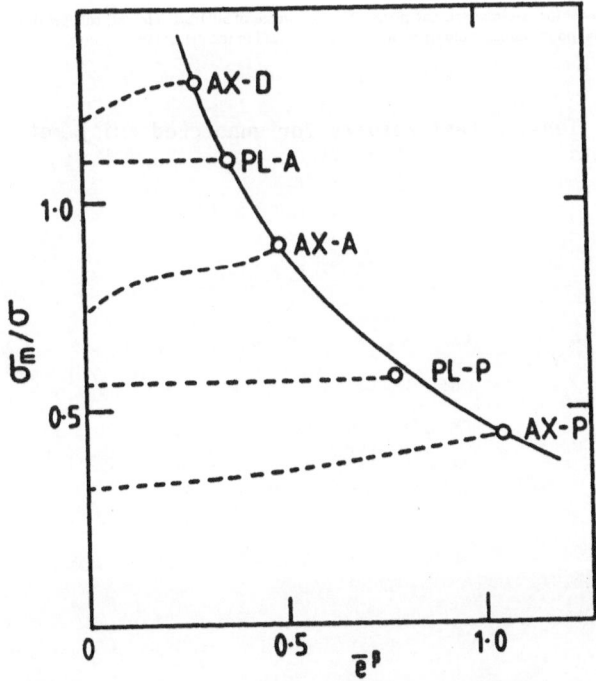

FIGURE 5. Tensile test results for notched axisymmetric and plane strain tests.

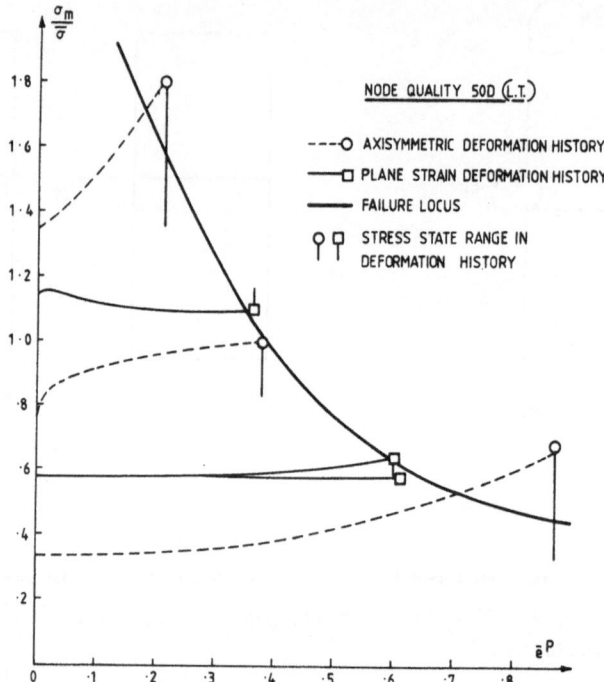

Failure locus for axisymmetric and plane strain specimens of 50D L.T. The vertical lines through the final points indicate the stress state range involved in the deformation history.

FIGURE 6. Tensile test results for unnotched axisymmetric and plane strain tests.

such as that by Rudnicki and Rice [7] address the problem of localisation into an infinitely thin shear band, the finite element calculations for models of periodic arrays of voids show a similar effect. Necking of the ligament which involves relatively diffuse strain concentrations is markedly favoured by plane strain conditions, and inhibited by the axisymmetric conditions. In contrast the experimental data shows that the ductility in both axisymmetric and plane states of strain lies on the same locus. A feature of real materials is that they contain random rather periodic arrays of particles which nucleate voids. The necessary consequence of this is that there exist regions within the material which contain more voids than average. Fracture seems certain to intiate in such statistical inhomogeneitives, and due to their arbitrary shape, the local strain state within them does not correspond to that remotely applied to the macroscopic body as a whole.

On this basis it is natural not to expect strain state effects in real materials to be as marked as those in model systems. Similarly comparisons of the ductilities of real materials with those determined from finite element calculations on periodic arrays of voids show better agreement with plane strain calculations than with axisymmetric ones because failure is expected to initiate first in statistical inhomogeneities with local conditions which approximate to plane strain irrespective of the remote strain state on the macroscpic body as a whole.

8. References

1. J R Rice and D Tracey (1969) J.Mech.Phys.Solids 17, 201.

2. B Budiansky, S Slutsky and J W Hutchinson (1981) 'Mechanics of Solids' Pergamon Press, Oxford.

3. P W Bridgman (1952) 'Studies in Large Plastic Flow and Fracture', McGraw-Hill, New York.

4. J W Hancock and A C Mackenzie (1976) J.Mech.Phys.Solids 14, 107.

5. A Needleman (1972) J.Mech.Phys.Solids 20, 111.

6. A Gurson (1977) Trans A.S.M.E. J.Eng.Mat.Tech. 99, 2.

7. J W Rudnicki and J R Rice (1969) J.Mech.Phys.Solids, 17, 201.

8. D P Clausing (1970) Int.J.Fracture.Mech. 6(1), 71.

9. J W Hancock and D K Brown (1983) J.Mech.Phys.Solids 31, 1

10. ABAQUS (1982) ABAQUS Users Manual, Hibbitt, Karlsson and Sorensen Inc. Providence, Rhode Island.

160

11. R D Thomson (1985) 'Ductile Fracture by Void Nucleation, Growth and Coalescense' Ph.D. Thesis, University of Glasgow.

9. Acknowledgement

The work was carried out in the Department of Mechanical Engineering of the University of Glasgow and acknowledgements are due to Professor B F Scott for the provision of laboratory facilities and to H.K.S. for access to ABAQUS.

Solid State Chemistry
and
Physics of Fracture

Session Chairmen: R. Thomson and D. Pettifor

THEORETICAL APPROACHES TO MATERIALS DESIGN:
INTERGRANULAR EMBRITTLEMENT

M. E. Eberhart[*] and D. D. Vvedensky[**]
Materials Science and Technology Division
Los Alamos National Laboratory
Los Alamos, New Mexico 87501

[*]Also at: Department of Materials Science and Engineering,
             Massachusetts Institute of Technology, Cambridge,
             Massachusetts 02139

[**]Permanent address: The Blackett Laboratory, Imperial College, London,
             SW72BZ, UK.

1. INTRODUCTION
     One of the ultimate objectives for the electronic structure theory
of solids is the first-principles design of materials. Major steps in
this direction have already been taken in the form of parameter-free
calculations, which are capable of yielding accurate descriptions of a
number of structural, electronic, and magnetic properties of metals,
semiconductors and even disordered alloys. Furthermore, extensions of
these approaches to point defects (substitutional impurities, interstitials, and vacancies) and to interfaces and clean and covered surfaces
are showing great promise. However, only recently has there been an
attempt to correlate the results of electronic structure calculations
with mechanical properties, and only in the past few years have the
specific features of electronic structure that could give rise, for
example, to brittle versus ductile behavior,[1-5] been addressed. Indeed,
despite the complex and manifold origins of mechanical behavior and the
relatively poor characterization of the pertinent structures at the
atomic level, general trends in certain mechanical properties may be
correlated with specific features of electronic structure. An interesting illustration is the control of mechanical properties of semiconductors by electrically-active impurities.[6] At relatively low temperatures
($\lesssim 500°C$) the dopants have been shown to affect yield stress and hardness
through their influence on dislocation velocities,[6-7] the effect being a
particularly strong function of dopant concentration in Si and Ge.[*]
     In this paper, we investigate one specific aspect of mechanical
behavior, intergranular fracture, by examining the influence of grain-
boundary-induced electronic structure on the tendency toward intergranular fracture in polycrystalline materials. We then explore the implications of our model for impurity embrittlement and consider explicitly
possible microscopic mechanisms of embrittlement by sulfur and by hydrogen.

164

For the case of _intrinsically_ brittle grain boundaries, our calculations were motivated by the working hypothesis that the extent to which a grain boundary disrupts the electronic structure of the parent crystal depends upon the ease with which s-hybridization can be enhanced to accommodate the oblique bonding geometries found near the grain boundary. Thus, for intergranularly brittle materials, enhanced s-hybridization is inhibited, and we identify the resulting appearance of localized grain-boundary electronic states near the Fermi energy as being indicative of a marked decrease in the mechanical stability of the interface relative to the parent crystal. Thus, the grain boundary is more responsive to an external stimulus and can accommodate greater strain than the parent crystal because the energy barriers to bond movement and charge redistribution have been lowered locally.[2] This leads to localization of strain and plastic deformation and subsequently to intergranular fracture. Recent scanning electron microscopy studies have indeed revealed localization of plastic flow accompanying intergranular fracture in Ni_3Al.[8-9]

The embrittlement of metals by impurities (extrinsic embrittlement) has been by far the most extensively studied aspect of intergranular fracture, having been addressed by many authors from a variety of perspective.[1-3] In fact, the first attempts to explain embrittlement at the atomic level were in systems believed to model the electronic structure of sulfur in nickel.[1-3] One complicating feature of impurity-induced embrittlement is the possibility of interactions among the segregants at the grain boundary. A single impurity can induce a qualitatively different charge distribution locally than interacting impurities. For example, as a single impurity sulfur interacts ionically with nickel, while if sulfur-sulfur interactions become important, then covalent sulfur-sulfur-nickel bonds are formed.[5] In fact the formation of metal-mediated covalent sulfur-sulfur bonds is quite a general feature of sulfur interactions with a metallic environment, having been observed not only for grain-boundary geometries, but also in iron-sulfur proteins,[10] and on nickel surfaces[11-12] where sulfur is a well-known poison of certain types of catalytic reactions. The overall effect of the interacting sulfurs at the grain boundary is to induce charge redistribution into bonds parallel to the grain boundary, thus catalyzing cleavage in nickel

The organization of our paper is as follows. In Section 2 we discuss the theoretical framework used for calculating electronic structures at grain boundaries, including the problem of structurally characterizing grain boundaries. In Section 3 we apply our model to study intrinsic embrittlement in polycrystalline $L1_2$ compounds (Cu_3Au structure). The wealth of theoretical and experimental work carried out on the $L1_2$ compounds, especially the identification of fracture modes, makes

---

*
This example also illustrates the particularly strong interplay between theory and experiment in manipulating mechanical behavior when the atomic structures are well-characterized (dislocations and substitutional impurities in otherwise perfect crystals) and the microscopic mechanisms responsible for macroscopic behavior have been identified (dislocation velocity).

these systems a convenient starting point for microscopic theories of intergranular embrittlement. In Section 4 we consider embrittlement by sulfur and hydrogen impurities. We discuss briefly of other microscopic approaches to impurity embrittlement, including the work of Messmer and Briant, which has been so important in developing interest in this problem. Finally, in Section 5 we summarize our work and outline future directions of research.

2. METHOD

To apply the general considerations presented in the introduction to specific materials, we use the multiple scattering-Xα (MS-Xα) cluster method[13] to calculate the density of states (DOS) of a polyhedron representative of local grain boundary structure[14] the cluster is first divided into three separate regions: (1) an atomic region, inside the atomic muffin-tins centered on the constituent atoms, (2) an intersphere region, between the atomic spheres and the surrounding outer sphere, and (3) extramolecular region, outside the sphere that surrounds the complex. The potential is constructed by spherically-averaging in regions 1 and 3, and by volume-averaging in region 2. Thus, solutions to the Schrödinger equation in regions 1 and 3 are derived by direct numerical integration of the radial equation, while in the intersphere region the constant potential facilitates an analytic rapidly-convergent multicenter partial-wave expansion of the wavefunction. The eigenstates of the cluster are then found by matching the logarithmic derivative of the intersphere wavefunction to that of the radial solutions at the sphere boundaries.

Although we have not performed a total energy analysis of our systems, Danese and Connolly[15] have demonstrated the feasibility of such calculations within the MS-Xα framework, even for diatomic molecules, by including non-muffin-tin corrections to the charge density. Moreover, Case, Cook, and Karplus[16] have used the MS-Xα wavefunctions for accurate calculations of a variety of one-electron properties, including dipole and quadruple moments, diamagnetic susceptibility, and nuclear quadruple coupling constants. .We have furthermore shown elsewhere[11] that MS-Xα eigenvalue spectrum compares remarkably well with the DOS obtained from the linearized augmented plane-wave (LAPW) calculations of Wimmer, et al.[17] for a clean Ni(100) surface and with various adsorbates. These results indicate that the MS-Xα wavefunctions are a suitable basis for analyzing local electronic interactions near grain boundaries.

Since the atomic structures of grain boundaries are not generally known, we base our studies upon the polyhedral models of grain boundaries.[14] Since the pertinent changes in local electronic structure arise quite generally from the lower symmetry, the decrease in coordination numbers, and the oblique bonding geometries at grain boundaries, we do not expect our findings to result from any particular choice of polyhedral model. Indeed, the general results presented in the following sections seem to be reproduced by any of the Bernal polyhedra in which there is a significant deviation from the cubic crystal structure. Thus, by comparing the DOS of a polyhedral model of grain boundary structure with the DOS of the corresponding single crystal (which is calculated within either the cluster or band structure framework) we can identify the states appearing as a direct result of grain boundary geometry. Of course, a rigorous identification of interface states requires a proper band structure calculation that considers the grain boundaries embedded in the parent crystal. However, at the moment, such calculations are not feasible for technical reasons, not the least of which is the number atoms in the two-dimensional unit cell. In fact, we hope that the work presented here will motivate others to calculate grain-boundary electronic structure at a more fundamental level.

3.   INTERGRANULAR EMBRITTLEMENT IN THE $L1_2$ INTERMETALLIC COMPOUNDS

The nickel aluminides are members of a class of intermetallic compounds that exhibit a number of attractive mechanical properties, including a high specific strength that increases with temperature. Ni_3Al, in particular, is ductile as s single crystal, showing an elongation of 20% before ductile failure. However, polycrystalline Ni_3Al shows elongations of less than 2% and fails intergranularly, which complicates the processing required for parts. Recently, Takasugi and Izumi[18] have noted that when boron is segregated to the grain boundaries of polycrystalline Ni_3Al the material is rendered ductile and the room-temperature elongation to failure is once again found to be near 20%. We first describe the electronic structure of clusters modelling crystalline Ni_3Al before turning our attention to the local electronic structure of grain boundaries.

Crystalline Ni_3Al was modelled by a nineteen-atom cubo-octahedral cluster representing the local environment of Al in the Ni_3Al lattice. The central Al atom was surrounded by twelve nearest-neighbor Ni atoms and six second-neighbor Al atoms. The Gaussian-broadened DOS for this cluster is shown in Fig. 1 and compared with the DOS obtained from a linearized muffin-tin orbital (LMTO) band calculation for Ni_3Al.[19] Note the excellent agreement in DOS profiles and band-widths. The Fermi level in both calculations passes through the top of the nickel d-bands, near a DOS minimum. This comparison lends further support to our contention that the polyhedral structural models for Ni_3Al grain boundaries should reproduce the gross general features of local grain-boundary electronic structure.

In Fig. 2(a) we compare the Gaussian-broadened DOS for a cluster believed to be a typical grain-boundary polyhedron (a trigonal prism of Ni atoms capped by three Al atoms) with that for the nineteen-atom cluster model of crystalline Ni_3Al. The most important changes in the electronic structure resulting from the local geometry of the grain-boundary are (1) an enhancement of the Ni-Al sp bonding, (2) the filling of the Ni 3d levels, and (3) the appearance near the Fermi level that show a strong Ni spd-Al sp mixing. These trends are entirely consistent with those found by Hackenbracht and Kubler[20] when comparing the electronic structures of Ni_3Al and NiAl. We can furthermore conclude from Ref. 20 that while the feature (1) is a general result of changes in local coordination at the grain boundary (and therefore sensitive to local fluctuations in stoichiometry), the features (1) and (3) occur because of the angular deviations from cubic symmetry in the polyhedral models[14] and are therefore intrinsic to the grain boundary geometry.

Similar results are obtained for the grain-boundary electronic structure of Ni_3Si (Fig. 3), which is also intergranularly brittle. There is again a markedly enhanced hybridization with s orbitals on both

Ni and Si throughout the bands, which is accompanied by the appearance of diffuse states near the Fermi energy. However, for Cu_3Pd, which fails in

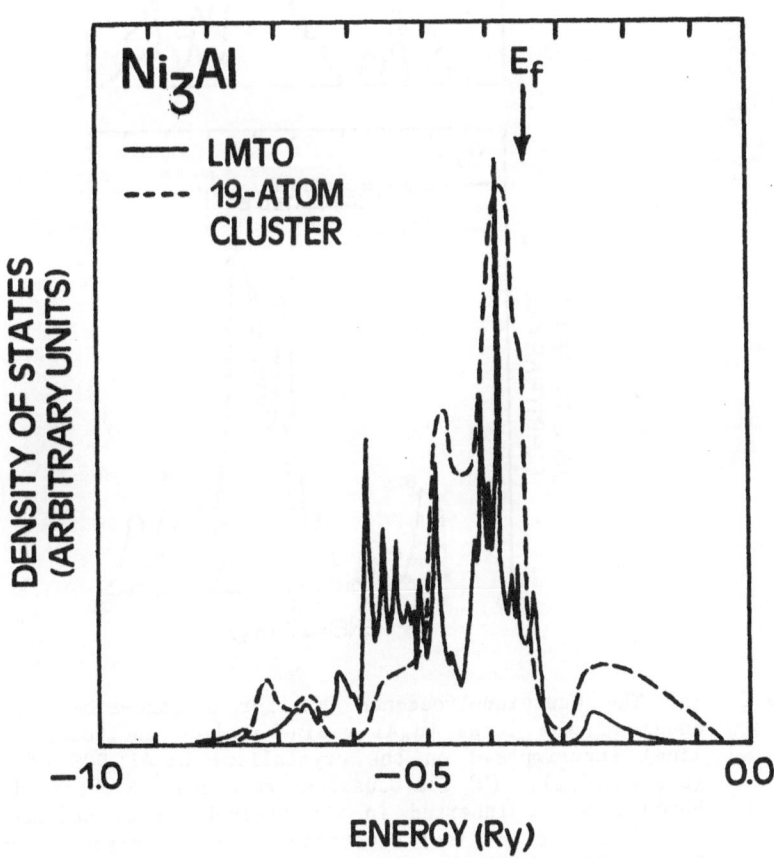

Figure 1. The DOS obtained from the Gaussian-broadening of the orbital energy levels for a nineteen-atom cluster representative of the local environment of crystalline Ni_3Al compared with the DOS obtained from an LMTO band calculation.

168

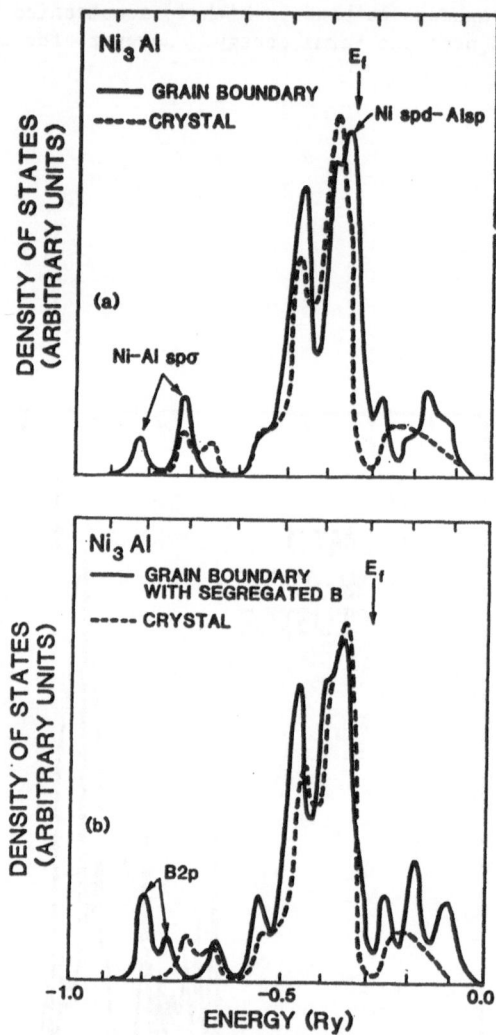

Figure 2. (a) The Gaussian-broadened DOS for a nine-atom polyhedral model of the local Ni_3Al grain-boundary environment (solid line) superimposed on the crystalline Ni_3Al DOS of Fig. 1 (broken line). (b) The Gaussian-broadened DOS obtained when a boron atom is inserted in the grain-boundary polyhedron of Fig. 2(a) (solid line) superimposed on the crystalline Ni_3Al DOS of Fig. 1 (broken line). The combination of the crystalline and grain boundary DOS provides a first-order approximation to the electronic structure of polycrystalline Ni_3Al with and without boron segregated to the grain boundaries.

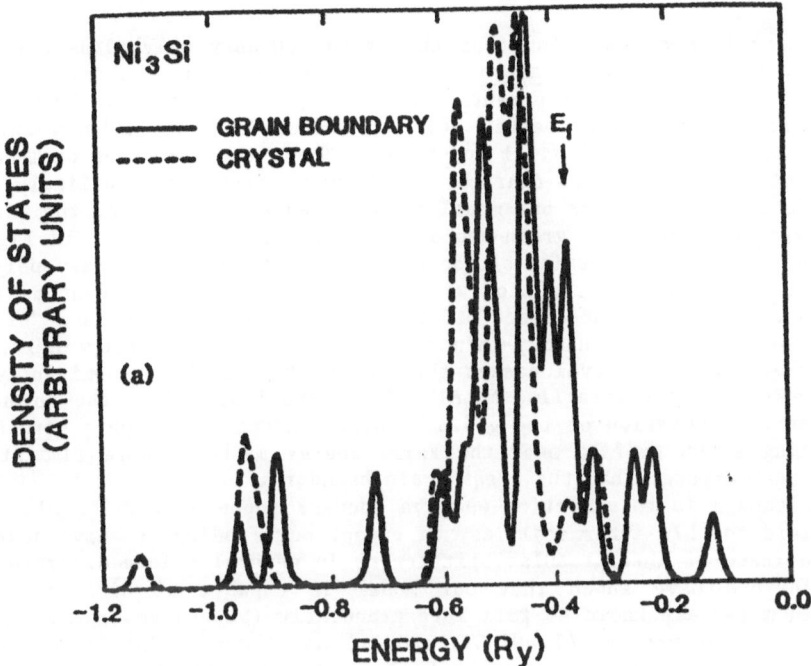

Figure 3. Same as Fig. 2 (a) for Ni$_3$Si.

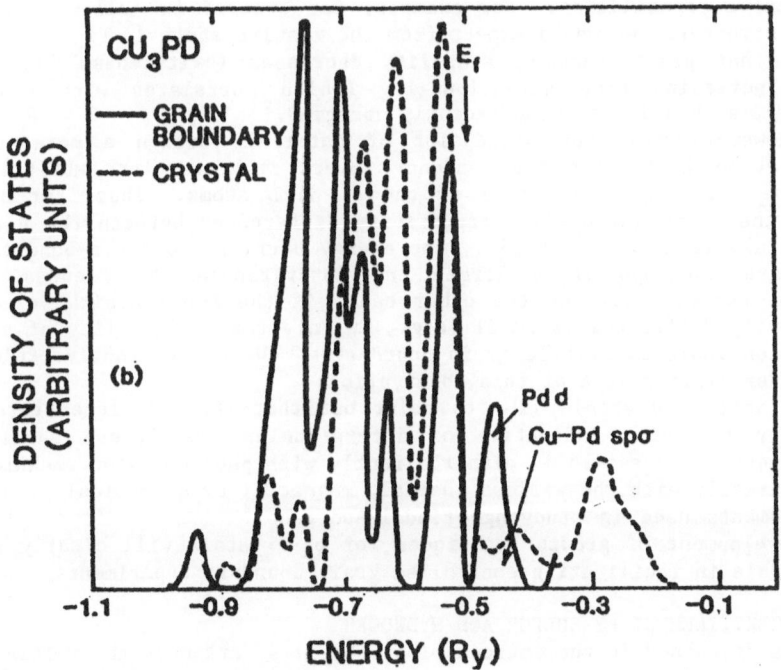

Figure 4. Same as Fig. 2 (a) for Cu$_3$Pd.

in a ductile mode, we find that the grain boundary polyhedron develops directional Pd d-levels <u>above</u> the Fermi energy, in contrast to the diffuse sp-levels of the parent crystal (Fig. 4). The grain boundaries of Cu_3Pd are evidently better able to accommodate the enhancement of s-bonding than those of Ni_3Al and Ni_3Si. Thus, the formation of localized grain-boundary states near the Fermi energy in Cu_3Pd is unlikely.

Further support for our model is provided by considering the effect of segregated boron on grain-boundary electronic structure. In Fig. 2 (b) we compare the Gaussian-broadened DOS for the grain-boundary polyhedron coordinating a boron atom with the DOS of the nineteen-atom Ni_3Al cluster. The prominant direct effect of the boron impurity is a "bridging"[1] of the grain-boundary-induced Ni-Al sp levels. Accordingly, we find that the Fermi level passes through the top of the predominantly Ni 3d levels, as in crystalline Ni_3Al. The overall effect of the boron is thus an accommodation of the grain-boundary-induced electronic structure producing a DOS profile near the Fermi energy markedly more similar to the parent crystal than the clean grain boundary.

Although in this section we have focussed upon a specific class of materials the $L1_2$ intermetallics, we expect our findings to have bearing on the issue of intergranular brittleness in general. Indeed, extensive calculations have shown that our model is capable of discriminating between materials known to fail intergranularly (Ni and Fe containing S, P, or H segregants, Ni_3Al, Ni_3Si, and Ir) and those exhibiting significant plastic deformation prior to failure (Ni, Fe, Al, Cu, Cu_3Pd, Ni_3Al with segregated B at grain boundaries, and Ir with segregated Th at grain boundaries). Moreover, despite the model being crude and strictly applicable only to incipient mechanical response, it does seem to provide predictive capabilities. For example, for ordered A_3B alloys with the Cu_3Au structure, we would expect from the results shown in Figs. 2(a), 3, and 4 that grain boundary stability decreases (brittleness increases) with increasing bond directionality (which correlates with ordering energy), as has indeed been recently observed.[21]

However, from the standpoint of materials design a more useful correlation is between the tendency toward intergranular embrittlement and the s-electronegativities of the A and B atoms. Thus, for alloys where there are large electronegativity differences between the A and B s-orbitals (e.g, Ni_3Al, Ni_3Si), the energy barriers to A-B s-bond formation are correspondingly large, and intergranular brittleness is a natural consequence. On the other hand, if the A-B s-orbital electronegativity difference is small (e.g., Cu_3Pd), the availability of s-bond formation inhibits brittle grain boundaries. We are currently exploring the wider implications of this observation.

Finally, we should like to point out that the interface states we identify as being indicative of intergranular brittleness should be experimentally observable, either directly with photoemission techniques, or indirectly with the various thermal, magnetic, or electrical transport measurements used in studying Friedel-Anderson states in dilute alloys. The development of growth techniques for bi-crystals will clearly prove invaluable in facilitating controlled grain-boundary experiments.

4.   EMBRITTLEMENT BY SULFUR AND HYDROGEN

As discussed in the introduction, the first attempts to explain embrittlement at the atomic level addressed the influence of sulfur on the electronic structure near nickel grain boundaries. We begin this section with a brief review of these first attempts at explaining sulfur-induced embrittlement of nickel.

The first microscopic theory of sulfur embrittlement was due to Losch,[3] based upon calculations by Walsh and Goddard[22] for sulfur chemisorption on Ni(100). Losch conjectured that segregated sulfur forms a layer of covalent sulfur-metal bonds along the grain boundary which causes a weakening of neighboring metal-metal bonds. Embrittlement results because (1) the covalent sulfur-metal bonds are themselves intrinsically brittle and (2) the weakening of the nearly metal-metal bond strength promotes decohesion.

Taking a similar point of view to that of Losch, Haydock[2] proposed that embrittling impurities induce a covalent character in the intergranular bond. The covalency thus inhibits charge mobility and the transfer of bonds, which leads to intergranular embrittlement. Haydock was also the first to point out the analogy between bond mobility at grain boundaries for embrittlement and bond mobility at surfaces for catalysis, a point to which we shall return below.

Messmer and Briant[1] performed the first quantitative study of the influence of sulfur on the electronic properties of nickel grain boundaries, using the MS-Xα method.[13] The authors showed with extensive calculations that when an embrittling element such as sulfur is placed in a polyhedron believed to be typical of grain boundary structure, ionic sulfur-metal bonds are formed and charge is withdrawn from the metalmetal bonds of the polyhedron. Messmer and Briant argued that the depletion of charge from metal-metal bonds represented a reduction in the cohesive energy of the grain boundary. The argument was supported by calculations which showed a correlation between the amount of charge drawn by an impurity and its tendency to embrittle, being greatest for sulfur and decreasing through the series S>P>C>B, with boron being a "cohesive enhancer" by not depleting charge from metal-metal bonds.

While the different approaches just discussed agree upon the ability of sulfur to withdraw charge from neighboring metal-metal bonds, they differ strongly concerning the nature of the sulfur-metal bond. Indeed, as discussed in the introduction, a single impurity can induce a qualitatively different charge distribution locally than that of interacting impurities, a factor which must be considered given the impurity concentrations of embrittled grain boundaries (10-20 atomic per cent).

In Fig. 5 we see explicitly how a sulfur-sulfur interaction becomes an important factor in determining the electronic structure near grain boundaries. The formation of covalent sulfur-sulfur bonds mediated by the nickel results in a material that is more covalent and directional than one in which the sulfur atoms behave as dilute, non-interacting impurities.

In view of the analogy drawn by Haydock between embrittlement and heterogeneous catalysis we examine the influence of interacting sulfur adsorbates on the electronic structure of a Ni(100)[12] surface. The cluster used for the calculation is shown in Fig. 6 and in Fig. 7 we show the formation of a nickel-mediated sulfur-sulfur bond, which has obvious similarities to Fig. 5. This effect is not confined to the top-layer nickel atoms, as we see in Fig. 8 the formation of bonds with the second-

Figure 5. One of the orbitals of a cluster representative of nickel with
a large grain-boundary concentration of sulfur. This orbital
shows the directional bond-formation between the sulfurs,
producing a nearly covalent sulfur-nickel bond.

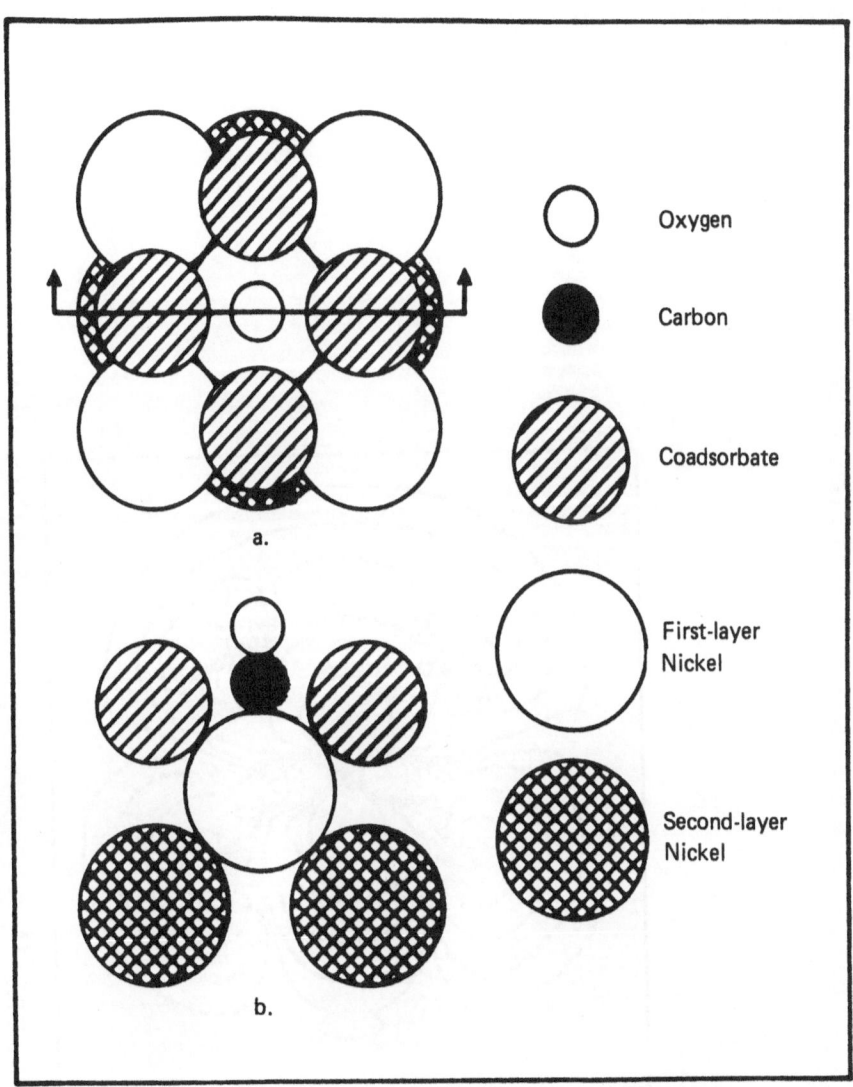

Oxygen

Carbon

Coadsorbate

First-layer
Nickel

Second-layer
Nickel

a.

b.

Figure 6. The cluster used for modelling the electronic structure of
sulfur adsorbed on Ni(100) and the cross-section used for
plotting the wavefunction contours in Figs. 7 and 8. The
adsorbed CO has not been included in these calculations.

174

layer atoms. The overall effect of the sulfur on the surface electronic structure is to decrease the ability of surface charge to respond to an external perturbation, producing a surface with a considerably diminished reactivity as compared with the clean surface.

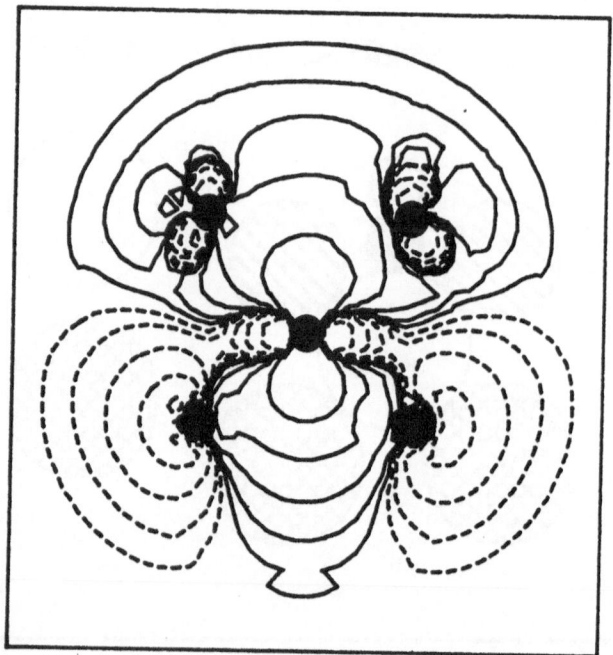

Figure 7. One of the orbitals of the cluster in Fig. 6 showing a nickel-mediated sulfur-sulfur covalent bond with a top layer nickel atom. Atomic positions are indicated by small darkened circles.

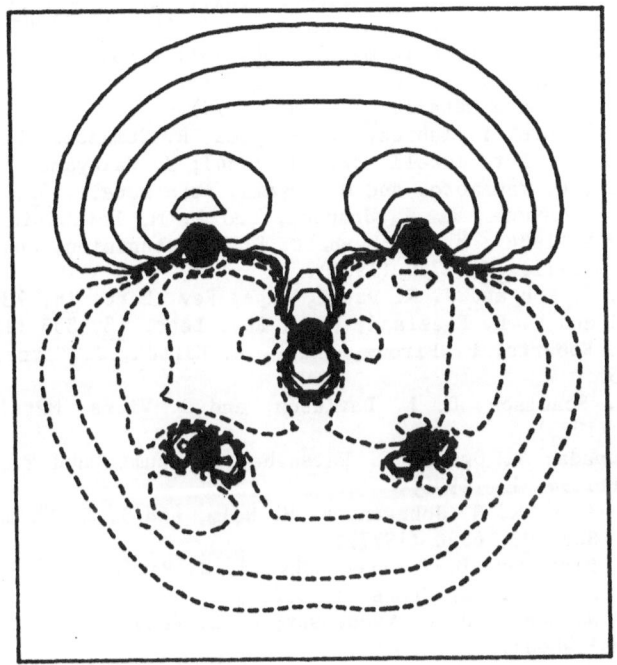

Figure 8. One of the orbitals of the cluster in Fig. 6 showing bond-
formation between the sulfur and the second-layer nickel
atoms.

Our conclusions for sulfur embrittlement are not dissimilar from
those of sulfur poisoning of surface catalytic reactions. Interacting
sulfur atoms form covalent bonds along the grain which are mediated by
nickel atoms. There is furthermore some secondary bond-formation (e.g,
Fig. 8) which causes a depletion of charge from neighboring metal-metal
bonds. Thus, charge along the grain boundary is localized in covalent
bonds, while bonds within the grains are weakened, with both effects
implying an embrittled grain boundary. Note that our conclusions are
not dissimilar from those of Losch, Haydock, and Messmer and Briant, but
that the mechanism requires the impurities to be interacting.

176

REFERENCES

1. C. L. Briant and R. P. Messmer, Phil. Mag. B 42, 569 (1980); R. P. Messmer and C. L. Briant, Acta metall. 30, 457 (1982); R. P. Messmer and C. L. Briant, Acta metall. 30, 1811 (1982).
2. R. Haydock, J. Phys. C 14, 3807 (1981).
3. W. Losch, Acta metall. 27, 1885 (1979).
4. M. Hashimoto, Y. Ishida, S. Wakayama, R. Yamamoto, M. Dcyama, and T. Fujiwara. Acta metall. 32, 13 (1984); S. Wakayama, M. Hashimoto, Y. Ishida, R. Yamamoto, and M. Doyama, Acta metall. 32, 21 (1984).
5. M. E. Eberhart, K. H. Johnson, and R. M. Latanision, Acta metall. 32, 955 (1984); M. E. Eberhart, R. M. Latanision, and K. H. Johnson, Acta metall. 33, 1769 (1985).
6. H. L. Frisch and J. R. Patel, Phys. Rev. Lett. 18, 784 (1967); J. R. Patel and P. E. Freeland, Phys. Rev. Lett. 18, 833 (1967).
7. S. G. Roberts, P. Pirouz, and P. B. Hirsch, J. Mater. Sci. 20, 1739 (1985).
8. E. M. Schulson, D. L. Davidson, and D. Viens, Metall. Trans. 14A, 1523 (1983).
9. S. Hanada, T. Ogura, S. Watanabe, O. Izumi, and T. Masumoto, Acta metall. 34, 13 (1986).
10. C. Y. Yang, K. H. Johnson, R. H. Holm, and J. G. Norman, Jr., J. Am. Chem. Soc. 97, 6596 (1977).
11. J. M. MacLaren, D. D. Vvedensky, J. B. Pendry, and R. W. Joyner, J. Chem. Soc. Faraday Trans (in press).
12. J. M. MacLaren, D. D. Vvedensky, J. B. Pendry, and R. W. Joyner, (to be published).
13. K. H. Johnson and F. C. Smith, Jr., Phys. Rev. B 5, 831 (1972); J. C. Slater and K. H. Johnson, Phys. Rev. B 5, 844 (1972).
14. M. F. Ashby, F. Spaepen, and S. Williams, Acta metall. 26, 1647 (1978).
15. J. B. Danese and J. W. D. Connally, J. Chem. Phys. 61, 3061 (1974).
16. D. A. Case, M. Cook, and M. Karplus, J. Chem. Phys. 73, 3294 (1980).
17. E. Wimmer, C. L. Fu, and A. J. Freeman, Phys. Rev. Lett. 55, 2618 (1985).
18. T. Takasugi and O. Izumi, Acta metall. 33, 1259 (1985).
19. R. C. Albers (unpublished). See also J. J. M. Buiting, J. Kubler, and F. M. Mueller, J. Phys. F 13, L179 (1983).
20. D. Hackenbracht and J. Kubler, J. Phys. F 10, 427 (1980).
21. T. Takasugi and O. Izumi, Acta metall. 33, 1247 (1985).
22. S. P. Walch and W. A. Goddard, III, Surf. Sci. 72, 645 (1978).

INTERATOMIC FORCES AND THE SIMULATION OF CRACKS

M. W. FINNIS

Theoretical Physics Division
Bldg. 424.4
A.E.R.E. Harwell
Oxfordshire OX11 ORA
United Kingdom

1. INTRODUCTION – WHAT CAN SOLID STATE PHYSICS CONTRIBUTE?

1.1 How many atoms?

What can solid state physics contribute to the understanding of fracture? If we leave aside the field of continuum mechanics, since the Workshop is concerned mainly with the atomistics of fracture, the current theoretical approaches have two extremes.

On the one hand we have the approach of theoretical chemistry, which as exemplified by the work of Eberhart, Johnson, Messmer and Briant[1-2] described at the previous Workshop can show us how the electronic energy levels and charge density of a cluster of atoms, with or without a foreign atom, depend on their configuration. It is now feasible to carry out such calculations on clusters of order 10–20 atoms to high accuracy. The results have been very suggestive of how embrittlement at grain boundaries might work. For example, an impurity such as sulphur or oxygen might draw electrons out of a neighbouring metal-metal bond, thereby weakening it.

On the other hand the techniques of atomistic modelling[3] have now been developed to a high level of sophistication to determine the atomic positions at dislocations[4-7], grain boundaries[8-9] and crack tips[3], taking into account the long range of the strain fields. Perhaps the greatest success of relevance to fracture has been in the description of dislocation core structure and its relationship to yield.

Let us consider briefly the characteristics of these opposite approaches. Chemical calculations on small clusters no longer need to make serious approximations in solving Schrödinger's equation. Refinements to the accuracy of energy levels, particularly for excited states, are still a topic of interest to the purists, but are not the main problem for applications. The limitation of such calculations is rather one of computing power, which restricts the number of atoms. Their geometry must be prescribed in advance and one has no way of including the effect of the matrix in which the cluster should be embedded either on the molecular wave functions or on the cluster geometry. While there is some prospect of developing practical embedding techniques for self-consistent cluster calculations appropriate to, say, a grain boundary, by analogy with a surface[10], the subsequent step of calculating the forces on the atoms and relaxing them to zero is further in the future. A more recent idea, which is to combine density functional theory and molecular dynamics in a unified Lagrangian equation of motion[11] may prove to be the way forward towards large scale, electronically self-consistent and atomistically relaxed simulations, but it is too early to tell how far in the future they lie.

The large scale atomistic simulations still have to use very simple empirically based force laws, and their conclusions are strongest when they are suggesting geometrical features, such as the possible structures of defects, or simulating mechanisms, such as point defect migration or dislocation motion under stress. As recently as five years ago the force laws used were, for metals, nearly all of the pair potential variety. According to the type of study, the pair potentials were either simple and illustrative (Lennard–Jones), fitted to bulk properties, or derived from perturbation theory on pseudopotentials in an electron gas. Ideal materials such as Lennard–Jonesium and Johnsonium have been popular and fruitful for general atomistic studies, while particular $s-p$ bonded metals have been modelled by pseudopotential-based pair potentials, which are oscillatory and of longer range. Empirical potentials have also been constructed for AB alloys[12,13] using the experimental heat of mixing to obtain an A-B pair potential, with reasonable results for grain boundary segregation. However, all the pair-potential models caused trouble at surfaces, e.g. in crack tip modelling, if it was required to reproduce experimental elastic constants for metals where $C_{12} \neq C_{44}$. For such cases a fictitious pressure, the Cauchy pressure, would have to be applied to the surfaces in order to obtain the observed volume per atom. While this could be done easily at external surfaces, simply by applying constant volume boundary conditions to the model crystallite, no satisfactory procedure was available for internal surfaces such as cracks, particularly if these surfaces were to be created or healed during the simulation. As a result, the simulation of cracks or cavities had to be carried out in the spirit of a qualitative study, with very idealised force laws, in which there was always a big question mark as to the role of non-central forces.

1.2 N–body potentials

This discomfort of atomistic modellers about internal surfaces has been relieved by the new generation of non-pairwise potentials based on the embedded atom concept[14-18], which gives a realistic representation of cohesion, elasticity and surface properties. The energy of an atom is written as the sum of a pairwise term, acting between atoms i and j, and a term which depends on the local density of atoms ρ_i :

$$U_i = f(\rho_i) + \tfrac{1}{2}\sum_j V_{ij}. \tag{1.1}$$

The local density ρ_i is itself expressed as a sum of atomic functions :

$$\rho_i = \sum_j \rho_{ij}. \tag{1.2}$$

V_{ij} and ρ_{ij} are function only of the interatomic distance $i-j$, and can be short ranged. Since they involve no significant extra computation compared to simple pairwise potentials, the embedded atom potentials can be applied to simulations of up to tens of thousands of atoms with the present generation of mainframe computers.

Although they are a big step forward from pair potentials, it is not yet clear whether embedded atom potentials can describe the changes in chemical bonding energy which are involved in embrittlement. The embedded atom potential is non-directional. That is to say, the energy of an atom depends on the number of neighbours in each coordination sphere in a non-linear way (unlike a pair potential energy), but does not depend on the relative bond-angles of those neighbours, which is where we might expect the orientation of atomic orbitals to make a difference. Compare for example the energy of an atom in the face-centred cubic (fcc) structure and one in the hexagonal close-packed (hcp) structure of ideal axial ratio and the same atomic volume. The first two shells of neighbours in each case contain 12 and 6 atoms respectively. They are at the same radial distances in each structure, but differently disposed. In particular, the second neighbours in hcp lie on the corners of a triangular prism, whereas in fcc they are on the corners of an octahedron. Thus in these simple pair potential or embedded atom models, the energy difference between the structures has to be provided by differences in the interaction between third or more distant neighbours. One suspects however that in reality the difference in angular disposition of the first and second nearest neighbours should make a significant difference to the bonding energy in a system of tightly bound d-electrons. This is taken into account by models at a stage further in sophistication, namely tight-binding models.

1.3 Tight-binding

The tight-binding theory of metallic cohesion has been extensively applied to describe the energies of transition metals and their alloys. The review articles by Heine et al[19] and Pettifor[20] describe the theory and a wide range of its applications. Earlier work calculated the energies of atoms in an arbitrary configuration, but this configuration had to be prescribed in advance. Progress in understanding embrittlement was made by Hashimoto et al[21], who calculated the local densities of electronic states at a grain boundary in iron with and without segregated phosphorus and boron. They first prescribed the atomic configurations by relaxing the grain boundary under the action of empirical Morse potentials. They concluded that phosphorus atoms segregated at the grain boundary produce strong bonding orbitals within clusters of Fe_9P below the original iron $3d$ band. Fe–Fe bonds outside these clusters are weakened by the corresponding enhancement of the anti-bonding density of states on neighbouring Fe atoms, thereby promoting intergranular embrittlement. Induced hydrostatic strains near the phosphorus also contribute to weakening the bonding. Boron on the other hand while also bonding strongly to the iron forms its bonding states within the Fe $3d$ band and does not thereby weaken the Fe–Fe bonds much. The authors also suggested that the interstitial position of B does not introduce the hydrostatic tension and consequent weakening of bonds compared to substitutional phosphorus.

Atomistic relaxation within the tight binding model became possible as computing power increased. At first, extended defects were treated for group IV semiconductors, which require a relatively small interaction matrix of four sp^3 hybrid orbitals per atom[22-23]. More recently, there have been a number of calculations of the energetics and configurations of extended defects in d–band metals in which atomic positions are relaxed by a variety of schemes.

Several applications of the tight-binding model have been made to the atomistic simulation of dislocations in metals. The approaches so far have been based on the band model (see below) to varying levels of approximation. Masuda and coworkers[24-25] worked with a model electronic density of states in which up to the fourth moment was calculated. This corresponds to including the effect of two shells of neighbours on the local density of states. The fourth moment approximation can predict the bcc-fcc structural energy difference, however the fcc-hcp energy difference is still problematic. In tight-binding a second-neighbour (fourth moment) model gives different energies for fcc and hcp crystal structures, but it is not adequate to predict the correct trend in crystal structure with band filling, for which higher moments are decisive[26-27].

To my knowledge the most sophisticated published application of tight-binding to atomistic relaxation is that of Legrand[28]. He used effectively six moments of the density of states, represented by a continued fraction obtained with the recursion method[19], and this was sufficiently accurate to obtain convergence of the stacking fault energy in hcp to within 10%. Relaxation of the $\frac{1}{3}\langle11\bar{2}0\rangle$ dislocation in titanium was performed. The forces used were approximations to the derivatives of the calculated tight-binding band energy, which would have been prohibitively expensive to calculate exactly[28-29]. The change in density of states on the atom moved was evaluated to obtain the force on that atom, call it atom i, which gave an energy change δE^i. Of course, when atom i is moved there is also a change in the density of states and therefore the energy of all the other atoms, which Legrand took into account in an approximate way as follows. The second moment contribution to the energy changes on all the atoms was calculated explicitly, but the contribution of moments 3–6 was assumed to be the same as on atom i. This assumption is motivated by the pair potential result; within a purely pair potential model, the energy change of the other atoms, $\sum_k \delta E^k$, would trivially equal δE^i.

To the forces due to the band energy were added the usual pairwise forces from the repulsive energy. The relaxed dislocation structure obtained was dissociated on the prism plane, explaining prismatic glide, which according to the author could not be achieved using pair potentials alone.

Surface relaxations and reconstructions have also been calculated within the tight binding model, assuming only a few layers are free to relax, with satisfactory results when compared to experimental LEED data. With few degrees of freedom, it is possible to calculate local densities of states to a higher number of exact moments (more levels in the continued fraction representation). Treglia et al[30-31] have pointed out the sensitivity of results on surface reconstruction to the tight-binding parameters used, and stressed the importance of self-consistency and charge neutrality. These points are dealt with approximately if the site energies ε_p^i are adjusted to maintain neutral atoms, which is an essential feature of the bond approach to be described. In metallic systems, and perhaps also some semiconductors, this approximation of perfect screening is good. Terakura et al[32] and Masuda–Jindo et al[33] have calculated surface reconstructions on W and Mo using both neutral atom and non-self-consistent formulations. They claim that the results are very similar; however this might be expected when the d–band is half full, since local band broadening would not lead to charge transfer. For general band fillings we believe that the neutral atom formulation will be different and more reliable than the completely non-self-consistent formulation.

This brings me back to the question with which I started. The contribution of solid state physics should be to try to bridge the gap between the chemical understanding of bonding and the capability which atomistic modelling already possesses to simulate the elementary processes of fracture such as crack tip opening or dislocation emission. We continually need to seek the simplest possible models, refining them as experience dictates and as computational power increases. Embedded atom models are an example of the fruit of such efforts and will be described further elsewhere. The remainder of this paper describes another recent approach, the tight-binding bond method, which is a variant of the tight-binding models referred to above. It is an extremely simplified description of the bond energies of d–band transition metals, but currently the best description that is at the same time computationally tractable for full scale atomistic modelling. We regard it as lying between the rapidly computable embedded atom models and the detailed quantum chemical models which are so far only tractable for relatively small clusters. The approach is particularly suited to calculating interatomic forces, and is similar to that of Moraitis et al[34] who calculated impurity-induced forces in a perfect lattice. Calculations of relaxed grain boundary structure using the method are in progress (Ohta, Finnis, Pettifor and Sutton, to be published). In this paper I describe the formalism and illustrate it with calculations of bond energies in close packed crystals.

2. TIGHT-BINDING BONDS

The definition and calculation of *bond* energies[35–37] represents a step forward in the solid-state physics approach compared to the previous tight-binding models which worked with local densities of states and *band* energies. The methods are closely related, as I will explain in this section. The bond model has recently proved its value by providing a qualitative account of the observed domains of stability of five of the transition metal-metalloid (p–d bonded) AB compounds, namely the NaCl, CsCl, NiAs, MnP and boride structures, leaving only the few FeSi structures and the ionic oxides unaccounted for[36].

The total bond energy of a material has the form :

$$U_{bond} = \sum_{i \neq j; pq} H_{pq}^{ij} \rho_{qp}^{ji}. \tag{2.1}$$

The i and j indices label atomic sites, as previously, and the p and q label localised atomic orbitals. We shall restrict the discussion here to d–band metals, for which p and q run from 1 to 5, and hybridisation with s and p electrons is ignored. These orbitals have the following symmetry :

Orbital	Symmetry
1	xy
2	yz
3	zx
4	$\frac{1}{2}(x^2 - y^2)$
5	$\frac{1}{2\sqrt{3}}(3z^2 - r^2)$

With the z–axis along the bond the matrix elements of the Hamiltonian H take the simple form[38] :

$$H_{11} = H_{44} = dd\delta,$$

$$H_{22} = H_{33} = dd\pi,$$
$$H_{55} = dd\sigma,$$
$$H_{pq} = 0 \;\; ; p \neq q, \tag{2.2}$$

and the bond energy, expression (2.1), becomes :

$$U_{bond} = \sum_{i \neq j} (dd\sigma \, \rho_{55}^{ij} + 2 \, dd\pi \, (\rho_{22}^{ij} + \rho_{33}^{ij}) + 2 \, dd\delta \, (\rho_{11}^{ij} + \rho_{44}^{ij})). \tag{2.3}$$

The matrix elements of H for a general orientation of the bond can be obtained in terms of $(dd\sigma, dd\pi, dd\delta)$ by the transformations given in Slater and Koster's paper[38].

Let us now discuss the meaning of the factors ρ_{pq}^{ij}. They are elements of the density matrix ρ. This formalism is much more familiar to chemists than to solid state physicists. A similar expression for the bond energy was used for example by Coulson[39], who wrote the energy as (in the original notation) :

$$2 \sum_{ij} p_{ij} \beta_{ij}. \tag{2.4}$$

The Hamiltonian matrix elements, or resonance integrals as they are sometimes called, were denoted β_{ij} by Coulson. He was concerned in particular with the contribution of the π-orbitals in hydrocarbons. Coulson called p_{ij} ($\frac{1}{2}\rho_{ij}^{pq}$ in our notation) the 'total order' of the bond, which was the sum of the 'partial mobile orders' of the occupied molecular π-orbitals, together with the contribution of the bonding σ-orbital, which was assumed to have the value 1 in all cases. By Coulson's definition :

$$p_{ij} = \sum_{n \; occ} p_{ij}^{(n)}. \tag{2.5}$$

The molecular orbitals were defined in terms of atomic orbitals :

$$\Psi^{(n)} = a_1^{(n)} \psi_1 + \dots a_k^{(n)} \psi_k. \tag{2.6}$$

If i and j are neighbours, the partial orders were defined by :

$$p_{ij}^{(n)} = a_i^{(n)} a_j^{(n)}, \tag{2.7}$$

for real coefficients a, or :

$$p_{ij}^{(n)} = \frac{1}{2}(a_i^{(n)} a_j^{(n)*} + a_i^{(n)*} a_j^{(n)}), \tag{2.8}$$

for complex coefficients. The idea was already familiar that the bond order in ethane was 1, in ethylene 2 and acetylene 3, and so too was the idea of resonance, that not all bonds are of integral order (e.g. Penney[40]). Coulson's contribution was to provide the first satisfactory quantitative definition of non-integral bond-orders in molecules.

Equations (2.5)-(2.8) can be generalised to extended systems by using the solid state concept of one-electron Green functions. The sums over occupied orbitals become integrals up to the Fermi energy ε_F. To make the connection to the solid state approach, we recall that the one-electron Green function is defined by :

$$G = (\varepsilon + i0 - H)^{-1}. \tag{2.9}$$

The matrix element G_{pq}^{ij} between orbital p on site i and orbital q on site j will sometimes be referred to for brevity as a Green function. For future reference we note that the local density of electron states on site i, projected onto orbital p is given by the diagonal Green function matrix element [19] :

$$n_p^i(\varepsilon) = -\frac{2}{\pi} \mathcal{I}mG_{pp}^{ii}(\varepsilon + i0).$$

(2.10)

where the factor 2 is for spin degeneracy, and $+i0$ indicates that the limit is to be taken as the real axis is approached from above.

The relations (2.9) and (2.10) assume that the orbitals form an orthonormal basis. We make this assumption for simplicity, to avoid introducing the matrix of overlaps S_{pq}^{ij}, but all the relations can be generalised for the case of non-orthogonal orbitals.

We are now in a position to define the density matrix ρ from the solid state viewpoint :

$$\rho = -\frac{2}{\pi} \int^{\varepsilon_F} \mathcal{I}mG(\varepsilon)d\varepsilon,$$

(2.11)

The usual expression for the band energy, or sum of occupied one-electron energies, is :

$$U_{band} = \sum_{ip} \int^{\varepsilon_F} \varepsilon n_p^i(\varepsilon)d\varepsilon.$$

(2.12)

This can be expressed in terms of the density matrix as follows. Firstly, we note from (2.9) that :

$$\varepsilon \mathcal{I}mG = H\mathcal{I}mG.$$

(2.13)

If we now replace the density of states in (2.12) by its Green function expression (2.10), and then insert (2.13) followed by (2.11), the band energy can be written in the compact form :

$$U_{band} = \mathcal{T}rH\rho.$$

(2.14)

To separate out the *bond* energy from (2.14) we simply have to subtract from the trace the terms corresponding to the diagonal elements of H. Let us call these elements ε_p^i for brevity, that is we introduce the notation $\varepsilon_p^i \equiv H_{pp}^{ii}$. They can be identified with atomic energy levels. Expression (2.14) becomes :

$$U_{band} = \sum_{ip} \varepsilon_p^i \rho_{pp}^{ii} + U_{bond}.$$

(2.15)

We have assumed, as in all practical cases, that there are no matrix elements of H between different orbitals on the same site. The factor ρ_{pp}^{ii} in the first term of (2.15) is, from (2.10) and (2.11), the total charge of both spins in orbital p on atom i. This term becomes the one-electron energy of isolated atoms as the bond lengths are increased, while the second term is the sum of bond energies we introduced at the start of this section. This completes the formal connection between the usual solid state description of band energy and the bond energy expression (2.1). To complete our model of the total energy, we add to U_{bond} a repulsive pairwise interaction to represent the ion-ion Coulomb repulsion and the remaining electrostatic energy of the electrons :

$$U_{tot} = U_{bond} + U_{rep}.$$

(2.16).

Thus the quantum mechanical binding is modelled by U_{bond}, and the remainder of the energy by an empirical pairwise potential, which is conveniently taken to be a Born–Mayer potential.

A formally identical expression for the tight-binding energy was used by e.g. Allan[41] although he did not refer to it as a bonding energy :

$$U_{bond} = \sum_{ip} \int^{\varepsilon_F} (\varepsilon - \varepsilon_p^i) n_p^i(\varepsilon) d\varepsilon. \tag{2.17}$$

For computational purposes, there are advantages and disadvantages in the form (2.17) for the bond energy compared to the explicit form (2.1). Calculation of the single site quantity n_p^i is a familiar operation, and details may be found elsewhere [19,42]. It involves the setting up of a cluster centred on i and containing all atoms within $LL-1$ hops of i. The matrix elements of H between all sites and orbitals must be specified. G_{pp}^{ii} is then obtained by the recursion method as a continued fraction

$$G_{pp}^{ii} = B2(1)/\varepsilon - A(1) - B2(2)/\varepsilon..../\varepsilon - A(LL) - T(\varepsilon). \tag{2.18}$$

There are various prescriptions for terminating the fraction with a function $T(\varepsilon)$ [19,43]. By comparison, to evaluate the bond energy sum requires the calculation of all the off-diagonal elements of G within the range of H. These are obtained by a similar method [19,35], but require in principle a larger cluster to obtain the same number of levels in the continued fractions, and furthermore two continued fractions must be calculated and subtracted for each element of G. Although it has these advantages, the form (2.17) does not give any physical insight into the energies or strengths of individual bonds. A related disadvantage of (2.17) is that it does not lead to a closely related expression for the interatomic forces, which are readily obtained from the energies of the corresponding bonds as follows.

3. THE BOND ENERGY AND INTERATOMIC FORCES

In this section we show how simple expressions for interatomic forces can be derived from the bond energy. The result represents a generalisation to extended systems of Lennard–Jones[44] and Coulsons'[39] result for tight-binding molecules.

We start by considering the effect on the bond energy of a perturbation in the Hamiltonion. To first order, from eqn.(2.1) :

$$\Delta U_{bond} = \sum_{i \neq j; pq} \Delta H_{pq}^{ij} \rho_{qp}^{ji} + \sum_{i \neq j; pq} H_{pq}^{ij} \Delta \rho_{qp}^{ji}. \tag{3.1}$$

We now show in outline how, with the assumptions of orthogonal orbitals and neutral atoms, the second term in (3.1) vanishes, leaving the simple result :

$$\Delta U_{bond} = \sum_{i \neq j; pq} \Delta H_{pq}^{ij} \rho_{qp}^{ji}, \tag{3.2}$$

from which interatomic forces are derived.

We proceed by writing :

$$\sum_{i \neq j; pq} H_{pq}^{ij} \Delta \rho_{qp}^{ji} = \mathcal{T}r \, H \Delta \rho - \sum_{ip} \varepsilon_p^i \Delta \rho_{pp}^{ii} . \tag{3.3}$$

Consider first the final term of (3.3). We recall that the quantity ρ_{pp}^{ii} is just the electronic charge in orbital p on atom i. In some applications, the site energy ε_p^i is assumed not to depend on the orbital p, e.g. for the five $d-$orbitals of the canonical theory, in which case the final term of (3.3) vanishes under the condition of atomic charge neutrality. More generally the final term of (3.3) vanishes under the assumption that there is no charge transfer between sets of orbitals having different self-energies. This is the assumption of our present model. It is well justified for $d-$band metals, in which there is nearly perfect screening, and was also applied by Pettifor and Podloucky[36] to the $pd-$bonded AB compounds. It will need to be reviewed when applications are considered with mixed s and p orbitals, where there may be charge transfer between them on a site, but for the present it is a very useful simplification.

Now consider the first term of (3.3). It has the form of the first variation of the energy with respect to charge density which we expect to vanish. We can demonstrate this by writing the trace in the basis of eigenfunctions of H :

$$\mathcal{T}r\, H\Delta\rho = \sum_n \varepsilon_n \Delta\rho_{nn} \ . \tag{3.4}$$

The quantity ρ_{nn} is the occupancy of the nth eigenfunction, as can be seen immediately from the relation[19] :

$$-\frac{1}{\pi}\mathcal{I}m G_{nn} = \delta(\varepsilon - \varepsilon_n) \ . \tag{3.5}$$

Because the total number of electrons is conserved the nth occupied eigenfunction remains occupied under a perturbation. At the same time the unoccupied eigenfunctions remain unoccupied and by definition they remain above the Fermi energy ε_F (which may in general changed by the perturbation). Hence for all n :

$$\Delta\rho_{nn} = 0 \tag{3.6}$$

and the right hand side of (3.4) vanishes.

This completes our derivation of (3.2). If the ΔH is due to the small displacement of atom i, (3.2) can be translated directly into a sum of pairwise interatomic forces, each given by :

$$\mathbf{F}_{ij} = \sum_{pq} \rho_{pq}^{ij} \nabla H_{qp}^{ji}, \tag{3.7}$$

The formula (3.7) is our central result for the bond model forces, and implicitly contains the perfect screening or neutral atom assumption. The gradient of H_{qp}^{ji} is its derivative with respect to the position of atom j and involves the dependence of H_{pq}^{ij} on the length R and orientation of the bond $i-j$. For the bond length dependence in d-band metals we shall assume the canonical R^{-5} variation as described in the following section. The angular dependence of the elements of H is given for each non-equivalent pair of orbitals qp by Slater and Koster[38]. A FORTRAN subroutine has been written to evaluate the first (and second) derivative matrices of H for the d-orbital case for an arbitrary orientation of the bond.

Although the bond energies and bond forces are formally pairwise, it should be clear that this provides no justification for empirical pair potential models, since the elements of the density matrix depend on the local atomic environment.

It is instructive to divide the force on atom i from atom j into radial and transverse components. The radial component of $\mathbf{F}_{ij}$ follows directly from (2.3) :

$$\mathbf{F}_{ij}^{radial} = -\frac{5}{R}\{dd\sigma\,\rho_{55}^{ij} + 2\,dd\pi\,(\rho_{22}^{ij} + \rho_{33}^{ij}) + 2\,dd\delta\,(\rho_{11}^{ij} + \rho_{44}^{ij})\}. \tag{3.8}$$

This force is directed to shortening the bond and has a positive sign. The influence of the local atomic environment enters through the factors ρ_{pp}. The interatomic forces also contain in general a 'bond-bending' component, due to the non-vanishing components of ∇H transverse to the bond direction. This will vanish in situations of special symmetry, such as in a regular lattice, but not in general.

4. THE ENERGY CHANGE TO SECOND ORDER

In order to obtain expressions for force constants within the tight-binding model, it is necessary to calculate the energy perturbed to second order in the change ΔH associated with moving a pair of atoms. This was carried out by Finnis et al[45], in order to investigate the relation between the force constants and phonon frequencies in group VB and VIB transition metals. However, the derivation of Finnis et al was based on the band model rather than the bond model, and did not consider any correction for self-consistency, which in the bond model means charge neutrality. In this section we show that the bulk-modulus obtained from the long wavelength limit of the phonon spectra calculated from band model force constants does not agree with its value calculated directly by homogeneous compression, or the bond model.

The important point is that when calculating quantities under homogeneous compression, no ingenuity is required to maintain charge neutrality of the atoms, since for simple lattices all sites remain equivalent. So, in the case of homogeneous compression, self-consistency in the sense of perfect screening can be obtained trivially simply by keeping the total band filling constant. This is not so for an inhomogeneous distortion such as a phonon in which there is automatically some charge transfer between atoms unless some adjustment of the matrix elements of H is made. In principle this charge transfer should be subject to some spacially localised charge conservation constraint, the simplest form of which is our neutral atom or perfect screening condition, achieved by adjustment of the diagonal site energies ε_p^i. The inconsistency of the usual band model in this respect can be understood as follows.

We consider first a general perturbation ΔH, which induces changes Δn and $\Delta\varepsilon_F$ in the density of states and the Fermi energy. This perturbation may either be localised, as when a single atom or pair of atoms is displaced, or it may be extended, as when the whole crystal is strained. There must rigourously be a correspondence between certain localised and extended perturbations, in the sense that the energy of phonons in the long wavelength limit is correctly described both by a force constant summation and by the elastic constants for homogeneous strain[46], but electronic models can easily be inconsistent in this respect unless handled with care, as the following derivation shows.

The change in band energy is given by :

$$
\begin{aligned}
\Delta U_{band} &= \Delta \int_{}^{\varepsilon_F} \varepsilon n d\varepsilon \\
&= \int_{}^{\varepsilon_F} \varepsilon \Delta n d\varepsilon + \int_{\varepsilon_F}^{\varepsilon_F + \Delta\varepsilon_F} \varepsilon(n + \Delta n) d\varepsilon \cdot
\end{aligned}
\tag{4.1}
$$

We now apply the condition of total charge conservation :

$$
\int_{}^{\varepsilon_F} \Delta n d\varepsilon + \int_{\varepsilon_F}^{\varepsilon_F + \Delta\varepsilon_F} (n + \Delta n) d\varepsilon = 0
\tag{4.2}
$$

to write (4.1) in the form :

$$
\Delta U_{band} = \int_{}^{\varepsilon_F} (\varepsilon - \varepsilon_F) \Delta n d\varepsilon + \int_{\varepsilon_F}^{\varepsilon_F + \Delta\varepsilon_F} (\varepsilon - \varepsilon_F)(n + \Delta n) d\varepsilon.
\tag{4.3}
$$

Equation (4.3) is exact; ΔH has not yet been assumed to be a small parameter. If we expand the second term and retain only second order terms we obtain :

$$
\Delta U_{band} = \int_{}^{\varepsilon_F} (\varepsilon - \varepsilon_F) \Delta n d\varepsilon + \tfrac{1}{2} n(\varepsilon_F)(\Delta\varepsilon_F)^2.
\tag{4.4}
$$

As far as we know, previous derivations of ΔU_{band} have obtained the first term of (4.4) but not the second. It is worth noting that the first term in (4.4) is not merely a first order term but is exact to all orders in ΔH; in particular it includes terms of second order in the atomic displacements which correspond to force constants.

Interatomic force constants would be calculated within the band model on the assumption that the movement of a pair of atoms does not involve a Fermi energy shift, or any shift in the diagonal elements ε_p^i (Finnis et al[45]). The energy changes described by force constants calculated in this way therefore correspond to only the first term of (4.4) and omit the contribution of the second term. We can regard the second term as arising from the condition of charge conservation, which is violated by the band model force constant calculation.

Let us now consider the simplest distortion, which is a homogeneous compression or expansion. It is easy to evaluate (4.4) within the canonical d–band model[47,48], in which the overlap parameters vary as the inverse fifth power of the interatomic distance R :

$$dd\sigma = -6W\left(\frac{2}{5}\right)\left(\frac{S}{R}\right)^5$$

$$dd\pi = +4W\left(\frac{2}{5}\right)\left(\frac{S}{R}\right)^5 \tag{4.5}$$

$$dd\delta = -1W\left(\frac{2}{5}\right)\left(\frac{S}{R}\right)^5 .$$

S is the Wigner–Seitz radius and W is an approximate bandwidth. When a crystal is dilated homogeneously, the band energy within this model simply scales as $(S/R)^5$, as does the Fermi energy ε_F, measured from the centre of the d–band ε_d. ε_d remains constant and is chosen as the zero of energy, so bond and band energies remain equal. It is therefore straightforward to derive expressions for the pressure and bulk modulus in terms of the band energy :

$$P_{bond} = (5/3)U_{band}/V$$

$$B_{bond} = (40/9)U_{band}/V \cdot \tag{4.6}$$

V is the total volume for which the band energy is defined. Referring to equation (4.4), we can consider B_{bond} as the sum of two parts :

$$B_{bond} = B_{bond}^{(1)} + B_{bond}^{(2)}. \tag{4.7}$$

corresponding to the first and second terms of (4.4). The second term of (4.4) gives the contribution :

$$B_{bond}^{(2)} = (25/9)n(\varepsilon_F)\varepsilon_F^2/V. \tag{4.8}$$

The contribution (4.8) represents a correction to the bulk modulus for charge neutrality, the condition which shifts the Fermi energy. It is automatically included in B_{bond} as given by the expression in (4.6). We see that for a half-filled band, the correction (4.8) vanishes, because $\varepsilon_F = 0$, but in general it is not negligible.

To illustrate the point, we require the density of states $n(\varepsilon)$, which we have calculated for a bcc crystal using the recursion method to 15 exact levels with the termination of Beer and Pettifor[43]. The integrations to the Fermi energy were carried out to high accuracy by quadratures on a rectangular contour in the imaginary half-plane, a procedure we developed mainly to avoid the singular behaviour of the density of states $n(\varepsilon)$ at the lower band edge. We have calculated $n(\varepsilon)$ for the perfect crystal at three volumes, with dilations of 0, +0.02 and −0.02. The contributions $B_{bond}^{(1)}$ and $B_{bond}^{(2)}$ from the second derivatives of the terms in (4.4) were then evaluated by numerical differencing. The resulting values of $B_{bond}^{(1)}$ and $B_{bond}^{(2)}$ are added together and compared with B_{bond} according to (4.6) in Fig.1. Also shown in Fig.1 is the bulk modulus as obtained by a force constant summation out to ten shells of neighbours, where the force constants were obtained by the method of Finnis et al[45]. As we anticipated, the force constant result corresponds to the part $B_{bond}^{(1)}$. The deviations in the force constant result are due to the truncation to tenth neighbours together with the restriction to fifteen levels in the continued fraction representation of the Green functions which enter the expressions for the force constants.

We see that the condition of charge conservation, expressed by the final term in (4.4), makes a very significant difference to the bulk modulus, away from the centre of the d-band. For a half filled band, there is no difference between the bond and band models, as mentioned previously in the discussion of surface relaxation. By inference, charge conservation is therefore very significant in the calculation of tight-binding force constants, and therefore also in the calculation of interatomic forces for atomistic relaxation. As discussed in the previous section, the simple expressions for the bond model forces automatically take care of this consistency condition because atomic charge neutrality is imposed in their derivation.

In leaving this section we note that expressions can be derived for the force constants within the bond model, but they appear more difficult to evaluate than the band model formulae, requiring the inversion of large response matrices, and work is still required to seek appropriate simplifications.

5. CALCULATIONS OF BOND ENERGY FOR CANONICAL d-BANDS

To illustrate the concept of bond energies we have evaluated them for the three close packed crystal structures bcc, fcc and hcp (ideal axial ratio) using the canonical d-band parameters given above. The expression (2.1) may be written for each bond $i-j$ as :

$$U_{bond}^{ij} = dd\sigma\, \rho_{55}^{ij} + 2\, dd\pi\, (\rho_{22}^{ij} + \rho_{33}^{ij}) + 2\, dd\delta\, (\rho_{11}^{ij} + \rho_{44}^{ij}). \tag{5.1}$$

The calculation of the elements of ρ in (5.1) requires off-diagonal elements of the Green function. Each of these we obtain from two recursion calculations, as described in Heine et al[19] and Finnis and Pettifor[35]. As in the calculations of diagonal elements, we used fifteen exact levels of the continued fraction with optimal square-root termination[43]. We describe the results for the three structures in turn.

For the bcc structure there are two contributing bonds, the first and second neighbours. In Fig.2a we have plotted the $dd\sigma$, $dd\pi$ and $dd\delta$ contributions to these bond energies, ie the three terms in (5.1), weighted by the number of neighbours, 6 and 8. It is clear that the nearest neighbour $\sigma-$ and $\pi-$bonds dominate the bonding energy, partly as a consequence of the R^{-5} fall off of the tight-binding parameters, which by second neighbours are at about half of their first neighbour values, and partly due to the smaller amplitude of the density matrix.

The fcc structure is simpler, because only a single set of three tight-binding parameters is required for the twelve equivalent nearest neighbours. The contributions are shown in Fig.2b, and as for bcc, the sigma and pi terms dominate.

The hcp case differs because, although all twelve neighbours are equidistant, and have the same tight-binding parameters, the basal plane neighbours are not equivalent by symmetry to the neighbours at $(a/2, \sqrt{3}a/6, c/2)$. Their density matrix elements must therefore be calculated separately. We see from Fig.2c that the sigma, pi and delta contributions from basal and non-basal planes are rather similar and not surprisingly resemble the fcc contributions. The differences between basal and non-basal bond energies are significant on the scale of the energy differences between structures. We also speculate that they explain the lower than ideal c/a ratio of the hcp transition metals Ti and Zr. For between band fillings of 1 and 2, the non-basal neighbours in the ideal structure have a lower $\pi-$bond energy than the basal neighbours (see Fig.2). Although the matrix element $dd\pi$ has the same value for each nearest neighbour bond, the $\pi-$bond charge $(\rho_{22}^{ij} + \rho_{33}^{ij})$ is larger between non-basal neighbours for the band fillings 1–2 and 5–7. From the results of Section 2, this means that the bond force pulling inwards the

188

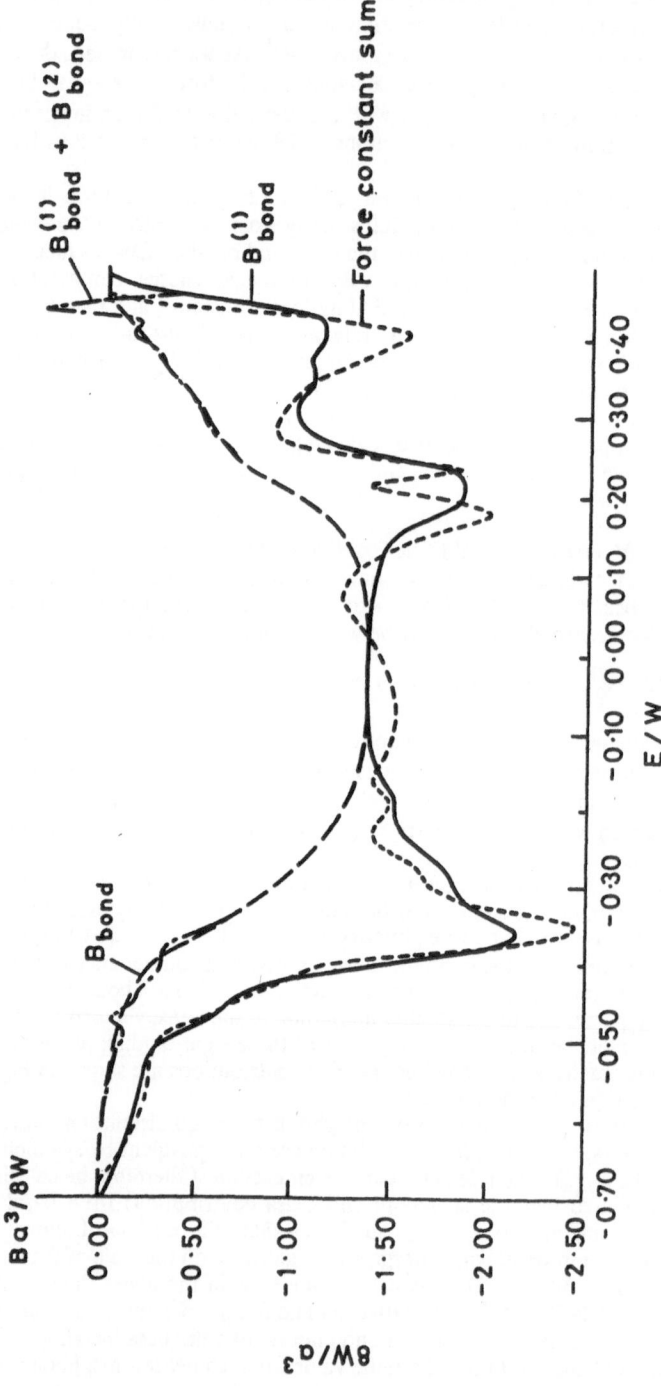

FIG. 1. BULK MODULUS IN BCC.

The bond energy contributions $B_{bond}^{(1)}$ and $B_{bond}^{(2)}$ calculated by homogeneous deformation and second differencing, compared to their theoretical total $B_{bond} = (40/9) U_{bond}/V$, for a bcc crystal with canonical tight-binding parameters. Also shown is the result of calculating B from the force constants out to ten shells of neighbours.

189

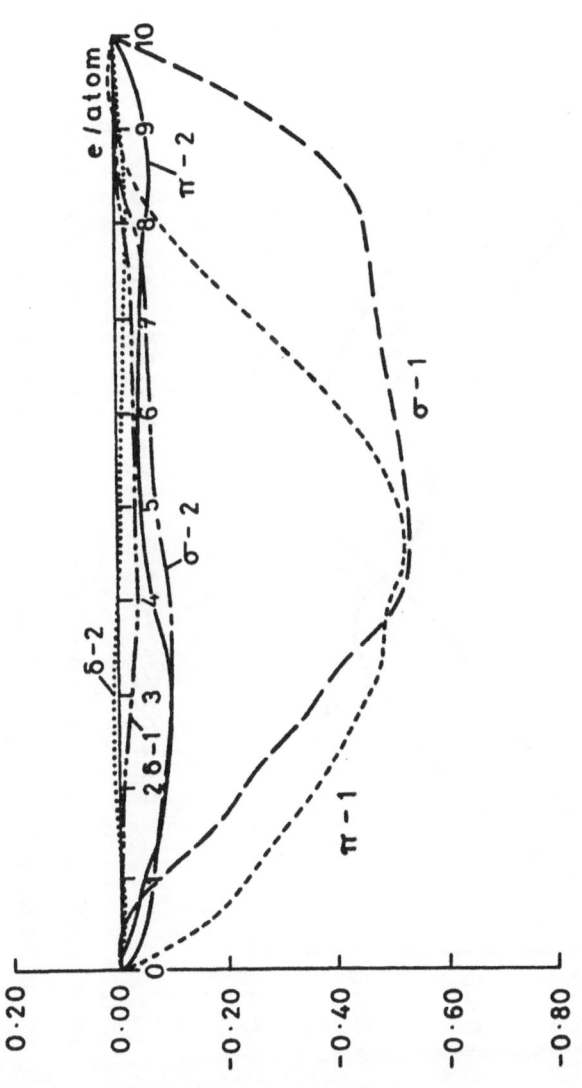

FIG. 2a. BOND ENERGIES IN BCC.

dd-sigma, dd-pi and dd-delta contributions to the bond energies, normalised by a factors 6 and 8 for first and second neighbours, labelled 1 and 2 respectively.

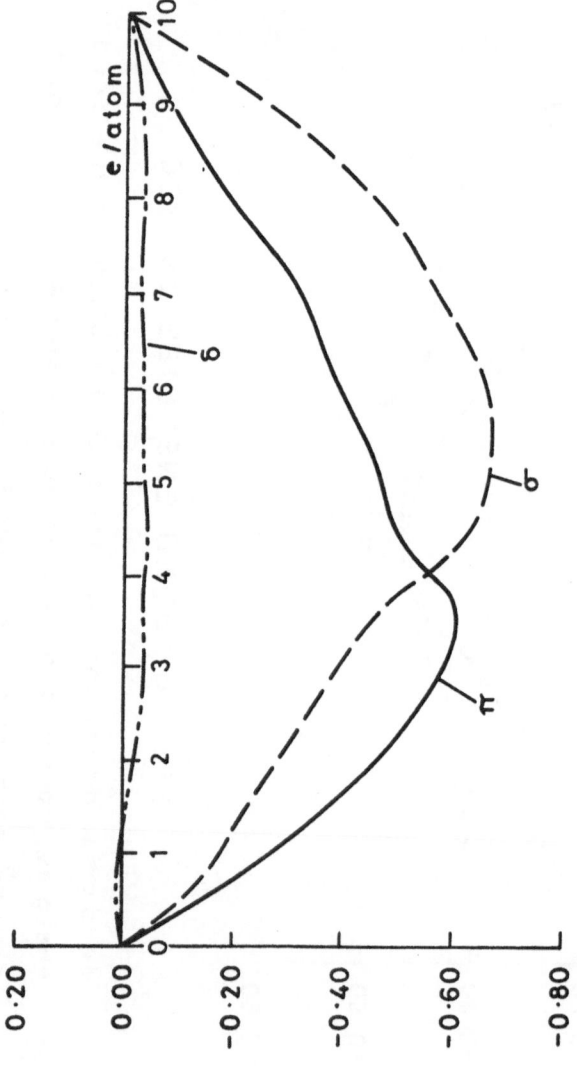

FIG. 2b. BOND ENERGIES IN FCC.

dd-sigma, dd-pi and dd-delta .contributions to the bond energies, normalised by a factor 12 so that their sum gives the energy/atom.

191

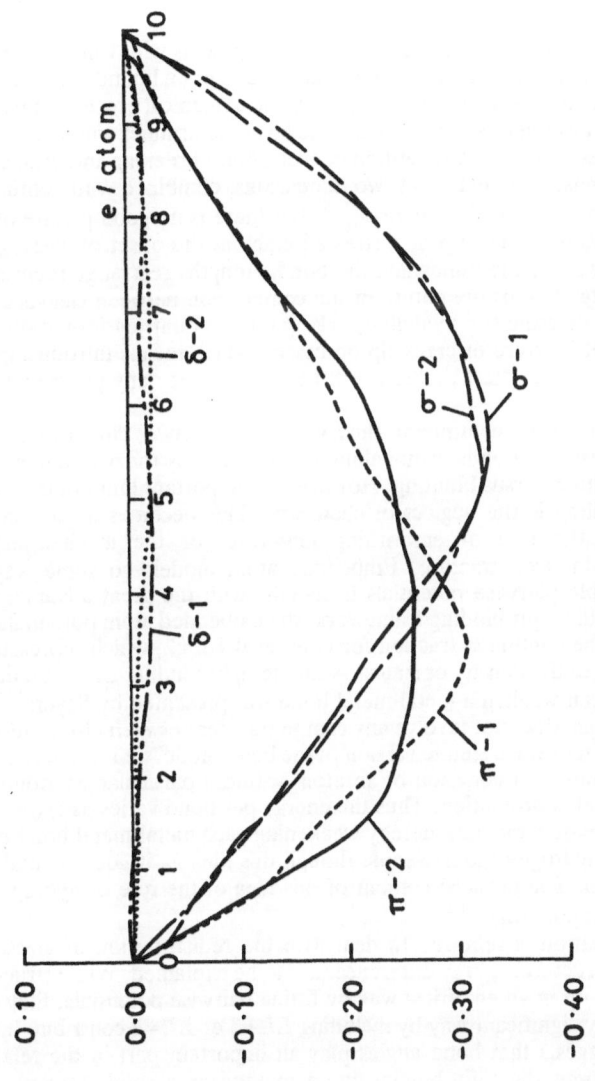

FIG. 2c. BOND ENERGIES IN IDEAL HCP.

dd-sigma , dd-pi and dd-delta contributions to the bond energies,
normalised by a factor 6 so that their sum gives the energy/atom.
1 and 2 denote neighbours at z = 0 (basal plane) and z=+-c/2.

non-basal neighbours is greater than that for the basal neighbours. If the repulsive term is truly central, it effects basal and non-basal neighbours alike. Thus due to the bond forces, the energy would be lowered by a reduction in c/a.

The bond energies shown here can be summed from a given site to give the conventional site energy (2.17), and if one was only interested in total energies (2.17) would be the simplest way to calculate them, not forgetting the additional pairwise repulsive term. However, the bond energy approach we have discussed relates more closely to ideas of chemical bonding and to the associated interatomic forces, which one needs to calculate in order to relax the atomic configuration.

6. DISCUSSION

It has always been the aim of theoretical metallurgists to find models with which to predict the atomic configurations and energies of crack tips, dislocations and grain boundaries. This is a prerequisite for describing the energies and therefore the possible dynamical processes involved in fracture. For example, the question of what, at an atomic level, determines whether a crack will propagate or blunt has focussed on the competition between bond-breaking and dislocation emission at the crack tip. Atomistic simulation in two dimensions, combined with continuum elasticity and dislocation theory, have made progress[49-51] but there is no good picture of the changes in bonding between atoms at the tip of a stressed crack at the onset of cleavage or blunting. A good description of how the bonds and the bond strengths rearrange themselves will be essential for an understanding of the nature of the competition between cleavage and dislocation emission. Larger scale atomistic modelling, taking into account jog formation, will also be necessary for a complete picture of crack tip processes, which means introducing the third dimension. Computationally very fast methods will be necessary for any progress along these lines.

The progression of models from continuum theory, to simple two dimensional and eventually perhaps three dimensional atomistic modelling is a natural way to try to understand the atomistics of crack propagation versus blunting. However, an important limitation of most contemporary atomistic modelling is the neglect of chemistry. This becomes most pressing when we try to understand the role of embrittling impurities or the mechanisms of environmentally enhanced or induced fracture. Embedded atom models go some way to meeting the objections to simple pairwise potentials in metals, with the great advantage of being easy to compute. Within the tight-binding framework, the embedded atom potentials are formally similar to truncating the continued fraction for G at level $LL=2$, which represents a second moment approximation to the density of states. A simple tight-binding $LL=2$ model of how the addition of hydrogen can weaken a metal-metal bond was presented by Sayers[52]. As stressed by Sayers, the mechanism does not involve any charge transfer between atoms, indeed this is explicitly excluded. It is therefore a simple version of the bond model we have described. The concept here is that the energy of cohesion of an atom within a particular environment varies as $\sqrt{z}$, where z is the local coordination. Thus the energy per bond varies as $1/\sqrt{z}$. The effect of the hydrogen is to increase z by one, thereby weakening each metal-metal bond from each of the neighbours of the hydrogen atom. Simple though this idea is, it goes beyond the pairwise additive approximation. The physical content of this idea of the role of hydrogen is captured by the embedded atom potentials[14,15].

A further level of sophistication is required to deal with the redistribution of electrons between bonds when detailed structural energy differences are to be explained. While strucural energy differences can be fixed up in an empirical way by fitting pairwise potentials, they can only be explained in a physically significant way by including $LL=3$ or $LL=4$ contributions to the density of states, which suggests that bond angles play an important part in the relative energies. We have presented here the tight-binding bond method as a solid state physics approach to including more chemistry into the modelling of bonds in metals. The energy of a bond in this picture is the sum of two parts. The chemical bonding is described by the first part which is the sum of products of bond charges ρ_{pq} and energy integrals H_{pq}. The bond charges are calculated from the local atomic configuration by using the tight-binding recursion method, starting from the parameters H_{pq}. The second term in the energy of a bond is a pairwise repulsion. It has recently become possible to justify this division into attractive, non-pairwise bonding energy and pairwise repulsive energy from first principles[53] and to calculate the Hamiltonian matrix elements from band theory[54], although it is more convenient in practice to use values fitted to properties of the perfect crystal.

We have estimated on the basis of preliminary calculations to relax a grain boundary (Ohta *et al*, to be published) that to include $LL=3$, that is the fifth moment of the Green functions, will be practical for atomistic relaxations using from a few tens of minutes to a few hours of the Cray computer, depending on the size and symmetry of the problem. Further work is in progress to improve the speed and convergence of the method.

REFERENCES

1. Eberhart ME, Johnson KH, Messmer RP and Briant CL, *Atomistics of Fracture*, pp 255–280, eds. R.M.Latanision and J.R.Pickens, Plenum, New York, (1983).
2. Briant CL and Messmer RP, Acta metall. **30**, 1811, (1982).
3. Sinclair JE, *Computer Simulation in Physical Metallurgy*, pp 107–128, ed. G. Jacucci, Reidel, Dordrecht, (1986).
4. Bacon DJ and Liang MH, *Interatomic Potentials and Crystalline Defects*, pp 181–200, ed. Jong K. Lee, The Metallurgical Society of AIME, (1981).
5. Takeuchi S, *Interatomic Potentials and Crystalline Defects*, pp 201–21, ed. Jong K. Lee, The Metallurgical Society of AIME, (1981).
6. Vitek V and Yamaguchi M, *Interatomic Potentials and Crystalline Defects*, pp 223–48, ed. Jong K. Lee, The Metallurgical Society of AIME, (1981).
7. Kuramoto E, Aono E and Tsutsumi T, Crystal Res. and Technol. **19**, 331, (1984).
8. Sutton AP, International Metals Reviews **29**, 377, (1984).
9. Hashimoto M, Ishida Y, Yamamoto R and Doyama M, Acta metall. **32**, 1, (1984).
10. Benesh GA and Inglesfield JE, J.Phys.C **17**, 1595, (1984).
11. Car R and Parrinello M, Phys.Rev.Lett. **55**, 2471, (1985).
12. Maeda K, Vitek V and Sutton AP, Acta metall. **30**, 2001, (1982).
13. Sutton AP and Vitek V, Acta metall. **30**, 2011, (1982).
14. Daw MS and Baskes MI, Phys.Rev.Lett. **50**, 1285, (1983).
15. Daw MS and Baskes MI, Phy.Rev.B **29**, 6443, (1984).
16. Foiles SM, Phys.Rev.B **32**, 7685, (1985).
17. Finnis MW and Sinclair JE, Phil.Mag.A **50**, 45, (1984). Erratum, Phil.Mag.A **53**, 161, (1986).
18. Matthai CL and Bacon DJ, Phil.Mag.A **52**, 1, (1985).
19. Heine V, Haydock R, Bullett DW and Kelly MJ, *Solid State Physics* **35**, eds. Ehrenreich, Seitz and Turnbull, Academic, New York, (1980).
20. Pettifor DG, *Physical Metallurgy*, Ch.3, eds. R.W.Cahn and P.Haasen, North Holland, Amsterdam, (1983).
21. Hashimoto M, Ishida Y, Wakayama S, Yamamoto R, Doyama M and Fujiwara T, Acta metall. **32**, 13, (1984).
22. Chadi DJ, J.Vac.Sci.Technol. **16**, 1290, (1979).
23. Chadi DJ, Phys.Rev.B **19**, 2074, (1979).
24. Masuda K and Sato A, Phil.Mag.A **44**, 799, (1981).
25. Masuda K, Yamamoto R and Doyama M, J.Phys.F, **13**, 1407, (1983).
26. Turchi P and Ducastelle F, *The Recursion Method and Its Applications*, pp 104–119, eds. D.G.Pettifor and D.L.Weaire, Springer Verlag, Berlin, (1985).
27. Beer NR, PhD Thesis, University of London, (1985).
28. Legrand B, Phil.Mag.A, **52**, 83, (1985).
29. Boswarva IM and Esterling DM, J.Phys.C, **15**, L729, (1982).
30. Treglia G, Desjonquères MC and Spanjaard D, J.Phy.C **16**, 2407, (1983).
31. Treglia G, Ducastelle F and Spanjaard D, J.Physique **41**, 281, (1980).
32. Terakura I, Terakura K and Hamada N, Surf.Sci. **111**, 479, (1981).
33. Masuda–Jindo K, Hamada N and Terakura K, J.Phys.C **17**, 1271, (1984).
34. Moraitis G, Stupfel B and Gautier F, Phil.Mag.B, **52**, 971, (1985).

35. Finnis MW and Pettifor DG, *The Recursion Method and Its Applications,* pp 120–131, eds. D.G.Pettifor and D.L.Weaire, Springer Verlag, Berlin, (1985).
36. Pettifor DG and Podloucky R, J.Phys.C **19**, 315, (1986).
37. Finnis MW, Sutton AP, Pettifor DG and Ohta Y, to be published.
38. Slater JC and Koster GF, Phys.Rev. **94**, 1498, (1954).
39. Coulson CA, Proc.Roy.Soc. **169 A**, 413, (1939).
40. Penney WG, Proc.Roy.Soc. **158 A**, 306, (1937).
41. Allan G, Ann.Phys. **5**, 169, (1970).
42. Nex CMMN, Computer Physics Communications **34**, 101, (1984).
43. Beer N and Pettifor DG, *Electronic Structure of Complex Systems,* Plenum, New York, (1985).
44. Lennard–Jones JE, Proc.Roy.Soc. **158 A**, 280, (1937).
45. Finnis MW, Kear KL and Pettifor DG, Phys.Rev.Lett. **52**, 291, (1984).
46. Wallace DC, *Thermodynamics of Crystals,* Wiley, New York, (1972).
47. Andersen OK, Phys.Rev.B **12**, 3060, (1975).
48. Pettifor DG, J.Phys.F **7**, 613, (1978).
49. Rice J and Thomson R, Phil.Mag. **29**, 73, (1974).
50. Sinclair JE and Finnis MW, *Atomistics of Fracture,* pp 1047–1051, eds. R.M.Latanision and J.R.Pickens, Plenum, New York, 1983.
51. Lin I.–H and Thomson R, Acta metall. **34**, 187, (1986).
52. Sayers CM, Phil.Mag.B **50**, 635, (1984).
53. Harris J, Phys.Rev.B **31**, 1770, (1985).
54. Andersen OK and Jepsen O, Phys.Rev.Lett. **53**, 2571, (1984).

DISCUSSION

Comment by B. D. Lichter:

If I understand you correctly, your model successfully predicts the HCP stability of Ti and Zr. Can you also predict the BCC stability of these elements which occurs at higher temperatures? More generally, how can your model take into account entropy effects, which may be as significant as energy (enthalpy) in determining lattice stability?

Reply:

The entropy effect on these lattices could be calculated by standard methods (in the harmonic approximation) from the phonon densities of states, which in turn could be calculated from the force constants. In principle, force constants can be calculated from our TBB model, as they were in the TB band model (Finnis, Kear and Pettifor, 1984), although the charge neutrality condition would make it a bigger calculation.

Comment by M. Daw:

Compared to the full tight-binding scheme of diagonalizing a large, cumbersome matrix, how is your approximate scheme going to work? Under what conditions will the approximations break down?

Reply:

In our approximate scheme, the elements of the Green Function G are calculated by the recursion method - the Lanczos algorithm - which does not require diagonalisation of the whole H matrix. In the recursion method, successive orders of approximation correspond to introducing the effects of successively further shells of neighbours to the bond under consideration. These successively further shells of neighbors contribute successively higher power moments of the functions ImG. We include only the effect of two shells of interactions (of order 100 atoms around a bond) in our TBB forces, equivalent to including up to the 5th moment of ImG. This is sufficient to resolve structural energy differences in a qualitative way, for predicting trends with bond filling. Other approximations include neglect of s-d hybridization in our d-band model and neglect of magnetic moments, which may both be important for quantitative predictions of energy.

Comment by C. Altstetter:

At what interatomic separations does your computational method begin to break down? Is complete separation (Cleavage) too difficult or inaccurate?

Reply:

I would expect our interatomic forces to be very inaccurate at separations of order >2X the equilibrium separation. We have no representation of Van der Waal's forces. However, most of the cleavage energy is taken care of within this separation. The models I have referred to predict reasonable surface energies.

APPLICATION OF THE EMBEDDED ATOM METHOD TO HYDROGEN EMBRITTLEMENT*

Murray S. Daw and Michael I. Baskes
Theoretical Division
Sandia National Laboratories in Livermore
Livermore, CA 94550

INTRODUCTION

This paper is concerned with atomistic simulations of the mechanical properties of metals and the effects of hydrogen. Figure 1 shows an example of such a simulation--a slab of Ni atoms which originally started with a small crack and a hydrogen atom at the crack tip. We performed molecular dynamics simulations on this slab using the Embedded Atom Method (1) to give us the forces for Newton's Law. External stress was applied and the result was that dislocations were emitted from the crack. The first dislocation originated near the hydrogen atom and subsequent dislocations came out from other areas. The same calculation without the hydrogen shows a lower dislocation emission rate. These calculations suggest that the process of hydrogen embrittlement may actually involve an enhancement of crack tip plasticity, as has been suggested by the work of Lynch (2) and Birnbaum and co-workers (3). In this paper, we will discuss how the calculations are done and what we conclude from them.

Many of the speakers at this Workshop have suggested that crack propagation is controlled by kinetics, rather than equilibrium energetics. Therefore, a knowledge of kinetics is of central interest, and this information is available from molecular dynamics. By molecular dynamics, we mean simply that we solve the coupled equations of motion of the atoms given the accelerations from Newton's Law, $a = F/m$. But how do we calculate the forces F? That is the point of this paper: We present here a way of calculating forces that is better than previously available.

We will first introduce you to the Embedded Atom Method, which represents a new advance in the calculation of energies for atomistic simulations. Then we will apply the method to basic parts of the fracture process: brittle fracture, dislocation dynamics, and crack tip plasticity.

Ultimately, we would like to be able to do the calculation illustrated in Fig. 2, where we have a macroscopic sample under applied stress, and we solve for the stresses and strains in most of the region by elasticity theory. Of course, there are regions, such as at the crack tip, that must be treated atomistically, perhaps involving on the order of 10^5 atoms. With this calculation, we would hope to attain a good understanding of dislocation emission and brittle fracture, and the effects of hydrogen on those processes.

In the past, workers in this field have used interatomic pair potentials to calculate the forces on the atoms (4-6). In this approximation (and we emphasize that it is an approximation), the total energy is given as a sum over pairs of atoms:

$$E = U(V) + \frac{1}{2} \sum_{ij} \phi(R_{ij}) \qquad (1)$$

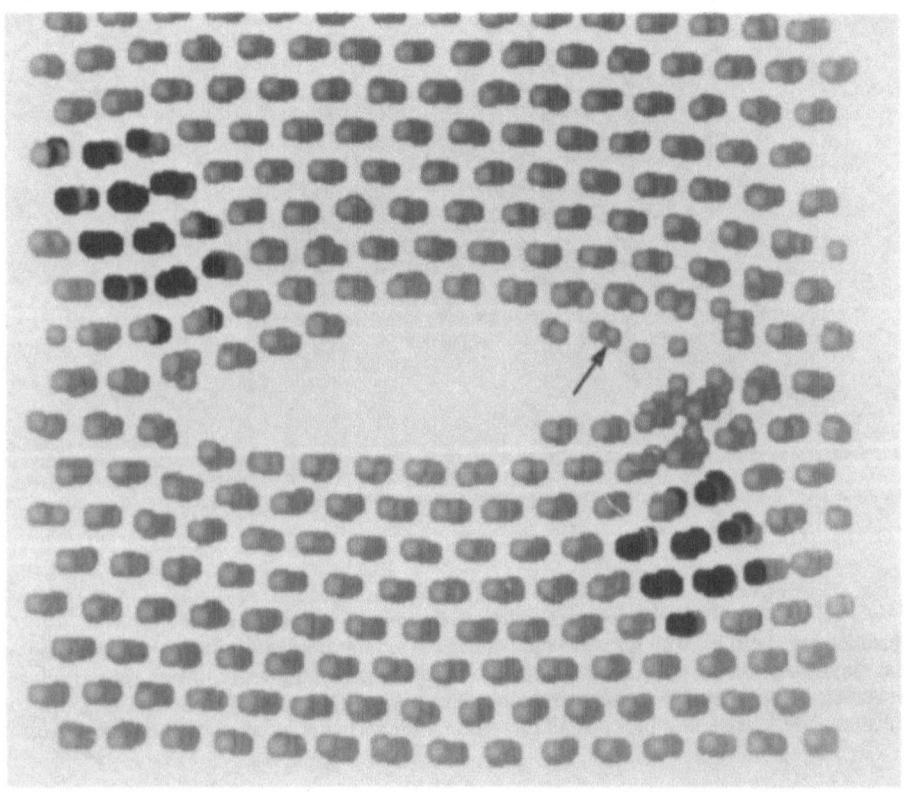

FIGURE 1. Fracture processes at a crack tip in nickel in the presence of hydrogen, as calculated using the Embedded Atom Method. The figure illustrates an intermediate stage in the process. The initial configuration consisted of a small, thin crack with one hydrogen atom present (indicated by pointer). The vertical axis is a [001] crystal direction, and the horizontal axis is a [1̄10] crystal direction. Forces were applied to the top and bottom surface atoms to create a tensile stress, and the motion of the atoms was solved at room temperature. The light gray atoms are nickel atoms in an f.c.c. lattice, and the dark gray atoms are nickel atoms near a dislocation core. The dislocations have been emitted from the crack tip and also the crack has advanced along the cleavage plane. Without hydrogen, the time to emit dislocations is longer.

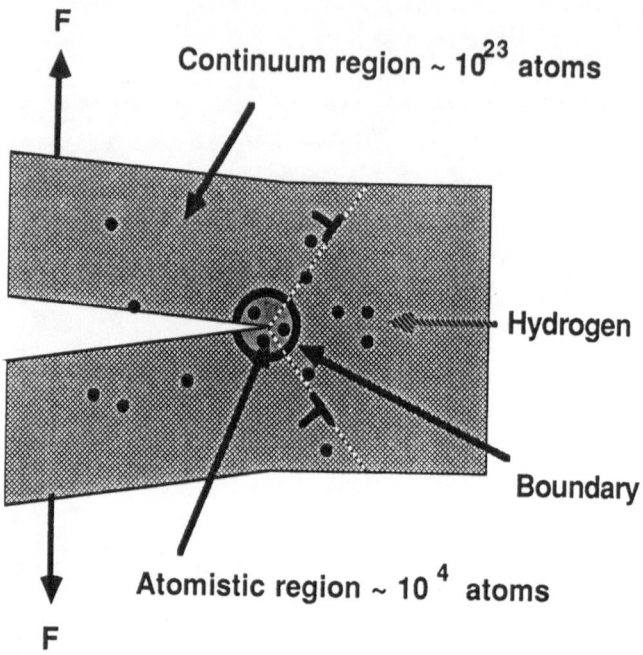

FIGURE 2. A goal of atomistic simulations: to model the fracture process of a macroscopic sample. The practical realization of this solution will involve the modelling of the majority of the sample by non-atomistic means, where certain non-linear zones such as the crack tip will involve atomistic regions. The dislocations created from the crack tip must be able to pass into the continuum region without being affected by the boundary.

The first term on the right hand side is a contribution to the total energy as a function of the total volume of the sample, and the second represents the pair-bond interactions in the solid. The pair potential approximation is reasonable for certain classes of problems in solid state physics. In the case of fracture and hydrogen embrittlement, however, there are two insurmountable problems. First, the volume-dependent energy presents a difficulty in that for atomistic calculations the volume is not always definable. The termination of the volume near a surface or near a nucleating or propagating crack is ambiguous. However, if we simply choose to ignore the volume-dependent energy (as numerous authors have done), the elastic properties of the material become unphysical. We will discuss this aspect later in the paper. The second major difficulty with the pair potential approximation is that metal-metal interactions are surely affected by the presence of hydrogen, yet the change in the metal-metal bond is neglected in the pair potential formulation (7). These two critical shortcomings can be avoided by using the Embedded Atom Method.

THEORY

In the Embedded Atom Method (or EAM), the main contribution to the total energy of a solid is the embedding energy, which is the energy to put an atom into an electron density created by all the other atoms in the solid. The embedding energy depends on the type of atom in question (the element), and the background electron density. Generally speaking, the embedding functions behave in one of two ways. An inert element, such as helium, is insoluble in everything, and so its embedding function rises monotonically from zero at zero background density. Chemically active elements, such as hydrogen or a transition metal, bond with the electron gas, and so the embedding energy has a negative slope at zero electron density. The embedding function for a given element is universal in the sense that it is independent of the source of the background density. We are thus suggesting that the cohesive energy of a metal, such as Ni, originates as a bonding of each Ni atom with the electron gas created by the other Ni atoms in the solid.

Two corrections to the simplest approximation are necessary. First, because the electron density in a real solid is not uniform, we approximate the correct electron density in the solid by a linear superposition of atomic electron densities. Second, the nuclei are not electrostatically isolated but do in fact interact with other nuclei. This electrostatic interaction can be represented by a pair interaction, which is small compared to the embedding energy and monotonically decreasing with distance. Again, the embedding energy is the main contribution to the total energy, and implicitly includes many-body interactions as will be discussed more fully below.

We can make these ideas more formal by approximations within density functional theory (8). Density functional theory states that the total energy of a system, E_s, is a functional of the electron density, ρ_s:

$$E_s = \int g(\rho_s(\mathbf{r}))\, d\mathbf{r} \; - \; \sum_i Z_i \int \frac{\rho_s(\mathbf{r})}{|\mathbf{r} - \mathbf{R}_i|}\, d\mathbf{r} \; + \; \frac{1}{2} \iint \frac{\rho_s(\mathbf{r}_1)\rho_s(\mathbf{r}_2)}{r_{12}}\, d\mathbf{r}_1\, d\mathbf{r}_2 \; + \; \frac{1}{2}\sum_{ij,\, i \neq j} \frac{Z_i Z_j}{R_{ij}} \qquad (2)$$

where we have separated out the electrostatic terms and always reference the energies to that of isolated atoms, so that we are always discussing cohesive energies. The functional g contains the kinetic, exchange, and correlation energies, and in this treatment we assume that this functional is semi-local, in that it can be written as an integral over a function of the electron density and successive derivatives:

$$G[\rho(\mathbf{r})] \approx \int g(\rho(\mathbf{r}), \underline{\nabla}\rho(\mathbf{r}), \nabla^2\rho(\mathbf{r}), \dots)\, d\mathbf{r} \qquad (3)$$

Many workers have attempted to establish the validity of this semi-local approximation (9), and we will not discuss it in detail here.

To lowest order, the electron density used in the expression for the total energy can be obtained by superposing atomic electron densities, so that

$$\rho_s(\mathbf{r}) = \sum_i \rho_i^a(\mathbf{r}) \qquad (4)$$

Given this assumption, we can rewrite the cohesive energy of the solid as:

$$E_s = \sum_i F_i(\rho_i) + \frac{1}{2}\sum_{ij,\, j \neq i} \phi_{ij}(R_{ij}) \qquad (5)$$

where the embedding energies F_i and the pair interactions ϕ are defined in terms of the kinetic, exchange, and correlation energy densities and the electrostatic interactions. The form in Eq. 5 is very similar to that obtained by others (10,11).

Having made a series of approximations to obtain the basic form for the EAM, we have obtained an expression which is an explicit function of the positions of the atoms in the solid. This expression is quite straightforward to use and is as computationally efficient to use as interatomic pair potentials. We can demonstrate this by outlining how the calculation is done in the EAM. To calculate the total energy in the solid, we group the terms in Eq. 5 so that it looks like a sum over the energy of each atom in the solid. Then

$$E_s = \sum_i E_i \tag{6}$$

where the energy for atom i is

$$E_i = F(\rho_i) + \frac{1}{2} \sum_{j \neq i} \phi(R_{ij}) \tag{7a}$$

$$\rho_i = \sum_{j \neq i} \rho^a(R_{ij}) \tag{7b}$$

First we sum $\phi(R)$ over the neighbors of i and this gives us the electrostatic contribution. Next we perform a similar sum of $\rho(R)$ over the neighbors of i and this gives us the total background electron density seen by atom i. We then take $F(\rho_i)$ to give us the embedding energy and add that to the electrostatic energy. In fact, the sum over $\rho(R)$ is analogous to the sum over $\phi(R)$, so that the EAM is no worse computationally than using pair potentials. But because we take a function of the ρ_i, the EAM contains more physical information than do pair potentials, as we shall see next.

In fact, it is interesting to note that pair potentials are a special case of the EAM (1,12). If we take a perfect, homonuclear, homogeneous, infinite, single crsytal, then every atom experiences the same background electron density, which we call $\bar{\rho}$, and the same electrostatic interaction, which we call $\bar{\phi}$. The total energy is then

$$\bar{E} = N\left(F(\bar{\rho}) + \frac{1}{2}\bar{\phi} \right) \tag{8}$$

If we now allow for slight distortions away from the perfect solid, the electron densities vary slightly, as do the electrostatic interactions. Expanding the total energy in a Taylor series in the deviations from the perfect case, we find that

$$\Delta E = \frac{1}{2} \sum_{ij} \psi(R_{ij}) + O(F'') \tag{9}$$

where we define an *effective* two-body interaction $\psi(R) = \phi(R) + 2F'(\bar{\rho})\rho(R)$. Higher order terms give interactions involving three bodies, four bodies, etc. This tells us when the pair potential approach breaks down: i.e., when higher derivatives of F are not negligible, or the variations are large.

From the derivation sketched in the preceding paragraphs, it is possible to determine the EAM functions from first principles (8). Functions so determined qualitatively reproduce transition metal properties. However, in order to get quantitative information, we have chosen instead to take a semi-empirical approach (1). In this scheme, we use our knowledge of the embedding function, electrostatic interactions, etc., to suggest parametric forms for these functions. We then determine the parameters in the functions by fitting to bulk empirical data, such as the lattice constant, the elastic constants, the sublimation energy, the vacancy formation energy the stacking fault energy, the dilute heats of alloying for intermetallic alloys, the hydrogen heat of solution, and the hydrogen migration energy (13).

The curvature of an embedding function is positive; that is, $F''>0$. The reason for this is quite simple. If one thinks about increasing electron density as being related to increasing bonding in some sense, then the slope of the embedding function represents a bond affinity for the element. As an atom becomes involved with an increasing amount of bonding, the affinity for additional bonds becomes less. Therefore the curvature should be positive. As we will see, this condition for positive curvature will occur again at two other places in this paper. We thus ask the reader to take note.

APPLICATIONS

The EAM is quite robust in that it describes surprisingly well the physical properties of metallic systems under a wide variety of conditions. To demonstrate this we have applied the method to transition metal phonons (14), transition metal liquids (12), intermetallic alloys (13,15), and metallic surface energy (1, 13), relaxations (1, 13), and reconstructions (16,17). We have also applied the method to the properties of H in metals, i.e., the ordered structures, critical temperatures, subsurface occupation (18,19), and molecular dissociation pathways (20) have been calculated on transition metal surfaces. Intermetallic alloy segregation to surfaces, interfaces, and defects (13) has been investigated. Most relevant to the topic at hand, the EAM has been applied to calculations involving the strength and ductility of materials (21), and these are the computations that will be discussed here.

At this point it is useful to notice that with the EAM we can calculate various quantities which come in as unknown parameters in continuum theory. For example, continuum theory cannot predict the surface energy of a material, but the surface energy is required in order to calculate the work of brittle fracture. Other such quantities include surface stresses, resistive force versus crack opening displacement at a crack tip, the formation energy of a ledge on a surface, and the dislocation core size and energy. Of course, the effects of H on all these quantities are also of interest and can be calculated using the EAM.

For example, we can calculate the binding of H to various defects in Ni. By placing the hydrogen atom near various defects, we find that the binding is determined by the local environment of the hydrogen atom, that is, those atoms within a few Ångstroms. We also find that H is bound to surfaces relative to the bulk, and also bound more strongly to ledges than to a flat surface. The binding energy to a ledge is about 0.1 eV relative to a flat surface. These facts will come into play later when we discuss the effects of hydrogen on crack tip processes.

Having now introduced and explained the utility of the EAM, we will now progress to an atomistic simulation of processes that occur in the fracture of materials. We begin by setting up a slab of material, illustrated in Fig. 3a, which is of finite thickness in one direction and periodic in the two directions within the plane of the slab. The slab is composed of an f.c.c. array of Ni atoms at some crystallographic orientation. In the first

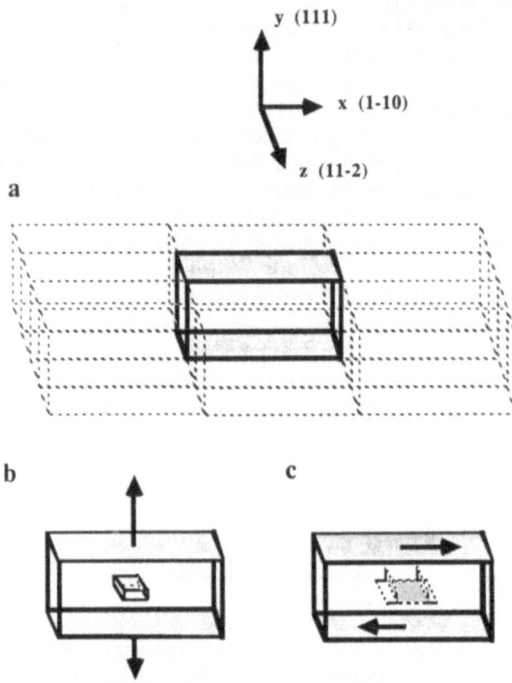

FIGURE 3. Illustration of the basic slab geometry used in the present calculations. The crystallographic directions are as drawn. a) The slab is finite in one direction and periodic in the other two. The finite direction is along the [111] axis, so that the two-dimensional, periodic plane is the slip plane in nickel. The unit periodic cell is outlined in dark, with some of its neighboring, identical periodic cells drawn in light. The top and bottom surfaces, shaded in the figure, are free (111) surfaces. The two periodic directions are chosen to be $[1\bar{1}0]$ and $[11\bar{2}]$. Thus, the x-direction is the slip direction and the z-direction is along the axis of dislocation cores. b) For the brittle fracture calculations, a crack may be introduced in the material and forces are applied normal to the free surfaces (normal to the slip planes). This produces a tensile stress. c) For the dislocation dynamics calculations, a dislocation is introduced, and the forces are applied to the free surfaces along the slip direction. This produces a shear stress.

set of calculations that we will show here, the slip plane is aligned parallel to the plane of the slab. The reason for this will become clearer as we proceed.

To this slab we can apply stresses by applying forces to the atoms on the top and bottom surfaces, which are otherwise free surfaces. We can introduce a crack in the middle of the slab by removing some atoms, as in Fig. 3b. In the case of the crack, let us suppose that we apply a tensile stress. We now follow the motion of all the atoms with molecular dynamics. That is, we solve Newton's equation for the motion of each atom. By increasing the stress gradually and watching for the crack to start propagating, we can calculate the stress required for brittle fracture of the pre-cracked slab. Because the slip planes are normal to the applied forces, the resolved shear stress is zero and the fracture is guaranteed to be brittle (unless another higher energy slip system becomes active).

The brittle fracture stress obtained is consistent with the Griffith criterion (21, 22). This criterion states that the brittle fracture stress is given by

$$\sigma = \sqrt{\frac{2\gamma K}{\pi c}} \tag{10}$$

where γ is the energy of the surface created by the fracture, c is the crack length, and K is the appropriate elastic strain modulus. For our calculations, we have $\gamma = 564$ ergs/cm^2, K=3.47x10^{12} erg/cm^3, and c=24.89Å, which gives $\sigma = 0.24$ eV/Å-atom, or 5.7% of C$_{44}$. This is in excellent agreement with the molecular dynamics results (21), which put the fracture stress at around 0.21 eV/Å-atom.

It is important in these calculations to apply the stresses gradually, in a ramped fashion. If the stresses are applied suddenly at the beginning, giving a step function in time, then two pulses are formed which travel inward from the surfaces and converge on the middle plane, and they cause undesirable effects such as premature fracture.

Note how the volume-dependence of a pair potential calculation would cause problems in the crack propagation problems. In this case, the volume is clearly ambiguous. The necessity of the volume-dependent term for the pair potential scheme is quite clear, however. For the energy of the system described by Eqs. 7, we can derive the elastic constants in a straightforward way (6). From this, we get

$$C_{ijkl} = B_{ijkl} + P_2\delta_{ij}\delta_{kl} - P_1(\delta_{ik}\delta_{jl} + \delta_{il}\delta_{jk}) \tag{11a}$$

where

$$B_{ijkl} = \frac{1}{2\Omega_0}\sum_m \frac{1}{(a^m)^2}(\phi''_m - \frac{\phi'_m}{a^m})a_i^m a_j^m a_k^m a_l^m \tag{11b}$$

and P_1 and P_2 are related to the volume-dependent energy. A simple example of the effect of the volume-dependent energy can be seen by forming the generalized Cauchy discrepancy, which is

$$C_{ijkl} - C_{ikjl} = (\delta_{ij}\delta_{kl} - \delta_{ik}\delta_{jl})(P_2 + P_1) \tag{12}$$

For cubic metals, this Cauchy discrepancy is C$_{12}$-C$_{44}$, which of course in general does not vanish. However, if we neglect the volume-dependence in the pair potential, then $P_1 = P_2$ =0 and the Cauchy relation (C$_{12}$-C$_{44}$=0 for cubic materials) holds, despite the fact that in real metals the Cauchy relation does not hold. Therefore, any atomistic simulation using pair potentials but neglecting the volume-dependent energy are simulating fracture in materials with unphysical elastic properties.

Because the EAM does not have a volume-dependent energy, the difficulty of the volume ambiguity is completely absent.. From Eq. 7, the elastic constants are given (1) by

$$C_{ijkl} = (B_{ijkl} + \bar{F(\rho)}'W_{ijkl} + \bar{F(\rho)}''V_{ij}V_{kl})/\Omega_0 \tag{13a}$$

where

$$B_{ijkl} = \frac{1}{2} \sum_m (\phi''_m - \frac{\phi'_m}{a^m}) \frac{a^m_i a^m_j a^m_k a^m_l}{(a^m)^2} \tag{13b}$$

$$W_{ijkl} = \sum_m (\rho''_m - \frac{\rho'_m}{a^m}) \frac{a^m_i a^m_j a^m_k a^m_l}{(a^m)^2} \tag{13c}$$

$$V_{ij} = \sum_m \rho'_m \frac{a^m_i a^m_j}{a^m} \tag{13d}$$

In this case the Cauchy relation is

$$C_{ijkl} - C_{ikjl} = (V_{ij}V_{kl} - V_{ik}V_{jl})F''(\bar\rho) \tag{14}$$

The sign of this relation for real metals requires that $F''>0$. Interestingly, this requirement on the curvature of the embedding function agrees with the previous argument about F', where it was pointed out that the positive curvature was related to the saturation of bonding affinity.

Now we add to the brittle crack propagation calculation the presence of some H atoms in the slab. We find that H atoms near the crack tip are particularly effective at reducing the fracture stress, by reducing the metal-metal bond strength near the crack tip (21).

The effects of hydrogen on the brittle fracture stress are again consistent with the Griffith criterion (22). Hydrogen near the crack tip can reduce the surface energy and thereby reduce the fracture stress. This is because, as we noted before, hydrogen is bound to the surface relative to the bulk, thus encouraging fracture. Away from the immediate crack tip vicinity, hydrogen atoms can further change the modulus of the material, again reducing the fracture stress. Thus it would seem that there are no surprises from the molecular dynamics/EAM calculations shown here.

It is again important to note that pair potentials could not correctly address the problem shown above, however. By their nature, pair potentials do not account for changes in metal-metal bonding because of the presence of hydrogen. This can be illustrated by the following example. Suppose we have two metal atoms. The energy of the two atoms is given, in the pair potential scheme, by

$$E = \phi_{M-M}(R_{M-M}) \tag{15}$$

where R_{M-M} is the distance between the two atoms. If we pull the two atoms apart, the force encountered in resistance to this is given by

$$f = \phi'_{M-M}(R_{M-M}) \tag{16}$$

Suppose now that we were to add a hydrogen atom to one side, so that only one metal atom experiences the effect of the hydrogen atom. Then the energy is given by

$$E = \phi_{M-M}(R_{M-M}) + \phi_{M-H}(R_{M-H}) \tag{17}$$

If we now increase R_{M-M} without changing R_{M-H}, the resisitive force is given by

$$f = \phi'_{M-M}(R_{M-M}) \tag{18}$$

That is, hydrogen had no effect on the resisitive force. This is not correct! Therefore, calculations which rely on pair potentials to describe hydrogen effects on fracture cannot correctly account for the effects of H on metal-metal bonding. And yet, surely this effect comes into play in hydrogen embrittlement!

However, the EAM does account for changes in M-M bonding caused by the presence of hydrogen. Repeating the simple heuristic argument of the preceeding paragraph, let us start with two metal atoms. In the EAM, the energy is given by

$$E = 2F_M(\rho_M(R_{M-M})) + \phi_{M-M}(R_{M-M}) \tag{19}$$

where ρ_M is the atomic density from a metal atom. Now the resistive force is given by

$$f = 2F'_M(\rho_M)\rho'_M(R_{M-M}) + \phi'_{M-M}(R_{M-M}) \tag{20}$$

In the presence of hydrogen, where only one metal atom experiences the effect of the hydrogen, the energy is given by

$$E = F_M(\rho_M(R_{M-M}) + \rho_H(R_{M-H})) + F_M(\rho_M(R_{M-M})) + F_H(\rho_M(R_{M-H}))$$
$$+ \phi_{M-M}(R_{M-M}) + \phi_{M-H}(R_{M-H}) \tag{21}$$

where now the embedding energy of one of the metal atoms arises from the electron density of the other metal atom *and* the hydrogen atom. The resistive force for changing the metal-metal bond distance, holding the hydrogen-metal bond fixed, is now given by

$$f = (F'_M(\rho_M + \rho_H) + F'_M(\rho_M))\rho'_M(R_{M-M}) + \phi'_{M-M}(R_{M-M}) \tag{22}$$

The change in the resistive force because of the presence of the hydrogen is now not zero, but

$$\Delta f = (F'_M(\rho_M + \rho_H) - F'_M(\rho_M))\rho'_M(R_{M-M}) \tag{23}$$

If we assume that the hydrogen electron density is small compared to that of the metal, we can expand this change in a Taylor's series, and this gives

$$\Delta f = F''_M\rho_H\rho'_M(R_{M-M}) \tag{24}$$

Therefore we see that if $F''>0$, then the metal-metal bond is weakened by the presence of the hydrogen (because ρ'_M is always negative). This is consistent with the previous argument about the curvature of the embedding function. In fact, it is the same argument, because we are saying that the metal-metal bond affinity is less because one of the metal atoms has now additional bonding.

The above argument about the weakening of the metal-metal bond by hydrogen is true when we are fracturing the bond, and it is equally true when we are shearing the bond. Therefore the conclusion that hydrogen weakens metal-metal bonds is generally true, both in fracture and in shear.

Let us step back for a moment to regain perspective. We want eventually to be able to do the problem shown in Fig. 2. We have done so far a part of that problem, namely the brittle fracture of a crack with and without hydrogen. Now let us consider another small part of the problem: namely, dislocation motion. Dislocations are important because we know that even what has been classically called brittle fracture does in fact involve a great deal of plasticity (2). Another reason why dislocations are important, is that several workers have observed a relationship between hydrogen and dislocations. For example, it has been observed that the introduction of hydrogen into a sample with a stressed crack cause dislocations to be produced at a faster rate and to travel at a faster rate than would be observed without the hydrogen (3). Also, many observations of transport of hydrogen by dislocations have been reported (23). So we will consider now the problem of a dislocation moving in an otherwise ideal material, and consider the effects of hydrogen. Eventually we will try to put all the pieces of the hydrogen embrittlement problem together, but that may be years away.

In Fig. 3c we show our same slab again, but this time instead of introducing a crack, we have introduced two extra half-planes (i.e., an edge dislocation). We can identify the dislocated atoms by an algorithm that compares the local atomistic environment of each atom with that in a perfect f.c.c. crystal. In effect, the algorithm tries to construct a Burger's loop, and the lack of closure identifies a dislocation (21). To be precise, we have adopted the formalism that Kroner developed from differential geometry to compute the dislocation density tensor on a lattice (24). The tensor can be contracted with itself to form a scalar dislocation density. This analysis is done independently of the molecular dynamics or energy minimization, merely to illustrate the calculations, and has no effects on the results. It simply gives us a quantitative tool for studying the dynamics of a dislocation.

The slab we illustrate in Fig 3c is constrained by the same periodic boundary conditions that were present in the brittle crack propagation calculations. That is, we are actually studying an infinite array of parallel edge dislocations. The reason for using periodic boundary conditions in this case is to avoid the interaction of dislocations with a surface (25). In previous calculations involving dislocations, workers have used several types of boundary conditions, falling into the two main categories of fixed strain condition or fixed stress condition. Free surface fall into the category of fixed (i.e., fixed at zero) stress, while a surface constrained to be undistorted falls into the fixed (i.e., at zero) strain category. Neither fixed strain nor fixed stress condition mimics the behavior of an infinite solid. The dislocation is attracted to a fixed stress surfaces by its image, and repelled from a fixed strain surface. By making the calculation periodic, we have avoided any surfaces perpendicular to the glide plane. We can therefore allow the dislocation to propagate forever, without interference from artificial boundaries.

Therefore, we can apply now a shear stress to the dislocation in Fig. 3c and watch the resulting dislocation motion. As one would expect, the dislocation starting at rest slowly accelerates to a terminal velocity, which is close to the speed of sound in the material. The actual limiting velocity is determined by the appropriate sound velocity for the finite slab, which is dominated by the propagation of surface waves. The calculations shown were performed at $T = 0$ K, but can also be performed at finite temperature. In Fig. 4, we show the velocity of a dislocation at three stresses: 0.015%, 0.06%, and 0.24% of C_{44}. The data points are from the molecular dynamics simulations. The solid lines are a comparison to a continuum calculation that will be described next.

The isotropic elastic continuum equation of motion for a dislocation under stress at $T = 0$ K is (26):

$$\frac{d}{dt} \left(\frac{m^*v}{[\, 1 - v^2/c^2 \,]^{1/2}} \right) = \sigma b \tag{25}$$

where m* is the effective mass, c is speed of sound, σ is the external stress, and b is the Burger's vector. The effective mass cannot be determined by continuum theory, but is a material parameter (like the speed of sound). The effective mass parameter is related to the energy of a dislocation at rest, and also to its dynamics through Eq. 25. We can, however, determine the effective mass from atomistic calculations, either by directly

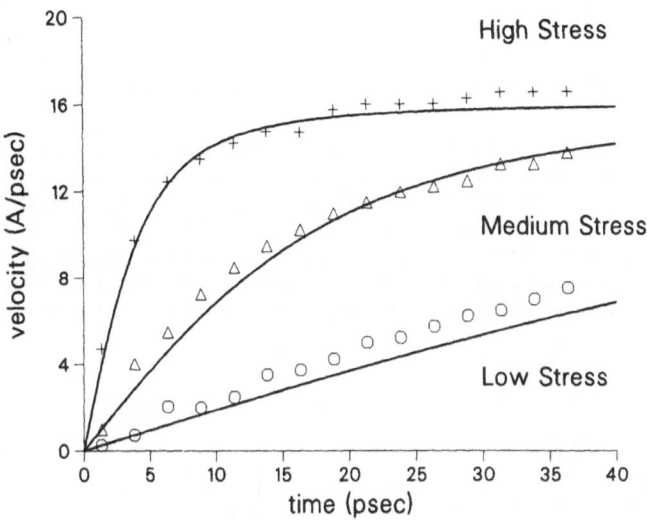

FIGURE 4. Edge dislocation velocity, in the slip direction [1 $\overline{1}$ 0], for the dislocation in Fig. 3c, under various shear stresses. The points are from the molecular dynamics calculations, and the solid lines are the fit from elastic continuum theory (see text). The low, medium, and higher stresses are 0.015%, 0.06%, and 0.24% of C_{44}. The dislocation accelerates to a terminal velocity, but in the lower stress cases the time illustrated in insufficient for the velocity to reach its terminal value.

calculating the energy of a dislocation at rest or by calculating the dynamics of the dislocation. The solution to Eq. 25 is

$$v = \frac{ct}{\sqrt{t^2 + \tau^2}} \qquad (26)$$

with $\tau = m^*c/\sigma b$. Now we can fit this solution to the results of the molecular dynamics simulations, where we have one parameter to fit to three functions. The fit determines the effective mass, (we get $m^* = 1.6 \times 10^{-3}$ eV-psec^2/Å^3) and results in the solid lines in Fig 4. The fact that the fit is good is quite encouraging, in that the atomistic and continuum calculations are consistent. The effective mass tells us that the inner cut-off on continuum theory (the point where local elastic continuum theory breaks down) is about 1 Burger's vector. A similar number is obtained by comparing to the rest energy of a dislocation, which gives $m^* = 3.9 \times 10^{-3}$ eV-psec^2/Å^3. In short, the continuum equation of dislocation motion (Eq. 25) is surprisingly valid.

    Now let us consider the effects of hydrogen on the dislocation. The first thing we try is placing the hydrogen at various positions around the dislocation. We find that the hydrogen is most strongly bound to the tensile region near a partial. The binding energy is relative weak, 0.05 eV. This number is consistent with the results of 0.05 eV from hybrid calculations of Baskes, Melius, and Wilson (27), and also with the 0.09 eV binding from thermal desorption measurements by Thomas (28). Relative to the migration energy of 0.41 eV, this agreement is quite good. The volume calculated for H by the present work is is 1 Å^3, consistent with length change measurements by Thomas and Drotning (29) of 1.4 Å^3. These values are consistent with the interpretation that the 0.05 eV binding is due to the purely elastic interaction of a 1 Å^3 dilatation center with the stress field of the dislocation.

We can make a more detailed comparison of the atomistic calculations to continuum theory by placing hydrogen at various points along a line intersecting one partial core. In Fig. 5, we compare the calculated atomistic energy to the continuum theory. Away from the core region, we can see that the continuum theory and atomistic theories agree quite well. The actual core region, where continuum theory becomes invalid, again appears to be on the order of one Burger's vector. EAM calculations show that the binding of hydrogen is a smooth function of position, having very simple behavior in the core.

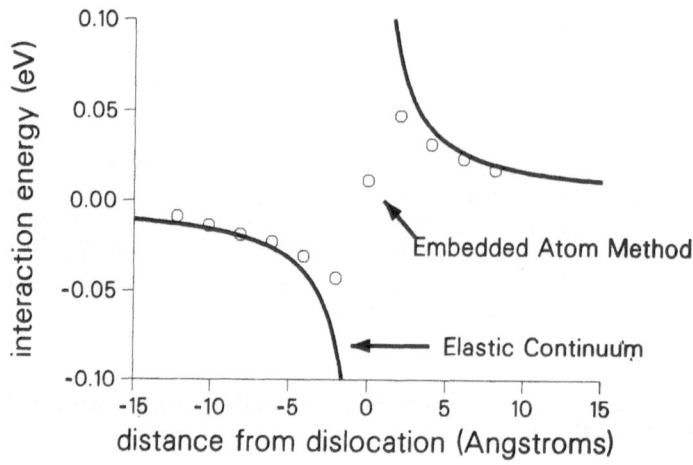

FIGURE 5. Binding energy of a hydrogen atom to an edge dislocation. The distance from the dislocation is taken along a line normal to the slip plane, passing through the core of one partial. The points are from atomistic calculations, and the solid line is the prediction of elastic continuum theory, given the dilatation volume of a hydrogen atom as calculated from atomistics.

To illustrate a contrast, we have also performed the calculation of the binding of He to a dislocation in Ni. Here we obtain a strong binding of 0.3 eV, which compares favorably to the hybrid calculations of Baskes, Melius, and Wilson (27), who obtain 0.3 eV. The volume of a He interstitial using the EAM is computed to be 4.5 Å^3 in Ni. Again, the binding energy is related to the volume, because the interaction of single He atom with a dislocation can be viewed in terms of a simple elastic interaction. (It has been known for some time that He will precipitate into bubbles in metals, and that it is very unlikely that a single He interstitial will exist for very long in the metal. But for illustrative purposes, we have considered on a single He atom in the current calculations.)

Therefore, to a dislocation, one helium atom looks like one big hydrogen atom, because the dislocation-impurity interaction is purely elastic. This is not true at higher concentrations of impurities!

We can now take the slab with the dislocation in it, insert an impurity such as H or He, and then apply a shear stress.

In the case of He, for stresses below values comparable to those used in the previous calculations for a free dislocation, the dislocation is pinned (the exact pinning stress depends on the concentration of He along the core). At the higher stresses where the dislocation is still pinned to the He atom, the dislocation bows out between pinning points. As the stress is increased beyond the breakaway value, the dislocation pulls away and accelerates. When a moving dislocation comes across a He atom, it is temporarily slowed down by the impurity, but the dislocation's inertia usually carries it beyond. The interaction of the He with the dislocation is entirely consistent with an elastic interaction between the two defects. The breakaway stress is consistent with the elastic binding energy over a region approximately one Burger's vector in size.

In practice, we find that He causes pronounced effects at the stresses which were used in the calculations shown in Fig. 4. Hydrogen at concentrations up to 10% at these same stresses appears to have no effect. We believe that this is because the elastic interaction of H with a dislocation is an order of magnitude weaker than that of He with a dislocation. We therefore expect the same effects for H, but at much lower stresses. We have observed the pinning of a dislocation in the case of 10% random H solution at very low stress (0.015% of C_{44}), but have not yet invested more computer time to the study of the breakway behavior.

We can make the following observations from the atomistic simulations:

1) We observe no drag of the interstitial by the dislocation. This is because the dislocation is moving in an otherwise perfect material, so that its velocity approaches the speed of sound in the material. The only means that a solute has to keep up with the glide motion of the dislocation is by diffusion, which even at room temperature is much too slow to match the high speed of a dislocation in a perfect material. Of course, in a real material, various imperfections pin the dislocation and lower its average velocity considerably, perhaps making it slow enough to carry along interstitials (23).

2) We observe no enhancement of velocity due to impurities. One possible explanation for the experimentally observed velocity enhancement (3) may be that hydrogen can elastically screen the dislocation from potential pinning points in the material, thus leading to a higher dislocation velocity. In our case, where we have no other pinning points present, this effect would not be expected.

3) The continuum theory of dislocation dynamics and dislocation-impurity interaction holds in Ni down to surprisingly short distances (i.e., about 1 Burger's vector).

In an attempt to carry the calculations one step further toward the goal of simulating the process illustrated in Fig. 2, we now consider a situation where crack tip plasticity is involved. Recall that the first calculations in this paper showed a geometry which was chosen to be brittle. This was accomplished by orienting the slip planes perpendicular to the applied forces, so that the shear stress resolved on the primary slip system was zero. In the following calculation, we rotate the crystal with respect to that slab geometry, so that now the slip planes are inclined at an angle to the applied forces and the resolved shear stress is near a maximum. This geometry then does not artificially constrain the fracture to be brittle. In fact, because the dislocations as they come out from the crack are affected by the free surfaces of the slab, it is likely that plasticity is facilitated.

Taking this new geometry and applying a tensile stress to the slab, we find that the crack begins emitting dislocations after an average time of about 1.5 psec. The stress is high (2% of C_{44}) and the temperature is 300 K.

Now we take the same slab, at the same temperature, apply the same stress for the same length of time, but add one hydrogen atom per 80Å of crack length near the crack tip. Again dislocations are emitted from the crack tip (Figure 1), but now in about half the time required without hydrogen. The first dislocation comes out near the hydrogen atom, but the successive ones have come out in other spots.

One should refrain at this point from making too hasty a conclusion about the effect of hydrogen on crack tip plasticity. Our primary concern is that if the dislocation emission is a thermally activated process, then the emission or lack or emission of dislocations is probabilistic in nature, and we have only one sample from each case. Therefore, we need to do many more runs to eliminate statistical variations. The results quoted in the previous paragraphs come from preliminary investigation of 10 runs in each case.

However, it does appear that hydrogen tends to make dislocation emission much easier, or at least faster. If indeed the process is activated, then the average time to emit a dislocation should follow an Arrhenius behavior

$$<t> = \tau\, e^{\frac{\Delta G}{kT}} \tag{27}$$

where ΔG is the energy barrier for dislocation emission. If hydrogen lowers this ledge energy, then indeed dislocation emission should be more abundant. This in fact brings us back to the calculations previously noted, where we saw that hydrogen atoms were indeed bound to ledges on surfaces relative to flat surfaces. By argument analogous to the one that led to the Griffith criterion, we can see that the reduction in ledge energy must make the work for dislocation emission lower, suggesting that the barrier is indeed lower.

We have seen from the present calculations that hydrogen can reduce the brittle fracture stress and the barrier for dislocation emission. It is not evident how these two effects compete in determining the overall, macroscopic appearance of hydrogen embrittlement. We do not yet understand, for example, the relationship between enhanced plasticity observed by Birnbaum and co-workers (3) and the occurrence of hydrogen embrittlement. This connection and other observations are discussed by Lynch (2).

CONCLUSIONS

In conclusion, we have shown that the EAM incorporates many physical properties of metallic systems that pair potentials cannot treat, and that the EAM is not significantly more complicated to use than pair potentials. Therefore, the EAM should replace the use of pair potentials in atomistic calculations of metallic systems.

We have also shown that hydrogen is bound to a surface relative to the bulk, leading to a reduction in the brittle fracture stress of a metal, in a manner very similar to local decohesion. We have shown that the hydrogen-dislocation interaction is elastic and weak, and that the relatively larger He interstitial can lead to dislocation pinning. We have further found some preliminary evidence that because hydrogen is bound to a ledge relative to a free surface, it facilitates dislocation emission from a crack tip.

It is true that the results of any single calculation in our review here can be strongly affected by boundary conditions. For example, the degree of plasticity of a crack tip propagating in a slab of the dimensions used in this work must be strongly influenced by the size of the slab. However, the *change* in the plasticity due to addition of hydrogen is a general tendency of hydrogen itself, related to its binding to surfaces and ledges. Therefore, the conclusions that hydrogen tends to weaken metal bonds and thereby lowers brittle fracture stress and facilitates dislocation emission are valid even for much larger, macroscopic systems.

REFERENCES

*Work supported by the Office of Basic Energy Science, Department of Energy.

1) M. S. Daw and M. I. Baskes, Phys. Rev. Lett. 50, 1285 (1983) and Phys. Rev. B29, 6443 (1984).
2) S. P. Lynch, J. Mater. Sci. 21, 692 (1986).
3) T. Matsumoto, J. Eastman, and H. K. Birnbaum, Scripta Met. 15, 1033 (1981). T. Tabata and H. K. Birnbaum, Scripta Met. 18, 231 (1984).
4) See papers contained in *Interatomic Potentials and Crystalline Defects*, edited by J. K. Lee (Metallurgical Society of AIME, New York, 1981).
5) W. A. Harrison, *Pseudoopotentials in the Theory of Metals*, (W. A. Benjamin, Inc., New York, 1966).
6) R. A. Johnson, Phys. Rev. B6, 2094 (1972).
7) M. I. Baskes, C. F. Melius, and W. D. Wilson, in *Interatomic Potentials and Crystalline Defects*, edited by J. K. Lee (Metallurgical Society of AIME, New York, 1981) (see Fig. 4 and discussion thereof).
8) M. S. Daw (in preparation).
9) See, for example, M. L. Plumer and M. J. Stott, J. Phys. C18, 4143 (1985).
10) K. W. Jacobsen, J. K. Nørskov, and M. J. Puska, (to be published).
11) M. Manninen, Phys. Rev. B (to be published).
12) S. M. Foiles, Phys. Rev. B32, 3409 (1985).
13) S. M. Foiles, M. I. Baskes, and M. S. Daw, Phys. Rev. B33, 7983 (1986).
14) M. S. Daw and R. L. Hatcher, Sol. State Comm. 56, 697 (1985).
15) S. M. Foiles and and M. S. Daw (submitted for publication).
16) M. S. Daw, Surface Sci. Lett. 166, L161 (1986).
17) S. M. Foiles and M. S. Daw, (in preparation).
18) T. E. Felter, S. M. Foiles, M. S. Daw, and R. H. Stulen, Surface Sci. Lett. 171 L379 (1986).
19) M. S. Daw and S. M. Foiles (submitted for publication).
20) S. M. Foiles, M. I. Baskes, C. F. Melius, and M. S. Daw, Proceedings of the International Symposium on the Properties and Applications of Metal Hydrides V, Maubuisson, France, May 25-30, 1986.
21) M. S. Daw, M. I. Baskes, C. L. Bisson, and W. G. Wolfer, *Modeling Environmental Effects on Crack Growth Processes*, ed. by R. H. Jones and W. W. Gerberich (Metallurgical Society of AIME, New York, 1986).
22) See, for a discussion, R. Thompson, "Physics of Fracture," in *Atomistics of Fracture*, ed. by R. M. Latanision and J. R. Pickens (NATO Advance Study Institute, Corsica, May, 1981, Plenum Press).
23) J. Donovan, Met. Trans. A7, 1677 (1976).
24) E. Kroner, in *Physics of Defects*, ed. R. Balian, M. Kleman, and J.-P. Poirier, (Les Houches 1980, Session XXXV, North-Holland, 1981).
25) See, for example, discussions and references in B. deCelis, A. S. Argon, and S. Yip, J. Appl. Phys. 54, 4864 (1983).
26) *Theory of Dislocations*, J. P. Hirth and J. Lothe (McGraw-Hill, New York, 1968).
27) M. I. Baskes, C. F. Melius, and W. D. Wilson, in *Hydrogen Effects in Metals*, ed. by I. M. Bernstein and A. W. Thompson, (The Metallurgical Society of AIME, 1980).
28) G J. Thomas, in *Hydrogen Effects in Metals*, ed. by I. M. Bernstein and A. W. Thompson, (The Metallurgical Society of AIME, 1980).
29) G. J. Thomas and W. D. Drotning, Met. Trans. A14, 1545 (1983).

DISCUSSION

Comment by H. Mughrabi:

Your calculations as presented here referred essentially to nickel. When I saw your results of the kinking of (partial) dislocations bowing out between pinning-points, it immediately occurred to me that similar calculations for metals with a bcc structure would be very interesting because of the special situation with regard to the different behavior of edge and screw dislocation components. Have such calculations been done? Are they feasible?

Reply:

We do have some functions representing bcc metals though we have concentrated on the fcc's. We have not looked at dislocation structure or dynamics in bcc's, however. The calculations are feasible; we just have not done them yet.

Comment by J. R. Rice:

In theoretical modelling of cleavage versus dislocation emission it is conventional to assume, conveniently, that the crack tip lies along the intersection of a slip plane with a crack plane. Do your calculations shed any light on whether cracks which are not initially oriented in that way tend to seek such orientations? For example, such effects might be seen for the case of a crack on (111), nominally advancing in the [110] direction, if there is a great enough repeat distance in the [112] direction (so there is enough flexibility for the crack tip to seek out a slip plane intersection).

Reply:

Nice idea. We will try it. In the meantime, please notice in our current calculations of crack tip plasticity, where the crack tip lies along the ideal intersection, that the periodicity in the [112] direction is rather long (~30Å). This is sufficient that the crack tip does not stay perfectly straight during the calculation. However, I have not sorted out the details, yet.

Comment by S. Altintas:

What computation time is required for a typical run (max. hr.)?
How many atoms can be taken into account?
What do you think about the size of the array (specimen) which would make your results comparable to real experiments? What results are size-independent?

Reply:

Currently, a simulation following 3,000 atoms for 10 picoseconds runs about one hour of CRAY-1 time. We are currently finding programming ways to speed this up, perhaps by as much as a factor of 5.
The time required is currently what limits our problem size. We may anticipate soon to be able to increase the size.

I think the only way to get around size effects is to border the atomistic region with a continuum of accurate response (say, a dynamical Green's function). The atomistic region for some problems may even then be required to be quite large if one wants to explain fully 3-D problems.

In our current calculations, we cannot conclude that real Ni should be ductile or brittle, because of size effects primarily. However, we <u>can</u> conclude that hydrogen lowers both the brittle fracture stress and the dislocation emission barrier.

Comment by K. Sieradzki:

When you orient a crack for slip in Ni, what happens when the system is shock loaded?

Reply:

I don't know. We have recently only loaded the system in a gradual way.

Comment by H. J. Engell:

1) What happens with the pair interaction if a hydrogen atom is in between the two pair atoms?

2) Hydrogen embrittlement happens normally at high surface concentrations of H. In your calculations, the H concentration was very low, even at the surface. Would an increase in H coverage qualitatively change your results?

Reply:

1) The argument presented was to demonstrate the general idea of metal-metal bond weakening by hydrogen. That is generally true regardless of the hydrogen position.

2) I expect no qualitative change in the results when the H coverage is raised. The weakening of M-M bonds still occurs.

Comment by D. J. Duquette:

While I'm sure that you cannot examine every binary system, the M-H system is one of the least understood. It would be interesting to examine M_a-M_b systems where M_a is a liquid metal which either embrittles, or does not embrittle M_b. This is especially true since a great deal of quantitative data relative to solubility, surface energy, etc., are known for binary metal systems.

Reply:

Stan Lynch has suggested that liquid metal embrittlement and hydrogen embrittlement are related by the fact that both proceed by enhanced dislocation generation due to a lowering of the ledge energy because of adsorption on the newly formed ledges. We do confirm the basic idea for the case of hydrogen. I agree that the question for liquids is also interesting.

Comment by I. M. Bernstein:

1. Can your calculations predict stress state changes in the vicinity of the crack?

2. Do you want to leave us with the impression that in brittle materials that embrittlement is a result (mechanistically) of a reduction in surface energy?

Reply:

1. Yes, but we have not looked at this yet.

2. I don't pretend to understand hydrogen embrittlement. I do believe that in totally brittle fracture (if such exists!), a reduction in fracture stress can be affected by a reduction in surface energy, as hydrogen can do. This is not a new idea.

Comment by D. D. Vvedensky:

One of the interesting aspects of your approach is that since you parameterize the calculation with various single-crystal bulk properties, you could identify those properties that most strongly influence the results. Therefore, having done this, you are in a position to predict mechanical behavior of a system by performing certain measurements on the corresponding single-crystal. Do you have any comments on this?

Reply:

This might be a good way of understanding certain tendencies in fracture. First, one must have a direct way of generating the functions from the material properties, rather than the indirect fashion that we now employ by fitting. R. Johnson at U. Virginia is doing exactly this first step. After that, it would be informative to modify individual material parameters to identify their affects on fracture. Perhaps this can one day lead to an improvement of the Rice-Thomson criterion.

Comment by D. D. Macdonald:

With respect to the dislocation/hydrogen interaction, what do you predict will happen when you replace H with D or T?

Reply:

Chemically, of course, the only difference is in zero-points. There is additionally (unrelated to the interaction with a dislocation) a difference in diffusion. We have estimated zero-point energies in Ni for H and they are about 0.05 eV in bulk, and in the binding energy, one is concerned with differences in zero-points for different sites, so I expect this to be a small correction for H and that H, D and T should have very similar binding energies to dislocations.

Comment by J. F. Knott:

You have drawn attention to the fact that both the work of fracture and the stress required to nucleate dislocations at a crack tip are reduced by the addition of hydrogen and that the net effect is therefore a "balance." It seems to me that any model which produces an <u>isotropic effect on bond</u> stiffness (i.e., reduces E and μ or C_{11} and C_{44} proportionately) is bound to give this result. What appealed to me about the Briant and Messmer model of S in nickel was that it suggested the possibility of <u>anisotropy</u>: strengthening planar p-orbitals in the S and weakening bonds normal to this plane. Is there any way in which similar anisotropy can be incorporated into the type of model that you have been using?

Reply:

I do not think that the weakening of M-M bonds by H, which leads to a reduction of brittle fracture stress and dislocation nucleation barrier, is unique to our calculation. I think the result is generally true of the M-H system. Please also be careful to draw the distinction between <u>moduli</u> and the forces given by very distorted bonds -- they are not necessarily tied together; this may be especially true for shear.

Comment by D. R. Baer:

We have now heard about pair potentials, embedded atom and tight binding methods of simulation. Some of the limitations of pair potentials have already been discussed. Can you compare strengths, weaknesses, computational speed, cluster or atom size and relation to reality of the different methods?

Reply M. Daw:

In general, the order in which you named the methods is also the ordering I would give for increasing reliability and decreasing speed. The way that the current methods stand, the size of the problem that is practical is limited by computer-time rather than memory size. The strength of the embedded atom method is that it is no more computationally intensive to use than pair potentials, but it incorporates important physics. The tight-binding bond method in turn incorporates more bond information than the embedded atom method, but at the cost of about (I believe) a factor of 100 in speed. Perhaps if this bond information is important, either the method can be speeded up, or the embedded atom method can be improved. I have yet to identify an area where the embedded atom scheme gives a clearly unreliable result (considering only metallic systems!). This is certainly something to watch for in our future work.

Reply by M. W. Finnis:

That is a wide question. Some of the limitations and comparisons are discussed in my written paper. Pair potentials now have little to recommend them since embedded atom potentials are about as fast to compute (or faster if the pair potential is a long range one). We expect embedded atom potentials to describe cohesive properties, but there may be situations where bond angles are significant, in which case more quantum mechanics, as in the TBB method, would be necessary. The TBB method itself would not work where there is interatomic charge transfer. Modeling energies are important, i.e., for ionic materials. We have not yet explored its accuracy very far, but there are probably limitations as we do not model orbital charge transfer on atoms. The cluster size limitations may be gotten around by elastic embedding procedures, but currently limit us to around 10,000 moveable atoms in the EAM and perhaps 100 in the TBB method, reflecting CPU constraints rather than storage. These numbers depend on the required number of iterations.

Comment by R. Thomson:

Your hydrogen enhanced dislocation emission result says that the hydrogen likes to produce corners. After a few (2-3) are made, the next is removed from the tip and the stress is lowered so that continued production of the dislocations should stop. Then I would imagine that cleavage would ensue. This suggests that the shape of the crack in hydrogen might be a mixed cleavage-emission one like that discussed by Neumann and Vehoff. Thus the effect of hydrogen may not simply be a shift of the brittle crack to a fully ductile one, but lead to a much more subtle result.

Reply:

Your observation is interesting, but I think that we have more work to do before making a conclusion about the effects of hydrogen on dislocation generation from a crack tip. The important conclusion now is that hydrogen can strongly affect the fracture process by the weakening of metal-metal bonds. What detailed path this takes, I am not yet sure.

Comment by H. K. Birnbaum:

Can you explain the observation that in the crack configuration which allows slip, additions of H seem to have a "non-local" effect? Addition of H seemed to cause increased dislocation generation at both ends of the crack.

Reply:

These results are very new and somewhat preliminary, so I cannot make a general statement about how dislocations are coming out of the crack. However, in the particular calculation shown, the first dislocation comes out near the hydrogen and subsequent dislocations come from both ends of the crack, which is about 10 Å long. Perhaps the first dislocation emission stresses the opposite end of the crack which then distorts as well. It is important to remember that plasticity occurs in our calculations even without hydrogen. Hydrogen seems to make the process occur sooner (with higher frequency?). But as I said, I hesitate to make general conclusions at this time because we must do a lot more work first.

Comment by R. M. McMeeking:

You have given us some insight into how good continuum theory is around dislocations in nickel. Can you make similar comments as to how good continuum theory is near the crack tip in your first (brittle) calculation?

Reply:

I haven't compared the atomistic crack tip to solutions from continuum theory. My guess would be that the non-linear region where continuum theory breaks down is rather small. I believe this is consistent with pair potential results of Sinclair and others.

Comment by A. S. Argon:

In your future calculations it would be useful to use longer cracks with proper border conditions prescribed by a continuum background field. In doing this, however, it is most desirable to use fraction border conditions which do not stifle the development of non-linear volume expansion near the crack tip. This benefit, I believe, overrides the artifact that a free surface will suck out a dislocation by its image effect. Displacement border conditions, on the other hand, very seriously stifle the development of non-linear processes near the crack tip.

Reply:

I do agree that displacement border conditions inhibit dislocation production, and we have avoided those. I also strongly agree that the question of boundary conditions is an important one and can influence the results of any single calculation. However, we believe that the effects of hydrogen can be demonstrated even though the boundary conditions may not be ideal. We believe that our conclusions about hydrogen are valid.

Comment by J. Hirth:

With regard to the influence of free surfaces, it would appear that their presence in the computation cell would greatly enhance dislocation blunting of a crack relative to the bulk case. This should appear to be so because a nascent dislocation would develop a nearby image that would attract the developing dislocation to the free surface.

Reply:

Your observation is correct. As I mentioned in the talk, I am not certain how to avoid the problem of dislocation-boundary interaction. But I do have some good ideas to try. In the meantime, I think the conclusions about the effects of hydrogen are valid anyway.

Comment by P. Neumann:

To get an answer whether dislocation emission or cleavage is occurring, care must be taken to provide an atomically sharp crack tip since the effect of H compared to other atoms, e.g., O, may depend strongly on the narrowness of the crack tip. Therefore, I would suggest to use an atomically sharp crack tip by starting with a reasonable distribution of atom positions of a crack and a finite stress. With approximately correct positions, the pulsing problem should be small.

Reply:

Sieradzki, Dienes, and Parkin (and co-workers) have suggested an alternative way of setting up the initial crack. At this point, we have not explored the crack-shape dependence on our results.

Comment by P. Neumann:

In the experiments by Vehoff and Klameth (Acta Met., 1985, p. 955) crack tip angle measurements as a function of H_2-pressure and temperature are evaluated to yield the binding energy and the pre-experimental factor of H bound to the crack tip. It is found that the energy is quite similar to that found by flash desorption from the free surface. The pre-exponential factor is, however, far off. Have you explanations for this?

Reply:

I will take a better look at your data. Off hand, I have no explanation.

Comment by S. M. Ohr:

Have you attempted to vary applied stress so as to see if there is a threshhold stress to emit dislocations from a crack tip?

Reply:

We have not yet had the time to make a systematic study of the dislocation emission as a function of stress and temperature and material properties. This is very important. One difficulty is that because the emission is a thermally activated process, there is always some finite probability for it to occur, but one may have to wait for impractically long times to detect the event. But we hope to overcome this difficulty. Also, we hope to understand the saddle point in more detail.

Comment by R. M. Latanision:

Regarding hydrogen in nickel, your calculations may be more relevant to the transmission electron microscopy observations of Howard Birnbaum -- in which crystallographic fracture is observed in nickel polycrystals which fails intergranularly when exposed to hydrogen. Can you treat intergranular fracture?

Reply:

Some work has been done on hydrogen at Ni grain boundaries, but we have not published this. This concentrated on finding the binding site and energy, but not on the intergranular fracture process. We do hope to pursue this more.

THEORY OF ENVIRONMENTAL EFFECTS ON TRANSGRANULAR FRACTURE*

K. SIERADZKI
Brookhaven National Laboratory
Upton, New York 11973

ABSTRACT

We discuss recent theoretical work on the transgranular stress-corrosion cracking of ductile metals. Computer simulations were used to study the behavior of cracks coated with thin solid films in the intrinsically ductile 2D triangular Lennard-Jones Solid. The key parameters studied were the elastic modulus mismatch and lattice parameter misfit between the film and substrate. It was found that elastically hard films could induce the nucleation of secondary brittle cracks in the normally ductile substrate. Films with a lattice parameter smaller than the substrate caused the crack to respond in a cleavage cracking mode, whereas films with a larger lattice parameter resulted in dislocation injection into the substrate. The concepts of film-induced cleavage and dynamic embrittlement are discussed. Molecular dynamic simulations were also used to study crack dynamics in the 2D triangular Johnson solid. Additionally we also present analytical calculations which explore the effects of thin film formation at crack tips on fracture. The parameters studied included elastic modulus mismatch, lattice parameter misfit and coherency. Suitable combinations of these parameters lead to film-induced micro-cleavage in ductile substrates. Finally crack dynamics are discussed and an estimation of film-induced micro-cleavage distances is presented.

I. INTRODUCTION

In this paper we consider mechanistic aspects of the phenomenon known as transgranular stress-corrosion cracking (SCC). In this phenomenon, owing to the action of an external environment, intrinsically ductile metals such as copper, α-brass, and austenitic stainless steels fail via a microscopic cleavage like fracture mode. In another paper in this volume Newman and Sieradzki[1] examine this and other experimental and electrochemical aspects of SCC. Here we shall review recent progress made in the understanding of this phenomenon based upon the concept of film-induced cleavage.

Over time scales relevant to the SCC process, most ambient temperature reactions between a metal and its environment are limited in extent to a surface or near surface effect. Hydrogen absorption is an obvious exception to this statement, however, in many systems hydrogen can easily be ruled out as a contributary cause to SCC.[2] The reaction product is usually in the form of a thin layer or film with differing chemical, structural, and mechanical properties with respect to the bulk metal

*This research was performed under the auspices of the U.S. Department of Energy, Division of Materials Sciences, Office of Basic Energy Sciences under Contract No. DE-AC02-76CH00016.

substrate. As examples, films may take the form of oxides, nitrides, chlorides or de-alloyed layers. De-alloying results when one element in an alloy is selectively removed leaving a porous layer enriched in the remaining constituents of the alloy. The size scale of the porosity is dependent upon many factors including alloy composition, corrosion potential (which provides the driving force for the selective dissolution process), surface energy, and de-alloyed layer thickness. Porosity resulting from selective dissolution can vary in size from 1 to greater than 10^3 nanometers. Alloys containing ~0.2 atomic fraction of less noble constituent form de-alloyed layers with nanometer scale porosity over relevant time scales (60 s) in SCC. The selective dissolution process may be understood within the context of the percolation model.[3,4] As discussed by Newman and Sieradzki,[1] the compositional dependence of SCC can be readily understood in terms of this model.

The major premise of this paper is that under appropriate circumstances thin films can induce brittle behavior on a microscopic scale of intrinsically ductile metals. We shall review results of recent molecular dynamic simulations[5,6] of film covered cracks which illustrate the strong manner in which films with suitable properties can modify behavior of intrinsically ductile materials. Film-induced cleavage is further supported by analytical calculations which define relationships which must exist between film and substrate in order for this mechanism of SCC to operate. The main film and interfacial parameters affecting material behavior are the following:

(a) lattice parameter misfit
(b) elastic modulus mismatch
(c) interfacial state of coherency
(d) degree of interfacial bonding
(e) geometric shape of film surrounding crack (determined by growth kinetics)
(f) porosity
(g) intrtrinsic brittleness or ductility of film.

We will show how suitable combinations of these parameters can induce cleavage like behavior of intrinsically ductile materials, i.e., a deformational mode transition. In this presentation we focus on parameters (a), (b), (c), and (g). Paskin et al.[5] have considered the effect of interfacial bond strength, and Sieradzki and Newman[2] have discussed the role of film geometry. The mechanical properties of porous materials and films are currently under active study in our laboratory.[7,8]

Following our discussion of microcleavage initiation we will consider the physical processes which allow the crack to propagate into the ductile substrate. Molecular dynamic simulations will serve to guide our discussion in this area. In this section we focus on crack arrest processes; other important aspects of the dynamic problem have been discussed in detail elsewhere.[2,6]

Finally the concepts of intrinsic brittleness and ductility require some elucidations. Conventionally these concepts are of an engineering origin and refer to some average measure of material performance. Examples of usage may involve partial or overall appearance of a fracture surface or strain to fracture in a tensile test. A related notion, cleavage, involves splitting along well-defined (usually low index) crystallographic planes. Cleavage behavior does not apriori preclude some degree of dislocation activity. For example Gilman[9] observed that considerable dislocation activity accompanied the cleavage of LiF crystals. In order to classify a material as intrinsically brittle or ductile in a self consistent manner we will employ the concepts of the analysis developed by Rice and Thomson.[10] Their analysis examines the stability of an atomically sharp crack against blunting by dislocation emission, and classifies a material as brittle if it can sustain an atomically sharp crack up to the Griffith load for that crack. Their analysis is essentially athermal and neglects dynamic effects. A material classified as intrinsically brittle by the Rice and Thomson analysis may display ductile behavior at some temperature above zero kelvin. Likewise any intrinsically ductile material would be expected to display brittle type behavior under severe enough dynamic loading conditions. Other parameters such as stress state may also be expected to influence intrinsic material behavior. Our usage of brittle or ductile film refers to the intrinsic behavior of the film material (isolated from the substrate).

II. COMPUTER SIMULATIONS

Here we review results of computer simulations directed toward investigating crack dynamics in a "brittle" solid and the behavior of cracks coated with a thin solid film in a "ductile" solid. The computer simulations were performed using the molecular dynamic technique described by Dienes and Paskin.[11]

In a series of simulations investigating film coated cracks,[5,6] the atoms are arranged in a two dimensional (2D) triangular lattice and interact via the 6-12 Lennard-Jones (L-J) potential

$$\phi_{ij} = \epsilon[(a/r_{ij})^{12} - 2(a/r_{ij})^{6}] \quad . \tag{1}$$

Here all energies are measured in units of the well depth ϵ and distances in units of the equilibrium spacing a. Forces (in units of ϵ/a) are calculated well beyond the distance 1.1a where the interatomic force is maximum and slightly less than the $\sqrt{3}$ a second nearest neighbor distance.

Crack dynamics in a nominally brittle solid was investigated using the Johnson potential[12] with the atoms arranged in a 2D triangular lattice.[13] Quantitative data on crack propagation will be reported on for the sample under either a constant load or a constant displacement boundary condition. Earlier work[14] reported on by our group at Brookhaven showed that a crack responds in a brittle manner in the 2D triangular Johnson solid. The Johnson potential is a polynomial potential up to and including a cubic term with the constants and ranges shows in Table I. The conversion factors and units for this potential and the 2D triangular solid are listed in Table II.

All calculations were performed at zero absolute temperature using the Verlet central difference method for solving Newton's equation of motion. The runs were performed on samples of central cracked geometry of various sizes. Runs on film coated cracks were performed using two

Table I. The Johnson Potential

Range (R_0)	Potential (ϵ_0)
$.7261224 \le r < .9172073$	$-8.7267265(2.6166387r-3.09791)^3+28.1179634-29.552166$
$.9172073 \le r < 1.1465091$	$-2.5402761(2.6166387r-3.115829)^3+4.969104r-6.2851$
$1.146509 \le r < 1.3147$	$-4.4311074(2.6166387r-3.066403)^3+4.8549901r-6.151563$

Unit of R_0 = 2.6166387 A.
Unit of ϵ_0 = 0.25163806 eV.

Table II. Conversion Factors and Units

1 Broken bond = $R_0/2$

1 Time step = $.02 \ R_0 \ (m/\epsilon_0)^{1/2}$

R_0 = 2.6166387 A

ϵ_0 = .25163806 eV

NBB = Number of broken bonds

L = Half crack length in units of R_0

L = 1/4 (2NBB + 1)

1 Broken bond/time step = $25 \ (\epsilon_0/m)^{1/2}$

M_0 = Young's modulus

ρ = Density of the material

Longitudinal sound velocity $v_L = (9M_0/8\rho)^{1/2}$

For our system (perfect-unstrained)

$M_0 = 60.13 \ \epsilon_0/R_0^2$

Mass per triangle = 1/2 m

Area of triangle = $(\sqrt{3}/4)R_0^2$

$\rho = 2/\sqrt{3} \ m/R_0^2$

$v_L = 7.6539774 \ (\epsilon_0/m)^{1/2}$

1 Broken bond/step = 3.266 v_L

different sized samples: one sample configuration consisted of 608 atoms arranged in 19 rows and the other configuration consisted of 2496 atoms arranged in 39 rows. The runs investigating crack dynamics were done using a sample consisting of 10704 atoms arranged in 79 rows.

The behavior of film coated cracks was investigated by examining the response of the system as a function of parameters characterizing the film-substrate composite. The parameters studied were: (a) elastic modulus misfit between film and substrate. (This is accomplished by varying the well depth in the L-J potential of the film with respect to the substrate.) (b) interfacial bond strength (This is accomplished by varying the well depth of the L-J interaction between film and substrate atoms at the interface.) and (c) lattice parameter misfit between film and substrate (This is accomplished by varying the equilibrium spacing of the film in the L-J potential with respect to the substrate.) Only the effect of parameters (a) and (c) are reported on herein. The interested reader is referred to the complete report[5] for a more detailed presentation of results. The results of these studies will be presented in the form of computer generated pictures of the lattice showing the near tip deformation and stress distributions.

The results of the crack dynamic studies on the Johnson solid are presented in the form of computer generated pictures of the lattice showing crack propogation and plots showing the change in crack length (number of new broken bonds, NBB) vs the time step (see Table II for conversion factors). A bond is considered broken when the bond length exceeds 1.3147 r/R_0 at which distance the potential to zero.

1.   Homogeneous solid

The results representing the behavior of the homogeneous 2D L-J triangular solid under constant load are displayed in Figs. 1a-1d. The material behavior is considered to be intrinsically ductile. Several dislocations have nucleated at the crack tip at time step 1100 (Fig. 1b) and have formed steps on the sample surface by time step 1650 (Fig. 1d). Similar results have been obtained on homogeneous L-J solids containing up to 10,000 atoms. The local force profile or distribution of "normal stresses" along the crack plane (row 9) developed during this run is shown in Fig. 2. The sigma up and down designation refers to the local forces acting on atoms - at equilibrium sigma up equals sigma down. During crack motion equilibrium is never attained so that the solid (σ_{up}) and dotted (σ_{down}) lines will in general not superimpose. The maximum force for a perfect triangular array of atoms interacting via the L-J potential is 4.66 ϵ/a. The maximum force (or stress) observed in this simulation is 4.1 ϵ/a at which point dislocation generation occurs. In the language of continuum mechanics, for the 2D L-J solid, the stress intensity for dislocation nucleation, k_d, (corresponding to a local force of 4.1 ϵ/a) is less than the stress intensity for brittle fracture k_G (corresponding to a local force of 4.66 ϵ/a). In the context of the R&T analysis the homogeneous solid would be classified as ductile. An interesting feature of the stress distribution shown in Fig. 2 is that even though the crack has widened by $\sqrt{3}/2a$, (by time step 1200, Fig. 1-c) the stress distribution at time step 1200 is similar to the stress distribution at time step 1000 which is prior to any dislocation nucleation and crack blunting. This interesting observation on the nature of geometric blunting formed the

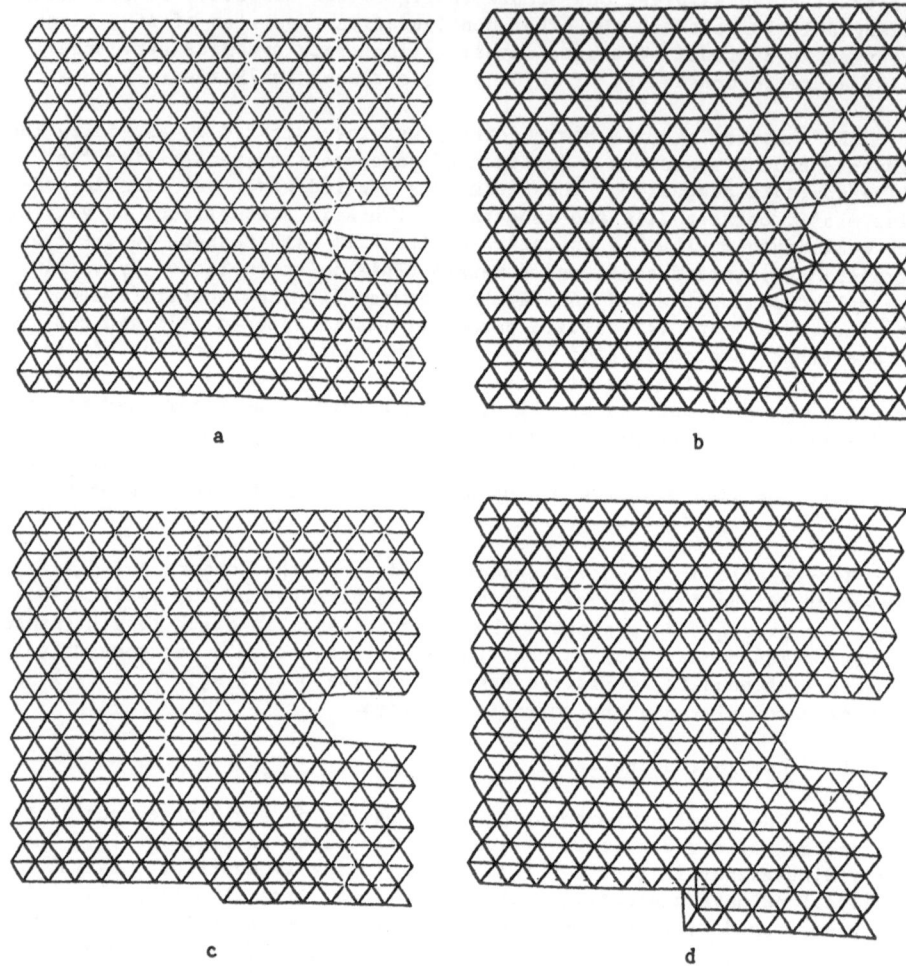

Fig. 1. Homogeneous sample (no film). Load raised from 1.90 to 2.09 at time step 1000. Time steps: (a) 1000; (b) 1100; (c) 1200; (d) 1650.

basis of a later study investigating the effect of atomic crack tip geometry on the near tip stress distribution.[15]

2. "Elastically hard" films

Figure 3 shows the results for a crack coated with an elastically hard film (film atoms are indicated by large circles). The film atoms interact with a well depth, ϵ_{film} such that $\epsilon_{film}/\epsilon_{substrate}$ = FF = 5. Dislocation generation and crack widening is completely prevented by the hard film. Crack nucleation and brittle propagation occurs in the matrix ahead of the film. The local force profile along the crack plane for this run is shown in Fig. 4. The theoretical maximum force that the crack tip bond can sustain is 5 (4.66)ϵ/a = 23.3 ϵ/a. As the applied

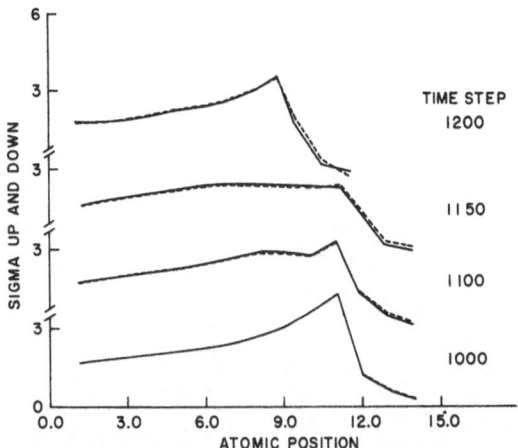

Fig. 2. Local force profiles for Fig. 1, the homogenous sample, on Row 9 at the time steps indicated.

load is increased the stress distribution is such that the theoretical maximum stress of the substrate (4.66 ϵ/a) is attained at the atomic position 10.0 and a crack is nucleated in this region. The double peak in the force profile apparent at time step 1525 is due to the secondary crack which has nucleated ahead of the main crack.

3. Films of negative misfit - $a_f > a_s$

The results for a crack coated with a film of three layers such that $a_f/a_s = 1.10$ is shown in Fig. 5. The misfit, f, is defined by

$$f = \frac{a_s - a_f}{a_f} . \qquad (2)$$

3a   3b

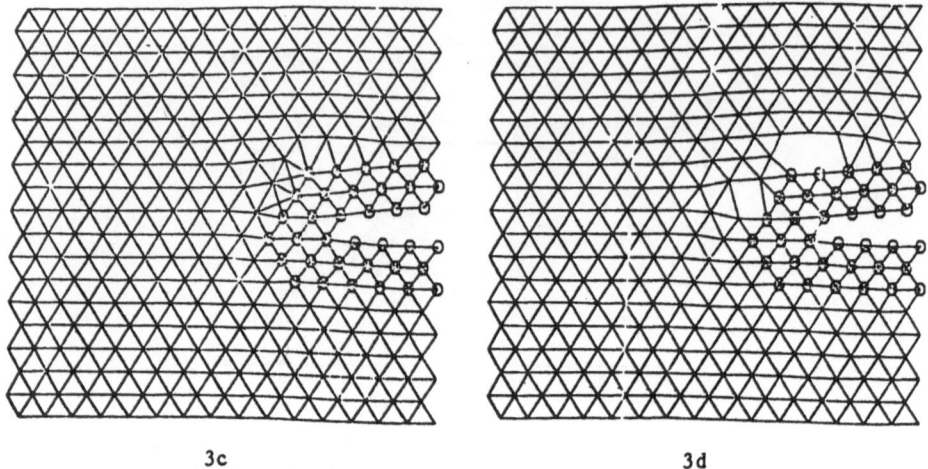

3c   3d

Fig. 3. Three layer film with FF = 5.0. Load raised
from 3.80 to 3.99 at time step 1300. Large circles
represent film atoms. Time steps: (a) 1475; (b) 1500;
(c) 1525; (d) 1550.

At an applied load of 1.5 (Fig. 5a) a dislocation has nucleated at the
film-substrate interface between rows 6 and 7. As the applied load is
increased other dislocations are generated. By Fig. 5d the dislocations
have moved to the surface of the sample. The increases in applied load
(Fig. 3b-3c) required for additional dislocation nucleation is a form of
strain hardening resulting from the crack-dislocation interaction. The

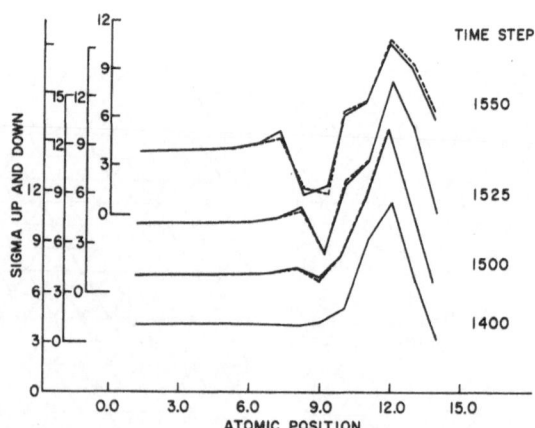

Fig. 4. Local force profiles on row 11 for
run shown in Fig. 3. The sample has a three
layer film (FF = 5.0).

dislocation motion is along a horizontal plane which is different from the usual slip plane in the solid (see Fig. 1). The plane of maximum shear has been altered owing to coherency stresses set up by the epitaxial film. This results in dislocation injection into the substrate at a lower value of applied load than that required for dislocation generation in the homogeneous solid (Fig. 1). The stress distribution for this run at an external load of 1.5 is shown in Fig. 6. Here the designations σ_{up} and σ_R refer to the normal stresses on planes parallel and perpendicular to the crack plane. The double peak in the profiles results from the crack tip and the "tip" located at atomic position 9 at the film-substrate interface. Owing to the coherency stresses induced by the epitaxial film,

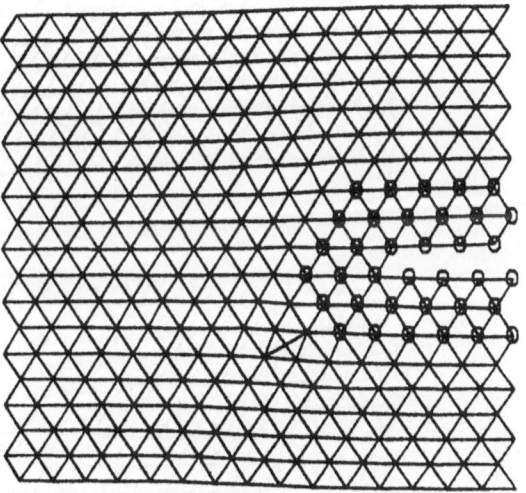

5a

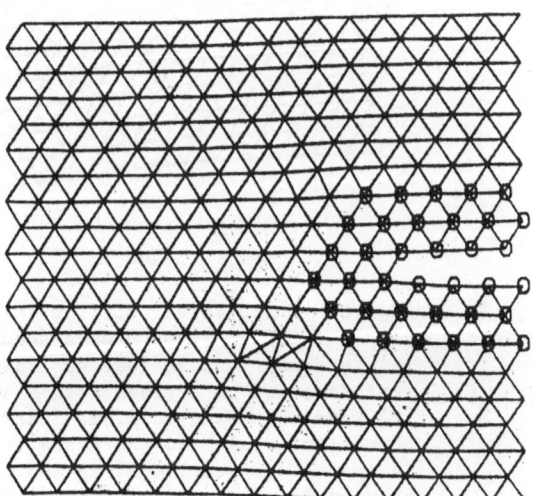

5b

228

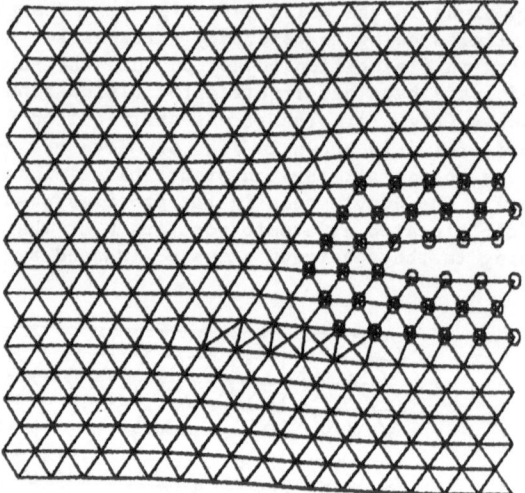

5c

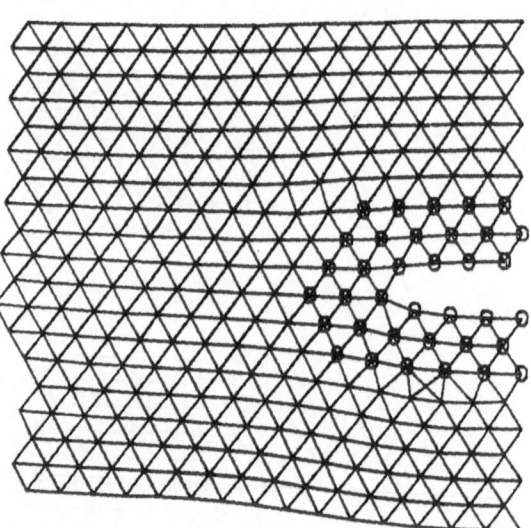

5d

Fig. 5. Three layers of film ($a_f=1.10\ a_s$)
with a crack of seven broken bonds. (a)
(a) External tensile load of 1.5. Incipient deformation between Rows 6 and 7.
(b) Load increased to 1.65. Increased deformation. (c) Load increased to 1.80.
Incipient dislocation along X-direction
between Rows 6 and 7. (d) Load increased
to 1.98. Dislocation moved out along X-
direction forming a jog.

the values of the normal stresses are significantly altered. σ_R is compressive while the stress normal to the crack plane, σ_{up}, is reduced. The difference between the normal stresses is proportional to the shear stress and is largest just on the substrate side of the film-substrate interface where dislocation nucleation occurs.

4. Films of positive misfit - $a_f < a_s$

The results for a crack coated with a film of three layers such that a_f/a_s = 0.95 is shown in Fig. 7. As the externally applied load is increased to 1.31 cleavage like crack propagation occurs. It must be emphasized that both film and substrate are composed of L-J material and are therefore intrinsically ductile as discussed earlier. The composite film-substrate system behaves in a brittle mode. Further studies have shown that the onset of brittleness is determined to a large extent by two parameters: (a) the degree of misfit, f, and (b) the film thickness. An interesting result of the simulation is that once cleavage is nucleated in the film, the crack continues to run in this mode through the substrate which under quasistatic loading shows (intrinsically) ductile behavior. Paskin et al.[6] have termed this effect as "dynamic embrittlement," and in the context of stress-corrosion Newman and Sieradzki[1] refer to it as "film-induced cleavage." In order to explore the response of the system (a_f/a_s = 0.95) to film thickness the sample was enlarged to a half sample size of 39 x 32 with a larger initial crack length of 15 broken bonds in the half sample. The results are summarized in Table III. All film coated cracks responded to applied loads by cleavage like propagation. The critical load for cracking decreased systematically as the number of layers is increased.

The force profile for the three layer film with a_f/a_s = 0.95 under no applied load is shown in Fig. 8. The dashed line is σ_R and the solid line is σ_{up} for atoms situated along the crack plane (row 9). The coherency stresses load the composite system such that the film is in a state of biaxial tension and the substrate is in compression. Figure 9 shows the stress distribution obtained for the situation described in

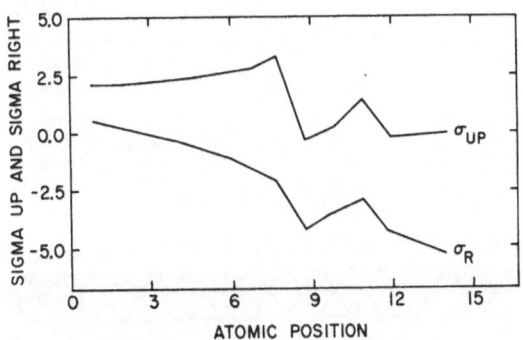

Fig. 6. Force distributions in the σ_{up} and σ_R directions along Row 9 for a sample containing three layers of film ($a_f = 1.10\ a_s$) and a crack under an external tensile load of 1.5.

230

Table III. Critical External Tensile Load Data

No. of Layers	Critical External Tensile Load (/d)
0	~2.1 (dislocation formation, not brittle)
1	1.76
2	1.65
3	1.30
4	1.21
5	1.18
8	1.05

Large sample - 39 x 32 for 1/2 sample
15 broken bonds in 1/2 sample
$a_f = 0.95\ a_s$

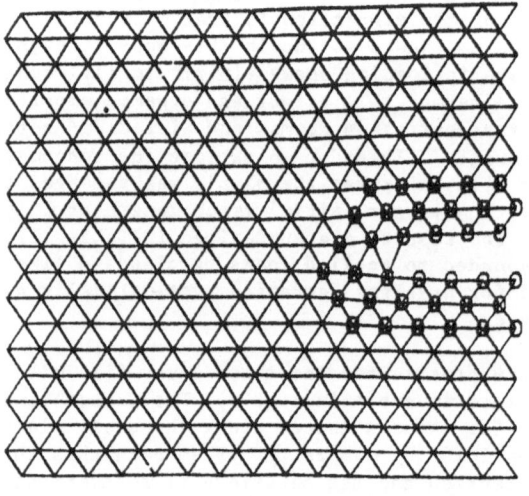

7a

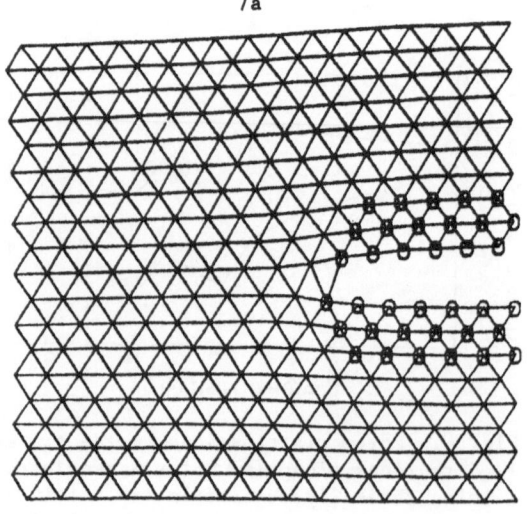

7b

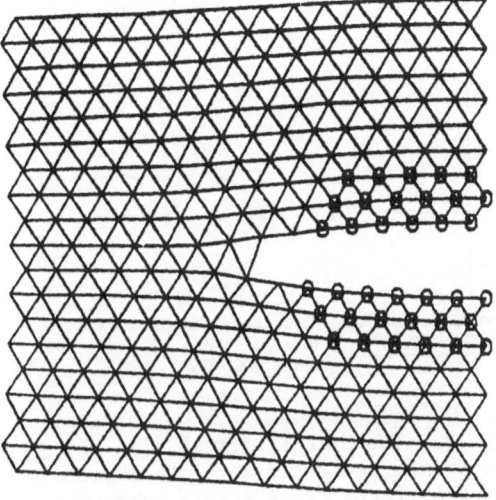

7c

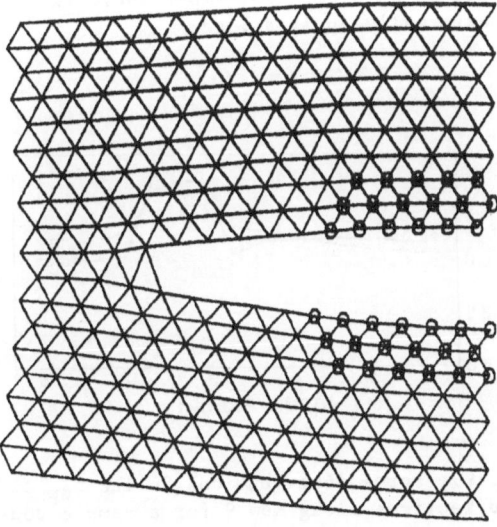

7d

Fig. 7. Brittle crack propagating with three
layers of film (a_f = 0.95 a_s). The external
tensile load was increased 1.28 to 1.31 at
at step number 400. Incipient breakage shown
in (a) at step number 600. The film bonds
are broken at step number 800. (b), cleavage propagation is shown at step number 1000
(c) and 1200 (d), with no indication of any
incipient dislocations.

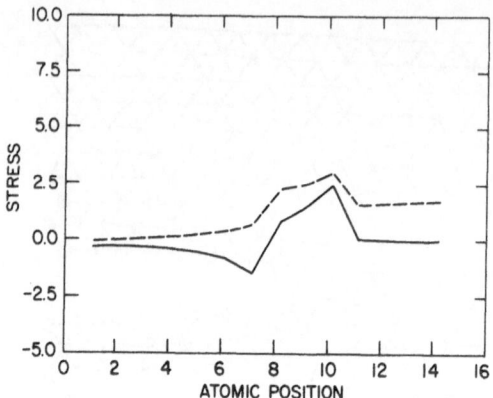

Fig. 8. The stress distribution along row 9 (the row defined by the lower crack surface) for a sample containing three layers of film $a_f = 0.95\ a_s$. The dashed line represents the horizontal stress and the solid line the vertical stress on an atom. Note that there is no externally applied load.

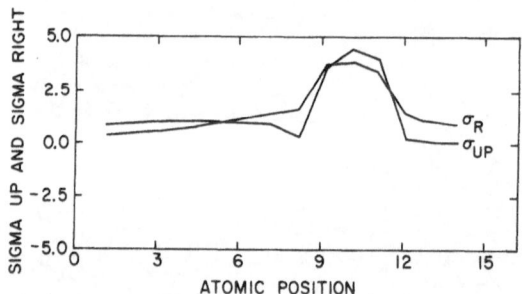

Fig. 9. Force distributions in the σ_{up} and σ_R directions along Row 9 for a sample containing three layers of film $a_f = 0.95\ a_s$ and a crack under an external tensile load 1.25.

Fig. 7 with an externally applied load of 1.25. There has been an elevation in the level of the normal stresses (compare to Fig. 8) to the point where the film bonds are about to break. At a slightly higher load (1.31) cleavage like crack growth is initiated. Note the low level of normal stresses in the substrate and shear stresses in the film.

All of the film covered crack simulation results are summarized in Table IV.

5. Crack dynamics

The dynamic crack velocity-crack length results obtained in the study for the Johnson solid are summarized in Fig. 10. The velocities are normalized to the longitudinal sound velocity, v_L, and the crack lengths normalized to the initial crack length, L_o. The constant displacement tests are designated by the level of nominal sample strain, e, and the constant load tests are designated by the level of the applied load, σ. "Fast" and "gradual" refer to the manner in which the samples were loaded. "Gradual" loading reflects a quasistatic loading procedure up to the critical Griffith value. In the "fast" loading proedure the applied load was increased from a value several percent below the Griffith load to a value several percent above the Griffith load. The resultant shock wave was allowed to damp out and then the crack was allowed to propagate. "Fast" load refers to a slightly overstressed system. Examination of Fig. 10 indicates that there is no significant difference in the curves either with respect to initial crack length or method of loading, i.e., constant load or constant displacement.

The "gradual" loading method resulted in a slower approach to the terminal velocity. In any case, the crack achieves terminal velocity by the time L/L_o is between 1.5 and 2. All data shown in Fig. 10 represents purely cleavage like propagation.

As crack propagation proceeded to values of L/L_o ~3 dislocation nucleation occurred at the tip of the propagating cracks studied. In Fig. 11, NBB is plotted against the time step for an initial crack of half length $L_o = 3.75$ propagating under a constant load of 3.37. At time

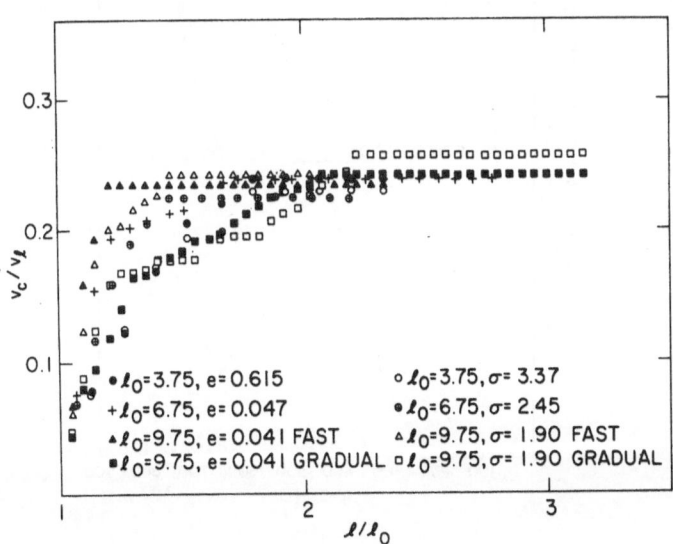

Fig. 10. Normalized crack velocity vs normalized crack length, v_L = longitudinal sound velocity, L_o = initial crack length (in half sample).

Table IV. Summary of Results of Molecular Dynamic Studies

	Ductile Behavior			Brittle Behavior	
(a)	$a_f = a_s$	Identical to homogeneous	(c)	$a_f < a_s$	"Cleavage" fracture
	$\epsilon_f = \epsilon_s$	L-J solid		$\epsilon_f = \epsilon_s$	
(b)	$a_f > a_s$	Dislocation nucleation in film-	(d)	$a_f = a_s$	Secondary crack nucleation
	$\epsilon_f = \epsilon_s$	substrate interface		$\epsilon_f > \epsilon_s$	

a_f - lattice parameter of film, a_s - lattice parameter of substrate
ϵ_f - elastic modulus of film, ϵ_s - elastic modulus of substrate

step 460, when 11 new bonds were broken ($L/L_o \approx 2.5$) a pair of dislocations was generated at the crack tip as shown pictorially in Fig. 12. At this instant the crack stops for about 200 time steps during which time the dislocations move away from the tip (Fig. 12 time steps 460 and 560). When the dislocations have moved far enough away the level of crack tip

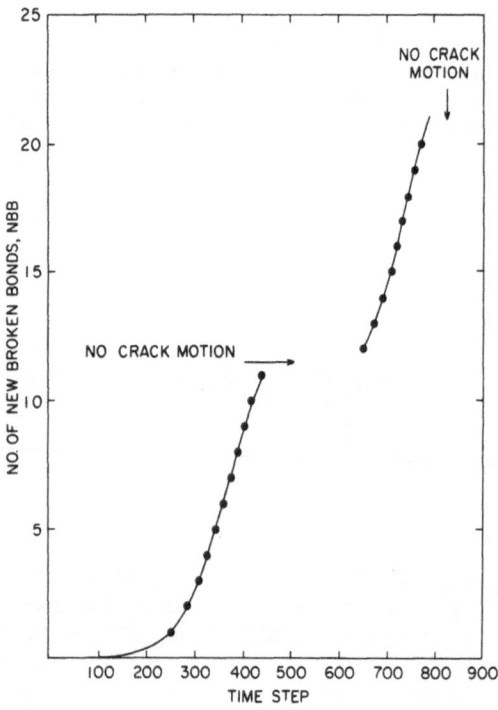

Fig. 11. Crack arrest caused by dislocation generation. Dislocation generated at time 460 , propagation renewed at 645, new dislocation generated at 801. Constant applied load of 3.37, $L_o = 3.75$.

shielding is reduced to a point so that by time step 645 crack propagation proceeds. Notice that the crack instantaneously adopts a velocity equal to the velocity it had just before it was arrested by the dislocation generation. Cleavage like propagation continues to NBB = 21, (time step 801) when a new pair of dislocations are generated and the sequence repeats.

These results will be discussed in conjunction with analytical calculations presented in a later section of this paper.

III. ANALYTICAL MODELING OF FILM EFFECTS

In this section we shall consider the manner in which thin films surrounding cracks can induce deformational mode transitions from the micromechanical viewpoint. The basis of our analysis is the approach developed by Rice and Thomson[10] (R&T) in their analysis of the stability of an atomically sharp crack in a solid against blunting by dislocation emission. Our approach is to examine the behavior of a crack which has been coated with a film of prescribed thickness. The film is character-ized by a set of parameters as discussed in the introduction. Here we consider the same parameters as those studied in the computer simulations discussed above.

1. Elastically hard films

An elastically hard film is simply defined as a film with shear modulus greater than that of the substrate. Consider a film free crack in a metal which under sufficient external loading is unstable to dislocation nucleation, i.e., the crack tip can spontaneously emit dislocations. If an elastically hard film surrounds the crack tip one expects the process of nucleating dislocations to be altered as discussed by Sieradzki.[16] According to the R&T analysis the following factors should enter into determining the nucleation criteria (a) the core cut off radius of a dis-location in the film, (b) the shear modulus of the film, (c) the surface energy of the film and (d) the surface or (interfacial) energy associated with fracture in the substrate. Here we shall assume that the film-metal interface is such so that it presents no barrier to dislocation transmis-sion. In the usual situation the interface will act as a nonrigid barrier to dislocation transmission and serve to strain harden the near crack tip region. This case is considered in section III-2 below.

As a simplification we consider a crack coated with a film loaded in longitudinal shear as illustrated in Fig. 13. The force per unit length of dislocation due to the applied load is

$$f_\sigma = \sigma_{\theta z} b_{screw} = \frac{K_{III} b}{(2\pi\rho)^{1/2}} \cos \theta/2 \sin\psi$$

$$= \left[\frac{2\mu_s \gamma_s b}{\pi \xi}\right]^{1/2} \frac{1}{\omega} \qquad (3)$$

where

$$G = 2\gamma_s = \frac{K_{III}^2}{2\mu_s}$$

$$\xi = \rho/b$$

and

$$\frac{1}{\omega} = \cos\theta/2 \sin\psi \quad .$$

Here $b_{screw} = b\sin\psi$, $2\gamma_s$ represents the work required to reversibly separate an internal interface, and μ is the shear modulus. The subscripts s
and f refer to substrate and film, respectively.

Rice and Thomson[10] showed that for the case of a crack, the
image force on a straight dislocation yielded the same result as for a
normal surface so we assume the situation is similar for the case of a
dislocation in a film surrounding a crack. The image force on a screw
dislocation in a film bounded by a substrate is given by

$$f_I = \frac{\mu_f b}{2\pi}\left(-\frac{1}{2\xi} + \frac{\xi}{h^2}\sum_{n=1}^{\infty}\frac{k^n}{n^2 - (\xi/h)^2}\right) \tag{4}$$

where

$$k = \frac{\mu_f - \mu_s}{\mu_f + \mu_s}$$

and h is the film thickness in units of b.

The force due to the ledge left behind during formation of the
dislocation results from the normal component of the Burgers vector to the
crack plane, and is given by

$$f_\ell = -\frac{2\gamma_f\alpha}{\pi\xi^2 + \alpha^2}\frac{1}{\omega'} \tag{5}$$

where

$$\alpha = e^{3/2} \, \xi_0/2 \quad,$$

$$\frac{1}{\omega'} = \sin\theta \, \cos\psi \quad,$$

and ξ_0 is the core cut off for the dislocation in the film. The critical distance, ξ_c, at which a straight dislocation is in unstable equilibrium is given by the condition $f_{total} = f_\sigma + f_I + f_\ell = 0$,

$$\frac{\mu_f b}{2\pi}\left[-\frac{1}{2\xi} + \frac{\xi}{h^2}\sum_{n=1}^{\infty}\frac{k^n}{n^2-(\xi/h)^2}\right] - \frac{2}{\pi\omega'}\frac{\gamma_f\alpha}{\xi^2+\alpha^2} + \left[\frac{2\mu_s\gamma_s b}{\pi\xi}\right]^{1/2}\frac{1}{\omega} = 0 \quad (6)$$

provided that $\xi_c < h$.

Figure 14 shows results of calculations for ξ_c using value appropriate for iron ($\gamma_{Fe} = 1.975$ J/m^2, $b_{Fe} = 2.49 \times 10^{-10}$ m, $\mu_{Fe} = 6.92 \times 10^{10}$ Pa) for different values of the ratio μ_f/μ_s and varying film thickness h. The core cut off distance chosen for the dislocation in the film was 2/3. The results indicate that for values of h greater than 2 and ratios of μ_f/μ_s greater than ~3.5 brittle behavior is predicted. As the value of $2\gamma_s$ is reduced the range of stability of an atomically sharp crack in the μ_f/μ_s vs h plot is increased. If weak interfaces exist in the solid in close proximity to the crack tip, elastically hard films would tend to promote fracture of the solid at such interfaces. This is the situation discussed by Sieradzki[16] for the case of intergranular embrittlement of a HSLA steel in a gaseous Cl$_2$ environment.

2. Ductile films

Suppose that the film which forms over a metal as a result of a corrosion process is intrinsically ductile. If such a film covers a crack surface, can the normally ductile deformation mode of the film be altered by the composite film-substrate system in such a way so as to result in the initiation of a cleavage cracking mode? The molecular dynamic simulations showed that under the right conditions this behavior is possible and here we examine the analytical basis for this. Our analysis indicates that under appropriate circumstances the film-substrate interface can serve to provide an extreme degree of strain hardening in the near crack tip region. The hardening results in a deformational mode transition, i.e., the crack undergoes a transition from an emitting crack to a cleaving crack. In general the film will have a lattice parameter different from that of the bulk substrate metal. The lattice misfit and film thickness combine to determine the nature of the interface and the coherency strains. In this section we ignore any effect of elastic modulus mismatch between film and substrate since these effects are probably relatively unimportant in materials not containing weak interfaces.

238

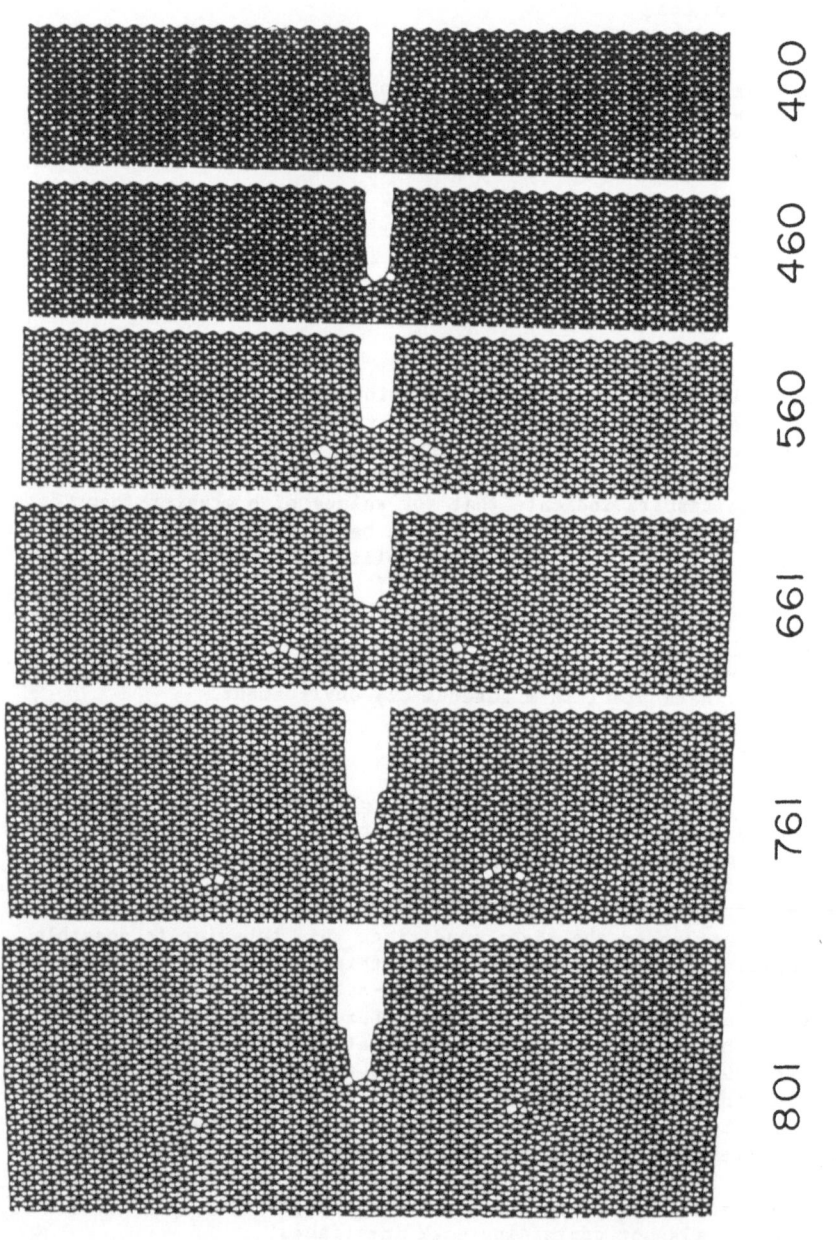

TIME STEP

400   460   560   661   761   801

Fig. 12. Series of simulation pictures (half samples shown) during the dynamic simulation of Fig. 11. The dislocation shows up as missing bonds as atoms move out of the short range of the potential.

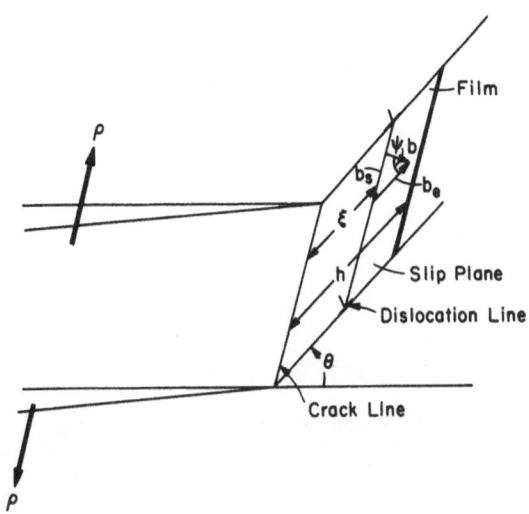

Fig. 13. Geometry of the dislocation crack
configuration loaded in longitudinal shear.

2a. Cleavage nucleation by coherent ductile films of positive misfit

Sieradzki and Newman[2] have discussed how a crack surface and a
planar surface differ in the manner in which coherency strains are accommodated. For $a_f < a_s$ coherency strains within the film are tensile while
on the substrate side of the interface the strains are compressive. This
situation allows the interface to serve as a barrier to dislocation motion. If the barrier is strong enough the emitting crack will be transformed to a cleaving crack. For coherent films of negative misfit, i.e.,
$a_f > a_s$, this situation is not possible, since the signs of the coherency
strains are reversed.

In general, coherent films of positive misfit will induce a
stress intensity, k_f, which will tend to open the crack. Sieradzki
and Newman[2] have shown that the induced stress intensity is given by

$$k_f = 2\mu_f \ (1-\nu^2) \ f \ t^{1/2} \tag{7}$$

where t is the film thickness. In order for the crack to experience
stresses induced by the film, t should be greater than the microscopic
crack tip opening displacement. The total stress intensity, k_T, acting
on the crack is given by

$$k_T = k_f + K + \mathcal{L} \tag{8}$$

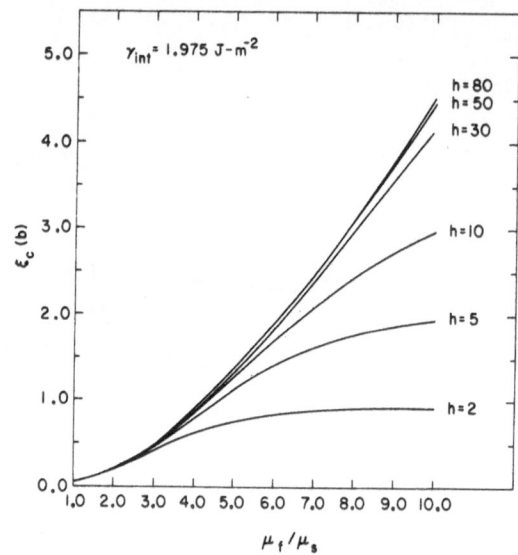

Fig. 14. Results of calculations for
$\gamma_{int}=1.975$ J/m^2. Values of ξ_c less
than 2/3 indicates ductile behavior.

where K is the stress intensity resulting from the applied loads and $\mathcal{K}$ is the stress intensity resulting from shielding effects. The fracture condition may be expressed as $k_T > k_G^f$ where k_G^f is the Griffith value of the stress intensity for the film. The main contribution to $\mathcal{K}$ comes from dislocation-crack interactions for the materials under consideration here. In the stress-corrosion problem there are four primary sources of dislocations:

(1) The grown in structure present in the annealed crystal. These dislocations are geometrically homogeneous and will not contribute to the shielding term.

(2) Dislocations produced at sources other than the crack tip. In general these dislocations will have shielding and antishielding components. If the sources are operated by crack tip stress fields, the antishielding portions of the dislocations are attracted towards the tip and the shielding portions are forced away from the tip. For sources located on the substrate side of the film substrate interface the antishielding dislocation will tend to agglomorate at distances on the order of twice the film thickness from the crack tip. The contribution of these to $\mathcal{K}$ is balanced in part by their shielding counterparts and in part by shielding dislocations as discussed in (3). Within the film itself the absolute number of active sources is low (owing to the thinness of the film in stress corrosion) so that this contribution to $\mathcal{K}$ can be neglected.

(3) In the next section we shall discuss the production of dislocations at the crack tip during the crack arrest process. These dislocations will dominate geometrically in the immediate region surrounding the crack which will become filmed owing to a corrosion process. These shielding dislocations will contribute most to $\mathcal{L}$ in eqn. (8). Their density is determined mainly by the dynamics of crack arrest and so their number will not vary substantially during static loading.

(4) The crack tip in the inherently ductile film serves as the dominant source of dislocations in the immediate vicinity of the tip.

We shall assume that the external loading is such so that k_T is maintained at a value equal to $k_G{}^f$ (eqn. (8)). The nucleated dislocations pileup at the barrier at a distance h (h = t/b) from the crack tip equal to the film thickness. The pileup exerts a force on an incipient dislocation opposing its nucleation. As the pileup grows it becomes increasingly more difficult for the crack tip to nucleate the next dislocation. Eventually the local k required to nucleate the N+1 dislocation is greater than $k_G{}^f$ and the emitting crack undergoes a transition to a cleaving crack. Lin and Thomson[17] have also discussed the occurrence of such a transition. Referring to Fig. 15 the number of discrete dislocations in the pileup, N, required to cause the deformational mode transition is given by the condition that the sum of the forces acting on the incipient N+1 dislocation is zero, i.e.,

$$F_{total} = 0$$

$$= k_G{}^f (b/8\pi\xi_j)^{1/2} \sin\phi \, \cos(\phi/2) \, \cos\psi \; - \mu b/4\pi\xi_j - \sigma_o b$$

$$+ \mu b/2\pi \sum_{\substack{i=1 \\ i \neq j}}^{N} (\xi_i/\xi_j)^{1/2} (\xi_j - \xi_i)^{-1} \; . \tag{9}$$

The equation is solved for N using an iterative alogarithm. The position ξ_1, (in units of b) that the first dislocation to be emitted from the crack takes up is equal to h, the film thickness unless friction is too great to allow the dislocation to reach the film-metal interface. The first term in eqn. (9) is the force on a dislocaiton at ξ_j resulting from the crack tip stress field. The second term is the image force on a dislocation at ξ_j and the third term is the frictional force. The fourth term is the force on a dislocation at ξ_j produced by all the dislocations at ξ_i, $i \neq j$. The force at $\xi_j = \xi_o$, the position of the incipient dislocation (assumed equal to the core cut-off distance in the film), is continually examined. When $F(\xi_o) = 0$ a dislocation is nucleated there and moves away eventually taking up a position determined by eqn. (9). Finally, after N dislocations have been nucleated $F(\xi_o)$ remains negative and no further nucleation is possible. In this analysis we have neglected

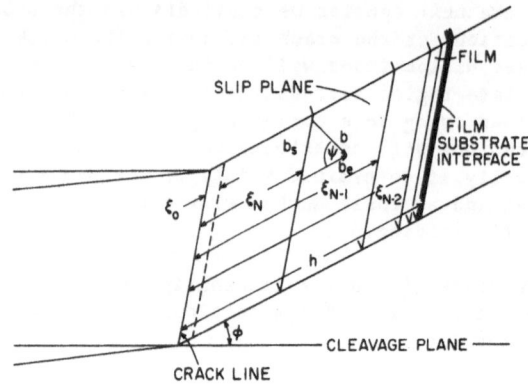

Fig. 15. The geometry of dislocation, crack
and film-substrate boundary.

the contribution of the ledge terms. Using values appropriate for α-brass
Sieradzki and Newman[2] determined as a function of h the number of dislocations, N, required to affect the deformational mode transition. The results are shown in Fig. 16. If N is greater than the interface barrier
can withstand, interface integrity breaks down and ductile behavior ensues. If the interface can support the pileup a cleavage crack is nucleated. The maximum number of dislocations that a film of thickness h and
misfit f can withstand is given approximately by[2]

$$N_{max} = f \, h \qquad\qquad\qquad\qquad (10)$$

Using eqn. (9) and (10) coherent film parameters required for microcleavage of α-brass are summarized in Fig. 17.

2b. Cleavage nucleation by incoherent films

Equilibrium considerations[18] indicate that the maintenance of
coherency becomes difficult as the product fh increases. Most films will
lose coherency with the substrate by forming a network of misfit dislocations. These misfit dislocations can harden the near tip region in a manner similar to dispersion hardening. This effect is possible for films of
positive or negative misfit. We shall examine the situation by assuming
that all the misfit strain is accommodated by a network of misfit dislocations in the interface.

The relevant question is whether or not a dislocation formed at
the crack tip can be transmitted through the dislocated interface a k values less than the critical Griffith value. The crack geometry is illustrated in Fig. 18. Our approach is to consider the balance between the
crack tip force and the line tension of the curved dislocation. We assume
that if the crack tip force is sufficient to allow a dislocation to reach
the saddle point (semicircular) geometry it will be nucleated and pass
into the substrate. Under these circumstances the interface may even
serve as a source of dislocations (via Frank-Read multiplication). On the

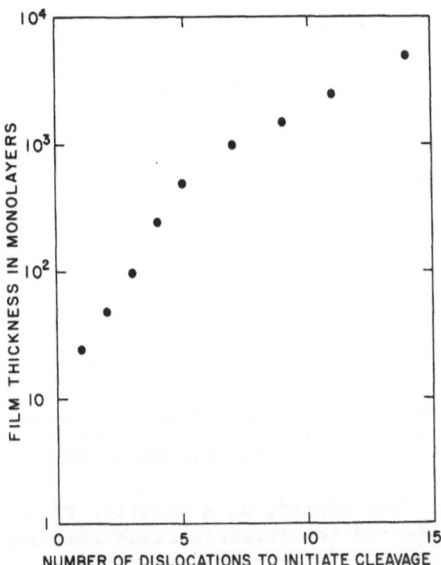

Fig. 16. The results of numerical calculations using
eqn. (9) and values appropriate for α-brass. The results indicate the number of dislocations required to
shift the balance from ductile to brittle behavior in
α-brass as a function of the thickness of a ductile
de-alloyed layer surrounding the crack tip.

other hand, if the value of k required to push the dislocation into the
saddle point configuration is greater than $k_G{}^f$ cleavage crack nucleation
is possible.

The crack is under mode I loading. The stress field produced by
the applied load yields a normal force on a dislocaton segment of length
ds given by

$$dF = (\sigma_{\rho\phi}b)ds$$

$$= \frac{1}{\beta} \frac{kb}{(8\pi\rho)^{1/2}} ds \quad , \tag{11}$$

where

$$\frac{1}{\beta} = \sin\phi \; \cos(\phi/2) \; \cos\psi \quad .$$

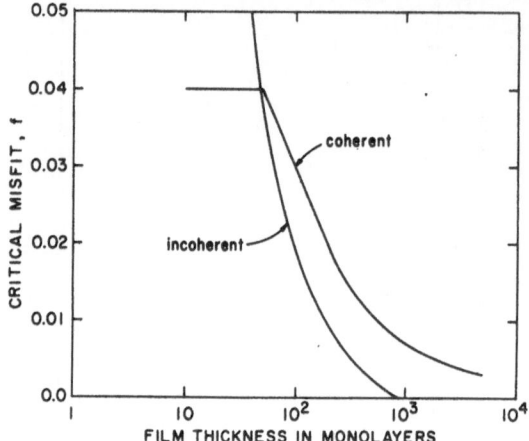

Fig. 17. The misfit, as a function of film thickness, required for cleavage crack initiation in coherent and incoherent ductile films.

The outwardly directed component of this force on the slip plane is

$$dF_\perp = \frac{1}{\beta} \frac{kb}{(8\pi\rho)^{1/2}} \; rd\theta \cos\theta \quad . \tag{12}$$

As shown in Fig. 18, $\rho = r \cos\theta + t$ and $r = \ell/2$ where ℓ is the spacing between misfit dislocations given by

$$\ell = \frac{a_f a_s}{a_f - a_s} \quad . \tag{13}$$

Using these substitutions eqn. (12) may be written as

$$dF_\perp = \frac{kb}{\beta} \left(\frac{\ell}{16\pi}\right)^{1/2} \frac{\cos\theta}{[\cos\theta + (2t/\ell)]^{1/2}} \; d\theta \quad . \tag{14}$$

Integrating we obtain

$$F_\perp = \frac{kb}{\beta} \left(\frac{\ell}{16\pi}\right)^{1/2} I(2t/\ell) \quad , \tag{15}$$

where the integral

$$I(2t/\ell) = \int\limits_{-\pi/2}^{+\pi/2} \frac{\cos\theta}{[\cos\theta + (2t/\ell)]^{1/2}} \, d\theta$$

is expressible in a standard form as a sum of elliptic integrals of the first and second kind.

The inwardly directed component of the line tension may be expressed as[19]

$$T = \frac{\mu b^2}{4\pi}\left(\frac{2-\nu}{1-\nu}\right)\ln\frac{8r}{e^2 r_0} \quad . \tag{16}$$

Combining eqn. (15) and (16) yields an expression for the value of k required to transmit a dislocation through a film of thickness t with an interfacial dislocation spacing ℓ:

$$k_D = \frac{\mu\beta}{I(2t/\ell)}\left(\frac{b}{\pi\lambda}\right)^{1/2}\left(\frac{2-\nu}{1-\nu}\right)\ln\frac{4\lambda}{e^2\xi_0} \quad . \tag{17}$$

Here $\lambda=\ell/b$. Owing to the $r^{-1/2}$ stress dependency, at fixed k and λ, the closer the interface is to the tip the easier it will be for the crack tip stresses to push a dislocation through the interface. Hardening is facilitated by small λ and large h. If $k_D > k_G{}^f$ cleavage crack nucleation is possible. We have evaluated eqn. (17) as a function of film thickness and misfit using values appropriate for α-brass and the results are shown in Fig. 17. Examination of this figure indicates that a very thin (<50 monolayers) ductile film on α-brass can initiate cleavage only if it remains coherent and has the required misfit. For thicker films a deformational mode transition may occur for proper combinations of misfit and film thickness regardless of the state of pseudomorphism between film and substrate. We assume that even if coherency is lost the integrity of the interface is such so that a high degree of bonding exists between film and substrate.

IV. DYNAMIC CONSIDERATIONS - CRACK ARREST

In previous sections we considered several ways in which films surrounding cracks could initiate cleavage like crack propagation in intrinsically ductile materials. In this section we consider the question, how far will the cleavage crack penetrate into the intrinsically ductile matrix? We believe that this question is inherently related to the ductile brittle transition which is known to occur for a number of materials as a function of temperature and strain rate. As a simplification we assume that the crack suffers no loss in velocity in traversing the interface.

246

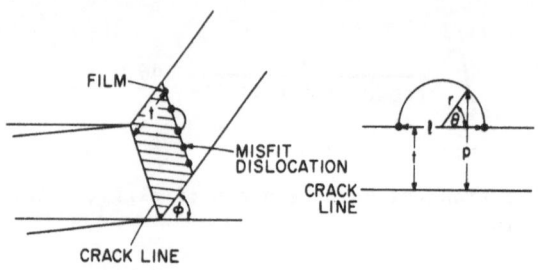

Fig. 18. The geometry of dislocation, crack and incoherent interface with misfit dislocations at a distance ℓ.

Consider the mode of cleavage crack propagation observed in the simulation illustrated in Fig. 12. Dislocation generation at the moving crack tip is concomitant with the cleavage process. Immediately following dislocation nucleation, the crack stops for an instant and as the dislocations move away the cleavage cracking continues. If this crack were loaded quasistatically it would continually emit dislocations and blunt. In the case of Fig. 12, cleavage cracking continues owing to purely dynamic effects. Even though the crack has stopped for an instant in time the local displacement rate of atoms in the vicinity of the crack remains extremely high - on the order of $0.2 v_L$. Thus, the crack is shock loaded and restarts once the dislocation-crack interactions permit. We believe the behavior described here may be generalized to any class of materials, i.e., under sufficient loading rates all materials should display brittle behavior.

Experimental attempts at observing such behavior may be fraught with difficulties. Consider a test sample configured for a high strain rate thought experiment. If the straining is inhomogeneous, the parts of the sample for which local displacement rates are much less than v_L can accumulate significant plastic damage. The plastic damage would tend to arrest any cleavage crack which may have nucleated in some other part of the sample. Any cleavage nucleated in intrinsically ductile materials via a bulk loading procedure would therefore be expected to be quite short lived. The observation of short cleavage cracks (<1 μm) in heavily deformed material seems an extremely difficult experimental task. In the case of film induced cleavage, only a small part of the sample is shock loaded (the material in the immediate vicinity of the tip) so that microcleavage may be more easily observed. Additionally, since the crack may be penetrating into low defect density material longer cleavage cracks may occur.

Crack arrest can occur for a variety of reasons:

  (a) restraining effect of unbroken ligaments
  (b) arrest at preexisting or deformation generated defects
  (c) crack bifurcation
  (d) exhaustion

The restraining effect of ligaments is generally accepted to be most significant in arresting macroscopic cleavage cracks.[20] For the micro-cleavage situation in SCC the effect of ligaments seems limited for two reasons:

(i) in some cases (e.g., α-brass) the ligaments are themselves extremely crystallographic and are only at most tens of nanometers in height.

(ii) Owing to time retardation effects, only a small portion of the ligament can assume a role in arresting the crack.

We expect that in some cases crack arrest will occur at defects such as slip bands. If this were the case, then as the defect density increases the cleavage or crack jump distance should decrease to an approximately constant distance limited by the maximum defect density attainable. Alternatively a crack traveling at high enough velocity may bifurcate, leading to rapid arrest.

As an upper limit to the crack jump distance in intrinsically ductile materials we have analyzed an arrest process resulting from exhaustion. Consider a dynamic crack which has just crossed the film substrate interface. It carries with it a kinetic energy,[21]

$$T = \left(\frac{\kappa \, \sigma_G^2}{2E^2}\right) \ell^2 v_c^2 \tag{18}$$

where ℓ is the crack length, ρ the density, v_c the crack velocity and κ an undetermined constant. The quasistatic assumption made by Mott in deriving this expression has been criticized by a number of researchers.[22] However, Freund[23] has remarked on the general validity of Mott's description - the only difficulty is in defining the value of the constant κ which determines the terminal velocity. Here we adopt a value of κ corresponding to the terminal velocity being the Rayleigh Wave velocity v_R. When the moving crack emits a dislocation it loses an amount of energy equal to[19]

$$E_D = \frac{Eb^2}{8\pi(1+\nu)} \, \frac{1}{\beta_2} \, \ell n \, \frac{r}{r_0} \tag{19}$$

where

$$\beta_2 = (1 - v_c^2/v_R^2)^{1/2} \; ,$$

r_0 is the core radius of the dislocation and r is the smallest relevant macroscopic sample dimension. Applying conservation of energy

$$\Delta(W - U) = \Delta(T + E_D + E_s) \tag{20}$$

where W is the work done during the crack jump by external loads, U is the strain energy stored in the system ($U = \pi\sigma_G^2\ell^2/E$) and E_s is the surface free energy term ($E_s = 2\gamma$). Henceforth we ignore E_s and also a force times distance work term done by the crack on the dislocation.[2] These terms contribute an energy per unit crack advance which will be negligible when compared to the contribution of eqn. (19).

We now argue that the work done during a microcleavage event by the external loading will always be identically zero. At the instant the crack starts to move in a cleavage mode elastic waves leave the tip and move out toward the external boundaries. The amount of time required for the external loads to sense crack motion is given by, $t = r/v_L \sim 10$ μs. The total crack jump time, t^*, is given by v_c/x where x is the crack jump distance. Assuming the microcleavage crack moves at speeds of at least $0.1\ v_L$, $t^* \sim 10$ ns for typical values of $x \sim 1$ μm. t^* is $\sim$3 orders of magnitude smaller than t which means that by definition W=0 for microcleavage. Similar arguements regarding microcleavage apply for the restraining effects of ligaments. Substitution into eqn. (20) yields the following expression for the change in crack velocity, δv, per charcteristic advance distance, $\delta\ell$:

$$\delta v_c = -\left[\left(\frac{v_R^2}{v_c} - v_c\right)\frac{\delta\ell}{\ell} + \frac{\mu b}{\gamma}\left(\frac{\pi b\ell_0}{(2\pi)^2}\frac{1}{\beta_2}\ln\frac{r}{r_0}\right)\frac{v_R^2}{\ell^2 v_c}\delta N\right] \quad (21)$$

where ℓ_0 is the original crack length and δN is the number of dislocations emitted during a crack advance of $\delta\ell$. We will assume that a single dislocaton is emitted per unit crack advance (equal to b). Equation (21) can be used to estimate the extent of a microcleavage event as a function of the initial crack velocity. We have solved this equation iteratively using values appropriate for α-brass, and the results are shown in Fig. 19 for an initial crack length $\ell_0 = 10^{-3}$ m. Equation (21) represents a correction to the previously published result[2] in which we inadvertently left out the U term in the energy balance (eqn. (20)).

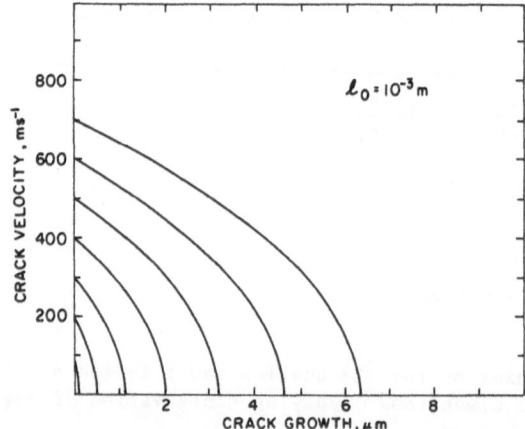

Fig. 19. The crack jump distance as a function of the velocity of the crack as it leaves the film. The distances we evaluated using eqn. (21).

REFERENCES

1, R. C. Newman and K. Sieradzki - this conference.

2. K. Sieradzki and R. C. Newman, Philos. Mag. 51:95 (1985).

3. K. Sieradzki and R. R. Corderman, in preparation.

4. R. R. Corderman, K. Sieradzki, and R. C. Newman, in preparation.

5. A. Pakin, K. Sieradzki, D. K. Som, and G. J. Dienes, Acta Metall. 30:1781 (1982).

6. A. Paskin, K. Sieradzki, D. K. Som, and G. J. Dienes, Acta Metall. 31:1253 (1983).

7. K. Sieradzki and Rong Li, Phys. Rev. Lett. 56:2509 (1986).

8. K. Sieradzki, J. Phys. C 18:L855 (1985).

9. J. J. Gilman, J. Appl. Phys. 27:1262 (1956); J. Metals 9:449 (1957).

10. J. R. Rice and R. Thomson, Philos. Mag. 29:73 (1974).

11. G. J. Dienes and A. Paskin in "Atomistics of Fracture," R. M. Latanison and J. R. Pickens, eds., Plenum, New York (1983), pp. 671-705.

12. R. A. Johnson, Phys. Rev. 134A:1329 (1964).

13. G. J. Dienes, K. Sieradzki, A. Paskin, and B. Massoumzadeh in "Modeling Environmental Effects on Crack Growth Processes," R. H. Jones and W. W. Gerberich, eds., Metall. Soc., Warrendale, PA (1986), pp. 85-98.

14. A. Paskin, B. Massoumzadeh, K. Sieradzki, and G. J. Dienes, Scripta Metall. 18:1135 (1984).

15. A. Paskin, B. Massoumzadeh, K. Shukla, K. Sieradzki, and G. J. Dienes, Acta Metall. 33:1987 (1985).

16. K. Sieradzki, Acta Metall. 30:973 (1982).

17. I. H. Lin and R. Thomson, Scripta Metall. 17:1031 (1983).

18. J. H. van der Merve, J. Appl. Phys. 34:117 (1963).

19. F. R. N. Nabarro, "Theory of Crystal Dislocations," Oxford Univ. Press, (1967).

20. G. T. Hahn, R. G. Hoagland, J. Lereim, A. J. Markworth, A. R. Rosenfield in "Crack Arrest Methology and Applications," G. T. Hahn and M. F. Kanninen, eds., ASTM STP 711-American Soc. for Testing and Materials, Philadelphia, PA, (1980) p 289.

21. N. F. Mott, Engineering 165:16 (1948).

22. F. Erdogen, "Fracture," H. Liebowitz, ed., Academic Press, New York (1968), p. 497.

23. L. B. Freund, J. Mech. Phys. of Solids 20:129 (1972), 20:141 (1972), 21:41 (1973).

DISCUSSION

Comment by D. J. Duquette:

I am concerned that you are overgeneralizing your model. For example, I understand the concept of dealloying of brass which you have proposed, but I cannot believe that every reaction product film is epitaxial, crystalline, and brittle. Even when they are, for example, for metals which show cracked oxides, at elevated temperatures, SCC does not generally occur.

Reply:

See the paper by Newman and Sieradzki in this volume.

Comment by W. W. Gerberich:

Following the dynamic pop-in from the film, the crack velocities you estimated for α-brass were about 200-700 m/s for crack jumps of 1 to 6 μm. Do the dislocations in this case pop back onto the free surface? Another way of asking this, is there a lower limiting velocity below which the dislocations will not be drawn back by the image force? It would seem that experimental probes of these surfaces, with regard to dislocation content, might help define the velocity at which the cracks ran into the matrix, if you could get such effects in a system with low bulk strains.

Reply:

I believe that many of the dislocations will be removed at the newly created crack surface by image forces. As the crack arrests, dislocations which run into the solid are produced in a cascade type of process. This dislocation distribution (structure) should be visible by various experimental techniques.

Comment by A. S. Argon:

As I understood your model, you are using the porous layer to inject a very fast crack into the copper. Yet a porous material is likely to have a rather low velocity of sound and therefore a low terminal crack speed which is likely to be too low to produce the "overshoot" effect that you are discussing. Can you comment on this?

Reply:

We are doing experiments to characterize the mechanical properties of the porous material (layer). That information will provide us with values of elastic modulus and fracture strength as a function of porosity. The low velocity of sound in the porous solid may not represent a severe limitation to the operation of dynamic embrittlement if the crack can reach a velocity ~0.2 of the Rayleigh velocity in the bulk solid.

Comment by R. H. Jones:

Your analysis of brittle film induced cleavage of ductile materials depended on the strain energy release rate of a solid film; however, your examples include cases with porous films. Is the strain energy release rate from these films sufficient to cause brittle cracking of a ductile material? It is very likely that the porous films are macroscopically brittle but each ligament is probably ductile and will fail with a small energy release rate.

Reply:

The rest of the solid (unfilmed material) will easily provide the strain energy to drive the crack. It is possible that on the scale of the ligaments (1-100 nm) the response may not reflect the behavior of the macroṣċopiċ solid. Each of the small ligaments may behave in a manner similar to a whisker.

Comment by M. O. Speidel:

Does your model give us some guidance as to which material-environment couples could be susceptible to transgranular stress corrosion cracking?

Would your model allow the calculation of a lower limit to the stress intensity (K_{ISCC}) which must be exceeded for stress corrosion cracking to occur?

Reply:

Dr. Newman is going to provide an answer to your question regarding material-environmental couples.

I'm not sure that there is a <u>measurable</u> K_{ISCC} in all systems. Minimum energy criteria must, of course, be met for crack growth; however, not all of that energy must come from applied loads.

Comment by R. M. Latanision:

I am concerned about treating the concept of dealloying/porous films on inducing SCC in ductile metals as a generality. I am prepared to accept that the concept may apply to Cu-Zn alloys, where dealloying is clearly observable. In the case of Fe-Cr-Ni alloys, however, is there observation of dealloying? Some time ago, Roger Staehle and I proposed a nickel-enrichment model of SCC in which we expected nickel to be enriched on a Fe-Cr-Ni surface due to selective dissolution of the more active Fe and Cr. We are never able to detect enrichment, either by atomic absorption analysis of the solution or Auger analysis of the solid surfaces. Do you consider this concept to be generally useful or specific to certain systems?

Reply:

Dr. Newman's group has recently obtained Auger data on Fe-Cr-Ni and Fe-Ni alloys showing selective dissolution of Fe (and Cr) and the development of Ni rich sponge. I believe that the concept is applicable to alloy systems composed of a more noble and less noble component elements.

Comment by B. D. Lichter:

Regarding the question of the brittle behavior of normally ductile porous solids, we offered a simple explanation in the case of gold sponges produced by reaction of a Cu-25Au single crystal in aqueous $FeCl_3$. This was based on SEM observation of the fracture surface of a failed sponge sample, which breaks at vanishingly small applied load. (A stereo pair will be shown in the presentation to be given in Monday's workshop). Observation of the fracture surface reveals that the fine gold fibers (~100 nm) have necked-down to fine points and have undergone characteristically ductile failure. Thus, failure of the sponge is "brittle" on a macroscopic scale, but it occurs by ductile overload of the fibers on a microscopic scale. As Roger Newman points out, the structure of the sponge near the alloy-sponge interface (~20 to 100 nm) may be quite different from that of a completely dealloyed sample, in that the sponge coarsens with time due to surface diffusion, or, alternatively, dealloying is incomplete at the interface, as our microprobe measurements indicate (B. D. Lichter, unpublished research). Thus, a crack propagating through such a structure may sharpen to atomic dimensions, allowing a large increase in stress concentration.

Solution Chemistry

Session Chairman: H. J. Engell

SURFACE CHEMISTRY IN AQUEOUS SOLUTIONS

Digby D. Macdonald
Chemistry Laboratory
SRI International
Menlo Park, CA 94025

ABSTRACT

A review is presented of the chemistry of surfaces in aqueous solutions, with emphasis on those process that contribute to stress corrosion cracking.   Accordingly, the thermodynamic criteria for passive film formation are discussed in order to identify conditions under which stable and metastable corrosion products form on the surface.   The electrical properties of both metal/electrolyte and oxide/electrolyte interfaces are also examined, and recent quantum mechanical modifications to the theories for the electrical double layer are discussed.   Passivity, passivity breakdown, and the kinetics of repassivation play important roles in determining the susceptibility of a metal or alloy to stress corrosion cracking, and these phenomena are discussed at some length, particularly with respect to recently developed deterministic models for chemically-induced breakdown.

INTRODUCTION

It is now widely accepted that stress corrosion cracking (SCC), corrosion fatigue (CF), and to a lesser extent hydrogen embrittlement (HE), are essentially electrochemical in nature.   In both SCC and CF, the cojoint action of an applied stress and an aggressive environment results in the propagation of a crack through a sensitized microstructure via dissolution of the matrix at the crack tip (1-7).   In the case of HE, atomic hydrogen enters the lattice via an electrochemical reaction (at least in the case of aqueous environments), is transported to the region of maximum hydrostatic stress in front of the crack tip, and subsequently embrittles the matrix resulting in discontinuous crack growth (8).

In all three cases, charge transfer reactions play an important role in determining the susceptibility of a metal to environment enhanced fracture (EHF). The importance of electrochemical reactions in crack propagation is readily gleaned from a consideration of the slip dissolution model that has been proposed by various authors (1-7) to describe SCC (Figure 1).   In this model, crack propagation occurs because of the cyclical occurrence of slip rupture of the film, dissolution, and repassivation. Additionally, the highly non-uniform current distribution that exists down the crack is maintained by the highly passive crack sides.   If this passivity is destroyed by, for example, the solution within the crack becoming too aggressive, then crack blunting will occur and the crack will cease to grow.

From these brief introductory remarks, it is clear that crack propagation involves a number of specific electrochemical and surface chemical issues. The most important of these include the dissolution of metals, the nucleation and growth of passive films, passivity breakdown, and the general properties of metal/electrolyte and oxide/electrolyte interfaces. The purpose of this review is to discuss these issues with respect to their implications for crack propagation. Because the topics are very broad, and have been subjected to extensive research for more than a century, the discussion presented here is not exhaustive, but is restricted to a consideration of general principles with reference to specific examples, as the need arises.

THERMODYNAMICS

Because of their very nature, metal/solution and oxide/solution interfaces exhibit discontinuities in many properties of interest in electrochemistry and corrosion science. The most important property, as far as charge transfer processes are concerned, is the electrical potential. This quantity, which is defined as the work done in bringing a unit positive charge from a field free point at infinity to the point of interest, provides the driving force for charge transfer reactions as well as defines conditions under which various reactions are spontaneous in the thermodynamic sense.

The role of potential in determing the chemical thermodynamic properties of a metal/solution interface is clearly evident from the potential/pH diagram for iron (Figure 2). This diagram is constructed by plotting equilibrium potentials for reactions between various species against the pH of the aqueous phase, thus defining regions in potential/pH space over which different phases are thermodynamically stable. These diagrams, which were first popularized by Pourbaix and his coworkers (9), have proven to be extremely valuable for interpreting experimental electrochemical data and for displaying the thermodynamic properties of complex metal/water systems in a relatively simple and convenient form. In recent years, E/pH diagrams for various metal/water systems (10-15) have been extended to elevated temperatures (25 < T < 300°C), in order to understand corrosion phenomena in water-cooled nuclear reactors, geothermal systems, and in geochemical environments. Many of these diagrams (e.g., Figure 3) have included complexing species, such as chloride and sulfide, frequently at high concentrations, in order to provide realistic descriptions of the environments of interest. Comparison of Figures 2 and 3 demonstrates that temperature has a significant effect on the thermodynamic properties of metal/water systems, such that metals tend to become more active as the temperature is increased, as reflected by the shift in the equilibrium potential for the Fe^{2+}/Fe reaction to more negative potentials.

As valuable as they may be, potential/pH diagrams provide little guidance as to how fast interfacial reactions occur. However, many authors (9) imply that the thermodynamic description does in fact contain kinetic information by labelling various regions of E/pH space as "corrosion" and "passivity". The use of this terminology should be discouraged. That this is so, is easily gleaned by noting that iron corrodes quite freely in concentrated hydrochloric acid but not in concentrated nitric acid, even though the pH of these two media are such that the systems are located in the stability regions of Fe^{2+} and Fe^{3+} (Figure 2). The explanation for this discrepancy in behavior lies in the fact that metastable oxide phases may form in the "corrosion" region, thereby passivating the surface. In the particular case of iron (Figure 2), Fe_3O_4 and Fe_2O_3 form metastabally at potentials above the extensions of lines 13 and 17, respectively. Thus, assuming that all of the phases that may form are included in the diagram, the unambiguous "corrosion" region is that bounded by line 23 and the extension of line 13 into the Fe^{2+} stability region. However, even in this case, the kinetic interpretation is not without difficulties, since adsorbed species (e.g, oxygen atoms or OH radicals) may confer passivity on the system without the formation of a bulk oxide or hyroxide phase.

With few exceptions (e.g., the noble metals), metals in contact with aqueous environments are thermodynamically unstable, particularly if the potential is relatively high due to the presence of oxygen or some other oxidizing species. For reasons evident in Figures 2 and 3, and because of the possible formation of metastable oxides, as discussed above, the surfaces of common engineering metals and alloys are covered by oxide, hydroxide (e.g., $Fe(OH)_2$), or oxyhydroxide (e.g., γ-FeOOH) films. These films, which frequently are no more than a few nanometers thick, are responsible for "passivity"; the condition whereby highly reactive metals exhibit kinetic stability in environments in which, thermodynamically, violent reaction should occur. It is the presence and the properties of these films that allow reactive metals, such as Fe, Cr, Ni, Al, Ti, etc., to be used in engineering structures, and it is interesting to note that our metals-based civilization rests on the continued durability of ultra-thin corrosion product films. The breakdown of passive films is discussed at some length later in this paper.

ELECTRICAL PROPERTIES OF INTERFACES

All corrosion processes, such as metal dissolution at the tip of an advancing crack or the discharge of hydrogen onto a metal surface, involve charge transfer reactions, in which electrons and/or ions move across regions of widely differing electrical potential. Since the driving force for such reactions is the potential difference or the potential gradient (i.e., electric field), it is evident that an understanding of the physico-chemical processes involved in charge transfer reactions relies heavily on a knowledge of the electrical properties of the interface. This subject has been researched for more than a century, and extensive reviews are available in the literature (16-25). Accordingly, only the very basic concepts will be presented here.

Models for the Electrified Interface

Because most experimental studies on the electrified interface have been carried out on noble metals, or at least on "film-free" metals such as mercury, the classical models of Helmholtz (Figure 4) and Gouy-Chapman-Stern (Figure 5) confine the entire potential drop across the interface to the solution side. Accordingly, the metal is regarded as being simply a "sea" of mobile electronic charge that is incapable of supporting any significant field. Such models predict that the electrical properties of a metal/solution interface are independent of the identity of the metal; however, this prediction is not bourne out by experiment so that the models depicted in Figures 4 and 5 must be regarded as gross oversimplifica-tions. Nevertheless, they have considerable merit because they describe in some detail the physico-chemical processes that occur on the solution side as the potential difference across the interface is varied.

The simple "double layer" model proposed by Helmholtz (Figure 4) is physically unrealistic, in that competing electrical and thermal effects are expected to establish a distribution of excess charge on the solution side of the interface. Accordingly, Gouy (27, 28) and Chapman (29), proposed a model that recognized the existence of a "diffuse" layer, in which the excess charge decreases with distance in a roughly exponential manner. The basis of this model is that the concentration of an ion at point x from the surface (see Figure 5) is described by Boltzman statistics

$$n_i = n_i^o \exp\left(-z_i \, e\phi/k_B T\right) \tag{1}$$

where z_i is the charge on the ion, e the electron charge, and ϕ the local electrical potential with respect to that in the bulk solution. The total charge per unit volume at point x is therefore given by

$$\rho(x) = \sum_i n_i z_i e \tag{2}$$

From Poisson's equation, the gradient in potential, i.e. the electric field, is derived as

$$\frac{d\phi}{dx} = \left\{\frac{2k_B T}{\varepsilon \varepsilon_o} \sum_i n_i^o \left[\exp\left(\frac{-z_i e\phi}{k_B T}\right) - 1\right]\right\}^{1/2} \tag{3}$$

For a symmetrical electrolyte (e.g., NaCl), equation (3) becomes

$$\frac{d\phi}{dx} = -\left\{\frac{8k_B T \, n^o}{\varepsilon \varepsilon_o}\right\}^{1/2} \sinh\left(\frac{ze\phi}{2k_B T}\right) \tag{4}$$

which shows that the electric field $d\phi/dx$ is related to the local potential by the hyperbolic sine function. Equation (4) is readily integrated to yield

$$\tanh\left(ze\phi/4k_B T\right) = \tanh\left(ze\phi^o/4k_B T\right) \exp(-\kappa x) \tag{5}$$

where

$$\kappa = \left(\frac{2n^o z^2 e^2}{\varepsilon \varepsilon_o k_B T}\right)^{1/2} \tag{6}$$

The parameter κ has units of reciprocal distance, and describes the rate with which the local electrical potential varies with x. For dilute solutions, and assuming that $\varepsilon = 78.49$ for water at 25°C, the characteristic distance is related to concentration by

$$\frac{1}{\kappa} = 3.04 \times 10^{-8}/zC \tag{7}$$

where C is the bulk electrolyte concentration in units of mol/ℓ. Equation (7) shows that, as the electrolyte becomes more dilute, the diffuse layer extends further and further into the solution.

The relationship between the charge density q^m (C/cm^2) on the metal surface and hence that in the solution phase, q^s, is readily derived as

$$q^m = -q^s = \left(8k_B T \varepsilon \varepsilon_o n^o\right)^{1/2} \sinh\left(ze\phi^o/2k_B T\right) \tag{8}$$

The differential capacitance, C_d, (F/cm^2) is also readily derived by differentiating q^m with respect to ϕ^o to yield

$$C_d = \left(\frac{2z^2 e^2 \varepsilon \varepsilon_o n^o}{k_B T}\right)^{1/2} \cosh\left(\frac{ze\phi^o}{2k_B T}\right) \tag{9}$$

The differential capacitance predicted by equation (9) passes through a minimum as a function of the potential ϕ^o. This behavior is qualitatively similar to that observed for a number of systems (see, for example, Figure 6) provided that $|\phi^o - \phi^o_m|$, where ϕ^o_m is the potential at the minimum, is not too large (< 200 mV). However, for larger potential excursions the simple V-shape predicted by equation (9) is no longer observed. Furthermore, the capacitances predicted by the Gouy-Chapman model generally are much higher than those observed experimentally. All of these observations suggest that the Gouy-Chapman model is not sufficiently sophiscated to account for many observed properties.

Stern (30) modified the Gouy-Chapman model to recognize that, over the dimensions of the diffuse layer, ions cannot be regarded as point charges. Ion size becomes even more important because real ions in solution are hydrated. Accordingly, the actual radii, which determine how close ions may approach the metal surface, are considerably larger than the crystallographic values listed in many sources.

The structure of a metal/electrolyte interface, as envisaged by the Gouy-Chapman-Stern model, is shown schematically in Figure 5. The inner or "compact" layer adjacent to the metal surface is defined by the plane of centers of hydrated ions which reside in contact with the surface; the region outside this layer corresponds to the classical Gouy-Chapman diffuse layer. Defining the local electrical potential at the compact layer as ϕ_2 and thickness of the layer as x_2, equations (3) to (6) may be modified in an appropriate manner to describe the electrical properties of the diffuse and compact layers as follows (26)

$$\tanh\left(ze\phi/4k_B T\right) = \tanh\left(ze\phi_2/4k_B T\right) \exp\left[-\kappa(x-x_2)\right] \tag{10}$$

$$\left(\frac{d\phi}{dx}\right)_{x=x_2} = -\left(\frac{8k_B T \, n^o}{\varepsilon \varepsilon_o}\right)^{1/2} \sinh\left(\frac{ze\phi_2}{2k_B T}\right) \tag{11}$$

$$\phi_o = \phi_2 - \left(\frac{d\phi}{dx}\right)_{x=x_2} \cdot x_2 \tag{12}$$

$$q^m = -q^s = -\varepsilon \varepsilon_o \left(\frac{d\phi}{dx}\right)_{x=x_2} = \left(8k_B T \varepsilon \varepsilon_o\right)^{1/2} \sinh\left(\frac{ze\phi_2}{2k_B T}\right) \tag{13}$$

$$C_d = \frac{\left(2\varepsilon\varepsilon_o z^2 e^2 n^o / k_B T\right)^{1/2} \cosh\left(ze\phi_2/2k_B T\right)}{1 + \left(x_2/\varepsilon\varepsilon_o\right)\left(2\varepsilon\varepsilon_o z^2 e^2 n^o / k_B T\right)^{1/2} \cosh\left(ze\phi_2/2k_B T\right)} \tag{14}$$

or more appropriately

$$\frac{1}{C_d} = \frac{x_2}{\varepsilon\varepsilon_o} + \frac{1}{\left(2\varepsilon\varepsilon_o z^2 e^2 n^o / k_B T\right)^{1/2} \cosh\left(ze\phi_2/2k_B T\right)} \tag{15}$$

Equation (12) assumes that the electrical potential varies linearly across the compact layer and that a sharp transition occurs in the ϕ vs x behavior at $x = x_2$. Neither of these assumptions is particularly well-justified. Nevertheless, they are made for lack of a more accurate description of how ϕ varies with distance over atomic and molecular dimensions. Furthermore, the point charge assumption is still retained in treating the diffuse layer, so that the model is not expected to be particularly valid at potentials close to ϕ_m, or at high electrolyte concentrations.

The form of equation (15) indicates that the ideally polarizable interface, (i.e., one at which no charge transfer occurs) may be represented by two capacitors in series

$$\frac{1}{C_d} = \frac{1}{C_H} + \frac{1}{C_{GC}} \tag{16}$$

where the first (C_H) corresponds to the classical Helmholtz layer and the second to the Gouy-Chapman diffuse layer, as expressed by equation (9). The capacitance C_H is independent of concentration and potential ϕ^o, but C_{GC} is clearly a function of both of these variables. The manner in which each component, and the total differential capacitance, varies with ϕ and concentration is shown schematically in Figure 7. Clearly, at high potentials (positive or negative with respect to ϕ_m) and/or concentrations the capacitance tends towards the Helmholtz limit, whereas at small values for $|\phi^o - \phi_m|$ and at low concentrations the differential capacitance is dominated by that of the diffuse layer, C_{GC}. Although the G-C-S model is successful in accounting for the deviation from the monotonic V-shape predicted by equation (9) alone, the fact that the limiting forms correspond to the Helmholtz and Gouy-Chapman models indicates that the Stern modification suffers from the same restrictions as do the two earlier models. This is particularly evident from the failure of the G-C-S model to account for the "fine structure" exhibited in Figure 6.

A number of papers (31-46) have appeared over the past twenty years, in which attempts have been made to account for the "fine structure" of differential capacitance vs voltage curves. Space does not permit a detailed discussion of these studies, but it is necessary to outline the general developments that have taken place. The reader is referred to the original literature for a detailed discussion of this subject.

Bockris, Devanathan, and Muller (31) have addressed the problem of the maximum in the capacitance vs potential curve that occurs on the positive side of the capacitance minimum, in terms of a solvent orientation model, which also allows for specific adsorption of anions at the interface, and for dielectric saturation of the solvent within the IHP. The general features of this model are shown in Figure 8. The assumptions made by these workers include the following: (1) Adsorbed cations with their primary hydration shells remain outside of the layer of strongly adsorbed solvent molecules. (2) The inner layer of solvent molecules exists under conditions of dielectric saturation, in which the dielectric coinstant approaches the Maxwell limit of n_r^2, where n_r is the refractive index. Accordingly, the dielectric constant of the inner layer is ~6. In the outer layer, which contains adsorbed cations, a somewhat higher dielectric constant (ca 30-40 compared with ~78 for bulk water) is proposed due to the partial orientation of water dipoles by the electric field. (3) Solvent molecules in the inner sheath (IHP) are assumed to be adsorbed in either one of two states; with the dipole pointing towards or away from the surface, depending upon the potential or charge on the metal. As the potential is changed, these dipoles may flip from one orientation to the other. This model has provided a semi-quantitative explanation for the following experimental observations: (1) The double layer capacitance at cathodic potentials is not sensitive to the size of the cation, at least for alkali and alkaline earth cations. This is so because the capacitance is dominated by the IHP, a region into which the cations are assumed not to penetrate. (2) The capacitance hump is attributed to effects that arise from the simultaneous change in the charge due to specifically adsorbed anions in the IHP and two-dimensional anion-anion repulsion. (3) Finally, the theory has been found to account generally for the shape of the capacitance vs potential (or charge) curve over a wide range of potential. However, the calculated behavior is sensitive to the values assumed for various parameters, and to the adsorption isotherm adopted for describing the specific adsorption of anions (and other species) at the interface. Nevertheless, the theory provides a physically realistic explanation for many of the anomalies that existed between previous theories and experimental observations, although it is apparent that other models (37-39) are also capable of explaining many of the same phenomena.

Over the past decade, a renaissance has occurred in the theoretical description of the electrified interface, particularly with respect to the physics of the inner Helmholtz layer region. The fundamental contribution of these more recent treatments is to recognize that the metal simple cannot be regarded as a two-dimensional "sea" of electrons. Instead, based on quantum and statistical mechanical arguements, various authors (40-46) have proposed that any comprehensive model for the electrical double layer must recognize the finite probability of finding electrons from the metal in the inner Helmholtz (IH) region, as indicated by the finite value for $\psi\psi^*d\tau$, where ψ is the electron wave function, ψ^* is its complex conjugate, and $d\tau$ is a volume element. This phenomenon is depicted schematically in Figure 9, in which the electron density (ρ), normalized to that in the bulk metal (ρ_m), is plotted as a function of distance away from the interface. The distribution of electrons is such that the metal surface is electron deficient, whereas the IH region is electron rich. In classical electrodynamics, this appears as a surface dipole oriented perpendicular to the surface, as shown in Figure 9. Because the penetration of electrons into the IH region increases as the potential of the metal is made more negative, the magnitude of the surface dipole is potential-dependent. According to Price and Halley (46), the capacitance of the compact layer (i.e., the IH region) is given by

$$\frac{1}{C_c} = 4\pi \ x_2(o) - \frac{32\pi^2 \kappa'}{\rho_m |e|} \cdot \sigma_m \tag{17}$$

where σ_m is the charge on the metal, $x_2(o)$ is the dimension of the IH region, and κ' is the characteristic penetration (units of reciprocal distance) of electrons into this region. They further argue that κ' can assume values of $1/8\pi$ or $1/3\pi$, depending upon whether or not a "slide" model or a "swing" model is assumed for the dependence of the electronic charge density on the interfacial field. Although the "jellium" model for a metal/vacuum interface indicates that this view is too simplistic (46), the assumptions apparently are qualitatively correct. Accordingly, the capacitance of the IH region becomes

$$\frac{1}{C_c} = \begin{cases} 4\pi \ x_2(o) - \dfrac{4\pi}{\rho_m |e|} \cdot \sigma_m & \text{(Slide Model)} \quad (18) \\[3ex] 4\pi \ x_2(o) - \dfrac{8}{3}\left(\dfrac{4\pi}{\rho_m |e|}\right) \cdot \sigma_m & \text{(Swing Model)} \quad (19) \end{cases}$$

A test (46) of these predictions, assuming that C_c dominates the interfacial capacitance in the neighborhood of the point of zero charge, is shown in Figure 10. Clearly, the model is in accordance with the experimental data for a number of metals in aqueous NaF solutions, at least at small charge densities. A more demanding test (46) is shown in Figure 11, in which the slope of $1/C_c$ vs σ_m is plotted against $4\pi/\rho_m$. In this case, the model is in only qualitative agreement with experiment. Nevertheless, the model successfully predicts that the interfacial capacitance depends upon the identity of the substrate metal and predicts the correct trends.

(2) Oxide/Solution Interface

For reasons noted above, most metals of engineering interest are covered by oxide or other corrosion product films when exposed to aqueous environments. Accordingly, any discussion of the interphase region must necessarily include an analysis of the corrosion product/solution interface. The physics of this interface are complicated by the fact that the corrosion product generally has electronic properties that differ markedly from those of a metal, and that the adsorption of "potential-determining" ions or the ionization of acidic surface groups play an important role in determining the electrical properties of the interface. Nevertheless, the principles outlined above remain valid and applicable to oxide/solution interfaces, but the models must be modified to take into account these additional phenomena.

The principal difference between a metal/electrolyte interface and an oxide/electrolyte interface, ignoring for the time being any semiconductor properties of the oxide itself, lies in the fact that an oxide surface generally is amphoteric in nature, and hence can adsorb potential-determining protons or hydroxyl ions, depending upon the pH of the aqueous medium. This adsorption process is best viewed as a surface acid/base reaction.

$$\underset{\overset{|}{\underset{OH_2^+}{|}}}{M} \quad \xleftarrow{ H^+ } \quad \underset{\overset{|}{\underset{OH}{|}}}{M} \quad \xrightarrow{ OH^- } \quad \underset{\overset{|}{\underset{O^-}{|}}}{M} \quad + \quad H_2O$$

Thus, in sufficiently acidic environments, the surface will be positively charged, whereas in sufficiently basic solutions a net negative charge will reside at the interface. At some pH between these extremes, the surface is electrically neutral; this pH is defined as the "point of zero charge (PZC)." It is apparent that the surface potential (ϕ^o), and hence the properties of the double layer, will depend to a large extent upon the thermodynamics of adsorption (or reaction) of the potential-determining ions with the surface. However, the potential ϕ^o may also depend upon the potential in the metal phase (ϕ^m), depending upon the electrical properties of the intervening oxide layer. In the case of highly resistive oxides, little dependence of ϕ^o on ϕ^m is expected, whereas in the case of a highly conductive oxide ϕ^o may depend principally upon ϕ^m.

Various models for the adsorption of potential-determining ions, notably H^+ and OH^-, at oxide-electrolyte interfaces have been reviewed by Wright and Hunter (47, 48) and by Hunter (49). The simplest, and the most commonly invoked, assumption is that first proposed by Gaudin and Fuerstenau (50), which states that the surface potential ϕ^o varies in a Nernstian manner with pH

$$\phi^o = (2.303 \ RT/F)(PZC - pH) \tag{20}$$

where PZC is the point (or pH) of zero charge. Berube and de Bruyn (51) pointed out that equation (20) is valid only when the chemical potentials of the potential-determining ions in the surface is constant. This condition may well hold for anionic oxides, but it is unlikely to be valid for a solid whose surface groups are only weakly dissociated at the point of zero charge, as noted by Wright and Hunter (47, 48). Levine and Smith (52) present a modification of equation (20), which takes into account the effect of surface charge, in addition to pH, in determining ϕ^o. These authors claim that their model yields reasonable simulations of experimental data for a number of oxides, but this is disputed by Wright and Hunter (47, 48), who point out that Levine and Smith did not specify several of the parameters used in their calculations.

A more thorough analysis of the surface potential at oxide/solution interfaces has been published by Wright and Hunter (47, 48), and the essential features of their treatment are given below. Consider the simplest case of a single ion adsorbing at specific surface sites, of which there are N_s per unit area. The electrochemical potential of the ion on the surface is given by

$$\tilde{\mu}_s = \mu_s^o + k_B T \ln[\{n/N_s - \Sigma n_i)\}\{(N_s - \Sigma n_i^o)/n^o\}] + ze\phi^o \tag{21}$$

where n is the number of adsorbed ions or molecules per unit area, Σn_i is the sum of all i species adsorbed, and ϕ^0 is the surface potential. Assuming ideal behavior in solution, the electrochemical potential of the ion in the bulk electrolyte phase is

$$\tilde{\mu}_b = \mu_b^0 + k_B T \ln(X/X^0) \tag{22}$$

where the electrical potential in the bulk has arbitrarily been set equal to zero, and the mole fraction of the solute in the bulk is designated by ^{v}. Selecting the standard state $\phi = 0$, $n = n^0$, and $X^0 = 1$, and equating $\tilde{\mu}_s$ and $\tilde{\mu}_b$ at equilibrium yields

$$n/(N_s - \Sigma n_i) = X \exp\left[-(ze\phi^0 + \delta)/k_B T\right] \tag{23}$$

where

$$\delta = \Delta\mu^0 - k_B T \ln\left[n^0/(N_s - \Sigma n_i^0)X^0\right] \tag{24}$$

Equation (23) is the Stern adsorption isotherm, that relates the number of ions adsorbed and the local potential at the surface. In the case of hydrogen ions, the isotherm becomes

$$\exp\left\{-(e\phi^0 + \delta_{H^+,s})/k_B T\right\} = (1/X_{H^+})\left\{n_{H^+}/(N_s - \Sigma n_i)\right\} \tag{25}$$

At the point of zero charge, equation (25) can be written as

$$\exp\left\{-(e\phi_z^0 + \delta_{H^+,s})/k_B T\right\} + (1/X_{H^+})\left\{n_{H^+,z}/(N_s - n_{H^+,z}\right\} \tag{26}$$

and if the parameter $\delta_{H^+,s}$ is assumed to be constant, rearrangement yields

$$\Delta\phi^0 = (2.303\ RT/F)\left\{pH_z - pH + \log\left(\frac{n_{H^+}}{n_{H^+,z}}\right) + \log\left(\frac{N_s - n_{H^+,z}}{N_s - n_{H^+}}\right)\right\} = \phi^0 - \phi_z^0 \tag{27}$$

where ϕ_z^0 is the surface potential at the point of zero charge. If only specific adsorption of H^+ and OH^- occurs in the compact layer, ϕ_z^0 is equal to zero. Furthermore, since $n_{H^+} \ll N_s$ equation (27) reduces to

$$\phi^0 = (2.303\ RT/F)\left\{pH_z - pH + \log\left(n_{H^+}/n_{H^+,z}\right)\right\} \tag{28a}$$

If a substantial fraction of the surface sites are occupied by hydrogen ions at the point of zero charge, then the variation of n_{H^+} with pH is small, and $n_{H^+} \sim n_{H^+,z}$. Accordingly, equation (28a) reduces to equation (20). Apparently, this is not a good assumption for weakly dissociated oxides, such as SiO_2 (47, 48), and presumably also the various iron oxides. More sophisticated models were developed by Wright and Hunter (47, 48) in which the number of occupied sites is a function of surface charge. These models are claimed to be more applicable than is equation (20) for the case of weakly dissociating oxides (47, 48).

Data for the point (or pH) of zero charge for various oxides are available in the literature. For example, Atkinson, Posner, and Quirk (53) report PZC values of 7.55 ± 0.15 for geothite (α-FeOOH) and 8.45 - 9.27 for hematite (α-Fe$_2$O$_3$), depending upon the method of preparation. Tewari and McLean (54) measured the PZC of magnetite (Fe$_3$O$_4$) in potassium nitrate solutions, as a function of temperature between 25°C and 90°C. The PZC was found to decrease from 6.55 at 25°C to 5.4 at 90°; a change of 1.55 units. One the other hand, the quantity 1/2 pK$_w$, where K$_w$ is the ionization product of water, changes by only 0.79 over the same temperature range. Accordingly, the variation in the PZC for Fe$_3$O$_4$ cannot be attributed to a change in the dissociation of water alone, contrary to the case for alumina (Al$_2$O$_3$), which was also studied by Tewari and McLean (54). Similarly, the change in the PZC of Ni(OH)$_2$, NiO, Co(OH)$_2$, and Co$_3$O$_4$ with temperature can be accounted for entirely by the change in pK$_w$, except when the PZC is measured from the primary equilibrium (i.e., from the change in pH upon suspension of the solids in the solution). In this case, the PZC decreases more with increasing temperature than does 1/2 pK$_w$. The fact that the PZCs for the three iron oxides (Fe$_3$O$_4$, α-Fe$_2$O$_3$, α-FeOOH) lie within the range of 6.55 (Fe$_3$O$_4$) to ~ 9 (α-Fe$_2$O$_3$) demonstrates that in acidic solutions the surfaces are positively charged, whereas in moderately basic solutions (pH > PZC) the interfaces are negatively charged.

The role played by surface charges in determining the adsorption of ions from aqueous solutions has been investigated by Tewari and coworkers (56, 57), and by others (58, 59). For example, the adsorption of Co^{2+} onto Fe$_3$O$_4$, Al$_2$O$_3$, and MnO$_2$ is found to increase markedly as the pH is increased through the PZC. However, the data do not support a simple adsorption or ion exchange model, but instead suggest that the adsorbing species is a hydrolysis product.

PASSIVITY

A comprehensive discussion of the growth of passive films on metal surfaces is beyond the scope of this paper, and several thorough treatments of the subject are available in the literature (60-62). Rather, we will discuss two important aspects of the problem that have a direct bearing on the phenomenon of stress corrosion cracking; transient growth or "repassivation" and chemical breakdown. The first is important because it describes events that occur subsequent to film rupture due to slip (Figure 1). Accordingly, the "repassivation" rate has an important influence on the crack propagation rate and crack morphology. The chemical breakdown of passive films is also a matter of considerable importance, because if the passive film on the crack walls spontaneously breaks down then crack blunting and hence crack arrest may occur. As noted above, a high aspect ratio crack will result only if dissolution is restricted to the crack tip and the sides remain passive in the acidic, chloride-containing environment that is known to exist within cracks propagating through metals in brines.

(1) Repassivation

The importance of the repassivation rate of a freshly generated surface in determining the crack propagation rate stems from the theoretical work of Hoar et al. (63, 64) and others (1-7). Subsequently, a number of authors have studied the electrochemistry of freshly generated metal surfaces, either by rupturing the existing passive film by rapid straining (65) or by scratching the surface with a hard stylus (66-75). The first of these techniques suffers from the fact that the exposed surface is poorly defined, so that it is difficult to quantify the results. However, the scratching technique also suffers from the fact that the trailing scratch is in a state of varying repassivation, so that a composite response is observed. Nevertheless, the extensive work reported by Burstein and coworkers (66-72) has provided great insight into the kinetics and mechanism(s) of repassivation of base metals and the subject is treated briefly below.

A typical current transient observed on scratching iron in 0.5 M $NaClO_4$, pH = 9.35, at E = -605 mV (NHE), as reported by Burstein and Ashley (66) is shown in Figure 12. Note that the stylus contacts the surface at t = 0 but is lifted at t = 0.8 ms. Accordingly, the decreasing transient observed for t > 0.8 ms is due only to repassivation of the base metal in the scratch. The decay in the scratch current, which reflects both dissolution processes and film formation, varies with accumulated charge, potential (Figure 13), electrolyte composition, and pH (Figure 14). The form of these plots indicates that the total current decays linearly with accumulated charge

$$i = k \left(1 - q/q_{max}\right) \tag{28b}$$

where k is a potential dependent rate constant of the form k = k' exp ($\beta FE/RT$). Burstein and Ashley (66) showed that at potentials below the Fe/Fe(II) reversible potential, the metal oxidizes by randomly forming FeOH across the surfaces to produce a monolayer of adsorbed OH radicals. At potentials above the Fe/Fe(II) reversible potential, a monolayer of Fe(II) oxide apparently forms; this too by random oxidation of iron atoms on the surface. An interesting and somewhat unexpected result of this work is that no evidence was found for separate nucleation and growth steps for the formation of corrosion product phases on the metal surface.

The transient described by equation (28b) is not universally observed. In fact, most transients appear to be of exponential or linear-log forms, sometimes involving powers of time (73). These somewhat complex empirical relationships no doubt reflect simultaneous dissolution and repassivation, which render a quantitative interpretation of the data difficult. Nevertheless, these studies have shown that the current density for the dissolution of a base metal can be very high (sometimes many A/cm^2), and that repassivation rates can vary over a very wide time scale. Both features have very important implications for stress corrosion cracking, since crack advancement is ultimately related to the dissolution charge passed per breakdown cycle.

(2) Chemical Breakdown

Chemically-induced passivity breakdown has been a subject of extensive research over the past century. Despite this effort, a complete understanding of the processes leading to breakdown has not been achieved, and a great deal remains to be discovered about this important phenomenon. The subject has been reviewed (76, 77) so that no attempt is made here to provide the reader with a comprehensive discussion of the subject. Instead, I will discuss the point defect model for passivity breakdown, which apparently provides the most quantitative interpretation of passivity breakdown that has been reported to date. This model, which was developed in our laboratory over the past five years (78-87), has now been extended to account for the effect of minor alloying elements, such as molybdenum, on passivity breakdown (84), and to describe the statistical nature of the breakdown event (85, 86).

The point defect model (78, 79) proposes that the breakdown of passivity occurs as the result of an enhanced flux of cation vacancies (J_{ca}) from the film/solution interface to the metal/film interface (Figure 15). This enhanced flux is envisaged to arise from the absorption of aggressive anions (e.g. Cl^-) into oxygen vacancies at the film/solution interface (Figure 15). If the flux J_{ca} exceeds the rate at which the cation vacancies are consumed at the metal/film interface (flux J_m, see Figure 15), the excess vacancies coalesce to form a cation vacancy condensate. When the condensate exceeds a critical areal size (ξ, mol vacancies/cm^2), the film becomes unstable and breakdown occurs. This site may or may not repassivate, depending upon the prevailing conditions.

Assuming quasi steady-state conditions, passivity breakdown (79) occurs when

$$(J_{ca} - J_m)(t_{ind} - \tau) > \xi \tag{29}$$

where τ is a constant. By expressing J_{ca} in terms of fundamental thermodynamic parameters, by assuming that J_m is constant, and by supposing that the enhanced concentration of cation vacancies is due to anion absorption/Schottky pair reaction at the film/solution interface, equation (29) leads to the following expressions (79) for the breakdown voltage (V_c) and the induction time (t_{ind}) for a single nucleation site

$$V_c = \frac{4.606\ RT}{\chi\alpha F} \log\left(\frac{J_m}{J^o\ u^{-\chi/2}}\right) - \frac{2.303\ RT}{\alpha F} \log(a_x) \tag{30}$$

$$t_{ind} = \xi'\left[\exp\left(\frac{\chi\alpha F\Delta V}{2RT}\right) - 1\right]^{-1} + \tau \tag{31}$$

where

$$J^o = \chi KD\left[N_v/\Omega\right]^{1+\chi/2} \exp(-\Delta G_s^o/RT) = \hat{a}\ D \tag{32}$$

$$K = F\epsilon/RT$$

$$u = \left(\frac{N_v}{\Omega}\right)\exp\left[\frac{\Delta G_{A-1}^o - F\phi_{f/s}^o}{RT}\right]\exp\left[\frac{-\beta F pH}{RT}\right] = \hat{b}\ \exp\left[\frac{-\beta F pH}{RT}\right] \tag{33}$$

$$\xi' = \xi / J^{\circ} \, \bar{u}^{\chi/2} \, (a_x)^{\chi/2} \, \exp \left(\frac{\chi \alpha F V_c}{2RT} \right) \tag{34}$$

and

$$\Delta V = V - V_c \tag{35}$$

D is the cation vacancy diffusivity, χ is the oxide stoichiometry ($MO_{\chi/2}$), N_v is Avogadro's number, Ω is the mole volume/cation of the film, ε is the electric field strength, ΔG_S^o is the change in standard Gibbs energy for the Schottky pair reaction

$$\text{Null} \rightleftharpoons V_M^{\chi'} + \left(\frac{\chi}{2} \right) V_O^{\cdot\cdot} \tag{36}$$

ΔG_{A-1}^o is the same quantity for the absorption of aggressive anions into oxygen ion vacancies at the film/solution interface

$$V_O^{\cdot\cdot} + X^- \rightleftharpoons X_O^{\cdot} \tag{37}$$

a_x is the activity of anion x in the solution, and α and β describe the dependence of the potential drop across the film/solution interface on the applied potential and pH, respectively (78)

$$\phi_{f/s} = \alpha V + \beta pH + \phi_{f/s}^o \tag{38}$$

Equation (30) shows that the critical voltage for a single breakdown site becomes more negative as the halide activity is increased, with a functional dependence of $V_c \propto \log a_x$. The proportionality constant is given by 2.303 RT/αF, which, since $\alpha < 1$, has a value that is greater than 0.05915V at 25°C, as observed experimentally (79). Furthermore, the induction time [equation (31)] increases with decreasing applied voltage but decreases with increasing halide activity. Indeed, Lin, Chao, and Macdonald (79) found equation (31) to be in excellent agreement with experimental data reported in the literature, even to the extent that it successfully predicts the induction time using a value of α derived from film growth and breakdown voltage data.

Any real metal surface, however, contains a large number of defect sites at which pits may nucleate, and hence the breakdown parameters tend to be distributed (88-102). These sites include ghost grain boundaries, inclusions, and emergent dislocations, and represent structural imperfections either inherent to the passive film or projected into the film from the underlying metal phase. Assuming that these imperfections differ in the diffusivity of cation vacancies (D), and that the breakdown sites are normally distributed in D, we may write the distribution functions for V_c and t_{ind} as follows (85, 86):

$$\frac{dN}{dV_c} = \frac{-b\gamma}{\sqrt{2\pi}\ \sigma_D\ a_x^{\chi/2}}\ \exp\left[-(e^{-\gamma V_c} - e^{-\gamma \bar{V}_c})^2\ b^2/2\ \sigma_D^2\ a_x^{\chi}\right]\ \exp\left(-\gamma V_c\right) \tag{39}$$

and

$$\frac{dN}{dt_{ind}} = -\frac{C\ \exp\left(-\gamma V\right)\ \exp\left[\{-e^{-\gamma V}\ g[1/t_{ind} - \tau)-1/(\bar{t}_{ind}-\tau)]\}^2\right]}{(t_{ind} - \tau)^2\ a_x^{\chi/2}\ \sqrt{2\pi}\ \sigma_D} \tag{40}$$

where

$$\gamma = \chi F\alpha/2RT \tag{41}$$

$$b = J_m\ u^{\chi/2}/\hat{a} \tag{42}$$

$$C = \xi\ u^{\chi/2}/\hat{a} \tag{43}$$

$$g = C/(a_x^{\chi/2}\ \sqrt{2}\ \sigma_D) \tag{44}$$

and σ_D is the standard deviation for the normal distribution of breakdown sites with respect to D. The parameter $\bar{t}_{ind}$ is defined as the characteristic induction time for breakdown. It is actually defined by equation (31) with $V_c = \bar{V}_c$ [equation (35)], where $\bar{V}_c$ is the breakdown voltage at the fiftienth percentile. Equations (39) and (40), and the parameters contained therein, shows that the distributions in the breakdown voltage and induction time are functions of temperature, pH, anion activity (a_x), and applied potential (dN/dt_{ind} only), in addition to being dependent upon the intrinsic parameters that describe the normal distribution in the cation vacancy diffusivity ($\bar{D}$, σ_D).

Cumulative probability distribution functions [$P(V_c)$] where

$$P(V_c) = 100 \int_{-\infty}^{V_c} \left(\frac{dN}{dV_c}\right)\ dV_c/ \int_{-\infty}^{\infty} \left(\frac{dN}{dV_c}\right)\ dV_c \tag{45}$$

are plotted in Figure 16 for four values of $\bar{D}$ and for other parameters selected in References (85) and (86), with the curve for $\bar{D} = 5 \times 10^{-20}$ cm^2/s coinciding with the experimental data of Shibata (88) for the breakdown of the passive film on Fe-17Cr in 3.5% NaCl solution at 30°C. Clearly, the cumulative probability is a sensitive function of the mean diffusivity ($\bar{D}$), with the curves shifting to the right (i.e. to more positive breakdown voltages) by $\sim$ 0.05 V for every decade decrease in the value of $\bar{D}$. This correlation suggests that the most breakdown-resistant passive films are those with the smallest cation vacancy diffusivity. We reached a similar conclusion in our recent work (84) on the role of minor alloying elements in the breakdown of passivity, in which we argue that the principal effect of molybdenum (for example) in the film is to decrease the cation vacancy diffusivity and the concentration of mobile vacancies.

Using the same values listed above for the various parameters, we plot in Figure 17 the differential cumulative probabilities for the induction time, $\Delta N]_{t_j}^{t_j+1}$, where

$$\Delta N]_{t_j}^{t_{j+1}} = \int_{t_j}^{t_{j+1}} \left(\frac{dN}{dt_{ind}}\right) dt_{ind} \tag{46}$$

These data were generated without additional approximations or assumptions, except for the assignment of a value for σ_D, so that comparison of the $\Delta N]_{t_j}^{t_{j+1}}$ values with experimental data from Shibata (88) provides a good test of the validity of the model. Clearly, the PDM accounts for the highly assymetric form by the distribution in induction time, and does so quantitatively for a standard deviation (σ_D) equal to $\bar{D}$ (5×10^{-20} cm^2/s). The need to assign a value for σ_D in equation (40), rather than in equation (39), arises from the finding that dN/dV_c is insensitive to values chosen for the standard deviation, whereas the distribution function for the induction time is sensitive to this parameter (85).

The predicted changes in the distributions in V_c and t_{ind} with varying halide activity are shown in Figures 18 and 19. Increasing halide activity is seen to move the distribution in V_c to more negative voltages and that in t_{ind} to shorter times. Both correlations are consistent with experimental findings (88-94, 103, 104), which show that passivity breakdown occurs more readily as the concentration of the aggressive anion is increased. From equation (2), the shift in the breakdown voltage at the fiftieth percentile with halide activity is described by

$$\bar{V}_c = \frac{1}{\gamma} \ln \left(\frac{b}{\bar{D}}\right) - \frac{2.303 \; RT}{\alpha F} \log a_x \tag{47}$$

which also provides a convenient method for determining α [see also reference (79)]. Mechanistically, the effect of increasing the activity, a_x, is to enhance the generation of cation vacancies at the film/solution interface according to the reactions shown in Figure 15.

The effects of pH on the distribution functions in V_c and t_{ind} are shown in Figures 20 and 21, respectively. Increasing pH is predicted to shift the distribution in V_c to more positive voltages and to move the distribution in the induction time to longer times. Accordingly, the passive film is predicted to be more resistant to breakdown. It is evident, however, that the predicted effect depends upon the value selected for β; the negative value adopted here is consistent with the calculations of Chao et al. (78) who found $\beta = -0.0144$ from Sato, Noda, and Kudo's (105) data for the growth of passive films on iron in phosphate buffer solutions.

CONCLUDING REMARKS

In this paper, I have attempted to review a number of surface chemical phenomena that have mechanistic significance for the stress corrosion cracking of metals and alloys in aqueous environments. The areas chosen for discussion are by no means the only ones of importance, but they demonstrate the enormous complexity of the chemical and electrochemical properties of metal and oxide interfaces in contact with aqueous solutions. Fortunately, the topics discussed continue to be areas of intense research, and it is likely that significant advances will be made within the next few years in developing more quantitative models for the various processes involved.

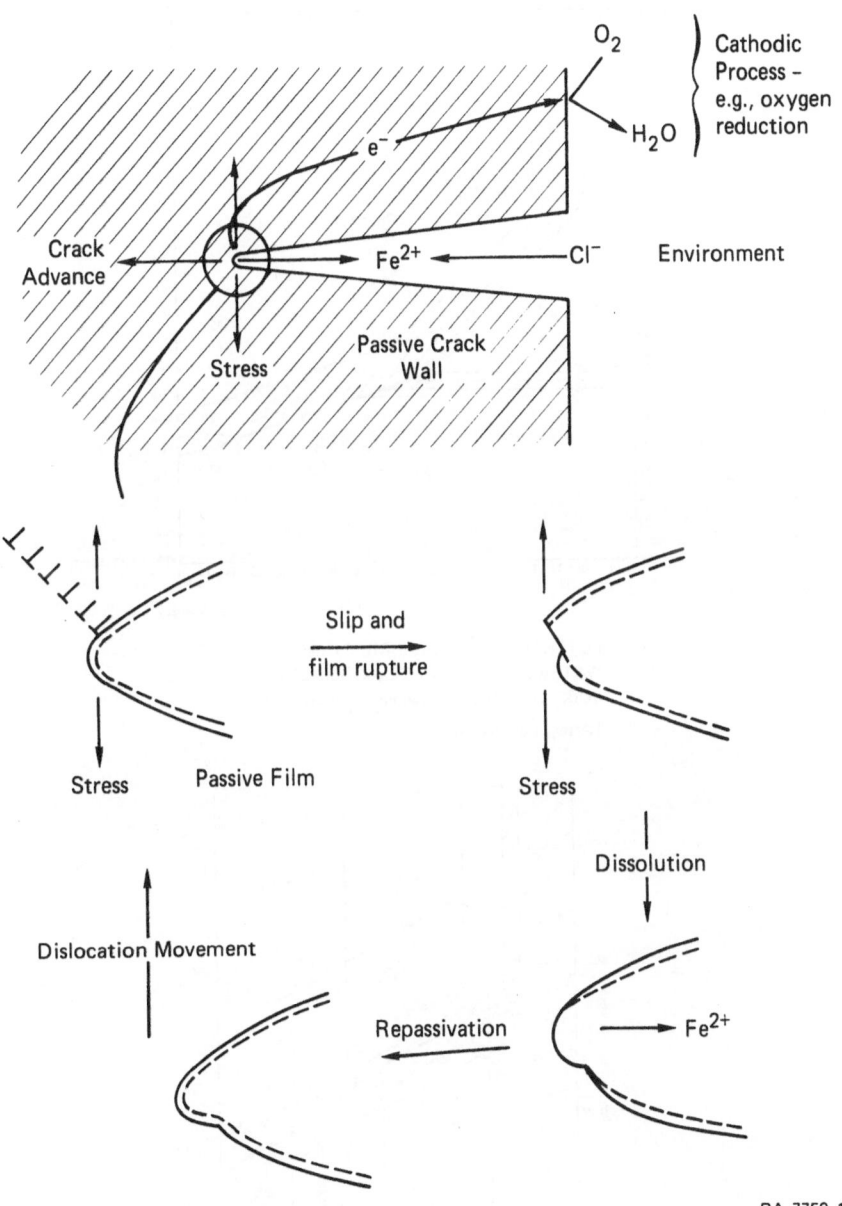

FIGURE 1 SCHEMATIC REPRESENTATION OF THE SLIP DISSOLUTION/
REPASSIVATION MODEL FOR STRESS CORROSION CRACKING

Note that crack advance is discontinuous, with the rate of advance
being determined by the relative rates of dissolution and repassivation
at the crack tip. Film rupture at the crack tip is envisaged to result
from slip due to the movement of dislocations in the stress field
that exists of the apex.

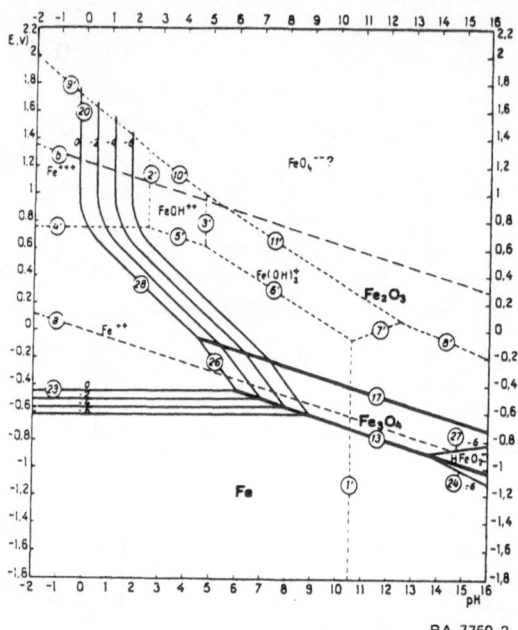

RA-7759-2

FIGURE 2 POTENTIAL-pH EQUILIBRIUM DIAGRAM FOR THE IRON-
WATER SYSTEM, AT 25°C (CONSIDERING AS SOLID
SUBSTANCES ONLY Fe, Fe_3O_4 AND Fe_2O_3)

[After Pourbaix (9)].

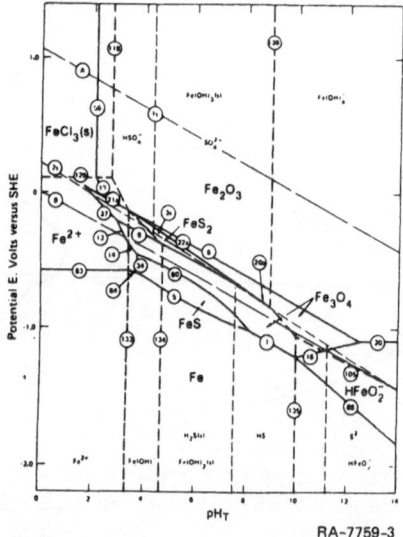

RA-7759-3

FIGURE 3 POTENTIAL-pH DIAGRAM FOR IRON IN HIGH-SALINITY
BRINE AT 250°C IN THE PRESENCE OF 10 ppm TOTAL
DISSOLVED SULFIDE ($H_2S + HS^- + S^{2-}$)

Activities of HSO_4^- and $SO_4^{2-} = 10^{-6}$ molal. Activities of
dissolved iron species = 10^{-4} molal. (S) indicates a soluble
molecular species. [After Macdonald (10)].

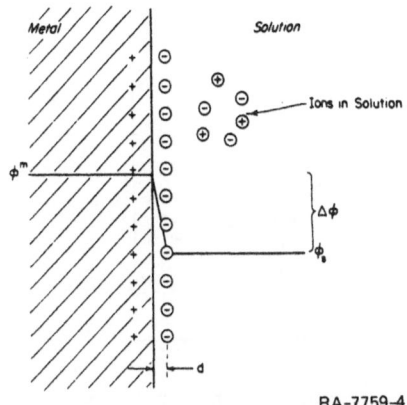

RA-7759-4

FIGURE 4   HELMHOLTZ (26) MODEL FOR THE ELECTRICAL DOUBLE
LAYER AT A METAL/ELECTROLYTE INTERFACE

ϕ_s and ϕ_m refer to the electrical potentials in the solution and metal
phases, respectively.

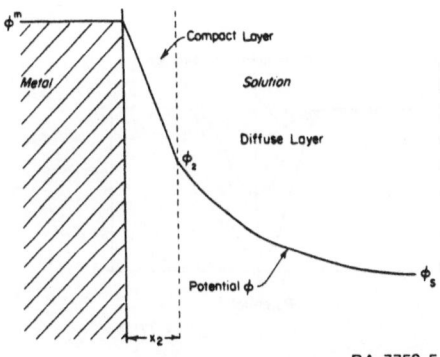

RA-7759-5

FIGURE 5   GOUY-CHAPMAN-STERN MODEL FOR THE ELECTRICAL
DOUBLE LAYER AT A METAL/SOLUTION INTERFACE
(27-30)

274

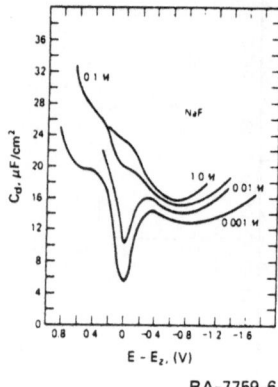

RA-7759-6

FIGURE 6   DIFFERENTIAL CAPACITANCE VERSUS POTENTIAL
FOR SODIUM FLUORIDE SOLUTIONS IN CONTACT
WITH MERCY AT 25°C

[After Bard and Faulkner (16)] .

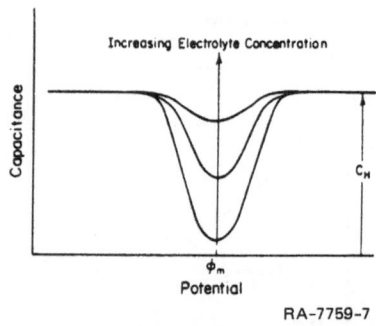

RA-7759-7

FIGURE 7   SCHEMATIC REPRESENTATION OF THE VARIATION
OF INTERFACIAL CAPACITANCE WITH POTENTIAL
AND ELECTROLYTE CONCENTRATION ACCORDING
TO THE GOUY-CHAPMAN-STERN MODEL (30)

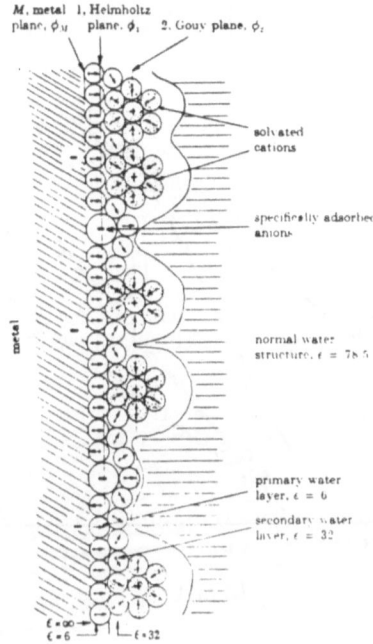

RA-7759-8

FIGURE 8   MODEL OF THE DOUBLE LAYER, ACCORDING
TO BOCKRIS et al. (31)

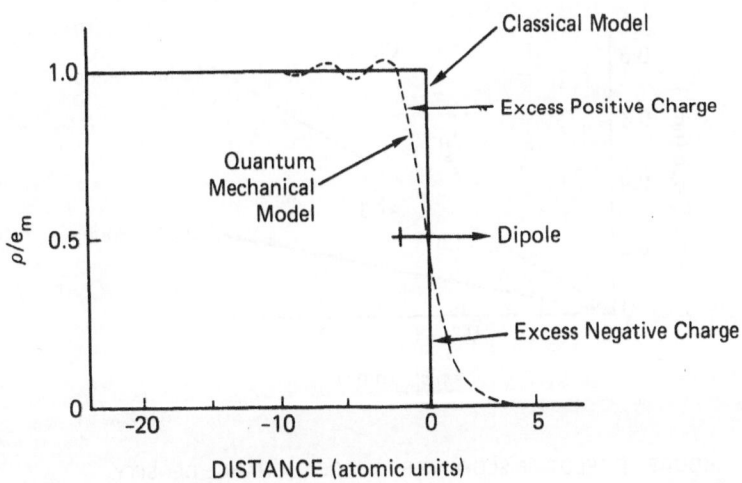

RA-7759-9

FIGURE 9   QUANTUM MECHANICAL PENETRATION OF ELECTRONS
INTO THE INNER HELMHOLTZ REGION RESULTING IN
A SURFACE DIPOLE

276

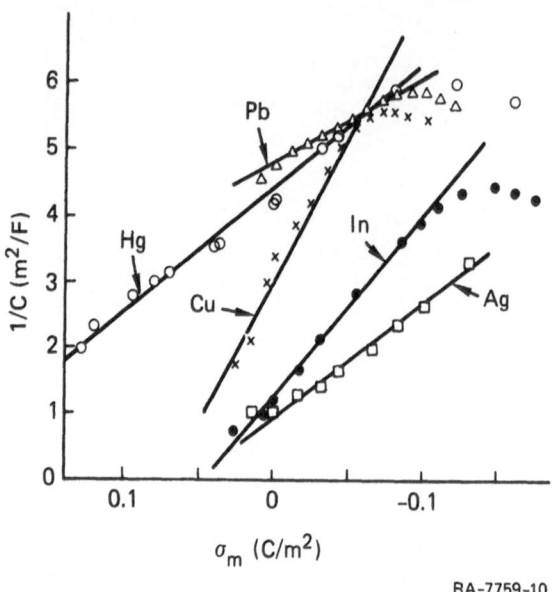

FIGURE 10   TEST OF EQUATION (17) FOR VARIOUS METALS
IN SODIUM FLUORIDE SOLUTIONS [AFTER PRICE
AND HALLEY (46)]

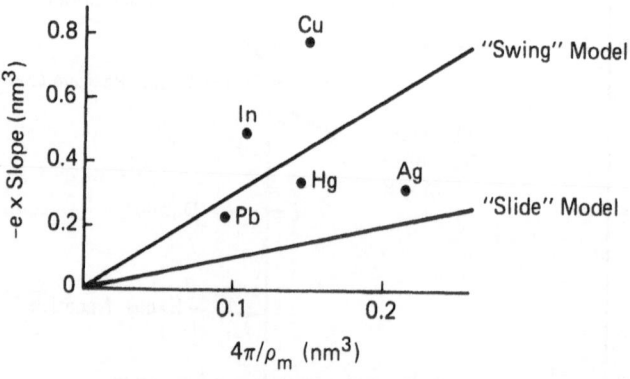

FIGURE 11   PLOT OF SLOPE VS RECIPROCAL CHARGE DENSITY
FOR THE "SWING" AND "SLIDE" MODELS ACCORDING
TO PRICE AND HALLEY (46)

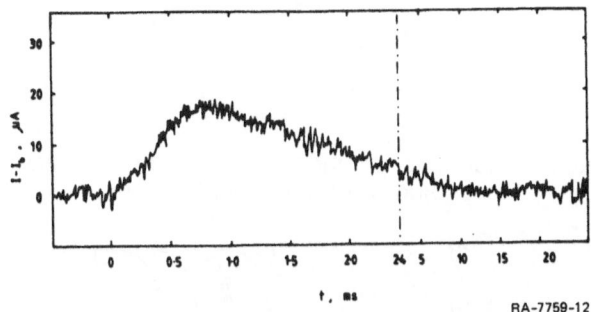

FIGURE 12   CURRENT TRANSIENT DUE TO SCRATCHING Fe AT E =
−605 mV(nhe) IN pH 9.35 ELECTROLYTE CONTAINING
0.5 M NaClO$_4$

Stylus contact is made at t = 0 and broken at t = 0.8 ms.
Note scale change at t = 2.4 ms. [After Burstein and Ashley (66)] .

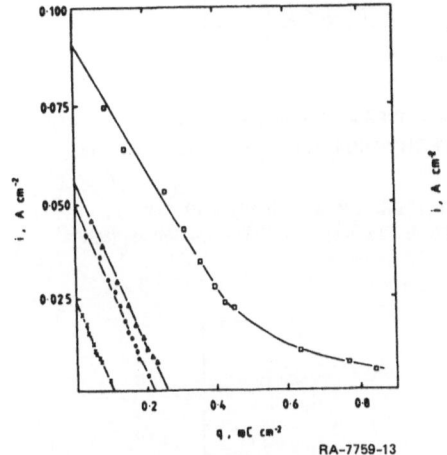

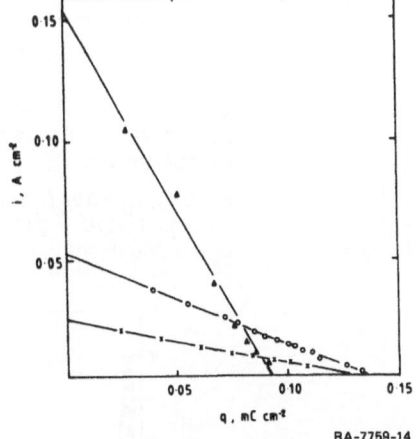

FIGURE 13   DECAY OF SCRATCH CURRENT DENSITY AS A
FUNCTION OF THE CHARGE DENSITY PASSED
FOR DIFFERENT POTENTIALS IN ELECTROLYTE
OF pH 9.35

E, mV(nhe): x = −580; o = −530; △ = −505; and □ = −255.
[After Burstein and Ashley (66)] .

FIGURE 14   DECAY OF SCRATCH CURRENT DENSITY AS A
FUNCTION OF THE CHARGE DENSITY PASSED
AT −580 mV(nhe) IN NEUTRAL AND ALKALINE
ELECTROLYTES

x: 0.1 M K$_2$B$_4$O$_7$, 0.2 M KOH, and pH 10.15.
o: 0.034 M Na$_2$B$_4$O$_7$, 0.021 M NaOH, 0.5 M
NaClO$_4$, and pH 9.35. △: 0.02 M Na$_2$B$_4$O$_7$,
0.2 M H$_3$BO$_3$, 1.0 M NaClO$_4$, and pH 7.85.
[After Burstein and Ashley (66)] .

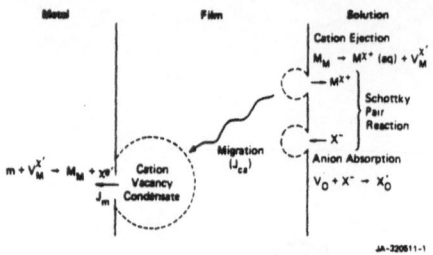

FIGURE 15   PROCESSES LEADING TO THE BREAKDOWN OF PASSIVE FILMS
ACCORDING TO THE POINT DEFECT MODEL

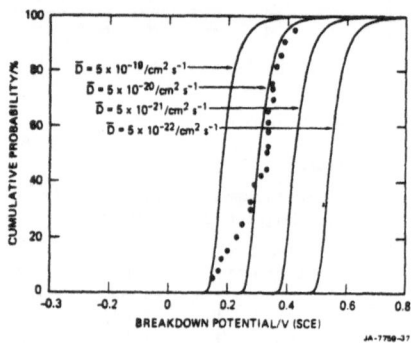

FIGURE 16   CUMULATIVE PROBABILITIES FOR THE BREAKDOWN VOLTAGE
AS A FUNCTION OF $\bar{D}$ FOR NORMAL DISTRIBUTIONS IN THE
DIFFUSIVITY D

β = -0.01, σ_D = 0.75 $\bar{D}$, T = 298.15 K, a_{Cl} = 0.402 x 10^{-3}, χ = 3.
*Data for Fe-17Cr in 3.5% NaCl solution at 30°C from Shibata (88),
$\bar{V}_c$ = -0.046 V (SCE).

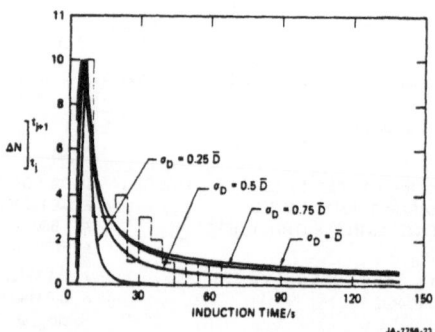

FIGURE 17   DIFFERENTIAL CUMULATIVE PROBABILITIES FOR THE INDUCTION
TIME AS A FUNCTION OF σ_D FOR NORMAL DISTRIBUTIONS IN D

$\bar{D}$ = 5 x 10^{-20}/cm^2 s^{-1}. (---) Data for Fe-17Cr in 3.5% NaCl solution at
30°C from Shibata (88). $\bar{V}_c$ = -0.046 V (SCE), V = 0.02 V (SCE), $\bar{t}$ = 7.5 s,
τ = 0. Other parameters as in Figure 16.

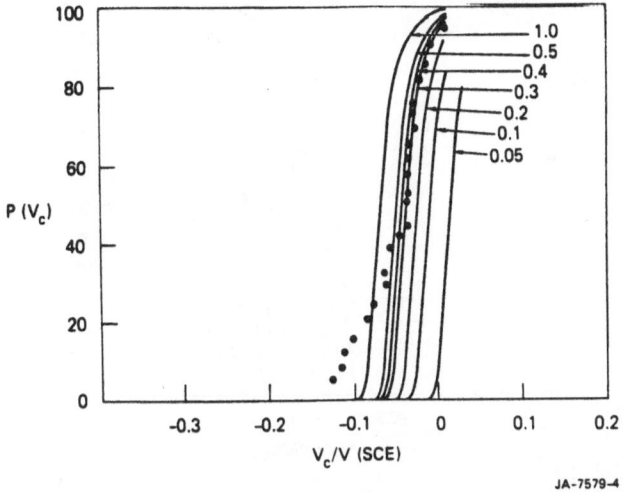

FIGURE 18   CALCULATED DISTRIBUTIONS FOR THE BREAKDOWN OF PASSIVE
FILMS AS A FUNCTION OF HALIDE CONCENTRATION (mol/dm^3)

β = –0.01, pH = 7, $\bar{D}$ = 5 x 10^{-20} cm^2/s, σ_D = 0.75 $\bar{D}$, T = 298.15 K, χ = 3.
● Data from Shibata (88) for Fe–17Cr in 3.5% NaCl solution at 30°C.

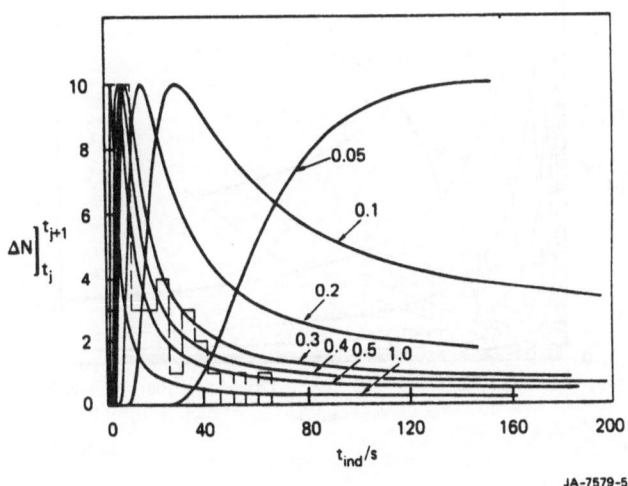

FIGURE 19   CALCULATED DISTRIBUTIONS IN THE INDUCTION TIME FOR THE
BREAKDOWN OF PASSIVE FILMS AS A FUNCTION OF HALIDE
CONCENTRATION

(– – –) Data from Shibata (88) for Fe-17Cr in 3.5% NaCl solution at 30°C.
All other parameters are as listed in Figure 18, with τ = 0, $\bar{t}$ = 7.5 s,
V = 0.02 V (SCE).

280

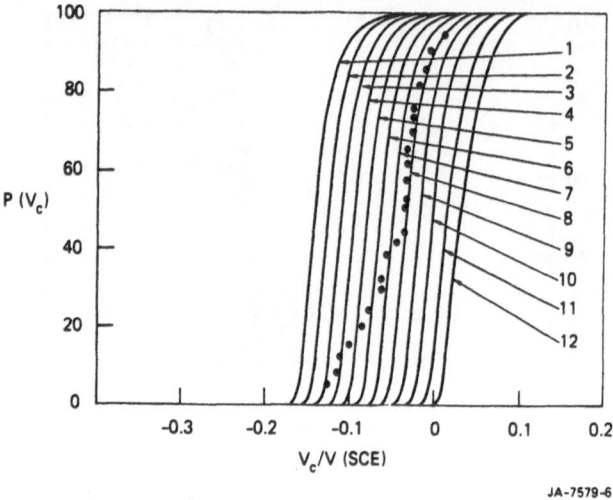

JA-7579-6

FIGURE 20   CALCULATED DISTRIBUTIONS IN THE BREAKDOWN VOLTAGE
AS A FUNCTION OF pH

All other parameters are as listed in Figure 18, $a_{Cl} = 0.402 \times 10^{-3}$.

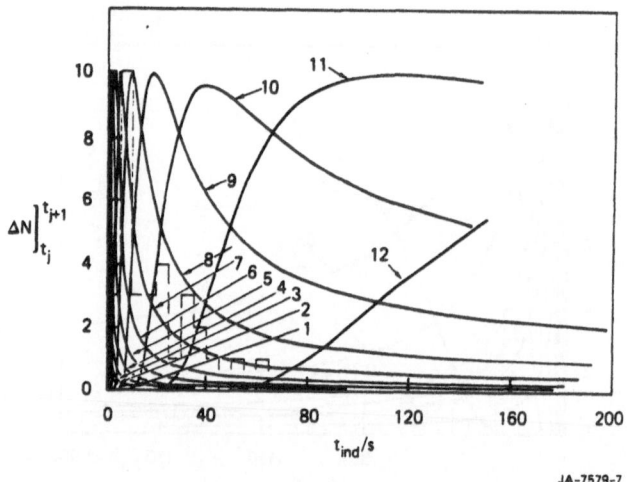

JA-7579-7

FIGURE 21   CALCULATED DISTRIBUTIONS IN THE INDUCTION TIME
AS A FUNCTION OF pH

(———) Data from Shibata (88) for the breakdown of passive
films on Fe–17Cr in 3.5% NaCl solution at 30°C. All other
parameters are as listed in Figure 19.

The importance of this quest is simply that any quantitative model for crack propagation via active path dissolution will require prior quantization of the component phenomena of film rupture, bare metal dissolution, repassivation, and chemically-induced passivity breakdown, together with more accurate chemical and electrochemical models for the crack environment. The stimulus for this work clearly is the enormous cost of stress corrosion cracking to our industrial societies; a problem that normally might be dealt with through engineering or "applied" science. However, the required advances are highly fundamental in nature, and every effort must be made to attract those researchers who deal with fundamental scientific problems to tackle the challanging and important subject of stress corrosion cracking.

ACKNOWLEDGEMENTS

The author gratefully acknowledges the partial support of this work by the Basic Energy Sciences Division, US Department of Energy, through Grant No. DE-DG03-84ER45164. Thanks are also due to the organizers of this conference for a travel grant to partially cover the cost of attending this meeting.

REFERENCES

1. Bursle, AF and Pugh, EN: An Evaluation of Current Models for the Propagation of Stress-Corrosion Cracks, in Environment-Sensitive Fracture of Engineering Materials, Foroulis, AZ(ed): Met. Soc. AIME, Warrendale, PA, 1979, p. 18.
2. Champion, FA: Simp. Internal Stresses in Metals and Alloys, Inst. Met., London, 1948, p. 468.
3. Logan, HL:, J. Res. Nat. Bur. Stand., $\underline{48}$: 99 (1952).
4. Staehle, RW: The Theory of Stress Corrosion Craacking Alloys, NATO, Brussels, 1971, p. 223.
5. Scully, JC: Corros. Sci., $\underline{15}$: 207 (1975).
6. Vermilyea, DA: J. Electrochem. Soc., $\underline{119}$: 405 (1972).
7. Ford, EP: Mechanism of Environment Sensitive Cracking of Materials, Met. Soc., London, 1977, p. 480.
8. Foroulis, ZA(ed): Papers in Environment-Sensitive Fracture of Engineering Materials, Met. Soc., AIME, Warrendale, PA, 1979.
9. Pourbaix, M: Atlas of Electrochemical Equiliria in Aqueous Solution, Pergamon, Oxford (1966).
10. Macdonald, DD: ASTM Sp. Publ. 717, ASTM, Philadelphia, Pa (1981).
11. Biernat, RJ and Robins, RG: Electrochim. Acta $\underline{17}$: 1261 (1972).
12. Townsend, HE: Corr. Sci. $\underline{10}$: 343 (1970).
13. Lewis, DJ: Inorg. Nucl. Chem. $\underline{33}$: 2121 (1971).
14. Ashworth, V and Bowden, PJ: Corr. Sci. $\underline{10}$: 709 (1970).
15. Macdonald, DD and Syrett, BC: Corrosion $\underline{35}$: 471 (1979).
16. Bard, AJ and Faulkner, LR: Electrochemical Methods, John Wiley and Sons, New York (1980).
17. Payne, R: Electroanal. Chem. 41: 277 (1973).
18. Grahame, DC: Chem. Rev. $\underline{41}$: 441 (1947).
19. Delahay, P: Double Layer and Electrode Kinetics, Interscience, New York (1965).
20. Hills, G: J. Phys. Chem. $\underline{73}$: 3591 (1969).
21. Parsons, R: Mod. Aspects Electrochem. $\underline{1}$: 103 (1954).

22. Conway, BE: Theory and Principles of Electrode Processes, Ronald, New York (1965).
23. Frumkin, AN and Damaskin, BB: Mod. Aspects Electro-Chem. $\underline{3}$: 149 (1964).
24. Mohilner, DM: Electroanal. Chem. $\underline{1}$: 241 (1966).
25. Payne, R: Techniques in Electrochemistry $\underline{1}$: 43 (1972).
26. Von Helmoltz, HLF: Ann. Physik. $\underline{7}$: 337 (1979).
27. Gouy, G: Compt. Rend. $\underline{149}$: 654 (1910).
28. Gouy, G: J. Phys. Radium $\underline{9}$: 457 (1910).
29. Chapman, DL: Phil. Mag. $\underline{25}$: 475 (1913).
30. Stern, O: Z. Elektrochem. $\underline{30}$: 508 (1924).
31. Bockris, JO'M, Devanathan, MAV and Muller, K: Proc. Roy. Soc. $\underline{A274}$: 55 (1963).
32. Devanathan, MAV and Tilak, BVKSRA: Chem. Rev. $\underline{65}$: 635 (1965).
33. Damaskin, BB: Elektrokhimiya $\underline{5}$: 71 (1969).
34. Parsons, R: Rev. Pure Appl. Chem. $\underline{18}$: 91 (1968).
35. Parsons, R: Croat. Chem. Acta $\underline{42}$: 390 (1970).
36. Payne, R: Adv. Electrochem. Electroanal. Eng. $\underline{7}$: 1 (1970).
37. Levine, S, Mingins, J and Bell, GM: J. Electroanal. Chem. $\underline{13}$: 280 (1967).
38. Levine, S, Bell, GM and Calvert, D: Can. J. Chem. $\underline{40}$: 518 (1962).
39. Levine, S, Bell, GM and Smith, AL: J. Phys. Chem. $\underline{73}$: 3534 (1969).
40. Schmickler, W: J. Electroanal. Chem., $\underline{150}$: 29 (1983).
41. Kornyshev, A, Schmickler, W, and Vorotyntsev, M: Phys. Rev. B, $\underline{25}$: 5244 (1982).
42. Badrali, JP, Goodisman, J, and Rosinberg, M: J. Electroanal. Chem., $\underline{130}$: 31 (1981); $\underline{143}$: 73 (1983).
43. Borkovec, M and Westall, J: J. Electroanal. Chem., $\underline{150}$: 325 (1983).
44. Liu, SH: J. Electroanal. Chem., $\underline{150}$: 305 (1983).
45. Henderson, D, Blum, L, and Lozado-Cassou, M: J. Electroanal. Chem., $\underline{150}$: 291 (1983).
46. Price, D and Haley, JW: J. Electroanal. Chem., $\underline{150}$: 347 (1983).
47. Wright, HJL and Hunter, RJ: Aust. J. Chem. $\underline{26}$: 1183 (1973).
48. Wright, HJL and Hunter, RJ: Aust. J. Chem. $\underline{26}$: 1191 (1973).
49. Hunter, RJ: Mod. Aspects Electrochem. $\underline{11}$: 33 (1975).
50. Gaudin, AM and Fuerstenau, DW: Ming. Eng. $\underline{7}$: 77 (1955).
51. Berube, YG and de Bruyn, PL: J. Colloid. Interface Sci. $\underline{28}$: 92 (1968).
52. Levine, S and Smith, AL: Disc. Faraday Soc. $\underline{52}$: 290 (1971).
53. Atkinson, RG, Posner, AM, and Quirk, JP: J. Phys. Chem. $\underline{71}$: 550 (1967).
54. Tewari, PH and McLean, AW: J. Colloid. Interface Sci. $\underline{40}$: 267 (1972).
55. Tewari, PH and Campbell, AB: J. Colloid. Interface Sci. $\underline{55}$: 531 (1976).
56. Tewari, PH, Campbell, AB, and Lee, W: Can. J. Chem. $\underline{50}$: 1642 (1972).
57. Tewari, P and Lee W: J. Colloid. Interface Sci. $\underline{52}$: 77 (1975).
58. Matijevic, E, Abramson, MB, Ottewill, R, Schultz, KF, and Kerker, M: J. Phys. Chem. $\underline{65}$: 1724 (1961).
59. Davies, JT and Rideal, EK: Interfacial Phenomena, Academic Press, New York (1963).
60. Young, L: Anodic Oxide Films, Academic Press, N.Y., 1961.
61. Kruger, J and Frankenthal, J: Passivity of Metals, ed. R. P. Frankenthal and J. Kruger, The Electrochem. Soc., Princeton, N.J., 1978.
62. Fromhold, AT, Jr.: Theory of Metal Oxidation, Vo. 1 – Fundamentals, North Holland, Amsterdam, 1976.

63. Hoar, TP and Jones, RW: Corros. Sci., 13: 725 (1973).
64. Hoar, TP and Galvele, JR: Corros. Sci., 10: 211 (1970).
65. Scully, JC: "The Role of Surface Films in Stress Corrosion Cracking and Corrosion Fatigue", in ref (1), p. 71.
66. Burstein, GT and Ashley, GW: Corrosion, 39: 241 (1983).
67. Burstein, GT and Newman, RC: J. Electrochem. Soc., 128: 2270 (1981).
68. Burstein, GT and Newman, RC: Electrochim. Acta, 25: 1009 (1980).
69. Burstein, GT and Kearns, MA: J. Electrochem. Soc., 131: 991 (1984).
70. Burstein, GT: J. Electrochem. Soc., 130: 2133 (1983).
71. Burstein, GT and Marshall, PI: Corros. Sci, 24: 449 (1984); 23: 125 (1983).
72. Burstein, GT and Ashley, GW: Corrosion, 40: 110 (1984).
73. Barbosa, M and Scully, JC: in ref (1), p. 91.
74. Pessall, N and Liu, C: Electrochim. Acta, 16: 1987 (1971).
75. Lizlovs, EA and Bond, AP: J. Electrochem. Soc., 122: 719 (1975).
76. Janik-Czachor, M: J. Electrochem. Soc., 129: 513C (1981).
77. Galvele, JR: Passivity of Metals, ed. P. Frankenthal and J. Kruger, Ed., The Electrochem. Soc., Princeton, N.J., 1978.
78. Chao, CY, Lin, LF, and Macdonald, DD: J. Electrochem. Soc., 1981, 128(6): 1187.
79. Lin, LF, Chao, CY, and Macdonald, DD: J. Electrochem. Soc., 1981, 128(6): 1194.
80. Chao, CY, Lin, LF, and Macdonald, DD: J. Electrochem. Soc., 1982, 129(9): 1874.
81. Silverman, S, Cragnolino, G, and Macdonald, DD: J. Electrochem. Soc., 1982, 129(11): 2419.
82. Chao, CY, Szklarska-Smialowska, Z, and Macdonald, DD: J. Electroanal. Chem., 1982, 131: 279.
83. Chao, CY, Szklarska-Smialowska, Z, and Macdonald, DD: J. Electroanal. Chem., 1982, 131: 289.
84. Urquidi, M and Macdonald, DD: J. Electrochem. Soc., 132: 555 (1985).
85. Macdonald, DD and Urquidi-Macdonald, M: "Distribution Functions for the Breakdown of Passive Films", Electrochim. Acta, in press (1986).
86. Macdonald, M and Macdonald, DD: "Theoretical Distribution Functions for the Breakdown of Passive Films", J. Electrochem. Soc., submitted for publication (1986).
87. Lenhart, SJ, Urquidi-Macdonald, M, and Macdonald, DD: "Photo Effects in Passivity Breakdown", J. Electrochem. Soc., submitted for publication, (1986).
88. Shibata, T: Trans. ISIJ, 23: 785 (1983).
89. Shibata, T and Takeyama: Proc. 8th Intl. Congr. Met. Corr., Dechema, Franfurt/Main, p. 146 (1981).
90. Shibata, T and Takeyama: Nature, 260: 315 (1976).
91. Shibata, T and Takeyama: Trans. Iron and Steel Institute of Japan, Research Article 785, 1983.
92. Shibata, T and Takeyama, T: J. Jap. Inst. Metals, 42: 743 (1978).
93. Shibata, T and Takeyama, T: J. Jap. Inst. Metals, 43: 270 (1979).
94. Shibata, T and Takeyama, T: Boshoku Gijutsu, 26: 25 (1977).
95. Williams, DE, Westcott, C, and Fleischmann, M: in "Passivity of Metals and Semiconductors", Edt. by M. Froment, Elsevier, Amsterdam, p. 217 (1983).
96. Williams, DE, Westcott, C, and Fleischmann, M: J. Electroanal. Chem., 180: 549 (1984).
97. Williams, DE and Westcott, C: in Proc. 9th Intl. Congr. Met. Corr., Nat. Res. Counc./Toronto, Canada, 4: 390 (1984).

98. Williams, DE, Westcott, C, and Fleischmann, M: J. Electrochem. Soc. <u>132:</u> 1796 (1985).
99. Doelling, R and Heusler, KE: in Proc. 9th Intl. Congr. Met. Corr., Nat. Res. Counc./Toronto, Canada, 2: 129 (1984).
100. Fratesi, R: Corrosion, 41: 114 (1985).
101. Evans, UR, Mears, RB, and Queneau, PE: Engineering, <u>136:</u> 689 (1933).
102. Evans, UR: "Corrosion and Oxidation of Metals", Edward Arnold, London, Ch. 22 (1960).
103. Heusler, KE, and Fischer, L: Wekst. Korros., <u>27:</u> 551 (1976).
104. Strehblow, HH: Werkst. Korros., <u>27:</u> 792 (1976).
105. Sato, N, Noda, T, and Kudo, K: Electrochim. Acta., <u>19:</u> 471 (1974).

DISCUSSION

Comment by P. Marcus:

Is your model for breakdown of passive film by vacancy diffusion to the metal/oxide interface applicable to passive oxides that can be only 2-3 atomic layers thick?

Reply:

Possibly not, except that the model really only requires reactions to occur at the metal/film and film/solution interfaces. Accordingly, provided that sufficient film is present to support the concept of surface vacancies (i.e., at least one or two monolayers); it is possible that this or a modified point defect model can be formulated to account for the breakdown of extremely thin passive films.

Comment by J. R. Rice:

How do considerations of stressing or deformation of the substrate enter your modelling of film breakdown?

Reply:

Substrate deformation is not considered in the point defect model. However, deformation will result in an increase in the tensile stress in the film and hence should result in film rupture at lower cation vacancy accumulations at the metal/film interface. Thus, straining of the metal should lower the breakdown voltage and decrease the induction time.

Comment by J. Lumsden:

1. Does your point defect model depend upon the film being crystalline? Most experimental results, particularly for stainless steels, indicate the films are noncrystalline.

2. On the point defect model, the breakdown potential is correlated with void formation at the film/metal interface, yet the same breakdown potential is obtained if the film is broken down mechanically in the scratch technique.

Reply:

1. While the original model was based upon concepts of crystalline defects (point defects in a regular array), it is not necessary that this idea be retained. For example, defects in glassy structures exist and can be used to describe the movement of charged species. Also, the concept of point defects is necessary only for specifying the reactions that occur at the metal/film and film/solution interfaces, and not for describing the properties of the "bulk" film.

2. In the point defect model, chemical breakdown occurs because of mechanical rupture of the passive film due to the formation of a critical void at the metal/film interface. Therefore, it is not surprising that a correlation exists between the "electrochemical" and "mechanical (i.e., scratch)" breakdown experiments.

Comment by M. O. Speidel:

What is the fundamental reason for the critical potential of stress corrosion cracking of steel in high-temperature water? What is the potential of the crack tip?

Reply:

We don't currently understand the fundamental reason for the critical potential, although it is clear that it is related to the electrochemical processes that occur at the crack tip. We have used transmission line electrical models to compute the change in potential at the crack tip upon changing the external potential, and it is apparent that the mechanism of shutting down a growing crack by switching the potential to a value below the critical potential is more complex than that for a simple dissolution mechanism.

Comment by K. Nisancioglu:

You discussed chemical breakdown of oxide films at length, which I think was very interesting, but did not say much about the significance of this on the surface chemistry of a crack tip.

Reply:

The real significance relates to two phenomena:
      (1) Maintenance of passive crack walls, and
      (2) Nucleation of cracks at pits.
In the first case, a high aspect ratio crack can form only if the crack walls do not break down, whereas in the second, breakdown is a necessary condition for the formation of a stress raiser of sufficient dimensions to nucleate a crack. With respect to the crack tip, I believe that breakdown is principally a mechanical effect, if for no other reason than that the less extensively strained crack walls adjacent to the strained crack tip remain passive in nominally the same environment.

Comment by M. Pourbaix:

I think that the significance of the two critical potentials between which SCC may occur may be easily explained when one knows the composition of the electrolyte inside the cracks: below the lower potential, SCC does not occur because the metal inside the crack is immune; and above the higher potential, SCC does not occur because the metal inside the cavity becomes passive. This has been mentioned during Latanision II in Calcataggio.

Reply:

I believe that the critical potential for SCC reflects a process that is more complex than the simple transition from the active to the immune state. The reason for this is that the change in potential at the crack tip should be less than that imposed externally. Since the transition from the active state to the immune state is described by a continuous current/voltage curve, it is difficult to rationalize why the critical potential should be so sharp. Also, in the case of Type 304 SS, the critical potential is substantially different from the equilibrium potentials for Fe/Fe^{2+}, Cr/Cr^{3+}, Ni/Ni^{2+}, even allowing for IR drop down the crack.

Comment by J. F. Knott:

With respect to current transients in a scratching electrode experiment, it is often observed that the initial form is exponential, followed by something characteristic of diffusive growth. I have two questions:
   a)   The difference between "patchy growth" and random placement of hydroxyl groups or whatever during exponential decay. Presumably, we always have bare metal during an exponential transient?
   b)   Even if growth occurs by diffusive transport, charge transfer must occur, so that the film is not passive. Unless the current transient drops completely to zero, is any film truly passive? How does one define a "passive" film in practice - is there a specified current density below which it can be defined as "passive"?

Reply:

   a)   It is not necessary to have entirely bare metal for an exponential decay, because various mechanisms for the growth of passive films predict transients of this form. Indeed, if the metal is bare and the dissolution reaction completely diffusion controlled, the current decays with $t^{\frac{1}{2}}$ (the Sand equation). Accordingly an exponential transient is not necessarily a good mechanistic aid.
   b)   The "passive" state does not imply a zero current. Because all passive films have a finite impedance (albeit very large, in many cases) the passive current must also be finite. This current may be carried by the movement of vacancies (or ions in the reverse direction) through the film or by electrons. In the latter case, the electrons must react with a redox couple at the film/solution interface in order for a finite dc electron current to exist.
   The definition of passivity is purely operational in the sense that it implies a current that is less than that expected from pure dissolution and one that does not change strongly with potential when the potential is above the primary passivation value.

CRACK-TIP ELECTROCHEMISTRY: RECENT DEVELOPMENTS

A. Turnbull

Division of Materials Applications,
National Physical Laboratory, Teddington, Middlesex TW11 0LW, UK

ABSTRACT

Over the last decade the understanding of crack-tip electrochemistry has advanced very considerably due principally to the development of mathematical models of the mass transport and electrochemistry within cracks.   In parallel with this development, experimental measurements have focussed on the kinetics of reactions on straining crack-tips and on the nature and influence of the films formed.
The important features of mass transport theory are described and the application demonstrated for steels in marine environments. The effect on crack chemistry of crack geometry and mechanical and environmental variables is examined and the interrelation between crack chemistry, crack-tip kinetics and crack growth is discussed.

INTRODUCTION

The rate of growth of stress corrosion and corrosion fatigue cracks results generally (an exception is bulk hydrogen charging) from the interaction between mechanically strained metal and the local environment at the crack tip. Crack advance may be due to anodic reactions such as dissolution or film formation (leading to film-induced cleavage) or due to the cathodic generation and absorption of hydrogen atoms. Hence, characterisation of the appropriate reaction kinetics at straining crack-tips is essential to the proper understanding and prediction of stress corrosion cracking and corrosion fatigue. The best approach to attaining that objective involves a combination of crack electrochemistry studies and measurement of transient kinetics on rapidly scraped or strained metal surfaces.   In the latter experiments attention has to be given to (a) the nature of the material, which should simulate crack tip metal (this is a problem when cracking is path sensitive) (b) the nature of the environment which should simulate crack-tip chemistry and (c) the mass transport conditions of the experiments.   If crack tip kinetics are transport dependent the last of these poses particular problems of experimental simulation.   Modelling may then be necessary to translate results from the transient current experiment to real crack tip conditions.
Determination of the local crack tip environment and potential has been undertaken using a variety of techniques (1) though at the basic level little advance on the pioneering methods of Brown (2) and also Davis (3) has been made during the last decade. Nevertheless, the quantity of data available has gradually increased, the techniques have been made more rigorous, and a reasonable body of reliable data exists for a number of metal/environment systems (1) but as yet with important

limitations with regard to shallow cracks and to corrosion fatigue cracks.

Although the establishment of such an experimental database has been important, the major advance in understanding crack electrochemistry during the past decade has originated in the development and application of mathematical models. These models embrace both mass transport and electrochemistry and consequently can describe the interaction between mechanical variables (eg stress), environmental variables (eg applied potential) and metallurgical variables (eg. alloying elements).

In this paper recent developments in our general understanding of the factors controlling crack chemistry are described, making considerable use of mass transport theory and modelling and giving emphasis to corrosion fatigue cracks. Specific examples of both modelling and experiment applied to structural steel in marine environments are discussed and finally the interrelation between chemistry, kinetics and crack growth is examined.

MASS TRANSPORT THEORY OF RELEVANCE TO CRACKS

Mass transport in electrolyte systems, whether related to cracks, pits or porous electrodes can be described by the same basic set of physical equations (4). For dilute solutions the important relationships are as follows.

The flux (J) of each dissolved species is given by:

$$J_i = C_i v - D_i \nabla C_i - z_i u_i F C_i \nabla \phi \tag{1}$$

in which the three terms represent the contribution from fluid flow, diffusion and ion migration respectively. C_i is the concentration of species i, D_i is the diffusion coefficient, z_i is the charge, u_i the mobility, F is Faraday's constant and ϕ is the potential.

The mass conservation of species is described by

$$\frac{\partial C_i}{\partial t} = - \nabla J_i + R_i \tag{2}$$

where $\partial C_i/\partial t$ is the rate of change of concentration with time and R_i represents the rate of production or depletion of a species by chemical reaction in solution eg hydrolysis reaction. In addition the solution can be considered to be electrically neutral to a good approximation, ie

$$\sum z_i C_i = 0 \tag{3}$$

The fluid velocity can be derived from solution of the hydrodynamic equations governing incompressible flow:

$$\frac{\partial v}{\partial t} + v \nabla v = - \frac{1}{\rho} \nabla P + \nu \nabla^2 v \tag{4}$$

and div v = 0 .

where P is the hydrodynamic pressure, ρ is the density and ν is the kinematic viscosity.

When the solution is concentrated, more complex relationships have to be established which makes mathematical solution more intractable and limited also in application because of the lack of relevant data describing interactions between different species (eg friction coefficients).

Within the constraints of dilute solution theory an important first step is to describe the fluid flow in relation to cracks.

Fluid flow in cracks

The fluid flow in cracks (5) can originate as a consequence of

(a) fluid motion of the bulk solution (externally induced flow);
(b) cyclic displacement of the crack walls (stress-driven flow).

Natural convection due to bubble formation or due to gravity effects resulting from density gradients in the crack solution is likely to be insigificant compared with forced convection. Similarly, flow induced by a rapidly moving crack has little practical significance.

(a) Externally induced flow

Flow of the bulk solution can have obvious implications in relation to the external electrode potential when the free corrosion potential is mass transport controlled and this is readily characterised.

Motion of the bulk solution will also induce movement of the fluid in a crack but the important question is whether the induced flow is going to be significant compared to diffusion and possible stress-driven flow.

The Reynolds number for the crack when the flow is externally induced is given by $U_c h/\nu$ where U_c is the characteristic velocity (see later) and h is half the crack opening. Since h is small the Reynolds number is generally small and the flow within the crack is laminar.

The orientation of the flow with respect to the crack orientation is important and three situations can be envisaged (Fig 1), viz. flow along or across the mouth of the crack (an example would be a small crack in a flat plate with the fluid flowing over the plate and essentially parallel to it; flow through the crack (an example would be an edge crack with the sides unsealed and the velocity vector orientated in the through-thickness direction; flow directed into a surface crack (in this case the surface crack is orientated perpendicular to the velocity vector).

When the flow is across or along the crack mouth it is most probable that flow separation will occur and a recirculatory system established as typified by Fig 2. Note that Fig 2 is somewhat artificial since there is no interchange with the bulk, nevertheless it is of qualitative interest. In Fig 2 the recirculation pattern (streamlines) is shown for cavities of varying depth but constant crack opening (6) (rectangular geometry with crack sides sealed). The flow velocity at any position is tangential to the streamline at that position, the arrows indicating direction. Actual velocities of relevance to cracks are unknown (although the recent work by Andresen et al (7) should

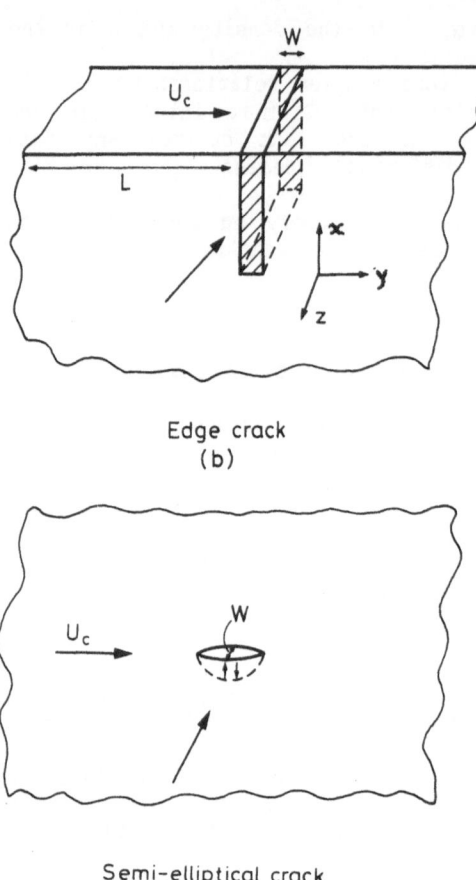

Edge crack
(b)

Semi-elliptical crack
(a)

Fig 1. Orientation of flow with
respect to orientation of edge
and semi-elliptical cracks.
The edge crack may be sealed
along its side.

be revealing in this respect) but the general trend is important and that is the decreasing magnitude of the velocity with distance from the crack mouth. However, in real cracks the crack mouth opening, and hence the mean crack width, increases with increasing crack depth (at constant ΔK) and the geometry is not necessarily rectangular. Recently, Alkire (8) has solved numerically the hydrodynamic equations for a semi-circular cavity, which is more relevant to cracks in service. The influence of varying Peclet number on the metal ion concentration in a crack of constant depth is shown in Fig. 3. The Peclet number represents the ratio of convective mass transport to molecular diffusion and is given by hU_C/D_N where h is half the crack opening, U_C is the characteristic velocity (defined by Alkire as the velocity at centre of cavity opening) and D_N is the characteristic diffusion coefficient. An increasing Peclet number can be construed as representing either an increasing crack width or increasing velocity. The effect of increasing Peclet number is seen in this figure as enhancing mass transport of ions (at saturation level, $C = 1$, at boundaries) out of the crack with a corresponding increase in concentration gradients (contours close together) at the highest value.

Conversely, decreasing the flow velocity or decreasing the crack width (eg decreasing stress intensity) should result in a diminution of the effect of flow on transport of ions compared to diffusion. Alkire's model runs into difficulties near the crack mouth in the limit of small Peclet numbers as in (a) of Fig 3 since conceptually it would be expected that if the velocity is maintained constant and the width decreased the concentration just outside the crack would show much steeper gradients than represented in (a). Nevertheless, the important factor is that movement of the bulk solution can significantly influence the solution composition within a crack. Decreasing the crack depth tends to enhance the role of convective motion but this is counteracted by the decreased crack opening. Hence specific predictions as to actual behaviour are still lacking for real cracks. A

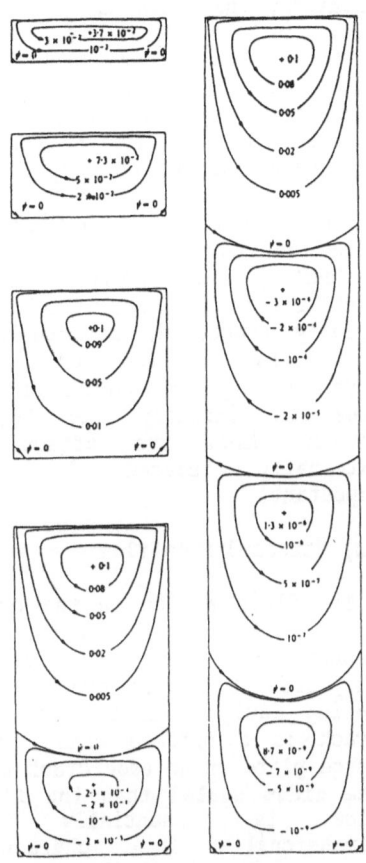

Fig 2. Streamline patterns in rectangular cavities, assuming creeping flow, for varying depth/width ratios.

comment that should be made is that if the flow is across the crack (ie parallel to the short transverse direction or into the plane of Fig 3) then the pressure drop is very much smaller and induced flow would be expected to be less significant.

In the above discussion the important flow velocity was not simply the free stream velocity but the velocity at the mouth of the cavity and this can be an order of magnitude less than the free stream velocity (5). When the flow is directed in the through-thickness direction of an edge crack specimen and the sides are unsealed then the characteristic velocity can be the free stream velocity and a considerably enhanced influence of convection on mass transport is predicted. Recently Andresen et al (7) have confirmed the significance of net mass transport in the through thickness direction on crack chemistry.

Similarly, when the fluid flow is orientated directly into the crack mouth the characteristic velocity is again the free stream velocity and significant convective motion is expected.

The emphasis given to externally induced flow within a crack arises because it is an area which has largely been neglected experimentally despite the well known influence of flow on pitting corrosion (9). The strong effects shown in the model cracks (physical models) of Andresen et al (7) and the observed effects on corrosion fatigue (10) and stress corrosion cracking (11,12) show the need to address this area more fully. In particular, tests should be done in which the crack sides are sealed since this is most practically relevant and avoids some of the artificially enhanced convective effects of the edge-cracked specimen. The absence of these seals may provide some explanation for the effect of flow on fatigue crack growth rates in low alloy steel in high temperature water observed by Scott (10). The reduced growth rate of cracks in constant extension rate testing of low-alloy steel in high temperature waters due to increasing flow (11,12) may be a direct consequence of removal of damaging sulphur compounds (dissolved from inclusions) from relatively shallow cracks.

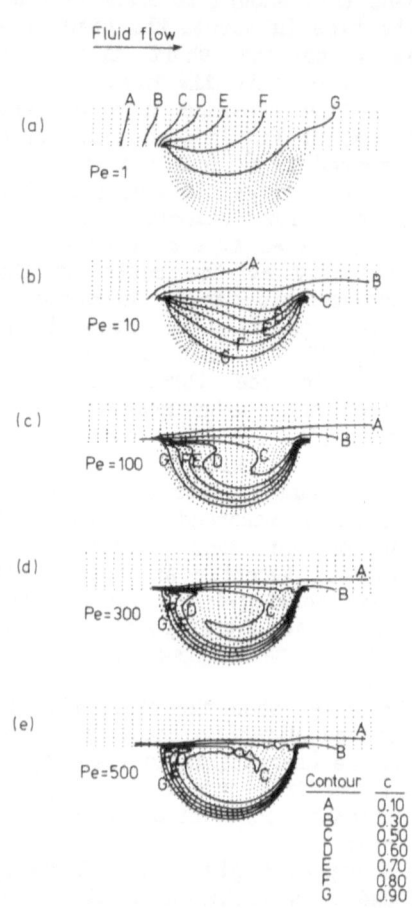

Fluid flow

(a) Pe = 1

(b) Pe = 10

(c) Pe = 100

(d) Pe = 300

(e) Pe = 500

Contour	c
A	0.10
B	0.30
C	0.50
D	0.60
E	0.70
F	0.80
G	0.90

Fig 3. Effect of Peclet number on metal ion concentration when the concentration at cavity wall is equal to 1 everywhere. Equiconcentration contours are shown for various fractions of the saturation value.

Modelling of crack electrochemistry when externally induced flow is important is rather difficult because the problem is at least two-dimensional in character. It is likely that solutions will be limited because of computer costs to very simple reaction schemes for the immediate future though with the rate of advance of computing technology relatively complex problems of this type will become readily tractable. Nevertheless, there is a need for more extensive experimentation in this area to identify the bounds within which the effect of externally induced flow is important.

(b) Stress-driven flow

The fluid motion in the crack in stress-driven flow is associated with the cyclic displacement of the crack walls during fatigue loading. If the crack is a surface crack in a large plate or an edge crack with the sides sealed then the fluid flow is essentially two-dimensional in character. However, Turnbull [13] demonstrated that because the width of the crack is very small it is possible realistically to average between the walls of the crack to produce an average one-dimensional velocity, described for a trapezoidal crack by

$$v(x,t) = -x \frac{\left[\dot{h}_o(t) + \frac{x}{2}\dot{\theta}(t)\right]}{h(x,t)} \tag{5}$$

where h_o is half the crack tip opening displacement, θ is the crack angle and $\dot{h}_o(t)$ and $\dot{\theta}(t)$ are respectively dh_o/dt and $d\theta/dt$.

Crack geometries (Fig 4) are often more complex than the simple model geometries assumed for mathematical analysis. For example, the crack is usually assumed to be a Mode I crack with no significant Mode II components. Derivation of the relevant fluid velocities when shearing

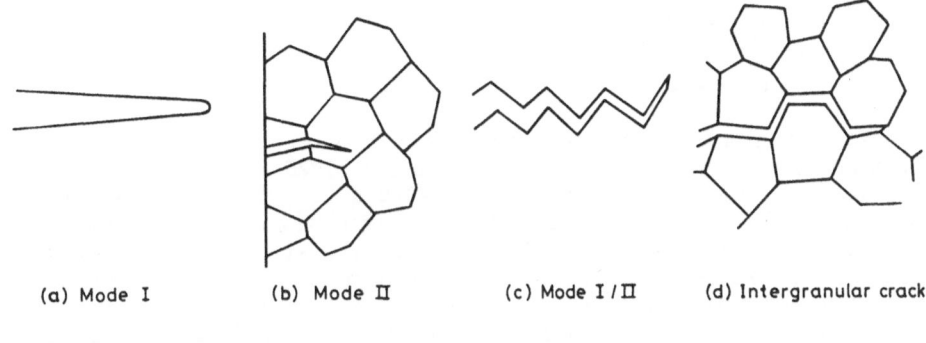

(a) Mode I (b) Mode II (c) Mode I / II (d) Intergranular crack

(e) Trapezoidal crack (f) Parallel-sided crack

Fig 4. Typical crack geometries (a to d) and geometries
adopted for model cracks (e, f).

stresses are important or when the surface roughness is of the order of
the crack opening, as in intergranular cracking in some conditions, has
not yet been attempted but is likely to be difficult.   However, an
important factor in relation to the fluid flow is the Reynolds number
for the crack and this is given for stress-induced flow by

$$Re_c = \frac{4\pi f \delta h l}{\nu} \tag{6}$$

where f is the cyclic frequency, δh is the amplitude of opening (the
opening of the mouth would give the maximum value).   Since δh is
small, the cavity Reynolds number will be small and the flow will
generally be slow viscous laminar flow.
With a knowledge of the fluid velocity it is possible to evaluate the
maximum extent of ingress into the crack (during a cycle) of fresh
solution from the crack mouth (or egress of solute species from the
crack tip), in the absence of diffusion.   For a trapezoidal or
wedge-shaped crack the position of maximum ingress (at peak stress) is
given by $x_{tr}/l \approx R^{1/2}$ and for a parallel-sided crack by $x_{tr}/l = R$ where
R is the stress ratio and x is measured from the crack tip.   (It is
useful to note that there is no dependence on ΔK which only affects the
volume/area ratio of the crack.) Thus, decreasing values of R give more
effective mixing of the crack solution with the bulk, ie greater
replenishment of the crack tip solution.   This replenishment is also
more effective for the parallel-sided crack.
Comparison of mass transport by stress-driven flow and by diffusion
yields an expression for the crack depth below which diffusion is

dominant and this is given by

$$l_{crit} \leqslant \frac{(D_N/f)^{1/2}}{(1 - R^\alpha)} \tag{7}$$

where α is geometry dependent and has values of about 0.5 for a wedge or trapezoidal crack and 1.0 for a parallel-sided crack; f is the cyclic frequency. Thus, for crack depths below l_{crit}, stress-driven flow can be neglected. For crack depths above this value the contribution of diffusion to mass transport will depend on the stress ratio, R, and will decrease in relative significance with decreasing R and increasing l. Detailed comparison between transport by ion migration and fluid flow (and diffusion) is more difficult but in general terms it is clearly going to depend on the potential gradient and if this is small ion migration can be neglected as a significant mechanism for transport of reacting species. If the potential gradient is several hundred millivolts per centimetre then ion migration would potentially be a highly significant transport mechanism for the movement of anions and cations in and out of the crack.

In the above analysis externally induced flow has been assumed to be insigificant. However, it is necessary to try and assess the relative contributions of both flow processes to transport with the crack. Without knowledge of the detailed fluid velocities in the crack associated with externally induced flow this is not readily possible but rough comparisons can be made by considering the maximum velocities.

For stress-driven flow a reasonable approximation to this is given by

$$U_{max} \approx \frac{2\pi f\delta hl}{h_m} \tag{8}$$

where δh and h_m reflect amplitude and mean openings at the crack mouth. Assuming a stress ratio of 0.5, $\Delta K = 20$ MPam$^{1/2}$, $f = 0.1$ Hz and $l = 1$ cm then $U_{max} \approx 0.2$ cms^{-1}. For externally induced flow, the velocity adjacent to the crack mouth (and hence close to the maximum value) will be given approximately by the friction velocity (5) and for laminar flow

$$U_c^l \approx 0.576 \; U_\infty Re_B^{-1/4} \tag{9}$$

and for turbulent flow

$$U_c^t \approx \left\{ 0.0225 \; U_\infty^2 Re_B^{-1/4} \right\}^{1/2} \tag{10}$$

where U_∞ is the free stream velocity and Re_B is the bulk Reynolds number.

Comparison of these different velocities can determine when stress-driven flow is dominant though the reverse condition is not readily quantified because of the unknown dependence of externally induced flow on crack depth. For laminar flow, with $U_\infty = 1.0$ cms^{-1} and $Re_B \approx 10^3$ the characteristic velocity is given by $U_c \approx 0.1$ cms^{-1}. Thus, under the conditions described above for stress-driven flow, externally induced flow could be neglected without undue error.

ELECTROCHEMICAL REACTIONS IN CRACKS

In describing crack chemistry it is necessary to consider anodic and cathodic processes both of which may be solution composition dependent and may occur at diferent rates on the crack tip and crack walls. Typical reactions to be considered in cracks are given in Table 1.

TABLE 1

Anodic dissolution

$$M \longrightarrow M^{n+} + ne^- \tag{11}$$

hydrolysis reaction

$$M^{n+} + H_2O \rightleftharpoons M(OH)^{(n-1)^+} + H^+ \tag{12}$$

cathodic reduction

$$H^+ + e^- \longrightarrow H \tag{13}$$

$$H_2O + e^- \longrightarrow H + OH^- \tag{14}$$

$$O_2 + 2H_2O + 4e^- \longrightarrow 4OH^- \tag{15}$$

$$M^{n+} + ne^- \longrightarrow M \tag{16}$$

When a supporting electrolyte is also involved the number of different species can be seven or more and relevant mass transport equations are necessary for each species. Further reactions to be considered are dissolution and hydrolysis of alloying elements and precipitation of species of limited solubility eg $Fe(OH)_2$. Treatment of differential reaction rates associated with the crack tip and crack walls has also to be considered carefully. For metals in the active state the influence of crack tip reactions (though enhanced compared to those on the walls) on crack chemistry can usually be ignored because of the small area of activity at the tip. When the walls are passive the situation can be more complex and the crack tip reactions can sometimes dominate. A useful rule of thumb (14) is that crack tip reactions can be ignored in terms of their effect on crack chemistry if the ratio of the average current density at the tip to that on the wall is < 10^3 for deep cracks ($\approx$ 10 mm) and < 10^2 for shallow cracks ($\approx$ 0.5 mm).

Combining the assumption of one-dimensionality for the crack with reaction processes the mass transport equations reduce to a form (15) typified by equations (17) to (19)

$$J_i = C_i v - D_i \frac{\partial C_i}{\partial x} - z_i \frac{D_i F}{R'T} C_i \frac{\partial \Phi}{\partial x} \tag{17}$$

$$\frac{\partial C_i}{\partial t} - C_i \frac{\partial V}{\partial x} = - \frac{\partial J_i}{\partial x} + k_1 C_k - k_{-1} C_{ij} - \frac{k}{h(x,t)} \left\{ C_i C_p^q - K_{So} \right\}^m$$

$$\text{solution reaction} \quad \text{precipitation reaction}$$

$$+ \frac{i}{nFh(x,t)} \exp\left\{\frac{\beta F \Phi}{RT}\right\} + \frac{\Theta(t)}{h(x,t)} D_i \frac{\partial C_i}{\partial x} + z_i \frac{D_i F}{R'T} \frac{\Theta(t)}{h(x,t)} C_i \frac{\partial \Phi}{\partial x} \qquad (18)$$

$$\text{electrode reaction}$$

and

$$\sum z_i C_i = 0 \ . \qquad (19)$$

Typical boundary conditions would involve a statement of the flux at the crack mouth or more usually the concentration of the particular species, and a statement of the flux at the crack tip in terms of the current densities at the tip.

The reciprocal of the crack width is clearly apparent in these equations and associated with reactions occurring on the crack walls. This term is the ratio of area of metal/volume of solution (A/V) and decreases with increasing ΔK (range of stress intensity factor) and R (stress ratio) which conceptually can be considered as effecting a dilution of the species in the crack. It is also an essential factor in describing crack depth effects because the A/V ratio decreases with increasing depth at constant ΔK. Hence any explanation of crack depth effects must embrace this parameter as well as the changing magnitudes of the mass transport processes (16). The terms directly involving the crack angle, Θ, in equation (18) arise because the normal to the crack walls of trapezoidal cracks is no longer perpendicular to the x direction and has to be resolved into its components.

PREDICTION AND EXPERIMENT IN SPECIFIC METAL/ENVIRONMENT SYSTEMS

The analysis of the previous section is essential to the proper understanding of mass transport processes in cracks and the influence of different variables such as crack size on their relative magnitude. However, to enable a more detailed assessment of crack-tip electrochemistry the mass transport and electrochemical equations have to be solved and combined with experimental data for the specific system of interest. Much of the discussion in this section will be based on research in structural steels in marine environments but the results will be presented with a view to the broader implications in relation to other systems. The parameters of interest are the potential, pH and local metal ion concentration at the crack tip since these are the parameters necessary to successfully guide the obtention and application of relevant transient electrode kinetic data. The effect of externally induced flow on these parameters is not discussed further in the absence of specific experimental data or modelling predictions.

Potential drop in cracks

The potential drop in cracks has been measured in a number of systems and the potential at the crack tip shows the characteristic dependence on external potential depicted schematically in Fig 5 and more

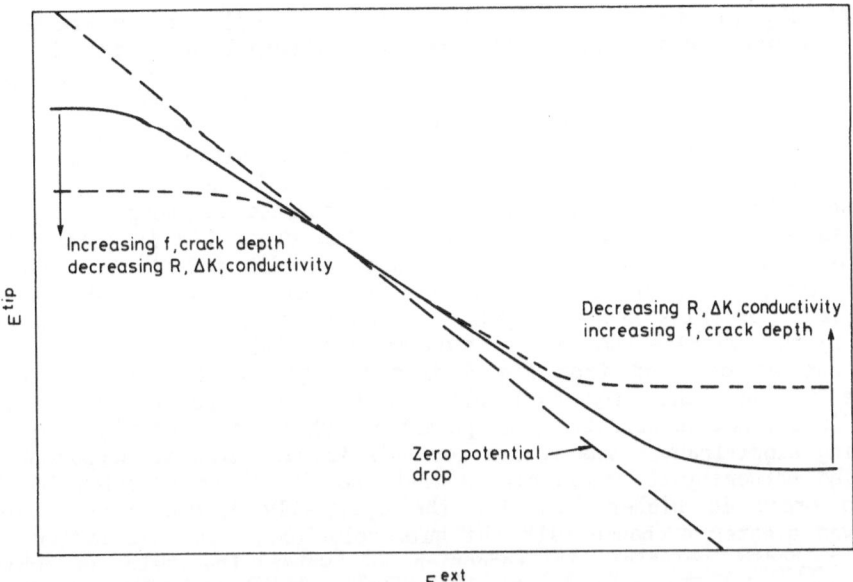

Fig 5. Schematic illustration of variation of crack tip potential with external potential showing limiting conditions and effect of experimental variables.

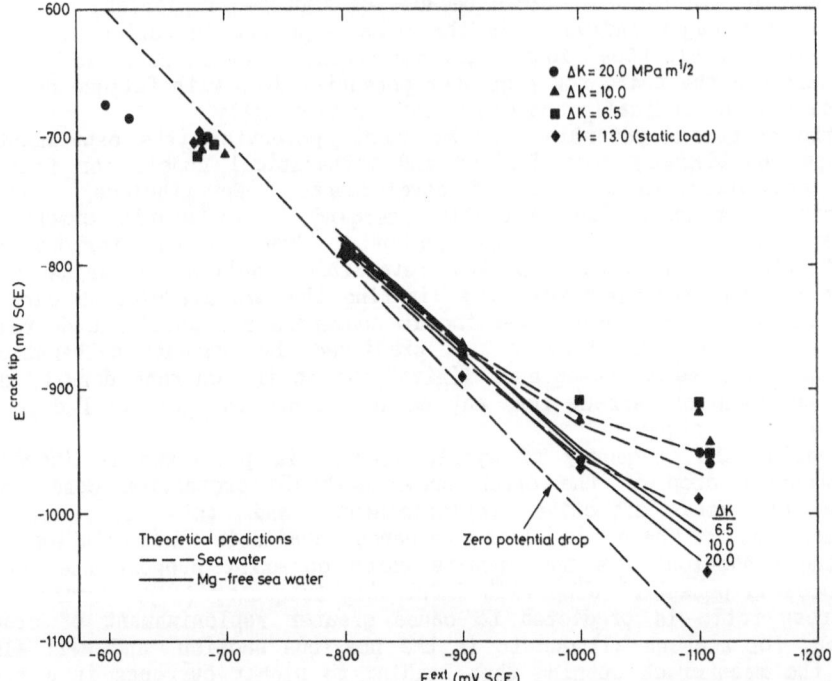

Fig 6. Comparison of predicted and experimentally measured crack-tip potentials as a function of potential at the crack mouth. R = 0.5; f = 0.1 Hz; l = 15 mm; a/W = 0.3; T = 5°C; BS 4360 50D steel in ASTM sea water.

specifically in Fig 6 for both static and cyclically loaded cracks in steel in sea water (17). The relevant external potential is the potential at the crack mouth which can then be correlated with potentials measured in service conditions. In laboratory tests using notched specimens the potential at the crack mouth may be significantly different (in excess of 50 mV in some circumstances) from that associated with the external steel surface and should be measured independently using a Luggin capillary inserted into the notch.

It is apparent from Figs 5 and 6 that the crack tip is polarisable only over a limited range of anodic and cathodic potentials. For a given metal/environment system the limiting potentials will depend on mechanical variables such as cyclic frequency, stress ratio and ΔK and environment variables such as solution conductivity.

It can be observed from Fig 6 that the potential drop in static cracks is less than for cyclically loaded cracks particularly under cathodic protection conditions at potentials where the potential drop is becoming significant. The reason for this trend, which is supported by detailed mathematical modelling, is that the pH of the solution in the static crack is higher than for the cyclically loaded crack (which involves greater exchange with the bulk solution). In this system the main cathodic reaction is reduction of water the rate of which diminishes with increasing pH. Thus, improved mixing with the bulk solution due to fatigue loading gives rise to lower pH values over some region of the crack and hence higher current. In NaCl solutions, oxygen reduction also tends to increase the potential drop and this is more important in fatigue cracks because of improved replenishment. In sea water the major influence is the reduced pH due to replenishment of buffer species, viz HCO_3^- ions, into the crack. Under cathodic protection therefore the trend for a greater potential drop with fatigue cycling is predicted theoretically and confirmed experimentally.

At the corrosion potential and at anodic potentials the experimental data are considerably more limited and mathematical models for fatigue cracks are still in a state of development. Nevertheless, results from both approaches are gradually emerging. In static cracks an important factor determining the potential drop is the ferrous ion concentration which tends to its reversible equilibrium value with respect to the iron electrode thus limiting the net dissolution current in the crack. When fatigue cycling is commenced a significant decrease in ferrous ion concentration is predicted by recent mathematical modelling (17). As a consequence limitations on the current density due to the build-up of ferrous ions may be absent and the potential drop is predicted to be higher.

Increasing the frequency of cyclic loading is predicted to increase the potential drop in the crack under cathodic protection conditions because of more effective replenishment, and this is observed experimentally. There is less evidence available in relation to corroding conditions but the results which do exist support the trend for increased potential drop with increasing frequency (17). Decreasing the stress ratio is predicted to cause greater replenishment of crack solution, for reasons alluded to in the previous section, and will also reduce the mean crack opening thus leading to higher currents in a more restrictive geometry. The former aspect is the more significant and experimental evidence in support of these predictions is shown in Fig 7. A similar trend of increased potential drop with decreasing stress ratio is experimentally observed under freely corroding conditions though the

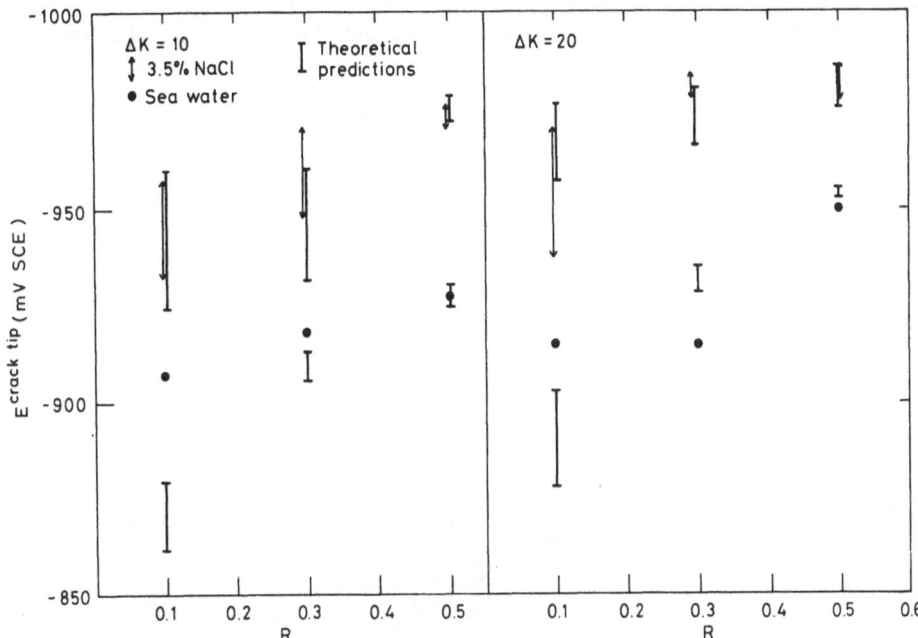

Fig 7. Variation of the crack-tip potential with R value and ΔK for
BS 4360 50D steel in sea water at 5 °C and in 3.5% NaCl at
20 °C; E^{ext} = - 1.0 V (SCE); l = 15 mm; f = 0.1 Hz, Units of
ΔK are $MPam^{1/2}$.

effect is less marked possibly because of the proximity to the zero-
potential drop condition and hence the small degree of overall
polarisation involved (18).

Increasing the stress amplitude should cause the crack to be more
open and the potential drop to be smaller as is indeed evident from Fig
6. Similarly, decreasing the crack depth will be expected to lower the
potential drop and this is observed experimentally in fatigue cracks
under cathodic protection conditions (17). In static cracks under
corrosion conditions the potential drop can tend to a limiting value
with increasing depth because of transport limited current. It is
anticipated that fatigue cycling would alter that behaviour but
modelling and experiment have not advanced to that stage.

<u>pH in cracks</u>

The pH obtained in a crack is dependent on a number of factors viz
the rate of production and hydrolysis of metal ions, salt concentration,
mass exchange with the bulk solution, buffering reactions, the rate of
cathodic reduction processes and possibly the reversibility of the
hydrogen electrode. Prediction of pH values under a particular set of
conditions will not be straightforward. Nevertheless, some general
trends can be established and modelling and experiment allow more
specific statements for a number of systems.

When cathodic protection is applied to the external metal surface the
potential in the crack will also go more negative, provided there is no
actual physical blocking of the crack, and typical crack tip potentials

were illustrated in Fig 6.   (Even if there is induced physical blocking
the driving force for dissolution from oxygen reduction on the external
surface will be negated and the rate of dissolution will diminish.) The
net current in the crack is cathodic and since this involves the
production of hydroxyl ions (Table I) the pH will increase and without
strong buffering the solution in the crack will become alkaline. This is
exemplified for cathodically protected structural steel in sea water (17)
in Fig 8.   Distinction between statically loaded and fatigue loaded

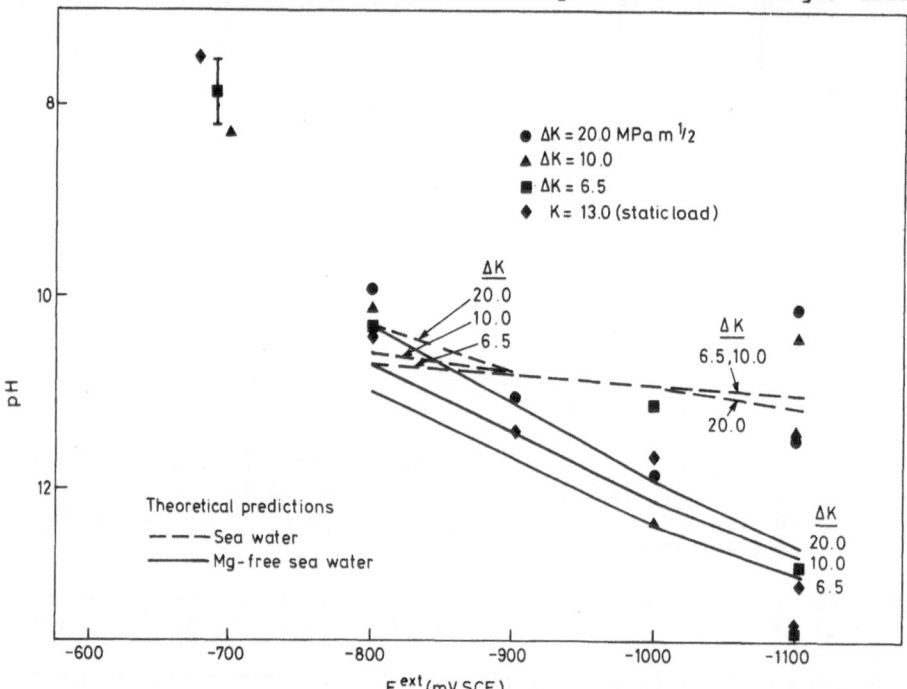

Fig 8.   Comparison predicted and experimentally measured crack-tip pH
         values as a function of potential at the crack mouth. R = 0.5;
         f = 0.1 Hz; l = 15 mm; a/W = 0.3; T = 5 °C; BS 4360 50D steel
         in ASTM sea water.

cracks (and effects of ΔK) is not readily apparent in these crack-tip pH
values.   Nevertheless, modelling does suggest a significant difference in
pH profile along the crack and in particular a much steeper decline in pH
towards the bulk value in the cyclically loaded crack due to improved
replenishment of the HCO_3^- buffer.   This is consistent with the trend for
a greater potential drop in the fatigue crack as discussed in the previous
section. Increasing frequency and decreasing stress ratio have the same
effect of steepening the pH profile, based on model calculations, but
experimental evidence is lacking as yet.
   The buffering reactions of sea water, including the impeding effect of
calcareous scale on oxygen reduction in the crack, tend to lead to lower
pH values than in 3.5% NaCl though the effect on crack tip values is
less pronounced than the effect on the pH distribution along the crack.
   The situation under freely corroding and anodic polarisation
conditions is much more complex and is critically dependent on the alloy
composition.   The main source of production of hydrogen atoms in the

crack is from hydrolysis of metal ions and the hydrolysis constants of
different ions range over several orders of magnitude as indicated in
Table 2. For structural steels the main hydrolysis reaction in the crack

---

TABLE 2. Hydrolysis of metal cations (19)

$$Fe^{2+} + H_2O \rightleftharpoons FeOH^+ + H^+ \qquad \log K = -9.5$$

$$Cr^{3+} + H_2O \rightleftharpoons CrOH^{2+} + H^+ \qquad \log K = -4.0$$

$$Ni^{2+} + H_2O \rightleftharpoons NiOH^+ + H^+ \qquad \log K = -9.86$$

$$Mn^{2+} + H_2O \rightleftharpoons MnOH^+ + H^+ \qquad \log K = -10.59$$

$$Al^{3+} + H_2O \rightleftharpoons AlOH^{2+} + H^+ \qquad \log K = -4.97$$

---

is hydrolysis of ferrous ions since even in fatigue cracks the crack is
usually oxygen free except in unusual loading circumstances (low stress
ratio, high frequency). The hydrolysis constant is small for this
process and when model calculations are made which take into account
cathodic reduction and mass transport the pH values are predicted to be
neutral to weakly alkaline. The significant reaction tending to increase
the pH is the cathodic process (mainly water reduction) which is often
neglected in crack pH calculations, on the basis that the cathodic current
is significantly less than the anodic current. However, whereas every
electron involved in reduction of water produces a hydroxyl ion the same
extent of production of hydrogen ions does not ensue from the dissolution
process because of the intermediate hydrolysis reaction. If full model
calculations are carried out including both anodic and cathodic processes
but based simply on ferrous ion hydrolysis as the source of H^+ atoms,
whether for static or fatigue cracks, the predicted pH can be considerably
alkaline (of order 10) particularly for deep cracks. Experimental pH
values tend to be in the range 7.0 to 8.5 at the free corrosion potential
(17,20) (= - 690 mV(SCE)), and for static cracks or crevices were not
significantly affected by crack or crevice depth (1). The explanation
for these pH values resides in the limited solubility of ferrous ions.
As the pH increases, ferrous hydroxide precipitates and this tends to
act as a buffer limiting the pH to a value associated with the ferrous
ion concentration through the solubility constant. In tests in sea
water under freely corroding conditions the pH tended to rise when
loading was changed from static to dynamic conditions. Recent model
calculations (18), although at a preliminary development stage, provide
an explanation in terms of the reduced concentration of ferrous ions
(and hence higher pH) in the cracks consequent upon cyclic loading.
   At the current time, insufficient data exist to predict behaviour over
a wider range of loading conditions, but a general description of the pH
as lying between 7.0 and 8.5 is likely to encompass most situations. The
lowest pH in static cracks is likely for anodic polarisation of shallow
cracks but previous predictions suggest that this value will still be
close to 7.0. Analysis for fatigue cracks is incomplete. Note that
the increase in the activity of the hydrogen ion which can occur as the
metal salt concentration increases in the crack will also lead to
enhanced cathodic reduction so that the net effect of salt concentration

on pH may be small for this system.

In order to obtain a lower pH in a crack (apart from adding a strong buffer) it is evident that this must arise from changes in material chemistry. Chromium is a common alloying element used in high strength steels such as AISI 4130, 4340 at the 1% (mass %) level. Provided that chromium-ion concentrations of a reasonable magnitude can be achieved in the crack then the much greater hydrolysis constant associated with this ion (Table 2) can in principle lead to much lower pH values. Thus, pH values as low as 4 at - 600 mV(SCE) were predicted mathematically for static cracks but only for relatively shallow cracks (21). As the crack depth increased the rate of metal dissolution became transport limited and hence the rate of production of chromium-ions diminished, with a resulting rise in pH. Measurements in deep crevices for 1% and 2% chromium steels in 3.5% NaCl confirmed these predictions of near neutral pH values (22). The effect of cylic loading on these predictions has yet to be examined. Nevertheless, in view of the dominance of diffusion in determining mass transport in very shallow cracks (see section on mass transport theory) the prediction of low pH values is likely to be sustained. The situation for deep cracks is less clear and speculation could be misleading.

As the chromium content of the steel is increased to the range associated with stainless steel, very low pH values of between 0 and 3 can be obtained - as reviewed previously (1). There have been some advances in current knowledge on crack pH since that review article and this applies also to aluminium alloy but the effect of fatigue cycling on crack tip pH values for these systems remains the challenge to the experimentalist.

Crack-tip kinetics and crack growth

Identification of the solution composition provides the basis for conducting separate electrochemical polarisation measurements to determine the kinetics of the anodic and cathodic process in that environment. The dynamic straining at the tips of cracks will lead to disruption of oxide or adsorbed films and it is necessary to simulate this process in controlled experiments. Methods of producing unfilmed material and of monitoring the electrochemical response have been critically discussed in a comprehensive review by Newman (23). Mechanical scraping provides the most generally effective method and numerous studies have been conducted for a range of metal/environment systems. Difficulties arise when the crack growth is path sensitive, as with integranular cracking, and simulating this process may require the development of alloys of appropriate composition. When the kinetics of reaction are mass transport limited, adequate simulation in a standard scraping test is particularly difficult and modelling may be a necessary complement when extrapolating to crack tip conditions.

Examples of maximum anodic and cathodic current densities determined using a scratching technique on structural steel in deaerated 3.5% NaCl (23) are shown in Figs 9 and 10. The anodic polarisation data at pH 8.5 are relevant to free corrosion conditions for this system under cyclic loading (for which the dissolution process is activation controlled since the ferrous ion concentration in significantly below the reversible value). Applying Faraday's law to estimate a maximum crack growth rate a value of $da/dt \approx 3.510^{-5}$ mms^{-1} is obtained at about - 690 mV(SCE). Maximum crack growth rates due to dissolution are

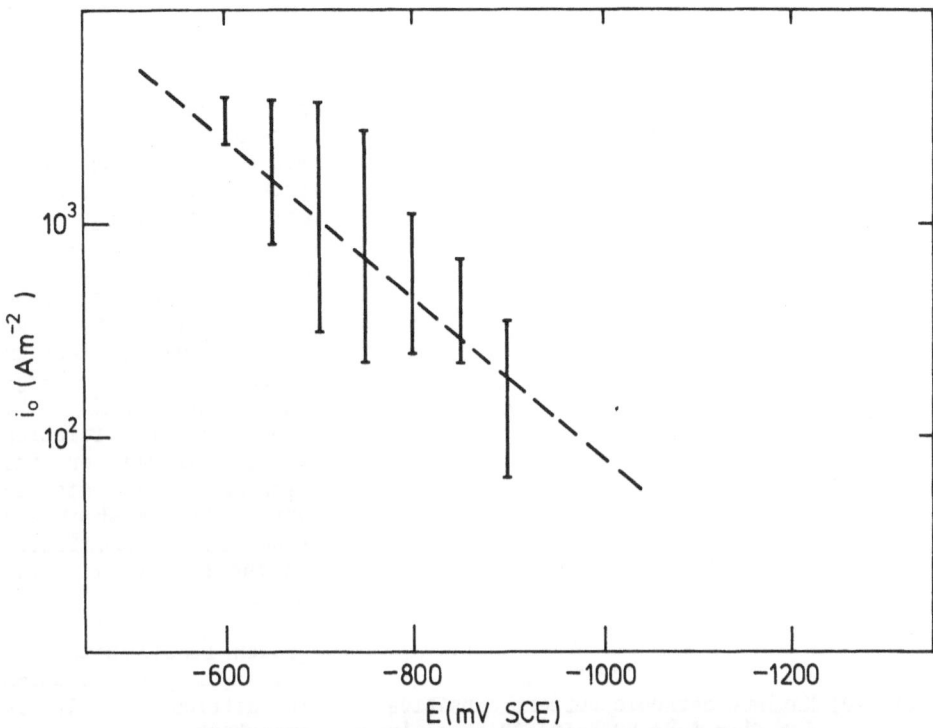

Fig 9. Maximum anodic current densities for film-free BS 4360
50D steel in deaerated 3.5% NaCl at 20 °C and pH 8.5.

only going to be sustainable at the higher frequencies where the rate
of crack advance is fast and bare surface is rapidly exposed.
However, Vosikovsky's fatigue data (25) for X-65 line-pipe steel in 3%
NaCl at 1 Hz indicates crack growth kinetics very considerably in excess
of this value for $\Delta K > 10$ MPam$^{1/2}$. A similar conclusion is derived
when compared with Appleton and Cowling's data for 4360 50D steel (as
used in the scratching tests) in 3.5% NaCl and sea water (26).
Consequently a mechanism of crack advance based on metal dissolution
does not seem sustainable for this system at the higher ΔK values and it
is necessary to involve absorbed hydrogen as the damaging species at the
free corrosion potential as well as at cathodic potentials.
Characterisation of the transient kinetics of water reduction (the
primary source of H atoms at neutral to alkaline pH values) at the
steady free corrosion potential is not directly possible because the
cathodic process on the scratched surface cannot be distinguished from
the much greater dissolution current density.   Conventionally,
polarisation curves are simply extrapolated and this is clearly feasible
with the maximum current densities.   Nevertheless, uncertainty remains
as to whether the current decay parameters can be similarly treated and
this is being investigated.
At the free corrosion potential the primary source of charging of the
steel is through the highly reactive crack tip but as the potential is
depressed the rate of water reduction on the external surface increases
with potential at a relatively greater rate compared to that at the

304

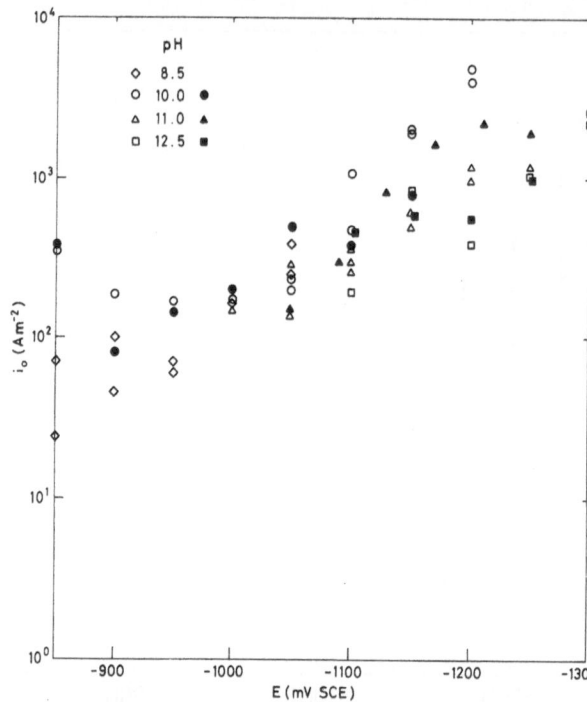

i_o (Am^{-2})

pH
◇ 8.5
○ 10.0 ●
△ 11.0 ▲
□ 12.5 ■

E (mV SCE)

Fig 10. Maximum cathodic current densities
for film-free BS 4360 50D steel in
deaerated 3.5% NaCl at 20 °C.

crack tip, (because of changes in pH and potential drop within the crack), and at - 1.0 V (SCE) exceeds the rate of generation within the crack. Hence, at these over-protection potentials crack chemistry has little relevance, except where calcareous plugging is important as in sea water. At intermediate protection potentials bulk charg-ing may still be important in relation to its effect on the hydrogen atom distri-bution in the steel and thus the concentration in the zone of triaxial stress.

Development of crack growth models based on the hydrogen mechanism is difficult. It is important to characterise (a) the electrode kinetics, which can vary between the crack and external surface, (b) the rate of hydrogen permeation into the steel, (c) hydrogen atom transport within the steel and (d) the response of the material to a given quantity of absorbed hydrogen. Combining these factors in relation to corrosion fatigue is a major challenge though steps towards this concept in static hydrogen embrittlement have already been undertaken (21).

Inevitably in this section it has not been possible to discuss adequately other metal/environment systems and their particular characteristics. However, having discussed hydrogen cracking mechanisms and anodic dissolution mechanisms it would be remiss not to mention briefly the film rupture mechanism of crack advance which can be of importance in a number of systems (eg α-brass in NaNO$_2$) (27). The important process at the crack tip in this case is the transient anodic reaction involving film formation. The particular characteristics of these films (which involve significant dealloying of the substrate) can lead to a type of brittle crack advance even in relatively ductile materials. Investigation of the nature of the films formed at crack tips in the local tip environment is an important aim.

CONCLUSIONS

Significant developments in the understanding and quantitative prediction of mass transport rates and electrochemistry in cracks have occurred over the last few years.

(1) Definition of the relative importance of diffusion and stress induced flow in corrosion fatigue cracks now exists which combined with awareness of the significance of volume of solution to metal area ratio enables a clearer understanding of the role of crack depth on crack electrochemistry.

(2) Experimental and mathematical modelling methods have been developed which enable the effect of mechanical loading parameters (R, ΔK, frequency) and environmental variables (potential, solution composition) on crack electrochemistry to be characterised.

(3) In relation to hydrogen assisted cracking prediction of the increased role of bulk charging on crack growth as the potential is depressed is an important advance.

FUTURE WORK

(1) The range of conditions under which externally induced flow is significant in relation to crack chemistry has to be identified experimentally and theoretically if proper simulation of service conditions is to be adopted in the laboratory.

(2) Extension of modelling and experimental measurement of crack-tip environments to other metal/environment systems is clearly of importance.

(3) Characterisation of transient kinetics is essential to the proper development of predictive models of crack growth and further investigation in simulated crack-tip environments is required.

(4) Evaluation of the time-dependent distribution of hydrogen atoms in stressed specimens taking into account the contribution to charging from both the crack and external surface is desirable.

(5) Modelling of crack growth based on anodic reactions is reasonably well advanced, with the important exception of film-induced cleavage, but further work to correlate the kinetics of hydrogen generation with the metal's response to absorbed hydrogen, and thus crack growth, is important.

REFERENCES

1. Turnbull A: The solution composition and electrode potential in pits, crevices and cracks, Corrosion Science, 23(8), (1983) pp 833-870.

2. Brown B F, Fujii C T and Dehlberg E P: Methods for studying the solution chemistry within stress corrosion cracks, J. Electrochem. Soc., 116 (1969) pp 218-219; Brown B F, Theory of Stress Corrosion Cracking in Alloys, NATO (1971).

3. Davis J A: Use of microelectrodes for study of aluminium alloys, pp 168-172 in Localized Corrosion, NACE, Houston, 1974.

4. Newman J: Electrochemical Systems, Prentice-Hall, N J, 1973.

5. Easthope P J and Turnbull A: Aspects of fluid flow of relevance to corosion fatigue and stress corrosion cracks, NPL Report DMA(A)105, (1985).

6. Pan F and Acrivos A: Steady flows in rectangular cavities, J. Fluid Mech., 28(4), (1967) pp 643-648. (Fig 2 reprinted by permission of Cambridge University Press).

7. Andresen P, Ballinger R, Marra M and Psaila, M: Modelling high temperature SCC based on crack-tip chemistry and mass transport

306

considerations in Modelling Environment Effects on Crack Growth Processes, AIME, Toronto, 1985 (to be published).

8. Alkire R C, Reiser D B and Sani R L: Effect of fluid flow on removal of dissolution products from small cavities, J. Electrochem. Soc., 131(12) (1984), pp 2795-2800. (Fig 3 reprinted by permission of the publisher, The Electrochemical Society, Inc.)

9. Beck T R: Effect of hydrodynamics on pitting, Corrosion, 33 (1977) pp 9-13.

10. Scott P M: AERE Harwell, Private Communication, (1985).

11. Choi H, Beck F H, Szklarska-Smialowska, Z and MacDonald D D: The effect of fluid flow on the stress corrosion cracking of ASTM A508 C12 steel and AISI Type 304 stainless steel in high temperature water, Corrosion, 38(2), (1982) pp 76-85.

12. Hurst P: Stress corosion behaviour of A533B and A508-III steels and weldments in high temperature water environments, 2nd Int. Symposium on Environmental degradation of Materials in Nuclear Power Systems - Water Reactors, Monterey, 1985.

13. Turnbull A: Theoretical analysis of the influence of crack dimensions and geometry on mass transport in corrosion fatigue cracks, Mat. Sci. Tech., 1, (1985) pp 700-710.

14. Turnbull A and Ferriss D H: Mathematical modelling of the electrochemistry in corrosion fatigue cracks. 1. Structural steel cathodically protected in 3.5% NaCl, in Corrosion Chemistry within Pits, Crevices and Cracks, HMSO (1986), to be published.

15. Turnbull A and Ferriss D H: Mathematical modelling of the electrochemistry in corrosion fatigue cracks in structural steel cathodically protected in sea water, accepted for publication in Corrosion Science, (1986).

16. Turnbull A and Newman R C: The influence of crack depth on crack electrochemistry and fatigue crack growth, in Small Fatigue Cracks, Santa Barbara, 1986 (to be published).

17. Turnbull A and Dolphin A S: Measurement of the pH and potential in corrosion fatigue cracks in structural steel in sea water, to be published.

18. Turnbull A and Ferriss D H: Work in progress.

19. Baes C F and Mesmer R F: The Hydrolysis of cations, John Wiley, New York (1976).

20. Hodgkiess T H, Cannon M J and McLachlan A: Electrochemical measurements within fatigue cracks in structural steel, in Corrosion Chemistry within Pits, Crevices and Cracks, HMSO (1986) to be published.

21. Gangloff R P and Turnbull A: Crack electrochemistry modelling and fracture mechanics measurement of the hydrogen embrittlement threshold, in Modelling Environmental Effect on Crack Growth Processes, AIME, Toronto (1985) (to be published).

22. Turnbull A and May A T: Cathodic protection of crevices in BS 4360 50D steel in 3.5% NaCl and in seawater, Materials Performance, 22 (10), (1983), pp 34-38.

23. Newman R: Measurement and interpretation of electrochemical kinetics on bare metal surfaces, in Corrosion Chemistry within Pits, Crevices and Cracks, HMSO (1986), to be published.

24. Herman R, Boomer D, Hamilton M and Turnbull A: manuscript in preparation.

25. Vosikovsky O: Fatigue crack growth in an X-65 line-pipe steel at low

cyclic frequencies in aqueous environments, <u>Trans AIME</u>, (1976), pp 298-304.

26. Appleton R J, Deans W F and Cowling M J: Corrosion fatigue mechanisms in a C-Mn steel, Glasgow Marine Technology Centre Report, ME-85-02, 1982.

27. Sieradzki K and Newman R C: Brittle behaviour of ductile metals during stress corrosion cracking, <u>Phil. Mag</u>. 51, (1986), pp 95-132.

DISCUSSION

Comment by M. Pourbaix:

We have been looking extensively at methods which would allow one to prepare solutions similar to those existing inside "occluded cells" (crevices, pits, cracks, etc.) in metals and alloys and to predict their behaviour in these solutions. This we have done notably with B. F. Brown, C. T. Fujii and L. Sathler. What we presently recommend is simply to add to oxygen-free water, up to saturation, increasing quantities of the less soluble salts likely to be formed in the cavity (for example, if chloride is present CuCl in the case of copper, and $FeCl_2$, 4 H_2O in the case of iron and unalloyed steels). In such solutions, in which one liter or more may be prepared, potentiostatic experiments make it easy to clarify all conditions of corrosion, immunity, passivation, hydrogen embrittlement, etc. (See proceedings of Latanision 2, p. 603).

Reply:

It is not realistic to assume that the crack tip solution will be saturated in ferrous chloride at the corrosion potential ($\approx$-690 mV(SCE)). Calculations indicate values considerably less and typically in the region of 10^{-3} and 10^{-1} moles litre^{-1} in corrosion fatigue cracks. The macro-cell method is useful in enabling relevant polarization measurements to be made, but it is first necessary to determine or predict the range of concentrations of metal ions in the crack and also the pH range.

Comment by C. J. Altstetter:

Your mathematical modeling of mass transport used very simple crack geometries; however, actual fatigue cracks (or SCC cracks) are very convoluted and branched. Does this cause a serious discrepancy in the comparison of computed and experimental results?

Reply:

The model has been developed initially for a Mode I crack and is applicable for a range of conditions but with the exceptions indicated in the text, e.g. intergranular cracking at low ΔK values where the crack opening may be of the same order as the surface roughness. In principle, it would be possible to model more complex geometries, but crack opening as a function of distance from the crack tip would have to be defined and these data are not available. The importance of crack geometry will depend on the particular metal/environment system. For structural steel in sea water and in 3.5% NaCl the effect would not appear to be too significant since reasonable agreement between experiment and theory is observed even at low R and ΔK values where crack closure and Mode II displacements would occur. However, I would not like to generalize at this stage.

Comment by R. M. Latanision:

Regarding the potential distribution in a crack tip, do you take into account the evolution of hydrogen gas (which may occur at bulk anodic or cathodic potentials) as H may affect both the resistance of the conductive path in the crack tip fluid and the convective term in your modeling?

Reply:

The evolution of gas in this metal/environment system will only occur at quite negative potentials of about -110mV (SCE) and this can be observed experimentally. It is certainly possible that this will disturb the convective motion and affect the transport path. In practice we have found that under static loading small potential fluctuations occur but this disappeared under fatigue loading, possibly because the opening and closing of the crack helped to sweep out the gas bubbles. Nevertheless, the effect would inevitably become important if the potential was depressed further. The important point is that for the range of potentials at which significant gas evolution is occurring the main source of hydrogen charging is from the external surface. Hence, the crack tip conditions are not so important except in relation to calcareous wedging.

Comment by R. H. Jones:

For your environment and material, is it known whether the transport and/or reaction rates within the crack electrolyte or the hydrogen transport and/or embrittlement processes within the material control ΔK_{TH} and da/dN.

Reply:

At the cathodic potentials where bulk charging is dominating, I believe that for a given potential crack growth will depend on the rate of accumulation of H atoms in the plastic zone and transport will be the rate limiting process. At other potentials there is less definitive evidence. In relation to threshhold behavior, the threshhold ΔK depends mainly on crack closure and the presence of the environment can lead to the same threshhold as in air but in some instances a high threshhold because of crack plugging; for example with calcareous scale.

Comment by A. W. Thompson:

Could you elaborate on the reasons for difficulty in comparing laboratory results to cracking in actual pipelines and structures under hydrogen conditions?

Reply:

In fracture mechanics specimens and also in tensile specimens, tests are usually conducted with the specimens fully immersed. When bulk charging is important (sea water at potentials less than -900 mV (SCE) and possibly brine/H_2S mixtures at the corrosion potential) the specimen is uniformly

charged. In a tubular steel as used in offshore structures or in pipelines, charging is from one surface only and hence a H atom concentration
gradient will exist through the thickness of the steel. As the crack
grows, it will see a varying hydrogen field. Hence, in relation to tubu-
lars, full immersion and uniform charging tests using standard specimen
geometries may not give realistic crack growth.

Comment by J. F. Knott:

You commented that the effect of frequency in corrosion fatigue was
only to alter the time in which the species in the crack reached a
distance, l_{crit}. Presumably, this referred to laminar flow? If the
frequency were high and low R ratios were employed, might not turbulent
flow be set up and a high proportion of the chemical reactions and
products be flushed out of the crack, each cycle?

Reply:

The calculation referred to laminar flow. Turbulent flow may exist
under the conditions you mentioned; and, indeed, for the combination of
low R and high frequency, considerable replenishment of crack solution
will occur each cycle.

Comment by K. Nisancioglu:

From what I gather from earlier talks, the crack width has atomic
(molecular) dimensions for an appreciable crack length starting from the
crack tip. Would your continuum model not break down in this "inner"
region? I think a natural extension of your work would be to develop an
atomistic model of mass and charge transport for this "inner" region and
match this with your continuum model for the "outer" region.

Reply:

In relation to cracking of ductile steels in sea water, the magnitude of
the crack tip opening is of the order of 1 μm, and this is very much
greater than the thickness of the double layer which is about 50 Å. Even
in the sharper crack geometries associated with the film induced cleavage
mechanism of stress corrosion cracking, the crack tip opening will be
about 500 to 1000 Å which is still an order of magnitude greater than the
thickness of the double layer. Hence I consider the continuum approach to
be sensibly based.

Comment by P. Marcus:

Is there any circumstance, high temperature water for example, where
surface diffusion along the crack walls could play a role in mass transport to the crack tip?

Reply:

I presume that you mean surface diffusion of hydrogen atoms. Because
the rate of generation of H atoms at the crack tip is greater than on the
walls, then it is likely that the activity gradient will lead to surface
diffusion away from the tip, if it is important at all.

Comment by M. Bernstein:

The role of Cr in high strength steels appears to be confusing. Your results suggest that it increases hydrogen ion concentration, but metallurgically it appears to improve resistance, likely by its effect on precipitates. Do you have suggestions as to a better alloying element to add to high strength steels?

Reply:

The two key factors are the electrochemical supply of H atoms and the materials response to that hydrogen. With regard to the first point, we haven't as yet predictions of the pH in fatigue crack in steels of low Cr concentration (1% by weight) but this will be done shortly. In relation to static cracks, previous modelling demonstrated that the presence of chromium in the steel at this concentration could lead to pH values of about 4 for shallow cracks (e.g. 0.5 mm); but for deep cracks (e.g. 1 cm) the pH was near neutral. The prediction was that crack growth would be enhanced in the shallow crack and this was subsequently confirmed (R. P. Gangloff and A. Turnbull in Modelling Environmental Effects on Crack Growth Processes, TMS-AIME, ed. R. H. Jones and W. Gerberich, 1986). The same sensitivity of crack growth to crack depth was also observed in corrosion fatigue cracking; and I anticipate that a similar variation in chemistry between shallow and deep cracks will be observed. With regard to the materials response, it should be pointed out that the sensitivity of K_{TH} to hydrogen content is not very high; hence the variation in pH in the crack does not necessarily have a dramatic response in that context. The dependence of crack growth on hydrogen content is likely to be more significant, but I have no specific information for the high strength steels. Also, I do not have any suggestions for better alloying elements.

ELECTROCHEMICAL THERMODYNAMICS AND KINETICS AND THEIR
APPLICATION TO THE STUDY OF STRESS-CORROSION CRACKING

MARCEL POURBAIX

Belgian Center for Corrosion Study, CEBELCOR
Avenue Paul Héger, gate 2. 1050 BRUSSELS, Belgium

ABSTRACT

After a short review of some fundamental concepts related to electrochemical thermodynamics and kinetics, a statement is made on the application of these concepts to the study of stress-corrosion cracking (SCC).
    The dangerous electrode potential range where stress corrosion cracking
may occur in a given solution may often be predicted by superimposing a
polarization curve determined in the bulk of the solution and polarization
curves determined in the solution existing inside the cracks. Above this
potential range SCC may stop due to passivation of the cracks; below this
potential range, SCC may stop due to immunity.
    Solutions similar to those existing inside cracks, which are often
"occluded corrosion cells" (OCC), may be prepared artificially, in any
quantity desirable for chemical and electrochemical studies, by simply
dissolving in oxygen-free water, and in the presence of the metal or alloy
in powder- or shavings-form, increasing quantities, up to saturation, of
the less soluble salt likely to be formed inside the crack; for instance,
$FeCl_2.4H_2O$ in the case of non-alloyed steel in the presence of chloride.
    Examples of the prediction of such SCC conditions are given for a non-
alloyed steel and for a 12% Cr chromium steel.

* * * * * * *

    Digby D. Macdonald (1) and A. Turnbull (2) have given us striking exam-
ples of the practical usefulness of electrochemistry and of the chemistry
of the crack-tip for the understanding and the mastering of stress-corrosion cracking.
    In the present talk I shall emphasize the interest of a joint consider-
ation of both electrochemical thermodynamics and electrochemical kinetics,
as well as of experimental potential-pH diagrams relating to "occluded
corrosion cells" (OCC), based on these two concepts. This talk will be
closely related to the one I gave during the NATO Workshop which was held
May 22-31, 1981 in Calcatoggio, Corsica (3).
    First, for those of you who would not be fully familiar with electro-
chemistry, I shall make a brief theoretical statement in addition to the
excellent introduction which was given to you yesterday by Digby D.
Macdonald. Details on these subjects can be found in treatises on electrochemistry and corrosion (4,5).

1. ELECTROCHEMICAL THERMODYNAMICS AND KINETICS

1.1 Chemical and electrochemical reactions. Equilibrium conditions. The first principle of electrochemical thermodynamics.

A <u>chemical reaction</u> (such as $Fe + 2H^+_{aq} = Fe^{++}_{aq} + H_2$) (Figure 1) in which several chemical species M (such as, Fe, H^+_{aq}, Fe^{++}_{aq} and H_2) take part, may be written

$$\Sigma \upsilon M = 0 \qquad (1)$$

The <u>equilibrium condition</u> of this reaction is

$$\Sigma \upsilon \mu = 0 \qquad (2)$$

where the μ are the chemical potentials (or free enthalpies of formation $g = h - TS$) of the reacting species M.

If one represents by (M) the values of the activities (or "corrected concentrations") of the dissolved substances and the fugacities (or "corrected partial pressures") of the gaseous substances, $\mu = \mu^\circ + RT \ln (M)$, where μ° is the "standard chemical potential" of the considered substance at the considered temperature, formula (2) may thus be written as follows:

$$\Sigma \upsilon \log (M) = \log K$$

where $\log K = - \dfrac{\Sigma \upsilon \mu^\circ}{4.575 \, T}$ $\qquad (3)$

K is the "<u>equilibrium constant</u>" of the chemical reaction.

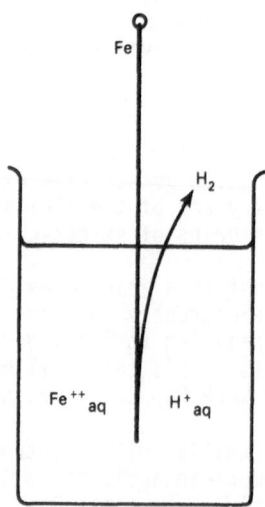

FIGURE 1. Chemical reaction.

The values of the "standard chemical potentials" (or "standard free enthalpies of formation") of the considered substances are given in thermodynamic tables. For instance, for the substances taking part in the reaction $Fe + 2 H^+_{aq} = Fe^{++}_{aq} + H_{2gas}$, these values are, at 25°C (298K):

$$\mu° \ Fe \quad = \quad 0 \ cal$$

$$\mu° \ H^+_{aq} \quad = \quad 0 \ cal$$

$$\mu° \ Fe^{++}_{aq} \quad = - \ 20300 \ cal$$

$$\mu° \ H_2 \quad = \quad 0 \ cal$$

and the equilibrium condition (Equation 3) is

$$\log \frac{(C_{Fe^{++}}) \times p_{H_2}}{(C_{H^+}{}^2)} = \frac{\mu°_{Fe^{++}_{aq}} \ H_2 - \mu°_{Fe} - 2\mu°_{H^+_{aq}}}{4.575 \times 298}$$

$$= \frac{20300}{1363} \quad = \ 14.9$$

Thus, according to this relationship, the thermodynamic equilibrium of iron in the presence of an aqueous solution where the activity of dissolved Fe^{++} ions is 1 mole per liter would be reached when the fugacity in gaseous H_2 is $10^{14.9}$ atmospheres.

An electrochemical reaction (such as $Fe = Fe^{++}_{aq} \ 2e^-$, or $2H^+_{aq} + 2e^- = H_2$) (Figure 2), in which chemical species and free electrons e^- take part, may be written

$$\Sigma\upsilon M + n(e^-) = 0 \tag{4}$$

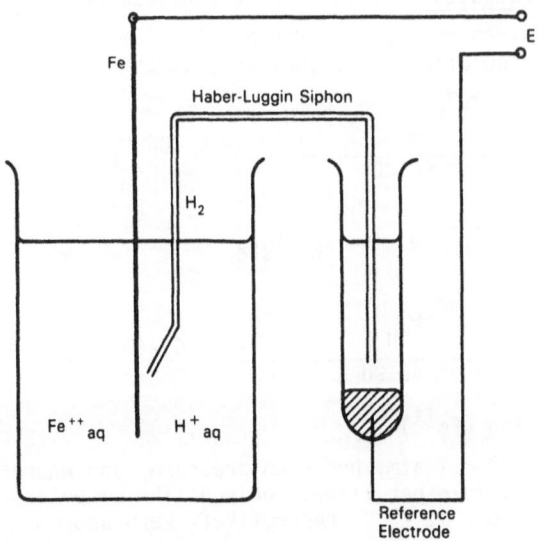

FIGURE 2. Electrochemical reactions.

and its <u>equilibrium condition</u> is

$$\Sigma \upsilon \mu - 23060 \, n \, E_o = 0 \qquad (5)$$

where the chemical potentials μ are expressed in calories per molar group and the electrode potential E_o is expressed in volts. This electrode potential E_o is to be measured versus a given reference electrode on which the state of equilibrium of a given reversible electrochemical reaction is obtained (for instance $H_2 = 2 \, H_{aq}^{++} + 2e^-$, where the hydrogen gas H_2 is under 1 atm. pressure and where the H_{aq}^{++} ions are at pH zero), and where the chemical potential μ are considered as equal to zero.
Formula (5) may be written

$$E_o = \frac{\Sigma \upsilon \mu}{23060 \, n} \qquad (6)$$

According to this formula (6), there is, for every electrochemical reaction, a given value of the electrode potential for which this reaction is in a state of equilibrium: no oxidation or reduction is possible. The value E of this <u>equilibrium electrode potential</u> only depends upon the values of the chemical potentials μ of the reacting chemical substances; i.e., upon the temperature, the activities (or "corrected concentrations" of the dissolved substances, and the fugacities (or "corrected partial pressures") of the gaseous substances. Formula (6), which is the equilibrium condition of an electrochemical reaction, expresses, in fact, <u>the first principle of electrochemical thermodynamics</u>.

The equilibrium conditions (Equation 5) may be written as follows:

$$E_o = E_o^\circ + \frac{0.0591}{n} \Sigma \upsilon \, \log (M)$$

where
$$\qquad (7)$$

$$E_o^\circ + \frac{\Sigma \upsilon \mu^\circ}{23060 \, n}$$

E_o° is the "<u>standard equilibrium potential</u>" of the electrochemical reaction.

According to the values of μ° given here above, the equilibrium conditions of the two considered electrochemical reactions are, at 25°C:

Reaction $H_2 = 2H_{aq} + 2e^-$: $E_o^\circ = \dfrac{2\mu^\circ_{H^+_{aq}} - \mu^\circ_{H_2}}{46120} = 0.000 \, V_{she}$

$$E_o = 0.000 - 0.0591 \, p_H - 0.0295 \, \log \, p_{H_2} \, V_{she}$$

Reaction $Fe = Fe^{++}_{aq} + 2e^-$: $E_o^\circ = \dfrac{\mu^\circ_{Fe^{++}_{aq}} - \mu^\circ_{Fe}}{46120} = 0.440 \, V_{she}$

$$E_o = -0.440 + 0.0295 \, \log \, (Fe^{++}) \, V_{she}$$

Thus, in a solution of pH zero under 1 atm. hydrogen pressure, and where the activity of dissolved Fe^{++} is 1 mole per liter, the equilibrium potentials of these two reactions would be, at 25°C, respectively zero and $-0.440 \, V_{she}$.

1.2 Electrochemical affinity (overpotential), direction and velocity of electrochemical reactions. The second principle of electrochemical thermodynamics.

When a given electrochemical reaction $\Sigma\mu M + n\ e^- = 0$ may occur on a metallic surface, the affinity A (or overpotential) of this reaction is equal to the difference $E - E_o$ between the actual electrode potential E of this metallic surface and the equilibrium electrode potential E_o of the reaction, both these electrode potentials being related to the same values of the chemical potentials μ of the reacting chemical substances (same temperature, same activities, same fugacities):

$$A = E - E_o \qquad (8)$$

As shown by Th. De Donder, the direction of a chemical reaction (indicated by the sign of its velocity v) is related to the sign of its affinity A by the relation

$$A \times v \geq 0 \qquad (9)$$

In the case of an electrochemical reaction, this formula of De Donder leads to the formula

$$(E - E_o) \times i \geq 0 \qquad (10)$$

where the "reaction current" i is considered as positive in the case of oxidations, and negative in the case of reductions. Thus:

o if $E > E_o$, $i \geq 0$ Only an oxidation is possible (or no reaction in case of irreversibility):

$$Fe \rightarrow Fe_{aq}^{++} + 2\ e^- \qquad \text{(corrosion of iron)}$$

$$H_2 \rightarrow 2H^+ + 2\ e^- \qquad \text{(oxidation of hydrogen gas)}$$

A reduction is impossible.

o if $E < E_o$, $i \leq 0$ Only a reduction is possible (or no reaction in case of irreversibility):

$$Fe_{aq}^{++} + 2\ e^- \rightarrow Fe \qquad \text{(electrodeposition of iron)}$$

$$2H_{aq}^+ + 2\ e^- \rightarrow H_2 \qquad \text{(evolution of hydrogen gas)}$$

An oxidation is impossible.

Thus, knowing the electrode potential E of a metallic surface in contact with an electrolyte, and the equilibrium potentials E_o of all the electrochemical reactions which might take place on the interface between the metal and the electrolyte, it is possible to predetermine in what direction (oxidation or reduction) the reactions are possible.

The relation

$$(E - E_o) \times i \geq 0 \qquad (10)$$

which relates the velocity i of an electrochemical reaction to its affinity $(E - E_o)$, expresses the second principle of electrochemical thermodynamics.

1.3 Polarization curves in stirred solutions. The Tafel rule.

The relationship between the affinity of an electrochemical reaction and its velocity may be often expressed experimentally by polarization curves (of the type $E = f(i)$ or $E = f(\log |i|)$) determined in stirred solutions,

316

under conditions where the values of the chemical potentials μ (notably
the concentrations of the dissolved reactants) remain constant during the
experiment (and equal to those existing in the bulk solution). The polarization curves then generally obey the Tafel rule, according to which the
logarithm of the reaction velocity is a linear function of the electrode
potential:

$$E = \alpha - \beta \log |i| \qquad \text{for an oxidation}$$
$$E = \alpha' - \beta' \log |i| \qquad \text{for a reduction}$$

Such Tafel lines are shown in Figures 3a and 4a.

In these figures, the crossing point of the two Tafel lines relates to
the equilibrium conditions of the considered reaction. The ordinate of
this crossing point is the equilibrium electrode potential E of the reaction, and its abscissa is its so-called "exchange current" i_o. In the case
of a reversible electrochemical reaction (Figures 3a and 3b), the exchange
current is relatively large (>10^{-8} to 10^{-6}A.cm^{-2}): any increase of the
electrode potential above its equilibrium value)which is the oxido-
reduction potential of the reaction) leads to an appreciable value of the
oxidation velocity, and any decrease of the electrode potential below its
equilibrium value leads to an appreciable value of the reduction velocity.
Contrarily, in the case of an irreversible electrochemical reaction
(Figures 4a and 4b), where the exchange current is relatively small (<10^{-8}
to 10^{-6} A.cm^{-2}), the occurrence of both the oxidation and the reduction
implies a more or less important overpotential: oxidation starts only above
a given "oxidation potential", and reduction starts only below a given
"reduction potential."

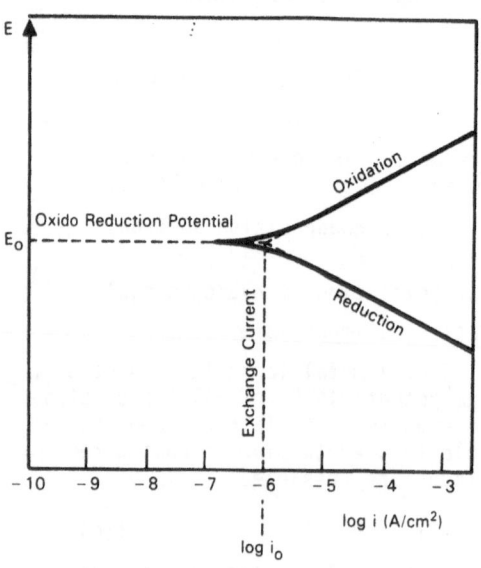

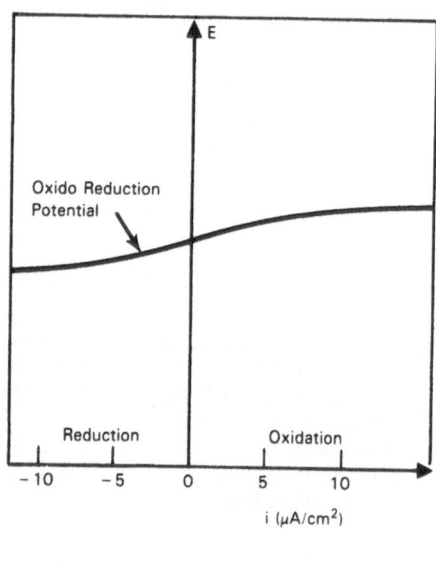

(a)   (b)

FIGURE 3.6 Polarization curve of a reversible electrochemical reaction
(i_o = 10^- A.cm^{-2}.
a. arithmetic current intensity scale E = f(i)
b. logarithmic current intensity scale E = f(log|i|).

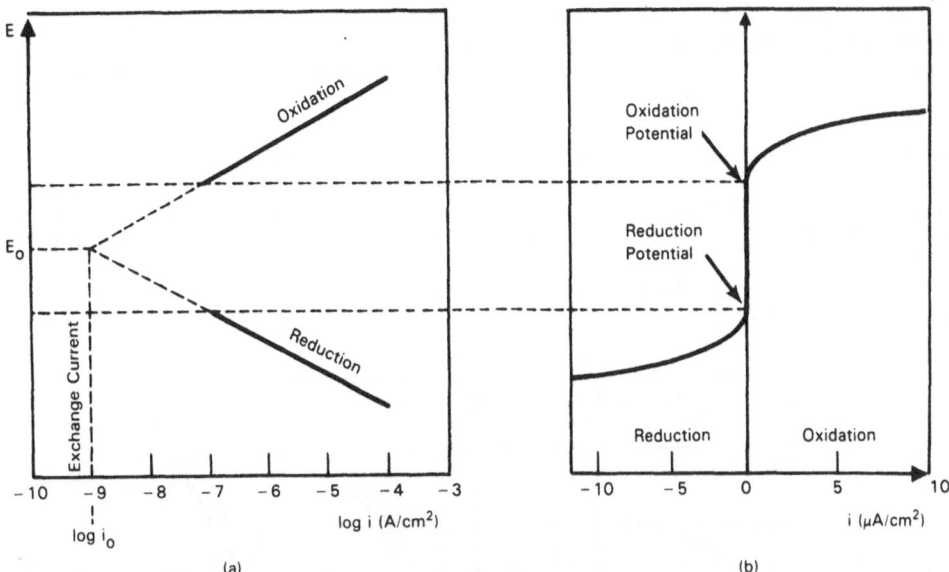

FIGURE 4. Polarization curve of an irreversible electrochemical reaction.
a. arithmetic current intensity scale $E = f(i)$
b. logarithmic current intensity scale $E = f(\log|i|)$.

1.4. The corrosion of iron with hydrogen evolution

Figures 5a and 5b relate to the application of the above mentioned concepts to the corrosion of iron with evolution of hydrogen, in a solution of pH zero containing 0.010 g.atom Fe^{++} per liter, at 25°C. The polarization lines marked a and c relate respectively to reactions $H_2 = 2H_{aq}^+ + 2e^-$ (a) and $Fe = Fe_{aq}^{++} + 2e^-$ (c), where the equilibrium conditions and the Tafel slopes are the following:

	reaction	
	$H_2 = 2H_{aq}^+ + 2e^-$	$Fe = Fe_{aq}^{++} + 2e^-$
Equilibrium potential $E_0(volts_{she})$	0.000	-0.500
Log exchange current i_0 (A.cm^{-2})	-5.85	-4.50
Tafel lines (volts per log unit)		
β ox	+0.123	+0.328
β red	-0.123	-0.328

In the absence of external currents, the total oxidation current $(Fe \rightarrow Fe_{aq}^{++} + 2e^-)$ and the total reduction current $(2H_{aq}^+ + 2e^- \rightarrow H_2)$ are equal, and correspond to the overall reaction

$$Fe + 2H_{aq}^+ \rightarrow Fe_{aq}^{++} + H_2$$

The electrode potential corresponding to this condition of zero net current is often called "corrosion potential," E_{corr}, and is -0.250 V$_{she}$ for the case of Figures 5a and 5b. The common value of the current at the zero current potential is often called "corrosion current," i_{corr}, and is in these figures 0.162 A. cm^{-2} (log i_{corr} = -3.79). Since 1 mA.cm^{-2} corresponds to the dissolution of 9.1 g iron per year and to a corresponding

318

thickness loss of 11 mm per year (or 44 mpy) for general corrosion, the corrosion current for Figures 5a and 5b corresponds to a loss of 1.47 g.cm⁻² per year, or to a thickness loss of 1.8 mm per year, or 0.072 mpy.

One also sees in these figures that, if the electrode potential of iron is being increased above -0.250 V$_{she}$, the corrosion velocity of iron will increase; the hydrogen evolution will decrease and will become zero at 0.000 V$_{she}$. If the electrode potential of iron is being deceased below -0.250 V$_{she}$, the corrosion velocity of iron will decrease and will become zero at -0.500 V$_{she}$, below which metallic iron will be electro-deposited on the metal.

Figures similar to Figures 5a and 5b have been used in drawing Figure 11 of the present report.

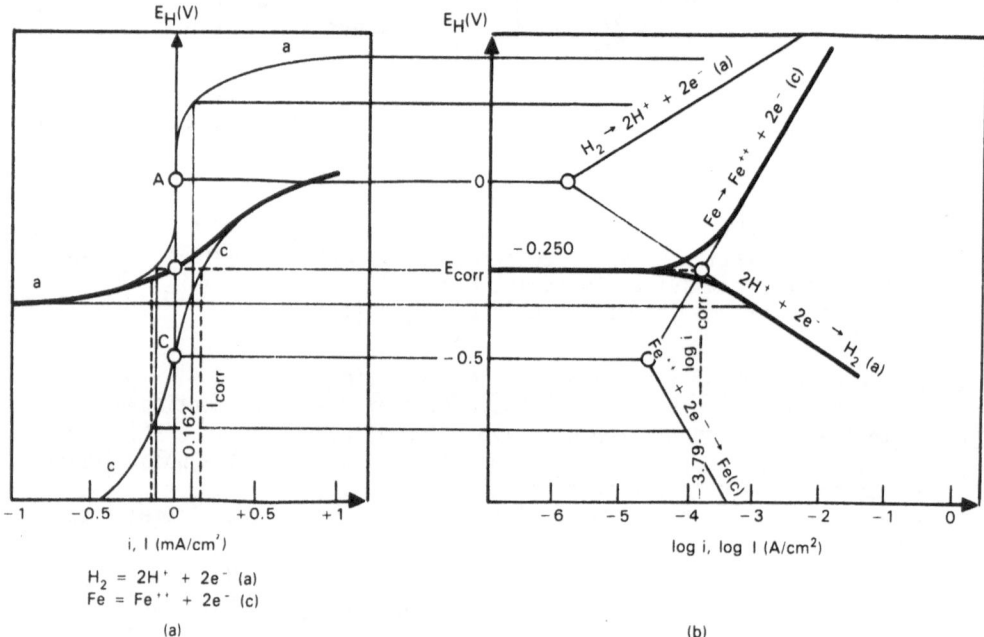

$$H_2 = 2H^+ + 2e^- \text{ (a)}$$
$$Fe = Fe^{++} + 2e^- \text{ (c)}$$

(a)   (b)

FIGURE 5. Corrosion of iron with hydrogen evolution, in solution of pH zero containing 0.010 g.atom Fe⁺⁺/liter. Anodic and cathodic polarization curves. a) arithmetic current intensity scale E = f(i)
         b) logarithmic current intensity scale E = f(log|i|).

2. ON THE ELECTROCHEMISTRY OF STRESS-CORROSION CRACKING

Thanks to work performed by H. H. Uhlig (6 to 14), R. W. Staehle (15), R. N. Parkins (16,17), P. E. Morris (18) and G. J. Theus (19), it is well known that, when SCC appears, it manifests itself in a certain range of electrode potentials (measured on the external surface of the metal) near the limit between passivity and general corrosion (Figures 6 and 7); i.e., according to R. W. Staehle's words, "in the potential range where the passivating film presents kinetic instability. This range of electrode potentials can be determined by using Parkins' "slow traction method" (Figure 8).

As mentioned during two meetings held respectively in 1978 in Rio de Janeiro (20,21) and in 1981 in Calcatoggio (22), it is probable that this range of dangerous electrode potentials, the precise significance of which does not seem to be currently known, must correspond to the circumstances in which corrosion occurs inside the "occluded corrosion cells" which the cracks in the metal constitute. If this is indeed the case, this dangerous potential range (measured on the outside surface of the metal) might be predetermined quantitatively by simply superimposing the polarization curves determined in the presence of the solution in which the metal is immersed (which show notably the conditions of general corrosion and passivity of the external surface), and the polarization curves determined in the OCCs (which show the conditions of corrosion, immunity and passivity of the internal surfaces; i.e., inside the OCCs as well as the conditions where hydrogen can evolve in these cavities). It should be understood that this superimposing must be done taking into account the existence of a diffusion potential (and of an eventual "ohmic drop") between the internal cavities and the external surfaces. This diffusion potential is most probably equal to the potential difference between two identical reference electrodes placed respectively inside and outside the OCC (Figure 9).

As previously shown (22 to 24), solutions similar to those existing inside stress-corrosion cracks (and other OCCs) may be prepared artificially, in any quantity desirable for chemical and electrochemical studies, by simply dissolving in oxygen-free water, up to saturation and in the presence of powder or shavings of the metal or alloy under study, increasing quantities of the less-soluble salt likely to be formed into the cracks or other OCCs. This salt may be, for instance, $Fe.CL_2.4H_2O$ in the case of iron or carbon steel at room temperature in the presence of chloride.

As shown by Figures 10a and 10b, which were obtained by Lucio Sathler in his Brussels thesis (25 to 27), the pH of the so-obtained solutions, which decreases by hydrolysis, stabilizes at 3.8. The concentration in dissolved $FeCl_2$ stabilizes at 4.6 M (i.e., 4.6 M in Fe and 9.2 M in Cl). The zero-current electrode potential of iron in this saturated solution is -0.33

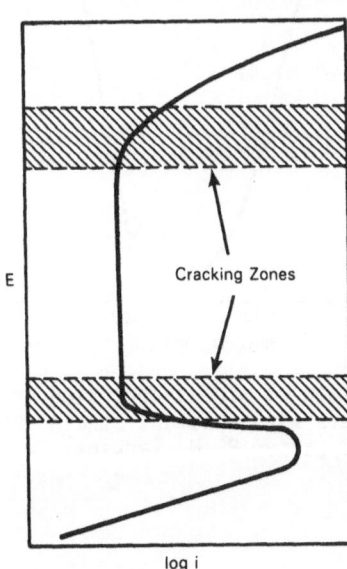

FIGURE 6. Potentiokinetic polarization curve and electrode potential values at which stress-corrosion cracking appears (schematic according to R. W. Staehle (15)).

320

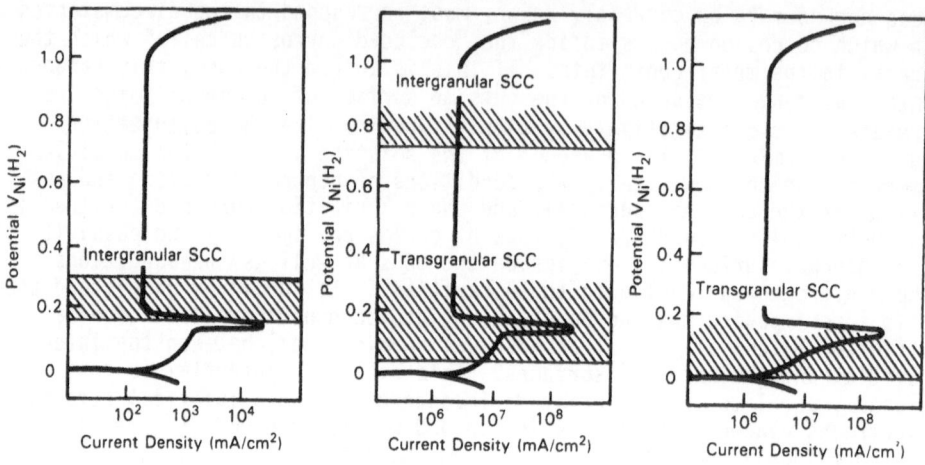

FIGURE 7. Potentiokinetic polarization curves and electrode potential values at which intergranular and transgranular stress-corrosion cracking appear, in a 10% NaOH solution at 288°C; a) 600 alloy (Inconel), b) 800 alloy (Incoloy), c) AISI 304 steel (G. J. Theus (19)).

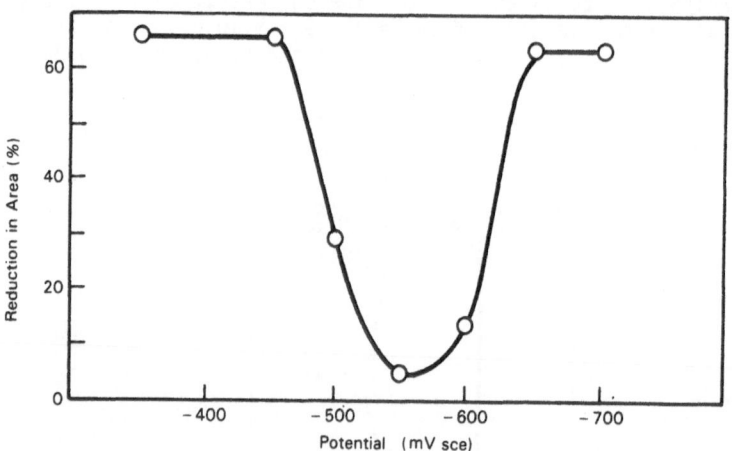

FIGURE 8. Determination of stress-corrosion cracking electrode potentials by potentiostatic tests with constant strain rate. Mild steel in solution 2N $(NH_4)_2CO_3$, at 75°C (after R. N. Parkins).

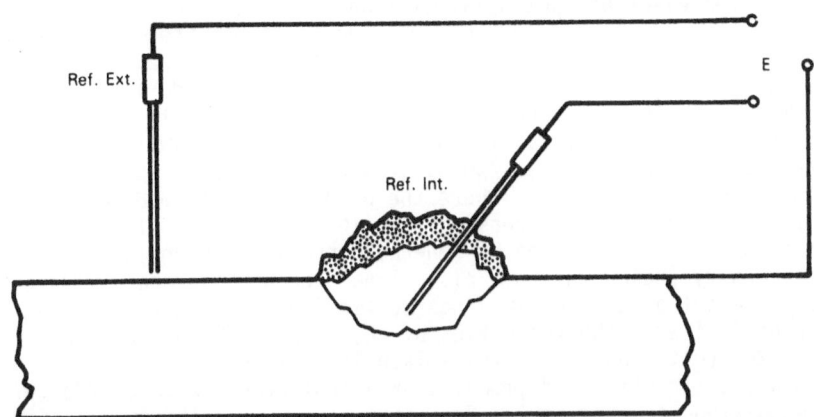

FIGURE 9. Measurement of electrode potentials inside and outside an occluded corrosion cell.

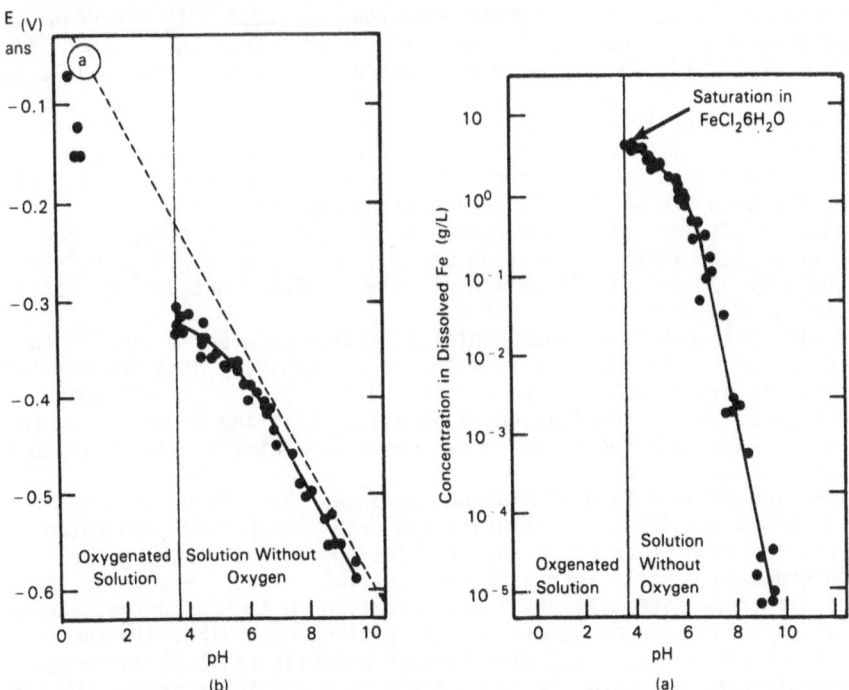

FIGURE 10. Solutions obtained by adding increasing quantities of ferrous chloride to pure oxygen free water, in the presence of iron powder, a) relation between pH and concentration in dissolved iron, b) relation between pH and electrode potential of iron.

V_{she}. Introduction of oxygen in this solution decreases the pH from 3.8 to 0.3 and raises the electrode potential of iron from -0.33 to about 0.15 to 0.7 V_{she}. L. Sathler used the ferrous chloride solutions thus obtained to study, by means of potentiokinetic and potentiostatic polarization experiments, the kinetic behavior of iron in OCCs which can form during localized corrosion of iron in the presence of solutions containing chloride.

The results of this work of Sathler have allowed us to draw Figure 11a, which is an <u>experimental</u> potential/pH diagram relating to the behavior of iron and unalloyed steel in OCCs, where the pH decreases from 8.0 to 3.8 (with chloride concentration increasing from 10^{-5} to 9.2 M). In Figure 11a, the dashed region is an area of general corrosion, which lies between an area of immunity (below the bold plain line) and areas of passivity and of pitting. The set of thin plain lines marked $\log v_{c,ox}$ drawn in the corrosion area indicates the velocities of the iron corrosion reaction Fe → Fe^{++} + 2e^{-} (c). The set of thin plain lines marked $\log v_{c,red}$ drawn in both the corrosion and immunity area indicates the velocities of the iron deposition reaction Fe^{++} + 2e^{-} → Fe. The set of the thin dotted lines marked $\log v_{a,red}$ drawn in both the corrosion and immunity areas indicates the velocities of hydrogen evolution 2H^{+} + 2e^{-} → H$_2$. The bold dotted line drawn in the corrosion area indicates the zero-current conditions of iron (where the velocity of iron corrosion is equal to the velocity of hydrogen corrosion, and where iron thus corrodes with hydrogen evolution according to the overall chemical reaction Fe + 2H^{+} → Fe^{++} + H$_2$). According to Figure 11a, when at zero-current in a dilute FeCl$_2$ solution of pH 8.0, iron has an electrode potential of about -0.5 V_{she}, and corrodes with a velocity of about $10^{-6.5}$ A.cm^{-2}. If anodically polarized at about -0.3 V_{she}, the corrosion velocity first increases up to about $10^{-3.0}$ A.cm^{-2} and then becomes zero, due to passivation of the metal. If the pH decreases up to about 7.0, some pitting appears. At lower pH, any polarization leads to an increased general corrosion, with perhaps some stifling at pH 3.8, when the solution becomes saturated in FeCl$_2$.4H$_2$O (due, as was shown by Sathler, to formation of film of γFeOOH between the metal and the FeCl$_2$.4H$_2$O deposit).

Figure 11b is a tentative experimental potential/pH diagram similar to the one drawn in Figure 11a, but relating to the behavior of a 12% Cr steel in chloride containing OCCs, of pH 8.0 to 2.0 (3,24). We are not aware of in-depth studies on the composition of solutions existing in OCCs of chromium steels. Figure 11b, which was drawn up as indicated previously, should be considered as provisional.

A comparison between Figures 11a and 11b shows that adding chromium to the steel raises its immunity potential and drops its passivation potential. This addition of chromium thus reduces the extent of the general corrosion area and greatly increases the passivation area which, for the values of electrode potential under consideration in these figures, already appears at pH values around 2, without any pitting corrosion. In the presence of the solution saturated with ferrous chloride (pH 3.8) which was already considered in Figure 11a in relation to non-alloyed steel, the corrosion of the steel with hydrogen evolution will take place, at zero current, at a velocity of $10^{-5.3}$ A.cm^{-2} (instead of $10^{-6.0}$) and an electrode potential of -.30 V_{she} (instead of -0.34). An anodic polarization current of around $10^{-4.5}$ A.cm^{-2} (which would raise the electrode potential of carbon steel to 0.29 V_{she} and would significantly increase the corrosion velocity) will raise the electrode potential of chromium steel above its passivation potential (-0.09 V_{she}) and thus cause passivation. It is currently admitted that localized corrosion phenomena which affect alloy steels can

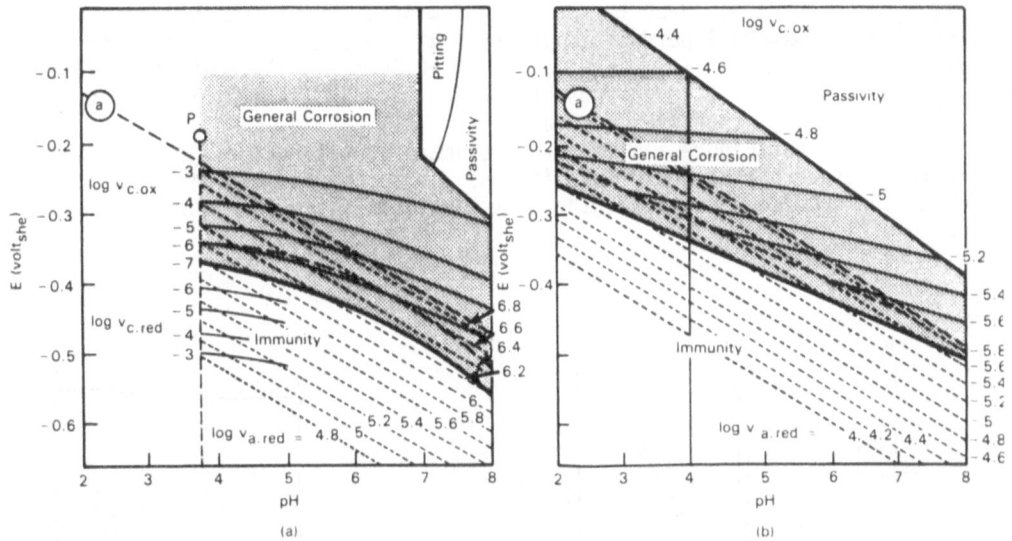

FIGURE 11. Experimental potential/pH diagrams relating to the behavior of steels in occluded corrosion cells, a) iron and unalloyed cells, b) 12% Cr steel.

disappear, either due to "repassivation" or to "deactivation." It is probable that these phenomena result from a change in the electrode potential (and possibly pH) of the OCCs in diagrams like that of Figure 11b: the repassivation probably comes from an increase in the electrode potential in the OCC above the passivation potential (-0.09 V_{she} in the case of Figure 11b pertaining to a 12% Cr steel). Deactivation is probably due to a drop of this potential below the immunity potential (about -0.33 V_{she}).

We now come back to the suggestion that the dangerous electrode potential range for SCC might be predicted by simply superimposing polarization curves relating to the behavior of external surfaces (in the bulk of the solution) and polarization curves relating to the behavior of internal surfaces (in the solutions existing inside the cracks).

Figures 12a, b, and c, which reproduce figures given previously (3), determine this dangerous zone of electrode potentials. Figure 12a shows the polarization curves determined on the external surface of the metal in the presence of pH 7.0, 0.1 M chloride solution. One can see notably the "passivation potential" P, and the "protection potential against the propagation of localized corrosion" plot (which is often not quite reproducible). Figure 12b shows four polarization curves determined in ferrous chloride solutions existing in the OCCs already considered in Figure 11a (pH 7.0 to 3.8, and 0.1 to 9.2 M in chloride). For four values of pH one can see the velocities of corrosion of the steel and the velocities of hydrogen evolution, along with the passivation potentials already shown in Figure 11b (-0.31 V_{she} at pH 6.3, -0.16 V_{she} at pH 4.9, and -0.09 V_{she} at pH 3.8). In Figure 12c, Figures 12a and 12b have been superimposed by shifting the lines in Figure 12b toward the top of a value equal to the diffusion potential (which we have assessed respectively at zero, 59, 100 and 115 mV for the four solutions under consideration). According to Figure 12c, corrosion in the OCC would start above an external electrode potential rising from -0.35 to -0.15 V_{she} as pH decreases from 6.3 to 3.8. And this corrosion would stop by "repassivation" above external potentials

324

rising from -0.21 to +0.30 V$_{she}$ as pH decreases. According to this exposé, the range of external potentials in which 12% Cr steel is sensitive to stress corrosion runs from about -0.21 to +0.30 V$_{she}$ (i.e., -0.46 to +0.05 V$_{she}$). Below -0.21 V$_{she}$ any possible cracks are in a state of immunity; above +0.30 V$_{she}$, they are passivated.

We recall that Figure 11b is an approximate provisional diagram. Complementary research that would make it possible to adjust this diagram, obtaining precise results for the approximations given in Figure 12, would be helpful.

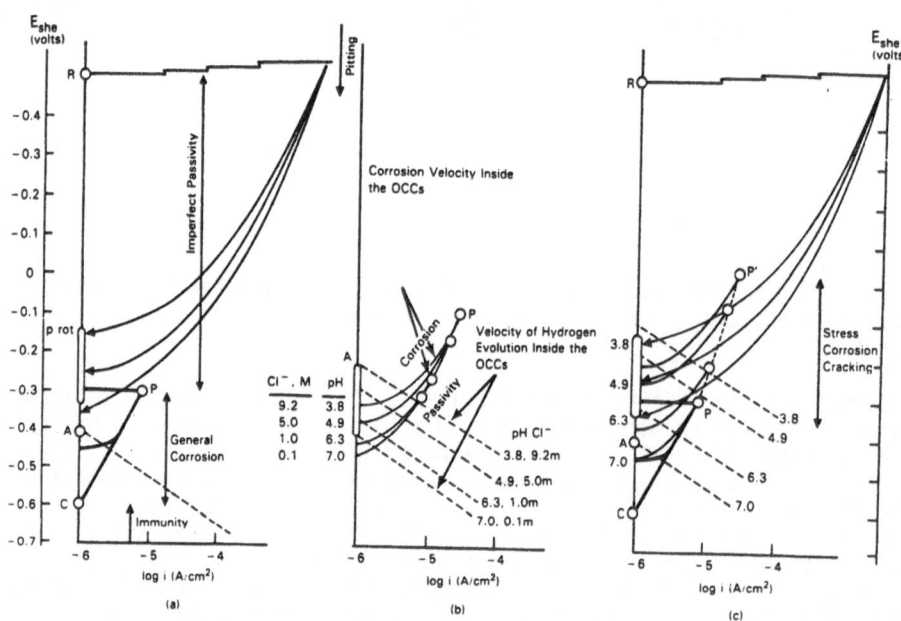

Figure 12. Electrochemical behavior of a 12% Cr steel in a 10^{-1} M Cl solution. Polarization curves in oxygen-free solutions (provisional schemes). a) external surface in the presence of the external solution (0.1 M Cl, ph 7.0); b) internal cavities in the presence of ferrous chloride solutions (pH 7.0 to 3.8); c) external and internal surfaces, and conditions of stress corrosion (as a function of the electrode potential of the external surface).

3. HYDROGEN EMBRITTLEMENT

As may be shown by Figures 12b and 12c, the method suggested here makes it possible also to predict the conditions where, as shown by B. F. Brown (28,29), hydrogen evolution may occur inside stress-corrosion cracks, thus leading to hydrogen embrittlement.

Before closing this statement, we wish to point out that hydrogen is not the only gas which can be formed in OCCs and which can therefore cause mechanical embrittlement of the metal. Methane (CH_4) can also be formed, and perhaps other hydrocarbons. In the presence of nitrates, nitrogen (N_2) and nitrogen oxides N_2O and NO can be formed by reduction of nitric acid, which can also have an embrittling effect.

REFERENCES

1. Macdonald, DD: Surface Chemistry in Aqueous Solutions, this volume,
2. Turnbull, A: Crack Tip Electrochemistry: Recent Developments, this volume.
3. Pourbaix, M: Occluded Corrosion Cells and Crack-Tip Chemistry. a) Proc. NATO Advanced Research Institute on Atomistics of Fracture, Calcatoggio, May 22-31, 1981, eds. R. M. Latanision and J. R. Pickens, Brussels, NATO, 1981, pp. 603-655. b) Rapports Techniques Cebelcor 141, RT.261, 1981.
4. Pourbaix, M: a) Lectures on Electrochemical Corrosion, 2nd edition, New York, Plenum Press, Brussels: Cebelcor, 1973. b) Lecons en corrosion electrochimique, 2e ed., Bruxelles: Cebelcor, 1975.
5. Kaesche, H: Die Korrosion der Metalle. Physikalisch-chemische Prinzipien und aktuelle Probleme, 2e Auflage, Spring, 1979.
6. Lee, H, Uhlig, HH: Effect of Nickel in Cr-Ni Stainless Steels on the Critical Potential for Stress Corrosion Cracking, J. Electr. Soc. 117:18-..., 1970.
7. Newberg, R, Uhlig, HH: Stress Corrosion Cracking of 18% Ferritic Stainless Steel, J. Electr. Soc. 119:981-..., 1972.
8. Mazille, H, Uhlig, HH: Effect of Temperature and Some Inhibitors on Stress Corrosion Cracking of Carbon Steels in Nitrate and Alkaline Solutions, Corrosion 28:427-..., 1972.
9. Newberg, R, Uhlig, HH: Stress Corrosion Cracking Behavior of Precracked 18-8 Stainless Steel, J. Electr. Soc. 120:1629, 1973.
10. Asphahani, A, Uhlig, HH: Stress Corrosion Cracking of 4140 High Strength Steel in Aqueous Solutions, J. Electr. Soc. 122:174, 1975.
11. Uhlig, HH, Gupka, K, Liange, W: Critical Potentials for Stress Corrosion Cracking of 63-37 Brass in Ammoniacal and Tartrate Solutions, J. Electrochem. Soc. 122:343, 1975.
12. Hikson, D, Uhlig, HH: Stress Corrosion Cracking of Mild Steel in Ammonium Carbonate Solutions, Corrosion 32:56, 1976.
13. Asphahani, A, Uhlig, HH: Stress Corrosion Cracking of High and Low Strength Carbon Steel in Nitrate Solutions, Corrosion 32:117, 1976.
14. Uhlig, HH: Applying Critical Potential Data to Avoid Stress Corrosion Cracking of Metals, J. Applied Electrochem., ...
15. Staehle, RW: Stress Corrosion Cracking on the Fe-Cr-Ni Alloy System, Proc. NATO Science Committee Research Evaluation Conference, "The Theory of Stress Corrosion Cracking in Alloy," Ericeira, 1971, pp. 223-228, ed. J.C. Scully, Published by NATO, Brussels, 1971.
16. Parkins, RN: Stress Corrosion Cracking of Low Strength Ferritic-Steels, Proc. NATO Science Committee Research Evaluation Conference, "The Theory of Stress Corrosion Cracking in Alloy," Ericeira, 1971, pp. 223-228, ed. J.C. Scully, Published by NATO, Brussels, 1971.
17. Parkins, RN: Prevention and Control of Stress Corrosion cracking, An Overview, Corrosion 1985, NACE, Paper 348.
18. Morris, PE: New Electrochemical Techniques for High Temperature Aqueous Environments, Proc. Corrosion 76, "Electrochemical Techniques for Corrosion," Houston, Texas, March 22-26, 1976, pp. 66-72, Publ. NACE, 1977.
19. Theus, GJ, Cels, JR: Slow Strain Rate Technique: Application to Caustic SCC Studies, Proc. ASTM Symposium on SCC "The Constant Strain Rate Technique," Toronto, May 2-5, 1977, pp. 76-31, Publ. RDTPA (Babcock & Wilcox).
20. Pourbaix, M: Electrochimie et Corrosion. Applications Pratiques Recentes, Rapports Techniques CEBELCOR, 134, RT. 244, 1978.

21. Pourbaix, M: Electrochemistry and Corrosion. Recent Practical Applications, Rapports Techniques CEBELCOR, 134, RT. 245, 1978.
22. Scully, JS(ed): Proc. NATO Science Committee Research Evaluation Conference "The Theory of Stress Corrosion Cracking in Alloys," Ericeira, NATO, Brussels, 1971.
23. Pourbaix, M: Electrochemical Aspects of Stress Corrosion Cracking, loc. cit. 22:17-63, 1971.
24, Pourbaix, M, Pourbaix, A: Analysis of Current Problems Caused by Localized Corrosion. Study of Their Industrial Impact and Proposals for Action. Rapports Techniques CEBELCOR, 152, RT. 287a, 1986.
25. Sathler, L: Contribution to the Study of the Electrochemical Behavior of Iron in the Presence of Solutions of Ferrous Chloride, with Reference to Localized Corrosion (in French), Thesis, Brussels, 1978.
26. Sathler, L, Van Muylder, J: On the Chemical and Electrochemical Nature of the Solutions Inside Occluded Corrosion Cells (in French), Rapports Techniques CEBELCOR, 126, RT. 224, 1975.
27. Sathler, L, Van Muylder, J, Winand, R, and Pourbaix, M: Electrochemical Behavior of Iron in Localized Corrosion Cells in the Presence of Chloride, Proc. 7th International Congress on Metallic Corrosion, Rio de Janeiro, 1978, 2:705-717, publ. ABRACO, 1979.
28. Brown, BF, Fujii, CT, Dahlberg, EPh: Methods for Studying the Solution Chemistry Within Stress Corrosion Cracking, J. Electrochem. Soc. 116:218-219, 1969.
29. Brown, BF: On the Electrochemistry of Stress Corrosion Cracking of High Strength Steels, Rapports Techniques CEBELCOR, E.76, 1975.

DISCUSSION

Comment by A. Turnbull:

The work done on the electrochemistry of iron in saturated ferrous chloride solutions was very informative, but do you think that the results can be applied directly to cracks. I am thinking particularly of the influence of crack size and environmental conditions on the salt concentration and of the influence of the cathodic reduction of hydrogen ions on the pH which are not simulated in the tests you referred to.

Reply:

I agree that the application to cracks of the method which I supported during the "Latanision 1 and 2 workshops" should be done with greatest care, for this implies, as far as possible, the knowledge of the local composition of the solution inside the cracks, and preferably on different spots of the cracks (crack tip, crack walls, crack mouth), as well as of the nature of the metallic surface (including the intergranular deposits) on their different sites. And I agree that the "future work" you are suggesting at the end of your paper may help to clear up the important influence of several factors in this respect.

As far as presently feasible, the influences of salt concentration and pH have been considered in Lucio Sathler's thesis and in figures 11a and 11b of my paper, and I hope that future work will soon make it possible to improve these two figures, which should be considered as provisional schemes. I think that more work relevant to the chemistry and electrochemistry of " occluded corrosion cells, o.c.c." is an urgent necessity.

Structure and Properties
of Interfaces

Session Chairmen: T. E. Fischer and J. Oudar

UNIVERSAL PROPERTIES OF BONDING AT METAL INTERFACES

JOHN R. SMITH
Physics Department, General Motors Research Laboratories
Warren, Michigan 48090-9055

JOHN FERRANTE and PASCAL VINET
National Aeronautics and Space Administration
Lewis Research Center, Cleveland, Ohio 44135

J. G. GAY and ROY RICHTER
Physics Department, General Motors Research Laboratories
Warren, Michigan 48090-9055

JAMES H. ROSE
Ames Laboratory—U.S. Department of Energy
Iowa State University, Ames, Iowa 50011

1. INTRODUCTION

Our knowledge of the fundamental properties of surfaces and interfacial bonds has improved considerably over the last decade. While the fracture process is complex, it is dependent in part on these properties. Thus it is of interest to consider current theoretical understanding of metal surfaces and bimetallic interfacial bonds.

We will not attempt to overview this field which has now become a relatively large one, but rather will be content with a discussion of selected properties we have analyzed which will hopefully be illustrative of the state of the art. We will find a complicated situation which will lead us to look for simplification. From studies of bimetallic interfaces we will discover a single energy relation for metallic adhesion, cohesion, and chemisorption as well as nuclear matter which greatly simplifies the problem.

Our approach will be that of the theoretical solid state physicist, and our conclusions will depend primarily on the results and capabilities of recent first principles methods aided and tested where possible with experimental results. As such our methods are more fundamental than others presented at this workshop. By the same token, our current results are limited to higher symmetry problems such as brittle fracture and monolayer adsorption effects. The field of fracture is inherently interdisciplinary, with solid state theory having played a relatively minor role up to now. The capabilities of the solid state theorist have improved in the last few years, however, so that one might now expect more contributions from that discipline in the future, including realistic models of the fracture process.

In Section 2 we will discuss some characteristics of metal surfaces and interfacial bonds. Surface energies or brittle fracture energies, surface states, interfacial electron density rearrangements and interfacial magnetics are examples of interface properties that are included. Some simplification of this complicated situation is found in parallels between molecular and bimetallic interfacial bonds.

The discovery of a universal relationship between total energy and distance between atoms is described in Section 3. Expected limitations and applicabilities are analyzed. As an example of simple relationships that can be derived from this universality, a relationship between surface energies and cohesive energies is found. A second example is the prediction of equations of state (pressure–volume relations) for metals. While the universal energy relations are limited to metals, covalent semiconductors and nuclear matter, equations of state (pressure–volume relations) are found to be of universal form in the absence of phase transitions for *all* classes of solids in compression. This universal equation of state allows for high temperature properties to be predicted, including high temperature thermal expansion and bulk modulus variation with temperature. Finally, Section 4 contains a summary.

2. SOME CHARACTERISTICS OF METAL SURFACE AND INTERFACIAL BONDS

2.1. Electrons and the Transition Metal Surface

2.1.1. Electronic Surface Rearrangement. We will want first to exhibit some results of our[1] self-consistent local orbital (SCLO) (see Section 2.2) calculations on transition metal surfaces. For a review of this work and related work in the field see Ref. 2. This discussion will illustrate the pertinent electronic phenomena and, in the process, the need for self consistency (Section 2.2). In Fig. 1 are our computed[3] electron density contours for the Ni(100) surface. As typical of transition metals, the electron density varies by a factor of ten in a distance of the order of one Å, primarily because of the presence of the *d* electrons. The *d* electrons are strongly hybridized with *s* and *p* orbitals, however, and when the metal is fractured to form a surface there is a spreading and smoothing of the electron density distributions into the vacuum as shown. Note also that the "surface" region is essentially one atomic layer as far as the electron density contours are concerned. That is, the electron density contours are virtually the same for all layers except the surface atomic layer. This is a manifestation of the short screening lengths which are typical of metals.

2.1.2. Surface States. We have found that transition metals can have large densities of surface states or surface resonances. A surface state has an electronic wave function which is localized to the surface. Surface resonances have wave functions in the surface region which have increased amplitude there because of resonances with the surface potential, but the wave functions are not localized to the surface. The combination of a large number of bulk states in resonance in the surface can have an effect in some experiments such as photoemission which is not unlike that of a surface state. Fig. 2 shows our computed[4] electron density amplitudes for surface states/resonances on a Rh(100) surface.

Ni(100) CHARGE DENSITY

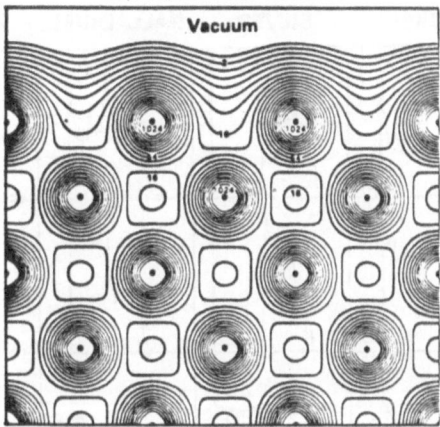

FIGURE 1. Electronic density contours from Ref. 3.

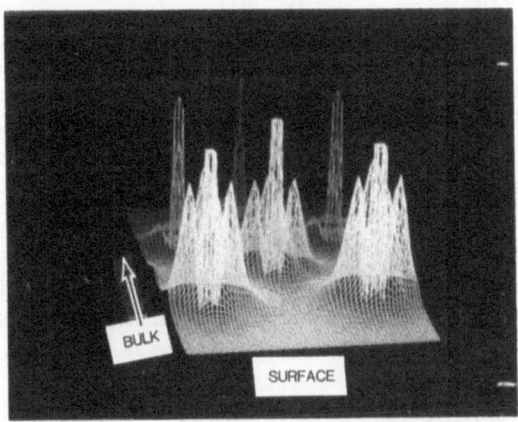

FIGURE 2. Surface state/resonance electron densities at a Rh(100) surface from Ref. 4.

It is clear from Fig. 2 that the surface state electron density decays rapidly in going from the surface into the bulk. What is not obvious from that figure is that the surface state electron density in the surface layer is a significant fraction of the total electron density in that layer. We can see that through Fig. 3. There the densities of electronic states (DOS) are plotted as a function of energy for (100) films of Cu, Ni, Ag, Pd, and Rh. The surface layer DOS (middle panel), are typically narrower than that of the bulk (top panel), primarily because surface atoms have a smaller number of near neighbors than do bulk atoms. The bottom panel is that part of the surface DOS which is due to surface states. It is clear that the bottom panel is a significant fraction of the middle panel. In fact, if we integrate the bottom and middle panels up to the Fermi level and then divide, we find the percentages to be: Cu, 36; Ni, 23; Rh, 27; Ag, 22; and Pd, 19. It should be noted finally that theory has progressed to the point that predicted surface state bands are pointing the way to subsequent experiments where the predictions are being confirmed[6].

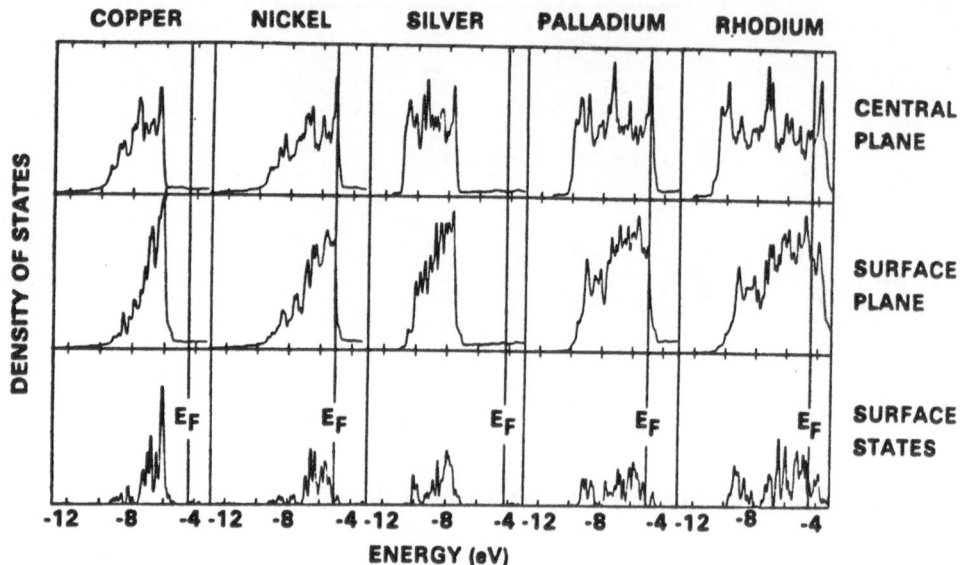

FIGURE 3. Projected densities of electronic states: top panel—central atomic plane; middle panel—surface plane; bottom panel—surface states in surface plane. Density of states scales scales are the same for all three panels of a given metal, but vary from metal to metal. See Ref. 5.

2.2. Self Consistency

We saw in Section 2.1 that the electron density distributions in the surface are considerably different than in the bulk so that it would be virtually impossible to guess a correct surface potential. This necessitates a self-consistent approach, in which the potential is determined from wave functions and vice versa. We begin with the Kohn-Sham equations[7]

$$[-\frac{1}{2}\nabla^2 + V(\vec{r})]\Psi_i(\vec{k}_{||},\vec{r}) = \epsilon_i(\vec{k}_{||})\Psi_i(\vec{k}_{||},\vec{r}) \tag{1}$$

$$\text{where} \quad V(\vec{r}) = \Phi(\vec{r}) + V_{\text{xc}}[\rho(\vec{r})] \quad , \tag{2}$$

$\Phi(\vec{r})$ is the electrostatic potential

$$\Phi(\vec{r}) = \int d\vec{r}' \frac{\rho(\vec{r})}{|\vec{r} - \vec{r}'|} - \sum_j \frac{Z_j}{|\vec{r} - \vec{R}_j|} \quad , \tag{3}$$

and $\rho(\vec{r})$ is the electronic charge density

$$\rho(\vec{r}) = \sum_{\text{occ.}} \left|\Psi_i(\vec{k}_{||},\vec{r})\right|^2 \quad . \tag{4}$$

Z_i is the nuclear charge of the atom at site $\vec{R}_i$, $\vec{k}_{||}$ is the component of the Bloch vector parallel to the surface, occ. refers to all the occupied states and $V_{\text{xc}}[\rho(\vec{r})]$ is the exchange-correlation potential. This potential is typically treated in the local density approximation[7], in which its value at any point $\vec{r}$ is given in terms of $\rho(\vec{r})$. Eqs.(1)–(4) when solved simultaneously assure that the solution is self consistent. This self consistency is a necessity for computation of electronic structure, as discussed in Section 2.1. For example, the question of the existence of surface states, much less their dependence on $\vec{k}_{\parallel}$ depends on the solution being fully self consistent.

In the SCLO method[1], the wave functions are expanded in terms of the localized orbitals $a_j(\vec{r} - \vec{R}_j)$:

$$\Psi_i(\vec{k}_{\parallel}, \vec{r}) = \sum_{i,j} C_{ij}(\vec{k}_{\parallel}, \vec{R}_j) a_j(\vec{r} - \vec{R}_j). \tag{5}$$

The $a_j(\vec{r} - \vec{R}_j)$ comprise all ground state atomic orbitals (including the core), plus other localized orbitals similar to the quantum chemist's double-zeta-plus-polarization basis sets[8]. Matrix elements of the Hamiltonian are then determined between local basis functions, and the electronic structure is obtained by solving the corresponding matrix eigenvalue problem.

In the adhesion (Section 2.5) and grain-boundary (Section 2.6) calculations for sp metals, Eqs.(1-4) are numerically integrated rather than being solved via Eq.(5).

2.3. Transition Metal Surface Energies

2.3.1. Thin Film Method.
In cleavage or brittle fracture of a crystalline material, periodicity is maintained parallel to the fracture plane. However, periodicity is lost in the direction perpendicular to that plane. As this loss of periodicity greatly increases the complexity of the calculation, we minimize this by dealing with thin, crystalline films. Because of the short screening lengths for metals mentioned earlier, thin films can be used to effectively model thick solids. The total energy of a film n atomic layers thick per surface atom is

$$E_n = nE_b + 2E_s, \tag{6}$$

where E_b is the bulk energy per film unit cell and E_s is the surface energy. It is also given by

$$E_n = \frac{1}{N}\left\{ \sum_{\substack{i,\vec{k}_{\parallel} \\ \text{occ.}}} \epsilon_i(\vec{k}_{\parallel}) - \frac{1}{2}\int d\vec{r}\,d\vec{r}\,' \frac{\rho_v(\vec{r})\rho_v(\vec{r}\,')}{|\vec{r} - \vec{r}\,'|} \right.$$
$$\left. + \frac{1}{2}\sum_{\substack{i,j \\ i \neq j}} \frac{Z_{c_i} Z_{c_j}}{|\vec{R}_i - \vec{R}_j|} + \int d\vec{r}\,\big(\rho(\vec{r})\epsilon_{\text{xc}}[\rho(\vec{r})] - \rho_v(\vec{r})V_{\text{xc}}[\rho(\vec{r})]\big) \right\}, \tag{7}$$

where N is the number of atoms per layer, $\rho_v(\vec{r})$ is the valence or conduction electron density, Z_{c_i} the net charge of the ion core at site R_j, ϵ_{xc} is the exchange-correlation energy per electron, and the rest of the quantities are defined in Eqs.(1-5).

334

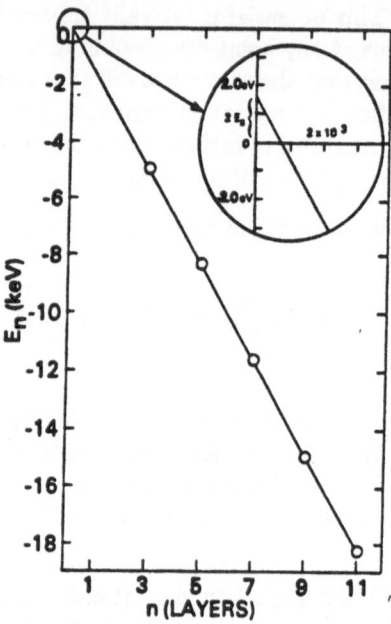

FIGURE 4. Total energy per surface unit cell E_n (circled points) plotted as a function of the number of atomic layers n in Cu(100) films. See Ref. 9.

The energy E_n has been determined[9] for Cu(100) films 3, 5, 7, 9, and 11 atomic layers thick. The results are shown in Fig. 4. The line is a plot of Eq.(6) while the circles are obtained from Eq.(7). Note first that the line runs right through the circles for films 3 atomic layers and thicker. This means that it would take the same amount of energy to cleave a block of copper into 3 layer thick films as into, say, 1 meter thick films. This is a manifestation of short screening lengths in metals mentioned several times earlier. Secondly, note that the bulk energy E_b is of the order of 10^3 eV, while E_s, which is of the intercept of the line with the ordinate, has to be magnified to be seen. It is shown in the inset as 0.93 eV per surface atom. This disparity between E_b and E_s means that, as is typical of total energy calculations, high accuracy is required in determining E_s.

The computed value of 0.93 eV per surface atom corresponds to 2300 ergs/cm², which compares well with the experimental value[10] of 2016 erg/cm². The experimental value is an estimate for polycrystalline Cu.

Now we would like to see if this sort of accuracy is maintained for series of transition metals. In the following, we describe the first self-consistent surface energy results for such a series. In Table 1 the computed surface energies[11] for the (100) surfaces of Cu, Ag, Ni and Fe are compared with experimental values for polycrystalline metals. First, let us look at the theoretical values in eV per surface atom. Note that Cu(100) and Ag(100) have nearly the same surface energies, which is consistent with them both being noble metals. For Ni(100) the d band is less than fully occupied and the surface energy is larger than that of the noble metals. Fe(100) has still fewer electrons in the d band and a still higher surface energy. This suggests a correlation of surface energy with d band filling, an idea which is not new but it is interesting that these first, self-consistent calculations confirm it. Now let us compare with experiment. The experimental values are by their nature in terms of energy per unit area so that the number of atoms per unit area enters, leading to Ag(100) having a somewhat smaller surface energy than Cu(100), consistent with the experimental values as shown.

TABLE 1. Comparison of predicted (100) surface energies with experimental values for polycrystalline metals.

Metal	Surface Energy		
	Theory[a]		Experiment[b]
	(eV/atom)	(erg/cm*2)	(erg/cm*2)
Cu	0.93	2300	2016
Ag	0.85	1650	1543
Ni(para)	1.22	3150	
Ni(ferro)	1.18	3050	2664
Fe(para)	1.88	3700	
Fe(ferro)	1.58	3100	2452

[a] Ref. (11)
[b] Ref. (12)

2.3.2. Effect of Spin Splitting. Now Ni and Fe are ferromagnetic metals, and one might inquire about the effect of spin splitting on their surface energies. The surface energy is the difference in energy between a final state which is a cleaved or brittlely fractured material and an initial state before fracture. Spin splitting will lower the energies of both the initial and final states of a ferromagnetic material. We[12-13] and others[14] have found that the magnetic moments of surface atoms in Ni(100) and Fe(100) are higher than those of the corresponding bulk atoms in these materials. This suggests that the final state energies may be lowered more than the initial state energy by spin splitting, thereby lowering the surface energies. This is in fact what happens, as can be seen from Table 1. The ferromagnetic values are lower, and this lowering is in the direction of the experimental values. The overall agreement with experiment is reasonable, especially considering that the comparison is between single crystal and polycrystalline values and that the error in the experimental values is perhaps as much as ± 20 percent. This suggests that the local density approximation[7] yields accurate surface energies for transition metals.

2.4. A Magnetic Overlayer

We have seen that spin splitting can be important in brittle fracture. One might then wish to investigate its effect in other interfacial phenomena. We first note that the isolated Fe atom has a magnetic moment of $4\mu_b$ (the orbital angular momentum is assumed to be quenched as it is in the solid), an Fe atom in an Fe(100) surface[13] has a moment of $2.9\mu_b$ and an Fe atom in bulk[13] Fe has a moment of $2.3\mu_b$. This prompts the question of whether the magnetic moment of atoms in an Fe overlayer on a nonmagnetic substrate could be larger than that found for bulk Fe atoms. On the one hand, the Fe atoms in the overlayer would have fewer Fe near neighbors than bulk Fe atoms have. The magnetic moments listed above would suggest that this would lead to a moment higher than in the bulk. On the other hand, the paramagnetic substrate will tend to quench the moment. Further, there is a large literature[15] on the question of the existence of magnetic *dead* layers at surfaces and interfaces.

We[13] chose an Fe monolayer on Ag(100) to investigate this question for several reasons. First, it is known[16] that a monolayer of Fe can be grown epitaxially on a Ag(100) substrate. This is important because it means that the system can be investigated experimentally. Secondly, the top of the Ag d band lies about 3.5 eV below the Fermi level. That, coupled with the expected narrowing of the Fe monolayer d band relative to bulk Fe may mean that the overlap of the Fe and Ag(100) d bands in energy space would be small. This could minimize the quenching of the Fe magnetic moment by the Ag(100). Finally, experimental results[15] suggest that Fe is less sensitive to demagnetization effects from substrate-overlayer *spd* hybridization than is, say, Ni.

Our results for the electron energy bands of seven atomic layers of Ag(100) with monolayers of Fe epitaxially adsorbed on either side of the film are shown in Fig. 5. The majority-spin bands are shown on the left and the minority- spin bands on the right. Note first that the Ag d bands, which lie between 3.5 eV and 11.0 eV below the Fermi level, exhibit very little spin-splitting so that the Fe overlayer provides little spin polarization to the Ag(100) substrate. The Fe overlayer d bands are contained essentially within the horizontal lines. Note first that the spin splitting is large: 2 to 3 eV compared to an average of about 1.5 eV in bulk Fe. Despite this large splitting, the Fe d bands do not overlap the Ag(100) d bands. This suggests that the quenching due to the substrate may in fact be small. In Fig. 6 are shown the majority and minority densities of states projected on the Fe overlayer.

There is clear separation of the majority and minority spin d bands, implying a large moment. In fact the magnetic moment of the Fe atoms in the overlayer is $3.0\mu_b$, 36 per cent larger than that of bulk Fe and certainly not a *dead* layer. This is the first example of an overlayer moment that is larger than that of the corresponding bulk atom. Enhanced overlayer moments have now been predicted[17] for a number of ferromagnetic elements on noble-metal substrates. This holds the promise of a new class of ferromagnetic materials in the form of superlattices.

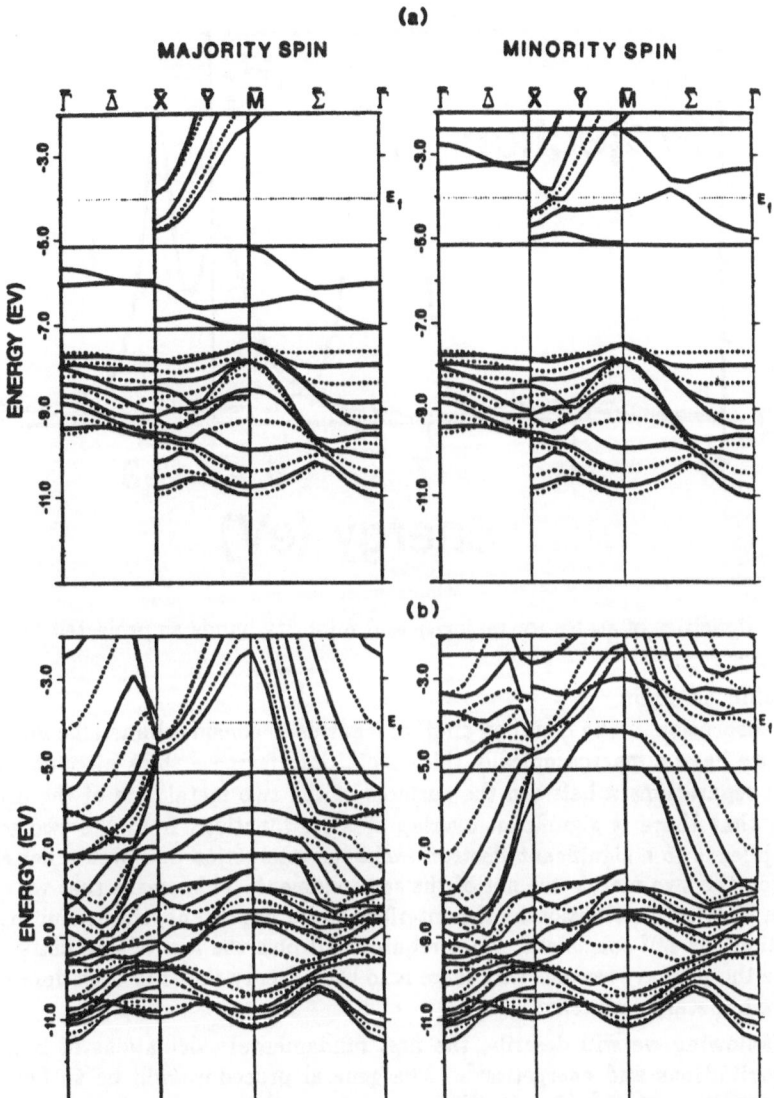

FIGURE 5. Electron energy bands along high-symmetry directions for a monolayer of Fe on Ag(100). Majority-spin bands are on the left and minority-spin bands are on the right. In (a) are those bands that are odd with respect to reflection in the mirror plane perpendicular to the surface and (b) exhibits the even bands. The Fe d bands lie essentially within the horizontal solid lines. The solid (dashed) energy bands correspond to wave functions symmetric (antisymmetric) with respect to reflection in the central atomic plane of the film. See Ref. 13.

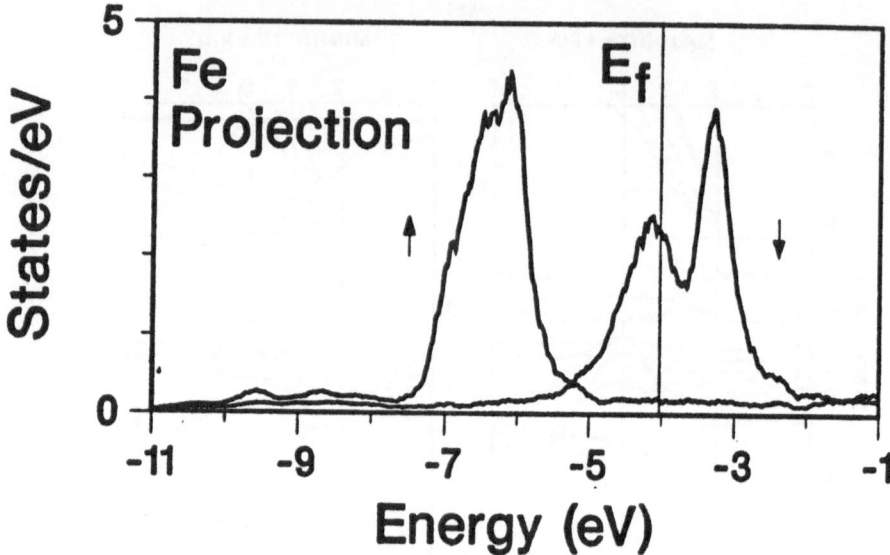

FIGURE 6. Densities of states for majority and minority bands as projected on the Fe monolayer. See Ref. 13.

2.5. Bimetallic Adhesion

Monolayer adsorption of one metal on another leads us to consider bimetallic interfaces, where now we have a macroscopically thick metal interfacing a thick piece of another metal. The separations a between the surfaces of the two metals are of the order of a few Å so that there is significant overlap of wave functions from the two metals. This overlap leads to a significant electron exchange interaction and even a net charge transfer when the two metals are not of the same element. This means that we expect the electron density distribution at the interface to be important, thus requiring that the calculations be self consistent. That requirement plus the loss of symmetry at the interface are the primary reasons why there is so little theory that has been done on the energetics of bimetallic adhesion.

In the following we will describe the first fundamental calculations of bimetallic electron distributions and energetics[18]. The general procedure will be to first solve the Kohn-Sham equations[7] (Eqs.(1-4)) for the electronic structure for an interfacial separation distance a. Next the total energy $E(a)$ will be computed and finally the adhesive energy E_{ad}, which is defined as

$$E_{ad}(a) = \frac{E(a) - E(\infty)}{2A},\tag{8}$$

where A is the cross-sectional area of the contact, is determined.

There are a number of questions that we would like to answer. We would like to determine the dominant energy component contributing to the bimetallic bond as well as the range of separations over which this bond is relatively strong. We would like to look for correlations between properties of bimetallic interfacial bonds and diatomic molecular bonds. We would like to compare contact potential properties with the actual electronic barriers that are found in the bimetallic interface. Finally, we would like to compare transition-metal interfaces with simple-metal interfaces.

Examples of simple metals are Al, Zn, Mg, and the alkali metals. We will consider these metals first. They are called simple metals because their electrons are nearly free in the bulk and so are well described for some purposes by a free-electron or jellium model. For the surfaces of these metals it has been found reliable[19] to terminate the jellium at a plane. The generalization of this to bimetallic adhesion can be seen in Fig. 7.

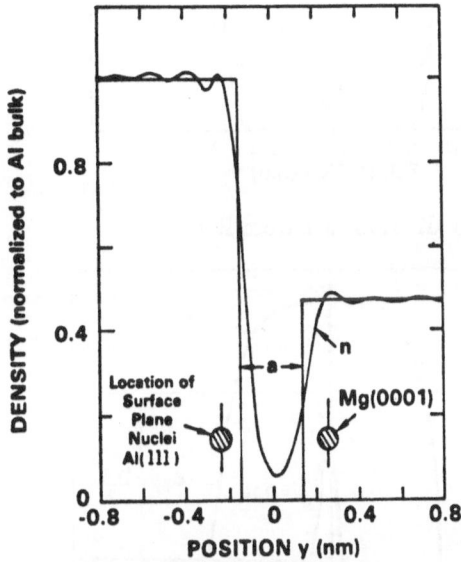

FIGURE 7. Electron number density n and jellium-ion charge density n_+ for an Al-Mg contact. When $a=0.0$, the distance between Mg and Al atomic planes is $(d_{Al} + d_{Mg})/2$, where d_{Al} and d_{Mg} are the respective bulk interplanar spacings.

So it is for this model that we solve for the electronic properties via Eqs.(1-4). Then we return the discrete atomic nature of the lattice via perturbation theory to the total energy as described in Ref. 18.

The electron densities of an Al-Mg contact for separations of 0, 0.16 and 1.6 nm are given in Fig. 8. For $a=0.0$, the electron density distribution falls smoothly between the Al and Mg. However, for a somewhat larger separation, $a=0.16$ nm, a significant dip in the electron density is evident. Thus one would have to conclude that the electron density distribution in the interface is sensitive to the interfacial spacing. For $a=1.6$ nm the electron density distribution is close to what one finds at a solid-vacuum interface. In Fig. 9 are shown the electronic potentials corresponding to the electron densities shown in Fig. 8.

For $a=0.0$ the potential rises smoothly from the Al to the Mg. At $a=0.16$ nm there arises a significant "bump" in the potential over 5 eV in height. Again this shows the sensitivity of the interfacial electronic properties to interfacial spacing. For $a=1.6$ nm the potential distributions are again close to solid-vacuum distributions.

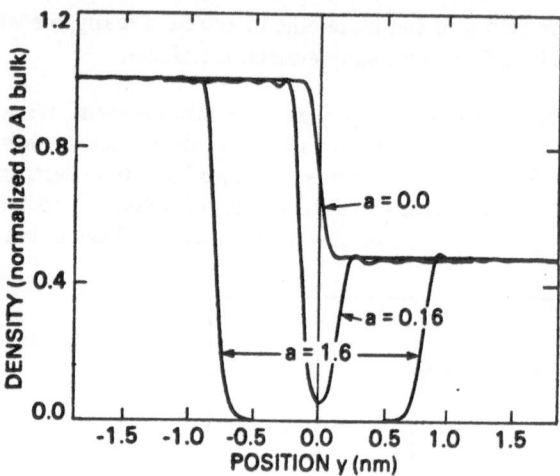

FIGURE 8. Electron density distributions from Ref. 18.

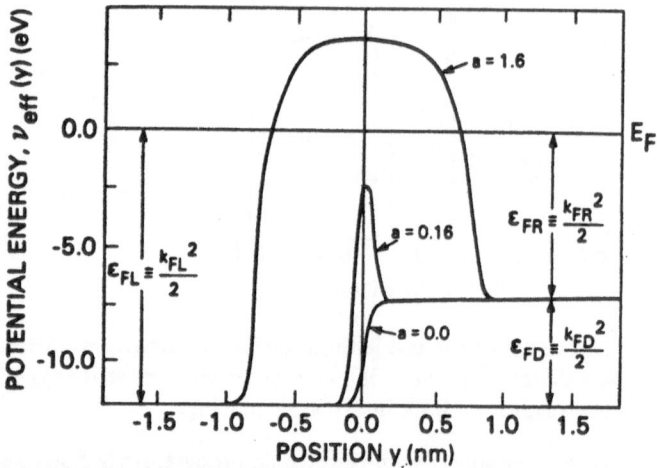

FIGURE 9. Electron potential distributions from Ref. 18.

One might well ask the question as to how much does the electron density distribution differ from what one would obtain by simply overlapping solid-vacuum distributions. These are compared for the Al-Na interface at $a=0.0$ in Fig. 10. It shows that simple overlap is a reasonably accurate procedure for the interfacial electron density distribution. But it is the interaction between the two metals in which we are primarily interested that causes the small difference between the self-consistent and the simple-overlap distributions. In Fig. 11 we plot this difference, self consistent minus simple overlap, again for the Al-Na interface at $a=0.0$. We see that the density difference shown in Fig. 11(a) exhibits quite a bit of structure. The potential difference that corresponds to this density difference is shown in Fig. 11(b). The well-known contact potential, ΔV, is shown there to be a rather small part of the potential difference. $\Delta V=0.81$ eV, which is rather small compared to the actual potential barriers shown in Fig. 9. In fact there is no obvious correlation between the contact potential and interfacial electronic barriers. Now let's increase the separation to $a=1.6$ nm and again view the difference distributions for the Al-Na contact as shown in Fig. 12.

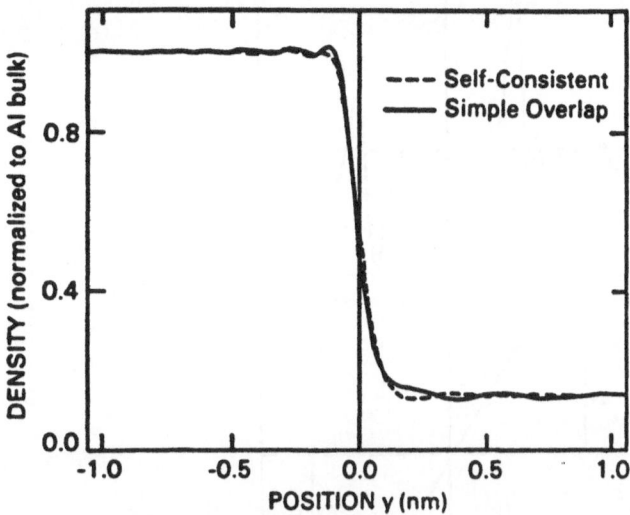

FIGURE 10. Electron density distributions in the Al-Na interface from Ref. 18.

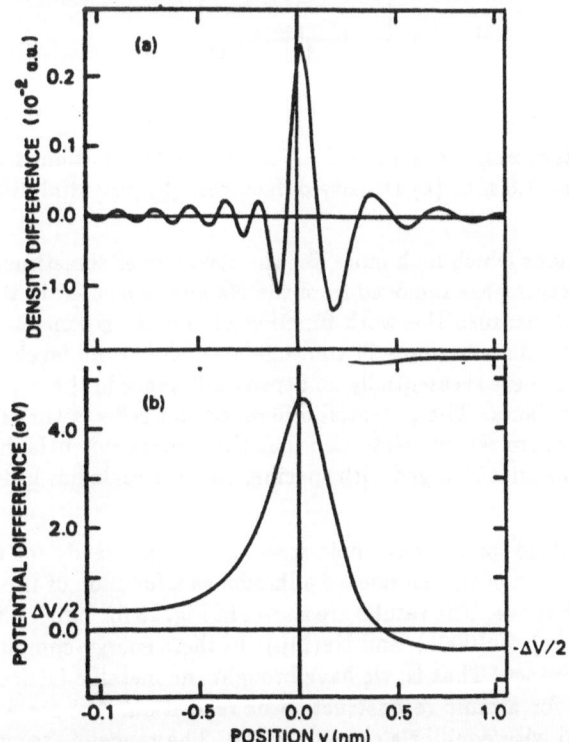

FIGURE 11. (a) Self-consistent electron density distribution minus overlapped solid-vacuum densities for $a=0.0$ in an Al-Na contact. (b) Self-consistent electron potentials minus overlapped solid-vacuum distributions from Ref. 18 for $a=0.0$ in an Al-Na contact. ΔV is the contact potential.

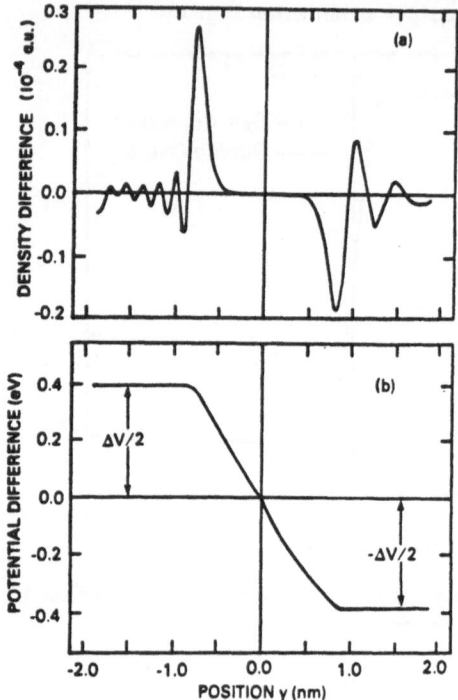

FIGURE 12. Self-consistent minus overlapped solid-vacuum distributions from Ref. 18 for an Al-Na contact at $a=1.6$ nm. (a) Density differences. (b) potential differences.

Now we see distributions which look more like the classic ones sometimes discussed in texts. We see that electrons are removed from the Na and deposited in the Al. One should expect that result because the work function of the Na is smaller than that of the Al and the charge redistribution will continue until the Fermi levels in the two metals are the same. Note there is essentially no density difference in the vacuum region between the two metal surfaces. The potential difference plot reflects that fact with a linear potential in the vacuum region. Note also that the contact potential has become the dominant barrier. It has not changed with spacing, only the scale has been changed.

Since we now have solved for the electronic structure, we are ready for the second step which is the computation of the energies of adhesion as a function of the interfacial separation a, as discussed above. The results are shown in Fig. 13 for bimetallic contacts involving Al(111), Mg(0001), Zn(0001), and Na(110). In these energy computations we have assumed a brittle contact. That is, we have brought the metallic lattices together rigidly, without allowing for atomic reconstruction or relaxation. We see a variety of shapes of these curves, all with equilibria close to $a=0.0$. The range of strong bonding, i.e., the range over which the slopes of the curves or the forces are relatively large, turns out to be approximately 0.2 nm or about one interplanar spacing. In the systematics of these curves we will find an important energy relationship which will be discussed in Section 3.

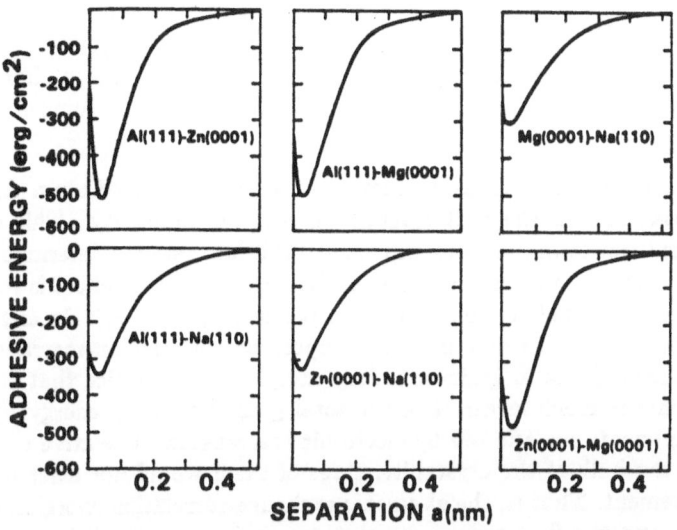

FIGURE 13. Adhesive binding energy versus separation a from Ref. 18.

2.6. Electrons and Grain Boundaries

The bimetallic interface is reminiscent of the grain boundary, although generally the grains are of the same element in the latter. We saw that electronic rearrangement, which is a quantum-mechanical effect, was of significance in the determination of energies of bimetallic adhesion. Grain boundary energies of metals have of necessity been determined via pair-potential approaches because a grain boundary has 10^2 to 10^4 atoms per unit cell. For that reason, fully self-consistent quantum-mechanical approaches have only recently been carried out for grain-boundary energies. It would be of interest to determine the importance of electronic or quantum-mechanical effects at metal grain boundaries nevertheless.

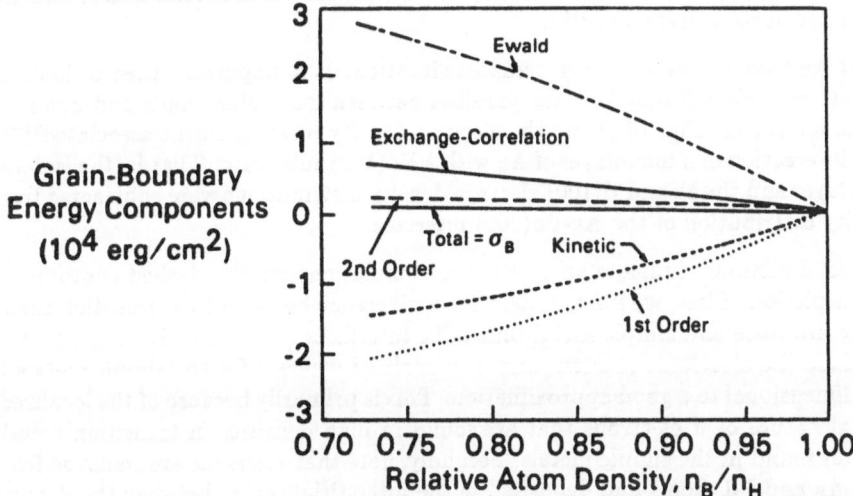

FIGURE 14. Grain-boundary energy components in aluminum as a function of the ratio of the volume-averaged boundary atom density n_B to the volume-averaged atom density of the host crystal, n_H. The width of the boundary is 10 atomic units. See

Ref. 20.

Recently, we[20] have carried out fully self-consistent, quantum-mechanical-grain-boundary-energy calculations for a series of grain boundaries in Al. Some of the results are shown in Fig. 14.

We have computed energies as a function of volume-averaged atom densities in the boundary, n_B, and in the host crystallites, n_H, for simplicity. This should be reasonably accurate for total energy estimates. If one is interested in determining atomic locations in the boundary via energy minimization then additional effort is required. A method is proposed for that in Ref. 20 (see, in particular, Eq.(4.3)). In any case, the results in Fig. 14 allow us to determine the importance of electron-rearrangement or quantum-mechanical effects in grain-boundary energies. First notice that each of the energy components is much larger than the total grain boundary energy σ_B. While all of these components are affected by electronic rearrangement relative to electronic distributions in the bulk of the crystallite, three of them would not exist were it not for this rearrangement. That is, the electronic exchange-correlation energies, electronic kinetic energies, and the first order perturbation energies would all be zero without electronic rearrangement at the boundaries. Note again that each of these terms is considerably larger than the total grain-boundary energies.

Thus one can make two conclusions. First, electronic rearrangement energies are essential in determining grain-boundary energies in metals. That is, one will have to go beyond pair potentials and do self-consistent, quantum-mechanical calculations of grain-boundary energies. Secondly, such calculations can now be done.

2.7. Some Parallels Between Adhesive and Molecular Bonds

In Section 2 we have seen a number of features of metal surface and interfacial bonding. Among the many effects are interfacial electron density rearrangement in cleavage, grain boundaries and at bimetallic interfaces; band-filling and spin-splitting effects in surface energies; adsorbate-substrate band overlap effects; interfacial magnetic effects and the effect of large surface-state densities.

What we have is clearly a very complex situation. It is important then to look for simplifications. We will first look for parallels between molecular bonds and bonds at bimetallic interfaces. Fig. 15 shows the electron density rearrangement associated[21-22] with the interaction of a monolayer of Ag with a Pd(100) substrate. That is, the isolated Ag monolayer and the clean Pd(100) electron density distributions were subtracted from the density distribution of the Ag-Pd(100) interface.

The solid contours indicate an electron accumulation and the dashed contours an electron depletion. First, we immediately see a difference between this transition metal bimetallic interface and simple metal bimetallic interfaces (compare with Figs. 11-12). That is, here the electronic rearrangement is localized whereas for the simple metals it was one dimensional to a good approximation. This is primarily because of the localized, directional nature of d electrons that are found in high densities in transition metals and are not found in the simple metals. Secondly, note that electrons are removed from both the Ag and Pd atoms and deposited in the interstitial region between the Ag and Pd atoms. This accumulation of bond charge is also found in covalent and metallic diatomic molecules. Thus we see a parallel between the bimetallic adhesive bond and the diatomic molecular bond.

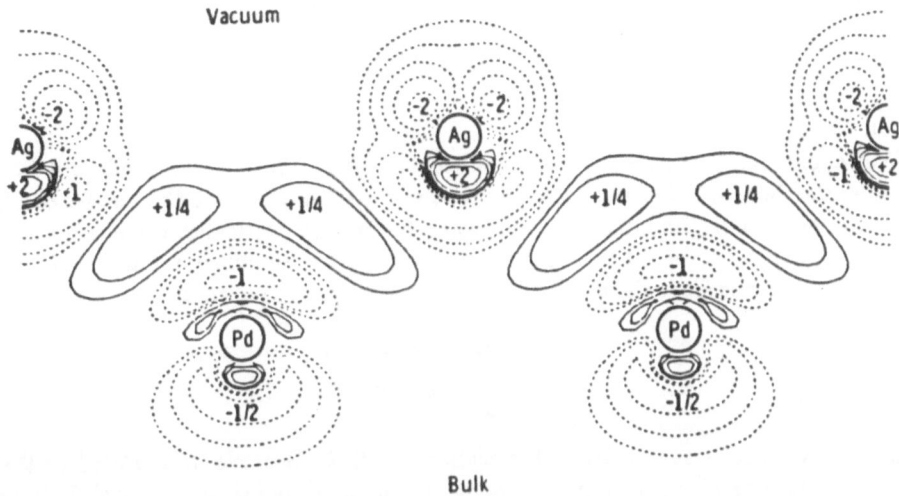

FIGURE 15. Electron density rearrangement associated with adsorption of a Ag monolayer on a Pd(100) substrate from Refs. 21-22.

Now let's examine the energy components of a bimetallic adhesive bond. In Fig. 16 we see a plot of the energy components of the Al(111)-Mg(0001) adhesive bond taken from Ref. 18. It is interesting that at large separations, the electronic kinetic energy is attractive. That is, the kinetic energy initiates the bond. At small separations, the kinetic energy forms the repulsive barrier. On the other hand, the electrostatic energy does just the opposite. It is repulsive at large separations and is attractive at small separations. This behavior is paralleled in molecular bonding[23] as shown for the diatomic molecule H_2^+. Note also that the dominant attractive energy component is the electron exchange-correlation energy.

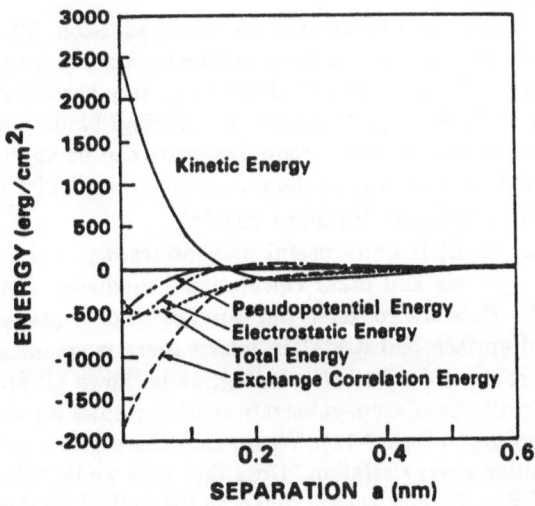

FIGURE 16. Self-consistent energy components of the binding energy for an Al(111)-

Mg(0001) contact as taken from Ref. 18.

3. A UNIVERSAL ENERGY RELATION

3.1. Bimetallic Adhesion

We've just seen that there are parallels between bimetallic adhesive bonds and diatomic molecular bonds. This encourages us to have a second look at the results of Fig. 13 for systematics in the bimetallic adhesion relations. In particular, we look for a simple scaling of those curves. Let us scale the energy by the equilibrium binding energy ΔE and the distance from the equilibrium separation a_m by the length l:

$$E(a) = \Delta E \, E^*(a^*) \tag{9}$$

$$a^* = \frac{a - a_m}{l}. \tag{10}$$

There are many ways to define the scaling length l. Because it is related to the size of the screening cloud of electrons about the ion core, we originally[24] took it to be the screening length. However, screening lengths are not well defined, e.g., molecules, and for transition metals where s, p, and d electrons hybridize in the screening cloud. It is convenient to define l in terms of an equilibrium quantity (as was done for ΔE), so that it can be more easily determined experimentally or theoretically. Thus we define l so that the second derivatives of the scaled curves are all the same at equilibrium. As this value of the second derivative is arbitrary, we take it to be equal to one for simplicity. The resultant expression for l is:

$$l = \sqrt{\frac{\Delta E}{[d^2 E(a)/da^2]_{a_m}}} \tag{11}$$

We[18,24,26-34] applied this scaling procedure to the data of Fig. 13 as well as to our adhesion results for the case where the two metals in the contact are identical[25] and the result is shown in Fig. 17.

We discovered that there is a universal equation of adhesion. That is, the adhesive energy relations for all 10 bimetallic contacts can be obtained from a simple scaling of a single, universal curve. Total energy calculations are very hard to do and this means that they don't have to be done provided the equilibrium binding energy, separation and second derivative are known. The second derivative can be approximated[25] by the elastic stiffness constant associated with the direction perpendicular to the interface.

3.2. Metallic Adhesion, Chemisorption and Cohesion

The generality of this result is quite useful and interesting, and one might wonder if it applies to other systems and other calculational methods. Let us turn next to chemisorption[26-28,35]. Here we consider the interaction of a gas atom (adatom) or molecule with a solid surface (substrate) in which there is significant wave-function overlap between adsorbate and substrate. In Fig. 18 is shown all first-principles, self-consistent total energy versus adatom-substrate spacing results we are aware of. These have been scaled according to Eqs.(9-11). We see that there is to a good approximation a universal chemisorption energy relation. Note that now we include elements beyond the simple metals of Fig. 17 and rather different theoretical methods are used here compared to those employed to carry out the adhesion calculations.

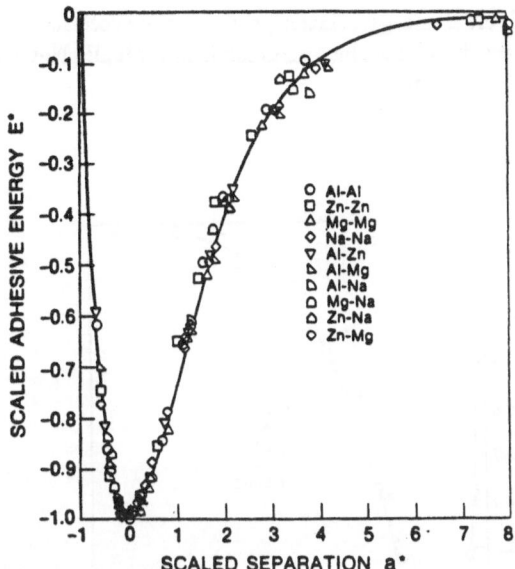

FIGURE 17. Adhesive energy results from Fig. 13 (Ref. 18) and Ref. 25 scaled using Eqs.(9-11).

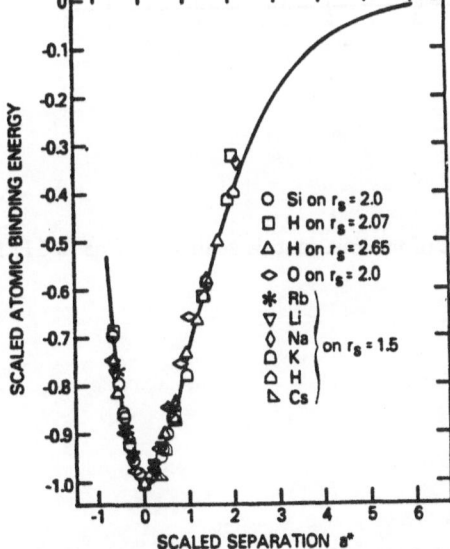

FIGURE 18. Atomic binding energy curves for chemisorption on jellium surfaces whose densities are indicated by the r_s values shown. The energy relations are scaled according to Eqs.(9-11). See Refs. 26-28.

A third field of solid energetics is found in cohesion. Here the crystal structure is maintained while the lattice constant is varied and corresponding total energies computed. Again these are difficult calculations and so the results are limited. The scaled cohesive energy curves are shown in Fig. 19. For reference, the equilibrium value is the heat of evaporation of the solid.

Again we find a universal energy relation, this time for cohesion. Note the variety of metals—noble, transition, band-overlap, rare-earth and alkali. Nevertheless, they obey an energy relation of the same form.

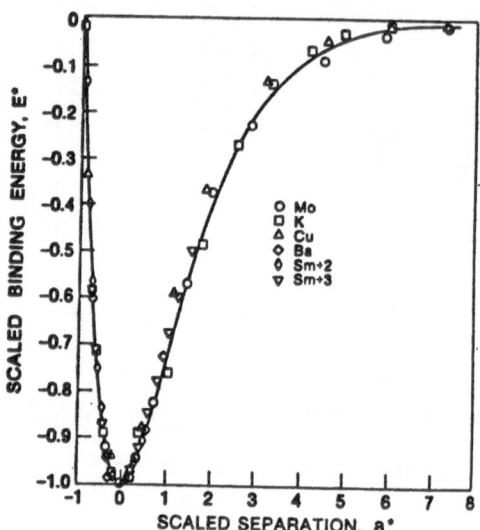

FIGURE 19. Bulk energies of various metals scaled via Eqs.(9-11) from Ref. 28.

Perhaps the most interesting result is that the universal relation for bimetallic adhesion is to a good approximation the same as the universal relation for cohesion which in turn is the same as the universal relation for chemisorption. To demonstrate this[27-28], in Fig. 20 we show scaled energy relations for representatives of each of those three categories as well as that for the H_2^+ diatomic molecule. These results reveal a fundamental relationship between the energetics of these various metallic configurations as well as an underlying simplicity of nature that had not been recognized before. It is this universal binding energy relation that Daw[36] et al. are already making use of in their computer simulations of fracture properties presented at this conference.

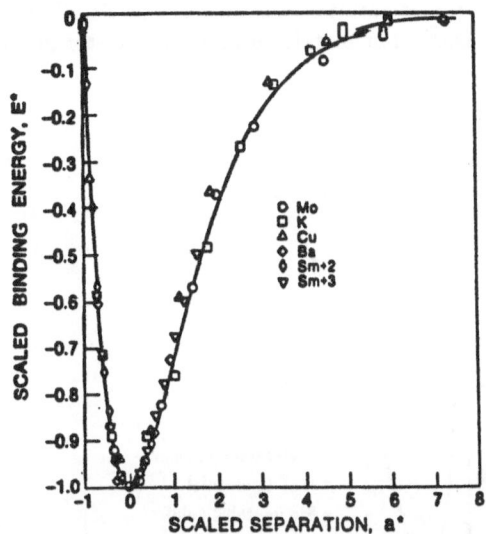

FIGURE 20. Binding energy as a function of interatomic separation for four systems as noted, scaled according to Eqs.(9-11). See Refs. 27-28.

3.3. Expected Limitations

We have found a single universality class which might include chemisorption, adhesion and cohesion of metals and elemental semiconductors. There are, of course, limitations to these universal relationships. First, ionic solids and inert gas solids do not fall in this universality class. Secondly, these universal relations apply only in the absence of phase transitions. That is, they may apply on either side of the phase transition but not through the transition. Finally, there is a limitation on the range in a. For a small enough so that the ion cores overlap, the shell structure of those cores plays a direct, important role in the energetics. Since this shell structure is specific to each element, it is doubtful that universality would extend into the region of ion-core overlap.

Now there are common characteristics of those particles and materials within the universality class. First, all particles are Fermions: electrons or holes. Secondly, the forces involved are exponential in nature. This exponential character comes from wave function overlap or from overlap of the ion-core-screening clouds. Agreed, asymptotic forms may not be exponential for metals and covalent semiconductors, but the behavior in and around the minimum is. It is this region of separations upon which we will concentrate. This character sets metals and covalent semiconductors apart from ionic or inert-gas solids.

3.4. Nucleons and Metals

There is another class of matter that is composed of Fermions and whose short-range forces are exponential in nature. That class is nuclear matter. Thus it would be of interest to determine whether nucleon-nucleon forces belong to the same universality class as forces between atoms of metals and covalent semiconductors. Recently Wiringa, Smith and Ainsworth[37] have empirically fit nucleon-nucleon data and have obtained a simple potential which yields a high-quality description of deuteron properties and nucleon-nucleon phase shifts below 330 MeV. We average the spin-isospin-dependent s- and p-wave components to obtain a form which should be representative of nuclear matter at moderate density. The scaled results are shown in Fig. 21, along with representative results from Fig. 20. We see[29] that metals and nuclear matter do obey a single energy form to a good approximation.

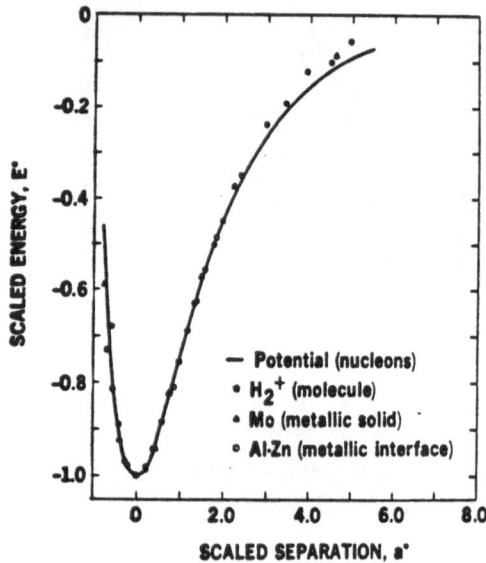

FIGURE 21. Comparison of two-nucleon potential with binding energy relations for the molecule H_2^+, the bulk metal Mo and the bimetallic interface Al-Zn, scaled according to Eqs.(9-11) (see Ref. 29).

9.5. Simple Relationships from Universality

9.5.1. Surface Versus Cohesive Energies.
It is a powerful result to be able to determine total energy versus interparticle spacing from a knowledge of only three equilibrium numbers. But there is more that can be done given the knowledge that there is a universal energy relation. That is, relationships between observables can be simply derived.

As the first example we consider the relationship between surface energies (otherwise known as cleavage or brittle-fracture energies—see Sections 2.3 and 3.1), and cohesive energies (see Section 3.2). The following relationship follows almost immediately[28] from a knowledge that binding energy relations are of universal form:

$$4\pi r_{ws}^2 \sigma \approx 0.82\, E_{coh} \tag{12}$$

Here r_{ws} is the Wigner-Seitz radius and σ is the surface energy. The predictions from Eq.(12) are compared with experimental data for metals in Fig. 22. One can see there that there is good agreement across the periodic table.

351

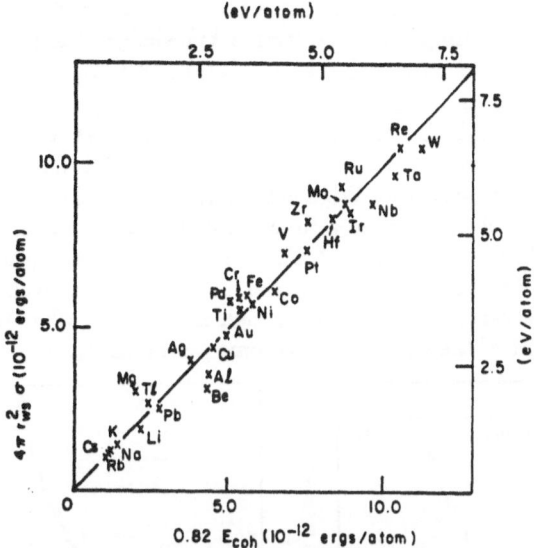

FIGURE 22. Plot of the left-hand side of Eq.(12) versus the right-hand side yielding the straight line. The experimental data are shown by the X's. See Ref. 28.

Now if nuclear matter, metals and covalent semiconductors belong to the same universality class (Section 3.4), then Eq.(12) ought to apply to nuclear matter also[29]. An experimental test of this is shown in Fig. 23. There one can see that Eq.(12) is accurate for the relationship between surface energies and cohesive energies not only for metals but also for nuclear matter. Note also that electron-hole liquids (EHL's) such as those found in Si and Ge also agree well with the predictions. Thus we have a relationship which applies over about 10 orders of magnitude in energy and about 21 orders of magnitude in density.

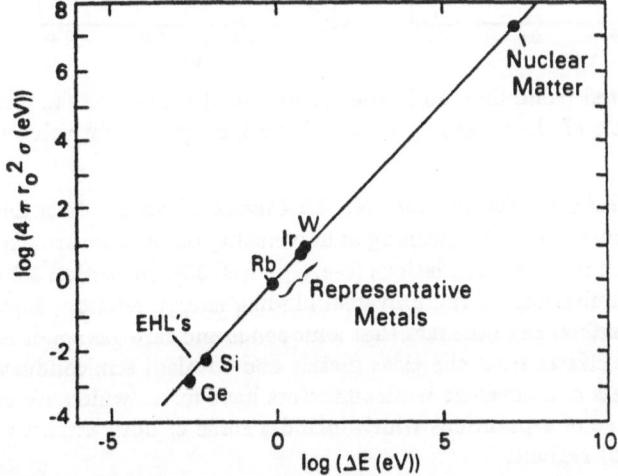

FIGURE 23. Predicted surface energies versus cohesive energies from Eq.(12) (straight line). Experimental data are given as points. See Ref. 29.

9.5.2. Equations of State for Metals. As a second example of relationships between observables which can be simply derived from a knowledge of universality, just taking a volume derivative of the universal energy relation (Eq.(9)), yields the pressure as a function of volume. This equation of state can be readily tested because there are considerable experimental high pressure data. Likewise cohesive energy and the zero-pressure bulk modulus data are well known for elemental solids and many alloys. Those two numbers, plus the zero-pressure Wigner-Seitz radius are all that we need[32] to define ΔE and l. Predictions of pressure as a function of volume are compared with experimental data in Fig. 24 for stainless steel, Rb, Cu and Li. These predictions, which are simple enough to be described as "back of the envelope", agree well with experiment. This is true while pressures obtained are as high as nearly 4 Mbar with compressions up to about 50 percent of the equilibrium volume per atom V_o.

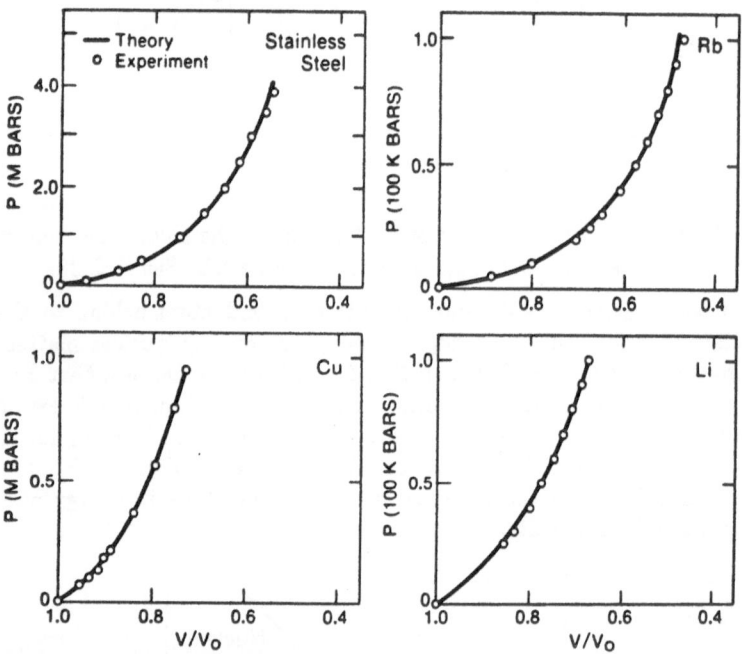

FIGURE 24. Predicted (solid line) and experimental (o's) results for the pressure P as a function of the ratio of the volume per atom V to the equilibrium volume per atom V_o. See Ref. 32.

9.5.3. A Universal Equation of State for All Classes of Solids. It is important to understand a little more about the meaning of universality because we are likely to see it used more and more in fracture calculations (see, e.g., Ref. 36). In Section 3.3 we pointed out some expected limitations of the universal binding energy relation, Eqs.(9-11). In particular, we made reference to the fact that ionic solids and rare-gas solids have energy relations in different classes from the class metals and covalent semiconductors are in. This is because metals and covalent semiconductors have forces which are exponential in nature over a range of separations which includes some of both attractive ($a^* > 0$) and repulsive ($a^* < 0$) regions.

However, if one is limited to the full repulsive region and to the expansive (attractive) region limited to near zero pressure, then wave-function-overlap or exponential effects can be significant for *all* classes of solids. In fact, one might expect[38] the shape of the pressure-volume curve to be dominated by overlap effects, even for ionic and rare-gas solids. That this is true is evidenced by Figs. 25 and 26.

Our first example is NaCl, an ionic solid whose high pressure behavior is generally thought of in terms rather different from those considered for a metal such as, e.g., Cu. However if we predict the pressure-volume relation for NaCl using the universal form (Eqs.(9-11) and Fig. 20) derived for metals, we obtain the solid curve of Fig. 25. One can see that it agrees well with the experimental data—apparently as well as for metals (Fig. 24). Also shown is a well known and extensively used pressure-volume relation for solids due to Murnaghan[39]. While good agreement with the data is obtained at low pressures, the Murnaghan expression deviates significantly at larger pressures presumably because of nonlinear pressure contributions accurately included in the universal relation.

Our second example is solid H_2, which is softer than NaCl so that it provides a more difficult test of nonlinear pressure effects. Solid H_2 below the metallic phase transition is a Van der Waals solid—certainly not metallic. Again we use the universal expression (Eqs.(9-11)), derived for metals and the result is shown in Fig. 26. The predicted pressure versus volume (solid curve) agrees well with the experimental data even though the pressures are so high that the change in volume per molecule is up to 80 percent of the zero-pressure value. Because of this relatively large change in volume, the Murnaghan relation is more in error than it was for NaCl.

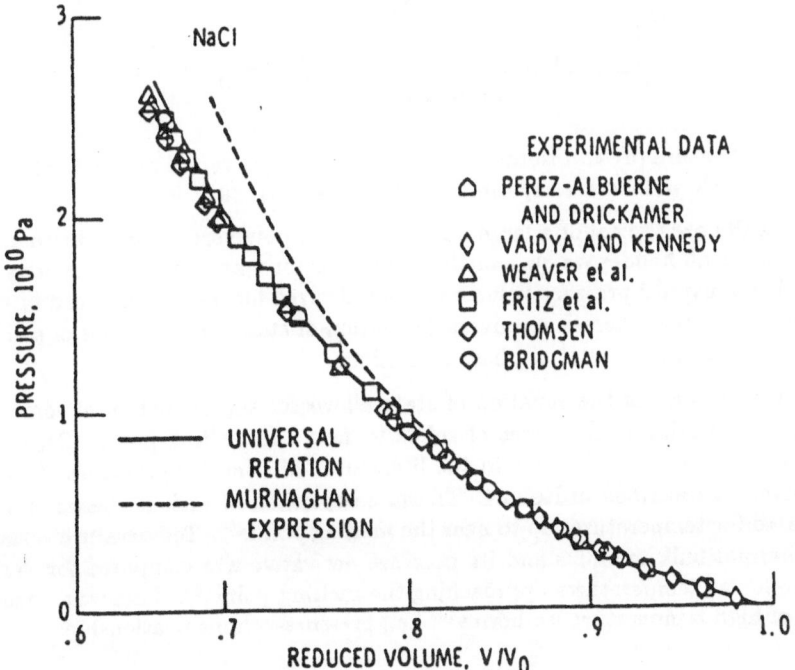

FIGURE 25. Pressure versus volume for NaCl, from Ref. 38.

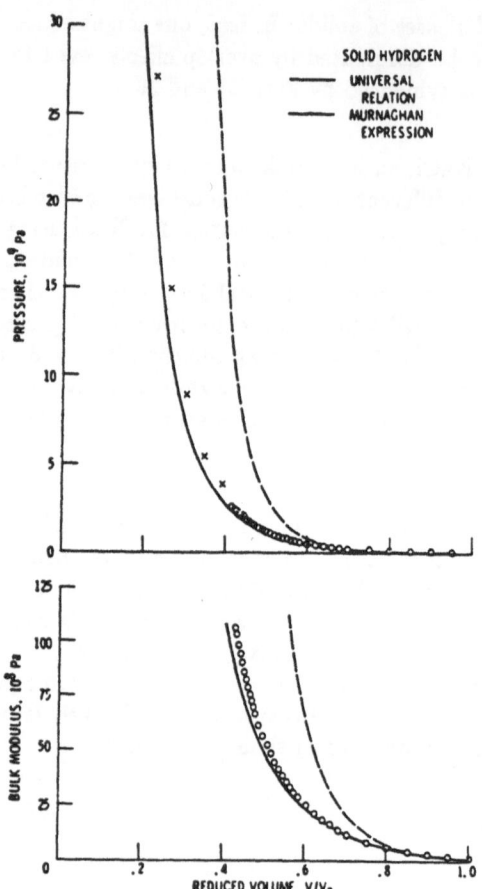

FIGURE 26. Pressure (a) and isothermal bulk modulus (b) versus reduced volume for solid hydrogen. x's and o's are experimental data. See Ref. 38.

These results are typical of a large number metals, polymers, semiconductors, rare gas solids and alkali halides we have made predictions for[38] and the accuracy is typical even though a universal pressure-volume relation derived for metals was used in every case. We conclude that there is a universal equation of state for *all* classes of solids in compression and in the absence of phase transitions.

The universal form of the equation of state allows for the prediction of some high temperature properties of all classes of solids to near the melting point. Because of space limitations, here we will refer to the literature. Melting temperatures of metals were predicted as described in Ref. 34. Thermal expansion of various classes of solids was calculated for temperatures up to near the melting points[40]. Temperature variation of the isothermal bulk modulus and its pressure derivative was computed for various classes of solids at temperatures approaching the melting points[40]. These included the prediction of high temperature isotherms[40], i.e., pressure-volume relationships.

4. SUMMARY

In this section we won't attempt to summarize all of the results of this review, but

355

will note representative ones. We first reviewed some characteristics of metal surfaces and interfacial bonds. Several properties of surface energies or brittle fracture energies of metals were shown to be important, incuding band filling, spin splitting and film thickness effects. Films as thin as three atomic layers were found to have surface energies close to those of macroscopically thick films which is due to the typically short screening lengths characteristic of metals. Spin splitting was shown to be an important factor in the surface energy of Fe(100). Interesting interfacial magnetic effects were found, including the possibility of large magnetic moments and a new class of superlattice magnetic materials. Adsorbate-substrate band overlap was found to play an important role in interfacial bonding. Large densities of surface states were found on all transition metal surfaces studied. In general, significant electronic rearrangement effects were found in surfaces, grain boundaries and bimetallic adhesion. This points out the need for fully self-consistent quantum-mechanical calculations to determine many properties of metallic interfaces. Several properties of bimetallic adhesion emerged. The range of the relatively strong bonding was found to be about one interplanar spacing. The kinetic energy was found to initiate the bond, and the exchange-correlation energy turned out to be the energy component which dominated the attractive components. Localized d orbitals in transition metals led to a more directional electronic rearrangement than we found in the simple metals. Electronic barriers at bimetallic interfaces were found to have little relation to the contact potential.

The large number and variety of characteristics of metal surfaces and interfacial bonds can be rather confusing. We sought some systematics in the various phenomena and found similarities between some properties of molecular bonding and bonding at bimetallic interfaces.

This led us to discuss the discovery of a single energy relation for metallic adhesion, cohesion and chemisorption—even for the energetics of nuclear matter. This at first suprising result provides a simple, powerful method for determining energy versus interparticle spacing.

Further, it was found that simple relationships between observables followed easily from the existence of a universal relationship between energy and distance. Some example relationships were given. A simple relationship between surface or brittle-fracture energies and cohesive energies was deduced and successfully tested with experiment. "Back of the envelope" equation of state (pressure-volume) calculations were made possible and results compared very well with experiment.

It was noted that the universal energy relationship did not extend to ionic solids or rare earth solids. However, a single equation of state was discovered to describe *all* classes of solids in compression and in the absence of phase transitions. With that, one can predict high temperature isotherms and high temperature behavior of thermal expansion, the bulk modulus and its pressure derivative.

It seems that we are entering a period of rapid progress in the area of theoretical modeling of fracture. Our understanding and calculational tools have developed to the point where one might expect realistic, quantum-mechanical theories of fracture. One can't help but look forward to the next five to ten years.

REFERENCES

1. Smith J. R., Gay J. G. and Arlinghaus F. J.: Self-Consistent Local-Orbital Method for Calculating Surface Electronic Structure: Application to Cu(100). Phys. Rev. B *21*, 2201 (1980).

2. Smith J. R.(ed): *Theory of Chemisorption*. Berlin, Heidelberg, New York: Topics in Current Physics, vol. 19: Springer-Verlag, (1980).

3. Arlinghaus F. J., Gay J. G. and Smith, J. R.: Ni(100) Surface Electronic Structure. Phys. Rev. B *21*, 2055 (1980). See also Zhu X. Y., Hermanson J., Arlinghaus F. J., Gay J. G., Richter R. and Smith J. R.: Electronic Structure of Ni(100) Films: Self-Consistent Local-Orbital Calculations. Phys. Rev. B *29*, 4426 (1984).

4. Gay J. G., Smith J. R. and Arlinghaus F. J.: Surface Electronic Structure of Rhodium(100). Phys. Rev. B *25*, 643 (1982).

5. Arlinghaus F. J., Gay J. G. and Smith J. R.: Surface States on *d*-Band Metals. Phys. Rev. B *23*, 5152 (1981).

6. Examples from our predictions are: Kevin S. D., Stoffel N. G. and Smith N. V.: Surface States on Low-Miller-Index Copper Surfaces. Phys. Rev. B *31*, 3348 (1985). Heimann P., Hermanson J. and Miosga H.: *d*-Like Surface-State Bands on Cu(100) and Cu(111) Observed in Angle-Resolved Photoemission Spectroscopy. Phys. Rev. B *20*, 3059 (1979). Kevan S. D. and D. A. Shirley: High-Resolution Angle-Resolved Photoemission Studies of the $\overline{M}$-Point Surface State on Cu(001). Phys. Rev. B *22*, 542 (1980). Westphal D. and Goldmann A.: Polarization Dependent Photoemission from it d-Like Surface States on Cu. Surf. Sci. *95*, L249 (1980). Goldmann A. and Bartels E.: High-Resolution Study of the $\overline{M}$ Surface State on Ag(100). Surf. Sci. *122*, L629 (1982). Koch E. E., Barth J., Fock J.-H., Goldmann A., and Otto A.: Surface Photoemission in the 4*d* Band from Polycrystalline Silver Surfaces. Solid State Commun. *42*, 897 (1982).

7. Kohn W. and Sham L.: Self-Consistent Equations Including Exchange and Correlation Effects. Phys. Rev. *140*, A1133 (1965).

8. Schaefer H.(ed.): *The Electronic Structure of Atoms and Molecules*. Reading, Mass.; Addison-Wesley, (1972).

9. Gay J. G., Smith J. R., Richter R., Arlinghaus F. J. and Wagoner R. H.: Surface Energies in *d*-Band Metals. J. Vac. Sci. Technol. *A2*, 931 (1984).

10. Wawra H. H.: The Surface Energy of Solid Materials as Measured by Ultrasonic and Conventional Test Methods. Z. Metallkd. *66*, 395 (1975).

11. Richter R., Smith J. R. and Gay J. G.: Total Energies and Atom Locations at Solid Surfaces. *The Structure of Solid Surfaces*, Springer Series in Surface Sciences, vol. 2 ed. by Van Hove M. A. and Tong S. Y., Berlin (1985), pp.35-40.

12. Zhu X. Y., Hermanson J., Arlinghaus F. J., Gay J. G., Richter R. and Smith J. R.: Electronic Structure and Magnetism of Ni(100) Films: Self-Consistent Local-Orbital Calculations. Phys. Rev. *29*, 4426 (1984). This trend of a larger magnetic moment for the surface was also found for 3, 5 and 7 layer Ni(100) films by Gay J. G., Richter R. and Smith J. R. (unpublished).

13. Richter R., Gay J. G. and Smith J. R.: Spin Separation in a Metal Overlayer. Phys. Rev. Letters *54*, 2704 (1985). This was first reported by Richter R., Gay J. G. and

Smith J. R.: An Iron Monolayer on Silver(100): Spin-Polarized Electronic Structure. J. Vac. Sci. Technol. *A3*, 1498 (1985). A surface magnetic moment larger than that of the bulk was also predicted for Fe(100) films of 3, 5 and 7 layer thicknesses by Gay J. G., Richter R. and Smith J. R. (unpublished).

14. See, e. g., Krakauer H., Freeman A. J. and Wimmer E.: Magnetism of the Ni(110) and Ni(100) Surfaces: Local-Spin-Density Functional Calculations Using the Thin-Slab Linearized Augmented Plane Wave Method. Phys. Rev. B *28*, 610 (1983). See also Ohnishi S., Freeman A. J. and Weinert M.: Surface Magnetism of Fe(001). Phys. Rev. B *28*, 6741 (1983).

15. See, e. g., Bergmann G.: Investigation of Magnetic Films by the Anomalous Hall Effect. J. Magn. Magn. Mater. *35*, 68 (1983), and references therein.

16. Smith G. C., Padmore H. A. and Norris C.: The Growth of Fe Overlayers on Ag(100). Surf. Sci. *119*, L287 (1982).

17. Fu C. L., Freeman A. J. and Oguchi T.: Prediction of Strongly Enhanced Two-Dimensional Ferromagnetic Moments on Metallic Overlayers, Interfaces, and Super-lattices. Phys. Rev. Letters *54*, 2700 (1985).

18. Ferrante J. and Smith J. R.: Theory of the Bimetallic Interface. Phys. Rev. B *31*, 3427 (1985).

19. Smith J. R.: Theory of Electronic Properties of Surfaces. *Interactions on Solid Surfaces*, Topics in Applied Physics, vol. 4, ed. by Gomer R., New York (1975), pp. 1-39.

20. Smith J. R. and Ferrante J.: Grain-Boundary Energies in Metals from Local-Electron-Density Distributions. Phys. Rev. B *34*, 2238 (1986).

21. Capehart T. W., Richter R., Gay J. G., Smith J. R., Buchholz J. C. and Arling-haus F. J.: Transition Metal Chemisorption on Transition Metals—Theoretical and Experimental Electronic Structure for Silver on Palladium(100).

22. Richter R. and Wilkins J. W.: Self-Consistent Surface Electronic Band Structure Calculations: Changes Upon Chemisorption. J. Vac. Sci. Technol. *A1*, 1089 (1983).

23. Feinberg M. J. and Ruedenberg K.: Paradoxical Role of the Kinetic-Energy Operator in the Formation of the Covalent Bond. J. Chem. Phys. *54*, 1495 (1971).

24. Rose J. H., Ferrante J. and Smith J. R.: Universal Binding Energy Curves for Metals and Bimetallic Interfaces. Phys. Rev. Letters *47*, 675 (1981).

25. Ferrante J. and Smith J. R.: Theory of Metallic Adhesion. Phys. Rev. B *19*, 3911 (1979).

26. Smith J. R., Ferrante J. and Rose J. H.: Universal Binding Energy Relation in Chemisorption. Phys. Rev. B *25*, 1419 (1982).

27. Ferrante J., Smith J. R. and Rose J. H.: Diatomic Molecules and Metallic Adhesion, Cohesion and Chemisorption: A Single Binding Energy Relation. Phys. Rev. Letters *50*, 1385 (1983).

28. Rose J. H., Smith J. R. and Ferrante J.: Universal Features of Bonding in Metals. Phys. Rev. B *28*, 1835 (1983).

29. Rose J. H., Vary J. P. and Smith J. R.: Nuclear Equation of State from Scaling Relations for Solids. Phys. Rev. Letters *53*, 344 (1984).

30. For a review, see Smith J. R., Rose J. H., Ferrante J. and Guinea F.: Universal Features of Binding Energy as a Function of Interatomic Spacing. *Many-Body Phenomena at Surfaces*, edited by Langreth D. and Suhl H., Academic Press (1984), pp. 159-174.

31. Smith J. R. and Ferrante J.: Metals in Intimate Contact. Materials Science Forum *4*, 21 (1985).

32. Rose J. H., Smith J. R., Guinea F. and Ferrante J.: Universal Features of the Equation of State of Metals. Phys. Rev. B *29*, 2963 (1984).

33. Smith J. R., Ferrante J. and Rose J. H.: Universal Binding Energy Relations for Bimetallic Interfaces and Related Systems. Journal de Physique, Colloque C4, supplement au no. 4, vol. *46*, (1985), p. C4-257.

34. Guinea F., Rose J. H., Smith J. R. and Ferrante J.: Scaling Relations in the Equation of State, Thermal Expansion and Melting of Metals. Appl. Phys. Lett. *44*, 53 (1984).

35. Perdew J. P. and Smith J. R.: Can Desorption Be Described by the Local Density Formalism? Surface Sci. *141*, L295 (1984).

36. Daw M. S., Baskes M. I., Bisson C. L. and Wolfer W. G.: Application of the Embedded Atom Method to Fracture, Dislocation Dynamics and Hydrogen Embrittlement. To be found elsewhere in this proceedings.

37. Wiringa R. B., Smith R. A. and Ainsworth: Realistic Nucleon-Nucleon Potentials with and without Δ(1232) Degrees of Freedom. Phys. Rev. A *29*, 1207 (1984).

38. Vinet P., Ferrante J., Smith J. R. and Rose J. H.: A Universal Equation of State for Solids. J. Phys. C: Solid State Phys. *19*, L467 (1986). ibid, Phys. Rev. B 15 (to be published).

39. Murnaghan F. D.: The Compressibility of Media Under Extreme Pressure. Proc. Nat. Acad. Sci. *30*, 244-247 (1944).

40. Vinet P., Smith J. R., Ferrante J. and Rose J. H.: Temperature Effects on the Universal Equation of State of Solids. Phys. Rev. B *15* (to be published).

DISCUSSION

Comment by C. J. Altstetter:

What universality have you been able to show when you consider the entropy changes?

Reply:

We have considered high temperature effects such as the temperature dependence of thermal expansion, the bulk modulus, and pressure derivatives of the bulk modulus. These are determined from a computed pressure-volume relation for temperatures up to the melting point. The pressure-volume relation is of universal form for all classes of solids which facilitates calculation of high temperature properties, as described in the text.

Comment by M. Bernstein:

Can your universality relations help predict which materials should be brittle or ductile in terms of your correlations between surface energy and cohesive energy?

Reply:

That is a very interesting question which I can't answer at this time. I will look forward to investigating brittle/ductile behavior in the light of the universal energy relation.

Comment by A. W. Thompson:

To follow up on Mel Bernstein's question, I would comment that there are two kinds of brittleness in materials. One is "intrinsic", such that it seems impossible to avoid brittleness in the material, and presumably the correct physics could explain why. The other could be called "extrinsic", and depends sensitively on microstructure, testing conditions, processing history, etc. The extrinsic kind seems to me unlikely to be explainable by physics in the foreseeable future.

Reply:

Thank you for that information.

Comment by D. G. Pettifor:

Your universal energy expression is applicable to metals. Are there any exceptions? For example, Cs under pressure undergoes structural phase transitions with discontinuities in the equation of state curve.

Reply:

I indicated that a limitation of the universal energy relation is that one should not expect it to apply through a phase transition. It can be applied separately to each phase, however.

Comment by M. W. Finnis:

Does the self-consistent local orbital method work for adhesion calculations on slabs of Al and Zn, the s-p bonded metals? If so, can you compare with the jellium results? I was wondering if it is okay to neglect the screening charges in the jellium calculations, i.e., the second order perturbation term.

Reply:

Yes, a self-consistent local orbital method should work well for Al, Zn and other sp-bonded metals. Your primary question seems to be about the size of the second order perturbation term. That term was investigated for surface energy calculations by J. Rose and J. F. Dobson, Sol. St. Commun. 37, 91 (1981). They found it to be small for the closest packed plane, which is the plane we always deal with in adhesion. Incidentally, we found the 2nd order term to be important for grain bounding energies, as discussed in reference.

Comment by M. Daw:

I think the physics community still has some skeptisism regarding the general universality of the binding curve. However, there is one important conclusion which is valid, and that is that the anharmonic part of the interaction is related to the harmonic part. This means that the P(V) is related to the bulk modulus, or the thermal expansion is related to the bulk modulus. Although this does not have to be true, it seems to be empirically valid. Do you have any comments?

Reply:

Yes, it is true that up to now we have primarily just tested the universality of the binding energy relation against experimental and theoretical results. These tests have in every case confirmed the universal nature to within experimental or theoretical accuracy. It should be remembered, however, that there are limitations. It applies only to metals, covalent semiconductors, and nuclear matter. Ionic solids and rare gas solids are included with these classes of matter only in equations of state in compression.

What remains to be provided is an explanation or a derivation of universality. We are working hard at that, and it will be forthcoming.

Comment by F. Nabarro:

Do you predict the Gruneisen constant to be a universal constant? What is its value?

Reply:

Yes, the Gruneisen parameter has a simple form in terms of the universal energy relation. See F. Guinea, J. H. Rose, J. R. Smith, and J. Ferrante, Appl. Phys. Lett. 44, 53 (1984), especially Eq. (4).

Comment by A. W. Thompson:

You alluded to surface energy values at 4 K, obtained from ultrasonic measurements of bulk modulus and the linear relation between bulk modulus and surface energy, and the comparison of these values to calculations. Have there been any elevated temperature calculations for surface energy to compare to what materials scientists would call "direct" surface energy measurements?

Reply:

We have not done elevated temperature calculations for the surface energy. The temperature derivative of the surface energy can be determined experimentally, however, so that perhaps it is reasonable to extrapolate higher temperature experimental surface energies to lower temperature.

Comment by J. F. Knott:

Two points on your density-of-states curves.
   a) Do these simply refer to d electrons? Is there not overlap of s and
      p states and will not this affect surface properties?
   b) Are the curves specific to a specific surface, e.g. (100)? Will not
      the form of the (d) surface states density-of-states distribution
      vary with the surface chosen?

Reply:

   a) No, s and p states are included in the density-of-states curves and
      they do affect surface properties.
   b) Yes, the density-of-states curves are all for the (100) surfaces.
      The surface density of states will depend on the surface chosen.

Comment by T. Watanabe:

1) You have mentioned the difference in surface energy between para- and
ferromagnetic nickel or iron. Since the surface energy depends on the
surface orientation, would you please tell us which orientation will give
the largest difference in the surface energy?
2) Can we expect similar magnetic effects on grain boundary phenomena,
such as segregation and even fracture, through the difference in boundary
energy between para- and ferromagnetic state? If so, how much can we
expect?

Reply:

1) I would speculate that the higher index planes would give a larger
difference in surface energy because they seem to have larger surface
moments.
2) Yes, in principle the magnetic effects would be present in grain
boundary phenomena. I suspect the difference in grain boundary energies,
paramagnetic to ferromagnetic, would be rather smaller than they are for
surface energies.

Comment by D. J. Duquette:

   You have shown significant effects of interbody separation for metals.
Can you comment on the possible effects of interacting electrochemical
double layers which can be expected to be present at the tips of growing
cracks in aqueous solutions -- particularly those which cause stress
corrosion cracking?

Reply:

   This is another very interesting question which I hope to deal with in
the near future. At the moment, there have been no first principles
calculations of the effects of impurity layers or adhesion in metals. We
have such calculations in progress, however. This has to be done before
we can deal with electrochemical double layers, a more complicated situation.

Comment by H. Mughrabi:

If you consider the anisotropy of the surface energy for one particular material, is this also encompassed in the universality relationship which you presented for different classes of materials?

Reply:

One cannot predict as yet the surface energy anisotropy from a universal binding energy relation because the latter does not apply through a phase transition (structural or otherwise). We are currently developing a method which overcomes this limitation.

STRUCTURE OF GRAIN BOUNDARIES AND INTERFACES

J. Th. M. DE HOSSON and V. VITEK*

Department of Applied Physics, Materials Science Centre, University of Groningen, Nijenborgh 18, 9747 AG Groningen, The Netherlands.
* On leave from the Department of Materials Science and Engineering, University of Pennsylvania, Philadelphia, PA 19104, U. S. A.

1. INTRODUCTION

Grain boundaries are the most common interfaces always present in materials if they are not in a single crystal form. Their presence affects a large variety of material properties since many important physical processes, such as diffusion, decohesion, segregation, cavitation, corrosion etc., occur preferentially at grain boundaries (for recent reviews see [1-3]). The property most strongly affected by grain boundary phenomena in metallic materials is their mechanical strength. It has been well established that segregation of various alloying elements and impurities to grain boundaries frequently makes the boundaries particularly suitable paths for brittle cracking at low temperatures ([4-12], for a number of reviews see ref. 13). Similarly , segregation of alloying elements has a pronounced influence, either detremental or beneficial, on fracture occurring during creep by cavitation at grain boundaries [14-17]. Both the effect of segregated impurities on cohesion and their fast diffusion in grain boundaries play the major role in the recently discovered, important phenomenon of brittle fracture at high temperatures [18-21]. Grain boundary cracking is also the main problem encountered in intermetallic compounds which would be otherwise very attractive structural materials for high-temperature applications (for reviews see [22, 23]). In crystallographically complex compounds the brittleness may be attributed to the insufficient number of slip systems so that an extensive plastic deformation cannot develop. However, in f.c.c. based $L1_2$ compounds, such as Ni_3Al, the available deformation modes are adequate and yet the intergranular fracture occurs readily. It has been argued by several authors [23, 24] that brittleness at grain boundaries is an intrinsic feature of these compounds. At the same time it has been found that ductility of these materials can be dramatically increased by dopping with boron which segregates to the grain boundaries [22 - 26].

All the grain boundary phenomena controlling the above mentioned fracture phenomena occur in a very narrow region, of the order of a few atomic spacings, where the two grains join. Hence, it is the atomic structure of grain boundaries which needs to be understood in order to establish physical mechanisms of various boundary phenomena. The structure of grain boundaries has, indeed, been studied extensively in the last decade (for reviews see [1-3, 27]). Direct observations of atomic configurations in grain boundaries have become possible only recently with the advent of the high resolution electron microscopy [28, 29] and use of intensive synchrotron sources of X-rays for diffraction from the boundary region [30, 31]. On the other hand crystallographic description of grain boundaries is now well established [27, 32 - 36] and tested experimentally. Many general features of the atomic structure of grain boundaries have been established on the basis of computer simulations of grain boundary structures which were instrumental in advancing our knowledge of boundary structures ([27, 37-40], and papers in [3 and 41]). These calculations have been done almost exclusively using pair-potentials to describe the interactions between the atoms. Such calculations are capable of revealing features common to certain classes of materials and in this paper we first review the most important results of these studies and summarize thus our present understanding of general aspects of the atomic structure of grain boundaries. However, when using pair-potentials it cannot be expected to obtain accurate results for specific materials. Furthermore, these methods cannot be applied when investigating cohesive properties as well as any chemical and electronic effects at interfaces (see e.g. [42, 43]). Hence, further progress in the atomistic studies of grain boundaries requires that more basic descriptions of interatomic forces are employed in future. We outline here possible future approaches and discuss in more detail the use of multi-body empirical potentials.

2. DEPENDENCE OF THE BOUNDARY STRUCTURE ON MISORIENTATION: STRUCTURAL UNIT MODEL

A grain boundary is characterized geometrically by five parameters: the orientation of the rotation axis, the angle of misorientation of the grains and the orientation of the boundary plane. The relative displacements of the grains and the position of the grain boundary (in non-primitive lattices) are sometimes regarded as additional four parameters but these are determined by energetics if the above five parameters are fixed. Interesting general questions are how does the atomic structure of boundaries vary with the geometrical parameters, what is the relationship, if any, between structures of boundaries with different parameters, and how sensitively the answers to these questions depend on the interatomic forces. To investigate the variation of the grain boundary structure with all five parameters is a very complex task and it is usual to vary only one of the parameters while the others are fixed. Detailed computer simulation studies using a number of different pair potentials were made to investigate the dependence on the misorientation of the grains. These calculations were performed for both tilt [44, 45] and twist [46, 47] boundaries. For a chosen rotation axis the misorientation is the only variable in the latter case but in the former case the orientation of the boundary plane can still be varied. Therefore, in the case of tilt boundaries the study was made for families of boundaries with the same mean boundary plane. This plane is defined such that its normal is parallel to the vector $n_1 + n_2$, where n_1 and n_2 are normals of the boundary plane in the upper and lower grains, respectively. Different boundaries possessing the same mean boundary plane differ only in misorientation. (Symmetrical tilt boundaries are obtained if the mean boundary plane is a mirror plane or if it contains a two-fold rotation axis, such as (001) and (110) planes in the cubic case.)

The general results of these atomistic studies which were found to be independent of the potential used form a basis for the structural unit model which relates the structures of boundaries corresponding to different misorientations. All tilt boundaries possessing the same mean boundary plane as well as the twist boundaries in a certain misorientation range are composed of mixtures of two structural elements identified as 'units' of two small period boundaries that delimit the misorientation range. (Additional filler units are present in twist boundaries.) The structure of each of the delimiting boundaries corresponds to the contiguous sequence of one of these units. If the unit of a delimiting boundary is not composed of units from other boundaries the corresponding delimiting boundary is called 'favoured', and this unit is then a fundamental structural element of nearby boundary structures [44].

As examples we show in figure 1 the Σ=353 (1780) 39.60° symmetrical tilt boundary, and in figure 2 the Σ = 57 [111] twist boundary. The structure of every unit cell of the Σ=353 boundary is composed of seven units of the Σ=5 (210) 36.87° boundary (marked by full lines) and one unit of the Σ=5 (310) 53.13° boundary (marked by broken lines). All other boundaries in the misorientation range 36.87° - 53.13° are composed of these units. The situation is more complex in the Σ = 57 boundary. The structure of this boundary could be assumed to contain units of the Σ=21, 21.79° boundary and units of the ideal crystal. The hexagons marked in Fig. 2 are, indeed, the units of the Σ = 21 boundary which are linked by filler units. However, when analyzing the stacking of the planes in the triangular regions, we can see that they correspond alternately to the ideal crystal and the (111) stacking fault. This is related to the dislocation content of this boundary as explained below.

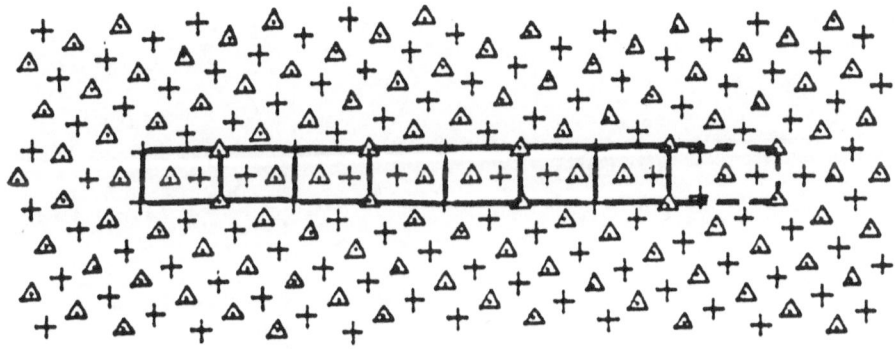

Fig. 1   Structure of the Σ=353 (1780) symmetrical tilt boundary. Triangles and crosses represent in this and all other figures of [001] tilt boundaries different (002) planes within the [001] period.

Geometrically, any large period boundary can, of course, be decomposed into strained units of smaller period boundaries, as first noted in [48]. However, it is not known a priori which boundary units to choose in such decomposition and whether the same units comprise boundaries nearby in the misorientation range or how distorted the units are. Atomistic studies revealed the continuity of the unit description and showed that relaxations minimized distortions of the units. Furthermore, calculations showed that the minority units represent regions of misorientation variation and are, therefore, sources of the elastic field of the boundary which can be interpreted as the field of network of DSC grain boundary dislocations superimposed on the reference structure composed of majority units. The structural unit model thus also provides a link between the dislocation model of grain boundaries and their atomic structure. In tilt boundaries rows of minority units parallel to the tilt axis can be identified with the cores of edge DSC dislocations and in twist boundaries the rows of filler units correspond to the cores of screw dislocations which then intersect at minority units. In figure 1 the rows of the Σ = 5 (310) units are the cores of 1/5 [210] edge dislocations. All these dislocations are complete DSC dislocations of the Σ=5 coincidence lattice.

In the Σ = 57 boundary the corresponding "complete DSC dislocations" would be 1/2 <1$\bar{1}$0> lattice dislocations since the ideal crystal is the reference structure.

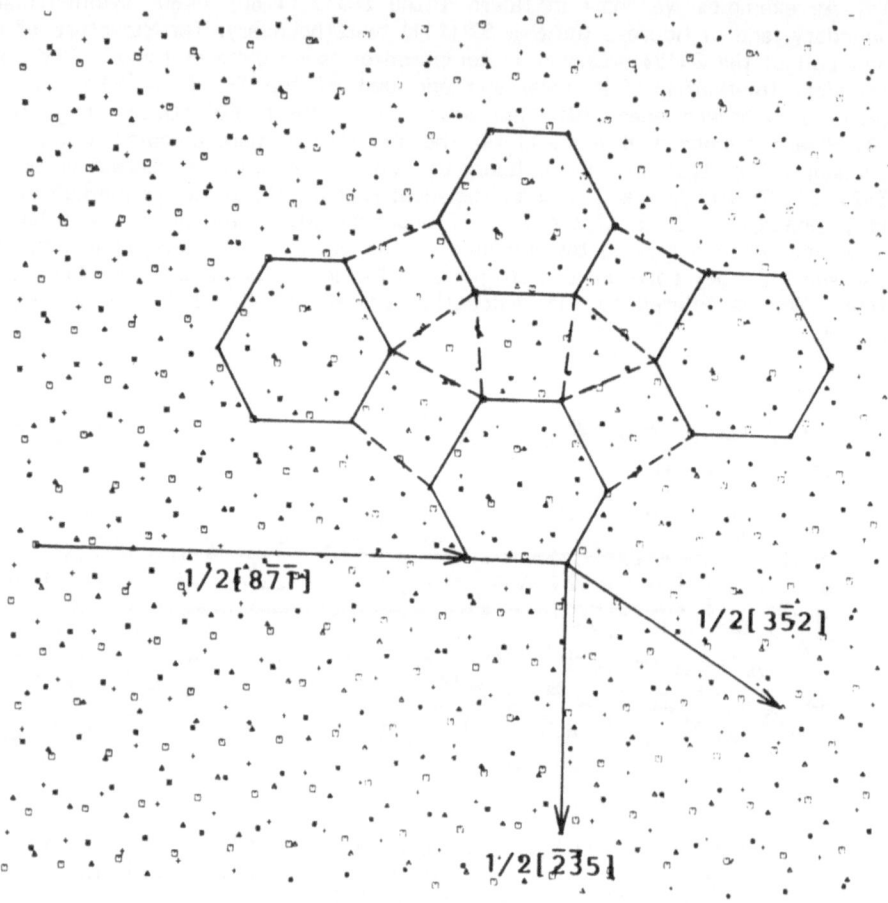

Fig. 2   Structure of the Σ=57 [111] twist boundary shown in [111] projection. The symbols in the sequence Δ + □ * represent atoms from four consecutive (111) planes. The geometrical boundary plane is positioned between layers marked + and . This notation is the same in all other figures of (111) twist boundaries.

However, it can easily be seen using Frank's formula [49], that the filler units represent cores of $1/6 <11\bar{2}>$ partial dislocations which is also consistent with the alternating regions of the ideal crystal and the stacking fault. Apparently, it is energetically favourable for every other node of $1/2 <1\bar{1}0>$ dislocations to dissociate and form thus a triangular network of $1/6<11\bar{2}>$ partials [50]. The situation is very much similar for lower angle grain boundaries ($\theta < 13.17°$) when considering the ideal crystal (together with the stacking fault) as a reference structure. However, in the case of the $\Sigma = 57$ ($\theta = 13.17°$) boundary the ratio of ideal crystal units and $\Sigma = 21$ units is 1:1 and therefore the $\Sigma = 21$ ($\theta = 21.78°$) can also be regarded as a reference structure and the dislocation content of the DSC dislocations relating to the Σ 21 coincidence. When moving to angles larger than $13.17°$ the units of the Σ 21 boundary are in majority and the ideal crystal or stacking fault will act as intersections of the corresponding DSC

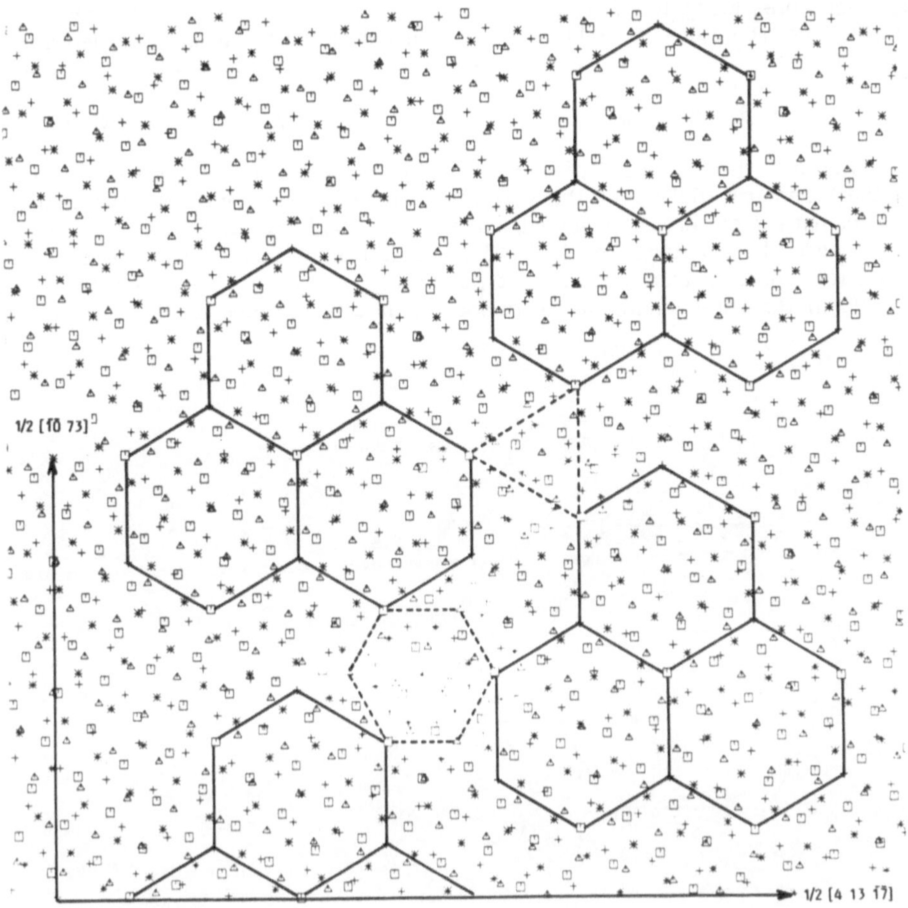

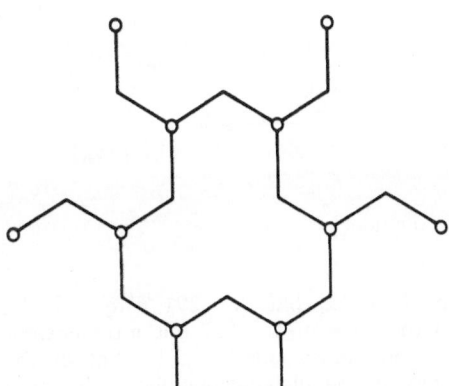

1/2 [1̄0 73]

1/2 [4 13 1̄7̄]

Fig. 3   (a) Structure of the Σ = 237 [111] twist boundary.

Fig. 3   (b) corresponding dislocation network; intersections
are indicated by O.

dislocations preserving the $\Sigma = 21$ structure. A similar picture is obtained when analyzing the structure of <111> twist boundaries with larger misorientations where other low Σ structures act as reference structures. To illustrate this point, grain boundary structures are presented for misorientations $\theta = 26.01$ ($\Sigma = 237$), $\theta = 27.80$ ($\Sigma = 13$), $\theta = 29.41$ ($\Sigma = 291$), $\theta = 30.59$ ($\Sigma = 97$), and $\theta = 32.20$ (Σ 39).

In figure 3 the structure of $\Sigma = 237$ [111] twist boundary is shown together with the schematic picture of the corresponding dislocation network. The regions, always composed of three (hexagonal) units of the $\Sigma=13$, $\theta=27.8°$ (Fig. 4) boundary, are surrounded by a network of $1/26$ <314> DSC dislocations relating to the Σ 13 coincidence. These dislocations are not pure screw but have edge components and cover $2/3$ of the total length of the period. Their intersections are $\Sigma = 21$ regions.

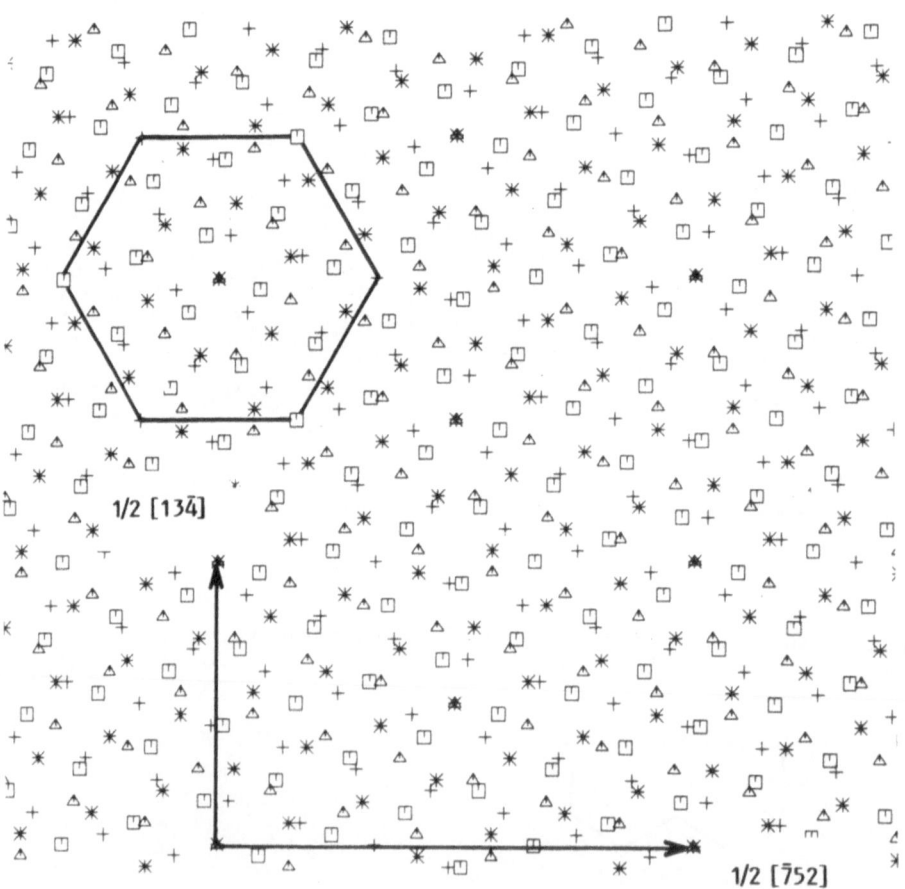

1/2 [13$\bar{4}$]

1/2 [$\bar{7}$52]

Fig. 4   Structure of the $\Sigma = 13$ [111] twist boundary.

When moving to angles larger than that of $\Sigma = 13$, like $\Sigma = 291$ (Fig. 5), the dislocation network is of the same type as in the case of $\Sigma = 237$ but intersections are now regions of the $\Sigma = 39$ boundary. For misorientations close to that of the $\Sigma = 39$, like $\Sigma = 97$ (Fig. 6) the Burgers vectors of the dislocations separating now

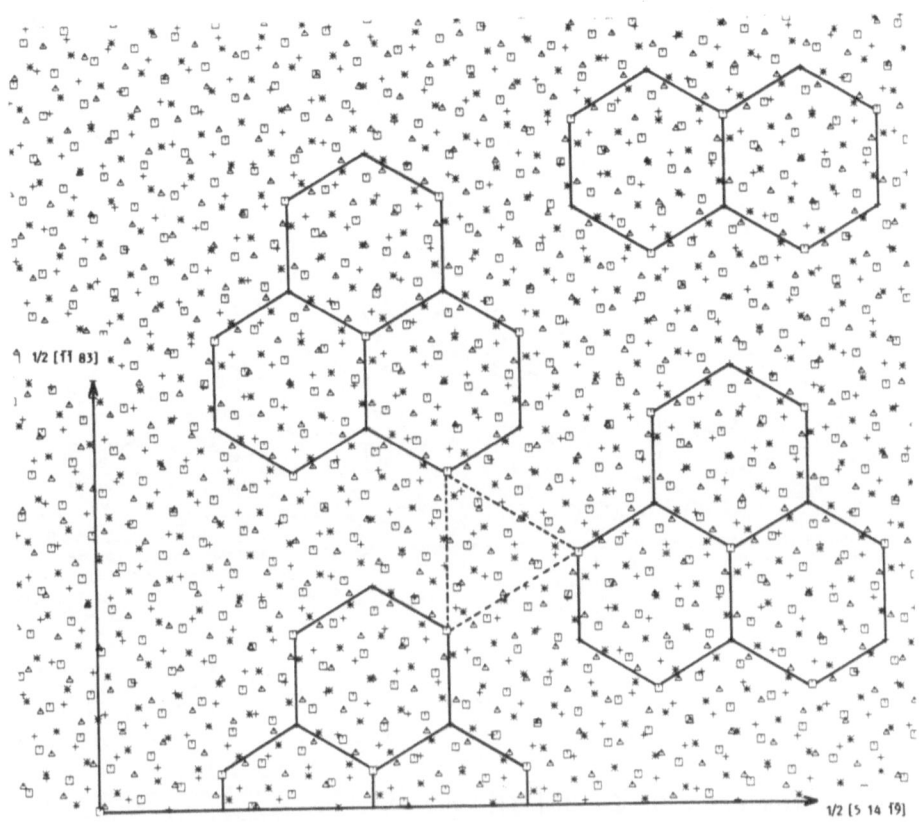

Fig. 5   Structure of the Σ = 291 [111] twist boundary.

Σ=39 units (Fig. 7) are again of the type 1/26 <314>. The network is still of the same type as in Σ = 237 but the intersections are now regions of Σ = 13.

The structural unit model is an example of the general rule governing the dependence of the boundary structure on geometrical parameters which was deduced on the basis of atomistic studies but is independent of interatomic forces employed. However, which of the delimiting boundaries are favoured may vary with the change of interatomic forces and, therefore, variation in boundaries favoured in different materials with the same crystal structure can be expected. Furthermore, although the structures of the same delimiting boundaries are often similar for different interatomic forces they are not identical and detailed positions of atoms in the units of the delimiting boundaries will vary with the potentials used and, presumably, from material to material.

370

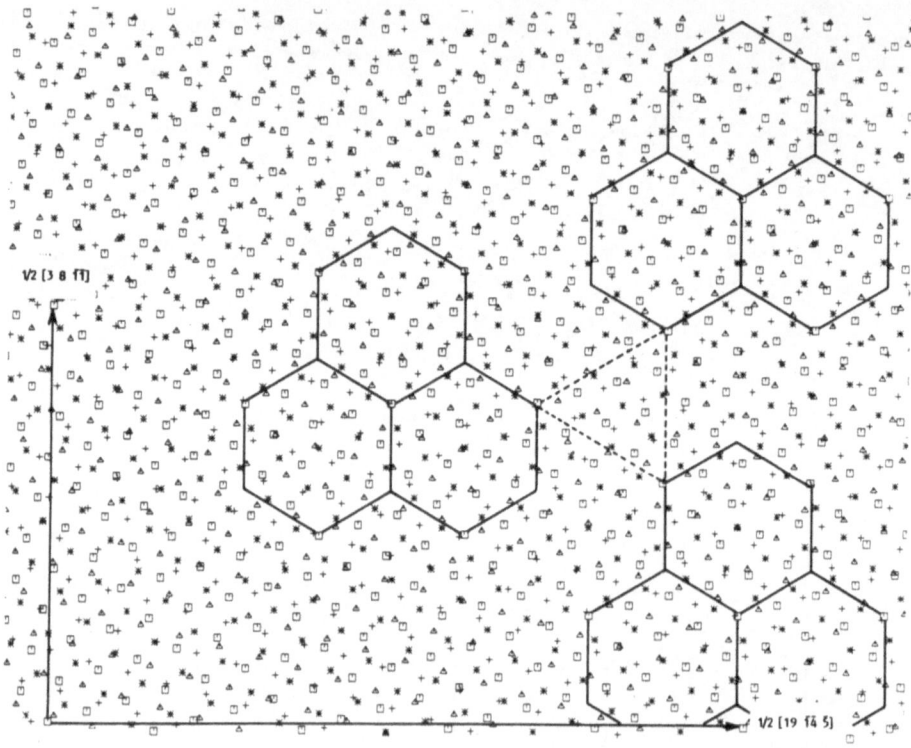

Fig. 6   Structure of the Σ = 97 [111] twist boundary.

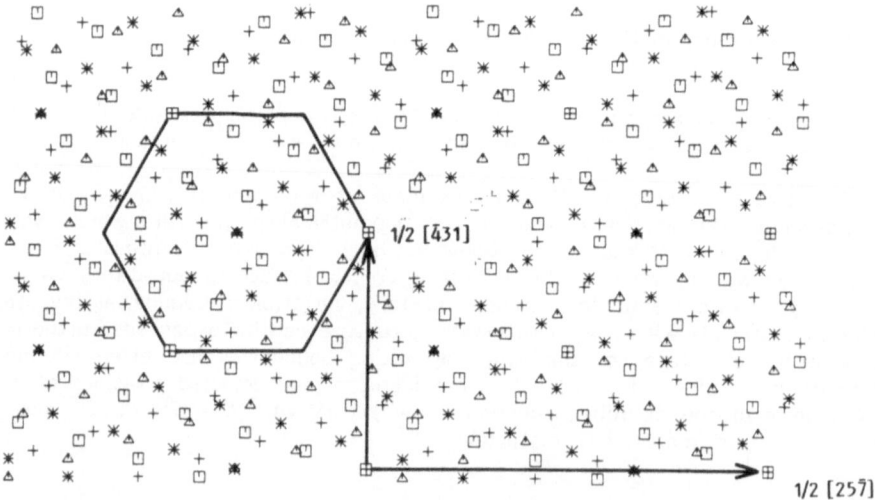

Fig. 7   Structure of the Σ = 39 [111] twist boundary.

3. MULTIPLICITY OF BOUNDARY STRUCTURES AND ITS PHYSICAL IMPLICATIONS

In a number of atomistic studies of grain boundaries several different meta-stable structures of a boundary characterized by the five geometrical parameters were found (e.g. [39, 45, 51, 52]). This structural multiplicity is not surprising and it would be of no particular significance if one of these structures possessed a much lower energy than the others. This structure would be then the only one which is physically important. However, already some of the early studies (e.g. [51]) indicated that several different structures might have very similar energies so that alternative structures may be of comparable importance. In particular a very extensive multiplicity of grain boundary structures may occur in general, large period boundaries. As explained in the previous section, structures of those boundaries can be regarded as composed of units of two short period delimiting boundaries. If the delimiting boundaries possess several alternative structures and units of all these structures participate in the structures of intervening boundaries, a very extensive structural multiplicity of the intervening boundaries follows. In principle, an infinite number of alternative structures may exist if the periodicity of the intervening boundary structures is allowed to become an indefinite multiple of the coincidence site lattice periodicity. An example of eight possible structures of the periodic $\Sigma=65$ (740)[001], 30.51° symmetrical tilt boundary is shown in Fig. 8.

Every period of this boundary consists of three units of the $\Sigma=5$ (210) boundary (marked B or B') and one unit of the ideal crystal (marked A). Two alternate structures of the $\Sigma=5$ (210) boundary, B and B', exist and their various combinations are found in different structures of the $\Sigma=65$ boundary. The lowest energy configuration corresponds to the unit sequence ABBB' and the highest energy configuration to the sequence AB'B'B. The energies of these two structures differ by 9% but much smaller differences exist for other alternatives. For example, the energy of the structure ABB'B' differs from that of the lowest energy structure by less than 1% [45].

The existence of alternative structures with similar energies suggests that at high temperatures transformations of the grain boundary structure could occur. These transformations may either be of the order-disorder type involving transition from a periodic structure to a non-periodic multiple structure, or the transitions from one alternative structure to another [52]. Such transformations of the grain boundary structure have been suggested earlier on thermodynamic grounds [53] and several experiments can be interpreted in terms of such transformations [54, 55]. Transitions between alternative multiple structures provide a basic physical model for such transformations the details of which will be different for different boundaries and different materials.

Another important feature of the multiple structures of tilt boundaries is that they may transform mutually by absorption or emission of vacancies. For example, structural unit B (Fig. 8) tranforms into unit B' if one of the atoms in its center is removed [45, 52]. Because the energies of these units are similar those transformations will be easy which explains the ability of grain boundaries to act as practically ideal sources and sinks of vacancies. Furthermore, the energetically most favourable structure of a vacancy in, for example, a boundary composed of units B corresponds to the isolated unit B' and vice versa [52]. Hence, the results of the simulation of alternative structures of grain boundaries also provide a general model for the structure of vacancies in grain boundaries.

372

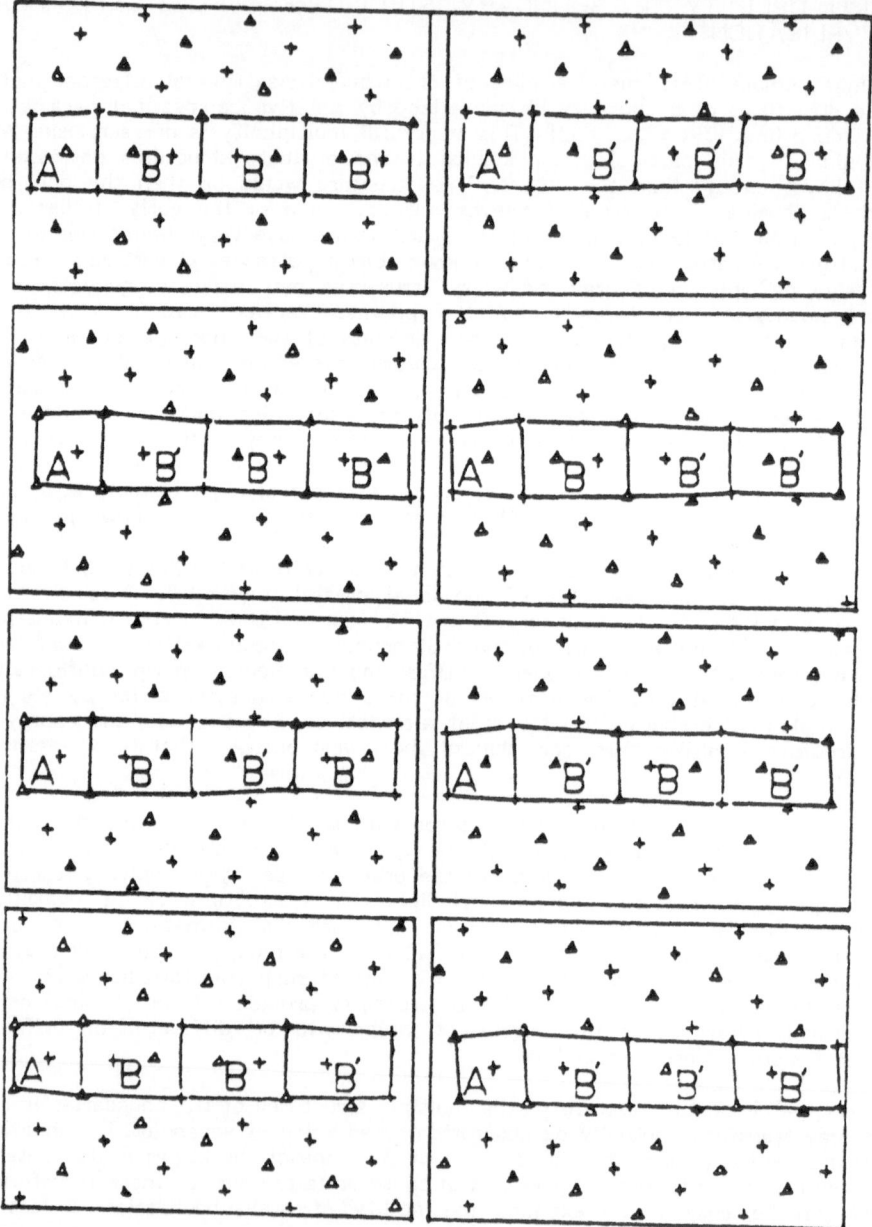

Fig. 8   Eight alternative periodic structures of the Σ=65 (740) boundary.

As an example of structural multiplicity which has been observed by transmission electron microscopy we discuss here the structure of (111) twist boundaries with misorientations close to that of the Σ = 3 twin [56].

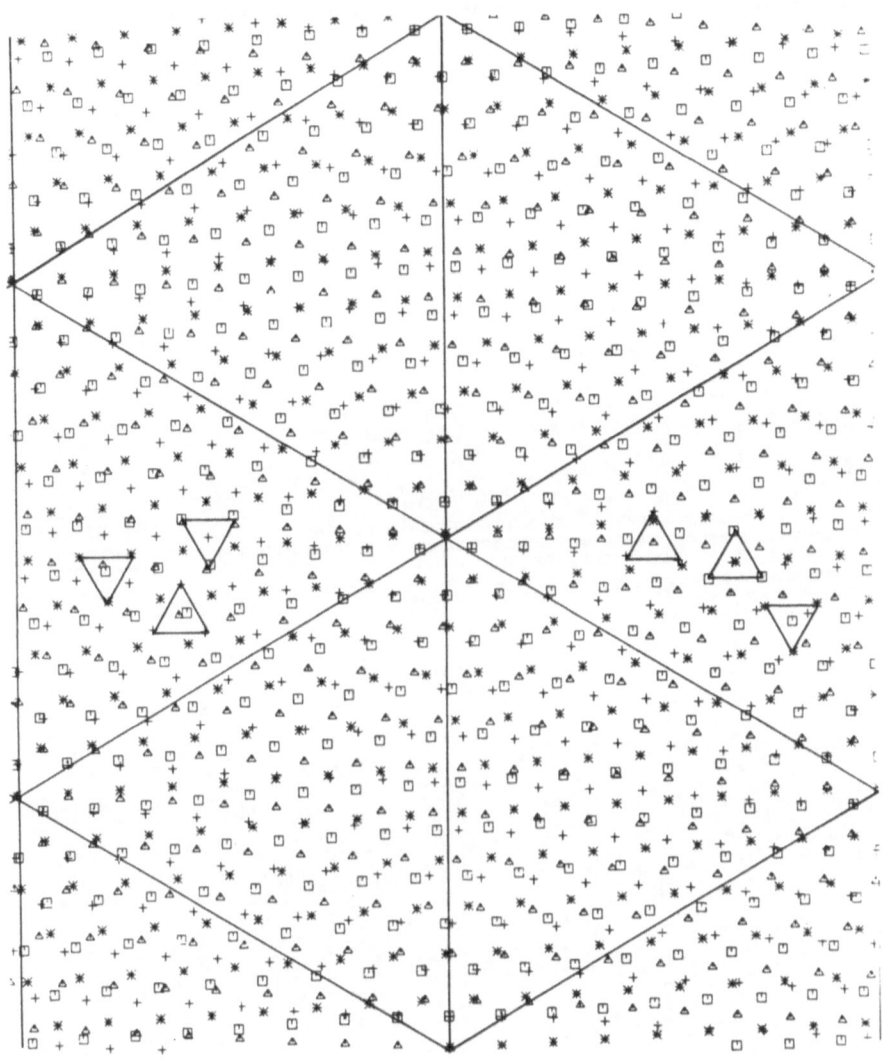

Fig. 9   Structure of the Σ=91 [111] twist boundary with the triangular pattern.

In Fig. 9 the calculated structure of the Σ = 91 [111] θ = 53.99° twist boundary is shown. In this configuration one finds that some atoms in the lattice planes adjacent to the boundary are at a distance closer than the nearest neighbour spacing in the f.c.c. lattice. Consequently, these atoms produce a large contribution to the boundary energy owing to the strong repulsion of atoms at these atomic separations. To search for another lower energy structure a different starting configuration has been adopted in which the close approach of atom pairs has been avoided. The lattice plane in the middle of the computational block was divided into two separate sections belonging to the upper and lower crystal, respectively, in such a way that no two atoms in different (111) planes are too close to each other. The initial density of this particular plane was taken to be the same as the density of the other (111) planes. After relaxation, structures different from that depicted in Fig. 9 were found depending on the interatomic potential used.

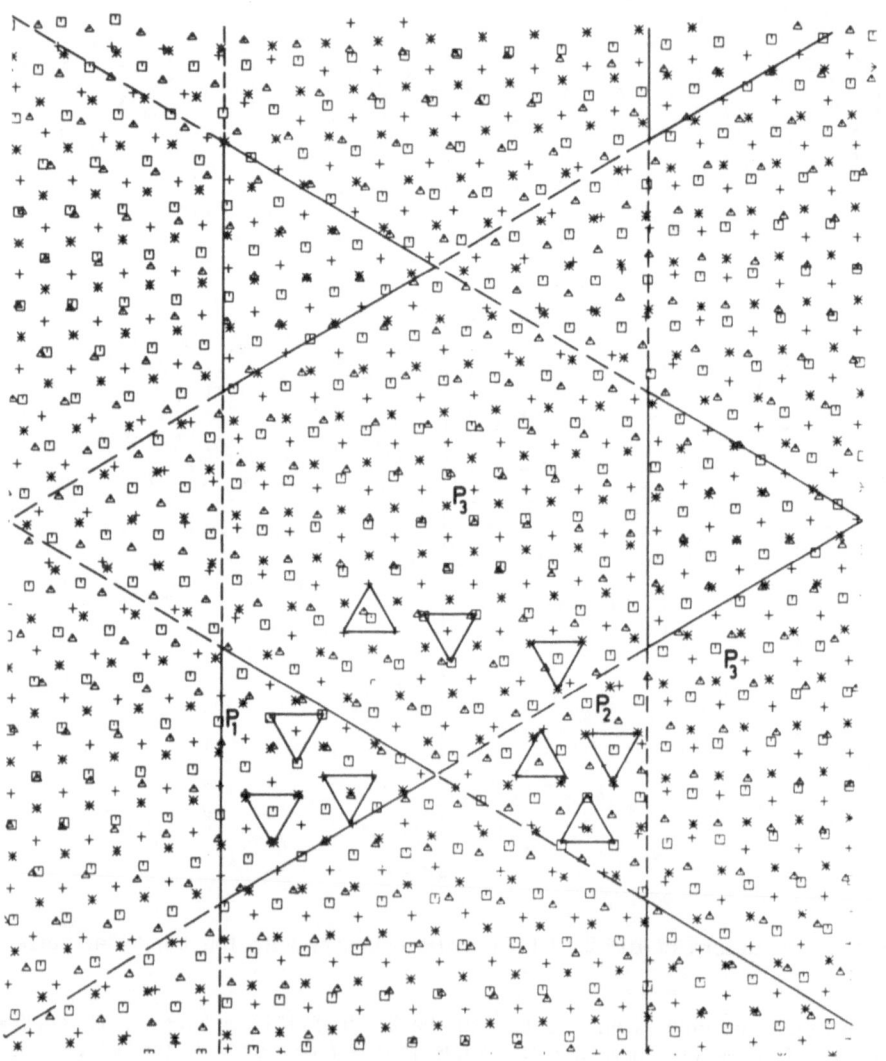

Fig. 10   Structure of the $\Sigma=91$ [111] twist boundary with the six-star pattern.

Figure 10 shows another calculated structure of the $\Sigma=91$ boundary which resembles closely the six-star pattern observed experimentally for boundaries with misorientation near the twin orientation (Fig. 11) [56]. The stacking sequences of (111) planes in a cross section along P_1P_3 in Fig. 10 is illustrated in Fig. 12 where the symbols I and II represent ABC... and CBA... stacking sequence of the (111) planes, respectively. Apparently, the boundary plane has a stepped character such that the double steps only occur at the corners of the triangles in Fig. 10, in agreement with the experimentally observed structure of the network.

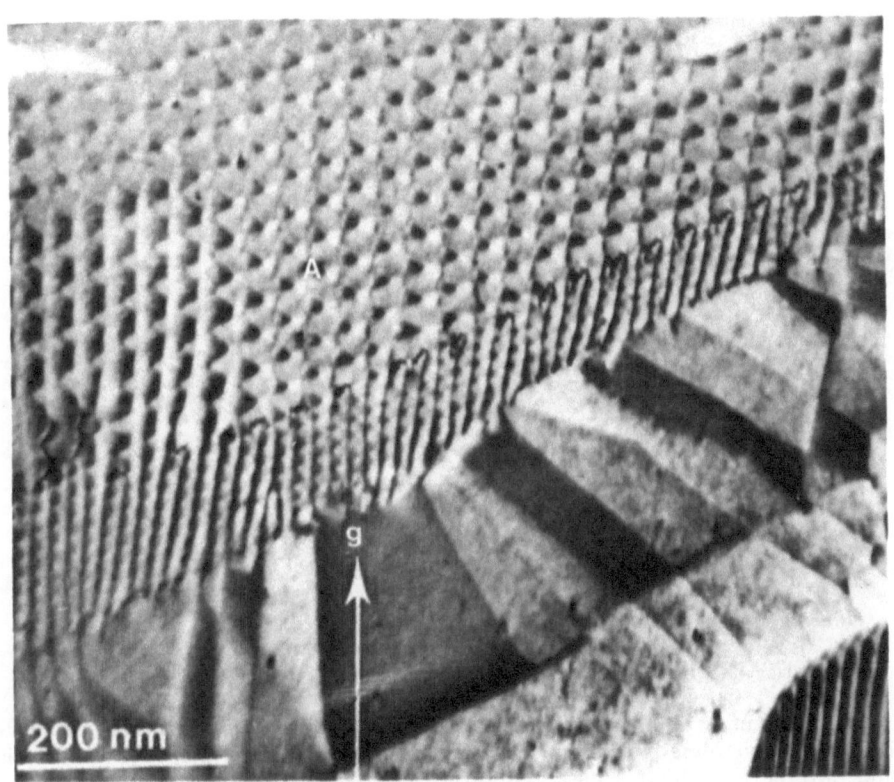

Fig. 11   Dark field image of a boundary the misorientation of which is 0.46° away from that of the Σ 3 twin (g = ($\bar{1}1\bar{1}$)$_m$ reflection). The steps caused by the dislocation network are seen as a black-grey-white contrast in area A.

Comparison of the structures with the triangular pattern (Fig. 9) and the six-star pattern (Fig. 10) shows that a shift of the dislocation lying vertically in Fig. 10 by half of the dislocation spacing transforms the configuration of Fig. 10 into that of Fig. 9. As a result of such a shift, the dashed triangle in Fig. 10 disappears. This implies that the triangular structure of Fig. 9 consists entirely of single steps in the boundary. Experimentally, a six-star pattern as well as a triangular pattern have been observed in near coherent twin boundaries [50, 56] in gold. This suggests that perhaps the energy difference, at least in Au, between the two alternative structures is not very large, as may have been anticipated from the previously discussed simple transformation of one structure into the other. This transformation preserves the dislocation density in the boundary plane. Both structures possess single atomic steps in the boundary, but the structure with the six-star pattern also contains double steps in the boundary, which increase the boundary energy [50].

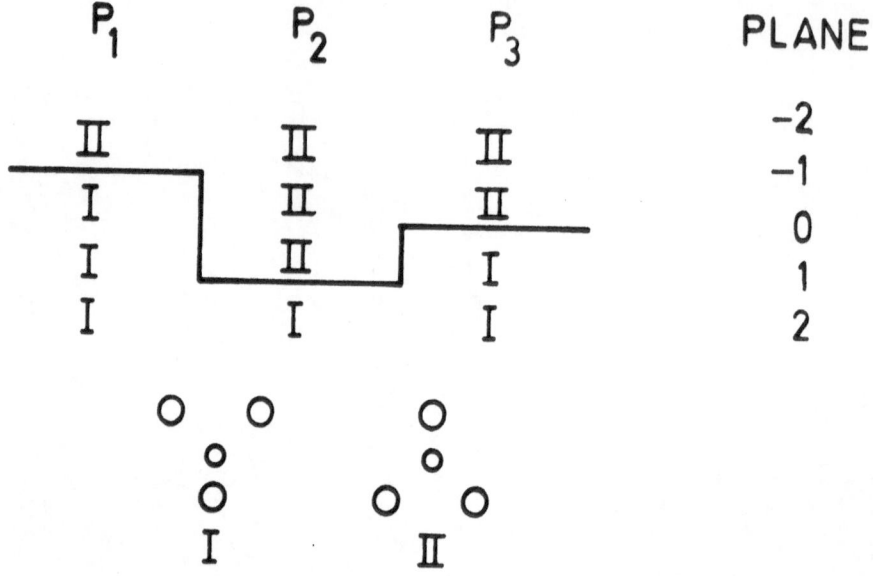

Fig. 12   Cross-section of the stacking of (111) planes in Figure 10 along P_1P_3.

4.   FUTURE ATOMISTIC STUDIES: BEYOND PAIR POTENTIALS

Previous studies of the atomic structure of boundaries, the principal results of which have been reviewed here, have been made using pair potentials to describe atomic interactions. In this case the total energy of the system of N particles is written as (e.g. [57])

$$E^{tot} = \frac{1}{2} \sum_{\substack{i,j=1 \\ i \neq j}}^{N} \phi^{ij} (r_{ij}) + U(\rho) \tag{1}$$

where ϕ^{ij} is the pair-potential, r_{ij} is the magnitude of the vector joining atoms i and j and U is the part of the energy dependent on the average density ρ. The pair-potential is also generally dependent on ρ. The term $U(\rho)$ is much larger than the pair-potential term and contains the bulk of the cohesive energy of the solid. Pair-potential then describes energy associated with rearrangement of atoms at constant volume [42, 57, 58] and papers in [59]). More generally, pair-potentials are justified if used to describe structural differences which do not lead to a drastic change of a reference state on the basis of which they have been derived [60, 61].

The pair-potential concept is well developed for simple metals and their alloys and it appears that it is valid even in transition metals when studying problems which do not involve large changes in the coordination number of atoms [57, 60,

62, 63]. It is, however, unfortunate that the pair-potentials are density dependent and the total energy contains a large term U(ρ) which contributes significantly to defect energies if there is a density change. Hence, the pair-potential formulation is not suitable for problems involving large local or total density changes, such as study of free surfaces. For the same reason the pair-potentials cannot be used to study problems of cohesion such as modelling of crack propagation since most of the cohesive energy is included in the U(ρ) term, although a different type of "bond breaking" potentials may be constructed [61]. On the other hand, structural features of those defects which do not involve large density and coordination variation (e.g. grain boundaries in metals) can be successfully investigated with the use of pair-potentials. This is the reason why pair-potentials have been adequate for studies of grain boundary structure but they are not sufficient for investigations of phenomena such as decohesion.

In recent years the most successful approach to calculations of the total energies of many-body systems has been based on the density fuctional theorem implemented in the local density approximation [64 - 66]. This formalism leads to an effective one-electron Schrödinger equation which in ab-initio calculations is solved self-consistently. Such calculations have been made in recent years to study simple lattice defects [67, 68] but their use in studies of extended defects which require inclusion of a large number of atoms and full structural relaxation is still not possible. One possible simplification, well suited for transition and noble metals, as well as for most non-metals, is based on expansion of the wave functions as linear combinations of atomic orbitals (LCAO). This is the tight-binding method first fully developed by Slater and Koster [69]. The total energy of a system can then be written as [70, 71]

$$E_{tot} = E_{bs} + U_r \tag{2}$$

$E_{bs} = \sum_l E_l$ is the band structure energy equal to the sum of all the single-particle energies over occupied states, and U_r is a short range repulsive term. At this stage an approximative empirical approach can be adopted which appreciably simplifies the calculations but preserves all the basic physical features of the general theory. U_r is approximated as $\frac{1}{2} \sum_{i,j} V(r_{ij})$ where V is a short range repulsive pair-potential fitted to obtain correct equilibrium density and compressibility of the ideal lattice [72]. To evaluate E_{bs} the tight-binding method is used which leads to the eigen-value problem

$$\sum_{\lambda'} H_{\lambda\lambda'} C_{\lambda'}^l = E_l C_{\lambda}^l \tag{3}$$

where λ and λ' mark all the atomic orbital states used, $H_{\lambda\lambda'}$ is the corresponding matrix element of the Hamiltonian H (called the hopping integral for λ ≠ λ') and C_{λ}^l are the coefficients of the LCAO expansion of the wave functions. The hopping integrals are chosen so as to reproduce, for example, the band structure or the density of states in the ideal lattice, or some other measurable quantities. The eigen-value problem can be solved using standard methods of band-structure calculations but periodic boundary conditions are then required. Hence, if studying the structure of defects we have to consider a periodic array of defects. Using the Helmann-Feynman theorem the forces on individual atoms can be derived in terms of gradients of the hopping integrals and coefficients C_λ^ℓ [73, 74]. The relaxation of the defect structure can then be carried out using standard techniques but the eigen-value problem need to be solved at every step to obtain the coefficients C_λ^ℓ. In order to evaluate the forces we also need to know the dependence of the hopping integrals on the separation of atoms and several forms of these dependences have been suggested [70 - 72].

Another approach to evaluate E_{bs} is to write the contribution of this energy to the energy of a defect as

$$\Delta E_{bs} = \sum_i \int^{E_f} (E-E_f) \, \Delta n_i(E)dE \tag{4}$$

Where E_f is the Fermi energy and Δn_i the change of the density of states at atom i. In the derivation of this equation the requirement of local charge neutrality was imposed [72, 75]. The density of states $n_i(E)$ can be conveniently evaluated by the method of Green's functions employing the recursion techniques of the Cambridge group [76 - 78]. This approach has been used for several defect studies in which relaxations have been carried out by evaluating the forces by numerical differentiation of the energy [79, 80]. Recently, it has been reformulated in ref. [81] to evaluate the forces directly which will enable much wider use of this method.

A more simplified evaluation of E_{bs} has been proposed in Refs. [72, 75]. If the details of the profile of the density of states $n_i(E)$ are neglected and it is regarded as a Gaussian, adjusted to its second moment,

$$E_{band} = E_o \sum_i (\sum_j H_{ij}^2)^{1/2} \tag{5}$$

where E_o is a constant and the squares of the hopping integrals, H_{ij}^2, are only functions of the distance between atoms i and j. The forces acting on atoms in a system of atoms can be easily evaluated.

Recently, it has been suggested by Finnis and Sinclair [82] to treat both the repulsive part and the band energy part semi-empirically and use eq. (5) only as a guideline for the functional dependence of E_{bs}. Hence, they suggest to write

$$E_{tot} = \frac{1}{2} \sum_{i,j} V(r_{ij}) - \sum_i (\sum_j \phi(r_{ij}))^{1/2} \tag{6}$$

where $V(r_{ij})$ is a pair-potential and $\phi(r_{ij})$ a short range pair-function. Both V and ϕ are adjusted empirically to fit various properties of a given material. This approach is philosophically the same as the embedded atom method proposed in ref. [83] and represents construction of empirical many-body potentials.

The many-body term of eq. (6) can be compared with the term $U(\rho)$ in eq. (1) but this part of the total energy is now expressed as a function of the positions of atoms rather than as a function of the material density. Hence, no problem arises when employing eq. (6) in problems involving density changes such as studies of free surfaces. This is of primary importance for investigation of fracture pheno-mena.

The empirical many-body potentials have been originally constructed for a number of b.c.c. transition metals [82] and a good agreement between calculated and measured vacancy formation energies and surface energies was obtained and very reasonable surface relaxation found [82, 84]. Recently, potentials of this type have also been constructed for noble metals Cu, Ag and Au [85]. In this case the pair potential extends up to the third neighbours and may have an attractive part. The properties fitted are: lattice parameter, cohesive energy, elastic constants and the stacking fault energy. These many-body potentials lead then to very reasonable values of vacancy formation energies and volumes. In the case of free surfaces the calculations show that in most cases the first atomic layer relaxes inwards rather than outwards as predicted when using pair-potential. The inwards relaxation is in agreement with a number of recent measurements of surface relaxation using LEED [86 - 88]. This demonstrates the ability of these many-body potentials to reproduce correctly some basic properties of surfaces and they can, therefore, be employed when studying both the grain boundaries and surfaces formed by splitting the material along a grain boundary. A preliminary study of this type has been carried out for the Σ 5 boundary in Au, the structure of which has been investigated using pair potentials by a number of authors.

A structure of the $\Sigma = 5$ (210) boundary calculated using the many-body potential for Au is shown in Fig. 13. This structure is very similar to the low energy structure (B) of this boundary found in previous studies using pair-potentials [52](see Fig. 8).

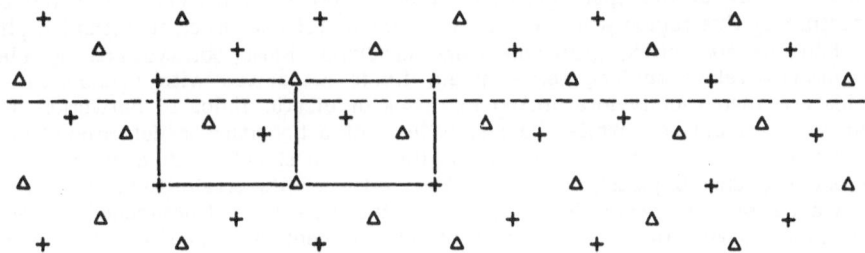

Fig. 13   The $\Sigma = 5$ (210) boundary calculated using the empirical many-body potential for Au.

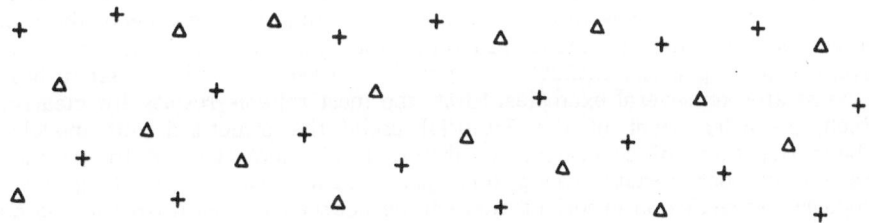

Fig. 14   Surface structure formed by splitting the bicrystal with the $\Sigma = 5$ (210) along the dashed line drawn in Fig. 13 (bottom part).

Two basic units of this boundary structure are delineated in the figure by full lines. The structure of the (210) surface was first studied by making a cut along a (210) plane in a single crystal and removing one part of it. Hence, in the starting configuration the atomic positions corresponded to those in the ideal lattice. The principal relaxation is in this case an inward motion of the uppermost layer by 0.03 of the lattice spacing. This is again in agreement with the observed trends of the surface relaxation [86 - 88]. The energy per unit area of this surface is 644 mJ/m^{-2}. However, a different surface structure has been found when using as a starting configuration the surface formed by splitting the bicrystal with the Σ 5 (210) boundary along this boundary. The separation of the two grains has been introduced along the dashed line drawn in Fig. 13 and the corresponding surface structure of the bottom part is shown in Fig. 14. The surface structure of the upper part is the same as when the cut has been made in the single crystal but in the surface structure of the bottom part the atomic configurations of the (210) boundary have been preserved. The energy per unit area of the surface with this structure is 860 mJ/m^{-2} and it is, therefore, a metastable structure of the (210) surface. However, it is this metastable surface which may be formed when fracturing along the (210) boundary rather than the fully relaxed low energy structure and, therefore, metastable higher energy surfaces rather than fully relaxed surfaces may be of importance when studying fracture phenomena.

5. CONCLUSIONS

The property strongly affected by grain boundary phenomena in both metallic and non-metallic materials is their mechanical strength. Grain boundary cracking occurs along the core of grain boundaries which is a very narrow region, of the order of a few atomic spacings, where the two grains join. Hence, it is the atomic structure of this region which needs to be understood in order to establish physical mechanisms controlling grain boundary cracking. Most observations by electron microscopy relate to long range strain fields associated with boundaries. These fields are then usually interpreted in terms of elastic fields of networks of grain boundary dislocations. While the strain field of a boundary is determined uniquely by the atomic structure of a boundary, the atomic structure of a boundary cannot be deduced unambiguously from the knowledge of its strain field, i.e. from the knowledge of the dislocation content of the boundary. Consequently, computer modelling proved to be an important complementary technique. However, the accuracy and the validity of the interatomic interaction functions used to describe the forces between atoms is essential to the success of any computer modelling study.

Pair potentials have been used to describe interatomic forces in most computer simulations of grain boundaries and other lattice defects. At the same time a complete description of interatomic forces in solids generally requires introduction of non-central many-body forces. The purpose of this paper has been, therefore, to discuss whether useful results regarding the grain boundary structure can be obtained with pair potentials. The positive answer to this question has been demonstrated on several examples. First, the most reliable results are clearly those which are independent of the potential used. The structural unit model which relates structures of boundaries for different misorientations of the grains is an example of such results. Using this model basic dependences of various grain boundary properties on misorientation can be deduced as shown here in the case of <111> twist boundaries. Their detailed studies for specific materials will require a more sophisticated description of interatomic forces. It is likely, however, that pair-potential calculations can similarly reveal dependences of the boundary structure on other geometrical parameters.

Further, general features of grain boundaries are the multiplicity of structures and related phenomena such as possible transformations of boundary structure, mechanisms of absorption and emission of vacancies and the structure of vacancies in boundaries. Again, detailed atomic positions in specific materials and related activation energies for the transformations and formation and diffusion of vacancies cannot be evaluated using pair potentials only. On the other hand, the present calculations are capable to reveal which structural alternatives can be expected. One example shown here are the [111] twist boundaries with misorientations near that of the $\Sigma=3$ twin for which two alternate structures and their relationship, were found.

The pair potential concept is well developed for simple metals. Unfortunately, pair potentials are density dependent and the total energy contains a large term $U(\rho)$ which does contribute to defect energies if there is a density change. Hence, the pair potential formalism is not suitable for problems involving substantial local or total density changes, such as in vacancies and their agglomerates or on free surfaces. For the same reason the pair potentials cannot be used to study problems of cohesion since most of the cohesive energy is included in $U(\rho)$. Structural features of free surfaces which do involve large density variation cannot be successfully investigated using pair potentials.

In this paper preliminary results of a computer modelling study are reported of both the $\Sigma = 5$ (210) grain boundary in Au and free surfaces formed by splitting the material along the grain boundary. The empirical many-body potential of the type suggested by Finnis and Sinclair [82, 85] has been employed. The many-body term is expressed as a function of the position of the atoms rather than as a function of the material density. This is of primary importance for investigations of fracture phenomena. An interesting result of this study is that a metastable surface may be formed when fracturing along the (210) boundary rather than the fully relaxed low energy structure. Consequently, higher energy surfaces rather than fully relaxed surfaces may be of importance when studying fracture phenomena. In particular, this result has to be kept in mind when interpreting experimental observations on embrittled materials. Most of these experiments are of the 'post-mortem' type, i.e. the structure of surfaces formed by fracturing samples along grain boundaries has been studied rather than the structure of corresponding grain boundaries. However, no clear link between the structure of the fracture surface and the corresponding grain boundary has been established.

Furthermore, grain boundary cracking is usually considered to be controlled by the cohesive strength. However, the process of cracking is, in general, complex and involves either competition between bond-breaking and dislocation emission [89, 90] or concomitant bond breaking and local dislocation activity [91, 92]. In alloys, not only cohesive strength but also resistance to shear may be affected by segregated impurities and both might be decreased. It has been suggested that the decrease of the shear resistance is halted at higher concentrations of impurities owing to their interactions while the decrease of cohesion continues [93, 94]. Therefore, it is essential for full understanding of the grain boundary brittleness to take into account the interplay between the local changes of cohesion and resistance to shear.

ACKNOWLEDGEMENTS

This research is a part of the research program of the Foundation for Fundamental Research on Matter (FOM-Utrecht) and has been made possible by financial support from the Netherlands Organization for the Advancement of Pure Research (ZWO - The Hague). This research was also supported in part by the U. S. National Science Foundation through the MRL Program Grant No. DMR-8216718.

REFERENCES

[1] Grain Boundary Structure and Kinetics, edited by R. W. Balluffi, ASM Publication, 1980.
[2] Structure and Properties of Grain Boundaries, edited by S. Hagege and G. Nouet, J. Physique Paris 43, C6, 1982.
[3] Structure and Properties of Internal Interfaces, edited by M. Rühle, R. W. Balluffi, H. Fischmeister and S. L. Sass, J. Physique Paris 46, C4, 1985.
[4] C. J. McMahon, Jr.: Temper Embrittlement in Steels, ASTM STP 407, p. 127, 1968.
[5] H. L. Marcus and N. E. Paton: Metall. Trans. 5, 2135, 1974.
[6] C. J. McMahon: Mat. Sci. Eng. 25, 233, 1976.
[7] M. P. Seah: Surf. Science 53, 168, 1975.
[8] C. L. Briant and S. K. Banerji: Int. Metal Reviews 23, 64, 1978.
[9] D. F. Stein and L. A. Heldt: in Interfacial Segregation, edited by W. C. Johnson and J. M. Blakely, ASM, Metals Park: Ohio, p. 239, 1978.
[10] J. P. Stark and H. L. Marcus: Metall. Trans. 8A, 1323, 1977.
[11] C. J. McMahon, V. Vitek and J. Kameda: in Developments in Fracture Mechanics -2, edited by G. G. Chell, Appl. Sci. Publishers, London, p. 193, 1981.
[12] M. P. Seah and E. D. Hondros: in Atomistics of Fracture, edited by R. M. Latanision and J. R. Pickens, Plenum Press: New York, p. 855, 1983.
[13] Atomistics of Fracture, edited by R. M. Latanision and J. R. Pickens, Plenum Press: New York, 1983.
[14] C. J. Middleton: Metal Sci. 15, 154, 1981.
[15] T. Takasugi and D. P. Pope: Metall. Trans. 13A, 1471, 1981.
[16] D. P. Pope and D. S. Wilkinson: in Creep and Fracture of Engineering Materials and Structures, edited by B. Wilshire and D. R. J. Owen, Pandridge Press, Swansea, p. 531, 1981.
[17] S. H. Chen, T. Takasugi and D. P. Pope: Metall. Trans. 17, 389, 1983.
[18] C. A. Hippsley, J. F. Knott and B. C. Edwards: Acta Metall. 30, 641, 1982.
[19] A. Kumar and B. L. Eyre: Proc. Roy. Soc. London A370, 431, 1980.
[20] J. Shin and C. J. McMahon, Jr.: Acta Metall. 1984.
[21] C. J. McMahon, Jr.: Z. für Metallkunde 75, 496, 1985.
[22] N. S. Stoloff: in High-Temperature Ordered Intermetallic Alloys, edited by C. C. Koch, C. T. Liu and N. S. Stoloff, MRS Symposia Proc., Vol. 39, p. 3, 1985.
[23] C. T. Liu and J. O. Stiegler: Science 26, 636, 1984.
[24] T. Takasugi, O. Izumi and N. Masahashi: Acta Metall. 33, 1247 and 1259, 1985.
[25] K. Aoki and O. Izumi: Acta Metall. 27, 807, 1979.
[26] C. T. Liu and C. L. White: in High-Temperature Ordered Intermetallic Alloys, edited by C. C. Koch, C. T. Liu and N. S. Stoloff, MRS Symposia Proc., Vol. 39, p. 365, 1985.
[27] A. P. Sutton: Int. Metals Reviews 29, 377, 1984.
[28] H. Ichinose and Y. Ishida: J. Physique Paris 46, C4 - 39, 1985.
[29] A. Bourret: in J. Physique Paris 46, C4 - 27, 1985.
[30] J. Budai and S. Sass: J. Physique Paris 43, C6 - 103, 1982.
[31] K. R. Milkove, P. Lamarre, P. Schmückle, M. D. Vaudin and S. L. Sass: J. Physique Paris 46, C4 - 71, 1985.
[32] W. Bollmann: Crystal Defects and Crystalline Interfaces, Springer-Verlag, Berlin, 1970.
[33] R. C. Pond and W. Bollmann: Phil. Trans. Roy. Soc. A 292, 449, 1979.
[34] R. C. Pond and D. S. Vlachavas: Proc. Roy. Soc. A 385, 95, 1983.
[35] D. Gratias and R. Portier: J. Physique Paris 43, C6 - 15, 1982.
[36] G. Kalonji and R. Portier: J. Physique Paris 43, C6 - 25, 1982.

[37] R. W. Balluffi: Metall. Trans. 13A, 2069, 1982.
[38] P. D. Bristowe: J. Physique Paris 43, C6 - 33, 1982.
[39] V. Vitek: in Dislocations 1984, edited by P. Veyssière, L. Kubin and J. Castaign, Editions CNRS: Paris, p. 435, 1984.
[40] P. D. Bristowe and R. W. Balluffi: in J. Physique Paris 46, C4 - 155, 1985.
[41] Computer Simulation in the Study of Solid-Solid Interfaces, edited by P. D. Bristowe and R. J. Harrison, Surface Science 144, 1984.
[42] A. M. Stoneham: Physica 131B, 69, 1985.
[43] V. Vitek and J. Th. M. De Hosson: in Proceedings of the MRS Symposium on Computer Modelling in Materials Science, to be published 1986.
[44] A. P. Sutton and V. Vitek: Phil. Trans. Roy. Soc. London A309, 37, 1983.
[45] G.-J. Wang, V. Vitek and A. P. Sutton: Acta Metall. 32, 1093, 1984.
[46] D. Schwartz, V. Vitek and A. P. Sutton: Phil. Mag. A51, 499, 1985.
[47] P. D. Bristowe and R. W. Baluffi: J. Physique 46, C4-155, 1985.
[48] G. H. Bishop and B. Chalmers: Scripta Metall. 2, 133, 1968, Phil. Mag. 29, 515, 1971.
[49] F. C. Frank: in Symp. on the Plastic Deformation of Crystalline Solids (Office of Naval Research, 1950), p. 150.
[50] R. F. Scott and P. J. Goodhew: Phil. Mag. A44, 373, 1981.
[51] P. D. Bristowe and A. G. Crocker: Phil. Mag. A38, 487, 1978.
[52] V. Vitek, Y. Minonishi and G.-J. Wang: J. Physique 46, C4-171, 1985.
[53] E. W. Hart: in Nature of Behaviour of Grain Boundaries, ed. Hsun Hu (Plenum Press, New York, 1972), p. 155.
[54] K. T. Aust: Can. Metall. Quart. 8, 155, 1972.
[55] T. Watanabe, S.-I. Kimura and S. Karashima: Phil. Mag. 44, 845, 1984.
[56] J. Th. M. De Hosson, F. W. Schapink, J. R. Heringa, J. J. C. Hamelink: Acta Metall. 34, 1051, 1986.
[57] R. Taylor: Physica B131, 103, 1985.
[58] V. Vitek, Y. Minonishi: Surf. Science 144, 196, 1984.
[59] Interatomic Potentials and Crystalline Defects, edited by J. K. Lee, TMS AIME, Warrendale, PA, 1981.
[60] A. E. Carlsson and N. W. Ashcroft: Phys. Rev. B27, 2101, 1983.
[61] A. E. Carlsson: Phys. Rev. B32, 4866, 1985.
[62] R. Taylor: in Interatomic Potentials and Crystalline Defects, edited by J. K. Lee, TMS AIME, Warrendale, PA, p. 71, 1981.
[63] A. H. Mac Donald and R. Taylor: Can. J. Phys. 62, 796, 1984.
[64] P. Hohenberg and W. Kohn: Phys. Rev. B136, 864, 1964.
[65] W. Kohn and L. J. Sham: Phys. Rev. B140, 1135, 1965.
[66] L. J. Sham and W. Kohn: Phys. Rev. B146, 561, 1966.
[67] J. R. Chelikowsky and J. C. H. Spence: Phys. Rev. B30, 694, 1984.
[68] M. Y. Chou, S. G. Louie and M. L. Cohen: Proc. 17th Int. Conf. Phys. Semiconductors, edited by D. J. Chadi and W. A. Harrison, Springer, New York, p. 43, 1985.
[69] J. C. Slater and G. F. Koster: Phys. Rev. 94, 1498, 1954.
[70] W. A. Harrison: Electronic Structure and Properties of Solids, Freeman, San Francisco, 1980.
[71] D. G. Pettifor: in Physical Metallurgy, ed. J. W. Cahn and P. Haasen, North Holland, Amsterdam, p. 73, 1983.
[72] F. Ducastelle: J. Physique Paris 31, 1055, 1970.
[73] D. J. Chadi: Phys. Rev. B19, 2074, 1979.
[74] R. M. Thompson and D. J. Chadi: Phys. Rev. B29, 889, 1984.
[75] F. Cyrot-Lackman: J. Phys. Chem. Solids 29, 1235, 1968.
[76] V. Heine: Solid State Physics, Vol. 36, 1, 1980.
[77] R. Haydock, V. Heine and M. J. Kelly: J. Phys. C8, 2591, 1975.

384

[78] C. M. Nex: Computer Physics Comm. 34, 101, 1984.
[79] K. Masuda, R. Yamamoto and M. Doyama: J. Phys. F: Metal Phys. 13, 1407, 1983.
[80] B. Legrand: Phil. Mag. B49, 171, 1984.
[81] D. G. Pettifor, M. F. Finnis and A. P. Sutton: to be published, 1986.
[82] M. W. Finnis and J. E. Sinclair: Phil. Mag. 50, 45, 1984.
[83] M. S. Daw and M. I. Baskes: Phys. Rev. B29, 6443, 1984.
[84] C. C. Matthai and D. J. Bacon: Phil. Mag. A52, 1, 1985.
[85] G. Tichy, G. Ackland, M. W. Finnis and V. Vitek: to be published, 1986.
[86] I. Stensgaard, R. Feidenhaus and J. E. Sorensen: Surf. Science 128, 281, 1983.
[87] J. W. M. Frenken, R. G. Sneek and J. F. van der Veen: Surf. Science 135, 147, 1983.
[88] J. Sokolov, F. Jona and P. M. Marcus: Solid State Coom. 49, 307, 1984.
[89] J. R. Rice and R. Thomson: Phil. Mag. 29, 73, 1984.
[90] R. Thomson: in Atomistics of Fracture, eds. R. M. Latanision and J. R. Pickens, Plenum Press: New York 1983, p. 167.
[91] C. J. Mc Mahon and V. Vitek: Acta Metall. 27, 507, 1979.
[92] M. L. Joke, V. Vitek and C. J. Mc Mahon: Acta Metall. 28, 2479, 1980.
[93] M. E. Eberhart, K. H. Johnson and R. M. Latanision: Acta Metall. 32, 955, 1984.
[94] M. E. Eberhart, R. M. Latanision and K. H. Johnson, Acta Metall. 33, 1769, 1985.

DISCUSSION

Comment by S. M. Ohr:

It is not clear at present if the fracture behavior of a grain boundary depends more on its structure or on the chemistry it possesses. Can you distinguish this by estimating the g.b. energy for different types of boundaries theoretically with or without impurity atoms?

Reply:

A recent paper by Maurer and Gleiter (Scripta Met. 19(1985):1009) calls special attention to this effect, i.e., the dependence of interfacial structure on the electronic configuration of the alloying elements. It was concluded that crystallographic parameters alone are not sufficient to characterize the grain boundary structures and consequently future atomistics studies should go beyond pair potentials. One approach which is underway is to employ the full-scale parameterised tight-binding method in which the electronic density of states can be conveniently evaluated by the method of Green's function recursion techniques. In that case it is possible to calculate the grain-boundary energy for various types of boundaries with and without impurities. Consequently, the effect of impurities on decohesion of grain boundaries can be studied. However, from such a study it cannot be concluded whether the fracture behavior of a particular grain boundary depends more on its structural aspects or on the chemistry it possesses, since the resistance to shear may be affected by the electronic structure of impurities as well.

Comment by A. S. Argon:

On the problem of intergranular fracture at high temperatures, it has been widely observed that the boundaries that produce fracture are often very rough and have precipitates on them. Do you feel the roughness in the form of serrations precede the precipitation or is a result of this precipitation of the particles?

Reply:

In my opinion there is not a clear cut answer to this question because it pretty much depends on the plasticity in the grain and the nucleation/ growth of second-phase particles at the grain boundary. If strain locali- zation does occur at high temperature deformation, it will lead to cracks at grain boundary precipitates and finally grain boundary fracture, inde- pendent of the roughness of the grain boundary by second-phase particles. On the other hand, precipitates may change the intrinsic grain boundary dislocation structure considerably. Calculations by Hashimoto et al (Acta Met. 32(1984):1) indicate that P segregation in Fe changes completely the initial structure of the grain boundary. If precipitates introduce imper- fect steps in a tilt boundary, i.e., net components of the Burgers vector of the intrinsic dislocations parallel to the boundary plane, they will affect grain boundary sliding. Grain boundary sliding is an important source of creep intergranular fracture in metals and ceramics under conditions where diffusional creep is relatively unimportant.

Comment by P. Neumann:

What significance do you attribute to the grain boundary plane inside the lattice in addition to Σ (Gleiter explained his results with inter- locking features on the GB which do depend on the orientation of the GB plane in the lattice).

Reply:

In general, we may conclude from Gleiter's experiment that the stronger embrittlement of grain boundaries with higher energy (or misorientation) is due to an increasing number of segregation sites with an increasing number of secondary dislocations in the interface. The grain boundary plane in these experiments is influenced by the [110] orientation of the single crystal on which the Cu-0.1 at %Bi spheres are deposited.
Experimentally it was shown by Watanabe (Res. Mechanica 11(1984):47) that low Σ boundaries are resistant to intergranular fracture whereas random boundaries are preferential sites for crack nucleation and propa- gation. It leads to the conclusion that the change in fracture mode between intergranular and transgranular fracture depends on Σ existing in part of a propagating crack. These results seem to be independent of the boundary plane. If the plastic work associated with the propagation of the microcrack is small, the fracture energy is determined by $2\gamma_s - \gamma_b$, where γ_s is the surface energy and γ_b the grain boundary energy. γ_b depends on the boundary plane. Calculations by Sutton and Vitek of $\Sigma=3$, $70.53°|[110]$ boundaries showed a variation between 22 mJ/m² (.001 mean boundary plane) and 2878 mJ/m² (111 mean boundary plane). Nevertheless, γ_s is roughly $3\gamma_b$. Consequently, one does not expect a strong dependence of the boundary plane on the fracture energy.

Comment by R. H. Jones:

There is some evidence that segregation to the grain boundary affects the Petch slope. Is there any evidence that the motion of grain boundary dislocations is a factor in intergranular fracture?

Reply:

To the best of my knowledge, there are no fracture experiments done on clean bicrystals focusing on the effect of moving grain boundary dislocations on intergranular fracture. It has been suggested that grain boundary sliding is affected by the movement of intrinsic grain boundary dislocations (Kegg et al, Phil. Mag. 27(1973):1041; Gates, Acta Met. 21(1973):855). In such a mechanism it is clear, of course, that there must be a net component of the Burger's vector of the intrinsic dislocations parallel to the boundary plane. If tilt boundaries are equilibrated, they will contain only perfect steps and grain boundary dislocations with the Burgers vectors normal to the boundary plane. Sliding can then occur only if extrinsic dislocations are introduced in the boundary by plastic deformation of the grains interiors. In the case of twist boundaries, sliding seems to be affected by both intrinsic screw grain boundary dislocations and the glide/climb of extrinsic grain boundary dislocations (Watanabe, Phil. Mag. 37A(1978):649).

Comment by T. Watanabe:

1. Do you think that there is a relationship between multiplicity of grain boundary structure and the stability of boundary?
2. How much can we apply our present knowledge of boundary structure for cubic crystal to other crystals such as HCP crystal?

Reply:

1. Yes. Besides the transmission electron microscopic observations of [111] twist boundaries in bicrystals of Au presented here, which support the idea of multiplicity of grain boundary structures, there exists other experimental and theoretical evidence. For instance, the $\Sigma=5$ [001] twist boundary in Au was studied using X-ray diffraction by Sass and co-workers (Acta Met. 31:699, 1983). A pronounced disagreement between experiment and theory is apparent if one takes one CSL structure to calculate the diffraction intensities. It has been found by Yoonsik Oh and Vasek Vitek that structures composed of Quasi-random mixtures of the units of CSL structure and the structure with the P42'2' symmetry, the boundary plane of which is positioned alternately below and above that of the CSL structure, may possess lower energy than any of the CSL periodic structures. Since randomness is also favoured for entropic reasons, this structure will be even more favoured at finite temperatures. An excellent agreement between calculated X-ray intensities and experiment is now obtained.
2. Presumably the message will be the same, i.e., in a certain misorientation range the grain boundary can be composed of mixtures of structural units that delimit the misorientation range. However, this has to be investigated for specific non-cubic materials in more detail. In this respect, it should be emphasized that even in the cubic materials investigated the favoured boundaries are not always associated with the lowest available Σ and that not the same boundaries are favoured with the same crystal structure.

Comment by T. Watanabe:

Regarding the effect of grain boundary inclination, I would like to make my comment. I suppose that there must be some relationship between the magnitude of the inclination effect and Σ value, possibly strong effect for low Σ boundaries since the size of CSL unit decreases with Σ. Do you agree with my comment?

Reply:

No. As pointed out by Sutton and Vitek (Phil. Trans. Roy. Soc. London A309(1983):37) a general tilt boundary can be classified by two parameters Σ and p. The parameter p is determined by the mean boundary plane. Since all boundaries in a p-system share the same mean boundary plane, p varies with the inclination of the two grains and not their misorientation. Σ varies only with the misorientation of the two grains and not their inclination. Therefore, Σ and p are independent parameters. Intrinsic secondary grain boundary dislocations stress fields accompany the introduction of boundary units belonging to the same p-system, i.e., p = constant, but Σ varies.

Comment by K. Sieradzki:

What fraction of gbs in a real material is high Σ?

Reply:

Recently, Watanabe et al. have undertaken a systematic investigation into the grain boundary character distributions in polycrystalline materials. The frequency of random boundaries never exceeded 65% in Al. According to this study, it has been made clear that the frequency of the three types of grain boundaries, i.e., low angle boundary, coincidence boundary and random boundary, strongly depends on the original orientation and the amount of prestrain in aluminum polycrystals produced from single crystals. The frequency of coincidence boundaries can be as high as 50% for recrystallized polycrystals. Finally, one should keep in mind that the frequency of coincidence boundaries in the recrystallization structure of zone-refined high purity aluminum is strongly affected by impurities (Bellus and Aust, Met. Trans. 6A:219, 1975).

SEGREGATION AT INTERFACES

H.J. GRABKE

Max-Planck-Institut für Eisenforschung GMBH
4000 Düsseldorf 1
Federal Republic of Germany

1. INTRODUCTION

Surfaces , grain boundaries and interfaces of inclusions or precipitates are a sink for elements dissolved in metals. The segregation of dissolved C, Si, P, Sn...to interfaces in Fe, Ni and other metals or alloys leads to an enrichment in the range of one or only a few monolayers at and near the interfaces. This equilibrium segregation is induced by the decrease of interfacial energy upon segregation, additionally a release of elastic energy occurs in case of the interstitial elements which are too large in size for the interstices, and in case of substitutional elements being larger than the atoms of the host metal. The equilibrium segregation

$$A(dissolved) = A(segregated) \qquad (1)$$

is exothermic in character, as adsorption, and the concentrations at interfaces decrease with increasing temperature. Surface segregation should lead to the same states and binding modes as adsorption of these elements from the gas phase.
The presence of adsorbed or segregated elements on a metal surface affects gas-metal reactions, corrosion, catalysis, surface diffusion, sintering, recrystallization, adhesion, friction, wear etc. Also the composition of grain boundaries and other interfaces is very important by affecting the fracture and corrosion behavior of metals. Several elements such as S, P, Sn...when segregated at grain boundaries of iron-base materials act embrittling, i.e. they induce intergranular fracture. Other elements such as C, N, B...are not embrittling, they even may enhance the cohesion of the grain boundaries and thus favour transgranular fracture.
The surface and grain boundary segregation of the elements A = C, Si, Sn, N, P, O, S...have been investigated systematically in the recent years at the Max-Planck-Institut für Eisenforschung. The surface concentrations were determined by AES in dependence on the bulk concentration x_A and on the temperature of equilibration, and ordered structures on the surface of single crystals were observed by LEED/1-7/. The binding modes of the segregated species were characterized by XPS, ARUPS and ELS/8-14/. Analysis of grain boundaries is possible after intergranular fracture of specimens inside the UHV-system, in case of embrittlement of the alloy by the segregating element the fracture is intergranular and the grain boundary concentration can be determined by AES/15-25/. All investigations of the surface and grain boundary segregation have been performed in the temperature- and concentration range of the α-solid solution, to avoid the formation of threedimensional, stable or metastable compounds.

2. SURFACE SEGREGATION

2.1. Carbon

The surface segregation of carbon has been investigated in the temperature range 400-800°C on single crystals with carbon concentrations between 10 to 100 ppm C /1-3/. On Fe(100) a c(2x2) structure (fig. 1) with 50 at%C is approached for high concentrations and low temperatures. This is the saturation coverage ($\theta = 1$). The degree of coverage θ, decreases with increasing temperature and decreasing bulk concentration, as expected from thermodynamics, see fig. 2.

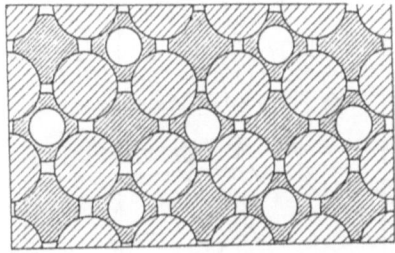

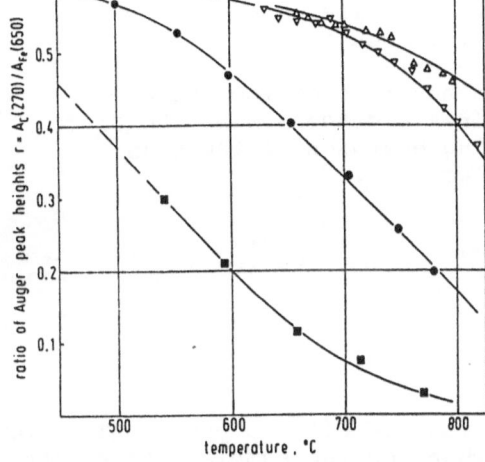

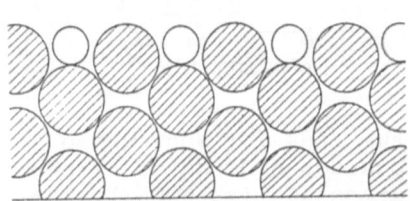

FIGURE 1. Model of the c(2x2) adsorption structure of carbon or nitrogen on the Fe(100) surface, a) top view, b) cross-section in (100) direction

FIGURE 2. AES-measurement of the equilibrium segregation of carbon on Fe(100), surface concentration of carbon in dependence on temperature and bulk concentration

These dependencies are described by the Langmuir-McLean equation:

$$\theta/(1-\theta) = x \exp(-\Delta G^o/RT) \qquad (2)$$

where θ is the degree of coverage, x is the bulk concentration and ΔG^o is the Gibbs' free energy of segregation. This equation can be rewritten according to

$$\ln \frac{\theta}{1-\theta} = -\frac{\Delta H^o}{RT} + \frac{\Delta S^{xs}}{R} + \ln x \qquad (3)$$

where ΔH^o is the enthalpy of segregation and ΔS^{xs} is the excess entropy of segregation. For the case of carbon on Fe(100) the segregation enthalpy is -85 kJ/mol.
For carbon segregation on other orientations of iron, the results are not so simple and clear /3/.
The binding mode of carbon on Fe(100) has been characterized by photoelectron-spectroscopy (XPS). Spectra from single crystal surfaces with segregated C have been taken and have been compared with spectra of graphite and cementite $(Fe,Cr)_3C$. In

order to obtain a thermodynamically stable cementite a Fe-2%Cr alloy has been carburized in CH_4-H_2. The photolines of the C 1s photo electrons of carbon are shown in fig. 3 /13/.

FIGURE 3. Investigation of the binding modes of carbon by photoelectron spectroscopy
(1) segregated carbon and graphite on Fe(100)
(2) only segregated carbon on Fe(100),
(3) graphite, and
(4) carbon in cementite $(Fe,Cr)_3C$.
Intensities of the photo lines plotted versus the binding energies of the C 1s electrons.

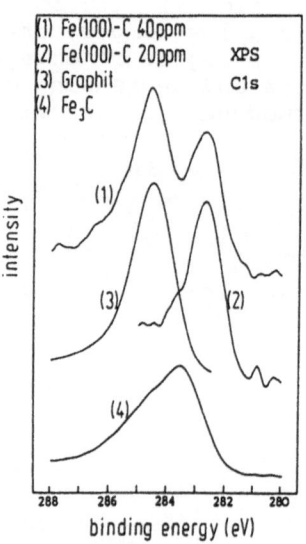

The sample with 20 ppm C only shows the peak for segregated carbon, on the sample with 40 ppm C besides segregated carbon there is graphite deposited by oversaturation (at 600°C). The energies of the C 1s electron levels are distinctly different from the energy level of C 1s in graphite and in the carbide. The segregated carbon is a special binding state of carbon. Related to the homopolar bond of carbon in graphite the Fe-C bond shows a strong polarity with electron transfer from iron to carbon atoms since the chemical shift of the C 1s level amounts to - 2.0 eV. For the carbide the chemical shift is half as high. Segregated carbon is a special state of carbon with relatively high negative charge.

2.2. Silicon

The surface segregation of Si has been studied on Fe-3%Si single crystals in the temperature range 450-900°C /6/. The kinetics of segregation has been measured on surfaces which had been cleaned before the sputtering. These measurements yielded after evaluation according to a solution of the diffusion equation given by McLean, diffusion coefficients of silicon in α-iron which are in good agreement with literature data. This is true for the orientation (100) but for the orientation (111) there seem to be kinetic obstacles, maybe in the transition of the silicon atoms through the surface.

The dependence of surface concentration on temperature has been measured for the equilibrium of silicon segregation on Fe(100) (see fig. 4.). This dependence can be described by the Langmuir-McLean equation (2) resp. (3). The result for the

Gibb's free energy of segregation is

$$\Delta G^O = -48000 - 15\ T \qquad J/mol \qquad\qquad (4)$$

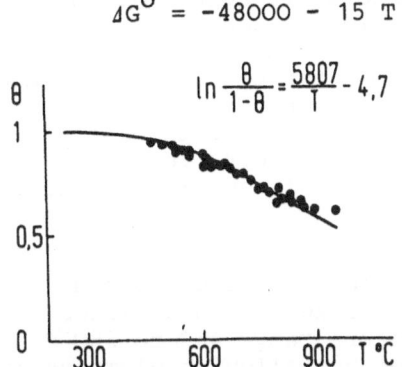

$$\ln \frac{\theta}{1-\theta} = \frac{5807}{T} - 4,7$$

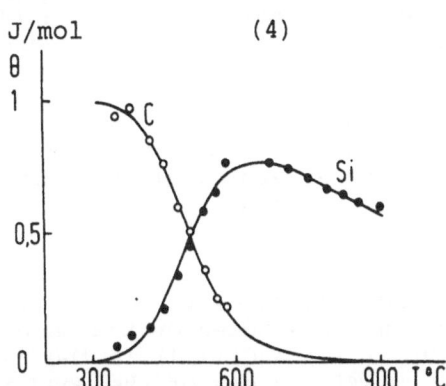

FIGURE 4. Surface segregation of silicon on Fe-3%Si (100), temperature dependence of the surface concentration

FIGURE 5. Temperature dependence of the displacement equilibrium in carbon and silicon segregation, Fe-3%Si with 40 ppm carbon

So the value of the segregation enthalpy of silicon on iron ΔH^O_{Si} = -48 kJ/mol is much lower than for carbon. Thus carbon can displace silicon from the surface on Fe-3%Si samples, which contain small concentrations of carbon:

$$C(diss.) + Si(segr.) = C(segr.) + Si(diss.) \qquad (5)$$

This equilibrium of mutual displacement has been measured in dependence on temperature for the segregation on Fe-3%Si(100), the carbon contents of the samples in these measurements were between 10 to 100 ppm, see fig. 5. According to the higher value for ΔH^O_C the C-segregation prevails at lower temperatures, for higher temperatures the carbon segregation decreases and silicon is able to segregate to the surface. This equilibrium of mutual displacement was described by relatively simple equations, which only consider the site competition of both segregating elements:

$$\theta_{Si}/(1-\theta_{Si}-\theta_C) = x_{Si}\ \exp(-\Delta G^O_{Si}/RT) \qquad (6)$$

$$\theta_C/(1-\theta_C-\theta_{Si}) = x_C\ \exp(-\Delta G^O_C/RT) \qquad (7)$$

Here the thermodynamic values for the Si-segregation stay unchanged, while the value for ΔH^O_C is increased by the presence of Si.

The binding mode of Si on Fe(100) was investigated by XPS. Shape, position and width of the Si photolines are independent of coverage, i.e. only one bonding state is observed in this case /9/. The Si 2p binding energy compared to pure silicon is lower by 0.1 eV. For the silicides Fe_3Si, FeSi and $FeSi_2$ the binding energy of the Si 2p electrons increases with Si content and is higher than for pure silicon.

In the valence band photoemission spectra of the Fe(100)-c(2x2)Si surface two additional lines corresponding to Si 3p and Si 3s states are observed /11/. Furthermore Si induces changes of the density of Fe 3d states in the range of 2 - 4 eV.

A detailed analysis of angular resolved photoemission using synchrotron radiation between 14 and 40 eV revealed the following results: a) Silicon segregates in the topmost iron layer of Fe(100) since Si enrichment induces a pure surface state. b) Si atoms occupy the fourfold sites in the center of a square of Fe atoms in the bcc Fe(100) surface since the Si 3p band reveals the same symmetry in k-space. c) There is a lateral interaction of Si atoms. d) The interaction of Si with Fe is low, this is inferred from the weak dispersion of the Si $3p_2$ state.

2.3. Tin

The segregation of tin on an iron single crystal containing 4wt% Sn was studied on the three low indexed surface planes (100), (110) and (111) in the temperature range 450 to 650°C /7/. Tin is one of the most interesting alloying elements in iron since it shows a different behaviour for each plane and even for different temperatures on a given plane.

On Fe(100) tin segregation up to a coverage of 50% of a monolayer leads to a c(2x2) structure as monitored by LEED. With this coverage the saturation is not yet attained, on Fe-Sn single crystals with tin concentrations in the range 0.036 - 4 wt% further segregation occurred leading to a degree of coverage of about 1.2. This observation has been made under conditions, where the solid solubility of Sn in iron was not exceeded.

The diffraction pattern of the overstructure vanishes when the coverage becomes greater than half a monolayer. At the higher coverage the Sn atoms are disordered. When the order/disorder transition takes place the binding energy of Sn 3d photoelectrons immediately rises by 0.15 eV towards the value of pure tin while the Fe 2p binding energy falls back to the value of a clean iron surface (Fig. 6). A tin surface plasmon appears as a new feature in the electron energy loss spectrum. A layer of pure Sn has developed with a predominant lateral Sn-Sn interaction. This layer is nearly closed-packed but disordered, and most probably has the character of a two-dimensional fluid.

FIGURE 6. Photoelectron spectroscopy on Fe-4%Sn(100), shifts of the photo-lines in the transition from the ordered c(2x2) structure to the disordered higher coverages

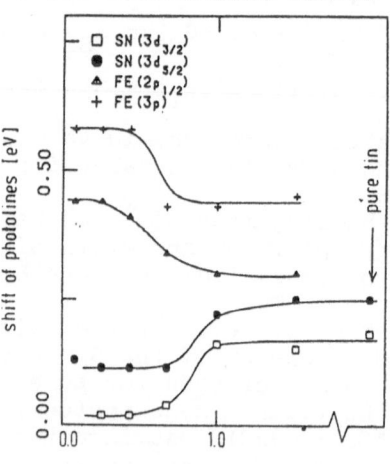

ratio of Auger peak heights A_{Sn}/A_{Fe}

The segregation on Fe(111) shows similar characteristics, a transition from the ordered p(1x1)structure to a disordered layer at higher coverage. On Fe(110) two different hexagonal structures are observed, the structure with lower degree of coverage is stable on Fe-0.019%Sn. At higher bulk contents up to 4%Sn a structure is formed which probably resembles the intermetallic compound FeSn, three layers of this structure being stabilized epitaxially on Fe(110).

So a wide variety of coverages and structures are observed in the system Fe-Sn. No segregation enthalpies could be derived since no Langmuir-Hinshelwood type temperature dependence was observed in any case. Even at low bulk concentrations always saturated structures were observed, so the value of the enthalpy for surface segregation of tin must be relatively high,

$\Delta H_{Sn} \leq -200$ kJ/Mol.

The high tendency for surface segregation of tin plays an important role in the creep rupture of steels. Impurities in heat resistant steels, such as tin lower the critical radius for formation of cavities

$$r_c = 2\gamma/\sigma$$

where γ is the surface energy of the cavity which is reduced by tin segregation, and σ is the uniaxial stress under creep conditions. Thus, tin in steels leads to premature failure upon creep at elevated temperatures, by favouring the void formation and cavity growth. This effect was demonstrated by creep experiments at 550°C, testing a 1%CrMoNiV-steel doped with different concentrations of tin: 0.004, 0.022, 0.061 and 0.12%Sn /35,36/.The rate of primary and secondary creep were not affected by tin, but the period of secondary creep was shortened with increasing tin content, the tertiary creep starts earlier and leads to premature failure of the tin-doped steels (fig. 7). Surface segregation of tin accelerates the nucleation and growth of cavities, the coalescence of the cavities causes early rupture.

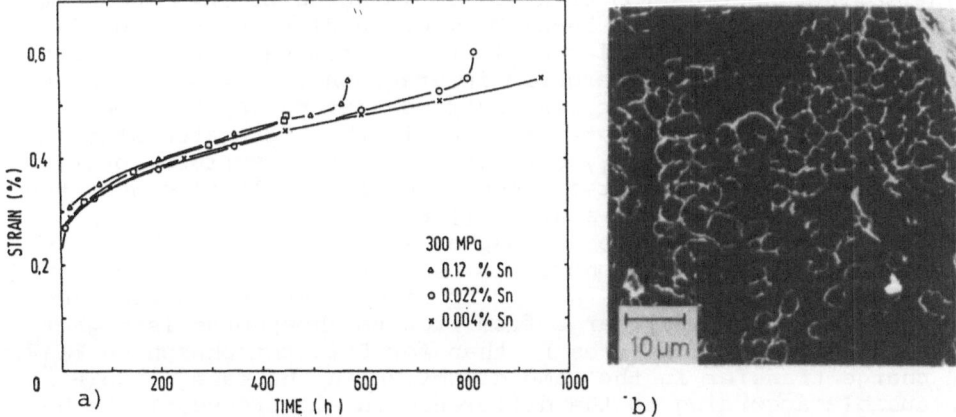

a)   b)

FIGURE 7. Creep of a 1%CrMoNiV-steel doped with different tin concentrations, at 550°C
a) creep curves b) cavities at a grain boundary after rupture

2.4. Nitrogen

According to LEED-investigations nitrogen forms the same segregation structure on Fe(100) as carbon, see fig. 1. This segregation equilibrium cannot be determined exactly, since at elevated temperatures there is a steady N_2-desorption:

$$N(diss.) = N(segr.) \rightarrow N_2(gas) \qquad (8)$$

However, segregation measurements with Fe-N alloys with 150 and 580 ppm N yielded the value $\Delta H_N^0 = -110$ kJ/mol for the segregation enthalpy of nitrogen, this is in good agreement with results from kinetic measurements of the nitrogen absorption and -desorption in iron. In this reaction the segregated nitrogen is an intermediate, and information about the segregation equilibrium can be derived /2,3,4/.

The N 1s binding energy for segregated nitrogen is lower than for the nitrides Fe_2N and Fe_4N by 0.9 eV /13/. This shift is comparable with the result for segregated carbon and the carbides. It is, however, more difficult to determine the chemical shift in relation to a homopolar nitrogen bond. Adsorbed N_2 on Fe(100) at low temperatures /26/ reveals a N 1s binding energy of 400.3 eV. Comparing binding energy with bond charge as proposed by Siegbahn /27/ gives an excess charge of 0.25 electrons at nitrogen atoms in the nitrides and half an electron for segregated nitrogen. No excess charge, i.e. a homopolar bond, is expected at 399.5 eV N 1s binding energy.

The bonding of nitrogen to iron surface atoms is therefore strongly heteropolar with electronic charge transferred from iron to nitrogen.

2.5. Phosphorus

Upon saturation of Fe(100)-surfaces with segregated phosphorus also a c(2x2) structure is obtained. This has been observed with Fe-1%P single crystals at temperatures 800°C. Under these conditions one has the α solid solution /8/.

The surface segregation enthalpy could be determined /37/ by measurements in the system Fe-Si-P (where a displacement equilibrium was observed similar as in the system Fe-Si-C, but with additional repulsive energetic interactions): $\Delta H_P^0 = -180$ kJ/mol. In XPS investigations it was observed that phosphorus does not segregate only on the outer surface in the segregated state which will be denoted P_I, but with increasing surface coverage there is also enrichment of phosphorus in the lattice near the surface in a segregated state P_{II}, Fig. 8.

The species of chemisorbed phosphorus denoted as P_I /8/ is distinguished from phosphorus in the three-dimensional compound Fe_3P by a difference in their P 2p binding energy values. The transfer of negative charge from iron to phosphorus is higher for the segregated species P_I than for P in the phosphide Fe_3P. A charge transfer in the same direction in the case of Fe_3P is plausible according to the difference in electronegativity between Fe and P, and obvious from a chemical shift of the P 2p photoline to lower binding energy in comparison to solid state phosphorus.

FIGURE 8. Photo-lines of phosphorus,
(1) in the compound Fe_3P,
(2) segregated on a single crystal Fe-1%P and
(3) on polycrystalline Fe-0.08%P upon saturation with segregated phosphorus. Both photolines (2) and (3) show a shoulder at higher binding energies, corresponding to the photoline for Fe_3P, but originating from P_{II} segregated in the subsurface lattice.

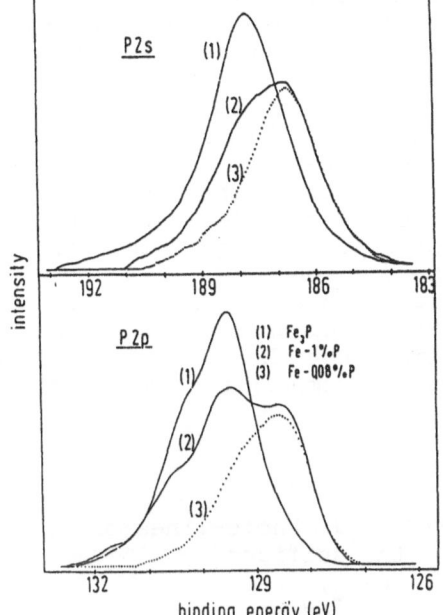

After long-term segregation further enrichment of phosphorus depending on the bulk content is observed trespassing the saturation level of a complete c(2x2) structure. At least two reasons support the idea that phosphorus is enriched in a different chemical state denoted $P_{II}/8/$ below the topmost iron lattice layer. Upon long-term segregation the shape of the P photolines changes in a way that a second line appears with a higher binding energy overlapping with the line corresponding to P_I. The subsurface location P_{II} enrichment can clearly be deduced from the progress of phosphorus signal intensity with increasing subsurface enrichment of phosphorus P_{II}, i.e. for rising bulk contents of phosphorus (refer to fig. 8). As the information depth for the P photolines is about 15 Å, it takes subsurface layers into account. With P_{II} enrichment the peak area of P 2p rises relative to Fe 2p, i.e. the ratio Fe 2p/P 2p decreases. Hardly any change, however, is observed in the Auger peak height ratio Fe(L₃VV)/P(LMM). Since the majority of P(LMM) Auger electrons comes from the first one or two layers of the sample, it can be excluded that P_{II} occupies surface sites. The P 2p binding energy of 129.5 eV for the subsurface P_{II} equals the value for Fe_3P as well as for phosphorus dissolved in the iron matrix. However, Fe_3P precipitation can be excluded from the fact that the LEED pattern is unchanged through the accumulation of P_{II}. Subsurface segregation obviously is a peculiarity of phosphorus in iron.

2.6. Sulfur
Sulfur is extremely 'surface active' on iron surfaces, even for very small bulk concentrations >1 ppm in the stability range of the α-phase up to 900°C always saturation of the iron surface with sulfur was observed after equilibration /2/. The presence of sulfur on the iron surface strongly affects surface reaction

kinetics in the case of carburization and nitrogenation of iron
/2, 4/ and also the surface diffusion is influenced by adsorbed
sulfur, the surface self-diffusivity of iron is enhanced in the
presence of adsorbed sulfur /28/. On Fe(100) the c(2x2) structure is obtained, up to high temperatures,a very distinct LEED
pattern of this structure can be observed. By measurements and
theoretical calculations of the LEED-intensity - energy curves
it was proved that sulfur occupies the fourfold-coordinated
sites of the (100)plane in the centre between four iron atoms
(as shown in fig. 1 for carbon). The radius approaches that of
the S^{2-} ion.
Fig. 9 shows the XP-spectra of sulfur segregated on the (100)
surface of an iron single crystal containing 20 ppm sulfur, for
comparison the sulfur photolines were measured on the surfaces
of the compounds of FeS and FeS_2.

FIGURE 9. Photo-lines of S,
in the sulfides
(1) FeS_2,
(2) FeS and
(3) segregated on iron with
dissolved sulfur (two different surface concentrations).

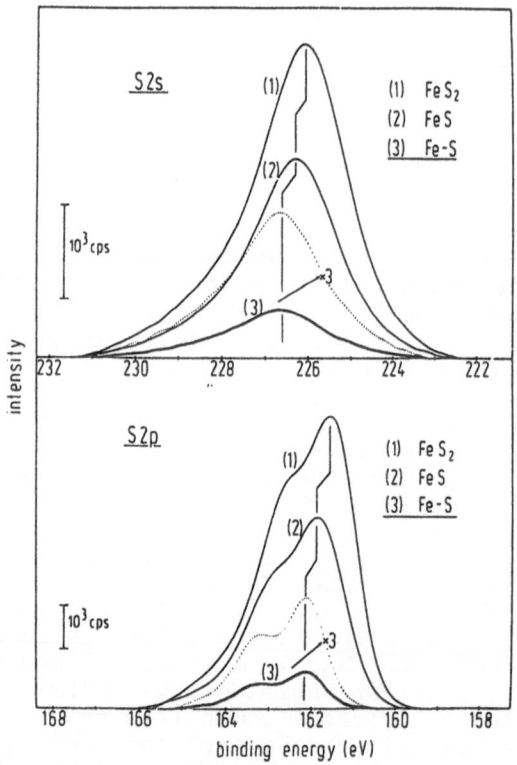

A S 2p binding energy of
164.0 eV is observed for S_8 with its homopolar bond. In relation to this value the observed binding energy of 162.2 eV for
segregated sulfur indicates a net gain of negative charge at
the sulfur atoms. In contrast the Fe 2p photoline shifts to
0.2 eV higher binding energy at sulfur saturation compared to
a clean iron surface. This shift can be anticipated from a loss
of charge from the surface iron atoms. A high degree of electron
transfer from iron to sulfur atoms is indicated by both line
shifts. This result agrees well with the Fe_4S cluster calculations of Briant and Messmer, who predicted for sulfur segregated at grain boundaries a nearly ionic bond with iron /29/.

2.7. Hydrogen adsorption on iron with segregated carbon, silicon or sulfur

The adsorption of hydrogen on iron surfaces is of great interest in relation to the embrittling effect of hydrogen in fracture of Fe,Ni and steels. Segregated impurities at grain boundaries may interact with hydrogen from the gas atmosphere upon fracture in hydrogen atmospheres. Furthermore, additional or synergetic effects are discussed of hydrogen atoms and embrittling segregants, such as S,P,Sb,Sn at grain boundaries /38-40/.

Therefore the hydrogen adsorption was investigated on a clean Fe(100)surface and on samples with the same orientation but presegregated with C,Si or S /41/. In each case the saturated c(2x2)structure was attained by heating single crystals containing C,Si or S at appropriate temperatures and quenching. The adsorption of hydrogen was conducted at low temperatures, e.g. 230 K, and the amount of adsorbed hydrogen atoms was determined by thermal desorption spectroscopy. Fig. 10a) shows such thermal desorption spectrum (t.d.s.) for adsorption on clean iron. The height of the desorption peaks gives the amount of adsorbed H, which increases with exposure in the adsorption period (measured in Langmuir = 10^{-6} Torr·sec). From the desorption temperature a value can be derived for the desorption energy, which is equal to the adsorption enthalpy ΔH in the case of iron, since the dissociative adsorption of hydrogen is nonactivated at about room temperature. For hydrogen on clean Fe(100) two desorption peaks are observed, one at about 400K which corresponds to an adsorption state β_2 and one at about 300K which corresponds to a state β_1 /42-44/. The state β_2 is occupied at first and has the higher adsorption enthalpy, about -100 kJ/mol H_2, the other state β_1 has the adsorption enthalpy -75 kJ/mol H_2, the coverage at saturation corresponds to half a monolayer for both states.

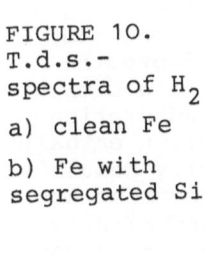

FIGURE 10.
T.d.s.-
spectra of H_2

a) clean Fe

b) Fe with
segregated Si

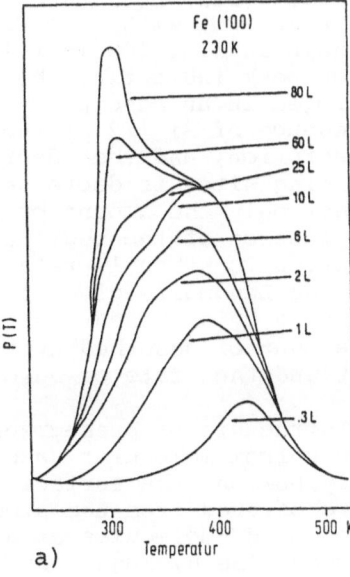

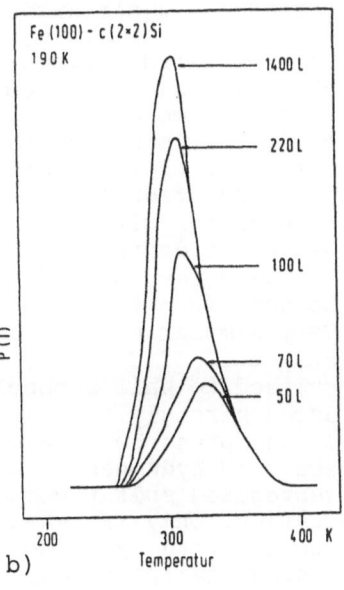

It may be assumed (see fig. 11) that upon saturation with β_2
a c(2x2)structure is formed by the hydrogen atoms, the occupation of the remaining adsorption sites leads to a p(1x1)
structure and is less exothermic according to repulsive interactions of the H-atoms. For adsorbed H the structure shown in
fig. 11a) cannot be confirmed by LEED and AES. However, this
discussion is supported by the study of hydrogen adsorption
on the presegregated surfaces /41/.

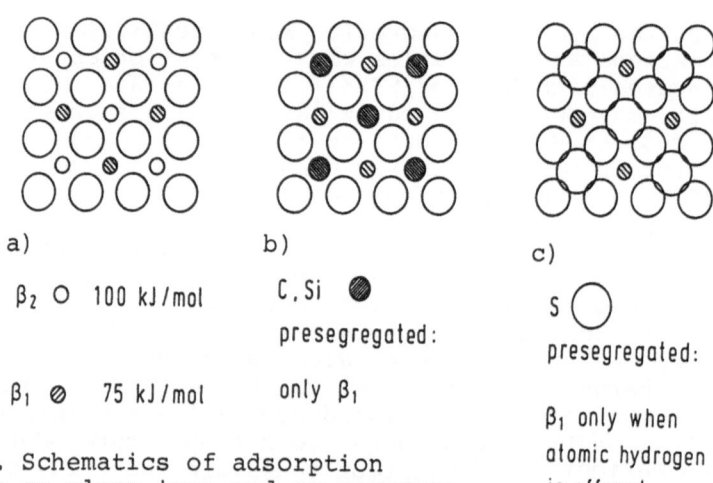

a)

β_2 O   100 kJ/mol

β_1 ⊘   75 kJ/mol

b)

C, Si ●

presegregated:

only β_1

c)

S ◯

presegregated:

β_1 only when
atomic hydrogen
is offered

FIGURE 11. Schematics of adsorption
structures on clean iron and on presegregated surfaces

In Fig. 11 it is assumed that the hydrogen atoms occupy the
same adsorption sites as the carbon, silicon and sulfur atoms,
namely the fourfold coordinated 'central sites' on the (100)
face. This is supported by the fact, that Si and C atoms block
just the number of sites for hydrogen adsorption which correspond to their number in the c(2x2) structure.
The t.d.s. for hydrogen on a Fe(100) with a c(2x2) structure
of segregated silicon is shown in fig. 10b. In this spectrum,
there is only one desorption peak indicating the adsorption
of half a monolayer of hydrogen in an adsorption state similar
to β_1. Obviously by the presence of Si 1/2 of the adsorption
sites are blocked for H-adsorption, and the adsorption enthalpy
for adsorption in the remaining sites is decreased by repulsive
interaction with silicon. Not only the amount of adsorbed hydrogen is decreased by the presence of the segregated Si-atoms
but also the rate of adsorption is strongly retarded - as can
be seen from the values of the Langmuirs given in fig. 10b).
Very similar effects are caused by a presegregated c(2x2)
structure of C-atoms, the amount of adsorbed hydrogen is decreased to half a monolayer and the rate of adsorption is reduced strongly.
In the presence of a c(2x2)structure of presegregated S, however, no hydrogen is adsorbed from molecular gaseous H_2. The
segregated sulfur completely poisons the surface for H_2-disso-
ciation. Only if H-atoms are offered from the atmosphere (which
are obtained by dissociation of H_2-molecules at a hot Pt-wire),
a slow adsorption was observed. The hydrogen can be adsorbed un

to a coverage of one half of a monolayer (fig. 11c) on surfaces
with a c(2x2)S, however, the adsorption is slow and the binding
force is low (small adsorption enthalpy) according to repulsive
interactions with the sulfur atoms. In spite of these effects
of sulfur on the hydrogen adsorption, obviously both elements
may act together enhancing embrittlement of Ni and of Fe and
steels. This is possible if the hydrogen enters the crack or
the grain boundary as an atom (proton), coming from the lattice
of the metal or discharged from H^+-ions coming from an aqueous
electrolyte.
The effects of segregated tin on hydrogen adsorption are very
similar to the effects of sulfur. Adsorption of molecular hydrogen at 200K was minimal. Only if atomic hydrogen was offered
to the sample with c(2x2)Sn a slow adsorption was observed at
a low adsorption enthalpy /41/. Also tin obviously hinders the
dissociation of H_2 and decreases the adsorption enthalpy of H
by repulsive interactions.

3.   GRAIN BOUNDARY SEGREGATION
3.1. <u>Phosphorus</u>
The equilibrium of grain boundary segregation has as yet been
studied in detail only for phosphorus in α-iron /16-20/. Fe-P-
samples with phosphorus contents between 0.003 to 0.3% have
been annealed at temperatures between 400 to 800°C till the
equilibrium was established. After that, as already described
in the introduction, the samples were fractured in the UHV-
system and the grain boundary concentrations were determined
by AES. Many measurements are performed on different spots of
different grain boundary faces. These measurements with the aid
of a calibration /19/lead to quantitative data on the grain boundary concentrations, which show some scatter on the different
grain boundary faces. However, from the average values steady
curves are obtained, see Fig. 12.

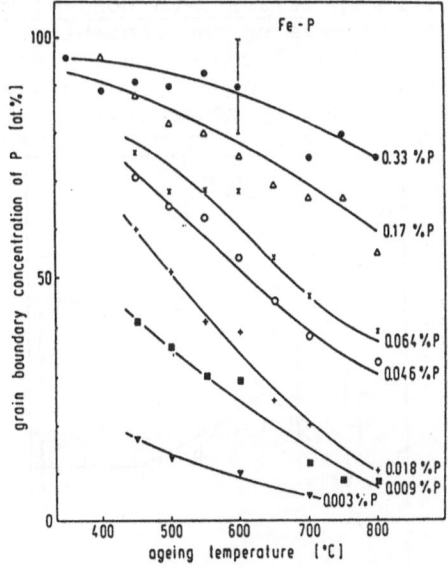

FIGURE 12. Dependence of the
grain boundary concentration
of phosphorus in Fe-P alloys
on the bulk concentration and
annealing temperatures.

The grain boundary concentrations decrease with increasing temperature and with decreasing bulk concentration as expected, for low temperature and high bulk concentrations a grain boundary coverage of 100 at% is approached, i.e. the phosphorus content of the grain boundaries corresponds to an atomic monolayer. It is well possible to describe the results with the Langmuir-McLean equation. For the Gibbs' free energy of segregation one obtains ΔG° = -34300 - 21.5 T J/mol. Thus the segregation enthalpy at grain boundaries is much lower than the value for the surface segregation of phosphorus: -180 kJ/mol /37/. The bonding state of the Fe-P chemical bond was therefore characterized for segregated phosphorus on the fracture surface as well as on Fe(100) and on the free surface of polycrystalline iron. A comparison was made to the Fe-P bond in Fe_3P /8/.

On the fracture surface of polycristalline iron containing 0.09at% phosphorus the P 2p spectrum reveals the presence of both states of segregated phosphorus with a slightly increased amount of P_{II}. Thus, phosphorus has segregated in the plane of fracture and in lattice planes adjacent to the grain boundary.
With increasing grain boundary concentration of phosphorus the part of intergranular fracture increases, phosphorus reduces the cohesion of the grain boundaries. This is the reason for the temper embrittlement and the long-term embrittlement of low alloy ferritic steels,which has been studied in detail for some turbine steels /21/. Phosphorus at grain boundaries also can cause intergranular corrosion, for example of iron in hot nitrate solutions at high potentials /25/, under certain conditions also stress corrosion cracking may be enhanced by the presence of phosphorus at grain boundaries /30/.

3.2 Phosphorus and Carbon, Nitrogen or Boron
Upon increasing the content of dissolved carbon in Fe-P alloys at constant P-concentration and constant annealing temperature (which can be done by carburization in CH_4-H_2), an increasing grain boundary concentration of carbon and a decreasing grain boundary concentration of phosphorus is observed, see fig. 13.

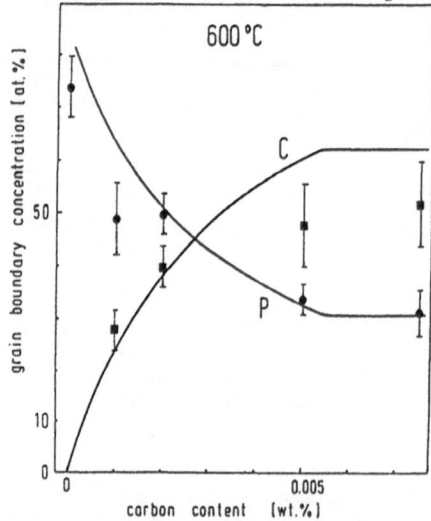

FIGURE 13. Grain boundary concentrations of phosphorus and of carbon in Fe-0.17%P. Dependence on the bulk concentration of C after annealing at 600°C.

Carbon is able to displace phosphorus from the grain boundaries. This is an equilibrium of mutual displacement, similar as already described for the surface segregation of silicon and carbon. For describing this equilibrium equations corresponding to the equations (6) and (7) can be used. In the description of this displacement equilibrium only the competition of both elements for the available sites is considered, no energetic interactions need to be taken into account /16,31/. With increasing carbon concentration in the bulk and at the grain boundaries the tendency to intergranular fracture decreases and more and more transgranular fracture occurs. For the higher carbon concentrations it is very difficult to find intergranular fracture faces, on which grain boundary concentrations can be determined. Thus AES studies of grain boundary segregation of carbon are possible only in ternary alloys Fe-P-C or Fe-S-C where simultaneous segregation of carbon and of the embrittling elements P or S occurs. Such studies have now been performed with Fe-0.043%P and Fe-0.16%P in which different carbon contents between 10 to 90 wt ppm C have been introduced by carburization in CH_4-H_2 and which had been annealed for sufficient time at temperatures between 500-700°C, till the segregation equilibria of phosphorus and carbon were established /32/. Such studies lead to the (preliminary) result for the grain boundary segregation of carbon in α-iron: ΔG_C = -57000 -21.5 T J/mol. In this expression it was assumed that the segregation entropy of carbon and of phosphorus are similar at grain boundaries.

The part of intergranular fracture is smaller for samples with equal grain boundary concentrations of phosphorus if they contain additional carbon. Thus, carbon has two effects:

(1) Upon increasing grain boundary concentration the phosphorus is displaced from the grain boundaries.

(2) Carbon decreases the tendency for intergranular fracture, obviously it increases the cohesion of the grain boundaries.

The equilibrium of mutual displacement can be written for carbon and phosphorus similarly as for carbon and silicon on surfaces, equ.(5):

C(dissolved) + P(segregated) = P(dissolved) + C(segregated)   (9)
  ↓ carbide-forming elements: Cr, Mn etc.
Fe, Mn, Cr-carbides

With the arrow indicating the effect of carbide-forming elements on this displacement equilibrium, the increase of P-segregation is explained which is observed upon alloying a steel with Cr or Mn. The concentration of carbon in solution and at the grain boundaries is decreased by such elements, therefore the grain boundary concentration of P is enhanced. By other authors /45,46/ this effect was explained by cosegregation, e.g. of Cr and P. Such cosegregation i.e. a mutually enhanced segregation by the attractive interaction of solute elements was observed in our studies as yet in no case. Chromium enhances P-segregation at grain boundaries of Fe-Cr-C-P alloys compared to Fe-C-P alloys, but not in Fe-Cr-P alloys compared to equivalent Fe-P alloys, see Fig. 14.So the effect

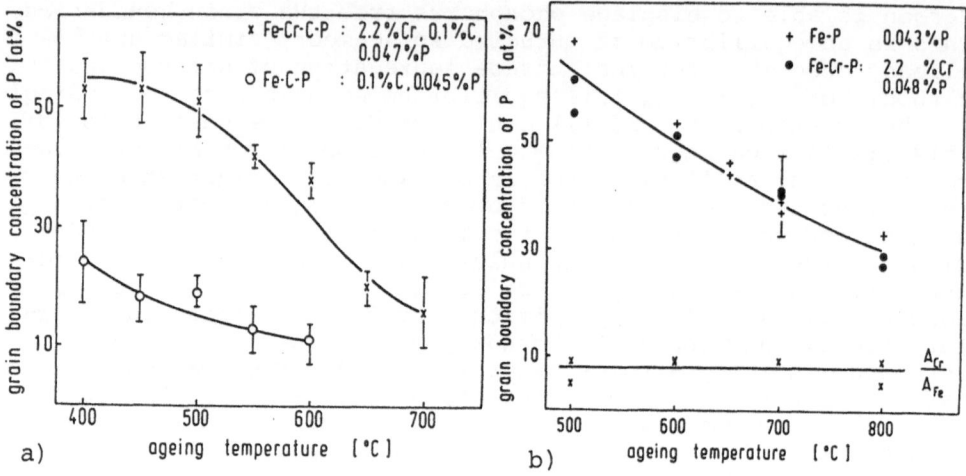

FIGURE 14. Effect of chromium on the grain boundary segregation of phosphorus, grain boundary concentration vs. equilibration temperature a) Fe-C-P and Fe-Cr-C-P alloys, b) Fe-P- and Fe-Cr-P alloys, with about equal concentrations of C and P.

of chromium is not caused by cosegregation of Cr and P, but by the decrease of free carbon in solution and at the grain boundaries upon formation of Cr, Fe-carbides. This decrease of carbon concentration causes a shift of the equilibrium equ.(9) to the side of phosphorus segregation.

Other alloying elements in steels also affect the grain boundary segregation of P. Ti and Nb show a strong chemical interaction with P, the phosphorus is tied up in the steel matrix by formation of phosphides, small precipitates or clusters. Therefore Ti and Nb can effectively decrease the grain boundary segregation of P/21,22/, for example in interstitial free steels, where these elements are not precipitated as carbides. The elements Mo and V, however, are scavenging phosphorus when they are present as carbides, Mo_2C and VC. By FIM and atom probe it was proved that P-atoms are enriched around small Mo_2C-particles /33/, in this way again phosphorus segregation is reduced, but not by formation of phosphides or clusters as presumed by other authors/47,48/. VC and also TiC to some extent act in a similar way as Mo_2C, we now suspect that the P-atoms are located in dislocations around the carbides which are stabilized by the fine precipitates.

Equilibria of mutual displacement also exist for phosphorus and nitrogen/17/ and for phosphorus and boron/34/. With increasing nitrogen content in Fe-P-N alloys the grain boundary concentration of N increases and the grain boundary segregation of P decreases. As in the case of carbon the grain boundary segregation of nitrogen does not embrittle iron, but rather increases grain boundary cohesion leading to transgranular fracture. Boron is very effective in displacing the P from grain boundaries, during heat treatments of steels in the stability range of austenite, even very small contents ≲ 25 ppm B suppress phosphorus segregation and diminish the tendency to intergranular fracture.

3.3 Tin

The grain boundary segregation of tin in iron was studied on nine melts with concentrations between 0.022 to 3.1%Sn. After equilibration at temperatures between 500° and 750°C for sufficiently long time the samples were fractured and analyzed in the usual way. The calibration of the grain boundary analysis was obtained from our surface segregation studies on single crystals /7/. The measurements showed a wide scatter of concentrations from grain boundary to grain boundary (Fig. 15), which may be caused by a strong dependence of the tin segregation on grain orientation /35,36/.

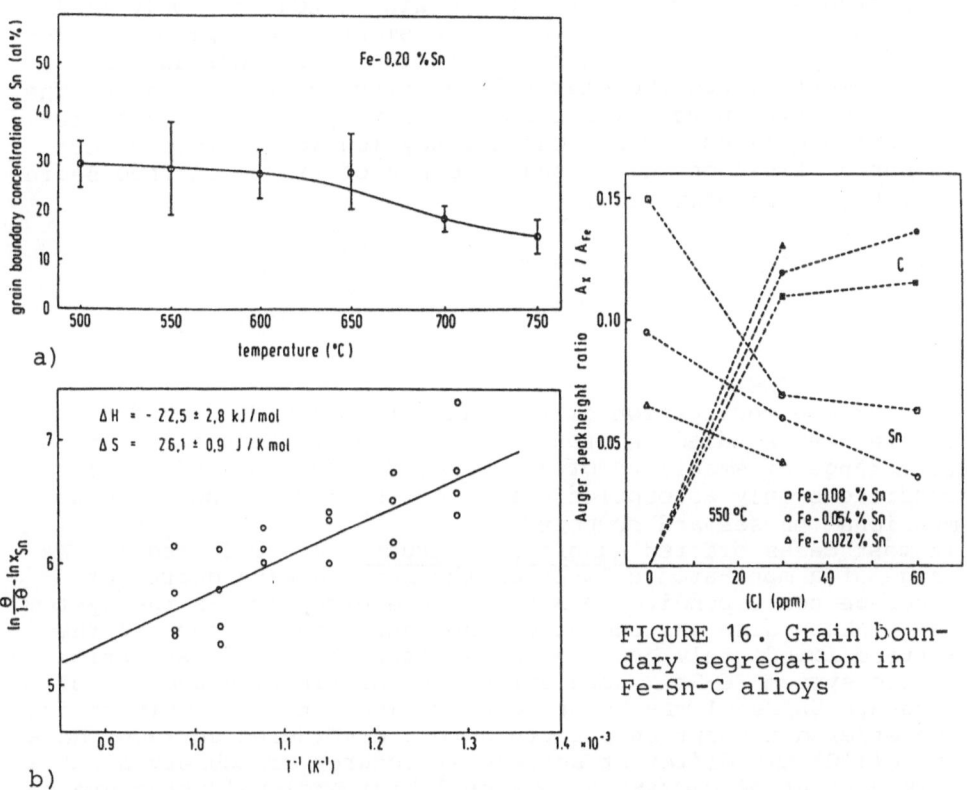

a)

b)

FIGURE 16. Grain boundary segregation in Fe-Sn-C alloys

FIGURE 15. Grain boundary segregation of tin in Fe-Sn alloys
a) grain boundary concentration vs temperature for Fe-0.20%Sn
b) evaluation for all investigated alloys according to the Langmuir-McLean equation.

In contrast to the surface segregation, the grain boundary concentrations are rather low, always much below one monolayer. The tendency to grain boundary segregation is much less than for surface segregation, for the Gibbs' free energy of segregation the expression is obtained: ΔG^{o} = -22500 -26.1 T J/mol The value for the enthalpy of segregation $\Delta H \cong$ -20 kJ/mol is very low compared to surface segregation and grain boundary segregation of other elements. The effect of carbon was studied on tin grain boundary segregation, samples with different concentrations of tin and carbon were annealed at 550°C up to

5000 h, somes samples were held in hydrogen at 550°C and completely decarburized. Similarly as in the case of phosphorus, carbon can displace tin from the grain boundaries (Fig. 16). With increasing carbon concentration, the grain boundary concentration of tin decreases and the grain boundary concentration of carbon increases, simultaneously the part of intergranular fracture decreases. Consequently the presence of tin in carbon steels is no great danger to their mechanical properties, the tendency of grain boundary segregation is low, the diffusion of tin to the grain boundaries is slow and furthermore concentration of dissolved and segregated carbon will be high enough to keep tin from the grain boundaries. Only if low alloy steels with carbide forming elements are applied at elevated temperatures slow tin segregation will occur and embrittlement, since the embrittling effect of tin is large. The effect of tin on void formation and cavity growth in heat resistant steels upon creep was already mentioned, this effect is based mainly on the strong tendency of tin to surface segregation (see chapter 2.3).

4. CONCLUSIONS

The surface segregation of the elements A = C,Si,Sn,N,P,O and S have been studied on iron in the concentration- and temperature range of stability of the α-solid solution. Under these conditions only adsorption phases can be formed, no three-dimensional phases and compounds.

In most cases ordered adsorption structures are formed in the range of a mono-atomic layer on the surface with degree of coverage corresponding to A/Fe <1. One exception is the system Fe-P, where also enrichment of phosphorus takes place in the lattice immediately beneath the surface. The other exception is the system Fe-Sn, where enrichment of tin with degrees of coverage Sn/Fe >1 was observed on Fe(100) and a transition from the ordered adsorption structure to a disordered surface phase. On Fe(110) two different surface structures are observed: At a bulk content of 0.02%Sn a hexagonal bidimensional structure of only tin atoms, is stable, which is related to the diamond structure of α-Sn. At higher bulk contents a surface compound with three layers composed of Fe and Sn atoms is formed epitaxially /7/.

The dependencies of the segregation equilibria

$$A(dissolved) = A(segregated)$$

on temperature and on bulk concentration can be described by the Langmuir-McLean equation

$$\ln(\Theta_A/(1 - \Theta_A)) = - \Delta H_A^{o}/RT + \Delta S_A^{xs}/R + \ln x_A \qquad (10)$$

From this equation the values are obtained for the enthalpy of segregation ΔH_A^{o}, which are given in table 1.

Table 1 Enthalpies of segregation ΔH_A (kJ/mol) on and in iron

Segregant	Surface segregation	Grain boundary segregation
C	- 85	- 57
Si	- 48	
Sn	$\leq$ -200	- 22
N	-110	
P	-180	- 34
S	-190	

In the simultaneous segregation of different elements the adsorbed atoms compete for the surface sites, also in the case
of substitutional and interstitial elements, such as carbon and
silicon. The surface concentrations are determined by the respective bulk concentration and the free energies of segregation according to

$$\theta_1/(1 - \theta_1 - \theta_2) = x_1 \exp(- \Delta G_1^o/RT) \tag{11}$$

$$\theta_2/(1 - \theta_1 - \theta_2) = x_2 \exp(- \Delta G_2^o/RT) \tag{12}$$

In the case of carbon = 1, silicon = 2 these equations which
only consider site competition are sufficient for describing
the experimental results, in the case of phosphorus = 1, silicon = 2 additional repulsive interactions must be considered by
introducing an interaction term $\alpha_{1,2}$ into the exponentials /37/.

$$\theta_1/(1 - \theta_1 - \theta_2) = x_1 \exp((- \Delta G_1^o - \alpha_{1,2}\theta_2)/RT) \tag{13}$$

$$\theta_2/(1 - \theta_1 - \theta_2) = x_2 \exp((- \Delta G_2^o - \alpha_{1,2}\theta_1)/RT) \tag{14}$$

Studies of the binding modes of segregated atoms (C,Si,Sn,N,P,S)
by photoelectron spectroscopy (XPS) always showed a more or
less distinct shift of the photo-lines to diminished binding
energies of the core electrons (in comparison to the binding
energies in the element), see table 2.

Table 2 Chemical shift of the binding energies of core
electrons of surface segregated elements

Segregant	Chemical shift (eV)	Effect on the grain boundary cohesion
S	- 2	
Sn	- 0.15	strongly embrittling
P	- 2	
Si	- 0.1	weakly embrittling
N	- 2	
C	- 2	cohesion enhancers
B	not determined	

The chemical shift of the photolines may be interpreted by an
electron transfer from iron to the segregated elements. The
chemical shift and the electron transfer are in the same direction but not equal as in the compounds with these elements:
carbides, silicides, phosphides, sulfides etc. The chemical
shift is rather high for carbon, nitrogen, phosphorus and
sulfur, so the segregated state of these elements has a relatively high negative charge. The chemical shift is less for
elements silicon and tin on iron, which obviously form a more
homopolar bond.

The thermodynamics of the grain boundary segregation of phosphorus in polycrystalline iron can be described well by the
Langmuir-McLean equation in spite of some scatter of the
values. The enthalpy of segregation is much less at the grain
boundary than on the surface, see table 1. The Langmuir-McLean
equation cannot be applied satisfactorily on the grain boundary segregation of tin, the scatter of the data is very great,
however, also a rather low value results for the segregation
enthalpy. In this case the segregation obviously is strongly
dependent on the grain boundary orientation, such orientation
dependence also has been observed in the studies of the surface segregation of tin /7/. The enthalpy of surface segregation could not be determined for tin, since even at low bulk
concentrations always saturated structures were observed and
no temperature dependence of the surface coverage could be
measured. However, according to the latter fact the segregation
enthalpy most probably is high.

Also in grain boundary segregation, displacement equilibria
are observed of elements, which compete for the sites in the
grain boundary. In this site competition substitutional and
interstitial elements have equal rights as shown for the elements phosphorus and carbon and for the elements sulfur and
nitrogen. The description of these displacement equilibria is
possible with the simple equations (11) and (12) without considering repulsive or attractive interaction energies, just assuming site competition.

A cosegregation, i.e. mutually enhanced segregation by the
attractive interaction of solute elements was observed in no
case. The cosegregation was assumed in the previous literature
in many systems, however, for several cases other explanations
could be given, as demonstrated for the system Fe-Cr-C-P.
Other cases of 'cosegregation' may be explained by the formation of thin layers of threedimensional compounds at grain
boundaries.

The elements phosphorus and sulfur reduce the cohesion of the
grain boundaries and in this way they embrittle iron, whereas
the elements carbon and nitrogen do not reduce the grain boundary cohesion, but rather increase the cohesion of the grains.
These contrary effects of the segregated elements cannot be
explained with different binding modes, as was proposed by
Messmer and Briant /29/, since in the studies of surface segregation by photoelectron spectroscopy the shifts of the
binding energies of electrons have been determined to be in
the same direction for all elements.

REFERENCES

1. H.J. Grabke, G. Tauber, H. Ciefhaus: Scripta Metallurg. 9 (1975) 1181
2. H.J. Grabke, W. Paulitschke, G. Tauber, H. Viefhaus: Surface Sci. 63 (1977) 377
3. H.J. Grabke, H. Viefhaus, G. Tauber: Archiv Eisenhüttenwes. 49 (1978) 391
4. H.J. Grabke: Mat. Sci. Engg. 42 (1980) 91
5. H. Viefhaus, H.J. Grabke: Surface Sci. 109 (1981) 1
6. H. de Rugy, H. Viefhaus: Surface Sci.
7. M. Rüsenberg, H. Viefhaus: Surface Sci. 159 (1985) 1-23
8. B. Egert, G. Panzner: Surface Sci. 118 (1982) 345
9. B. Egert, G. Panzner: Physical Rev. B 29 (1984) 2091
10. G. Panzner, B. Egert: Surface Sci. 144 (1984) 651
11. B. Egert, H.J. Grabke, Y. Sakisaka, T.N. Rhodin: Surface Sci. 141 (1984) 397-408
12. G. Panzner, W. Diekmann: Surface Sci. 160 (1985) 253-70
13. W. Diekmann, G. Panzner, H.J. Grabke: Surface Sci., in press
14. G. Panzner, D. Mueller, T.N. Rhodin: Physical Review B 32 (1985) 3472
15. G. Tauber, H.J. Grabke: Ber. Bunsenges. Phys. Chem. 82 (1978) 298
16. H. Erhart, H.J. Grabke: Metals Sci. 15 (1981) 401
17. H. Erhart, H.J. Grabke: Scripta Met. 15 (1981) 531
18. H. Erhart, H.J. Grabke, R. Möller: Archiv f. Eisenhüttenwes. 52 (1981) 451
19. H. Viefhaus, R. Möller, H. Erhart, H.J. Grabke: Scripta Met. 17 (1983) 165
20. H.J. Grabke, H. Erhart, R. Möller: Microchimica Acta Wien Suppl. 10 (1983) 119
21. H. Erhart, H.J. Grabke, R. Möller: Archiv f. Eisenhüttenwes. 54 (1983) 285
22. R. Möller, H.J. Grabke: Scripta Met 18 (1984) 527
23. R. Möller, H. Erhart, H.J. Grabke: Archiv f. Eisenhüttenwes. 55 (1984) 543-548
24. M. Paju, R. Möller: Scripta Met. 18 (1984) 813
25. J. Küpper, H. Erhart, H.J. Grabke: Corrosion Sci. 21 (1981) 227-238
26. D.W. Johnson, M.W. Roberts: Surf. Sci. 87 (1979) L255
27. K. Siegbahn et al.,"ESCA, Atomic, Molecular and Solid State Structure studied by Means of Electron Spectroscopy (Almquist and Wiksells, Uppsala, 1967)
28. H.J. Grabke, E.M. Petersen, S.R. Srinivasan: Surface Sci. 67 (1977) 501-16
29. R.P. Messmer, C.L. Briant: Acta Met. 30 (1982) 1811
30. H.J. Krautschick, K. Bohnenkamp, H.J. Grabke: in Vorbereitung.
31. R.R. de Avillez, P.R. Rios: Scripta Metallurg 17 (1983) 677-680
32. H. Hänsel, B. Bennett, H.J. Grabke: in Vorbereitung
33. R. Möller, S.S. Brenner, H.J. Grabke: Scripta Metallurg. 20 (1986) 587-592
34. M. Paju, H. Viefhaus, H.J. Grabke: to be published

35. W. Jäger, H.J. Grabke, Jin Yu: Proc. 2nd Int Conf. on Conf. on Creep and Fracture of Engineering Materials and Structures", April 1984, Swansea. Ed. B. Wilshire, D.R.J. Owen. Pineridge Press, Swansea U.K. 1984, II. S. 649/659
36. W. Jäger, H.J. Grabke, R. Möller: in Proceed. Int. Conf. 'Residuals and trace elements in iron and steel', Portoroz, Okt. 1985
37. H. Viefhaus: to be published
38. C.J. McMahon Jr. C.L. Briant, S.K. Banerji: in 'Fracture 1977' Proceed. 4th Int. Conf. on Fracture, Waterloo, Canada 1977
39. Jun Kameda, C.J. McMahon Jr.: Metallurg. Trans. 14A (1983) 903-11
40. G.W. Simmons, P.S. Pao, R.P. Wei: Metallurg. Trans. 9A (1978) 1147-58
41. A.K. Birchenall, B. Egert, H.J. Grabke, H. Viefhaus: in "Wasserstoff in Metallen", Ergebnisse eines Schwerpunktprogramms, Deutsche Forschungsgem. Mai 1986, p. 231-243
42. F. Bozso, G. Ertl, M. Grunze, M. Weiss: Appl. Surface Sci. 1 (1977) 103
43. J.B. Benzinger, R.J. Madix: Surface Sci. 94 (1980) 119
44. G. Wedler, K.P. Geuss, K.G. Colb, G. Mc Elhiney: Appl. Surf. Sci. 1 (1978) 471
45. M. Guttmann: Materials Sci. Engg. 42 (1980) 227-232
46. M. Guttmann, Ph. Dumoulin, M. Wagman: Metallurg. Trans. 13A (1982) 1693
47. C.J. McMahon Jr., A.K. Cianelli, H.C. Feng: Met. Trans. 8A (1977) 1055
48. Jin Yu, C.J. McMahon Jr.: Met. Trans. 11A (1980) 277

DISCUSSION

Comment by J. Chene:

In the measurements of H adsorption on iron by thermal desorption, is there any contribution of H absorbed species at 230K? Is there evidence for a subsurface state?

Reply:

In the desorption of hydrogen from an iron sample, H-atoms coming from solution in the bulk must pass through the adsorbed state β_1 or β_2. So it cannot be discerned if a certain amount of hydrogen seen in the desorption spectra comes from adsorbed, absorbed or subsurface states.

Comment by H. Birnbaum:

1. Have you measured the thermodynamics of B segregation in Fe?
2. Your heats of segregation would suggest that all of the measured impurities including C would be embrittling species according to the theory of Rice and Hirth. This leaves only B as a solute which is supposed to strengthen the grain boundaries. We have looked for segregation of B at grain boundaries in Ni and at surfaces in Ni but have not observed such segregation. We do see precipitates of nickel borides at the surfaces and

at grain boundaries or at surfaces where the segregation is true segregation rather than precipitation.

Reply:

We have looked for surface segregation of boron on bcc iron single crystals, doped with B. But, we never observed segregation of boron alone. In the presence of oxygen, B_2O_3 is formed on the surface and in the presence of N (dissolved), a layer of BN.

Grain boundary segregation of B was observed in fcc iron. After annealing of iron containing some B and P within the temperature range 900-1100°C, the material showed intergranular fracture caused by grain boundary segregation of P. With increasing concentration of B (in the range 10-30 ppm) the grain boundary concentration of P decreases and the part of intergranular fracture is diminished. This indicates true segregation of B at austenite grain boundaries (besides also nonequilibrium segregation is observed). In and on ferrite no true segregation of B was observed.

Comment by J. R. Rice:

Let ΔW be the reduction, due to segregation, in the theoretical work of separation of a grain boundary at constant composition. Then, at low temperature (neglect entropy terms) the thermodynamics of metastable interfaces, out of equilibrium with the bulk, give an expression which reduces (assuming Lagmuir-McLean conditions) to an

$$\Delta W = [(\Delta H)_{gb} - (\Delta H)_s] \, \Gamma_{gb}$$

Here Γ_{gb} is the grain boundary segregant coverage; and it is assumed that $(\Delta H)_f$, the segregant enthalpy on the freshly fractured surfaces, is approximately equal to $(\Delta H)_s$, the enthalpy for a surface at segregation equilibrium with coverage $\Gamma_{gb}^s/2$. If the latter condition is not met, we can nevertheless assert from energy minimization, at low temperature, that $(\Delta H)_f \geq (\Delta H)_s$, and hence assert that the above ΔW then provides at least an upper bound to the reduction in work of separation.

Using the data for C, P and Sn that you present, we then conclude that all three reduce the theoretical work of separation of a gb in Fe, but for a given coverage Γ_{gb}, the reduction caused by P and Sn is 5 and 6 times, respectively, that caused by C.

Thus, I raise the following question:

Could it be the case that C seems to be a "cohesion enhancer" not because it actually increases the work of separation, but, rather, because it displaces elements like P and S which have a far more deleterious effect on the work of separation?

If such is the case, then one may speculate that the hallmarks of "cohesion enhancers" are as follows:

1. $[(\Delta H)_{gb} - (\Delta H)_s]$ is small (or even negative!) so that the segregant does not itself much embrittle (or even increases the work of separation);

and

2. $[- (\Delta H)_{gb}]$ is relatively large so that the segregant readily displaces others.

Reply:

In our studies of grain boundary segregation in binary and ternary iron alloys, the effect of the segregants on grain boundary cohesion was classified according to the part of intergranular fracture, shown by specimens fractured at ~-100°C by impact (for AES-analysis of the grain boundaries). With increasing P- or S-concentration in the grain boundaries, the part of intergranular fracture increases in relation to the part of transgranular fracture. The presence of carbon causes displacement of P from the grain boundaries and decrease of intergranular fracture. But if two ferrite samples had the same P-concentration at the grain boundaries, the one with additional C-segregation showed less intergranular fracture. So we concluded that C not only displaces P but also increases grain boundary cohesion. In contrast, we observed some embrittlement with increasing C-content in Fe-C-P alloys quenched from annealing in the austenite range.

One main point of my paper was to show that the conclusions of Briant and Messmer[1] are not applicable -- they assumed that strong embrittling elements draw charge from the neighbouring metal atoms and thereby weaken the grain boundaries whereas the cohesive enhancers do not draw charge off the metal atoms. Our XPS-studies demonstrate that all segregants, embrittling and non-embrittling, draw charge from the iron atoms, P and S as well as C and N are negatively charged in the segregated state.

Certainly your thermodynamic analysis of the fracture process and the effects of the segregants is right, and the conclusions are evident for P, S, and Sn in iron. The effects of carbon may somewhat depend on other influences on fracture, e.g. the relation of strength of the material in the grain and at the grain boundaries which is affected by the distribution of C and microstructure.

1. C. L. Briant, R. P. Messmer: Acta Metall. 30:(1982)1811.

Comment by R. M. Latanision:

Your work on hydrogen adsorption/absorption on alloyed Fe(100) is very interesting indeed, especially in terms of the issue of whether the source is molecular or atomic hydrogen. You have reported that sulfur inhibits absorption from both kinds of hydrogen sources. One would expect that surface segregated Pt might increase hydrogen absorption from a molecular source, since platinum is a good dissociation catalyst. Have you performed any experiments on Fe-Pt or Fe-Rh, for example?

Reply:

You propose to study the effect of dissociation promoters on the adsorption of hydrogen on iron. However, iron itself already is a very good catalyst for hydrogen dissociation. On pure iron the adsorption of molecular hydrogen and dissociation is a very fast non-activated process at room temperature and even lower temperature. This is known from many studies of several authors with different techniques. According to this fact, the adsorption enthalpy can be derived from the desorption energy observed in thermal desorption studies. So there is no great chance to see promotion of hydrogen adsorption on iron by Pt, Rh, or else.

Comment by J. P. Fidelle:

You said you had found no cosegregation instances. But are not the cases of S and H in steels and nickel instances of cosegregation? As shown by several authors, suitable segregation of S in nickel enhances sensitivity to H embrittlement and a shift from transgranular to intergranular for fracture surfaces, the same way as an increase in severity of the H embrittlement condition does.

Reply:

Cosegregation or 'synergetic cosegregation' is according to Guttmann[1] the mutually enhanced segregation of two elements 2 and 3 in a solvent 1. Precondition for such cosegregation is the fact that the balance between all the interactions between atoms results in an attractive interaction between the two solutes 2 and 3 with respect to the solvent 1, then both segregations should enhance each other.

This phenomenon may well be possible. We have never observed such cosegregation - and I think that this situation should rather lead to cluster formation or precipitation in the grain. What I wanted to show is that many observations have been misinterpreted as cosegregation, e.g. the increase of P-segregation in grain boundaries of steels upon alloying with Cr or Mn. Thereby the carbon solubility and activity is decreased and also the grain boundary concentration of C, so that the grain boundary concentration of P may increase. The simultaneous occurrence of Cr at the grain boundaries seen by AES is caused by Cr-containing carbides, precipitated at the grain boundaries.

Your case of S and H in nickel is certainly not cosegregation since the interaction of these elements is repulsive. The enhancement of embrittlement in the presence of both elements usually is interpreted as an additional effect, not by a mutual enhancement of segregation.

1. M. Guttmann: Surface Science 53(1975):213-227.

Comment by H. Mughrabi:

Could it be that, in addition to the factors discussed, the structure of the grain boundary plus segregated atoms could play a decisive role with respect to enhancing (or reducing) cohesion through the formation of a strong amorphous (metallic glass) interface on the grain boundary in some cases? For example, it is known that the metalloids boron and carbon promote the formation of metastable amorphous iron-based metallic glasses.

Reply:

Effects of the segregants on the structure of the grain boundary certainly may play a role in embrittlement. You propose that segregation of the nonembrittling elements C or B leads to formation of an amorphous layer which acts as a glue holding the grains together. But as yet there is no evidence for such amorphous layer. On the other hand, there is some evidence that embrittling elements, such as P, S and Te can cause a reconstruction and formation of low-indexed faces, most probably by preferential formation of certain low energy segregation structures. For the system Fe-Te and Cu-Bi even facetting of the grain boundaries was observed, and in the case of Fe-Te the grain boundaries even crack to form open surfaces during annealing and Te-segregation.

Comment by W. Losch:

Did you measure the H-desorption with mass analyzer; and, if yes, did you see a difference (in H/H_2 rate) in the presence of Si and S compared to a clean Fe surface?

Reply:

Yes, we have recorded the desorption with a quadrupoll-mass analyzer. But as far as I know, there was no difference in the ratio H/H_2 for desorption from clean iron, and iron with segregated Si or S. I supposed to get more H-desorption from the surfaces with segregation coverage and asked the student to look for the atomic hydrogen, but he did not find a higher H/H_2 ratio.

Comment by J. Smith:

The results of adhesion experiments would seem to bear on the question of C as an embrittling agent. As I recall, Don Buckley, Stephen Peppper et al., NASA Lewis have investigated the effect of fractional monolayers of C on Fe surfaces. They brought the Fe surfaces together in UHV and found that C decreased adhesion. This seems consistent with Jim Rice's comment that C is an embrittler, but just displaces P which may be a stronger embrittler.

Reply:

In my answer to Prof. Rice I have explained why we think that C enhances the cohesion of ferrite grain boundaries. Concerning adhesion measurements I may remind of own earlier work[1] where we brought iron samples in contact in an UHV-apparatus and measured the work of separation. The iron samples were covered with different amounts of segregants or adsorbants (controlled by AES). All studied elements, C, P, S, N, O, even H, increased the adhesion. At that time this result occurred to be reasonable to us, since all these elements (except hydrogen) could lead to additional chemical bonding upon contact of the iron surfaces. So these results are not in agreement with the study you cited, however, I think the relation to grain boundary cohesion cannot be established easily.

1. W. G. Hartweck, H. J. Grabke: Acta Metallurg. 29(1981):1237.

Comment by T. Watanabe:

You have mentioned that you observed strong boundary misorientation dependence of segregation in iron-tin alloys. How did you treat the measurements of the amount of segregation for your quantitative discussion? I would suggest that simple averaging of the data is dangerous because such averaging can mask the effects of boundary structure and of the boundary character distribution in polycrystals, related to processing.

Reply:

I agree, the scatter of the data on grain boundary concentrations in the case of our study on the system Fe-Sn was too great to allow the process of averaging. In the case of our studies on the system Fe-P, the scatter was much less. The strong scatter of the grain boundary concentrations in Fe-Sn alloys is in agreement with the large variety of coverages and structures observed in our studies of Sn surface segregation on different orientations of Fe-Sn single crystals[1].

The quantitative evaluation of our data on grain boundary segregation of tin may be only justified to get the important information that the enthalpy of grain boundary segregation is <u>rather low</u> for Sn in iron, let me say about -20± kJ/mol, and probably very high for surface segregation.

1. H. Viefhaus, M. Rusenberg: Surface Sci. 159(1985):1-23.

Comment by A. W. Thompson:

Would you expect that the cohesion-enhancing elements like C, N, and B could also be effective at the interfaces of particles like carbides? We know that P, for example, decreases cohesion of Fe_3C in steel, and it would be interesting if B or N could reduce that effect.

Reply:

We have no information on the segregation at the interfaces of carbides in ferrite. Recently we did a field-ion microscope and -mass spectrometer (atom probe) study[1] in the system Fe-Mo-C-P and found P enriched near Mo_2C particles, not in the interface but most probably in dislocations around the carbide -- this leads to the well-known effect of Mo decreasing the grain boundary segregation of P. Further atom probe studies should be done on segregation at carbide interfaces -- with AES such studies are not well possible. So we do not know about B and N at carbide interfaces.

1. R. Moller, S. S. Brenner, H. J. Grabke: Scripta Met. 20(1986):587-592.

Comment by J. Oudar:

Did you have any indication on the effect of the grain boundary orientation on the critical amount of phosphorous which induces the intergranular fracture in iron?

Reply:

As yet we have no results on the orientation dependence of grain boundary segregation. We have started such studies, but at first we met a difficulty -- electron channelling induced effects in grain boundary AES measurements[1].

The critical amount of phosphorus which induces intergranular fracture of iron is rather low, in the range of some at% of a monolayer, but it is not known if this amount is distributed equally in the grain boundaries.

1. B. Bennett, H. Viefhaus: Aurf. Interface Analysis 8(1986):127-132.

Comment by M. Pourbaix:

Would you agree that the aqueous intergranular corrosion embrittlement and stress-corrosion cracking of steels may be enhanced by the segregation of species which may form volatile hydrides (S, P, Sn, As...) when in contact with the acid and reducing solutions which exist inside cracks and other "occluded corrosion cells" formed in the presence of chlorides?

414

Reply:

In the study[1] which we conducted about the effect of phosphorus on the intergranular stress corrosion cracking of steels in nitrates, we found a certain enhancement of i.g.s.c.c. with increasing P-concentration at the grain boundaries. The effects were observed at rather low (external) potentials, -300 mV$_H$ to -50 mV$_H$ in NH_4NO_3 and -80 mV$_H$ to -50 mV$_H$ in $Ca(NO_3)_2$, so that formation of PH_3 may be possible, but I do not know if this is of significance for the process. Under similar conditions Hunt and Seah at NPL found no effects of Sn, As and other impurities except perhaps sulfur.

1. H. J. Krautschick, K. Bohnenkamp, H.J. Grabke: to be published in Corr. Sci.

Comment by M. Pourbaix:

It seems to me that the stress corrosion cracking which is being observed when steels are being tested in nitrate solutions may be related to the evolution of gaseous nitrogen and nitrogen oxides in the solutions of nitric acid which probably exists inside the cracks. Would you agree at such a mechanism and has research already been conducted in this direction?

Reply:

During intergranular stress corrosion cracking of steels in nitrates the evolution of nitrogen is observed[1,2]. At corrosion potential the reaction is:

I am not sure if this gas evolution is of importance for the process. We think the active dissolution at the crack tip and a slow oxide layer formation at the walls are the main features of this i.g.s.c.c. The active dissolution at the crack tip, i.e. in the grain boundary, can be enhanced by the presence of phosphorus.

1. A. Baumel, H. J. Engell: Archiv Eisenhuttenwes. 32(1961):379.
2. H. J. Engell, K. Bohnenkamp, A. Baumel: Archiv Eisenhuttenwes. 33(1962):285.

Comment by M. W. Finnis:

Is it possible that the role of embrittling or cohesion enhancing elements has nothing to do with their effect on bond strengths, but reflects their ability to block or to smooth the passage of dislocations?

Reply:

I would not say that the role of embrittling or cohesion-enhancing elements has <u>nothing</u> to do with their chemical interaction with the host metal. But certainly their effect on dislocation emission and mobility plays a role in fracture. I will not elaborate on this effect since I am a chemist. The third point which is important is the effect of the segregants on grain boundary structures.

Comment by K. Sieradzki:

In the case of C and Si where is the charge drawn from (Fe core levels or Fe valence bands - and how much from each)? Can we see some Fe peaks and corresponding shifts?

Reply:

Certainly the charge is drawn from the Fe valence bands. The XPS results and further UPS and ELS investigations concerning your question are presented in detail in the papers of B. Egert, G. Panzner and. W. Diekmann, cited in my paper.

Workshop Session 1:
Novel Aspects of Fracture

Session Chairman: P. Haasen

NOVEL TECHNIQUES AS APPLIED TO FRACTURE PROCESS ZONE THEORY

W. W. GERBERICH
Department of Chemical Engineering and Materials Science
151 Amundson Hall
University of Minnesota, Minneapolis, MN 55455

1. INTRODUCTION

A number of relatively new techniques utilizing in situ high voltage electron microscopy (HVEM),[1,2] synchrotron radiation,[3] selected area electron channeling patterns (SACP's),[4,5] acoustic emission (AE)[6,7] and flaw reconstruction[8-10] are being applied to improve the fundamental understanding of fracture processes. Only a few of these will be discussed in detail, specifically the AE and SACP techniques, and how these are contributing understanding to the details of sub-critical crack growth.

Recognition that very complex environmental/microstructural interactions control the crack growth process has been a driving force to apply all possible techniques to promote understanding. Besides the typical compositional and segregation influences on electrochemical attack and grain boundary separation, there are micromechanical factors which have only recently been recognized as having a major role in subcritical crack growth. The need to address the micromechanical aspects is illustrated in Figure 1. This shows a crack growth rate (velocity, v) versus applied mode I stress intensity, K_I, for three different materials. The top two are 4340 steel heat treated to room temperature yield strengths of 1620 and 1340 MPa while the bottom one is a body-centered cubic Ti-30Mo alloy with 485 ppm hydrogen having a yield strength of 900 MPa at 200 K. One notes both differences in magnitude and shape of the v-K_I curves. The effect of test temperature is not an important consideration here since these predominantly stage II crack velocities were nearly maximum for each material condition at each of the three test temperatures shown. Thus, the crack velocity in the 4340 steel decreasing with increasing tempering temperature was microstructurally related. The important feature here is that the velocity curve for the highest strength steel is smooth, v for the lower strength condition is fluctuating by a factor of 10, and v for the lowest strength Ti-30Mo alloy is varying by about a factor of 100. Also note for the 450°C tempered steel that one period of fluctuation occurs in about 1 MPa-m$^{1/2}$ while for the Ti-30Mo, one period occurs in about 10 MPa-m$^{1/2}$.[†]

Recent investigations[11-13] of the micromechanical influences have emphasized a process zone in which a microcracked region in front of the main crack changes the local stress magnitude and distribution. A schematic of such a process zone and the crack-tip opening angle[13] associated with it is shown in Figure 2. It would appear that the above order of magnitude scalings in crack velocity variability and periods of fluctuation are related to

---

† Note that these were constant load tests so that an increase in K_I scales with crack extension.

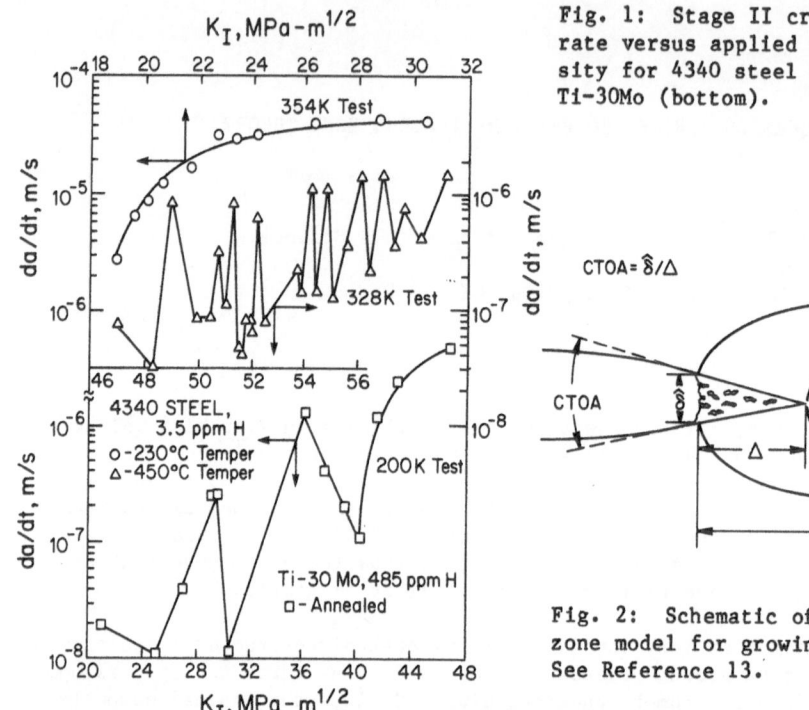

Fig. 1: Stage II crack growth rate versus applied stress intensity for 4340 steel (top) and Ti-30Mo (bottom).

Fig. 2: Schematic of the process zone model for growing cracks. See Reference 13.

the process zones in these three materials which, from top to bottom in Figure 1, are about 10 µm, 100 µm and 1000 µm, resepctively. The purpose of the present paper is to apply several of the above novel techniques to understand more about the process zone. While the understanding of crack growth fluctuations might seem to be a second order effect in many cases, we propose that detailed understanding will be a key factor in microstructural design for good resistance to environmental degradation.

2. NOVEL TECHNIQUES

There are a number of novel techniques for probing critical questions associated with microcrack formation at a macroscopic crack front. Some of these are outlined in Table 1. Others not mentioned but in use for some time are more standard SEM, TEM and synchrotron radiation techniques. Note that one of the overriding aspects is whether cracking is continuous or not at the atomistic level and, if discontinuous, when and where in the process zone is it occurring? It may not be possible to have well-characterized stress or strain states to pin down the question of discontinuous or continuous cracking using HVEM. However, slightly thicker films on the order of 1 mm thick are accessible to synchrotron radiation sources. In principle, this technique could easily characterize the stress state. In situ SEM studies on materials under well characterized stress, strain and/or displacement states have been accomplished by Vehoff and Neumann[14] and Davidson, et al.[15] However, these have generally been under displacement rate or fatigue cycling conditions and the time duration between microcrack events was not delineated. Again, few if any sustained load cracking studies have been performed in the electron microscope.

Table 1: Novel Techniques for Probing Critical Questions
Associated with Microcrack Formation

General Questions and Techniques	Information Sought		References
When occurring? - In situ SEM, HVEM - Acoustic Emission (A.E.)	K_I t dependence T	continuous? discontinuous?	(1,4,14-15) (16-19)
Where occurring? - Ultrasonic Probes - A.E. Source Location - Tritium Autoradiography - Selected Area Electron Channeling Patterns (SACP)	contiguous to crack? in advance of crack? nucleated at weak interfaces? in which of several fracture modes?		(9,10) (6,7) (20) (4)
How occurring? - Tritium Autoradiography - Fracture Reconstruction - SACP's	binding energies - which interfaces? process zone size effects?		(20) (8) (4,5)

The continuous/discontinuous question has mostly been addressed using
acoustic emission techniques.[16,17] Some have used the AE event as a measure
of the "brittle" crack event[16] while others[18] have suggested that the AE may
be measuring fracture events in the remaining ligaments after the majority
of the material has cracked on an atomistic scale. In a given material/
environment combination either one of these views could be correct but both
would require a process zone. That is, even if in a given material the
acoustic events were the ductile ligaments tearing after the main crack
advance, this would require invoking a process zone concept unless the ligament area fractions were negligible. This requires a good calibration of
acoustic emission amplitude or energy versus fracture area and stress at the
acoustic source. Such studies have been underway for some time[6] with more
precise quantification using velocity inversion and signal processing
methods close at hand.[19]

In addressing the where question, weak interfaces are being probed by
tritium autoradiography in terms of hydrogen embrittlement[20] while ultrasonic probes, selected area channeling patterns (SACP's) and acoustic emission are being tuned to source location. For example, using ultrasonic
angular scans, flaw shape, orientation and location may be obtained.[9,10]
Selected area channeling may be used both in situ or after the fact to
gather information on the crystallography and strain level associated with
environmentally induced microcracking.[21] Although this will be detailed in
the application sections the quality of the strain information possible is
illustrated in Figure 3. Here, the effect of 20 percent deformation on line
broadening is given, along with a calibration of line-width degradation versus plastic strain. The increased scatter with large strain appears to be
associated with the finest-line method as much as it is with inhomogeneous
deformation. AE techniques have located microcrack sources in several ways.
One is to simply fatigue precrack a sample, load it until an acoustic event

422

Fig. 3(a): Effect of
prior plastic strain on
SACP quality in Ti-30Mo,
a BCC alloy.

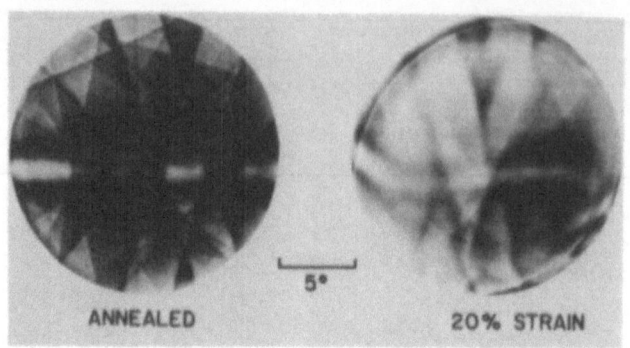

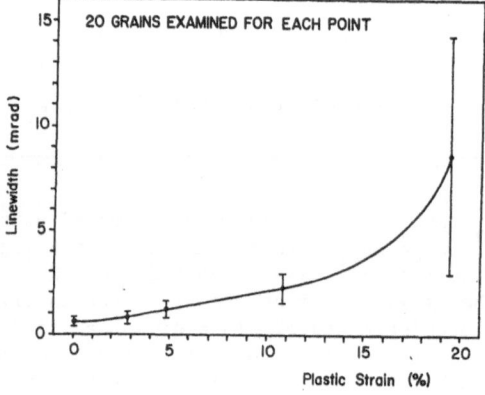

Fig. 3(b): Channeling
line width as a function
of prior plastic strain
for Ti-30Mo.

occurs, unload it and then surround the microcrack event by ductile fatigue
striations formed at an intermediate ΔK. Such a technique, illustrated in
Figure 4, has been successful for isolating cleavage microcracks formed at
temperatures below the ductile-brittle transition (DBTT) in both 10 µm and
100 µm grain size steels. The relative sharpness of the SACP from this
microcleavage facet also implies little surface plasticity associated with
this event. Such studies are making the case that process zone effects are
extremely important to understanding toughness in the DBTT range as well as
fracture resistance associated with other discontinuous processes. A more
direct in situ AE technique is to source locate using multiple transducers.
Ohira and Pao[7] have applied this technique to detecting microcracks at MnS
inclusions in steel as shown in Figure 5. Most of the microcracks formed by
the time J_I ~ 85 kJ/m² which gave a local normal stress ranging between 400
to 800 MPa. None of the above mentioned techniques have been extensively
applied to locating environmentally-induced microcracks.

As an overview, the important question is how this process occurs in
terms of time, location and continuity of the responsible microconstitu-
ent(s) which fractures. That is, a complete history of microcrack develop-

ment with time would be of great benefit. Although there have been claims that fracture reconstruction is possible using topological mapping techniques,[8] these are necessarily deformation history dependent.[22] Nevertheless, in situ reconstruction of internal flaws from inversion of data from various types of electromagnetic fields is possible in principle.[9,10]

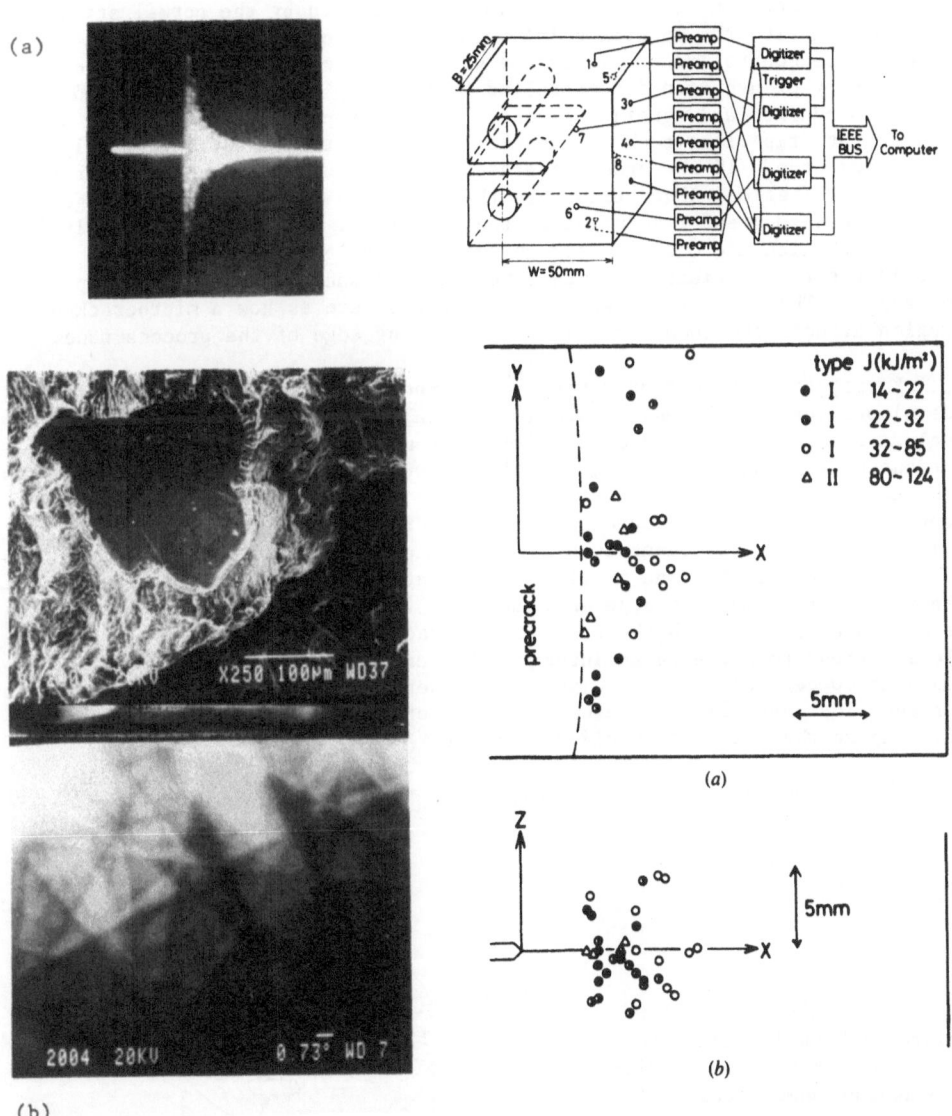

(a)

(b)

(a)

(b)

Fig. 4: Oscilloscope trace of AE event in (a). In (b), isolated cleavage facet in HSLA steel tested at -80°C (top); SACP from this grain showing little surfaxe plasticity (bottom).

Fig. 5: Experimental set-up for AE source location. Top (a) and side views (b) of microcrack sources. Courtesy of Ohira and Pao, Ref. 7, Met. Transactions A.

3. PROCESS ZONE MODELS

As alluded to above, there are a number of microstructural factors which control environmentally-induced crack growth. For internal hydrogen embrittlement of some steels, these are schematically shown in Figure 6. First, the amount of macroscopically-available hydrogen for diffusion and trapping in the dilatant zone at the crack tip is controlled by the normal stress interaction with the partial molal volume of hydrogen $\sigma_{11}\overline{V}_H$, giving a temperature-controlled step. On the way to the crack site, the apparent diffusivity of hydrogen is controlled by trapping at interfaces such as grain boundaries and carbides giving a second step. Finally, it then segregates at trap sites such as oxysulfides making this a third thermally-activated step. If the cracking process involves a fracture stress or de-cohesion criterion, then stress has the dual role of collecting hydrogen in the process zone and fracturing at the trap site. Given the relatively low stress at which microcracks form at sulfides in A533B steel, Figure 5, it should not be surprising if such sites would induce fracture with trapped hydrogen. The major point we wish to address here is how a microcracked region affects the local stress at the leading edge of the process zone.

First, consider one of several[11-13] theoretical models which have been developed for such zones. There are two major features to such zones, a volume fraction of microcracks and hence a microstructurally-dependent strength, σ_{sc}, and a size, Δ, dependent upon a quasi-static equilibrium with the applied stress. Considering the first, there will nearly always be a tendency for preferred microcracking whether this is a two-phase material with a ductile constituent such as a ceramic/metal composite or whether this is a pure polycrystalline metal. As the later is more germane, consider the hexagonal array as indicated in Figure 7(b). Note that there is a twist misorientation, ψ, with the macroscopic crack plane in Figure 7(a). For the local stress to nucleate an intergranular crack in this array, the worst case is 60 degrees or $\pi/3$ radians. Elsewhere, it is shown that if the twist orientation controls, the volume fraction of grain boundary microcracks forming at the crack tip would be given by[24]

$$f_v^c = \frac{3}{\pi} \sec^{-1}\left(K_I/K_{I_0\circ}\right)^{1/2} \quad ; \quad K_{I_0\circ} < K_I < 4K_{I_0\circ} \tag{1}$$

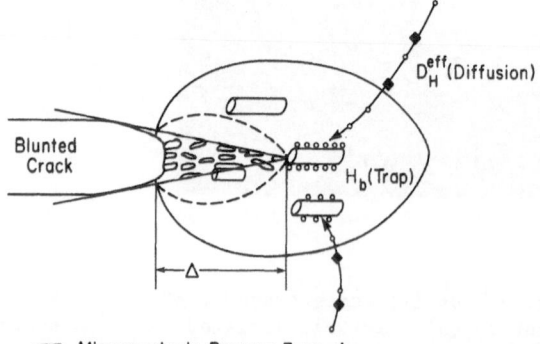

Fig. 6: Probable traps and microcrack nucleation sites for hydrogen-induced embrittlement of some steels.

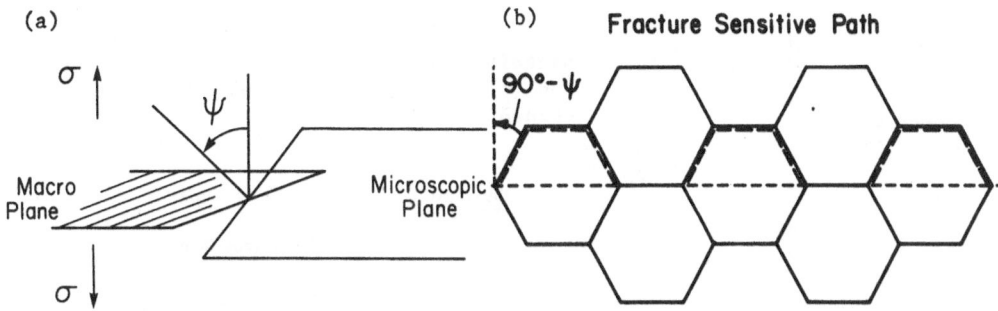

(a) (b) Fracture Sensitive Path

Fig. 7: Schematic of (a) the twist orientation of a fracture sensitive path and (b) a hexagonal array of such paths.

where K_I is the mode I applied stress intensity and K_{I_0} is the applied K_I value for a coplanar microcrack nucleating. Considering that such a micro-crack zone forms, the equilibrium of the three zones depicted in Figure 8 are given by

$$\Sigma\, F_y = \sigma\left(a/\pi\right)^{1/2} \int_{-a}^{a} f(x)dx - 2\sigma_{ys} \int_{b}^{a} f(x)dx - 2\sigma_{sc} \int_{c}^{b} f(x)dx = 0 \qquad (2)$$

Here σ_{sc} is the semi-cohesive strength in the microcracked process zone. In the absence of any such process zone, the first two terms revert to the classic Dugdale model with $f(x)$ given by $(a^2-x^2)^{-1/2}$. The semi-cohesive strength in an elastic-perfectly plastic material is given by

$$\sigma_{sc} \approx \sigma_{ys}(1 - f_v^c) \qquad (3)$$

It is seen that Eqs. (1)-(3) result in

$$K_I(\pi/c)^{1/2} = 2\sigma_{ys}\left[1 - \frac{3}{\pi} \sec^{-1}\left(K_I/K_{I_0}\right)^{1/2}\right]\cos^{-1}(c/a)$$

$$+ 2\sigma_{ys}\left[\frac{3}{\pi} \sec^{-1}\left(K_I/K_{I_0}\right)^{1/2}\right]\cos^{-1}(b/a) . \qquad (4)$$

This has been solved for a number of types of process zones noting that $b = c + \Delta$ and $a = c + \Delta + R_{P_{Io}}$ where $R_{P_{Io}}$ is the plastic zone size for the continuation of microcracking in the plastic zone at the leading edge of the process zone. If $R_{P_I} < R_{Io}$ this is equivalent to saying that no microcrack-ing can take place. As depicted in Figure 9, this model has been applied, successfully to a number of general cases as discussed elsewhere.[13] Since Eq. (4) is somewhat unwieldy, a simplification is in order based upon some plausible limits. At one extreme, assume there is almost no plasticity so that $R_{P_{Io}} \to 0$. Here both b and a would become $c + \Delta$ and with f_v^c unspecified one finds

$$K_I = 2\sigma_{ys}\left(\frac{c}{\pi}\right)^{1/2} \left[f_v^c \cos^{-1}\left(\frac{c+\Delta}{c+\Delta}\right) + (1 - f_v^c)\cos^{-1}\left(\frac{c}{c+\Delta}\right)\right] \qquad (5)$$

Since the first trigonometric term is zero and the second is accurately given by $(2\Delta/c)^{1/2}$, Eq. (5) conveniently reduces to

$$K_I = 2\sigma_{ys}\left(\frac{2\Delta}{\pi}\right)^{1/2}\left[1 - f_v^c\right] \qquad (6)$$

(a)

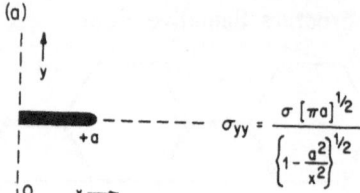

$$\sigma_{yy} = \frac{\sigma\,[\pi a]^{1/2}}{\left\{1 - \frac{a^2}{x^2}\right\}^{1/2}}$$

Fig. 8: Schematic of (a) elastic stress distribution function; (b) traction forces and (c) three-dimensional depiction with shaded ligaments.

Illustration of Process Zone Types

(b)

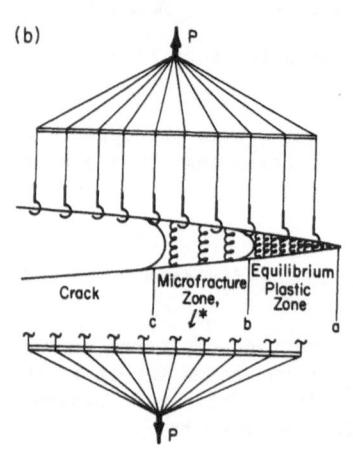

(a) Ductile

Fixed Zone Size:

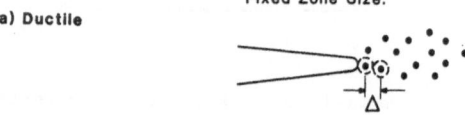

(b) Brittle

Variable Zone Size:

"Stress-Control"

(c) Semi-brittle

Variable Zone Size:

"Strain-Control"

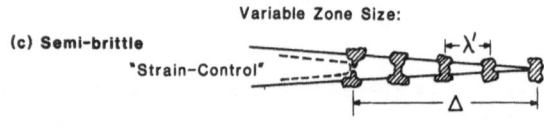

(c)

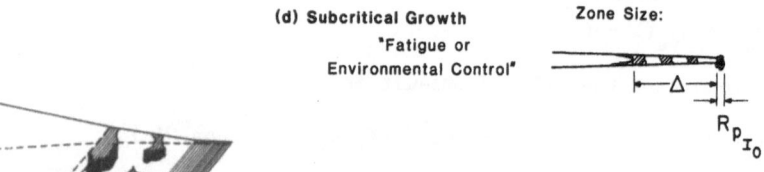

(d) Subcritical Growth

Fixed or Variable Zone Size:

"Fatigue or Environmental Control"

$$R_{p_{I_0}}$$

Fig. 9: Process zone, Δ, size variations as a function of different controlling micro-structural units.

This says that for equilibrium the higher the yield strength or the larger the process zone the larger the applied K_I field may be. Conversely, the greater the microcrack density within the process zone, the smaller K_I would be for equilibrium. At another reasonable extreme, assume that there is a large contiguous plastic zone such that $R_{p_{Io}} \simeq \Delta$. Using this and the good approximations that $\cos^{-1}[(c+\Delta)/(c+2\Delta)]$ and $\cos^{-1}[(c/(c+2\Delta)]$ are $(2\Delta/c)^{1/2}$ and $(4\Delta/c)^{1/2}$ respectively, Eq. (4) becomes

$$K_I = 2\sqrt{2}\ \sigma_{ys}\left(\frac{2\Delta}{\pi}\right)^{1/2}\left[1 + \left(\frac{1}{\sqrt{2}} - 1\right)f_v^c\right] \tag{7}$$

Note that the general form is the same as Eq. (6) so that these become

$$0 < R_{p_{Io}} < \Delta$$

$$K_I = \alpha\sigma_{ys}\left(\frac{2\Delta}{\pi}\right)^{1/2}\left[1 - \beta f_v^c\right]\ ;\qquad 2 < \alpha < 2\sqrt{2} \tag{8}$$

$$1 > \beta > 0.293$$

How does such a process zone model apply to environmentally assisted cracking? If the process zone changes with time, then the local driving force in terms of a K_I or a local stress also changes.

In a quasi-static sense, the stage II crack growth rate driving force resulting in the lower two curves of Figure 1 must be fluctuating. At a given moment for the 4340 steel tempered at 450°C, the process zone may be relatively large and the volume fraction of intergranular facets within it small, making the tolerable K_I at quasi-static equilibrium very high. With time, however, f_v^c would increase and the tolerable K_I according to Eq. (8) would decrease. This would be reflected in a larger crack velocity. To fit this into a kinetic model, one that has been developed for internal hydrogen embrittlement of high strength steels is utilized. For stage II crack growth rate, $(da/dt)_{II}$, a previous derivation has shown[25]

$$\left(\frac{da}{dt}\right)_{II} \simeq \frac{4D^{eff}}{X_{cr}} \left\{ \frac{\alpha_H^* C_o \exp[(H_b+\beta)/RT] + (1 + (\pi/2))\sigma_{ys} - \sigma_{fo}^*}{\alpha^* C_o \exp[(H_b+\beta)/RT]} \right\}^2 \tag{9}$$

Here, C_o is the initial uniform concentration of hydrogen, H_b is the binding energy at the fracture nucleation site, β is $((1/3)\sigma_{ii}\overline{V}_H)$, σ_{fo}^* is the macroscopic[†] fracture stress in the absence of hydrogen, D^{eff} is the trapped diffusivity of hydrogen in Figure 6 and X_{cr} is the critical distance the crack advances. Note that this is derived from Fick's second law of diffusion with an incubation time, t_i, being required to collect hydrogen at the binding site such that the fracture stress is degraded to $\sigma_{fo}^* - \alpha C_x'$. The steady state crack growth rate is then a repeat of X_{cr}/t_i. In this context, if the macroscopic criterion, σ_{fo}^*, in the absence of hydrogen changes due to the evolving character of the process zone, then $(da/dt)_{II}$ will fluctuate.

First, it is assumed that the process zone may allow a local increase or decrease in the stress required for intergranular separation. Consider the cracking at some N+1 step after the N_{th} step. If the stress scales with stress intensity then Eq. (8), with $\alpha\sigma_{ys}$ a constant, gives

$$\sigma_{f_{N+1}}^* = \sigma_{f_N}^* \left(\frac{\Delta_{N+1}}{\Delta_N}\right)^{1/2} \left[\frac{1 - \beta f_{v_{N+1}}^c}{1 - \beta f_{v_N}^c}\right] \tag{10}$$

The picture in Figure 10 is then invoked. Assume first that a process zone

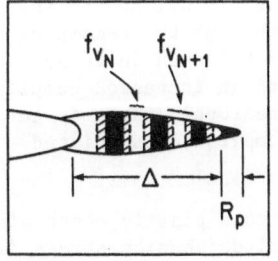

(a) $\Delta \sim$ constant

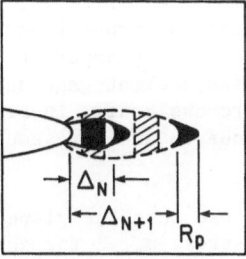

(b) $f_v \sim$ constant

Fig. 10: Schematic of process zone variations due to (a) an increase in microcrack volume fraction; (b) an increase in the process zone size.

has formed as in Figure 10(a) due to the most favorably oriented grains cracking. With $\Delta \sim$ constant, the local stress required according to Eq. (10) would <u>decrease</u> as the less favorably oriented grains crack with time. That is, as more hydrogen collects in the process zone, $f_v^c{}_{N+1}$ increases and there would be a reduced area of ligaments within the process zone holding on. This reduced local fracture stress requirement would cause an increase in $(da/dt)_{II}$ according to Eq. (9). As the area of the ligaments continues to get smaller the crack accelerates until it jumps through the remaining ligaments to the end of the process zone. At this point, there is an assumed straight crack front with no microcracks and no ligaments. In a very short time, however, hydrogen collects and traps resulting in fracture of the most favorably oriented grains. The first ones to fracture are in the highest stress region closer to the crack tip. As further out facets fracture, the process zone increases in size. This is depicted in Figure 10(b) and is modelled assuming that $f_v^c \sim$ constant during the stage where the process zone size increases. As the zone increases, the macroscopic fracture stress required would increase according to Eq. (10). This then produces a decreased crack growth rate from Eq. (9). With increasing Δ, the local stress R_p at the advancing process zone becomes so small that no further microcracking at the leading edge takes place or it evolves so slowly that further cracking is within the process zone. At this point, the mechanism depicted in Figure 10(a) takes over and one full fluctuation takes place. During stage II growth, the fluctuations within this "steady-state" regime are thus given by:

<u>Increasing da/dt:</u> $\Delta \sim$ constant
f_v^c increases
crack increase $\propto f_v^c$

$$\sigma_{f_{N+1}}^* = \sigma_{f_N}^* \left[\frac{1 - \beta f_{v_{N+1}}^c}{1 - \beta f_{v_N}^c} \right] ; \quad \beta = 0.6 \tag{10a}$$

<u>Decreasing da/dt:</u> $f_v^c \sim$ constant
Δ increases
crack increase $\propto \Delta$

$$\sigma_{f_{N+1}}^* = \sigma_{f_N}^* \left[\frac{\Delta_{N+1}}{\Delta_N} \right]^{1/2} ; \quad \Delta \text{ increases until } da/dt \to 0 \tag{10b}$$

Note that in both of these regimes the "crack increase" is all due to the changing process zone except where the front breaks through the remaining ligaments. In Figure 10(a), there is an apparent crack length increase detected at the crack-opening displacement gage due to an increased compliance associated with a weaker process zone. In Figure 10(b), there is an apparent crack length increase due to an increased compliance associated with a larger process zone.

---

† This is macroscopic in the sense that the plane strain plastic constraint raises σ_{yy} to this fracture stres, σ_{fo}^*. The actual decohesion stress is much higher as elevated by microscopic stress concentrations such as dislocation pile-ups.

4. APPLICATION TO Ti-30Mo DATA

One of the novel techniques, selected area channeling patterns, and process zone modeling was applied to internal hydrogen embrittlement of a body-centered-cubic Ti-30MMo alloy.[21] For 500 wt. ppm hydrogen and a test temperature of 260 K, the threshold stress intensity was 37.5 MPa-m$^{1/2}$. With a yield strength of 980 MPa corresponding to this test temperature and hydrogen content, the calculated process zone size from Eq. (8) is 1200 µm. This is reasonably consistent with the microcracked zone at the tip of a sectioned crack tip in Figure 11.

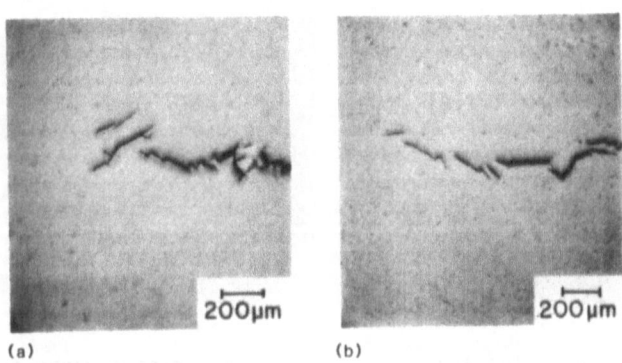

(a)   (b)

Fig. 11: Microscopic evidence of a microbranched region in Ti-30Mo. (a) 0.74 cm. from the surface of a 1.8 cm thick sample. (b) 0.46 cm from the surfae in a 1.8 cm thick sample.

To verify this, electron channeling was accomplished on fracture surfaces after the fact. As shown in Figure 12(a), multiple grain facets have cleaved in a sample containing 3200 ppm H and tested at 300 K. Similar features were found for samples containing 500 ppm H. Channeling was not possible on the as cleaved surface but after removing 0.4 µm from the surface, a sharp ECP indicating a low level of plastic strain resulted as shown in Figure 12(b). For another location, the strain was slightly higher as shown in Figure 12(c). A similar procedure for intergranular facets demonstrated that an order of magnitude more surface had to be chemically removed to achieve the same clarity of ECP. As shown in Figures 13(a) and (b), the growth was step-wise down the intergranular facet. Because of the curved nature of these lines, they are not emerging slip traces. The ECP in Figures 13(c) and (d) are still degraded after 0.4 microns are removed but approach the clarity of Figures 12(c) and (d) after 4.4 µm has been removed. Given that strain falls off as approximately 1/r from the fracture surface, it became clear that at the same distance from the original crack tip the strains were nearly an order of magnitude greater for the intergranular facets versus the cleavage facets. From the surface strains, it was determined that such facets were initially fracturing on the order of 260 to 600 µm from the original crack tip. Of course as the process zone grows, these distances could easily double. From such evaluations, the fracture sequence in the process zone appears to be that schematically shown in Figure 14.

430

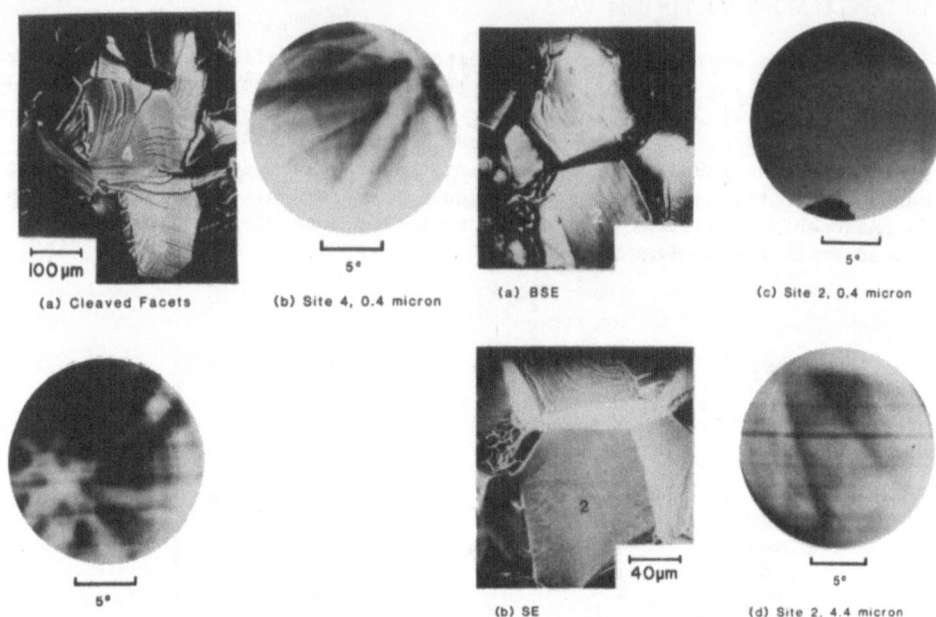

(a) Cleaved Facets   (b) Site 4, 0.4 micron   (a) BSE   (c) Site 2, 0.4 micron

100 µm   5°

(c) Site 5, 0.9 micron   (b) SE   (d) Site 2, 4.4 micron

40µm

Fig. 12: Fractography (a) and SACP's in (b), (c) at depths below cleavage facets of Ti-30Mo, 3200 ppm H at 300 K.

Fig. 13: Fractography (a), (b) and SACP's in (c), (d) at depths below intergranular facets of Ti-30Mo, 500 ppm H at 300 K.

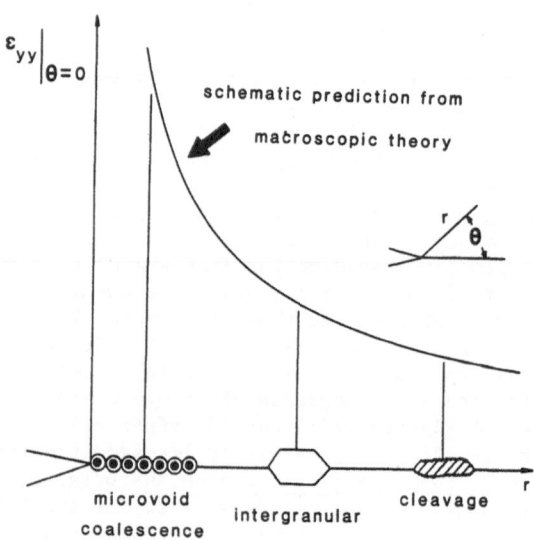

$\varepsilon_{yy}\big|_{\theta=0}$

schematic prediction from macroscopic theory

r θ

microvoid coalescence   intergranular   cleavage

r

Fig. 14: Strain distribution from macroscopic theory and associated distances from the crack tip at which microscopic fracture modes occur.

As established in independent fracture stress tests using 3 pt. notch-bend samples, $\sigma_{f_o}^* = 1900$ MPa. Also measured diffusivities allowed a value for $4D^{eff}/X_{cr} \approx 2 \times 10^{-7}$ s^{-1} to be established for 260 K. With these values it was possible to compare a calculated da/dt to a measured one. To start the calculation, an initial $\Delta \simeq 1200$ µm with f_v^c equal to 0.5 was allowed. The crack was then allowed to "grow" by having ligaments crack which increased f_v^c. This increased compliance was translated into an effective crack length. For example, if the original crack length was 10,000 µm and the process zone was 1200 µm with a f_v^c of 0.5, this was taken as an effective crack length of 10,600 µm. If the f_v^c was increased to 0.7, this was taken as an effective crack length of 10,840 µm. In this way, a new K_I could be determined from each effective a/W. The theoretical da/dt was determined from Eq. (9) based upon each σ_f^* as f_v^c was allowed to increase. Once f_v^c approached unity, within 10 percent, this process was terminated. The next phase, corresponding to Figure 10(b) was to allow Δ to grow at a constant f_v^c of 0.5. The only assumption here was that the initial value of Δ was 600 µm. The process zone was now allowed to increase until da/dt → 0. This occurred at a higher $K_I = 43.1$ MPa-m$^{1/2}$ and it is notable that the required process zone is now 2100 µm, nearly twice as large as the initial value at threshold. This whole process was repeated several times, the results being compared to the actual data in Figure 15. Although not totally reproducible, the growth rate range and approximate stress intensity period of the fluctuations is realistic. Still, there are a sufficient number of ways that the crack could be advancing in such a complex process that this can only be represented as one possibility from a large set.

STAGE II KINETICS

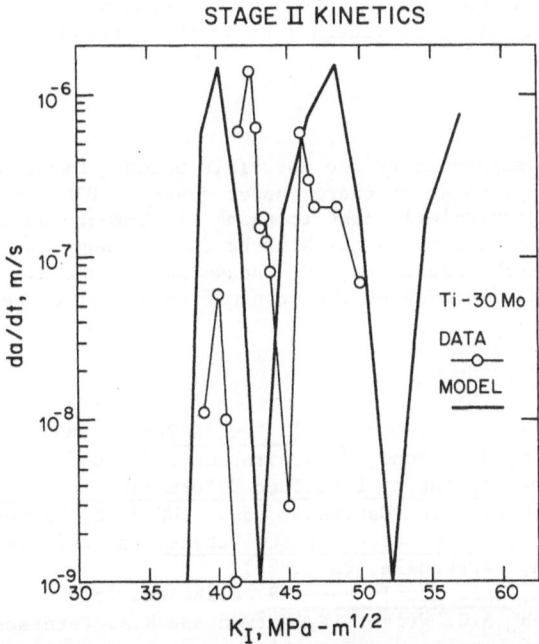

Fig. 15: Predicted versus observed hydrogen-induced crack growth rates in Ti-30Mo.

432

The important point to be emphasized here is that this material cracks by several microscopic fracture modes even though it is single phase. The extent of the process zone associated with the coarse microstructure is large. Because of this large Δ ameliorating the local stress field, cracks often stop. On a very gross scale this has been known to happen in stress corrosion cracking of polymers due to solvent-crazing where centimeter long crazes have been arrested.[26] In principal, such understanding may lead to better design of single and multiphase microstructures which posses better resistance to environmentally-induced cracking.

5. CONCLUSIONS

1. Several old as well as new and novel techniques are being applied to study the details of environmentally-assisted crack growth processes. To mention a few, there are *in situ* HVEM, SEM and acoustic emission techniques as well as post-fracture analysis using electron channeling.

2. Such studies have led to the formtion of process zone concepts which may be applied to both equilibrium threshold as well as quasi-static analysis of slow crack growth.

3. Application of one such process zone model has been made to the stage II crack velocities of hydrogen-induced cracking in a BCC Ti-30Mo alloy. This reasonably reproduces the observed orders of magnitude fluctuations of crack growth rate in this "steady-state" regime.

6. ACKNOWLEDGEMENTS

This work was supported by the Materials Science Division, Basic Energy Sciences of the Department of energy under contract DOE/DE-AC02-79ER 10433. The author would particularly like to thank his former and current Graduate Students, Dr. Ken Peterson and Mr. Xing-Fu Chen, along with his Post-Doctoral and research associates, Dr. Chong-Soo Lee and Mietek Kaczorowski and Ms. Tova Livne, for much of the information contained herein.

7. REFERENCES

1. S. Kobayashi and S. M. Ohr, Phil. Mag. A42(6) (1980) pp. 763-772.
2. I.M. Robertson, G.M. Bond, H.K. Birnbaum and H.G.F. Wilsdorf, in Fracture: Measurement of Localized Deformation by Novel Techniques, W.W. Gerberich and D.L. Davidson, eds., TMS-AIME, Warrendale, PA (1985).
3. J.C. Bilello, in Mech. Prop. of BCC Metals, pp. 207-216, M. Meshii, ed., TMS-AIME, Warrendale, PA (1982).
4. D.L. Davidson, Int. Metals Rev. 29, (1984) p. 75.
5. W.W. Gerberich, A.G. Wright, E. Kurman and K.A. Peterson, in Fracture: Measurement of Localized Deformation by Novel Techniques, pp. 59-74, W.W. Gerberich and D.L. Davidson, eds., TMS-AIME, Warrendale, PA (1985).
6. H.N.G. Wadley, C.B. Scruby and G. Shrimpton, Acta Met. 29 (1981) pp. 399-414.
7. T. Ohira and Y.-H. Pao, Met. Trans. A 17A (1986) pp. 843-852.

8. T. Kobayashi and J.H. Giovanola, "Fractographic and Topographic Analyses of MMCT Specimens 7 and 8," EPRI Project No. RP 2455-7, SRI International, Menlo Park, CA, December 1985.

9. J.H. Rose, "Reconstruction of the Electromagnetic Field From Scattering Data," in Rev. of Progress in Quant. NDE, Univ. of Calif. La Jolla, August (1986), Center for NDE, Iowa State University.

10. J.C. Moulder, P.J. Shull and T.E. Capobianco, "Uniform Field Eddy Current Probes: Experiments and Inverison for Realistic Flaws," ibid., National Bureau of Standards, Boulder.

11. W.W. Gerberich and N. Moody, "A Review of Fatigue Fracture Topology Effects on Threshold and Growth Mechanisms," pp. 292-341 in Fatigue Mechanisms, J. Fong, ed., ASTM, Philadelphia (1979).

12. E. Smith, Res. Mechanica 4 (1982) pp. 151-157.

13. W.W. Gerberich, "Interaction of Microstructure and Mechanism in Defining K_{Ic}, K_{Iscc} or ΔK_{th} Values," in Fracture and Interactions of Microstructure, Mechanisms and Mechanics, pp. 49-74, J.M. Wells and J.D. Landes, eds., TMS-AIME, Warrendale, PA (1985).

14. H. Vehoff and P. Neumann, in Hydrogen Degradation of Ferrous Alloys, pp. 686-711, R.A. Oriani, J.P. Hirth, M. Smialowski, eds., Noyes Publications, Park Ridge, NJ (1985).

15. D.L. Davidson, in Fatigue Mechanisms, pp. 254-275, J.T. Fong, ed., ASTM STP 675, Am. Soc. for Test. and Materials (1979).

16. W.W. Gerberich and C.E. Hartbower, in Fundamental Aspects of Stress Corrosion Cracking, pp. 420-438, Nat. Assoc. Corros. Engrs., Houston, TX (1969).

17. P. Padmanabhan, N. Suriyayothin and W.E. Wood, Met. Trans. 14A (1983) pp. 2357-2362.

18. R.A. Oriani, private communication, University of Minnesota, 1984.

19. R.H. Jones, P.H. Hutton, M.A. Friesel and S.M. Wolf, "Observation and Identification of Crack Growth Modes in Reactor Steels Using Acoustic Emission," Workshop Proceedings, Pacific Northwest Laboratory Report SA-12656, Richland, WA, December (1984).

20. N.R. Moody and S. Robinson, private communication, Sandia National Laboratories, Livermore, CA (1986).

21. K.A. Peterson, Ph.D. Thesis, University of Minnesota, Minneapolis (1983).

22. C.S. Lee, T. Livne and W.W. Gerberich, "The Acoustic Emission Measurement of Cleavage Initiation Near the Ductile Brittle Transition Temperature in Steel," submitted for publication (1986).

23. Discussion, this conference.

24. W.W. Gerberich and A.G. Wright, in Environmental Degradation of Engineering materials in Hydrogen, M.R. Louthan, Jr., R.P. McNitt and R.D. Sisson, Jr., eds., pp. 186-206, Virginia Tech Printing, Blacksburg, VA (1981).

25. W.W. Gerberich, T. Livne and X. Chen, "A Transient Model for Subcritical Cracking in BCC Alloys," in Modeling Environmental Effects on Crack Growth Processes, R.H. Jones and W.W. Gerberich, eds., TMS-AIME, Warrendale, PA (1986).

26. S.J. Israel, C.S. Kantamneni and W.W. Gerberich, " A Dugdale-Barenblatt Equilibrium Model for Crazes in Glassy Polymers," in ICM 3, Vol. 3, K.J. Miller and R.F. Smith, eds., pp. 393-402, Pergamon Press, NY (1979).

DISCUSSION

Comment by J. F. Knott:

Two points concerning the process zone:
1. The threshold stress intensity for (macroscopic) <u>crack growth</u>, K_{ISCC}, may be used as an engineering design parameter. Do I understand you to say that microcracking in the process zone may occur well below K_{ISCC}, as defined in engineering terms? (Following your model, first microcracking might occur at $K_{ISCC}/\sec^2 60° = K_{ISCC}/4$). Is this stable microcracking, or would a safety factor be put on K_{ISCC} for engineering design?
2. Above the (macroscopic) K_{ISCC}, i.e., when the crack is growing as a function of K, does the process zone remain constant in size or does it vary with K?

Reply:

1. In fact this model does say that microcracking of a few facets at $K_{ISCC}/4$ would occur but immediately arrest. Also, this would be for an ideal geometrically-arranged set of ideal boundaries. Unless one had large regions of the microstructure such that the preferred orientations were coplanar with a macroscopic crack, such microcracks would be stable. Thus, no safety factor would be needed for engineering design with K_{ISCC} for a randomly oriented microstructure. In severely oriented microstructures, engineers inherently put a safety factor on stress corrosion cracking by measuring K_{ISCC} in the most susceptible orientation and use that as design parameters.
2. For $K_I > K_{ISCC}$, this would be a regime where the process zone would increase with K_I to a "maximum" value in stage II. It is then assumed for the modeling that the process zone fluctuates between this maximum and minimum as dictated by the alternate microcracking and total crack front advance processes. In reality, even the maximum size of this process zone should increase with K_I for quasi-static equilibrium; but for the calculation we have tried to keep it simple.

Comment by T. Watanabe:

You have pointed out the importance of twist component of the boundary misorientation, only from the geometrical consideration of crack propagation across the boundary. How much do you take into account the effectiveness of the boundary as barrier to crack propagation, which may itself be strongly dependent on boundary structure?

Reply:

We recognize that only part of a very complicated problem has been attacked here, that of the geometrical effect on local stress. The problem you suggest brings in the additional feature that there are at least two probability distributions for a set of grains in the advance of a growing crack, that distribution associated with boundary macro-crack misorientation and that distribution associated with intrinsic grain boundary strength variation. Studies involving both distributions would be very useful and perhaps careful SACP (selected area channeling) measurements of opposing g.b. facets might be a way to sort this out.

Comment by C. J. Altstetter:

When stage II crack propagation has an "alpine" (peak/valley, start/ stop) character have you made any load adjustments; that is, have you found it necessary to increment the load to restart the crack motion?

Reply:

The data which were presented during the talk were generated under sustained load. In fact, in several cases, particularly for thinner specimens where K_{th} and hence the process zone is greater, we did increment the load to restart crack motion. The result was that in several cases da/dt actually varied by four orders of magnitude during stage II growth. In those cases where the crack arrested, several long hold times were attempted, one being 48 hours, to see if the crack would restart. In the 8.2 cm thick samples, only load incrementing could restart the crack whereas in the thicker sampler, time was often sufficient.

Comment by J. P. Fidelle:

Generally, and it is true for hydrogen embrittlement (HE), intergranular fracture corresponds to a more severe embrittlement than cleavage or quasi-cleavage (QC).
e.g.; in a given hydrogenated environment, fracture of a Q + T steel drifts from QC to IG when the steel is made more HE sensitive by tempering at lower temperatures.
- or the fracture of a given Q + T steel can shift from QC to TG as its H content is increased and/or if the environment aggressivity grows.
- in case of decreasing K fracture mechanics specimens, fracture at first generally displays dimples then QC and finally IG or vice versa in case of increasing K specimens.
Then, could you explain any further why between cleavage and MVC areas you find IG areas even if the intermediary zone does not seem to be completely intergranular?
P.S. If the question is already answered in the full manuscript, you can leave it.

Reply:

In both the 4340 steels and the Ti-30 Mo alloys, we found several modes of failure that were interspersed. In the steels, at the high strength levels used, the intergranular was more favored over cleavage. Thus, the transgranular regions tended to fracture after much of the intergranular process occurred. Depending on test temperature, the transgranular process was either quasi-cleavage of martensite lath (favored at high temperatures) or microvoid coalescence (favored at low temperature). In the Ti-30 Mo, the cleavage of whole grains was favored over the intergranular process. The reason that these secondary fracture modes are found interspersed is because of the local fracture criterion (e.g., cleavage plane orientation) is not satisfied everywhere while another mode is satisfied.

Comment by P. Haasen:

Is there a quantitative way to evaluate channeling patterns for strain (besides "pattern resolution")?

One is reminded often of early x-ray line broadening studies to measure dislocation densities which were superseded later by more quantitative methods.

Reply:

We have been reasonably successful in determining a correlation between plastic strain and hence dislocation density and line with degradation for deformation observed in single grains (see Scripta Met., 1986). The problem arises due to inhomogeneous slip or slip plane anisotropy which provides a greater scatter band to this relationship. Dave Davidson, of Southwest Research Institute, and I have a joint NSF contract where we are considering various Fourier transform and/or signal differentiation techniques. These would take all of the information in a channelling pattern and compare it to observed plastic strains and in this way narrow the scatter band.

Comment by A. W. Thompson:

Can you indicate to what extent physical observations of process zone sizes and crack volume fractions agree with the numbers used in the modeling? Have there been many such physical observations?

Reply:

In the present modeling of the slow crack growth of the Ti-30 Mo, we inferred the volume fractions of microcracks in the process zone from compliance changes during increasing da/dt. In other cases such as in fatigue (Moody, Sandia Livermore) cleavage, direct observations of the microcrack volume fraction is probably ± 30% volume fraction off of the predicted; but at this stage both model and measurements are relatively crude. As to the size of the process zone, this can be quantitatively inferred from the electron channeling measurements of plastic strain and this agrees within about a factor of two.

Comment by H. J. Engell:

In your kinetic modelling, you introduced as a linear factor the effective diffusivity of hydrogen. Experiments we did with the same steel at the same strength by quenching and tempering and by cold deformation did not give evidence of a direct correlation between diffusivity and/or trapping of hydrogen with crack growth rate.

Reply:

In our kinetic model, there are really three thermally activated processes, all of which contribute to the crack growth rate. One of these is effective diffusivity which can be affected by one set of traps, another is the binding energy at the crack nucleation site which can be affected by a different type of trap, and the third is the local stress state. Even if the strength level is the same, the crack nucleation site could be different because of a deformation or segregated/weakened interface and the process zone which formed could change the local driving force for continued growth. Thus, the trapped diffusivity only addresses part of the process which controls kinetics.

IN SITU TEM STUDIES OF CRACK TIP DEFORMATION IN MOLYBDENUM*

C. G. Park
Solid State Division, Oak Ridge National Laboratory
Oak Ridge, TN 37831, U.S.A.

S. M. Ohr
Department of Materials Science and Engineering
SUNY at Stony Brook, Stony Brook, NY 11794, U.S.A.

INTRODUCTION

The dislocation model of plasticity ahead of a propagating crack was first treated by Bilby, Cottrell, and Swinden (BCS) (1). They considered a plastic zone consisting of an inverse pileup of dislocations with an infinite density of dislocations at the immediate tip of a shear crack. Physically, this density is not possible since dislocations cannot be located less than a unit Burgers vector apart (2). A series of direct observations by TEM (3-5) have shown that crack propagation in many materials occurs through the emission of dislocations from the crack tip. These dislocations are driven rapidly out of the crack tip area and pile up in a plastic zone, creating a dislocation-free zone (DFZ) between the crack and the plastic zone. This crack geometry has been treated theoretically by extending the BCS model (6-8). The results show that the DFZ is a manifestation of the difficulty associated with the emission of dislocations from the crack tip, and this can be expressed as a critical stress intensity factor for dislocation emission "ke". ke is zero in the BCS model. The DFZ model has since been shown to be very useful in understanding the ductile vs. brittle behavior of various materials.

Although the mode of crack propagation has been shown to be closely related to the behavior of crack tip dislocations and the loading conditions, the details of their relationship are not well understood at present. This paper describes the effects of the loading condition on the distribution of dislocations, the DFZ, and the crack tip slip systems observed during in situ TEM deformation of molybdenum. The purpose of this study was to find parameters that are responsible for determining the geometry of crack tip deformation.

*Research sponsored by the Division of Materials Sciences, U.S. Department of Energy.

EXPERIMENTAL RESULTS

Direct observation of crack tip deformation was made in an electron microscope operated at an accelerating voltage between 120 and 200 KV. Depending on the specimen orientation and loading geometry, all three modes of crack tip deformation were observed. Figure 1 is an electron micrograph showing a typical plastic zone ahead of an antiplane shear crack of mode III type. The plastic zone consists of a linear array of approximately 130 dislocations on the (101) slip plane which is nearly coplanar with the crack. In the micrograph, the presence of a DFZ (about 2.5 µm long) can clearly be seen between the crack and the plastic zone. It was found, by contrast analysis, that the dislocations were very close to screw type with the Burgers vector $a/2[11\bar{1}]$. The slip system indicated represents the system of maximum resolved shear stress (4). It was noted that the distribution of dislocations in the plastic zone was not in the form of an inverse pileup, as was predicted by the BCS theory. That is, the dislocation density was not maximum at the crack tip; instead the maximum was located at a distance approximately 7 µm away from the crack tip (5).

Figure 2 shows the distribution of crack-tip-generated dislocations under the mode II loading condition. In this case, approximately 90 edge dislocations were emitted from the crack tip. These dislocations moved quickly along the slip plane and were piled up against a grain boundary, and as a consequence, the crack flanks were displaced along the direction parallel to that of crack propagation. In this specimen, a long DFZ, approximately 10 µm in length, was observed. It was noted, however, that the motion of the dislocations was impeded by the presence of the grain boundary, and thus the length of the DFZ observed could be shorter than the equilibrium value. We have also observed crack tip deformation under the mode I loading condition. In this loading geometry, crack tip deformation usually was caused by the emission of edge dislocations along slip planes that are inclined to the plane of the crack. As a consequence, the crack, that was originally sharp, became either blunted or wedge-shaped due to the slide-off process that occurred at the crack tip. The details of these observations are reported elsewhere (3-5).

Table I summarizes the observed configurations of crack tip deformation including the lengths of the crack, the DFZ, and the plastic zone and the number of dislocations (N) in the plastic zone. In all of the cases studied, the absence of dislocations in the DFZ was confirmed by imaging the area through 4-5 different diffraction conditions and hence it was possible to exclude a casual invisibility caused by local bending at the crack tip (9). Because of the thin foil geometry used in this experiment, these data cannot be compared directly with the predictions of the dislocation theory of fracture (3). However, an analysis of the present data indicates that the local stress intensity factor estimated is in surprisingly good agreement with the critical

stress intensity factor for dislocation emission ke predicted
by the theory. The data also indicate that ke for mode I is
greater than that for mode II and this quantity, in turn, is
greater than that for mode III. This result is again in good
agreement with the predictions of the theory.

DISCUSSION
The results of this study have demonstrated that cracks in
molybdenum propagate primarily through the emission of dislocations from the crack tip and the behavior of these dislocations depends on many parameters. The presence of the DFZ was
confirmed for all three modes of crack propagation. Throughout the experiments, dislocations were observed to originate
from the crack tip under an applied stress and these dislocations were strongly repelled from the crack tip.
The present results are in good agreement with the dislocation model of fracture (6-8). The length of the DFZ depends
on the loading condition as shown in Table I. It is shortest
for mode III cracks, indicating that crack tip deformation
under the mode III loading condition is the easiest between
three possible modes. The length of the DFZ decreases with
the increasing amount of crack tip deformation, which is
indicated by the number of dislocations N in the plastic zone.
The distribution of dislocations in the plastic zone is not in
the form of an inverse pileup. It is close to the distribution predicted by the DFZ theory, which shows a maximum near
the beginning of the plastic zone some distance away from the
crack tip. The choice of the crack tip slip system is
dictated primarily by two factors, i.e. the dislocation
emission condition and the crack tip Schmid factor. In all of
the cases studied, the deformation mode activated had the
maximum value of the ratio between the applied stress intensity factor and the critical stress intensity factor for
dislocation emission. Within each mode of deformation, the
slip system having the maximum crack tip Schmid factor is the
one activated at the tip of moving cracks.
It has been suggested (9) that the observation of the DFZ
may depend on the thickness of the specimens used and the DFZ
may be a property only of thin foils. In order to test this
possibility, in situ fracture experiments were carried out in
thick single crystal specimens of iron and copper using a 1 MV
high voltage electron microscope located at Nagoya University
(10). These experiments demonstrated the presence of the DFZ
in thick specimens, thereby confirming the earlier results
obtained from conventional electron microscopes. A recent
study by Chia and Burns (2) in LiF single crystals during mode
I deformation has shown that a DFZ is also present in bulk
specimens under various loading conditions.

440

REFERENCES

1. B.A. Bilby, A.H. Cottrell, and K.H. Swinden, Proc. R. Soc. London Ser. A, 272, 304 (1963).
2. K.Y. Chia and S.J. Burns, Scripta Met. 18, 467 (1984).
3. S.M. Ohr, Mater. Sci. Eng. 72, 1 (1985) and the references contained in this paper.
4. C.G. Park and S.M. Ohr, p. 105 in Fracture: Measurement of Localized Deformation by Novel Techniques, W.W. Gerberich and D.L. Davidson eds., AIME, New York, 1985.
5. C.G. Park and S.M. Ohr, to be published.
6. S.J. Chang and S.M. Ohr, J. Appl. Phys. 52, 7174 (1981).
7. S.M. Ohr and S.J. Chang, J. Appl. Phys. 53, 5645 (1982).
8. R.H. Zhao, S.H. Dai, and J.C.M. Li, Int. J. Frac. 29, 3 (1985).
9. I.M. Robertson, G.M. Bond, and H.K. Birnbaum, p. 158 in Second Int. Conf. Fundamentals of Fracture, ORNL/TM-9783, Oak Ridge, TN (1985).
10. H. Saka, Y. Zhu, T. Imura, and S.M. Ohr, to be published.

Table I. Geometry of crack tip deformation observed in molybdenum

Mode	Crack Length	DFZ	Plastic Zone	N
I	0.6 μm	8.8 μm	16 μm	420
	0.1	9.0	8	160
	0.3	5.5	8	250
II	0.9	8.8	12	100
	0.8	2.3	17	300
	0.4	12.0	8	120
	0.4	9.9	7	90
III	2.1	1.4	12	350
	0.6	3.6	15	400
	1.3	2.7	28	300
	0.1	2.5	21	130
	1.6	3.0	15	250
	2.8	2.5	20	300

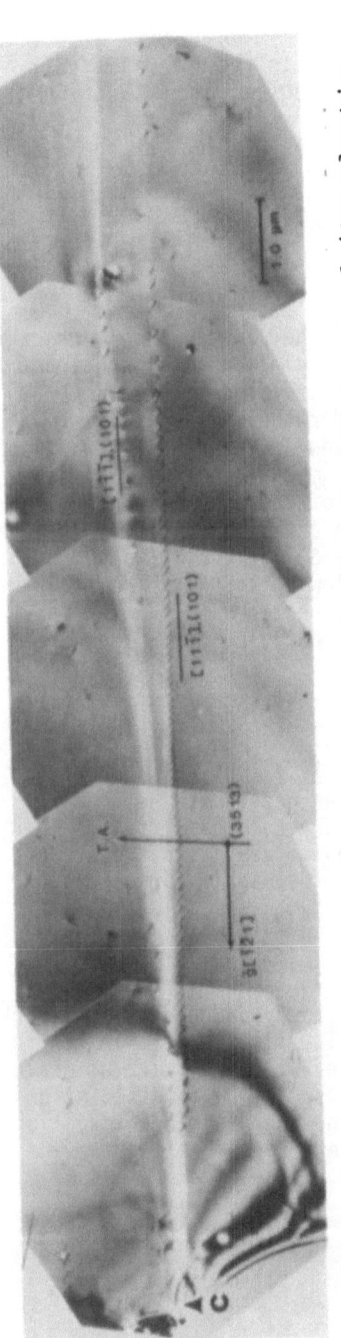

Fig.1 Electron micrograph showing a shear crack of mode III type and its plastic zone in molybdenum. A dislocation-free zone (DFZ) can be seen between the crack tip and the plastic zone. The plastic zone consists of linear arrays of screw dislocations.

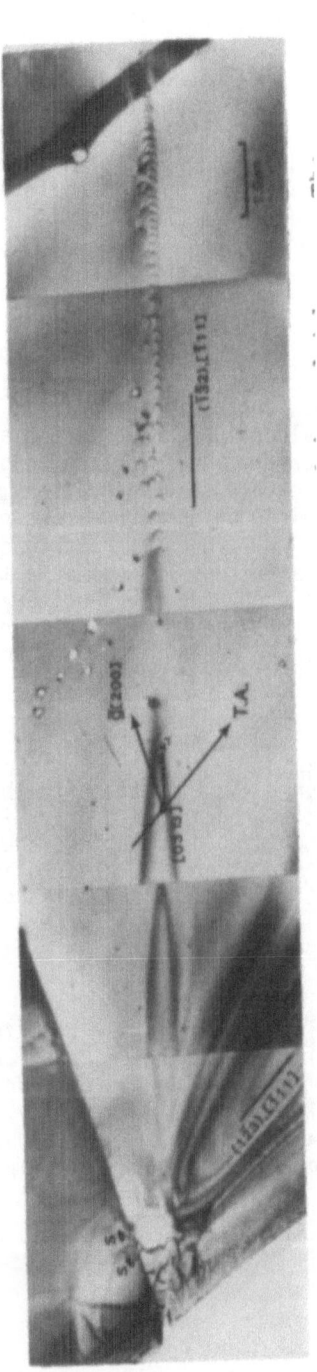

Fig.2 A mode II shear crack and the plastic zone observed in molybdenum. The length of the DFZ is 10 μm and the edge dislocations in the plastic zone are piled up against a grain boundary.

DISCUSSION

Comment by H. K. Birnbaum:

As you know, we have made similar observations in the TEM in mode I, but we do not observe any dislocation free zones in bcc Fe, MgO, or in a number of fcc metals. We observe dislocations very close to the crack tips in all cases. Could the difference in our observations be due to the fact that we always propagate our cracks for appreciable distances while you generally observe short cracks? Is the concept of a DFZ valid for a crack which propagates into a developed plastic zone?

Reply:

It is possible that the size of the DFZ could be reduced when a crack approaches a well developed plastic zone because dislocations cannot get away from the tangles in the plastic zone. We have also seen, however, that some dislocation tangles are dissolved under the stress field of an approaching crack. An important point to be made is that the interaction between a crack and the dislocations emitted from the crack is repulsive so that the dislocations are always pushed out of the crack tip area. When a DFZ is not present, the shielding of the crack tip is complete and hence the local stress intensity factor k is zero, i.e., all crack tip activity must stop. We see, however, that following crack tip deformation, cracks continue to propagate and emit dislocations indicating that the local k is not zero. This, we believe, is partly owing to the presence of DFZ at the crack tip.

Comment by H. Mughrabi:

You mentioned the limitation to thin foils. It is my impression that in your first micrograph of a crack in copper the nature of the planar array of dissociated dislocations ahead of the crack tip is itself a thin film effect, since normally cross slip occurs quite easily in copper and the dissociation is usually not recognizable at the magnification of your micrograph.

Reply:

I must point out that these stacking faults observed ahead of the crack tip are overlapping and hence represent a twin lamella. A similar configuration was also found in nickel. We have looked at the effects of crack tip stress fields, as well as purity, on the dissociation of dislocations. We believe that the stress-induced nucleation and the elastic interaction between the partials on parallel slip planes are responsible for the dissociation.

QUANTITATIVE STUDY OF STRIATIONS IN STAGE II FATIGUE CRACK GROWTH

H. MUGHRABI, R. PRASS, H.-J. CHRIST and D. PUPPEL

Institut für Werkstoffwissenschaften I, Universität Erlangen-Nürnberg, FRG

1. INTRODUCTION

Fatigue crack growth in stage II is characterized by the formation of so-called fatigue striations on the fracture surface (1,2). The most accepted mechanism of stage II fatigue crack growth by striation formation is that of alternating shear at the crack tip on at least two slip systems (2-5) which may be complemented by tensile decohesion of the blunted crack tip (6). The observed variety of fatigue striation profiles, cf. (1,2,7,8) suggests that no unique stage II fatigue crack growth mechanism exists that can explain all observed details.

The aim of the present study was to gain further insight into the mechanisms responsible for stage II fatigue crack growth by quantitative fractography, using scanning electron microscopy (SEM), and by transmission electron microscopy (TEM) of the dislocation substructure formed beneath stage II fracture surfaces (9,10).

2. EXPERIMENTAL

The materials investigated were an Al-1%Mg-1%Si alloy in the overaged state, an age-hardened 7075 aluminium alloy and oxygen-free copper (99.99%). Stage II fatigue cracks were introduced in notched specimens with a square cross section of 8 mm × 8 mm in load-controlled fatigue tests conducted in laboratory air. The fracture surfaces were studied by SEM, using the contamination line technique (11), compare Sect. 3. The dislocation substructure beneath the fracture surface was investigated by TEM, as in earlier studies by other authors (7,8,12).

The results to be reported below concern mainly Al-1%Mg-1%Si which was studied in most detail. In order to investigate the influence of certain fatigue parameters on the shapes of the fatigue striations, tests were performed at different stress intensity ranges ΔK, R-ratios (minimum to maximum stress) and frequencies f. The parameters of the fatigue tests on Al-1%Mg-1%Si to be discussed here are listed in Table 1.

3. RESULTS AND DISCUSSION

Figs. 1a to 1c show examples of typical fatigue striations observed on the fracture surfaces of fatigued Al-1%Mg-1%Si specimens. The profiles are clearly revealed by the sets of parallel contamination lines. The technique is as follows. The contamination lines are deposited in a direction perpendicular to the fatigue striations by exposure in the line-scan mode of the SEM for some minutes. Typically, sets of 3 to 5 parallel contamination lines were deposited. Subsequently, the specimen is rotated around the axis of the contamination lines, i.e. around the fatigue crack growth direction FCGD, by an angle of 45° (in the present study). The (somewhat distorted) profiles of the fatigue striations then become clearly recognizable. These profiles can be digitized and, by taking the mean of the available 3 to 5 profiles, a higher

specimen	min. load [kN]	max. load [kN]	f in Hz	R
1	0.28	7.00	2	0.04
6	0.40	10.00	2	0.04
10	0.20	5.00	2	0.04
17	-3.00	6.00	2	-0.50
19	0.30	7.50	0.1	0.04
20	0.30	7.50	20	0.04
26	-1.55	6.40	2	-0.24

TABLE 1. Test parameters of fatigued Al-1%Mg-1%Si specimens.

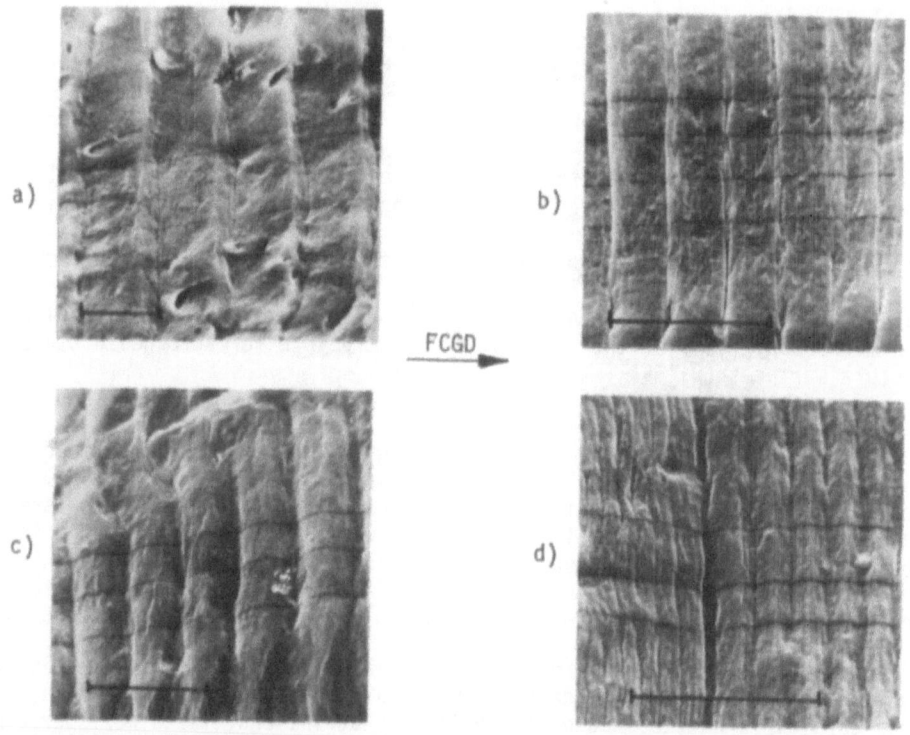

FIGURE 1. Typical fatigue striation profiles observed on Al-1%Mg-1%Si.
a) Saw-tooth profile, type A. b) Flat (squashed) profile, type B. c) Sinu-soidal profile, type D. d) Transition to ascent of fracture surface. Note transition from type B to type A profile. Scale marks represent ten microns.

accuracy corresponding to a more representative profile is obtained. Finally, the digitized profiles are rectified by correcting for projection effects. The true undistorted profiles so obtained can then be analyzed in more detail, as discussed below.

The fatigue striations observed in Al-1%Mg-1%Si material exhibited a vari-ety of profiles which are classified schematically in Fig. 2. Profiles of type E were only observed for $R < 0$. Among the remaining profiles, the saw-tooth-shaped profile A and the rather flat profile B played a special role in the following manner. It was noted that after the crack had propagated well into the specimen, a stage was reached when more or less unconstrained slip processes occurred at the crack tip, traversing the remaining cross section and causing an ascent of the lower (and a descent of the upper) fracture surface as in the example shown in Fig. 1d. The flat type B profile was generally found on the fracture surface region preceding the described ascent, whereas the saw-tooth profile A was typical of the ascending part of the fracture surface, as is evident from Fig. 1d. It was concluded that the original profile formed during crack growth probably was more similar to type A and that type B was a result of subsequent damaging deformation by compression. A peculiarity of the saw-tooth type A profiles was that, for FCGD from left to right, the rising flank of the fatigue striations on the lower fracture surface was smooth and featureless, whereas the falling flank (on the right) exhibited marked traces along the striation direction, as illustrated clearly in the example in Fig. 3 for wide striations.

One reason for the damage suffered by the striations due to compressive deformation was seen in the fact that crack growth occurred in a markedly crystallographic fashion with abrupt changes of fracture surfaces and FCGDs at the grain boundaries, as noted previously (4). Thus, matching difficulties could be expected during crack closure. In addition, however, as explained below, opposite fracture surfaces were related in a peak-peak, valley-valley fashion in the interior as in the crystallographic models of Pelloux (4,5) and Bowles and Bowles and Broek (7) rather than that of Neumann (3), suggesting peak-peak contact occurred during unloading. (It appears worthwhile to note that at the free surface the fracture surface mating was peak-valley.)

In order to minimize modification of the originally formed striation profiles by subsequent compression, most of the detailed tests were then performed without going into compression ($R > 0$), and the profiles selected for detailed evaluation were chosen preferentially from the early part of the ascending fracture surface region.

The evaluated profiles were approximated as saw-tooth profiles and could be characterized by the parameters defined in Fig. 4. Here a and b are the ascending and descending striation flanks, projected onto a plane perpendicular to the stress axis, corresponding to the original fracture surface before the transition to the ascent occurred. The angle δ was found to lie in the approximate range of $30°$ to $60°$, frequently being close to $\approx 36°$, as expected for crack growth on $\{110\}$ in the FCGD <001>, or to $\approx 55°$, corresponding to crack growth on $\{001\}$ in the FCGD <110> (4,5,7). The ratios of the projected flank lengths a and b and of the projected striation spacing W and the height h are illustrated in Figs. 5 and 6 as a function of the projected striation spacings W which range from about one to several ten microns. For specimens with $R > 0$, both sets of data are in accord with the qualitative dependence

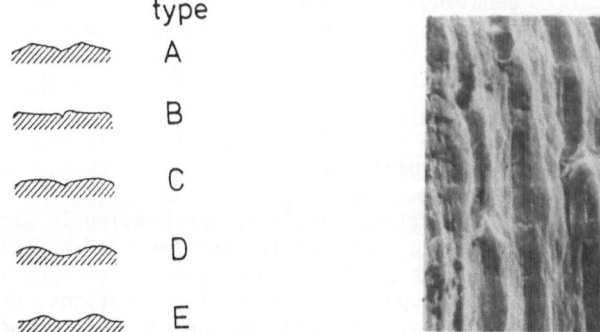

FIGURE 2. Types of fatigue striation profiles in Al-1%Mg-1%Si.

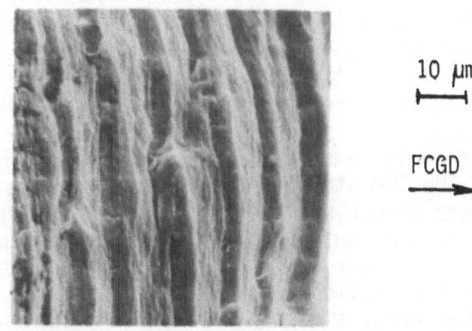

FIGURE 3. Fine structure of falling flanks of striations (Al-1%Mg-1%Si).

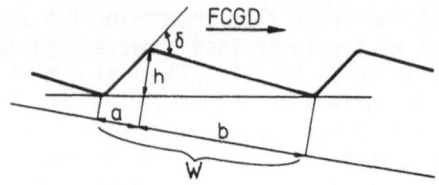

FIGURE 4. Definition of parameters used to characterize striation profiles.

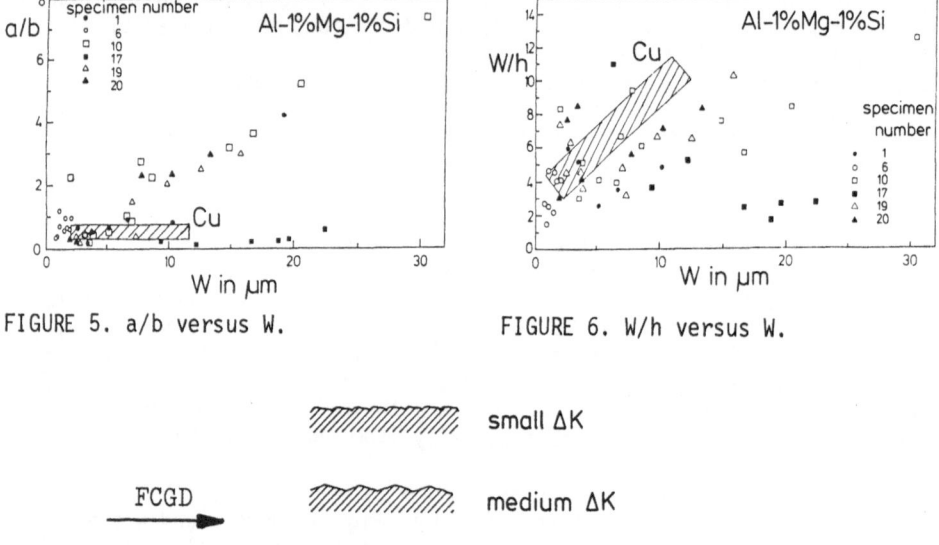

FIGURE 5. a/b versus W.   FIGURE 6. W/h versus W.

FCGD

       small ΔK

       medium ΔK

       large ΔK

FIGURE 7. Qualitative dependence of striation profiles on stress intensity ΔK

of the (saw-tooth) profiles on ΔK, as illustrated in Fig. 7. A similar tendency has already been reported by Meyn (13) for fatigued 2024 aluminium alloy. Aside from the significant scatter in the range of small W towards larger values of a/b and W/h, it is found that both a/b and W/h show a tendency
to increase with increasing W irrespective of frequency and load amplitude
(and hence crack length). This behaviour is roughly consistent with the picture that with increasing ΔK (increasing W) the flank a becomes larger, while
flank b remains more or less constant. The mentioned scatter for small W is
in accord with and suggested to be due to preferential squashing of the small
striations. For negative R-ratios (specimens 17,26), a/b and W/h do not exhibit an increase with increasing W. This behaviour is attributed to enhanced
damage in compression, rendering the profiles rather flat (more details in (9)).
    The TEM studies were made on specimens from regions of the fracture sur-
faces preceding the ascent. The observations confirmed the previously mentioned crystallographic type of fracture and revealed the existence of a dislocation cell structure beneath the fracture surface. There was no evidence that
the dislocation density varied with the periodicity of the striations. On the
other hand, a distinct periodic misorientation contrast due to lattice bending around an axis corresponding to the direction of the striations was observed. This is illustrated in Fig. 8 by the two dark-field micrographs of

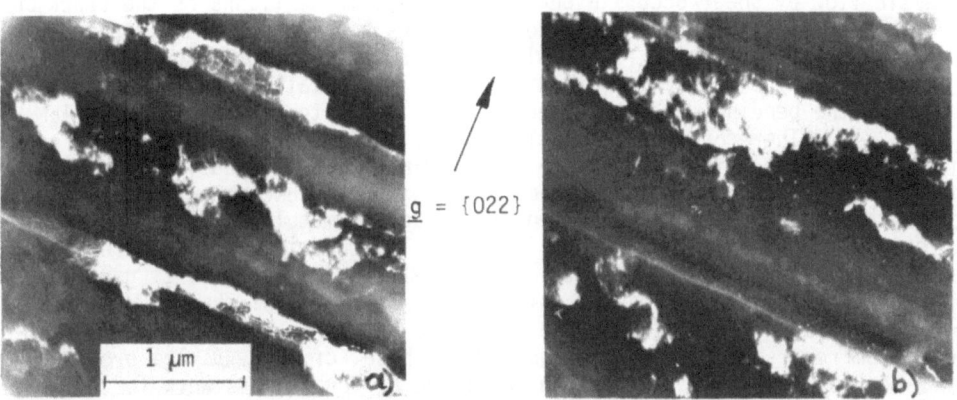

FIGURE 8. TEM misorientation contrast of striations. a) 0° tilt. b) 4° tilt.

the same area of a specimen (Al-1%Mg-1%Si) taken with a $\underline{g}$ = {022} diffraction vector parallel to the FCGD in the initial position (Fig. 8a) and after a 4° tilt (Fig. 8b) around an axis parallel to the direction of the striations. Typically, the total angle of misorientation was ≈8°(±4°).

Careful analysis of the misorientations with regard to their sign led to the model shown in Fig. 9. According to this model the peak-peak contact of the two fracture surfaces (Fig. 9a) gives rise to a mutual compressive deformation of the striation peaks (indicated by traces of slip planes in fig. 9b) in a process occurring after the striations have been formed. The net effect of this deformation which takes place during unloading (away from the crack tip) is to convert the type A saw-tooth profile into a flat type B profile and to introduce surplus edge dislocations of alternating sign beneath the valleys and the peaks of the striations, causing the lattice plane bending indicated in Fig. 9c.

Finally, Fig. 10 shows in a simple model the mechanism of formation of type A saw-tooth striations. When the load increases, the flanks a are formed by alternating slip at the crack tip (Fig. 10b). During unloading reverse

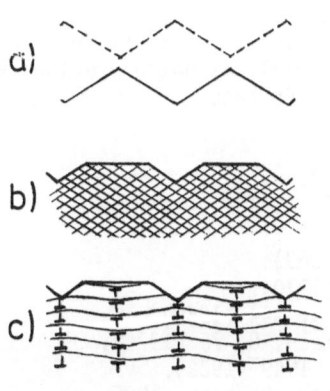

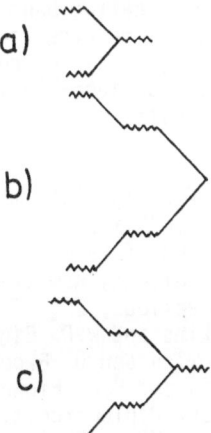

FIGURE 9. Model of deformation of saw-tooth striations by peak-peak contact during unloading. See text for details.

FIGURE 10. Model of crack growth. a) and c) correspond to minimum load, b) to maximum load.

448

plastic flow by shear occurs along a portion of these flanks as the crack tip resharpens, leading to the formation of flanks b which lie roughly perpendicular to the stress axis. (It should be noted that, if during unloading the resharpening of the crack tip occurred (partly) by lattice plane bending, the sense of the bending would be expected to be opposite to that found by TEM (Figs. 7,8). Therefore, it is concluded that the crack tip processes do not give rise to appreciable lattice plane bending.) With increasing ΔK, flanks a become larger, whereas flanks b remain approximately unchanged (possibly because the elastic strain energy density is approximately constant once the ascent of the fracture surface has set in). The described model corresponds to Pelloux' model (4,5), as extended by Bowles and Broek (7). It gives a fair description of the results obtained for Al-1%Mg-1%Si shown in Figs. 5, 6 and 7 and also explains the fine structure on flanks b as slip lines stemming from reverse sliding in the unloading phase, cf. Fig. 3. The proposed peak-peak and valley-valley mating could be proven in this work for Al-1%Mg--1%Si, as mentioned previously, in programmed loading experiments (9).

4. COMPARISON WITH OTHER EXPERIMENTS, CONCLUDING REMARKS

Similar, less detailed studies on an age-hardened 7075 aluminium alloy yielded results that were quite different from those described above (9), indicating that, even for similar classes of alloys, details of the fatigue fracture surface can differ significantly.

Yet other observations were made on fatigued polycrystalline copper. In this case no evidence for crystallographic crack growth mechanisms was obtained. The macroscopic fracture surface was perpendicular to the load axis, and the fatigue striations crossed the grain boundaries without disturbance, suggesting that the fracture mechanism was dictated by semi-macroscopic continuum effects, as in Laird's plastic blunting model (2). However, even in fully reversed fatigue tests, the peaks of the striations seemed to have suffered little damage and, in agreement with this impression, TEM revealed only dislocation cells not related to the striations and no periodic misorientations. Possible explanations could be that the fracture surfaces are related by a more favourable peak-valley mating and that matching is facilitated by the fact that the striations are not affected by the grain boundaries. The striation profiles were mostly of type C for W < 3 μm and of type A for W > > 5 μm. The ratios a/b and W/h depended on W as indicated in Figs. 5 and 6 by the shaded scatter bands, showing that while W/h increases, a/b remains constant. These results require further attention.

In summary, it can be concluded that the details of stage II fracture surfaces differ considerably from material to material, as noted previously (1,2), and that specific models must be developed in each case.

REFERENCES
 1. P.J.E. Forsyth, Acta metall. 11, 703 (1963).
 2. C. Laird, ASTM STP 415, Philadelphia, 1967, p. 131.
 3. P. Neumann, Acta metall. 17, 1219 (1969).
 4. R.M.N. Pelloux, ASM Trans. Quart. 62, 1 (1969).
 5. R.M.N. Pelloux, Eng. Fract. Mech. I, 697 (1970).
 6. B. Tomkins and W.D. Biggs, J. Mater. Sci. 4, 544 (1969).
 7. C.Q. Bowles and D. Broek, Int. J. Fract. Mech. 8, 75 (1972).
 8. K.J. Nix and H.M. Flower, Acta metall. 30, 1549 (1982).
 9. R. Prass, Diplomarbeit, Universität Erlangen-Nürnberg, 1986.
10. R. Prass, H.-J. Christ and H. Mughrabi, DVM Symposium, Erlangen, 1986.
11. R. Wang, B. Bauer and H. Mughrabi, Z. Metallk. 73, 30 (1982).
12. J.C. Grosskreutz and G.G. Shaw, Acta metall. 20, 523 (1972).
13. D.A. Meyn, ASM Trans. Quart. 61, 42 (1968).

THE BRITTLE-DUCTILE TRANSITION OF SILICON

M. Brede and P. Haasen
Institut für Metallphysik, Universität Göttingen, FRG

1. INTRODUCTION

At the preceding workshop 1981 on Corsika one of the authors analysed the early measurements of St. John (1) on the b/d transition of pre-cleaved Si single crystals. The analysis was based on the assumption that dislocations can be easily produced at the crack-tip in an otherwise perfect crystal. The condition for ductile behaviour was that the dislocation could blunt the crack fast enough in comparison with the built-up of the stress intensity at a constant crack opening rate $\dot{\delta}$ so that the critical K_{Ic} would never be reached. The theory of the b/d transition temperature T_c made use of the linear high stress-dislocation velocity behaviour that is well documented in the literature but contained two adjustible parameters: One is the distance Δ between dislocations emitted in sequence from the crack tip (the width of the "dislocation-free zone"). This is related to the "friction stress" τ^*, to the level of which the stress ahead of the crack is reduced by the shielding of the dislocations emitted ($\tau^* = Gb/\Delta$, G = shear modulus). The second parameter is K_{Ic} at T_c, taken to be several times the low T value $K_{co} = 0.9$ MPa $\sqrt{m}$ K_{Ic} according to the experiments (1,2). The rate-of-blunting -and shielding criterion for the b/d transition rather than that of dislocation formation at the crack tip is in contradiction to the classical Rice-Thomson theory (3) which calculates a very high activation energy of dislocation creation. Later estimates (see (1)) reduced this energy especially by considering partial dislocations as are observed during in-situ TEM crack opening experiments on metals of low SFE by Ohr (4). Recent work at Göttingen (2) and Nancy (5) concentrates on experiments under varying $\dot{\delta}$ and doping as well as oxygen contents of the silicon used. A full report of our work will be published elsewhere soon.

2. EXPERIMENTAL

Trapezoid-shaped specimen of maximum dimensions 30x18x1 mm³ were pre-cracked at room temperature. Due to the increasing width of the specimen with increasing crack length $\underline{a}$ the stress intensity K_I was almost independent of $\underline{a}$ for a given force pulling the crack open at constant rate $\dot{\delta}$. The crystallographic orientation of the crack front relative to the {111} slip and fracture planes is shown in fig. 1. This orientation proved to give the smoothest cracks on mode I deformation although the slip planes don't intersect the crack plane parallel to the crack front but rather in a zig-zag fashion. The specimen is heated under a load that keeps the pre-crack open, to a temperature near 800° C controlled to 0.1° C under high vacuum. The crack then is loaded at constant rate $\dot{\delta}$ in a computer-controlled Zwick machine measuring load, crack opening δ, T vs. time t.

Below the b/d transition temperature the crack opens at $K_{CO} = (2\gamma E)^{1/2} =$
0.9 MPa $\sqrt{m}$ independent of $\dot{\delta}$ and doping of the material. From this value
and Young's modulus E a surface free energy $\gamma = (2.6 \pm 0.3)$ ergs $/m^2$ is
calculated in agreement with values in the literature. The different
materials investigated so far are listed in tab 1.

No.	N_{dop}/cm^3	growth	oxygen/cm^3	source	ref.
1	$1.6 \cdot 10^{18}$ P	FZ	10^{15}	Wacker	
2	$6.5 \cdot 10^{18}$ P	FZ	10^{15}	Wacker	
3	$2.9 \cdot 10^{19}$ As	CZ	$6.7 \cdot 10^{17}$	Wacker	
4	$1.2 \cdot 10^{18}$ P	CZ	10^{17}	Siltec	
5	10^{16} P	FZ	10^{15}	Wacker	(5)
6	10^{16} P	CZ	$3.5 \cdot 10^{17}$	Wacker	(5)
7	$3.0 \cdot 10^{14}$ B	?	?	?	(1)

Tab. 1: Materials investigated

3. RESULTS

Michot (5) finds by X-ray topography some dislocations at the crack
front in initially dislocation-free material already at 1/3 K_{CO} in
specimens loaded very slowly. We have etched crack planes of crystals
fractured at T < Tc and find dislocation pits not visible at this magnification near the crack arrest position 2 in fig. 2. Position 1 marks the
crack front as introduced at room temperature. Position 2 was reached
at high temperature under K_{CO} as the crack advanced by Δa before stopping
again temporarily (Δa is also recorded by an intermediate load drop). The
final crack advance in the presence of dislocations then occurred at a
slightly higher $K_{IC} \approx 1.7$ K_{CO} and typically showed up in the pattern 4 and
5 on the crack plane (instead of pattern 3). Here we think the crack
advanced from the specimen sides, 4, towards the middle, 5 rather than
perpendicular to 2.

Fig. 3 shows K_{IC} vs. temperature for a particular $\dot{\delta}$ and type of material
(no. 1). At $T_c = 1077.2$ K the stress intensity increases sharply by a
factor 8 to a value corresponding to $\sigma_{uy} \cdot \sqrt{a}$, where σ_{uy} (T) is the upper
yield stress for gross plastic deformation. One notices for some specimens
(open squares), which showed an intermediate load drop and crack advance,
slightly increased values $K_{IC} > K_{CO}$ near T_c.

Similar K_{IC} (T) curves were measured for different $\dot{\delta}$ (betweeen 5 and 2500
$\mu m/min$) and materials (tab. 1) and the resulting T_c^{-1} vs. ln $\dot{\delta}$ were plotted
following an Arrhenius relation. This fitted well with the exception of
the very highest $\dot{\delta}$ which gave a too small T_c, Fig. 4.

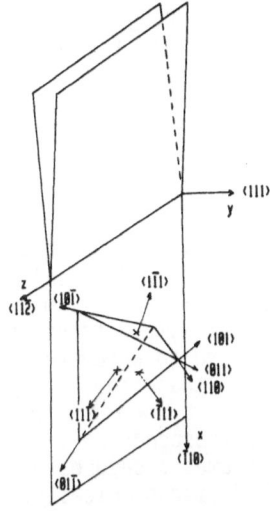

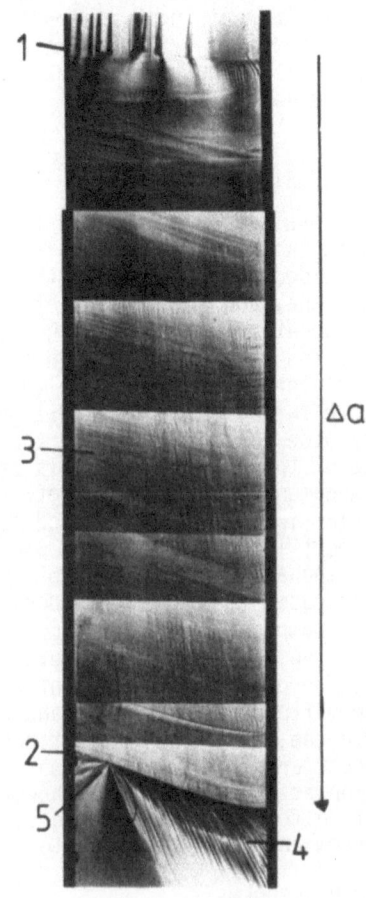

Fig.1   Slip systems and crack
plane/front

Fig.2   Crack propagating from
top to bottom (width 1 mm)

Fig.3 K_{Ic} vs. temperature
for material 1

Fig.4   Arrhenius plot of crack
opening rate $\dot{\delta}$ vs.
transition temperature T_C.

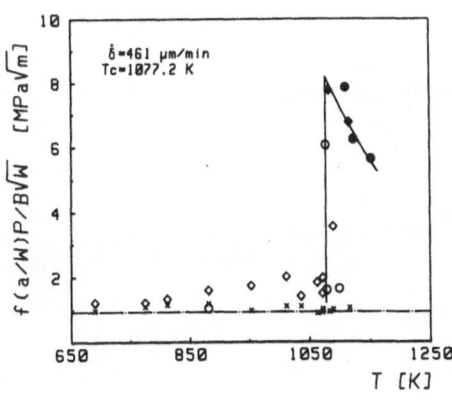

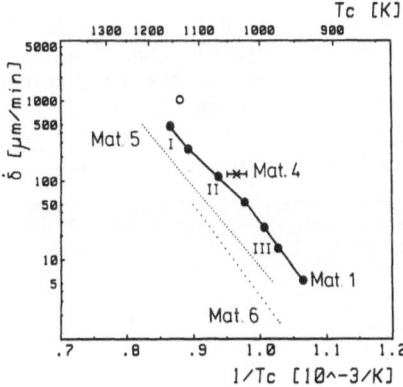

In the extrinsic temperature range, II, of doping the activation energy
(1.6 ± 0.1) eV is definitely smaller than in the intrinsic range, I,
U=2.0 eV. Surprisingly at low temperatures, III, there is again a steep
drop in ln δ $(1/T_C)$ also in materials 2 and 3. The high oxygen
CZ material 4 shows a less sharp b/d transition at lower T_C (c.f. however
also materials 5 and 6).

4. DISCUSSION

The evaluation of the b/d transition data follows the theory of ref.
(1) with two new points which are discussed in detail in ref. (2). (a) The
not-shielded stress ahead of the crack in the plastic zone is really small,
$\tau * \leqslant 8$ MPa, and independent of temperature. It is the minimum stress (for
any dislocation movement in the particular material) close to which the
dislocation velocity/stress in fact becomes much steeper than linear. This
can explain the steep drop in the ln δ $(1/T_C)$ dependence at low temperatures
as shown in ref (2). The small τ^* on the other hand corresponds to a
rather wide dislocation free zone, $\Delta \geqslant 10$ µm if one assumes a decohesion
stress of the order $E/10$. This still has to be studied microscopically.
(b) Arguments are given in ref. (2) that the free parameter of the theory,
(1), K_{Ic} (T_C) = K_{co}. Then good quantitative agreement between theory and
experiment is obtained in the whole range of temperatures and materials.
Indeed the activation energy of the velocity in intrinsic Si is near 2 eV
and decreases to 1.6 eV on n-doping (as in materials 1 and 3) for the extrinsic temperature range. Hopes to make low temperature ductile Silicon
by n-doping are however shattered by the discovery of the minimum stress
effect for dislocation movement. This also explains the oxygen influence
or the difference between FZ and CZ materials. There is no other way to
produce the necessary dislocations for the b/d transition but at the crack
tip, contrary to the conclusions of refs. (3) and (5). Dislocation
creation at the crack tip is however only a necessary but not a sufficient
condition for ductility - the dislocations also have to move fast enough
away from and into the crack to avoid brittle conditions.

5. ACKNOWLEDGEMENT

The DFG and Thyssen-Foundation supported this work. Dr. S. K. Hahn
of Siltec and Dr. H. Jacob of Wacker kindly supplied materials.

REFERENCES

1. St. John, C: Phil.Mag. 32,73 (1975)
2. Brede, M.: diploma thesis Göttingen 1983 and
             ph. d. thesis Göttingen 1986, in press
3. Rice, J., Thomson, R.: Phil. Mag. 29, 73 (1974)
4. Ohr, S.M.: Mat. Sci. Engg. 72, 1 (1985)
5. Michot, G., George, A.: Acta Met. to be publ.

DISCUSSION

Comment by M. Daw:

You say that the dislocations are originating from the crack. Are you sure that there are no other sources, such as intrinsic or extrinsic defects near the crack tip?

Reply:

Not as far as we know. Silicon, especially FZ material, is perhaps the most perfect crystal known. Inhomogeneities should have a rather course scale to be able to nucleate dislocations (i.e., nucleation at large oxygen precipitates which were artificially produced in CZ material).

Comment by A. Argon:

Have you done also experiments on predeformed silicon to separate the role of background initiated plasticity from that of crack tip initiated plasticity?

Reply:

Not so far. But we intend to, although it is experimentally not easy to homogeneously deform the large single crystals needed to cut the fracture specimens from.

Comment by P. Neumann:

The reason for the complicated crack-slip geometry in Si is due to the {111} nature of cleavage planes in Si. If the crack front is made parallel to a <110> direction in the {111} cleavage plane, there is only one inclined {111} slip plane cutting the <110> oriented crack front. This asymmetry leads to non-planar fracture surface.

Reply:

I agree. For a <112> crack front segments of 2 {111} slip planes meet along <110>.

WORKSHOP SUMMARY: NOVEL ASPECTS IN APPLICATIONS AND FRACTURE THEORY

PREPARED BY: P. Haasen
DISCUSSION LEADERS: M. Ohr, J. Rice, M. Daw, P. Haasen
RECORDERS: R. Thomson, W. Gerberich, K. Sieradzki, R. Newman

The following topics were discussed:
1. Recent understanding of relative role of plasticity and decohesion processes.
2. New experimental techniques and observations for monitoring/ evaluating fracture.
    a. In-situ TEM
    b. Fracture reconstruction
    c. Acoustic emission fracture relationships
3. Critical experiments
4. Development of linkage between continuum and atomistic aspects of fracture.
5. Can environmental effects be modeled using continuum theory?
6. Fracture of new materials-composites, polymers, ceramics, etc.
7. Relation of in-situ TEM observations to bulk fracture phenomena.
8. Is the relation of crack angle to crack growth rate in Fe-Si universal?

The results of the discussion in four separate groups are summarized as follows (referring loosely to the question period above).

The main issue in the plasticity and decohesion discussion centered about the role of plasticity in the fracture process or cleavage tendency in crystals. The following question came up: "Is there any material which cleaves without displaying measurable plasticity?". This led to a discussion of a proper working definition of cleavage. Cleavage was reasonably defined as an atom by atom separation across a close-packed plane. Silicon was suggested as a system which at low T shows apparently good agreement between measured work to fracture and the thermodynamic surface free energy. One difficulty arose in this regard as to how the surface free energy in Si is obtained. Another matter discussed considered the role of ductile processes such as pore formation in nucleating cleavage cracking. This process is observed in ductile-brittle transitions in steels. The mechanism of porosity-induced cleavage seemed widely appreciated but not at all understood.

The discussion centered mainly on recent observations by Birnbaum et al. on hydrogen effects in various metals and alloys. The hydrogen-induced fracture involves a narrowly confined microvoid coalescence, localized within 1-2 μm of the crack tip, leading to fracture surfaces which are macroscopically oriented parallel to slip planes. Other processes which localize strain in a similar way include fatigue. The effect of hydrogen is to make the material locally more responsive, probably by an internal effect although surface effects as described by Lynch may also be operating. (But note that brittle modes are hindered in TEM-type specimens.) The continued use of the term "hydrogen embrittlement" is acceptable if one regards embrittlement as the reduction in macroscopic fracture energy.

Any process which localizes plasticity, and makes the material more responsive plastically, will give a similar "brittle" appearance. Considering the extensive plasticity associated with intergranular, hydrogen-induced fracture, it is perhaps surprising that segregants are found in high concentration on the surfaces. The opinion was expressed that many measurements of this type may be giving concentrations that are too low. The "cleavage" observed (or rather deduced) by Neumann in Fe-Si/H$_2$ was felt to be plastic process.

Work on simple systems which can corroborate the theory is emphasized.

Silicon is obviously such a system. It seemed to us that the richness of the phenomena already briefly described there suggests that both new theoretical and experimental insights are to be expected. For example, can we be more certain of the sources of the observed dislocations? In other brittle systems, it is clear that sources very near to but not on the crack are numerous, but not actually visible as inhomogeneities of any kind in the electron microscope observations. Silicon is also a model material for the exploration of chemical effects because so much is known about surface adsorption on Si, both theoretically and experimentally.

The new theoretical developments range from more fundamental to more empirical (John Smith to Embedded Atoms) with ease of application running in the other direction. At this time, the embedded atom methods are limited to metals. In the semiconductors the tight-binding methods may be used, but the appearance of interband states make them more complex. It is expected to have in the near future a description of bonding at interfaces in metals, including the effect of impurities. Both size effects and charge redistribution will be taken into account with tight-binding. There is a need to establish rules for how adequate the various approximate methods are for various types of calculations by comparisons between them and experiments or more complex first principle calculations in simple systems. Some of this has been done for surfaces, with work on surface relaxation in W.

These theories should have their greatest impact in the predictions of new effects, such as that if adsorption on corners leads to enhanced dislocation activity. However, they will be difficult to apply in many cases of interest such as effects involving many dislocations. Dynamic problems, however, should provide an interesting field of application.

The interface between QM treating small numbers of atoms, and the need to embed these phenomena into a larger crystal with interactions over larger distances will provide focus anew on those mathematical techniques which have been developed to bridge these scales.

- The "process zone" concept used by Gerberich emphasizes the neglect of the third dimension in fracture studies. TEM samples do not have this kind of behaviour. Some "process zone" morphologies may be a useful area of application of percolation theory.

But on a different scale, Gerberich's attempt to model the bridging effects of plastic ligaments will probably be applied to a broad class of cases. We felt that such plastic bridging calculations will probably be most successful, however, in simpler situations such as the stress corrosion crack arrest problems. But bridging effects are wide spread, from concrete, to ceramics, to polymers as well as metals. A particularly successful attempt has been made to explain the toughness of cermets by Budiansky and Evans, regarding the bridging of the metal particles between the crack in the ceramic.

There was an expressed need to develop experimental results on simple systems, so theories on all relevant scales can be adequately tested. Otherwise the attempt to explain more complex phenomena will be on poor ground.

New Experimental Tools

Some discussion took place about the use of <u>acoustic emission</u> for fracture studies. The theory has already developed powerful tools for use in the interpretation of simple events in solids (i.e. Shear vs tension in time), and there is hope that it will see use at NBS in the interpretation of the stress corrosion cracking events shortly. The technique however will always be limited in spatial resolution to distances much larger than those of interest for atomic processes at crack tips. Gerberich's ability to correlate a specific AE event with the particular cleavage facet formed is considered an experimental tour de force.

The discussion on experimental techniques focused on two issues,

a) the lateral resolution of techniques which "measure" dislocation densities

b) the fracture reconstruction scheme introduced by researchers at SRI.

The difficulty of using the e^- channeling technique owing to calibration complexities was discussed. Tritium autoradiography was mentioned as a technique which could be used to study plastic damage. Mr. Chene claimed 150 nm lateral resolution for this technique. The <u>fracture reconstruction</u> scheme was generally considered to be a new technique fraught with difficulty. It appears to be of limited use because of the local distortions due to plasticity in the final stages of fracture.

Deformation processes such as rotation during shear were considered not to be reproduced by reconstruction. Nevertheless, no one seemed familiar enough with the Kobayashi technique to make absolute statements. Agreement was reached that this technique should be described in the open literature.

The contamination line technique is important in several applications, although it cannot deal with internal fracture. <u>SIMS</u> of hydrogen and nuclear profiling using deuterium were mentioned as techniques with potential for near-tip hydrogen studies. It is necessary to use dynamic (destructive) SIMS to get good measurements on hydrogen. SIMS is also useful for very low concentrations of grain boundary segregants (but are these relevant?).

The acoustic microscope should be watched, as lens developments may lead to sub-μm resolution. Acoustic emission is not used enough, and many studies are on complicated systems with inclusions etc. More studies should be done on "clear" systems, particularly with large-grained material where brittle events are occurring. In SCC, electrochemical transients are a useful confirmation of the source of AE, as shown by Newman and Sieradzki.

The development of fine, intense X-ray beams from synchrotron sources has many implications. X-ray line profile analysis can detect μm-scale hydrogen profiles and EXAFS can reveal impurity binding. X-ray topography has an inherent spatial resolution limit of about 5 μm. The load-pulsing method of Pugh (to mark the position of the crack front) should be used more. X-ray Kikuchi lines are better than conventional electron channelling for measurement of orientation of surfaces with plastic strain.

We spent time discussing the two camps represented by those who believe in DFZ's and those who do not. The consensus was that, in those materials that have sources or large initial dislocation densities, DFZ concepts will not apply, but that in those materials which were of low dislocation density and plasticity was only associated with crack-tip emission, they might. As an idealized system the latter experiments might be useful in probing the connection between atomistic and continuum models. On the other hand, that material which seems to provide the saviour (P. Haasen's dislocation-free silicon) provided results which did not agree with the Rice-Thomson model. That is, activation energies of 1/2 to 1 eV were measured for the brittle-ductile transition rather than the 100 eV prediction, and crack growth appeared to be controlled by the rate of dislocations leaving the crack tip.

The relationship between atomistic and continuum mechanics is closer than might be apparent, and some evidence (e.g. the simulations of Daw) suggests that the continuum equations may be valid down to a few Å. It is difficult to test this experimentally.

The group generally felt that there was little hope in obtaining a useful linkage that contained microstructure and interfaces in the near future. Nevertheless, in ideal systems, an approach to such a linkage may be possible. It seems currently possible to model most of the material with finite element of a dynamical Green's function, and only the nonlinear core with atomistic simulations. The group thought that the best way to probe atomistic and continuum models was with well-designed experiments which might be single-phase, single crystals or multi-phase but well oriented polycrystalline arrays.

Experiments on a "simple" material such as Si could be compared with predictions based upon rigorous atomistic simulation. Other areas were touched upon such as geometric blunting of cracks. In this regard we discussed the appropriate scale of size for the transition of atomic to continuum calculations. Little progress was made on this difficult problem.

Environmental effects can be incorporated into continuum studies via constitutive relations, e.g. the work of McMeeking incorporating hydrogen effects on dislocation velocity via a flow stress effect.

A comment was made that analytic continuum and finite element approaches would be very useful in understanding the nature of corrosion film induced stresses.

Ceramics phenomena were felt to be missing at the workshop.

Workshop Session 2:
Intergranular Embrittlement

Session Chairman: T. Watanabe

INTERGRANULAR FRACTURE

Wolfgang Losch
COPPE/Universidade Federal do Rio de Janeiro
Programa de Engenharia Metalurgica e de Materiais
Rio de Janeiro, Brasil

ABSTRACT

A review is given on the development of understanding atomistics of intergranular fracture behaviour in metals during the last five years. This involves both, new experimental results as well as theoretical modeling the possible mechanisms of atomic decohesion.

Experimental work has been focussed on mainly two aspects of brittle fracture. First: to obtain a clearer picture of the chemical influence of diverse impurity elements at the grain boundaries in metals and alloys. Second: the influence of grain boundary structure and orientation on fracture modes, including reconstruction of grain boundary structure due to impurity segregation. The review also includes embrittling effects of impurities in the presence of hydrogen. Not discussed in this paper is the complex field of hydrogen embrittlement.

Theoretical models on the atomistic mechanism of fracture may be divided basically into two approaches. One starting from quantum mechanical calculations, to show chemical bond modifications due to impurity segregation. The other approach is more related to structural considerations at the grain boundaries applying pair potential calculation methods. Both this approaches can explain a series of experimental facts and represent a considerable progress in understanding details of intergranular fracture, although a comprehensive description is still missed. Finally, fundamental processes on an atomically sharp crack tip, whether it will proceed in a brittle or ductile mode of fracture, are discussed.

1. INTRODUCTION

Application of modern physical methods has helped, successfully, to clear up our picture of the old metallurgical problem of intergranular fracture. It was by means of Auger electron spectroscopy (AES) that, at the beginning of the seventies, impurity segregation to grain boundaries (GB) could be proved to provoke embrittlement. Extended and systematic studies on many types of metals and impurities in the following years supported qualitatively and quantitatively these findings resulting in a well funded appreciation of GB segregation as the origin of embrittlement. Thermodynamical concepts have then been used to find a theoretical background in order to understand atomistic mechanisms of brittle fracture. These considerations resulted in atomic hard sphere models explaining embrittlement by atomic size effects. Until the change of the decade in 1980 most of the works and reviews are within the framework of these models [1-4]. However, the way fracture, i.e. atomic decohesion really may occur could not be explained in these treatments, since a detailed picture of atomic bonding was simply ignored.

It was again with the aid of physical considerations that new methods have been introduced in the treatment of intergranular fracture problems. By looking to atomic binding on the fundamental basis of quantum mechanic wave functions, the discussion of fracture problems changed drastically in its way of interpretation. A first attempt was presented by Losch [5] by transferring results from quantum mechanical calculations of free surface adsorption problems to the similar problem of interfacial segregation. The model could qualitatively explain that impurity segregation may result in reduction of bond strength along the GB, thus giving a first indication to the importance of inspecting the nature of atomic bonds in fracture problems. Briant and Messmer [6] came to similar results from quantum mechanical calculations of metal - impurity clusters, considered as units for a segregated GB model. A different method has been applied by Hashimoto et al. [7] in order to calculate metal - impurity interaction and again a reduction of bond strength along GBs was predicted.

It is intelligible that the nature of chemical bond has to be considered for any profound discussion of fracture problems; as a matter of fact bond disruption is involved in brittle fracture and also ductil failure proceeds via dislocation movement by breaking and reformation of chemical bonds.

However, the discussion cannot be limited to chemical effects only: structural influences have to be included which becomes evident when looking to the bond orientation relative to the fracture path. This all the more as GBs represent perturbed crystalline regions where localization of bonds rather than extended "smeared out" metallic bonds should be considered. Furthermore, fracture is a dynamic process, which means, in terms of electronic wave functions, reconstruction of atomic orbitals (from formerly molecular ones) on the freshly created fracture surface. It becomes evident from this that the simple mechanical fact of breaking a metal involves a complex world of fundamental problems; and our knowledge is quite distant from a complete and detailed understanding for a comprehensive description of the fundamental mechanisms.

The topic of this review is to discuss the present state-of-the-art of understanding the atomistic mechanism of intergranular fracture. In the following sections we will therefore first outline results of recent experimental work related to the chemistry and structure of GBs including the simultaneous effects of hydrogen and then pass to a presentation of suggested models of atomic mechanism of intergranular fracture. Consequently, the emphasis of our discussion of experimental work will be more related to exemplary investigations on the basic understanding of atomic fracture mechanisms, leaving apart important results from specific systems of engineering practice which are less illuminating in the present context.

2. BASIC INSIGHTS FROM EXPERIMENTAL WORK

2.1. Grain boundary chemistry

As already mentioned a great number of experiments on GB segregation have been carried out during the last decade principally by application of

AES, and excellent reviews have been presented on this subject [1,2]. AES is distinguished by fast analysis and high lateral resolution. The analysis depends on several normally uncontrolled parameters, such as angle of incidence, sputter conditions etc. Therefore quantitative calibration is difficult and information on chemical interaction of the impurities is poor. Some of the basic questions related to AES applications have been addressed in a careful work by Viefhaus et al. [8]. Several points of interest have been studied: i) concentration of the impurities in one or more atomic layers, ii) distribution on two corresponding fracture surfaces and iii) others to be discussed later. In case of P segregation in Fe after a careful calibration using single crystals and low energy electron diffraction (LEED) techniques they were able to show that segregation is restricted to the first atomic layer and that two corresponding fracture surfaces show equal concentrations of segregated impurities, thus justifying the common practice of analyzing one surface on one side of the specimen only.

The problem of lacking chemical information from AES seems inresolved. Surface methods highly sensitive to chemical states such as photoelectron and secondary ion mass spectroscopy (ESCA, SIMS) suffer by absence of high lateral resolution or by delicate interpretation respectively. Attempts for applications of ESCA have failed to give basic new information [9]. For detection of hydrogen, SIMS is actually the most recommended instrumentation and has been used to study segregation of deuterium [10]. Other methods, such as electron energy loss spectroscopy have also been applied [11,12] but interpretation is difficult and generally requires comparison with standards. Thus, absence of detailed information on the chemical state of the impurities at the GBs remains one of the serious and challenging problems of experimental work on the chemistry of grain boundaries.

In this context an observation about one more important result from the work of Viefhaus et al. [8] is of interest. The authors suggest possible position changes of P atoms from a second atomic layer with matrix atoms at the top atomic layer immediately after fracture, even at low temperatures. This is a well known process and has been observed specially when reactive elements with low surface energy are involved, as, for instance, during the formation of silicides by depositing thin metallic films on a silicon substrate [13,14]. Such fact should be taken into account in any chemical or structural study of a fracture surface: the freshly created surface does not necessarily represent the original GB state, surface reconstruction may have taken place.

In an extended work Loier and Boos [15] investigated the influence of the sulphur bulk concentration on the fracture mode in pure Ni. Interesting for us are their results for pure Ni which are shown in figure 1. The ultimate tensile strength R is taken as a measure of GB brittleness and plotted as a function of bulk sulphur content C_s after 20 days of heating to 1200°C.

464

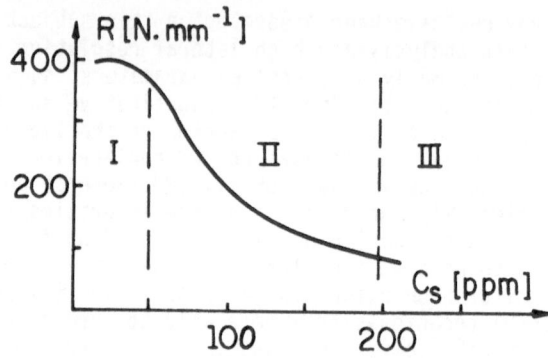

Figure 1 - Tensile strength R as a function of volume impurity S content
(after ref. [15]).

The curve may be divided into three regions. I:C_s < 50 wppm, with little
dependence of fracture mode on C_s; II: for 50 < C_s < 200 wppm, the fracture mode strongly depends on sulphur bulk concentration, reaching a
saturation in region III: with C_s > 200 wppm. This implies that in pure Ni
a certain minimum level of sulphur bulk concentration is necessary to provoke brittleness. Heating 35 days did not change substantially this result. The authors estimate the GB concentration of S by AES and find that
specimens with C_s = 70 wppm and 100 wppm present an intergranular concentration of 10 at.% and 14,5 at.% respectively. One may therefore conclude
from fig. 1 that a certain GB concentration (~ 10 at.%) is necessary to
initiate brittle fracture in Ni and the material becomes more brittle as
the intergranular concentration of S increases.

A somewhat similar work has been carried out by Briant [16] for the system Fe-S. Quantitative results of GB enrichment by S are not given. By
increasing the aging temperature from 400° to 550°C an increasing level of
unprecipitated segregated S was found. Special attention is called to the
fact that for sufficiently high S concentrations the density of sulfides
on the GBs makes AES analysis difficult.

The influence of oxygen on the brittleness of iron has been investigated
by Kumar and Raman [17]. Specimens containing less than 2 wppm of C, S, Si
and 70-125 wppm of O have been tested. It is shown that the percentage of
intergranular fracture is increased from zero to 100% when increasing the
heating temperature from 500° to 800°C. The authors state, however, that,
to observe Fe embrittlement due to O, two conditions must be satisfied:
i) segregation of O to the GBs is a necessary but not a sufficient condition, ii) sufficient O must be retained in solution to give a high yield
strength, in order to avoid blunting of the crack tip. This seems an important observation and the negligence of the second condition might have
been the origin for many controversies in former oxygen induced brittle
fracture research.

Embrittlement power is commonly attributed to the elements of the 3rd to
5th period of group IV to VI in the periodic system [2]. Some experimental
works have clearly shown that embrittling potence is not restricted to
those elements. Fukushima, Birnbaum [18] and Woodfort, Bricknell [19]
report on Ni embrittlement by chlorine. Even metals may act as embrittler

in other metals as shown by Allegra et al. [20], White et al. [21] and Nieh [22] for Zn in steel, Th in Ir and Ni in W, respectively.

Looking to embrittlement from the point of view of chemical interaction it seems surprising that even the noble element He can provoke intergranular fracture. Several authors report on this effect in steels as for instance, Caskey et al. [23] and Lane, Goodhew [24]. This type of embrittlement is of major importance especially in reactor engineering where radiation occurs. The evaluation of this effect must take into account that it is a somewhat artificial process: the material is energetically highly disturbed since the He is implanted into the matrix by radiation. The embrittling effect is certainly related to the kinetic implantation energy of the He atoms which are imbedded in the matrix or, at high energies, may destroy the lattice and create amorphous regions. Relaxation of thereby formed high stress fields results in segregation to GBs in such a level that their bond strength is reduced. Computer simulations, as for instance from Baskes and Vitek [25], have shown that high angle as well as dislocation edges at low angle GBs serve as strong sinks for He promoting the formation of He clusters and bubbles.

Grain boundary fracture in several sytems without any detectable segregated impurity has been reported by various researchers. Ogura et al. [26] have studied the intermetallic compound Ni_3Al and could not find any amount of impurity segregation at the intergranular interfaces. The fracture surfaces presented a plastically strained layer. The same entire absence of segregated impurities and hence an intrinsic brittleness was found for Ni_3Al and Ni_3Si by Takasugi et al. [27]. In certain conditions, even steels may become intrinsically brittle as reported by Lee and Morris [28] in a Fe-12 Mn steel. Even by carefully analyzing the fracture surface they could not detect segregated impurities. A possible influence of carbide could not be definitely excluded, however. These studies suffer from a real weakness of the AES technique. Not only the sensitivity of AES is rather low (about 0.1 % of a monolayer); the electron beam may also produce carbon adsorption on the surface, by cracking residual gases as CO or C_nH_m. Additional methods such as static SIMS could help to solve these problems.

Impurity segregation may also have beneficial effects on GB strength. This has been observed for boron by several authors. Liu et al. [29] showed that additions of 0.4 % B to Ni_3Al increased dramatically the ductility, and fracture mode changed from intergranular to transgranular. Boron segregated strongly to the GBs. The results suggest that alloy stoichiometry at the GBs strongly affects their cohesion. Also in case of the system Fe-12 Mn, which has been discussed above [28], the intergranular fracture type could be prevented by addition of small amounts of B [30].

A remedial effect, also of carbon, has been reported by Erhart and Grabke [31] in case of P segregation in Fe. The authors attribute this to site competition between C and P at the GBs. Lee et al. [32] could show in a systematic study that additions of Mo or W reduces brittleness of the Fe-P system whereas Mn enhances intergranular fracture.

2.2. Grain boundary structure

Experimental work on GB structure suffers from the fact of difficult accessibility of analytical methods such as LEED or ion scattering [33], which provide information on an atomic scale. Several simulation methods have, therefore, been developed in order to approach the atomistic structure of GBs by hard sphere models [34,35] or relaxation methods [36,37]. It is specially the latter one which, by the use of appropriate interatomic potentials, has found a lot of application in atomistic GB structure simulations related to intergranular fracture.

Several investigators, however, have published experimental work in order to overcome lack of structural information from GBs and trying to link GB chemistry to interfacial structure; with the aid of these ideas it has been possible to show the unquestionable importance of structural effects in intergranular fracture problems.

The degree of GB embrittlement as a function of GB misorientation has been studied by Roy et al. [38]. By means of a special sintering technique at 1050°C in hydrogen atmosphere a great number of isolated Cu + 0.1 at.% Bi spheres (diameter ~ 60 μm) have been deposited on a [110] single crystal surface.

During the annealing, for 30 min., GBs are formed between the crystal surface and the randomly orientated spheres (as checked by X-ray diffraction). Simultaneously, Bi segregated to the interfaces causing embrittlement. Subsequently GB fracture was initiated by vibrating the specimens ultrasonically, see figure 2.

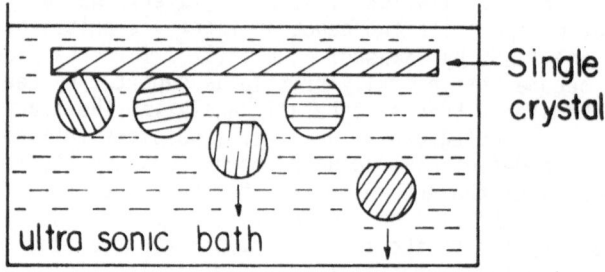

Fig. 2 - Scheme of the experiment of Roy et al. [38]

The vibrations caused the fracture of the embrittled interfaces and the spheres started to fall down. When checking by X-rays the orientational distribution of the spheres still stacking on the surface the initially randomly distributed texture changed to a selected one, with those spheres cracked off first which showed highest boundary energy (highest misorientation). Figure 3 shows a sequence of X-ray texture measurements from a) before and b)c) after 5 and 10 min. ultrasonic treatment respectively.

When annealing other specimens for 50 h at 1050°C the small spheres were able to rotate their orientation into energetically favourable positions which are those of low GB energy. In figures 3d and 3e, X-ray spectra of a set of spheres from a 30 min. sintering and subsequent ultrasonic treatment (d) are compared to those from spheres treated for 50 h without ultrasonic treatment (e). The spectra are identical leading to the conclusion that boundaries of least embrittlement (d) are boundaries of low energy. The stronger embrittlement of GBs with higher energy (or misorientation) is explained by increasing number a segregation sites due to increasing number of secondary dislocations in the interface.

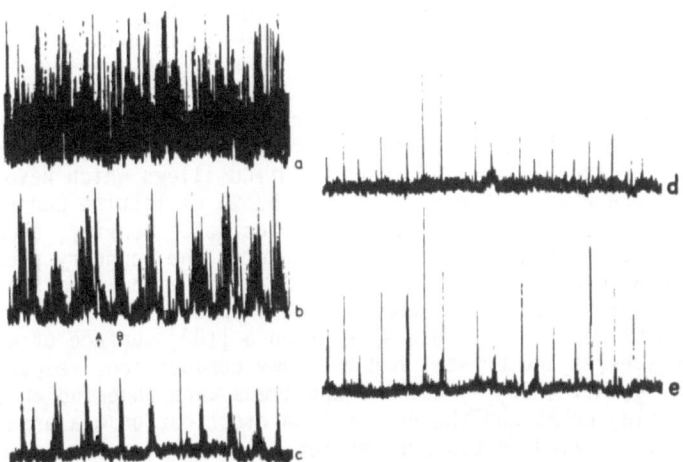

Fig. 3 - X-ray texture spectra a) before and after b) 5 min. and
c) 10 min. of ultrasonic treatment; d) 30 min. sintering and extended ultrasonic treatment, e) 50 h sintering without treatment
(from ref. [38]).

Another important question related to interfacial structure has been addressed in an experiment by Sickafus and Sass [39]: do impurities change
the GB structure after segregation? For this end they examined the structure of the same boundary both in the absence and in the presence of a segregated solute. Using the "hot pressing technique" bicrystals containing
a [001] twist boundary misoriented by $\theta \simeq 1.5°$ were produced of pure Fe
and of Fe + 0.18 at.% Au. Annealing temperature was 350°C to allow Au segregation to the GBs which was checked by Rutherford backscattering technique and energy dispersive X-ray spectrometry. As shown in figure 4
transmission electron micrographs indicate a different dislocation network, one aligned along <110> directions in pure Fe (fig. 3a) and another
with dislocation network alignment along <100> direction (fig. 3b) in case
of the interface with Au segregation. From this is it concluded that the
segregation of Au to the small angle twist boundary has changed its structure.

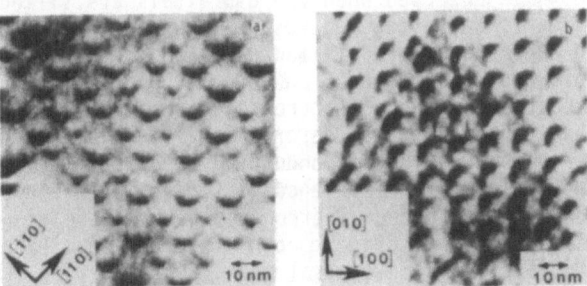

Fig. 4 - Transmission electron micrographs of $\theta \simeq 1.5°$ [001] twist
boundary in (a) pure Fe, (b) Fe + segregated Au, (from ref.
[39]).

The results of these two experiments cannot provide detailed structure information on an atomic scale; however, they strongly indicate the importance of interface structure considerations when modeling embrittling mechanisms. Other investigations addressing the problem of embrittlement and interface structure have been reported [40-44] resulting in the same conclusion reported above.

A recent paper calls attention to a special effect, i.e. the dependence of interface structure on the electronic configuration of the alloying elements. Maurer and Gleiter [45] studied Ni-Cu alloys which have fcc lattice structure with little variation (2.53 %) of lattice constant. Hence, if crystallographic parameters as coincidence density, boundary periodicity etc. are the controlling parameters, the boundary properties should be comparable in all Ni-Cu alloys. By using a modified "sphere rotation technique" (see [38]) samples (small spheres of ~ 7 μm diameter) were prepared by depositing these spheres on a [100] surface of a single crystal. The spheres and crystal had the same composition, respectively: Ni, $Ni_{2/3} Cu_{1/3}$, $Ni_{1/3} Cu_{2/3}$, Cu. The specimens were annealed at 50°C below the melting point and the orientation distribution measured, by X-ray methods, as applied in texture studies. Initially, during annealing, the spheres rotated into low energy orientational positions. After 50 h annealing rotation was no longer observed. The orientations of the spheres were now determined and the corresponding low energy GB type $\Sigma = 1$, $\Sigma = 3$, $\Sigma = 5$ ... up to $\Sigma = 35$ was calculated. The total number of possible crystallographic equivalents for this series of Σ is $n_p = 255$. The observed total number of equivalents showed strong variations depending on the alloy composition. It was found $n(Ni) = 7$; $n(Ni_{2/3} Cu_{1/3}) = 43$; $n(Ni_{1/3} Cu_{2/3}) = 99$ and $n(Cu) = 124$. Hence by adding Cu to Ni the finding was an increase in number of Σ equivalents.

Since the only significantly changing parameter by adding Cu is the density of occupied electronic states, it is concluded that crystallographic parameters alone are not sufficient for characterization of GB structures, and the authors consequently suggest that simulation methods based on pair potentials [36,37] should include a factor describing electronic configuration.

We would like to mention an additional observation: when following the alloying sequence Ni - $Ni_{2/3} Cu_{1/3}$ - $Ni_{1/3} Cu_{2/3}$ - Cu the difference Δn in observed total numbers Σ is $n(Ni_{2/3} Cu_{1/3})-n(Ni) = 36$, $n(Ni_{1/3} Cu_{2/3}) - n(Ni_{2/3} Cu_{1/3}) = 55$ and $n(Cu) - n(Ni_{2/3} Cu_{2/3}) = 26$. That means the most significant change Δn is observed when the d-orbitals are filled and the subsequent occupation of s-orbitals causes less number of orientational variations. To our understanding, this indicates the better adequacy of localized electronic orbitals rather than extended band states (as used by the authors) to describe interface electronic states. In other words: if for an interface the regular lattice description of solid state physics, i.e. extended electron states in the conduction band would still be valid, the atoms should not "feel" big differences in energy when forming an interface, since the free extended electron gas between the ion cores would serve as "lubricant" providing "smooth" interactions. Localized states, however, in particular, d-orbital are structured, i.e. they have preferential orientations in space. In Cu all d-states are occupied, bonding is mainly due to s-electrons and these have spherical orbitals, hence bonds are formed easily independent of orientational problems, which results in a great number of low energy interfaces. Whereas in Ni d-electrons participate in bonding and their structured orbitals permit only a reduced number of low energy interfaces. In summary: two conclusions may be drawn from this experiment of Maurer and Gleiter: electronic configuration may influence the geometrical structure of GBs and the more adequate description of GBs seems to be localized orbitals rather than extended electron states.

Some other work related to structural problems have been published. For instance, the dependence of impurity segregation on the GB structure was studied by Briant [46] and a relationship between structure and chemical bonds of the segregants is generally discussed unfortunately on a very qualitative basis and restricted to some GB units, as suggested by Ashby [34], therefore no guiding conclusions could be drawn off.

Other effects such as GB motion [47] or bulk dislocation interaction with GBs [48,49] may also influence the segregation level and consequently structure and fracture behaviour.

2.3 Effects of impurities and hydrogen

A very important aspect of intergranular failure is related to the simultaneous segregation of hydrogen and metalloid impurities.

This deleterious effect of impurities and hydrogen has been demonstrated clearly by several investigators. For iron, Jones et al. [50,51] studied effects of S and P as a function of cathodic potential in 1n H_2SO_4 solutions. GB chemistry was determined by AES on separated specimens by common vacuum fracturing. They observed a transition in the fracture mode from ductile to 100 % intergranular, when a GB sulphur concentration of 0.13 monolayers was achieved and when tested at -0.6 V and a strain rate of 7.10^{-5} sec.$^{-1}$. A cathodic potential less than - 0.6 V, i.e. a lower hydrogen pressure, did not provoke 100% intergranular fracture. Phosphorus was found to be less detrimental than sulphur. In contrast to sulphur 0.2 monolayers of phosphorus was not sufficient to cause intergranular fracture in straining electrode specimens tested at cathodic potentials of -0.5 to -1.0 V.

Although there is evidence that impurities and hydrogen may act simultaneously, little is known on the mechanism of this interaction. Two models have been suggested, one based on an additive effect [52] the other assumes a synergetic behaviour [53]. The additive model assumes an independent action of both types of segregants, while in the synergetic or adsorption model a cooperative effect is assumed, for instance, by first formation of hydrides, which, in turn, provoke embrittlement. This question was addressed in a work of Shin and Meshii [54]. Their results indicate that the effects of both hydrogen and sulphur in iron is to reduce the GB strength, that is, independently and additively. On the other hand, a relationship between sulphur content in steel and reversible and irreversible hydrogen trapping in the matrix volume has been reported by Iino [55]. The observed increase in H trapping with increasing S content indicates an interaction which might support a synergetic embrittling behaviour.

Some thorough studies in Ni also resulted in not unambiguous conclusions and it seems that the question whether an additive or synergetic (or combined) action exists, is a very difficult task requiring more detailed in-

470

formation. Bruemmer et al. [56] investigated the influence of S, P and Sb on intergranular hydrogen embrittlement of Ni. Using straining electrode· tests in 1n H_2SO_4 solutions at varying cathodic potentials they could show that increasing hydrogen pressure (potential -0.3 to -0.72 V) reduced the critical S fracture surface coverage from 0.2 to 0.05 monolayers. This is illustrated in figure 5.

It is seen that 100% intergranular fracture is observed at ~ 0.08 monolayer GB coverage of S when cathodic voltage is -0.72 V whereas -0.3 V gives 60% intergranular fracture at ~0.2 monolayer coverage. By estimating the embrittling potencies of the studied impurities the authors found the sequence S>Sb>P.

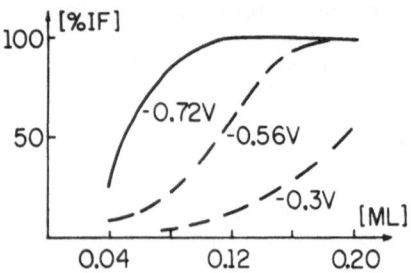

Fig. 5 - Percentage of intergranular fracture of Ni as a function of S
content in monolayers at GBs with varying cathodic test potentials (from ref. [56]).

The obtained results are discussed with respect to the additive and syn-ergetic models and it is argued that the behaviour illustrated in figure 5 may be explained in both. However, to our understanding, when less avail-able S requires more H to achieve intergranular fracture than this fact suggests an additional effect of S and H which is in agreement with the general conclusions of the authors who suggest support to the independent and additive effect of metalloids and hydrogen.

As mentioned already the technique SIMS has also been applied in order to find more insight in the impurity-hydrogen process in Ni [10]. Although evidence was found in the course of the experiments for strong interaction between substitutional S and interstitial ^2H (deuterium) solutes, no direct evidence for hydride formation at surfaces could be detected.

Since SIMS is a method very sensitive to chemical information this seems a strong argument against a combined action of metalloid segregants with hydrogen on the basis of hydrid formation. On summarizing: more results seem to support an additive process but the overall impression remains that the mechanism of impurity assisted hydrogen embrittlement is still an open question waiting for more and new suggestions and interpretations.

3. MODELS OF FRACTURE MECHANISM

As mentioned already in section 1 the theoretical description of inter-granular fracture has changed completely. Instead of a model drawn from thermodynamics currently different new approaches are available which pro-vide quite a detailed insight in the way how decohesion may proceed on an atomic scale. This gain in atomistic understanding still did not result in an overall picture permitting a quantitative description using macroscopic

parameters. No doubt, however, the significant progress lies in the way
how to view atomic bonding.

Physics, i.e. quantum mechanics describes electrons by their wave functions $\psi(\bar{r})$ and bonding may be represented by overlapping of these functions between the ion cores. The description of bonds by quantum mechanical wave functions in different environments has led to fracture models
which we will name chemical. Yet a rigorous treatment of fracture has to
take into account the dynamic process of decohesion as a result of applied
stress. As was seen in section 2.2, GB structure plays an important role
and approaches describing these effects we will denominate structural.

3.1 Chemical effects

Conducted by results from surface physics and by the conviction that any
atomistic discussion of fracture should focus on that what really will be
disrupted, that is the chemical bond, Losch [5] suggested a model which
could explain that impurity (I) segregation to GBs might result in reduction of bond strength along the GB. When the fracture path follows the GB
interface then because bonding is modified in this area.

Qualitatively this was explained in the following way: conventional embrittling elements belong to the 3rd to 5th period of groups IV to VI of
the periodic system. These elements have unfilled np^X shells and it is
known that p-electrons tend to form localized covalent bonds. By extract-
ing results from quantum mechanical calculations of chemisorption of S on
a Ni free surface and transferring them to the similar problem of a GB
Losch suggested that reduction of bond strength along the GB may be caused
by charge transfer from the metal to the impurity atom. The S atoms segregated to the GB of Ni may form hybridized bonds involving S (3p) and
Ni(4s) electrons. Since this covalent bond type is strong the involved 4s
electrons are now localized between the S and Ni atoms and do not longer
participate in the Ni-Ni metallic bonds which normally are formed by 3d
and 4s electrons. The result is a reduction of the Ni-Ni bond strength.

This effect of covalent bond formation along GBs is illustrated qualitatively in figure 6. It represents a two dimensional 53° GB segregated
with substitutional impurities. The strong I-M bonds (M = metal) are indicated by double lines, normal M-M bonds by dashed and weakened M-M bonds
by dotted lines.

From this picture it is suggested that bond strength is reduced in the
immediate neighbourhood of the impurity layer and fracture may occur preferentially along the GB, however, not by breaking M-I bonds across the
segregated layer, but by disruption of weakened M-M bonds adjacent to the
segregation layer. Due to irregularities the fracture path will certainly
not follow one side of the GB only but may cross the I-layer and continue
on the other side. Such a possible fracture path is indicated in figure 6.

Two other characteristics of impurity segregated GBs may contribute to
make the GBs a preferential fracture path. As mentioned already typical
embrittlers have external p-orbitals and these bonds are known to be highly localized and directed in space resulting in a very rigid character of
the bondings. Plastic deformation in materials formed by p-bondings is
more difficult and those materials are generally known to be brittle. This
fact may contribute to the brittle character of a segregated GB. A third

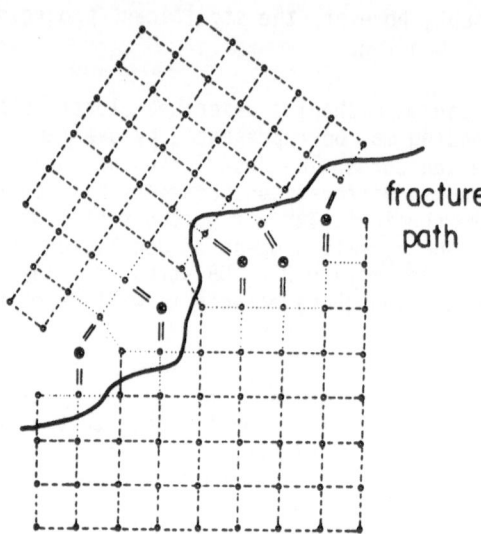

Fig. 6 - The 53° grain boundary after substitutional impurity segregation.
Dotted lines indicate modified M-M bonds. A possible fracture
path is indicated (after Losch [5]).

effect may result from a possible mutual interaction of the impurities it
self, i.e. bond modifications would again be the consequence. This point,
however, is not discussed more profoundly in [5].

All three effects presented by Losch are discussed in a very qualitative
manner as a first attempt to conduct fracture mechanism treatments in a
more physical based direction, i.e. to visualize the nature of the
chemical bonds. In section 2.1 we have discussed the work of Viefhaus et
al. [8] who have tried to demonstrate experimentally this suggested M-M
bond breaking. The authors wanted to show that impurities on a fracture
surface, as may be seen from figure 7, would generally be covered by a
layer of metal atoms. As discussed in 2.1 the failure of their experiment
does not indicate necessarily an alternative mechanism but may point to
another important process, that is fracture surface reconstruction immediately after fracture even at low temperatures [13,14].

In order to give the ideas of Losch's model a more quantitative basis
some attempts have been published trying to calculate the reduction of M-M
bond strength by application of methods from solid state physics. Anda et
al. [57] as well as Sayers [58] use the tight-binding approximation for
calculating a linear chain and a cluster model respectively with one impurity in a substitutional position. Both treatments come to the conclusion that really the M-M bond strength is reduced when an impurity forms a
strong I-M bond.

Certainly the most extended and well funded work on the influence of chemical effects in intergranular fracture has been conducted by Briant and Messmer in the course of the last years. The theoretical method they use to investigate the electronic structure of the local environment at a GB consists of two parts [6]: the first is to use a cluster to represent the local environment and the second is to use molecular orbital theory to solve for the electronic structure of the cluster. The choice of an appropriate cluster is simplified by the work of Ashby et al. [34] who have suggested that the structure of a GB can be described in terms of characteristic units. In case of a fcc solid a tetragonal dodecahedron has been chosen. This cluster constructed by 8 M atoms is shown in figure 7.

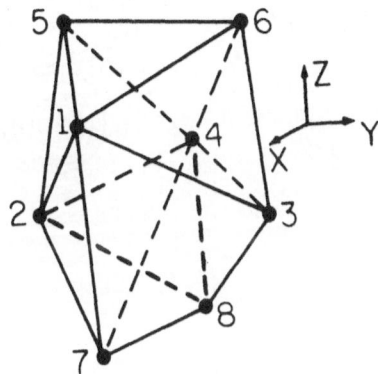

Fig. 7 - Tetragonal dodecahedron representing a structural GB unit (from ref. [6,60]).

Atoms 1 to 4 may be considered to represent the GB and atoms 5,6 and 7,8 are in the first layers adjacent to the GB. In order to investigate the electronic effects of sulphur in a Ni GB the authors calculate first the electronic structure of the cluster composed by 8 Ni atoms. For a second calculation a S atom is placed in the center of the cluster, i.e. in an interstitial site which an impurity may choose at a GB. Application of molecular orbital theory permits calculation of the energy levels of the diverse electronic orbitals as well as the spatial structure of the latters. More instructive, however, for our ends is a counter plot of the total valence charge density which is obtained by summing the charge densities over all the occupied valence orbitals. Such charge counter plots are shown in figure 8.

It is represented the x-z plane which contains the Ni atoms numbered 1,4,7,8 and the S atom in the Ni_8S cluster. The counters are logarithmic with increasing number representing increasing charge density. Counter line 4 differs most drastically between the two clusters and demonstrates clearly the influence of the S atom: high charge density between S and Ni_1 and Ni_4 means strong orbital overlapping thus strong bonding between these atoms. Consequently charge density between Ni_1-Ni_7 and Ni_4-Ni_8 is reduced. This means strong NiS bonding weakens the close Ni-Ni bonds a result which agrees completely with the suggestions of Losch [5]. For fracturing a GB with weakened M-M bonds less energy would be required hence giving a clear indication of a preferential fracture path.

It is seen that also the bond between 8 and atom 9 (figure 8b) is affected. This means that charge is transferred from atoms 8, 9 to the S atom.

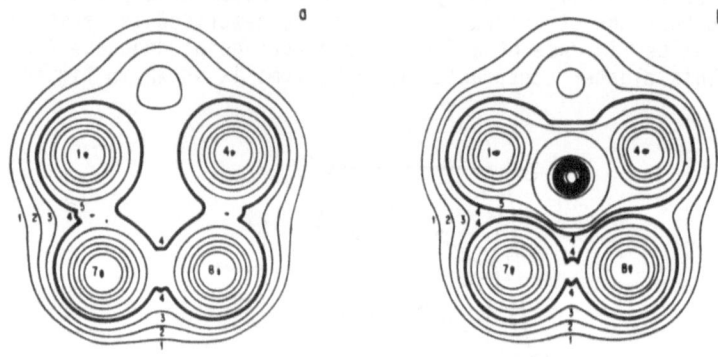

Fig. 8 - Charge counter plots in the x-z plane containing Ni atoms 1,4,7,8 in Ni_8 and the additional S atom in Ni_8S. Counters are logarithmic, counter 4 is drawn more darkly (after ref. [60]).

Although this cluster approximation suffers principally from two weaknesses, namely it has to work with a predetermined structure and it does not consider influences from next nearest neighbours, which may result in unrealistic charge distribution inside the cluster, this method has been most successful in demonstrating directly the weakening of M-M bonds and consequently suggests a clear picture of an electronic mechanism by which GB embrittlement could occur.

The problems of atomic arrangement in the cluster calculations has been treated in another work by Briant and Messmer [59]. The question is how far geometrical structure of the clusters and their number of metallic atoms could influence the charge transfer and consequently the M-M bond weakening. Again cluster structures suggested by Ashby et al. [34] are chosen which include the tetrahedron, the tetragonal dodecahedron, the capped trigonal prism and the Archimedian antiprism. The number of atoms is Ni_8S, Ni_9S and $Ni_{10}S$. The clusters are constructed in that way that the S has an increasing number of Ni neighbours. One would therefore expect less charge to be drawn off with increasing number of Ni atoms and consequently less M-M bond weakening. This picture is consistent with the results: the atomic arrangement does affect the amount of charge transfer and hence the degree of embrittlement. An increase in the number of nearest neighbours around the impurity decreases the embrittling potency. In general the important conclusion is: the reduction in M-M bond strength depends not alone on concentration but on geometrical structure of the environment as well.

The different embrittling potency of the elements S,P,C, B is demonstrated by Messmer and Briant [60] by substituting the diverse impurities in the center of a tetrahedron formed of 4 metal atoms. Again the results may best be represented in counter plots of valence charge density which in this case is defined by

$$\rho_{I-M} (\vec{r}) = \sum_{I-M} n_j \, \varphi_j^* (\vec{r}) \, \varphi_j(\vec{r})$$

where $\varphi_j (\vec{r})$ are the molecular orbitals of the system and n_j is the occupation of the jth orbital. By this formula only those orbitals are considered which contribute to the I-M bond. Figure 9 shows the plots for the clusters Fe_4S, Ni_4S, Fe_4P, Fe_4C, Ni_4B, respectively.

In the plots of figure 9 the impurity atom is placed in the center and it is seen that for the Fe_4S and Ni_4S cluster charge density is greater on the impurity atom. In Fe_4C and Ni_4B charge is distributed much more equally between the impurity and the host atom. The figure thus agrees with the experimentally proved sequence of embrittling potency S>P>C>B where the latter element B already serves as a cohesive enhancer. Again we would like to call the attention to the fact that only one possible atomic arrangement is considered and that atomic distances are maintained constant in the clusters what certainly does not correspond to reality.

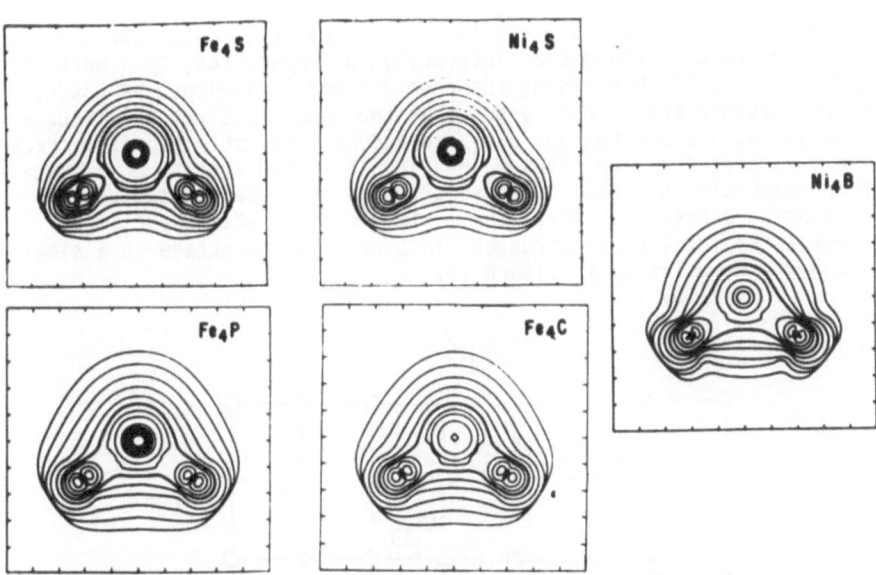

Fig. 9 – Valence charge density $\rho_{I-M} (\vec{r})$ including only those orbitals which contain a contribution from the impurity atom (from ref. [60]).

Consequently proceeding in their model calculations Briant and Messmer

476

investigate further the variation of metallic host atoms M = Fe, Cr, Mn, Ni for the impurity P [61] as well as the behaviour of Sb in Fe, Ni, Cr and their alloys [62]. The general result of Briant and Messmer's work is that the electronegative impurities draw off charge from the metal atoms and consequently weaken M-M bonds. In [61] charge distribution is calculated quantitatively and it is concluded that P is less electronegative with respect to Ni than to Fe. Therefore, Ni will not enhance P induced embrittlement but should counteract it. On the other hand, P is found to be more electronegative to Mn and Cr than to Fe thus enhancing embrittlement. Comparing Cr and Mn, P is found to be more electronegative with respect to Cr than to Mn in contradiction to Pauling's scale. The authors therefore recommend caution by applying Pauling's scale to metallurgical problems. In our understanding this appears exaggerated. Due to Pauling the difference P-Mn and P-Cr is small: 0.1, and it seems questionable whether the cluster method represents such a degree of accuracy.

Due to the inherent disadvantages of the cluster method (no next nearest neighbours, no relaxation) it is difficult to estimate the value of the quantitative results of Briant and Messmer given in [62] for a multicomponent system. Certainly the results may be considered as guidelines and useful help in planning future experiments as stated by the authors. There is no doubt, that in multi-component alloys chemistry is very complex and the embrittling power of an impurity will clearly depend on all elements present. No doubt also that the work of Briant and Messmer has contributed most substantially to the understanding of intergranular fracture problems.

Attention is called to another effect, which had been mentioned already by Losch [5], that is the mutual interaction of impurities, by a work of Eberhart et al. [63]. Discussing the chemical model developed by Losch, Briant and Messmer the authors argue that the impurities not only reduce cohesive strength along the GBs but also reduce shear strength at a crack tip. In a work of Rice and Thomson [64] it was stated that at an atomically sharp crack tip the ideal cohesive strength is probed by the concentrated tensile stress $\sigma_{\theta\theta}$, while the maximum concentrated shear stress τ_{max} probes the ideal shear strength. This we will illustrate in a simplified manner schematically in figure 10.

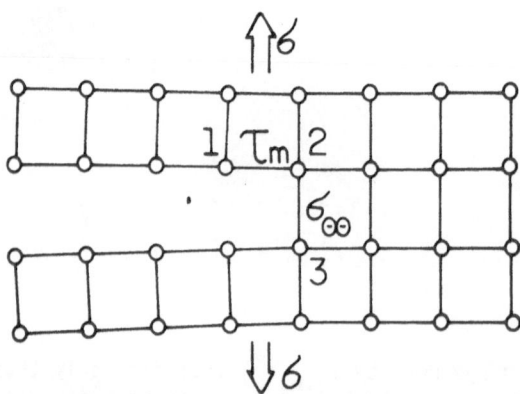

Fig.10 - Concentrated tensile stress $\sigma_{\theta\theta}$ and shear stress τ_{max} at an atomically sharp crack tip.

The tensile stress $\sigma_{\theta\theta}$ is represented on the bond between atoms 2,3 while τ_{max} stresses the bond between atoms 1,2. Bond weakening by impurities may affect tensile stressed bonds as well as shear stressed bonds. The latter case would result in emission of a dislocation at the crack tip which thus will be blunted and, in principle, behave in a ductile manner. The authors propose that a complete theory of intergranular embrittlement must deal not only with the way the cohesive strength changes but also the effect the segregation has on the shear strength. The experimental work of Loier and Boos [15], discussed in section 2.1, found that fracture mode changes significantly from ductile to brittle manner when reaching a. GB impurity concentration of ~10 at %, i.e. when mutual I-I interaction would start. Eberhart et al. claim that at this critical concentration the I-I interaction strengthens the bond within the GB plane resulting in an increase of shear strength while the cohesive strength across the GB plane continues to diminish. And these are the conditions for brittle fracture. They support their argumentation with molecular orbital calculations on a Ni_3S_2 cluster which show indeed attractive S-S bonding though less strong than Ni-S bonding. Fracture theories not considering I-I interaction are therefore said to be restricted to dilute concentrations [63].

3.2 Dynamic Model

Very recently a new aspect has been brought into the discussion by a work of Eberhart, Latanision and Johnson (ELJ) [65]. What they suggest is basically a dynamical consideration of electronic orbitals during stress with the argument that atomic movement may result in charge redistribution affecting chemical bonds. Thus, their basic question is how the charge density around an atom changes when subjected to stress.

This is, of course, a very difficult task and very little exists in literature on this subject. All the more, as in our context, the interest is focussed on the GB which is a defect area, not well understood even without applied stress.

In an extended overview ELJ first discuss chemical bonding and mechanical properties and define bond directionality and charge polarizability. The first is thought to describe how tightly packed the electron density is between bound atoms, i.e. it is directly related to the description of atomic orbitals as known from quantum mechanics, for instance, the tight and directed p and d bonds or the spherical s orbitals. And the ease with which charge redistributes as a result of the movement of neighbouring atoms is called charge polarizability. Both concepts may be considered to be inversely proportional since an increase in directionality reduces the ease of charge redistribution (overlapping of orbitals).

The fundamental statement of ELJ is the following: "when a system is subjected to an axial strain, the axial distance between atoms must increase. Because the nuclei of these atoms are moving apart the potential well between these nuclei is becoming less deep. This will result in a flow of electron density from the bonds parallel to the direction of applied stress to the bonds perpendicular to the direction of applied stress". The justification of this statement is seen in the success in explaining the trend in Poisson's ratio by considerations of the electronic configuration of the elements. Since the origin of binding energy is ultimately electrostatic in nature an increase in electronic charge between nuclei may result in a stronger attraction. Therefore, by increasing the atomic distance parallel to stress charge flows to bonds perpendicular to stress resulting in an increase in strength of these latter bonds. This is the way that ELJ explain the problem how cleavage can occur; we will come back to this question in section 4. Accumulation of charge between atoms perpendicular to stress will cause a contraction in this direction and, of course, the ratio of perpendicular to axial strain gives Poisson's ratio.

On applying this model to cleavage of two atomic planes it is clear that stress across the cleavage plane should cause charge flow to bonds perpendicular to the stress resulting in a strengthening of bonds parallel to the cleavage plane. This charge flow is related to charge polarizability and the suggestion is, therefore, that an "impurity will facilitate brittle fracture if it increases the charge polarizability of the atoms across a cleavage plane". In case of S at a Ni GB this means the following: as has been shown by Briant and Messmer [59-61] the S-Ni interaction is of an ionic type. From their former work [63] ELJ conclude that a simultaneous S-S interaction makes the S-Ni bond of more covalent, i.e. directional type by drawing charge off from the S atoms. "Therefore, when stress is applied perpendicular to the GB, the S atoms on either side of the GB are moving apart becoming less interacting. Hence the S-Ni bonds crossing the GB are becoming more ionic". As schematically shown in figure 11 the atoms parallel to the GB gain in charge and become, therefore, more covalent and more directed. This is the condition to promote bond breaking: charge is removed from the bonds across the GB resulting in strengthening and increasing directionality of bonds parallel to the GB, the latters thus being more resistent against shear. The overall result of this dynamical consideration will be fracture along the GB, i.e. an embrittlement.

ELJ try to extend their model to the problem of hydrogen embrittlement. Since they assume a possible participation of sulphur impurities we will, in continuation of section 2.3, briefly discuss these suggestions here. The authors propose that high hydrogen concentration is a condition for embrittlement. When the hydrogen concentration reaches a level where hydrogen atoms begin to interact highly polarizable orbitals are produced. Under stress a charge flow process will be induced similar to that suggested for the S-Ni system. In this case, however, by the dissociation of hydrogen molecules parallel to the applied stress and a recombination of hydrogen perpendicular to the stress.

It is tried to support this model by calculations of Eberhart et al. [66] on hydrogen assisted bond stabilization in amorphous silicon (a-Si). This work suggests that in regions of large lattice strains hydrogen can stabilize the Si-Si bond by forming a region of weakly interacting hydrogen atoms, i.e. the incorporation of molecular hydrogen in the a-Si matrix is assumed. In case of a metal ELJ further suggest that hydrogen rich regions should form in the presence of other defects, such as sulphur, by formation of H_2S - like molecules which acts as traps for additional hydrogen atoms. Hence, as discussed in section 2.3, this model of ELJ is following the suggestions of a synergetic effect between H and impurities.

3.3. Structural effects
Another extensively studied approach for illumination of GB problems is by simulation of GB structures using relaxation methods. The great importance of this simulations lies in the fact that currently no experimental

method exists for the direct observation of the GB structure on an atomic scale. Recently, progress could be noted in the direct observation of GB structures by application of high resolution electron microscopy. Due to problems in sample preparation this method is still limited to semi-conduc-

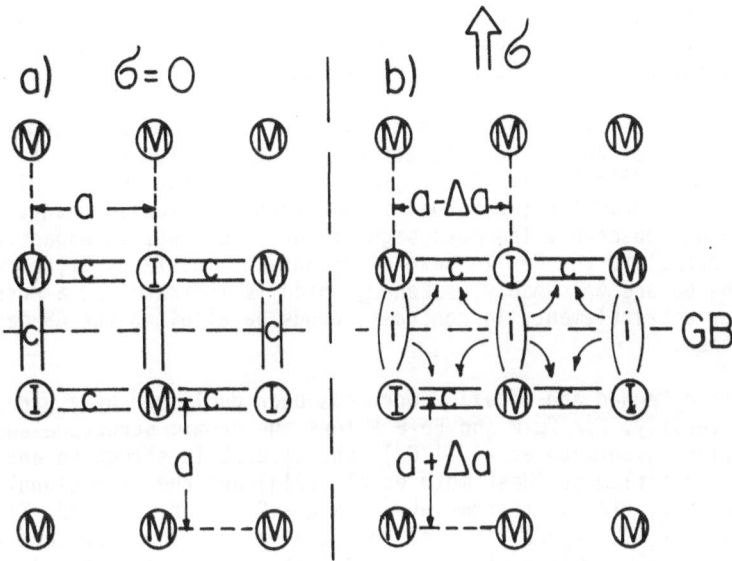

Fig.11 - A segregated GB without a) and with stress σ b). Bonds crossing the GB become more ionic due to the increase Δa in atomic distance by stress σ. (c: covalent, i: ionic).

tors and some special metallic GBs only. It should be mentioned that it is not strictly correct to classify the relaxation method, as done here, by "structural" effects only. As a matter of fact, the results of these investigations arrive to chemical conclusions quite similar to those discussed in section 3.1, thus giving support to the models based on "chemical" effects. As in section 3.1 we will not describe and not discuss in detail the relaxation methods, or their advantages and disadvantages, our subject is restricted to their results in terms of atomistic mechanism for embrittlement.

Hashimoto et al [7,67] for the first time applied this method to calculate the electronic interaction of P at a Σ = 5 tilt GB in Fe. The obtained results of the local density of states (LDOS) led to the conclusion that strong bonds are formed between P and Fe atoms, while bonds between these Fe atoms and their neighbours are weakened. This is in total agreement with the suggestions of Losch [5] and Briant and Messmer [6]. The origin of the bond weakening is seen in both chemical effects and lattice distortions due to segregated P. Furthermore they found that dilute P segregation does not affect the GB structure remarkably. However, a monolayer segregation causes a new structure. This result shows the importance of the relaxation methods when compared to cluster calculations of section 3.1 which cannot account for structure modifications.

The basic problem in relaxation methods is to find a potential which would describe the atomic interactions in a most appropriate manner. As stated by Maeda et al. [68] central force pair potentials are not sufficient to describe atomic interaction in metals. Therefore, one basic problem is to determine methods which fit the theoretical potentials to measured quantities. Furthermore, pair potential calculations do not consider directional nature of chemical bond. Certainly, the results have to be seen with certain caution as an approach to get insight into atomic arrangements at GBs where experimental methods are not available.

Sutton and Vitek [69] calculate tilt GBs of Cu and Au with segregated substitutional Bi and/or Ag. The structure and energies of the high angle GBs ($\Sigma = 5$, $36.87°$; $\Sigma = 17$, $28.07°$) are studied. Since the impurity concentration was held low (no mutual interaction) no drastic change in GB structure was observed. The most significant change was an expansion/contraction normal to the GB. In case of expansion, mostly by Bi, bonds across the GB are presumably weakened, which is indicated as a possible reason for embrittlement. In contrast, bonds parallel to the GB may be strengthened.

The most extended and detailed work has been published by a group from Tokyo University. For Fe-P and Fe-B alloys the atomic structure and stress distribution (Hashimoto et al. [70]), the electronic structure and intergranular embrittlement (Hashimoto et al. [71]) and the vibrational states (Wakayama et al. [72]) of atoms at $\Sigma=5$ and $\Sigma=9$ GBs are studied. The system is composed of 100 layers parallel to the GB plane. The structure is relaxed iteratively until a minimum in internal energy is achieved, while the total volume is maintained constant. We will limit our discussion here to results from the $\Sigma = 5$ tilt boundary.

Figure 12 shows a sequence of the $\Sigma=5$ GB before and after segregation. Without impurities, figure 12a, the GB of α-Fe is constructed by a periodic stack of capped trigonal prims, circles and crosses indicate atoms in different planes.

P and B atoms were then placed in substitutional and interstitial positions respectively of this prims. After the relaxation process, the structure did not change in the case of B segregation. The B atoms maintained their interstitial place in the center of the prims, figure 12b. Contrary to this, P segregation changed completely the initial structure of the GBs. P atoms after relaxation moved from their initially substitutional positions to interstitial ones. Two types of P segregated GBs are observed, figure 12c and 12d, and the local atomic environment surrounding the P atoms is similar to that of the crystalline $Fe_3 P$ compound. Hence it is concluded that P segregation induced a kind of structural transformation in the host GB. Furthermore, stresses have been calculated and are shown in figure 13. Arrows pointing to the right represent tension and those to the left indicate compression.

It may be seen that also a clean GB is not free of stress and that B did not change very much the stress distribution at the GB (fig. 13b), while P induced large hydrostatic tension and compression, figures 13c and 13d.

In order to elucidate the binding character of the segregated GBs in relation to the atomic structure the LDOS N(E) are calculated [71]. This is an advantage of the relaxation method over the molecular orbital cluster method, that it may include GB structures in its calculations. In figure 14 the density of states N(E) are shown for different atomic sites (see fig. 12) before and after segregation. First it is seen in figure 14a that an Fe atom in site A (GB) has a reduced N (E) at the strong bonding levels (at the bottom of the 3d band) compared to a bulk atom, site F. The effect of P and B have gained in strong bonding levels below and at the bottom of the band. Interesting is the result for atoms at site C. For boron segregation there is still an increase in strong bonding levels, while for phosphorus segregation the number of bonding states is reduced and that of non-bonding states (near zero level) is increased.

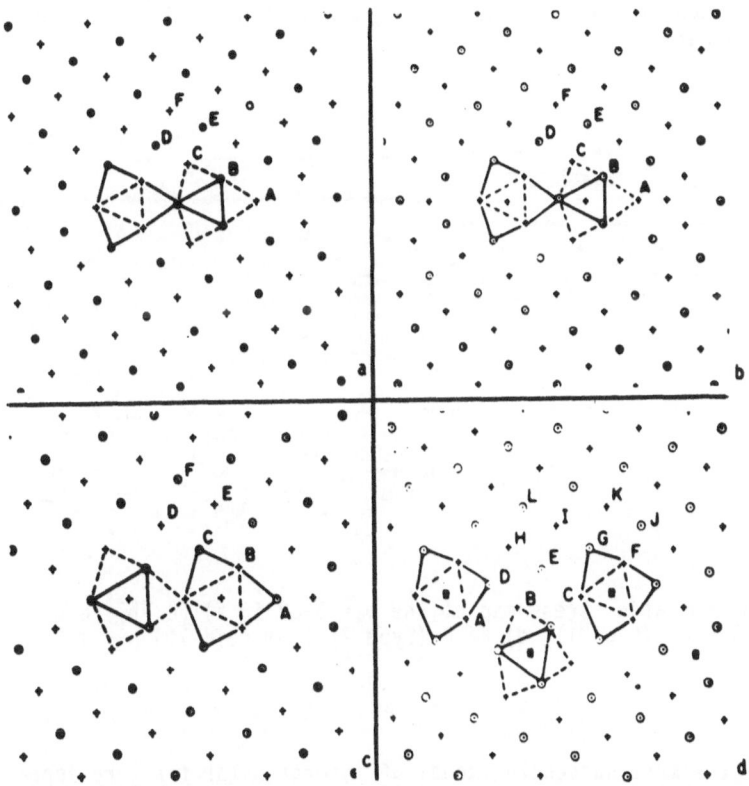

Fig.12 - Structure of the Σ = 5 tilt GB in α-iron: a) without impurities; b) with segregated B atoms; c) with segregated P atoms (type 1); d) with segregated P atoms (type 2) (after ref. [70-72].

The conclusion is that P and B segregation produced strong bondings within the Fe_9 X cluster, (X = P,B), while bonds between the cluster and surrounding Fe atom are weakened in case of P segregation which is not observed in case of B segregation. It is therefore suggested that after P segregation fracture will occur along the plane of these weakened bonds and B acts as an GB strength enhancer in agreement with former results [60]. A scheme of such GBs is represented in figure 15 with a possible fracture line indicated.

In section 2.1 we mentioned observations of brittle fracture in alloys such as Ni_3Al and Ni_3Si [26,27] without any detectable impurity segregation. A detailed phenomenological investigation of this type of A_3B alloys has been published by Takasugi and Izumu [73], using structural coincidence site lattice (CSL) and chemical concepts. The authors propose that both, structural and chemical effects may cause embrittlement. Alloys with large differences in valency $\Delta z > 2$ (Ni_3Ge, Ni_3Si, Ni_3Ga, Ni_3Al, Fe_3Ga, Co_3Ti) are found to be brittle, while alloys with a small difference $\Delta z < 2$ (Cu_3Pd, Ni_3Mn, Ni_3Fe) present ductile fracture. This chemical effect is explained by formation of a covalent bond type between A and B again resulting in a reduction of A-A (metal-metal) bond strength. This effect together with possible structural influences are shown in figure 16. Cavities due to structural and/or chemical effects may cross the GB vertical or nearly parallel, suggesting a more brittle behaviour of the latter structure.

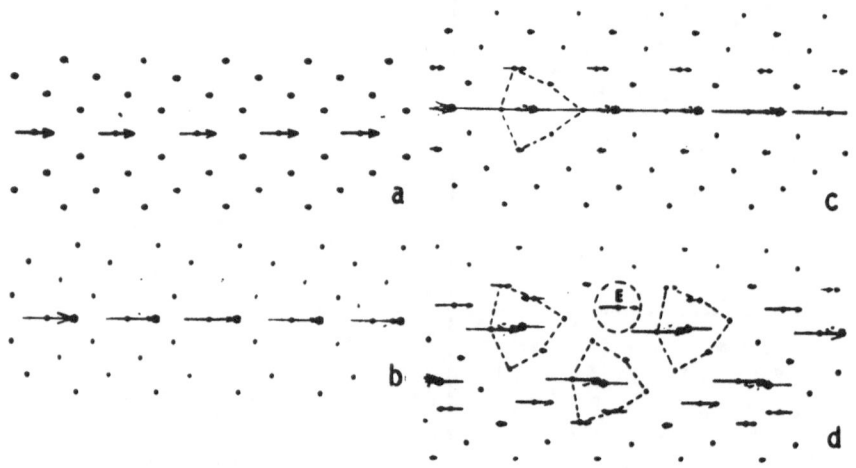

Fig.13 - Hydrostatic stress map of the $\sigma = 5$ GB in a) Fe, b) Fe-B, c) Fe-P (type 1), d) Fe-P (type 2) from ref. 70)

4. OUTLOOK

Actually the most successful model of intergranular fracture appears to be that developed in the work of Briant and Messmer (B+M) [6,59-62]. This chemical model is able to explain sequences of embrittling potencies, the beneficial effects of boron as well as influences from alloying elements. Also the conclusion drawn from the work of Loier and Boos [15], that a certain GB concentration is necessary to change drastically the fracture mode can be explained by results from B+M. As formerly discussed, in [59] B+M have shown that an increase in number of nearest M neighbours around an impurity decreases its embrittling potency. Hence, increasing the impurity level up to a starting I-I interaction reduces the number of M atoms around the Is and, consequently, increases I-M bond strength while simultaneously reduces M-M bond strength of next neighbours. This explanaion is different from that of Eberhart et al. [63] who try to explain the results from [15] by an I-I interaction.

The contribution of the structural models [67-73] mainly lies in the gain of information on atomistic arrangements at GBs. Of special importance is the result that GB can be constructed using structural units (see fig. 12) involving a periodicity. Since no experimental method is available to prove the results, it is difficult to judge the accuracy of these models. It is interesting to note that chemical models as well as structural considerations come to the conclusion that reduction of M-M bond strength is the origin of GB fracture caused by typical embrittling elements and difference in electronegativity seems to be the dominating parameter. In this case of embrittlement, structural effects seem to be of secondary importance. This is not so in those intergranular fractures where no impurities could be observed here structural influences seem to dominate. In this context, attention should be called to the role of carbon. To our understanding, C needs more and special investigations due to both its not unambiguous behaviour and detectability, respectively. In all studies, attention should be paid to the possible and probable fact of surface reconstruction after fracture.

The dynamical model of Eberhart et al. (ELJ) [65] has brought the important aspect to consider the bonds during mechanical stress. ELJ raise the basic question that a "complete theory of intergranular embrittlement must deal not only with the way the cohesive strength changes as a result of segregation of impurities, but also the effect the segregation has on the shear strength" [63]. In other words, an impurity may not only reduce M-M bond strength across a GB but also parallel to it, i.e. when at an atomically sharp crack tip the bonds parallel and perpendicular to applied stress σ are equal in strength, what makes a fracture brittle or ductile

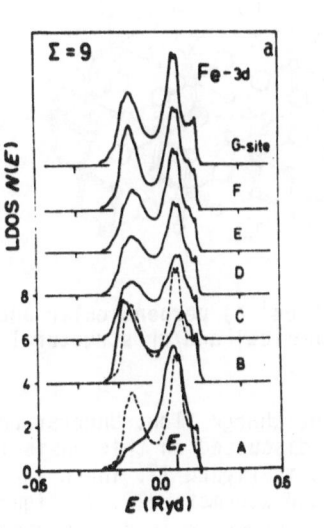
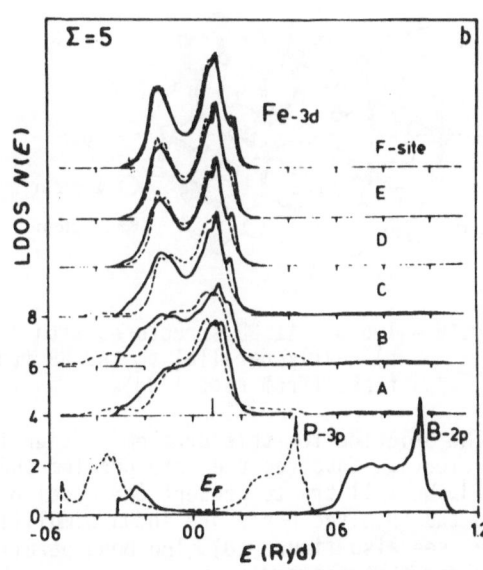

Fig.14 - a) LDOS of the Fe-3d band without segregation, b) after P(dashed) and B (full lines) segregation (from ref. [71]).

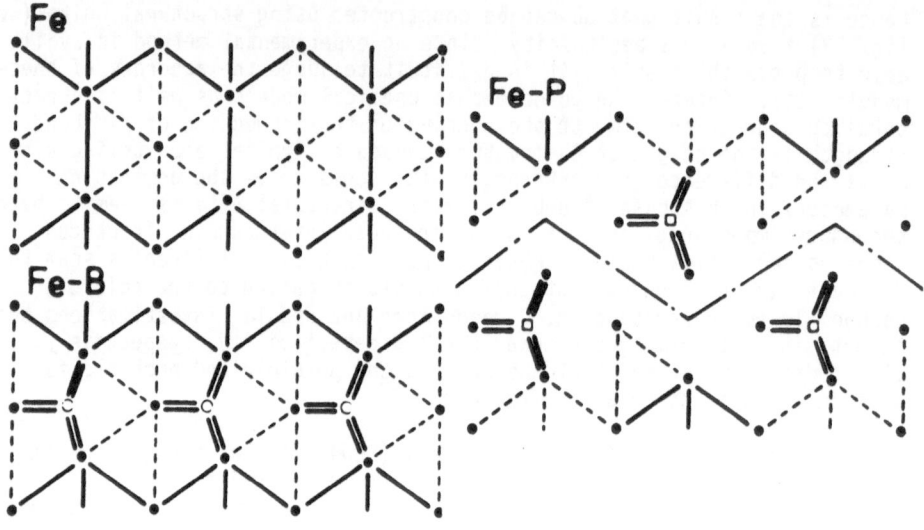

Fig.15 - Schematic representation of the effect of B amd P on the bonding
states at GB of Fe. Double, single and broken lines represent
strong, normal and weak bonds (from ref. [71]).

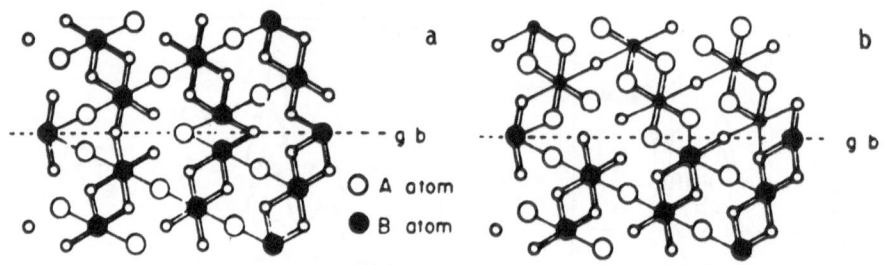

Fig.16 - Two $\Sigma = 11$ GB structures with "cavities" a) perpendicular and
b) nearly parallel to the GB from chemical and/or structural effects (from ref. [73]).

The solution for this problem is seen in the charge flow. Understanding
of cleavage involves the same problem and is discussed on this basis in
[65]. We will try to present the model of ELJ in figure 17. The crack tip
and the involved bonds are shown simplified and schematically in figure
17a (see also figure 10). The bond parallel to stress σ is represented by
$U_{\parallel}$ and that perpendicular to stress by $U_{\perp}$. Without stress both are equal
and their potential may be represented by $U_{\parallel}^0 = U_{\perp}^0 = U^0$ in fig. 17b.
Application of stres σ causes charge flow e, (fig. 17a) strengthening $U_{\perp}$
and weakening $U_{\parallel}$ with the corresponding changed potentials $U_{\perp}^0$, $U_{\parallel}^0$ shown in
17b. The force diagram $\partial U/\partial r = F(r)$ shows that fracture will occur at that

distance r where the force $\partial U/\partial r$ is a maximum and the restoring force is becoming smaller than the applied force. In this case, due to the charge flow, $F_\parallel^{max}$ is smaller than $F_\perp^{max}$ (although $r_\parallel > r_\perp$) and $U_\parallel$ will fail. This is the way how cleavage is explained by ELJ.

This is an interesting hypothesis and the justification is seen in electronic trends of Poisson's ratio. Furthermore, for amorphous silicon (a-Si) we have found theoretical [74] and experimental [75] evidence for such a negative charge accumulation of ~1% under compression of also ~1%.

However, in case of the interpretation of hydrogen embrittlement ELJ refer to calculations on Si-H-H-Si bonds in a-Si from Eberhart et al. [66]. As reported by Chabal and Patel [76] there seems no evidence for the incorporation of molecular hydrogen in the a-Si matrix. Instead H_2 is found to accumulate in micro-voids and bubbles.

To tackle the question whether the perpendicular $U_\perp$ or the parallel bond $U_\parallel$ will yield under stress different mechanisms may be considered and we would like to suggest an alternative process [77]. Generally, atomic interaction, i.e. a chemical bond is represented by a pair potential as shown, for instance, in figure 17b. The potential represents the interaction energy as a function of atomic distance when subjected to strain along the bond axis. Our question is, does the same potential also represent the interaction energy when the two atoms are subjected to shear stress? Probably not and we propose the following: the interaction potentials $U(r)$ depend on the direction of the applied force relative to the bond axis; or, in other words, the potentials $U(r)$ are functions of the spatial distribution of electronic charge (orbital overlapping) between the ion cores: $U(\rho(r))$.

Having this in mind we will return to the problem of a crack tip with bonds perpendicular and parallel to the applied stress σ. Although the more correct term for our discussion would be the overlapping integral we will use here the more common energy-displacement potentials, as used already in figure 17b, wich here are considered to represent overlapping of electronic orbitals. From figure 18c we suggest that the potential $U_t(y)$, which means displacement along bond axis, is different from the potential $U_s(y)$, representing displacement perpendicular to the bond axis. In case of a ductil material this seems evident. As shown in figure 18a $U_t(y)$ represents the common cohesive energy until rupte by tensile force. However, along the line indicated by s, in figure 18c, we have the extended metallic electron states still with strong overlapping at $y = a/2$. It is seen that less force and work of fracture is necessary for shearing the bond than for stretching it:

$$\int_0^{a/2} \frac{\partial U}{\partial y}\, dy \bigg|_s = \int_0^{a/2} F_s\, dy < \int_0^{a/2} F_t\, dy = \int_0^{a/2} \frac{\partial U}{\partial y}\, dy \bigg|_t \tag{1}$$

the crack will be blunted by emission of a dislocation, the fracture type is ductile.

In case of highly localized bonds, figure 18b, the situation is different. Not only at $y = a/2$ there will be no overlapping, also the potential $U_s(y)$ (this is our new suggestion) will be steeper than $U_t(y)$ resulting in less force F_t^{max} to break by stretching than F_s^{max} by shearing.

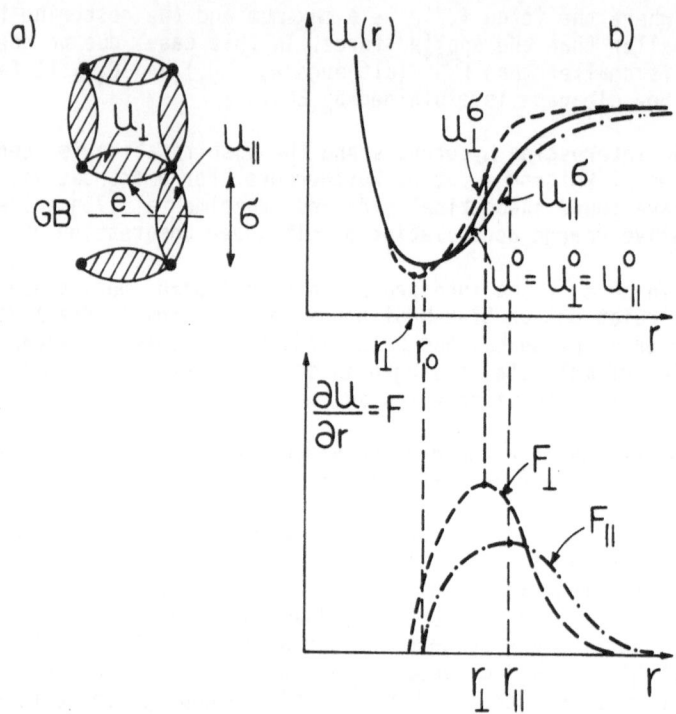

Fig.17 - Bonds at an atomically sharp crack tip. a) charge flow e from the bond parallel to stress $U_{\parallel}$ to the bond perpendicular to stress $U_{\perp}$. b) the corresponding potential and force diagrams.

Since there is no overlapping at a/2 anymore the total work necessary to break the bonds will be equal (contrary to ductile materials (1))

$$\int_{0}^{a/2} F_t \, dy = \int_{0}^{a/2} F_s \, dy \qquad (2)$$

but

$$F_t^{max} < F_s^{max} \qquad (3) \, .$$

From (3) we see that the fracture type will be brittle.

If this model is correct one may define the following general criterion whether a crack tip will behave in a ductile or brittle manner: that bond will yield which needs less initial work $\partial U/\partial y \, \delta y = \delta W$, or

$$\delta W_t \gtrless \delta W_s \begin{cases} \text{ductile} \\ \text{brittle} \end{cases} \qquad (4)$$

Of course, the situation at a real crack tip will not be so simple and ideal as that of figures 10 and 18: In order to calculate δW one will have to take an average over several bonds and bond orientations. Naturally, the two discussed cases represent extreme configurations and there will be all types of potentials between, depending on spatial charge distribution (overlapping orbitals) and on orientation of stress relative to bond axis.

This latter dependence emphasizes the influence of GB structure on fracture.

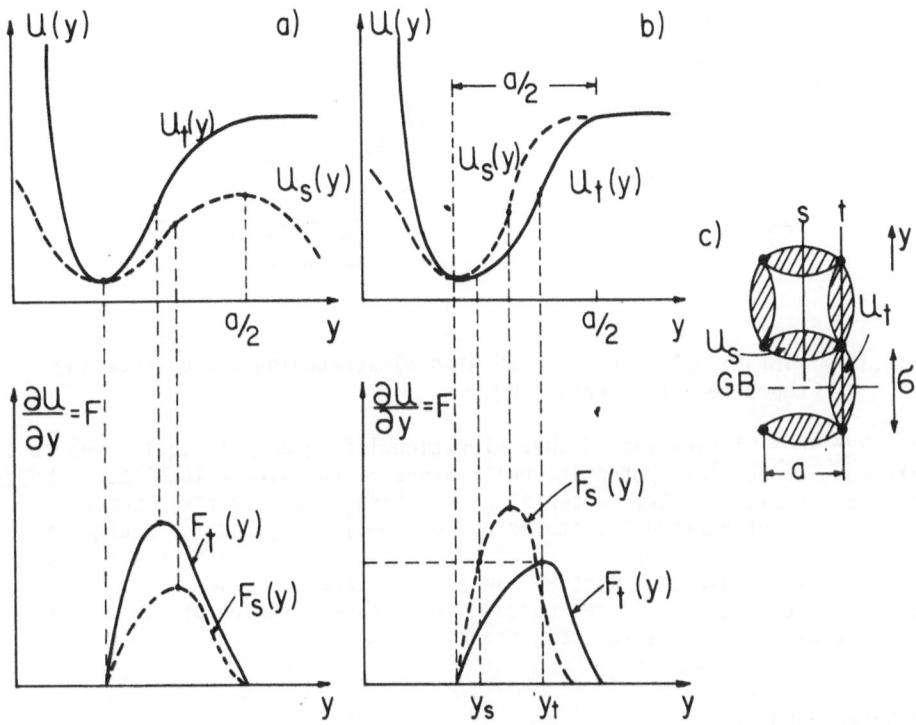

Fig.18 - a) shear U_s and tensile U_t potentials for metallic bond type with corresponding force diagram, b) the same for highly localized bond types, c) crack tip bonds under tensile stress along line t and under shear stress along line s.

How to justify this model? The ductile potential $U_s(y)$ of figure 18a is relatively easily explained by the metallic bond type from the extended electron states, even considering a certain localization of the orbitals. The strongest and unproved statement of our model is that the width of $U_s(y)$ is smaller than that of $U_t(y)$ in case of highly localized bonds.

In order to find support for our statement we will look for experiments which test individual chemical bonds. This, for instance, is done in research on vibrational states of molecules. Lucovsky and Pollard [78] report on vibrational resonance modes of Si_4H. It has been found theoretically and experimentally that the stretch-type motion of Si_4H, seen in figure 19a, has the lower frequency $\nu = 265$ cm^{-1} than the shear-type motion with

488

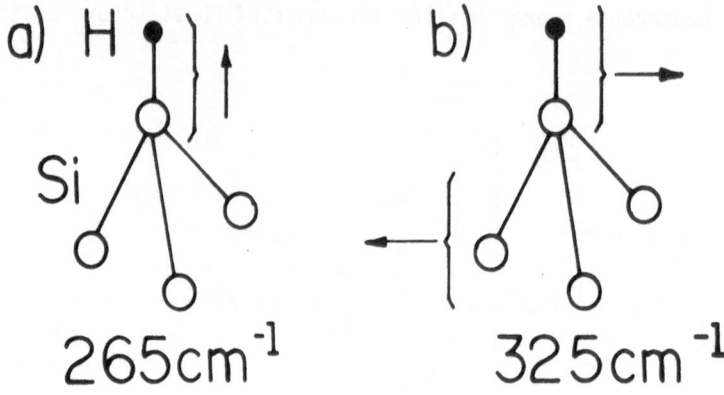

Fig.19 - Vibrational modes of Si_4H with a) stretching and b) shearing
character (from ref. [78]).

$\nu = 325$ cm^{-1}, figure 18b. Higher vibrational frequency is equivalent to a
steeper, or narrower potential well. Since Si has highly localized p-bonds
and is a typical brittle material we can take this as a first support for
our model. Furthermore, as stated by Coulson [79], generally charge distribution in molecules is more extended along bond axis than perpendicular to it. The decision whether the ideas of ELJ or those proposed here
(or both together) are more adequate for the description of the discussed
subject will need more investigations.

ACKNOWLEDGEMENT

The author is grateful to colleagues at COPPE/UFRJ and IGV/KFA for clari-
fying discussions. This work was supported by FINEP and CNPq, Brazil and
KFA Jülich, W.-Germany.

REFERENCES

1. E.D. Hondros, M.P. Seah; Int. Met. Rev. 22 (1977) 262.
2. C.L. Briant, S.K. Banerji; Int. Met. Rev. 23 (1978) 164.
3. M.L. Jokl, J. Kameda, C.J. McMahon Jr. V. Vitek; Metal Sience 14
   (1980) 375.
4. M.P. Seah, E.D. Hondros; in "Atomistics of fracture", R.M. Latanision,
   J.R. Pickens, Eds.; Plenum Press 1983.
5. W. Losch; Acta Metall. 27 (1979) 1885.
6. C.L. Briant, R.P. Messmer; Phil. Mag. B42 (1980) 569.
7. M. Hashimoto. Y. Ishida, R. Yamamoto, M. Doyama, T. Fujiwara; "V. Conference on Point Defects and Defect Interactions in Metals", Kyoto,
   Japan, 1981.
8. H. Viefhaus, R. Möller, H. Erhart, H.J. Grabke; Scripta Met. 17 (1983)
   165.
9. M.B. Hintz, L.A. Heldt, S.P. Clough, J.F. Moulder; Scripta Met. 17
   (1983) 1415.

10. H. Fukushima, H.K. Birnbaum; Acta Metall. 32 (1984) 851.
11. W. Losch, Acta Metall. 27 (1979) 567.
12. V.M. Bermudez; Appl. Surf. Sci. 17 (1983) 12.
13. G.W. Rubloff; Surf. Sci. 132 (1983) 268.
14. C. Achete, H. Niehus, W. Losch; Jour. Vac. Sci. Technol. B3 (1985) 1327.
15. C. Loier, J.Y. Boos; Metall. Trans. 12A (1981) 1223.
16. C.L. Briant; Acta Metall. 33 (1985) 1241.
17. A. Kumar, V. Raman; Acta Metall. 29 (1981) 1131.
18. H. Fukushima, H.K. Birnbaum: Scripta Metal. 16 (1982) 753.
19. D.A. Woodford, R.H. Bricknell; Scripta Met. 17 (1983) 1341.
20. L. Allegra, R.G. Hart, H.E. Townsend; Metall. Trans. 14A (1983) 401.
21. C.L. White, L. Heatherly, R.A. Padgett; Acta Metall. 31 (1983) 111
22. T.G. Nieh; Scripta Met. 18 (1984) 1279.
23. G.R. Caskey Jr., D.E. Rawl Jr., D.A. Mezzanotte Jr.; Scripta Met. 16 (1982) 969.
24. P.L. Lane, P.J. Goodhew; Phil. Mag. 48 (1983) 965.
25. M.I. Baskes, V. Vitek; Metall. Trans. 16A (1985) 1625.
26. T. Ogura, S. Hanada. T. Masumoto, O. Izumi; Metall. Trans. 16A (1985) 441.
27. T. Takasugi, E.P. George, D.D. Pope, O. Izumi; Scripta Met. 19 (1985) 551.
28. H.J. Lee, J.W. Morris Jr.; Metall. Trans. 14A (1983) 913.
29. C.T. Liu, C.L. White, J.A. Horton; Acta Metall. 33 (1985) 213.
30. S.K. Hwang, J.W. Morris Jr.; Metall. Trans. 11A (1980) 1197.
31. H. Erhardt, H.J. Grabke; Metal Science 15 (1981) 401.
32. D.Y. Lee, E.V. Barrera, J.P. Stark, H.L. Marcus; Metall. Trans. 15A (1984) 1415.
33. H. Niehus, G. Comsa; Surf. Sci. 140 (1984) 18.
34. M.F. Ashby, F. Spaepen, S. Williams; Acta Metall. 26 (1978) 1647.
35. H.J: Frost, F. Spaepen; Jour. Physique C6 (1982) 73.
36. K. Maeda, V. Vitek, A.P. Sutton; Acta Metall. 30 (1982) 2001.
37. T. Takasugi, O. Izumi; Acta Metall. 31 (1983) 1187.
38. A. Roy, U. Erb, H. Gleiter; Acta Metall. 30 (1982) 1847.
39. K. Sickafuss, S.L. Sass; Scripta Met. 18 (1984) 165.
40. T. Watanabe, S. Kitamura, S. Karashima; Acta. Metall. 28 (1980) 455.
41. T. H. Chuang, W. Gust, L.A. Heldt; M.B. Hinta, S. Hofmann, R. Lucic, B. Predel; Scripta Met. 16 (1982) 1437.
42. J.D. Russell, A.T. Winter; Scripta Met. 19 (1985) 575.
43. S. Hamada. T. Ogura, S. Watanabe, O. Izumi, T. Masumoto; Acta Metall. 34 (1986) 13.
44. A. Greenberg, Y. Komen, C.L. Bauer; Scripta Metl. 17 (1983) 405.
45. R. Maurer, H. Gleiter; Scripta Met. 19 (1985) 1009
46. C.L. Briant; Acta Metall. 31 (1983) 257.
47. M.B. Kasen; Acta Metall. 31 (1983) 489.
48. D.J. Dingley, R.C. Pond; Acta Metall. 27 (1979) 667.
49. D.A. Smith; Jour. Physique C6 (1982) 225.
50. R.H. Jones, S.M. Bruemmer, M.T. Thomas, D.R. Baer; Metall. Trans. 12A (1981) 1621.
51. R.H. Jones, S.M. Bruemmer, M.T. Thomas, D.R. Baer; Scripta Met. 16 (1982) 615.
52. K. Yoshino, C.J. Mc Mahon Jr.; Metall. Trans. 5 (1974) 363.
53. R.M. Latanision, H. Opperhauser Jr.; Metall. Trans 5 (1974) 483.
54. K.S. Shin, M. Meshii; Acta Metall. 31 (1983) 1559.
55. M. Iino; Metall. Trans. 16A (1985) 401.

56. S.M. Bruemmer, R.H. Jones, M.T. Thomas, D.R. Baer; Metall. Trans. 14A (1983) 223.
57. E. Anda, W. Losch, N. Majlis, J.E. Ure; Acta Metall. 30 (1982) 611.
58. C.M. Sayers; Phil. Mag. B50 (1984) 635.
59. C.L. Briant, R.P. Messmer; Jour. Physique C6 (1982) 255.
60. R.P. Messmer, C.L. Briant; Acta Metall. 30 (1982) 457.
61. C.L. Briant, R.P. Messmer; Acta Metall. 30 (1982) 1811.
62. C.L. Briant, R.P. Messmer; Acta Metall. 32 (1982) 2043.
63. M.E. Eberhart, K.H. Johnson, R.M. Latanision; Acta Metall. 32 (1984) 955.
64. J.R. Rice, R. Thompson; Phil. Mag. 29 (1974) 73.
65. M.E. Eberhart, R.M. Latanision, K.H. Johnson; Acta Metall. 33 (1985) 1769.
66. M.E. Eberhart, K.H. Johnson, D. Adler; Phys. Rev. B26 (1982) 3138.
67. M. Hashimoto, Y. Ishida, R. Yamamoto, M. Doyama, T. Fujiwara; Scripta Met. 16 (1982) 267.
68. K. Maeda, V. Vitek, A.P. Sutton; Acta Metall. 30 (1982) 2001.
69. A.P. Sutton, V. Vitek; Acta Metall. 30 (1982) 2011.
70. M. Hashimoto, Y. Ishida, R. Yamamoto, M. Doyama; Acta Metall. 32 (1984) 1.
71. M. Hashimoto, Y. Ishida, S. Wakayama, R. Yamamoto, M. Doyama, T. Fujiwara; Acta Metall. 32 (1984) 13.
72. S. Wakayama, M. Hashimoto, Y. Ishida, R. Yamamoto, M. Doyama; Acta Metall. 32 (1984) 21.
73. T. Takasugi, O. Izumi; Acta Metall. 33 (1985) 1247.
74. L. Guttman, W.Y. Ching, J. Rath; Phys. Rev. Lett. 44 (1980) 1513.
75. L. Ley, J. Reichardt, R.L. Johnson; Phys. Rev. Lett. 49 (1982) 1664.
76. Y.J. Chabal. C.K.N. Patel; Jour. Non-Cryst. Solids 77,78 (1985) 201.
77. W. Losch, submitted for publication.
78. G. Lucovsky, W.B. Pollard; in "Hydrogenated Amorphous Silicon II", Eds. J.D. Joannopoulos and G. Lucovsky; Springer Verlag 1984.
79. C.A. Coulson, "Valence", Oxford Univ. Press. (1963).

DISCUSSION

Comment by M. Daw:

1. The Si_4H example involves H on a surface. Do you have a more applicable example?
2. The harmonic part of the energy curve is related to the modulus but the anharmonic part is related to bond fracture. It may be true that the anharmonic part is related to the harmonic part for a bond tension, but maybe not for shear. I think you have assumed a relationship between the harmonic and anharmonic curves for shear. Is this true?

Reply:

1. No, unfortunately I do not.
2. When discussing bond fracture work due to shear or tension, I am not thinking in this relationship. However, in the example Si_4H vibration modes this relation may be of importance.

Comment by R. Thomson:

Do you feel that inhomogeneities in the gb can have a strong role to play in I-G failure? Most of our ideas have derived from smooth homogeneous boundaries, when actual boundaries are not homogeneous either in structural or chemical composition.

Reply:

I could believe that this depends on the segregation level. If at any point a certain minimum level is present, an inhomogeneity should not be important.

Comment by J. R. Rice:

Intergranular fracture is the conversion of a coherent grain boundary to a pair of free surfaces (not necessarily at composition/reconstruction equilibrium). Thus, the work of separation is given by the difference in energy between the pair of surfaces and the coherent g.b. For this reason, is it not incomplete in principle to focus, as in your discussion, only on binding in the gb and to ignore that in the resulting pair of surfaces?

Reply:

Yes, I agree, but my consideration is related rather to the work of breaking a bond only.

Comment by R. Thomson:

If your proposal that shear should change the fracture criterion, I believe you have to be careful about the details of the core of the crack and dislocation where the atoms separate. In looking at this, we have concluded (Lin and Thomson, Acta Met...) that shear is a minor component in the fracture criterion.

Reply:

My idea was simply to discuss the process of shearing or stretching an isolated bond not necessarily related to a general fracture criterion.

Comment by R. Neumann:

In Fe 3% Si bicrystals containing segregated S and P on the grain boundary we find indeed on 45° assymetric (100) tilt boundaries that S and P are found predominantly on one side only. More specifically if the grain boundary is a 100 plane in one crystal and a 110 plane in the other crystal, then S and P stick preferentially to the (110) plane side after interfacial fracture (Stenzel, Vehoff, Neumann, this meeting session 6/14).

Reply:

First of all I would see this as an indication of the good structural homogeneity of your grain boundary. Whether the fracture was by M-M bond rupture cannot be decided from this due to possible surface reconstruction.

STRUCTURE-DEPENDENT INTERGRANULAR FRACTURE AND THE CONTROL OF EMBRITTLEMENT OF POLYCRYSTALS

TADAO WATANABE
Department of Materials Science, Faculty of Engineering, Tohoku University
Sendai, Japan.

1. INTRODUCTION

Embrittlement of engineering materials gives rise to various problems in fabrication or in service of the materials (1). In most cases, embrittlement of polycrystalline materials is caused by intergranular fracture, increasing its magnitude with the percentage of intergranular fracture. Accordingly, it seems feasible that the control of embrittlement can be achieved by the control of intergranular fracture occurring in polycrystals. As has been shown by recent studies of the relationship between the structure and properties of grain boundaries,   it is indispensable to clarify the effect of grain boundary structure on intergranular fracture in order to control grain boundary embrittlement. This paper briefly reviews recent experimental studies of the effect of grain boundary structure on intergranular fracture associated with different types of embrittlement occurring in metallic materials (2). The importance of the grain boundary character distribution for polycrystals is emphasized. The possibility of the control of embrittlement of polycrystalline materials by controlling the grain boundary character distribution is discussed.

2. STRUCTURE-DEPENDENT INTERGRANULAR FRACTURE IN BICRYSTALS AND POLYCRYSTALS

During the past decade, basic research of the effect of grain boundary structure on intergranular fracture has been made extensively on metal bicrystals having well-characterized grain boundary. Our understanding of the effect on intergranular fracture has greatly advanced (2-5). It has been made clear that the resistance to intergranular fracture depends strongly on the type and structure of grain boundary, showing much higher resistance for low-energy special boundaries (low-angle boundary, low Σ coincidence boundary) than that for high-energy general, so called random, boundaries irrespective of material, test condition and environment.
As a good illustration of structure-dependent intergranular fracture, Fig.1 shows successive observations on fracture occurring during SEM in-situ deformation on a copper-bismuth alloy polycrystal in which bismuth embrittles grain boundaries. Intergranular fracture occurred preferentially at grain boundaries more perpendicular to the stress axis (horizontal direction). The most important finding was that even on the same boundary line, crack nucleation and propagation never took place at the parts (A and B) where twins met grain boundaries. The boundary structure of the parts would have been modified by the interaction with twin into lower-enrgy and stronger one than that of the original boundary. It is amazing to see how strongly the parts of the boundaries (A and B) stand deformation and fracture. Roy et al. (6) have also found that low-energy boundaries in copper-bismuth alloy are very resistent to fracture. So far quantitative studies of misorientation dependence of intergranular fracture at low temperature have been made on bicrystals of metals and alloys (refer to 2,3 and 5).

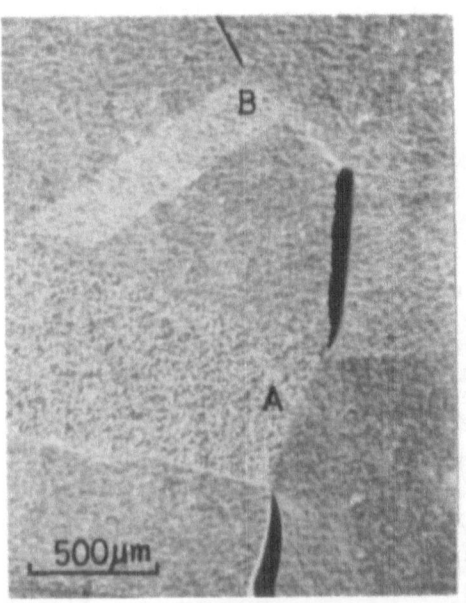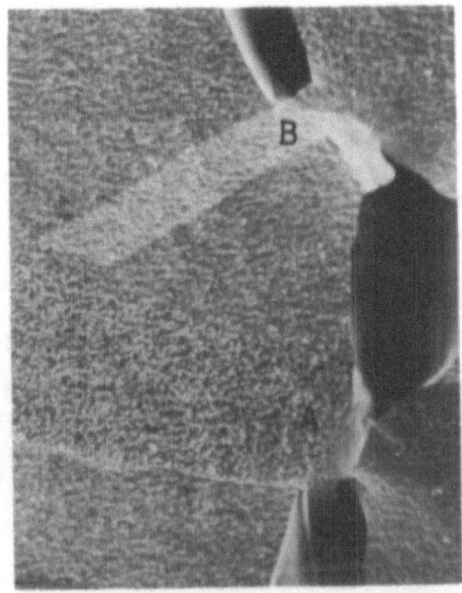

Fig.1 Successive observations of intergranular fracture at an SEM tensile
       stage in a copper-bismuth alloy polycrystal. Note the parts A and B.

It has been shown that low-angle boundaries and high-angle coincidence boundaries with relatively low Σ have higher fracture stress than that of random
boundaries.

The effect of boundary structure on high temperature creep intergranular
fracture was studied probably first by Watanabe and Davies (7) on orienta-
tion-controlled copper bicrystals. It was found that there is a close relationship between the amount of grain boundary sliding and the propensity to
creep intergranular fracture. Grain boundaries which can slide easily break
readily, through the generation of cavities at boundary irregularities such
as deformation ledges due to stress concentration caused by sliding.
Similar structural effects on creep intergranular fracture have been observed
on iron-tin alloy polycrystals (4). As clearly seen from Fig.2, high-angle
random boundaries (labeled by R) can slide and break easily while coincidence boundaries (labeled by Σ and numerals) are very resistent to fracture.
Of particular interest is that the $\Sigma 3$ coincidence boundary which was almost
perpendicular to the stress axis (horizontal direction) would not break at
all. This suggests that such special boundary has strong resistance to creep
fracture by vacancy condensation mechanism as well as sliding-assisted fracture mechanism probably because of poor effectiveness as vacancy sink or
source and diffusion path. More recently Lim and Raj (8) have studied the
effect of boundary structure on slip-induced intergranular cavitation in high
temperature-low cycle fatigue in nickel polycrystals. They found that the
propensity to cavitation for coincidence boundaries depends on Σ value and
the degree of the deviation angle from the exact coincidence misorientation,
$\Delta\Theta$. The coherent $\Sigma 3$ twin boundaries were not cavitated at all.
In relation to structural effect on high temperature intergranular fracture,
the effect of grain boundary structural transformation, as observed on grain
boundary sliding (9), should be taken into account in discussing fracture
behaviour over a wide range of temperature.

494

Kargol and Albright are probably the first who made a systematic study of the effect of boundary structure on liquid metal-induced intergranular fracture (10). They found that crack extension force was very high for low-angle boundaries and some high-angle boundaries with Σ3 and Σ11 coincidence orientations in <110>tilt aluminium bicrystals embrittled by liquid metal of Hg-3at. pct.Ga. Watanabe et al.(11) have studied the misorientation dependence of fracture stress and strain on <10$\bar{1}$0>tilt and twist zinc bicrystals embrittled by gallium. One of the results is shown in Fig.3. The intergranular fracture stress is greater at misorientation angles <20° and of near Σ9 coincidence orientation for both tilt and twist bicrystals. When fracture stress is great, fracture strain is also large. Accordingly it can be said that these special boundaries are strong and ductile in comparison with random boundaries. Watanabe et al.(12) have also studied structural effects on liquid metal-induced intergranular fracture in coarse-grained beta brass polycrystals embrittled by gallium. The character of all the boundaries contained in specimens was determined by the electron channelling pattern (ECP) technique. It was found that the initiation of intergranular fracture took place preferentially at random boundaries contacted by the liquid metal. The fracture path was not always the grain boundary or the grain interior.

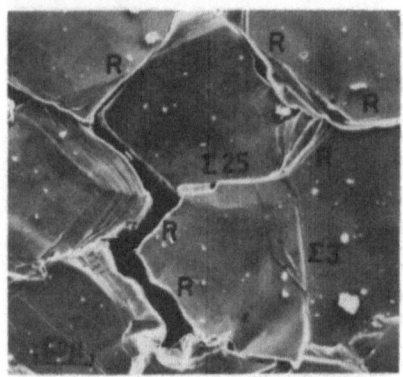

Fig.2 Creep fracture at random boundaries(R) in Fe-0.8 at.% Sn alloy at 973K and 29.4 MPa (4).

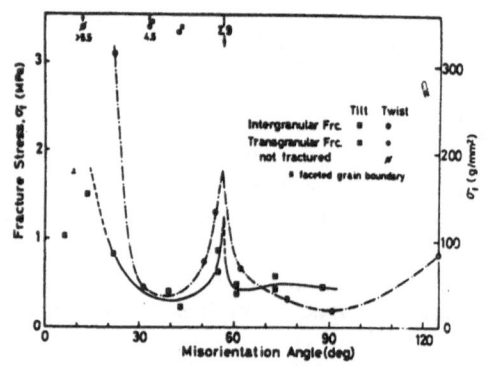

Fig.3 Misorientation dependence of fracture stress for <10$\bar{1}$0>tilt and twist zinc bicrystals embrittled by Ga. (11).

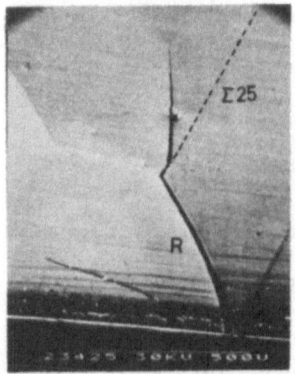

Fig.4 Observation of crack propagation in gallium-induced fracture of beta brass polycrystal(12)

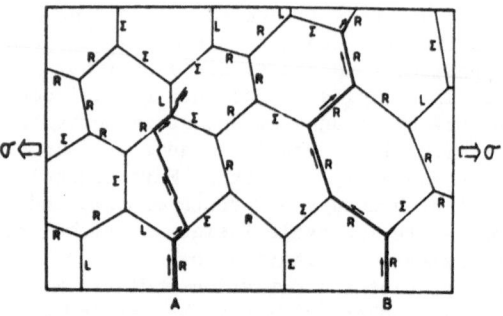

Fig.5 Schematic representation of grain boundary structure-dependent fracture processes in polycrystal (2).

The fracture mode changed from intergranular to transgranular, or vice versa during crack propagation, depending on the type of grain boundary which the propagating crack meets. Fig.4 clearly shows structure-dependent fracture processes in polycrystal. After nucleated at the random boundary at the bottom, crack propagated along the boundary passing a triple point to the Σ25 coincidence boundary. Probably because of high fracture resistance and the preference of stress condition, the crack entered the grain interior changing fracture mode from intergranular to transgranular (cleavage) fracture. However, after having passed through the grain, the transgranular crack was taken into another random boundary lying in front of the crack. Thus it has been made clear that the crack path and fracture mode can change between intergranular and transgranular in a polycrystal depending on whether weak random boundaries exist in front of a propagating crack. Fracture processes in a polycrystal are shown schematically in Fig.5 on the basis of the observations. Now we can predict that an increase in the number of inter-connecting random boundaries may lead to the predominance of intergranular fracture and embrittlement of polycrystals caused by intergranular fracture. On the contrary, it is expected that an increase in the frequency of strong (low-energy) boundaries will enhance the ductility of polycrystals. In fact, this can be a key to the control of intergranular fracture-induced embrittle-ment.

3. THE POTENTIAL FOR THE CONTROL OF INTERGRANULAR EMBRITTLEMENT

This section discusses the relationship between the grain boundary chara-cter distribution (3) and fracture processes in polycrystals in more detail, paying particular attention to the configuration of low- and high-energy boundaries. For simplicity, we classify grain boundaries into two groups: low-energy boundary and high-energy boundary. We assume that intergranular cracks nucleate and propagate only at high-energy boundaries and never at low-energy boundaries. Let us consider the effect of boundary configuration on fracture processes in polycrystals with particular boundary character dis-tributions defined by the frequency of grain boundaries of different types. Fig.6(A)-(F) show schematically possible fracture processes in polycrystals consisting of hexagonal networks of low-energy boundaries (denoted by Σ after coincidence boundary) and high-energy boundaries (denoted by R after random boundary). When the frequency of random boundaries is 2/3 of existing grain boundaries,Figs.6(A)(B), cracks propagate in ideally intergranular manner along only random boundaries at low and high temperatures. It is worth noting that when the configuration of the two types of boundary is characterized by specific boundary direction and position, there exist an ordered (regular) configuration. In this case crack propagation proceeds in a specific direc-tion associated with the directions in which random boundaries lie.
For the configuration with a lower frequency of random boundaries (1/3), cracks nucleated cannot propagate and remain inactive,Fig.6(C).
Fig.6(D) shows an interesting fracture behaviour for the configuration with the same frequency of random boundaries (1/3) as that of Fig.6(C), but with a partial modification in boundary sequence around the grains A,B,C and D (a degenerate configuration). In this case the crack can propagate in a limited region. Finally we consider the case where elongated grains make up polycrystals, Fig.6(E) and (F). In this case the anisotropy of fracture is expected depending on the type of transverse or longitudinal boundaries. Such situations seem to be realized in real polycrystalline materials which have fibrous grain structures or textures produced by drawing or rolling
Now we may conclude that grain boundary embrittlement can be controlled by manipulating the grain boundary character distribution and the boundary configuration in polycrystals containing low-energy and high-energy boundaries. This is a promising future subject of fracture control.

496

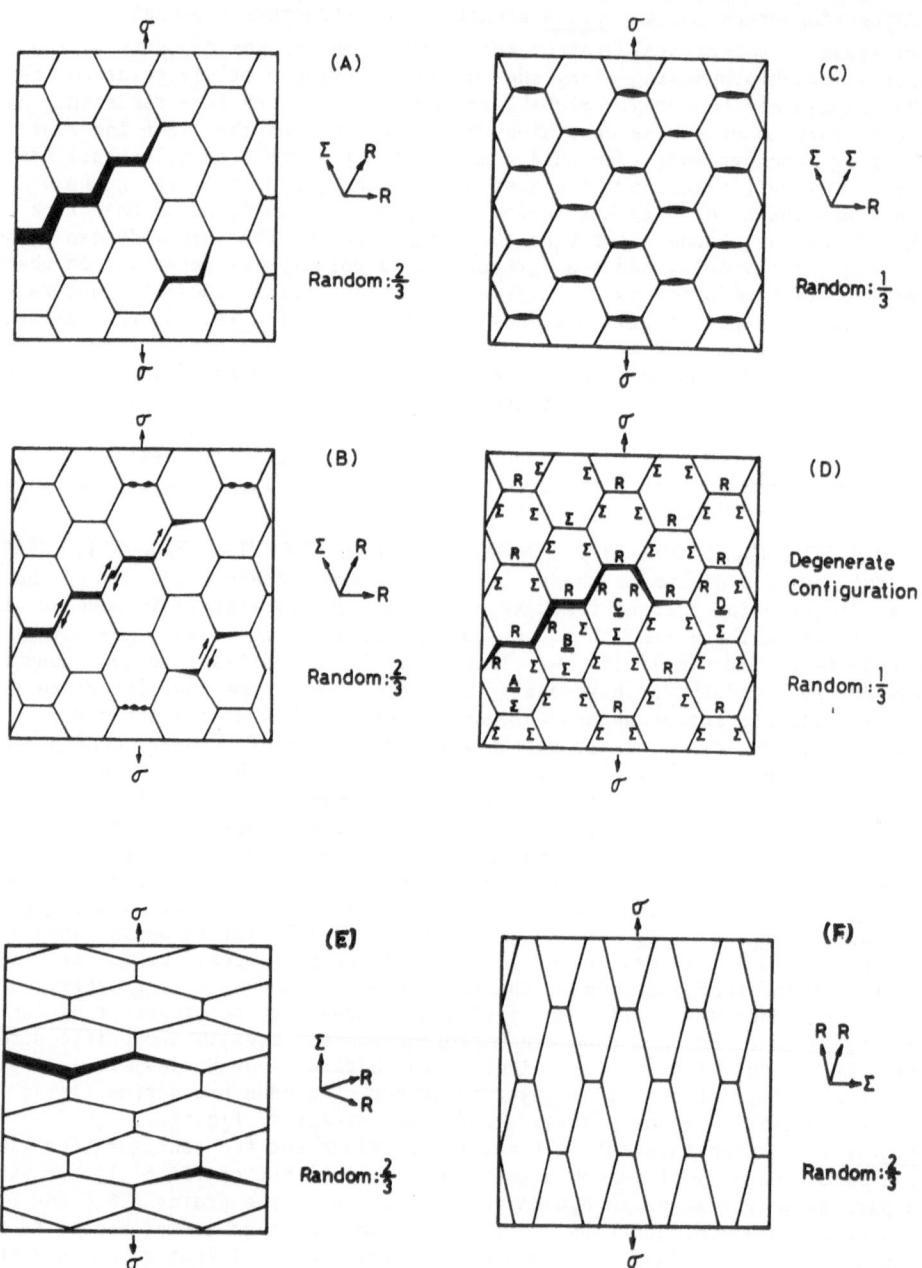

Fig.6 (A)-(F). Fracture processes in polycrystals with different fractions of random boundary (R) and boundary configurations.

REFERENCES

1. Briant, C. L. and Banerji, S. K.(ed): Embrittlement of Engineering Alloys, Academic Press (1983).
2. Watanabe, T: Res. Mechanica 11(1):47 (1984).
3. Biscondi, M.: J. de Phys. 43:c6-293 (1982).
4. Watanabe, T.: Met. trans. 14A:531 (1983).
5. Watanabe, T.: J. de Phys. 46:c4-555 (1985).
6. Roy, A., Erb, U. and Gleiter, H.: Acta Met. 30:1947 (1982).
7. Watanabe, T. and Davies, P.W.: Phil. Mag. 37:649 (1978).
8. Lim, L. C. and Raj, R.: Acta Met. 32:1183 (1984).
9. Watanabe, T., Kumura, S. and Karashima, S.: Phil. Mag. A49:845 (1984).
10. Kargol, J. and Albright, D.: Met. Trans. 8A:27 (1977).
11. Watanabe, T., Shima, S. and Karashima, S.: Embrittlement by Liquid and Solid Metals, AIME (1984):161.
12. Watanabe, T., Tanaka, M. and Karashima, S.: ibid. 183.

DISCUSSION

Comment by A. S. Argon:

I think for the purpose of high temperature creep fracture the specific nature of the boundary, i.e., whether it is random, low or high sigma is less important than the consequences of this structure in the high temperature precipitation of grain boundary particles. It is these particles which primarily result in cavity initiation by obstructing grain boundary sliding.

Reply:

As you point out, the presence of particle must be important to cavity initiation at sliding grain boundaries. However, in the absence of grain boundary particles, the difference in the propensity to fracture of different types of grain boundaries should be taken into account in discussing creep fracture processes in polycrystals (T. Watanabe, Met. Trans. 14A(1983):131, "Grain Boundary Sliding and Stress Concentration During Creep").

Comment by I. M. Bernstein:

Have you considered that this geometrical effect is primarily a result of relative differences in solute segregation such that random boundaries will have more solute and/or particles than low energy boundaries?

Reply:

Yes, we have. Some kind of grain boundary configuration which may result from relative difference of solute segregation between different type of grain boundaries seems very possible. We have observed that the distribution of dihedral angles at triple point tends to localize sharply around 120° with increasing tin concentration for recrystallized iron-tin polycrystals (tin concentration was from 0.16 at% to 1.08 at%). (T. Watanabe, to appear in Proceedings of JIMIS-4 on Grain Boundary Structure and Related Phenomena, 1986.) This may suggest the possibility of segregation-induced grain boundary configuration in polycrystals.

Comment by A. W. Thompson:

It appears that only strongly extended textured or very fine-grain materials can have a majority of coincidence or low-angle boundaries. If there is not a majority of such boundaries, a significant effect of those boundaries on yielding or fracture would require a "strongest link" instead of a "weakest link" approach to yielding or fracture. Grabke made the same point in his work on aluminum, that special boundaries only affect very fine-grained material.

Reply:

I completely agree with you regarding a "strongest link" approach to yielding and fracture. Strong grain boundaries can be the most important microstructural component in polycrystalline materials. Therefore, the control of the frequency of strong boundaries is of engineering importance in order to control strength and embrittlement of polycrystalline materials as discussed recently (T. Watanabe, Res. Mechanica 11(1985):47, "Grain Boundary Design for Strong and Ductile Polycrystals) and at this conference.

WORKSHOP SUMMARY: INTERGRANULAR EMBRITTLEMENT

PREPARED BY: T. Watanabe
DISCUSSION LEADERS: I. M. Bernstein, M. E. Eberhart, J. P. Hirth,
                    T. Watanabe
RECORDERS: W. Losch, H. E. Haenninen, P. Marcus, H. K. Birnbaum

The workshop was started by an introductory lecture by Losch on intergranular fracture. He reviewed recent experimental and theoretical work performed during the last five years, particularly focussing on two aspects of intergranular brittle fracture: One is the effect of grain boundary segregation of solutes or impurities in metals and alloys. The other is the effect of grain boundary structure and orientation on fracture modes. Also theoretical models of atomistic mechanism of fracture were discussed on the basis of quantum mechanic calculations which show chemical bond modification due to impurity segregation, and of pair potential calculations more related to structural considerations at the grain boundaries.

Watanabe discussed structure-dependent intergranular fracture in various environments, referring to recent experimental work on bicrystals and polycrystals of metals and alloys. He pointed out the possibility of the control of intergranular embrittlement of polycrystalline materials, by manipulating the relative density of special low-energy grain boundaries which can be preferential sites for fracture, and by controlling the configuration of different types of grain boundaries.

Thomson discussed theoretical bases of brittle-ductile transition behavior of materials.

In this workshop, the following topics drew particular interest of the participants and were discussed in depth:

1. the description of real grain boundaries
2. fracture stress of clean grain boundaries
3. effect of solutes on fracture stress of grain boundaries
4. synergisms between solutes and hydrogen in intergranular embrittlement
5. effect of solutes on grain boundary structure
6. the possibility of cohesive enhancers
7. electronic concepts of intergranular embrittlement, particularly relating to the effect of segregation on intergranular fracture.

Many stimulating and constructive discussions and comments made on the above topics are summarized as follows.

A. Experimental Aspects of Intergranular Embrittlement Studies

Recent studies of grain boundary structure have shown that the grain boundaries have a wide variety of atomic structures, dependent on the boundary misorientation and inclination, and other boundary parameters such as rigid body displacement. However, it was felt that the characterization and description of real grain boundaries are possible only for simply structured tilt or twist type boundaries, at present, and not for more general mixed type boundaries which contain both tilt and twist components of the boundary misorientation. Unfortunately there is no established theory that can well describe general grain boundaries, which may be the major part of real boundaries. Further development of the theory of grain boundary structure is needed.

More recently the characterization of grain boundaries in real materials has become much easier than before, after the advent of the modern electron channelling pattern (ECP) technique. This technique has been increasingly used and will become more popular in the field of grain boundary research, particularly in studies of the relationship between grain boundary structure and properties including fracture.

As for experimental techniques for analysis of grain boundary chemistry, Auger electron spectroscopy (AES) was thought to be the most appropriate analytical technique, and more reliable than ESCA and SIMS, in quantitative studies of chemical composition at fracture surfaces. However, there was a general recognition that experimental measurements of solute distributions at grain boundaries should be improved. For instance, determinations should be made of the solute distributions on both sides of an intergranular fracture, as recently performed by Viefhaus et al. [Scripta Met., 17(1983), 165].

Strong influence of grain boundary structure on intergranular segregation and fracture has been found in metals and alloys. Quantitative studies are needed on bicrystals which contain well characterized grain boundary and with controlled impurity content in the bulk and in the boundary. Only a few experimental studies have been performed so far on molybdenum-oxygen [Brosse and Biscondi; Proc. 10th Plansee-Seminar, (1981), p. 205], copper-bismuth [Franczkiewicz and Biscondi: J. de Phys., 46 (1985), C4-497] and Fe-Si and Fe-Sn systems [Watanabe et al. Scripta Met., 12 (1978), 361, Acta Met., 28 (1980), 455]. Particularly molybdenum and iron-silicon were seen as possible materials to study regarding the effects of boundary structure on fracture stress and segregation, as well as the electronic properties of grain boundaries. Also, it was thought that this kind of experimental work should concentrate on the grain boundaries that do not fracture in various materials. These boundaries may play a key role in the control of intergranular embrittlement of polycrystalline materials.

It was generally agreed that the distribution of grain boundary types in polycrystalline materials should be studied. Low-energy boundaries like low Σ coincidence boundaries resistant to fracture seem to occur much more frequently in real materials than expected from a random distribution of grain orientations. Although the reasons for this have not been known, one factor may be the presence of grain textures.

The correlation of the frequency of coincidence boundaries with grain size for recrystallized polycrystals reported by Watanabe, should be studied in various materials. For experimental determination of the grain boundary character distribution, the ECP technique is recommended to be the most appropriate and powerful tool with good accuracy.

Regarding the determination of the fracture stress of grain boundaries, bicrystals were recommended. It was suggested that K_{Ic} can be
measured as previously performed by Kargol and Albright on liquid metal-
induced intergranular fracture of aluminum bicrystals [J. Test. Eval., 3
(1975), 173]. There was a consensus of "no" as to whether the cohesive
energy can be measured. Nevertheless, it was suggested that one possible
experiment is the fracture of grain boundaries in thin films formed by
epitaxial growth on rigid substrates. If the two grains are sufficiently
thin, dislocation generation may be avoided so that the boundary cohesion
energy will be able to be determined without plastic blunting at the crack
tip.

It was felt that there are no demonstrated "cohesive enhancers."
Reported improvement in intergranular embrittlement seems to be explainable by site competition, bulk trapping and precipitate formation. The
most-cited "cohesive enhancer," boron, has been found to produce intergranular fracture at prior austenitic grain boundaries in quenched and
tempered low carbon steel when present together with hydrogen. Up to now,
reliable basic information is extremely scarce on synergisms between
solutes and impurities including hydrogen in intergranular embrittlement.
Experimental and theoretical work is needed to clarify the complex but
important effects of coexisting segregants.

With regard to another interesting subject, that is, the effect of
solute segregation on grain boundary structure, the question was raised
whether segregation-induced grain boundary structural transformation or
restructuring [as observed by Sickafus and Sass; Scripta met., 18 (1984),
165, on iron twist boundaries] is important and can affect the fracture
process. This may be a key issue to discrimination between intrinsic
fracture stress of clean grain boundary and the effect of solute segregation on the fracture stress. In order to clarify this, fracture experiments should be performed on bicrystals of absolutely pure material such
as silicon, with a well characterized boundary containing well defined
amount of a given segregant.

B. Theoretical Aspects of Intergranular Embrittlement Studies

It was pointed out that the original cluster calculations of Briant
and Messmer, particularly concerning the effect of solute on electron
distributions, were flawed as they are unable to calculate bond energies
or reflect crystal symmetry. The cluster calculations have a limited
applicability as the charge distribution, which they can handle, are of
limited interest in determining the energy of grain boundaries. Therefore, difficulties in handling total energies limit the utility of the
cluster calculation. Other limitations are the small number of atoms
handled, and the fact that the results are sensitive to the symmetry of
the cluster. Nevertheless, the concept is important and should be pursued
further on large clusters with a variety of symmetries and relaxations.

The suggestion was made that grain boundary dislocation glide and
climb had been demonstrated and could be an ameliorating factor for
decohesion. Peierls-stress like temperature dependence and impurity
pinning of such dislocation are possible and could contribute to a
ductile-brittle transition. In fact, quite recently, it has been reported
that boron can improve the intergranular brittleness of Ni_3Al and it

appears to increase the mobility of grain boundary dislocations and to enhance the accommodation of slip through the movement of the boundary dislocations [Schulson et al., Acta Met., 34 (1986), 1395].

It was generally agreed that the goal of the intergranular fracture research is "fracture prevention." For this purpose there is a strong need for the interaction between physicists, chemists and metallurgists in large research programs, through better understanding of mechanisms of intergranular fracture, primarily to control embrittlements of polycrystalline materials.

Workshop Session 3:
Hydrogen Embrittlement

Session Chairmen: I. M. Bernstein and H. Birnbaum

THE ROLE OF HYDROGEN TRANSPORT IN HYDROGEN EMBRITTLEMENT

M. HASHIMOTO and *R.M. LATANISION

R&D Laboratories-I, Nippon Steel Corporation
1618 Ida, Nakahara-ku, Kawasaki, 211   JAPAN

* Massachusetts Institute of Technology
Cambridge, MA   02139, U.S.A.

1. INTRODUCTION

The effect of hydrogen on the mechanical behavior of metallic materials has been extensively studied by numerous investigators. In recent years, considerable efforts have been directed toward understanding the fundamental aspects of hydrogen embrittlement. For a quantitative analysis of hydrogen embrittlement processes, the following steps must be clarified; (i) localization or accumulation of hydrogen at critical locations, (ii) microcrack formation, the nature of which depends upon the local hydrogen concentration and stress state, and (iii) hydrogen redistribution after microcrack formation. Repetition of the above steps (i)-(iii); i.e., the propagation of microcracks, continues until final fracture.

The above picture has been drawn by a number of investigators and used, for example, to explain the discontinuous nature of crack propagation[1], and the strong strain rate dependence of hydrogen embrittlement[2]. In order to obtain useful information on the atomistics of hydrogen embrittlement, for example, the critical hydrogen concentration necessary to initiate microcracks[3], quantitative determination of the local hydrogen concentration at the crack nucleation sites is required. Since the crack nucleation process and the local hydrogen concentration at these sites are interdependent, the local hydrogen concentration has to be determined under dynamic conditions.

The main characteristics of plastic deformation or fracture in crystalline metals represented by dislocation behavior suggests that hydrogen transport during these dynamic deformation processes is affected by two types of defect interactions: hydrogen trapping by newly generated dislocations and hydrogen transport by mobile dislocations.

There exist several models which consider the effect of hydrogen transport by mobile dislocations[4,5,6]. There exist, however, no general model which provides a quantitative expression for this phenomenon. Part of the difficulty stems from the fact that during plastic deformation there is always some contribution of the trapping effect by newly generated dislocations. Another problem is that since hydrogen dragged by mobile dislocations can interact with diffusing hydrogen, there must be an interdependence of the magnitude of the hydrogen flux due to moving dislocations and the local hydrogen concentration in lattice sites.

In this paper, a model describing general aspects of hydrogen transport during plastic deformation and experimental verification of the model will be discussed first. Then, the role of hydrogen transport in hydrogen embrittlement processes will be discussed.

2. HYDROGEN-DISLOCATION INTERACTIONS

It is well known that as elastic tensile stresses increase, and as elastic compressive stresses decrease, the chemical potential of an interstitial solute decreases(7). These elastic field may be due to externally applied stresses or residual stresses about defects in metals such as dislocations. Since the stress field around an edge dislocation has a non-vanishing hydrostatic component, σ_h, dissolved hydrogen can achieve an inhomogeneous distribution called a Cottrell atmosphere in this stress field.

By applying the linear isotropic elasticity theory to an edge dislocation (8), σ_h can be determined from

$$\sigma_h = \frac{\mu b(1+\nu)}{3\pi(1-\nu)} \frac{\sin\theta}{r} \tag{1}$$

where μ = shear modulus, b = Burgers length,
ν = Poisson's ratio, r = distance, and
θ = angle between glide plane and the point of interest.

As generally recognized in recent years, the equilibrium hydrogen concentration, p, in the vicinity of a defect must be expressed in terms of Fermi-Dirac statistics described by the following expression (9).

$$\frac{p}{1-p} = \frac{p_0}{1-p_0} \exp(\frac{H_B}{k_B T}) \tag{2}$$

where H_B = binding enthalpy to the defect
p_0 = fractional concentration in a region where H_B=0.
Substitution of equation (1) into (2), considering that H_B in this case is $\sigma_h \Delta V$ where ΔV is the volume expansion due to an interstitial solute, yields the following expression for the excess number of hydrogen atoms around an edge dislocation per unit length of dislocation (10).

$$\frac{N}{L}(\# H/m) = N_{lattice} \int_{r_0}^{R} rdr \int_0^{2\pi} d\theta \ (p-p_0) \tag{3}$$

$$= N_{lattice} \ p_0 \int_{r_0}^{R} rdr \int_0^{2\pi} d\theta \ (\frac{1+\xi^{-1}}{1+\xi^{-1}\exp(-\sigma_h \Delta V/k_B T)} - 1)$$

where r_0 = inner cutoff parameter
R = outer cutoff parameter
$\xi = p_0/1-p_0$
Nlattice = interstitial site density.
The equation (3) has been numerically integrated in this study using the Patterson method (11) by evaluating typically ten

thousand function points, and the results for the Fe-H system are shown in Figure 1. The values used in this calculation are μ=86 GPa, b=2.48x10^{-10} m, ν=0.29, and T= 303 K. The density of tetrahedral interstitial sites, Nlattice, corrected in order to account for the fact that no iron hydride is formed under these conditions, was taken to be 8.5x10^{28} sites/m^3 (10). Here three values of ΔV; 2.66, 2.0, 1.22 cc/mol, were used to cover the experimental scatter. Figure 1 shows that the excess number of hydrogen atoms around an edge dislocation strongly depends upon the inner and outer cutoff parameters and the volume expansion due to an interstitial solute, ΔV. For the lowest value of ΔV, the more convenient Boltzmann statistics gives a result similar to that of Fermi-Dirac statistics. For the higher ΔV values, however, there is a difference and the Fermi-Dirac statistics must be used. This is expected since a large ΔV indicates that there is a strong interaction of hydrogen with the edge dislocation and that the dislocation as a trap will saturate. Only Fermi-Dirac statistics allow saturation. Figure 1.b indicates that even for a Cottrell atmosphere, most of the excess hydrogen atoms are concentrated in the region a few Burger's distance from the dislocation core.

Since linear elasticity theory cannot be used near and in the core region of dislocations, it is necessary to evaluate excess hydrogen at the core separately. The excess number of hydrogen atoms at the core site may be estimated by applying equation (2) assuming that the site density in the core, N_{core}, does not depend upon the occupancy by hydrogen.

$$N/L(core) = N_{core} \frac{1}{1 + \xi^{-1}exp(-E_B/k_BT)} \qquad (4)$$

where E_B= the binding energy to the dislocation core
N_{core}= the site density in the core (sites/m).
Ncore has not been determined accurately and the proposed values are 1 - 100 sites per repeat distance along a dislocation (12). Therefore, we may consider

$$N_{core} = \gamma/b \quad where \quad \gamma=1-100 \qquad (5)$$

It is clear from equation (4) that when the binding energy is greater than about 0.4 eV and the fractional hydrogen concentration in the lattice is greater than 10^{-6}, the core is almost saturated with hydrogen.

By comparing equation (3) and (4), the boundaries separating where core or Cottrell atmospheres are dominant can be determined for various values of γ and ΔV, Figure 2. In the upper left region enclosed by a given boundary, core atmospheres contain more hydrogen than Cottrell atmospheres while in the region to the right and below a boundary Cottrell atmospheres are dominant. For reasonable values of γ, ΔV, E_B, and typical values of hydrogen concentration, it can be concluded that most of the excess hydrogen is concentrated in the core region. This suggests that the core interaction is more important than the elastic stress interaction in the Fe-H

508

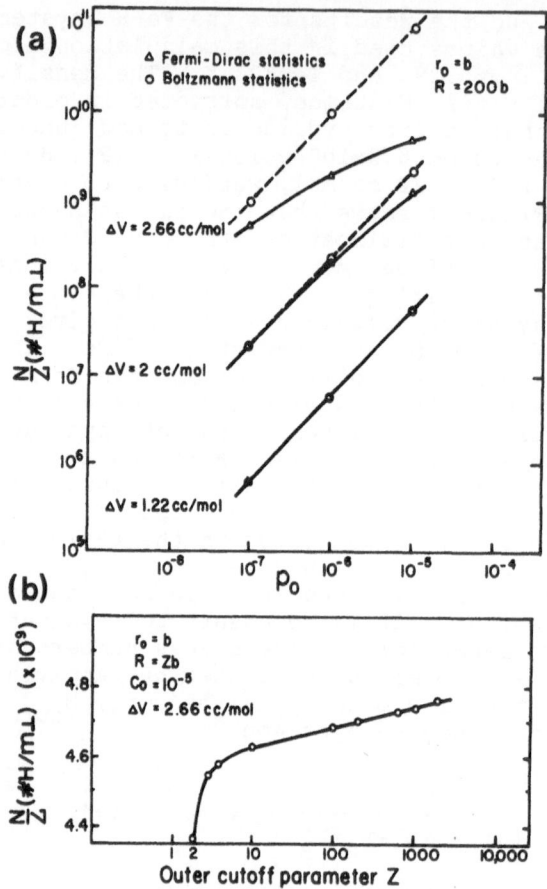

FIGURE 1. The excess hydrogen around an edge dislocation due to the elastic interaction.

(a) as a function of lattice concentration,
(b) as a function of outer cutoff parameter Z

FIGURE 2. A plot of boundaries separating regions where core or Cottrell atmospheres are dominant for given value of γ and ΔV.

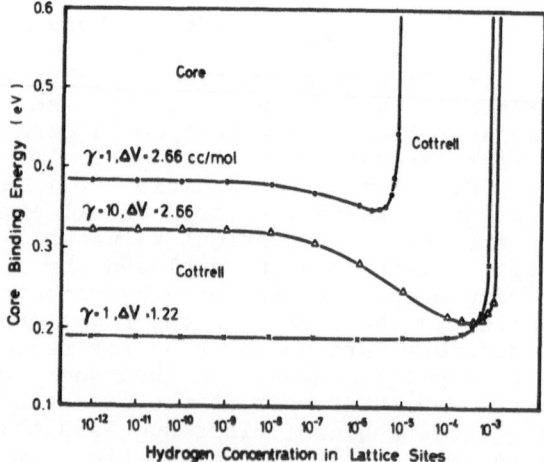

system under normal conditions(10).

The above discussion shows that the presence of dislocations enhances the solubility of hydrogen in metals, and hydrogen atoms are trapped around dislocations forming a highly concentrated atmosphere. Since diffusing hydrogen should fill these dislocations while diffusing through lattice sites, hydrogen-dislocation interactions hinder the lattice diffusion of hydrogen. This is known as the trapping effect and will be discussed in the next section.

3. STATIC TRAPPING EFFECT
3.1. Theoretical basis

Among the several models that have been developed to describe this trapping effect, the McNabb-Foster equations (13) are most suitable to describe saturable traps. McNabb and Foster (13) attempted to express the diffusion and the trapping process by a modification of Fick's equation. Their mathematical analysis of trapping describes diffusion with the relations;

$$\frac{\partial u_1}{\partial \tau} + \frac{\partial u_t}{\partial \tau} = \frac{\partial^2 u_1}{\partial x^2} \tag{6}$$

$$\frac{\partial u_t}{\partial \tau} = \lambda u_1 - \mu u_t - \nu u_1 u_t \tag{7}$$

where $u_1 = C_{lattice}/C_0$, $u_t = nN_{trap}/C_0$, $\tau = Dt/a^2$, $x = X/a$,

$\lambda = N_{trap}ka^2/D$, $\mu = Pa^2/D$, $\nu = kC_0a^2/D$

C_0 = some reference volume concentration
a = some characteristic length of the system
D = true diffusivity of H in the perfect lattice
$C_{lattice}$ = volume hydrogen concentration in the lattice
C_{trap} = volume hydrogen concentration in trap=nN_{trap}
n = fraction of trap sites filled with hydrogen
N_{trap} = trap site density
k = trapping coefficient
P = release coefficient.

McNabb and Foster (13) analyzed diffusion through a plate bounded by surfaces at X=0 and X=a and initially free of hydrogen. The concentration of diffusing hydrogen is maintained at C_0 on the input surface X=0 and zero on the exit surface X=a. These initial and boundary conditions simulate the hydrogen permeation technique such as developed by Devanathan and Stachurski (14). In the presence of trapping, the time lag was obtained by McNabb and Foster (13) as follows;

$$t_{lag} = \frac{a^2}{D} \{ \frac{1}{6} + \frac{\alpha}{2\beta} + \frac{\alpha}{\beta^2} - \frac{\alpha}{\beta^3}(1+\beta)\ln(1+\beta) \} \tag{8}$$

where $\alpha = \lambda/\mu$ and $\beta = \nu/\mu$.

The trapping effect is not only an interesting phenomenon but is also a useful tool to obtain information concerning hydrogen-defect interactions. Kumnick and Johnson (15) used this idea and determined the binding energy of hydrogen to a dislocation and the trap site density.

When the trap sites are almost saturated by hydrogen (n=1), the time lag becomes (15)

$$t_{lag} = \frac{a^2}{6D}(1 + 3N_{trap}/C_0) \tag{9}.$$

When the trap sites are almost empty (n<<1),

$$t_{lag} = \frac{a^2}{6D}(1 + kN_{trap}/P) \tag{10}.$$

In the intermediate region, all terms of equation (8) contribute to the time lag. Thus, by comparing the time lag data with equation (8), it is possible to experimentally determine several parameters used in the McNabb-Foster equations.

3.2. Experimental results

In order to characterize the nature of traps in Armco iron, hydrogen permeation experiments were carried out using electrochemical technique (14). The hydrogen charging solution used in the following experiment was 0.1 N NaOH in order to achieve a low hydrogen fugacity so that the trapping effect became evident.

By applying equation (8) and using the technique by Kumnick and Johnson (15), it may be possible to determine the binding energy and also the total trap site density. Figure 3 shows the time lag of annealed Armco iron as a function of 1/C, where C is the hydrogen concentration in the lattice at the input surface. It can be seen from this figure that traps in this condition are almost saturated with hydrogen since the time lag is very similar to that predicted from equation (9). Using equation (8), the parameters to make a best fit to experimental data were obtained as shown in Figure 3. The average value of four independent runs becomes E_B = 0.53 eV and N_{trap} = 6×10^{15} sites/cm^3.

The same analysis was performed on Armco iron deformed to various plastic strains before the hydrogen permeation experiments. Figure 4 shows a typical example of the final decay curves in a predeformed specimen. It can be seen from this figure that the time scale for the decay transient of the 12% predeformed specimen is more than two order of magnitude larger than that of the annealed specimen, which clearly shows the strong trapping effect in predeformed Armco iron.

In the case of predeformed specimen, an extremely long time is required to measure the each transient curve. During this period, the hydrogen concentration at the input side changes probably due to changes of the surface properties. Since the rate of flux increase due to changes in the surface is competitive with the time lag of deformed specimens, the

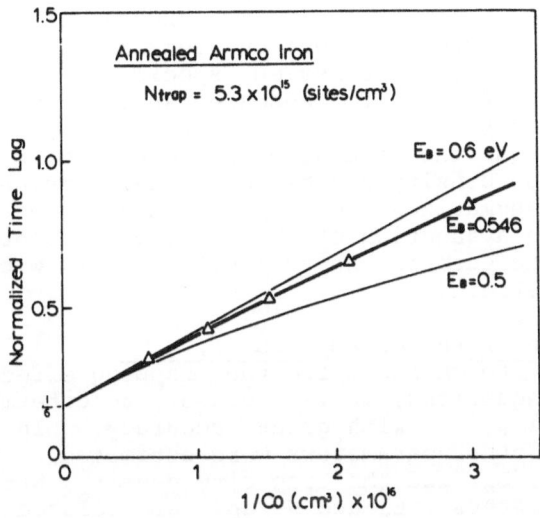

FIGURE 3.   Normalized time lag as a function of 1/C.

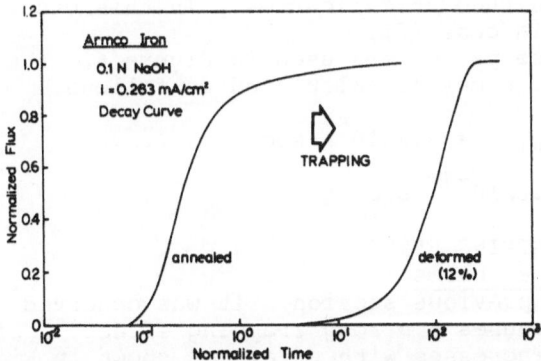

FIGURE 4.   Comparison of final decay curve between annealed and 12% predeformed specimen.

time lag could not be measured accurately. Therefore, the
binding energy for these predeformed specimen could not be
determined.

However, the experimental results by Kumnick and Johnson
(15) suggest that the binding energy of hydrogen to
dislocations in deformed specimens remains constant, and is
also equal to that for annealed specimen. Thus, the same
binding energy 0.53 eV for annealed specimen was assumed for
deformed specimens.

Even though the binding energy could not be determined,
the trap site density can be determined assuming equation
(9). Figure 5 summarizes the trap site density estimated by the
above method as a function of the total plastic strain. There
is a steady increase in the trap site density with increasing
prior plastic strain.

3.3. Estimation of the parameters λ, μ, ν

In order to characterize the trapping effect using the
McNabb-Foster equations, it is necessary to determine the three
parameters , λ μ ν , with great accuracy. In the previous
section, two other parameters were estimated. These are the
binding energy, E_B, and the trap site density, Ntrap. Note that
a set of parameters, E_B and Ntrap, are related to a set of
parameters, α and β. In other words, the ratio λ: μ: ν can be
obtained from these estimated parameters, E_B and Ntrap.
The third parameter, in this case the absolute values of λ: μ
: ν , may be estimated from the best fit of the theoretical
transient curve obtained by solving the McNabb-Foster equations
to the experimental data. In order to estimate the third
parameter, the following procedure was used. For the case of
the 2% predeformed specimen, Figure 6 shows the experimental
data of the final decay curve and the three theoretical decay
curves. By using the estimated values of E_B and Ntrap, α and β
can be estimated as 5350 and 109 respectively for the 2%
predeformed specimen. In other words, λ: μ: ν= 5350 : 1: 109.
The sets of parameters as shown in Figure 6 were used to
calculate the theoretical curves. In this case, the best fit
was obtained in case (2).

From these parameters used in Figure 6, the true physical
constant k and P may be calculated as follows.

$$kN_{lattice} = 3.3 \times 10^5 \text{ /sec} \tag{11}$$

$$P = 5.0 \times 10^{-4} \text{ /sec} \tag{12}$$

4. DYNAMIC TRAPPING EFFECT
4.1. Theoretical basis

In the previous section, it was observed that plastic
deformation causes a strong trapping effect. Since the trap
site density increases with strain as shown in Figure 5, it is
expected that the trapping effect becomes stronger and stronger
during plastic deformation. Therefore, even if the initial
trap sites are all saturated by hydrogen, plastic deformation
generates new trap sites and causes a further trapping effect.
This situation was named the dynamic trapping effect since the

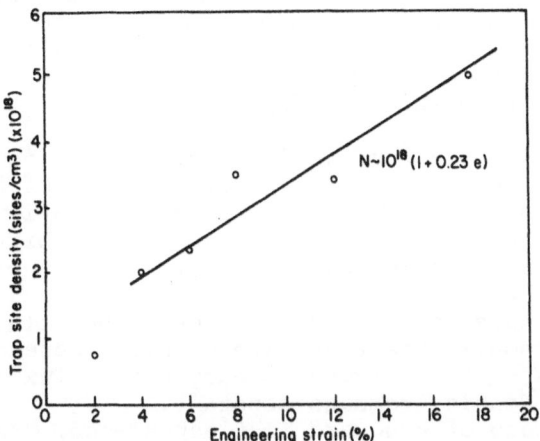

FIGURE 5. The trap site density as a function of total plastic strain.

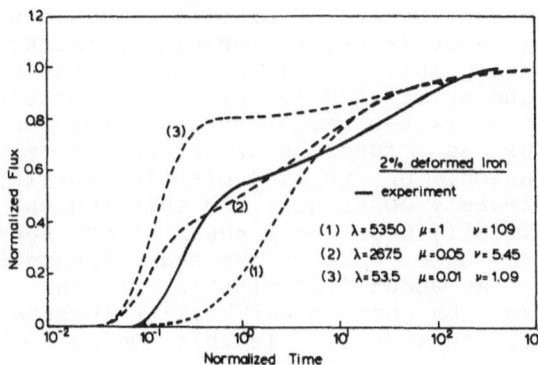

FIGURE 6. Procedure to estimate the parameters.

new trap sites are created during the diffusion of hydrogen resulting in dynamical interactions between diffusing hydrogen and the newly created traps. In order to model this situation, the McNabb-Foster equations, which describe a static trapping effect (Ntrap=constant), have been modified to include a time dependent trap sites density, Ntrap(t). It was found that the modification was done by simply replacing the λ parameter in equation (9) with the time dependent $\lambda(\tau)$ in the normalized equations (16).

One example of a numerical calculation using the modified equations is shown in Figure 7 for the case of the effect of dynamic trapping effect on the hydrogen permeation flux through a plate. Note that in this case the arbitrarily chosen parameters were used to demonstrate the nature of the modified equations.

In the case of a step function, the flux drastically decreases as soon as the trap site density increases. The flux then recovers to the original steady state flux level because the trap site density remains constant.

In the case of a square root dependence, the flux also first decreases and then starts increasing again. This suggests that the rate that hydrogen fills the trap sites exceeds the trap creation rate.

In the case of a linear increase in the trap site density, there seems to be a dynamical steady state, which suggests that the trap filling rate is dynamically balanced by the trap creation rate.

In the next section, the experimental data will be shown and then these data will be compared with the model described above.

4.2. Experimental results

The experiment of hydrogen permeation to steady state followed by deformation was performed for an Armco iron tensile plate specimen in order to examine this dynamic trapping effect. In this experiment, a specimen of thickness 0.2 cm was predeformed by 2% plastic strain to assure that the main defect population in the starting material was dislocations.

Figure 8 shows the stress strain curve and the hydrogen permeation flux as a function of strain or time. In this specimen dimension, up to 1.5% plastic strain, the stress remains approximately constant. In this region, Luder's bands are formed initially near the corners of the tensile specimen, and then, these Luder's bands propagate toward the center of the specimen. At about 1.5% plastic strain, Luder's bands propagating from the both corner combine together and uniform plastic deformation starts. In this region, the flow stress keeps increasing up to about 25 - 30% plastic strain. Beyond 30% plastic strain, macroscopic necking occurs and at about 41% this tensile specimen failures.

In the elastic region, the permeation flux increased slightly. Then, the permeation flux decreased from the steady state value as soon as the Luder's band passed the measuring area. After some recovery, it then started decreasing again when uniform plastic deformation begun.

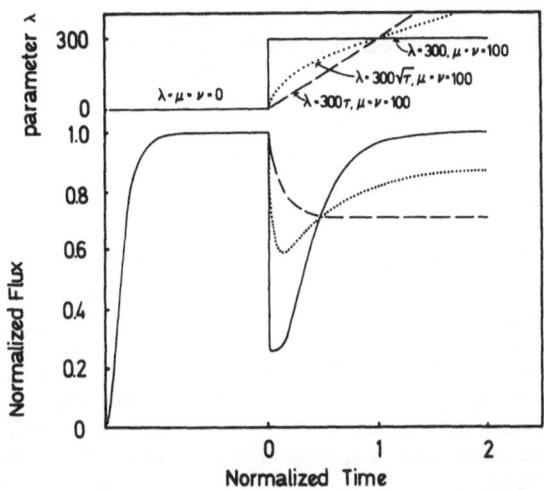

FIGURE 7. The result of the numerical calculation using the modified McNabb-Foster equations to describe the dynamic trapping effect.

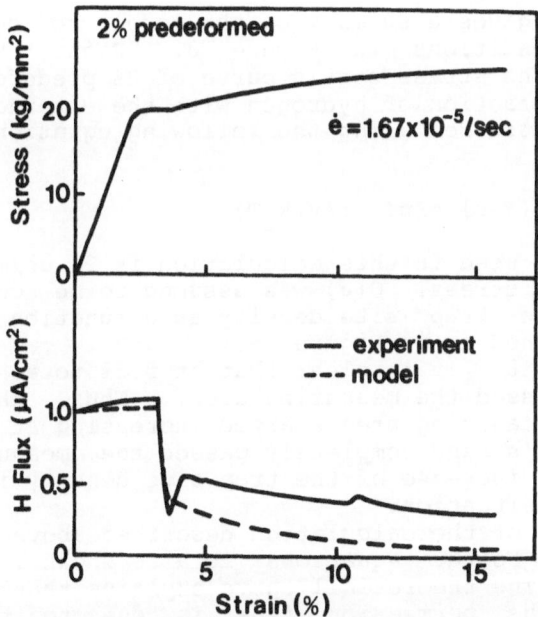

FIGURE 8. Experimental and theoretical results of hydrogen permeation through Armco iron during plastic deformation.

By comparing with the calculation described earlier (Figure 7), we may conclude that Luder's band deformation results in a rapid increase in the trap site density in the measuring area, similar to a step function. Also, no dynamic recovery was seen during uniform elongation, suggesting that the trap site density increases linearly in this region. This is in agreement with the predeformed static permeation experiment, Figure 5, where a continuously increasing trap site density was observed.

4.3. Comparison between model and experiment

In the section 3, the three parameters, λ, μ, ν, in the McNabb-Foster equations were estimated even though there is not complete agreement of the theoretical curves to the experimental data. The trap site density, Ntrap, was also estimated as a function of plastic strain. Thus, by considering the deformation pattern of Armco iron, it may be possible to simulate the experimental situation described in the previous section.

In the case of Armco iron, the Luder's bands propagation was observed at small strains, hence, the time dependent trap site density should be determined taking this localized deformation characteristic into account. The external stress also causes an increase in the hydrogen solubility through $\sigma \Delta V$ interaction, which should also be taken into account to describe the hydrogen permeation experiment during plastic deformation.

Figure 9 gives a summary of the model to describe the experimental conditions in Figure 8. Figure 9a shows a schematic of the stress-strain curve of 2% predeformed Armco iron. The interaction of hydrogen with the externally applied stress was considered using the following equation and shown in Figure 9b.

$$C(\sigma) = C(\sigma=0) \exp(\quad \sigma \Delta V/k_B T) \tag{13}$$

The value of ΔV used in this calculation is 2 cc/mol. After the lower yield stress, $C(\sigma)$ was assumed to be constant for simplicity. The trap site density as a function of plastic strain was obtained as in Figure 5.

From Figure 8, it is clear that at 3.2% total strain, the Luder's bands passed the measuring area. Thus, the trap site density in the measuring area started increasing at this strain until the Luder's band completely passed the measuring area. The first steep increase of the trap site density in Figure 9c describes this situation.

The result of the calculation described above using the modified McNabb-Foster equations is also shown in Figure 8 (dashed line). The theoretical curve explains that the initial increase in the permeation flux is due to the elastic interaction as shown in Figure 9b. The sharp decrease in the flux after 3.2% total strain can be described by this model fairly well, however, no recovery in the flux was obtained by this model. The experimental observation of this recovery at around 4% total strain may be due to the small time interval,

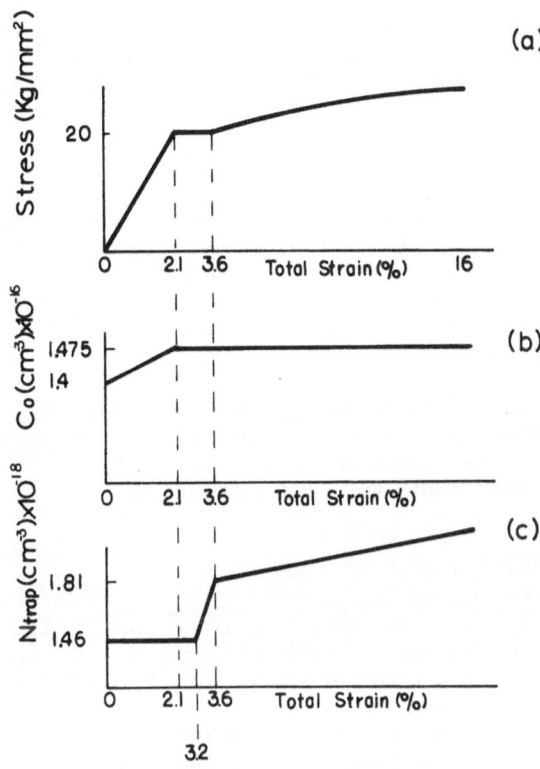

(a)

(b)

(c)

FIGURE 9. Parameters used in modeling the dynamic trapping effect.

FIGURE 10. Imaginary lattice and core structure used in an atomistic model.

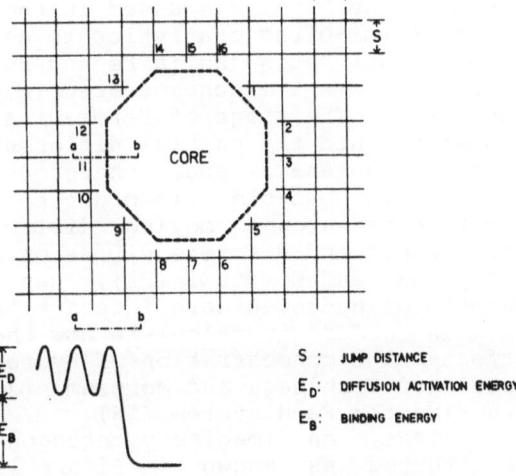

S : JUMP DISTANCE

E_D : DIFFUSION ACTIVATION ENERGY

E_B : BINDING ENERGY

during which there is no significant increase of the trap site density in the measuring area. The measuring area is only a small portion of the gauge section, therefore, any non-uniform deformation causes a non-continuous increase of the trap site density in the measuring area. After the second peak, both theoretical and experimental data show steady decrease in the flux, even though the magnitudes do not agree. The permeation flux during plastic deformation is very sensitive to the deformation pattern and the parameters. The above disagreement may be due to the error in the estimation of parameters and the modelling of exact deformation pattern in the measuring area. In any case, it was shown that the model describing the dynamic trapping effect is valid although the determination of the more accurate parameters and the more careful experiment have to be done.

In these experiments, no evidence of hydrogen transport by mobile dislocations was observed because dislocation transport was totally dominated by the strong trapping effect. A discussion of dislocation transport will be presented in the next section.

5. HYDROGEN TRANSPORT BY MOBILE DISLOCATIONS
5.1. Microscopic and discrete diffusion model

Hydrogen transport by mobile dislocations was first suggested by Bestien and Azou (17). Frank (18) reported the first experimental evidence of enhanced outgassing of hydrogen in mild steels due to plastic deformation. Tritium release (19,20), and slip plane decoration measurements (21) have suggested the possibility of hydrogen transport by mobile dislocations in a number of materials. Recently, Frankel (22) utilized the electrochemical hydrogen permeation technique with the addition of plastic deformation and clearly showed the increase in hydrogen permeation flux in the easy glide region of single crystalline nickel due to mobile dislocations.

It is well known that the amount of excess hydrogen dragged by moving dislocations depends upon the velocity of dislocations. As discussed in the section 2, it is essential to use Fermi-Dirac statistics to describe strong hydrogen-core interactions such as in Fe and concluded that most of excess hydrogen atoms are concentrated in a few Burger's distance from the core. The usage of Fermi-Dirac statistics introduces non-linearity into the partial differential equations that describe these phenomena, and, therefore, in general no analytic solution is possible. In order to describe such local hydrogen atmospheres around moving dislocations, the use of the microscopic and discrete diffusion model, such as developed by Yoshinaga and Morozumi (23), has a great advantage since the details of hydrogen-core interactions could be easily included.

In order to calculate how the motion of a dislocation affects the concentration of hydrogen associated with it, the model of Yoshinaga and Morozumi (23) was further modified to describe the Fe-H system (16).

First, an imaginary tetragonal interstitial lattice was constructed as shown in Figure 10. This lattice does not correspond to the real interstitial site structure. Therefore, it is called an imaginary lattice (23). The true interstitial site of hydrogen in Fe is a tetrahedral site (7), thus, in order to treat this problem rigorously, it is necessary to use a three-dimensional lattice. However, since this requires an extraordinary long execution time, the above simple two-dimensional model was used in this study.

The lattice constant of an imaginary lattice, s, was taken as the mean tetrahedral interstitial spacing in Fe, 1.25 A, since this gives the most reasonable physical picture of the diffusion process. This s corresponds to the jump distance in the diffusion process.

Then, a dislocation core was imbeded at the center of this imaginary lattice, the size of which is again one Burger's length, b, as used in the section 2. Since the exact potential profile in the core region is not known, the simple square well potential was assumed. The binding energy to the core, E_B, is, therefore, uniform through the core region.

The above two-dimensional structure was assumed to be repeated with the distance s in the direction perpendicular to the two-dimensional lattice.

Thus, for a given structure and the potential profile, the probability of finding a hydrogen atom at a particular site (i,j) can be calculated as follows.

$$dp(i,j)/dt = p(i-1,j) \{1-p(i,j)\} \ t\{i-1,j->i,j\}$$

$$-p(i,j) \{1-p(i-1,j)\} \ t(i,j->i-1,j)$$

+ repeated for the rest of
  the surrounding sites

$$+ V/s(p(i+1,j)-p(i,j)) \qquad (14)$$

where $p(i,j)$= the probability of finding a hydrogen atom
              at a site (i,j)
        $t(k,l->m,n)$=the transient rate of hydrogen from
              site (k,l) to site (m,n)
        V= the velocity of a dislocation.
Note that the above equation uses Fermi-Dirac statistics as can be seen from the (1-p) terms in the equation. Thus, the fractional concentration is bounded by the maximum value unity, or saturation.

In general, the transient rate can be expressed as follows.

$$t(k,l->m,n) = \kappa \nu \ exp(-ww(k,l->m,n)/k_B T) \qquad (15)$$

where ν= vibrational frequency of H atom at the lattice site
        κ= geometric factor
        E_D = the activation energy for diffusion
        $ww(k,l->m,n)$=the activation from site(k,l) to (m,n)
              $=E_D-(w(k,l)-w(m,n))/2$
        $w(i,j)$= the potential at site(i,j).

The dislocation core in this case was divided into two parts. The elastic stress field results in a potential in the top half of the core (including the extra half plane) that is so high that it was assumed that no hydrogen can enter this region. Therefore, hydrogen in the sites 1-2 and 12-16 in Figure 10 can not enter the core region either by a diffusional jump or by the core motion. Thus, there is no hydrogen in this half core region. The bottom half of the core was treated as described in equation (15). The hydrogen in this core region can interact with the lattice sites 3-11.

Making the above corrections to equation (14), a set of ordinary differential equations that describe diffusion around a moving edge dislocation can be obtained (16).

The effect of dislocation motion on the hydrogen concentration profile was then examined. In Figure 11, a dislocation which has an equilibrium atmosphere around it starts moving to the right at the velocity of 7.2 m/sec. At a short time, the atmosphere around the moving edge dislocation looks very complicated because both the core and Cottrell atmospheres are affected by the dislocation motion.

The steady state H atmospheres around the moving dislocation are shown in Figure 11.b. Since the velocity used in this calculation is very high, some of trapped H atoms escape from the potential field created by this edge dislocation. However, some of the excess H atoms are still dragged by this dislocation. Figure 12 shows the dependence of the core concentration at steady state upon the velocity. For comparison, the relationship for the core of no elastic stress field is given by the solid line.

The inclusion of the elastic stress field results in a higher efficiency of hydrogen dragging by the core. The reason for this is that the Cottrell atmosphere which develops in response to the elastic stress field acts as a supply of hydrogen for the core. It should be emphasized here that even in this case, the core drags most of the total excess hydrogen.

The characteristic velocity, $V10\%$, for the case of core and Cottrell interactions in Figure 12 is $v=10^{-1}$ which corresponds to 7.2 m/sec. This is larger than for the case of only core interaction. Therefore, mixed or edge dislocations can drag hydrogen up to higher velocities than screw dislocations.

In summary, the results obtained in this section are as follows. First, an atomistic model has been developed to examine the hydrogen concentration profile around a moving dislocation. From this model, the characteristic velocity, $V10\%$, was estimated for both a screw and an edge dislocation. The results of numerical calculations indicate that under the typical slow strain rate test, most of excess hydrogen atoms can be dragged by mobile dislocations. Even in the case of moving edge dislocations, most of dragged hydrogen atoms are still concentrated in the core region.

The atomistic technique described in this section is valid to use in studying the local hydrogen atmosphere around a moving dislocation. However, it cannot be used in systems which contain many dislocations, which requires a large execution time and huge storage spaces. Therefore, it is necessary to develop a model which describes the macroscopic

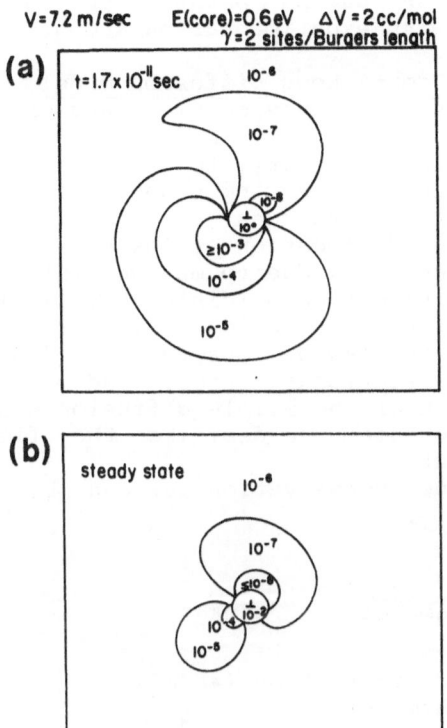

FIGURE 11. Effect of dislocation motion on hydrogen atmosphere around an edge dislocation in iron.

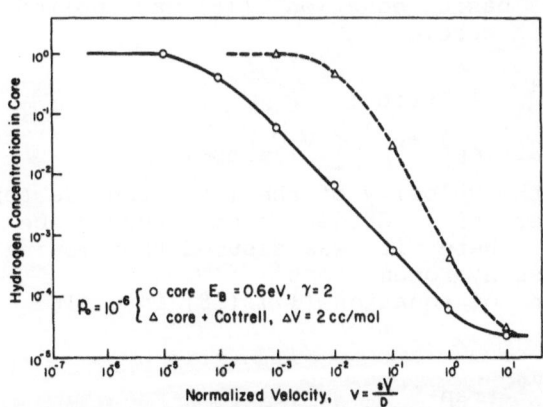

FIGURE 12. Hydrogen concentration in the core at steady state as a function of velocity.

hydrogen concentration during plastic deformation so that rigorous comparison with experimental data could be possible.

5.2. Macroscopic and continuous diffusion model

When trap sites themselves can move and drag some excess hydrogen atoms with them, i.e., hydrogen transport by mobile dislocations, it is necessary to take into account the additional flux due to these moving traps. As discussed in the section 5.1, all of the hydrogen around a dislocation can be dragged by the moving dislocation at low strain rates. As a result, the hydrogen flux due to moving dislocations can be approximated as J(dislocation)=C(dislocation)xVelocity.

The princip. of the derivation of modified McNabb-Foster equations is to use the law of conservation of hydrogen atoms for total hydrogen, stationary traps, and moving traps. As used in the derivation of the simple diffusion equation, Fick's first law was also used to describe the flux due to a concentration gradient.

In general, the conservation law can be expressed as follows;

$$\frac{\partial C}{\partial t} + \text{div } J = I_{source} - I_{sink} \tag{16}$$

where C = volume H concentration ($\#H/m^3$)
J = H flux ($\#H/m^2$-sec)
I = source or sink for H ($\#H/m^3$-sec).
This basic equation (16) was applied for total, and stationary and/or moving traps.

Note that the k parameter in the original McNabb-Foster equations does not have the dimension of rate (1/sec), even though the P parameter has that dimension. Here the K parameter (1/sec) will be introduced, which is related to the k parameter through the relationship; K=kNlattice.

First, the basic equation (16) was applied for total hydrogen. In this case,

$$C = C_{lattice} + \sum_i C_{trap-i} \tag{17}$$

$$J = -Dgrad(C_{lattice}) + \sum_i V_i C_{trap-i} \tag{18}$$

where V_i refers the velocity of the i-th trap (vector). Since the total number of hydrogen atoms should be conserved, Isink=Isource=0. Here it was assumed that moving traps can drag all of excess hydrogen atoms.

Substitution of equation (17) (18) into (16) yields the first equation.

$$\frac{\partial C_{lattice}}{\partial t} + \frac{\partial C_{trap-i}}{\partial t} = \text{div}(Dgrad\ C_{lattice})$$

$$- \sum_i \text{div}(V_i C_{trap-i}) \tag{19}$$

The same type of equation can be derived for the i-th trap

moving at the velocity V_i. In this case,

$$J(traps) = V_i\, C_{trap-i} \tag{20}$$

$$C(traps) = C_{trap-i} \tag{21}$$

$$I_{source} = H \text{ entry from lattice to traps}$$

$$= N_{trap-i}\, K\, \frac{C_{lattice}}{N_{lattice}}\left(1 - \frac{C_{trap-i}}{N_{trap-i}}\right) \tag{22}$$

$$I_{sink} = H \text{ escape from traps to lattice}$$

$$= N_{trap-i}\, P\frac{C_{trap-i}}{N_{trap-i}}\left(1 - \frac{C_{lattice}}{N_{lattice}}\right) \tag{23}.$$

Substitution of the above equations into equation (16) yields the second equation to describe the kinetics of the trapping effect.

$$\frac{\partial C_{trap-i}}{\partial t} = N_{trap-i}\left\{ K\,\frac{C_{lattice}}{N_{lattice}}\left(1 - \frac{C_{trap-i}}{N_{trap-i}}\right) \right. \tag{24}$$

$$\left. - P\,\frac{C_{trap-i}}{N_{trap-i}}\left(1 - \frac{C_{lattice}}{N_{lattice}}\right)\right\} - div(V_i C_{trap-i})$$

Thus, equation (19) and (24) are the equations to describe the general aspect of hydrogen transport during plastic deformation. These equations can be further reduced to the following normalized forms for a one-dimensional problem.

$$\frac{\partial u_1}{\partial \tau} + \frac{\partial u^{m+}}{\partial \tau} + \frac{\partial u^{m-}}{\partial \tau} + \frac{\partial u^s}{\partial \tau} =$$

$$\frac{\partial}{\partial x}\left(\frac{\partial u_1}{\partial x}\right) - v^+\frac{\partial u^{m+}}{\partial x} + v^-\frac{\partial u^{m-}}{\partial x} \tag{25}$$

$$\frac{\partial u^{m+}}{\partial \tau} = \lambda^+ u_1 - \mu u^{m+} - \nu u_1 u^{m+} - v^+\frac{\partial u^{m+}}{\partial x} \tag{26}$$

$$\frac{\partial u^{m-}}{\partial \tau} = \lambda^- u_1 - \mu u^{m-} - \nu u_1 u^{m-} + v^-\frac{\partial u^{m-}}{\partial x} \tag{27}$$

$$\frac{\partial u^s}{\partial \tau} = \lambda^s u_1 - \mu u_1 - \nu u_1 u^s \tag{28}$$

where u_1 = normalized H concentration in lattice

ui = normalized H concentration in trap
superscript s = stationary traps
m+ = moving traps (+ direction)
m- = moving traps (- direction)
v = normalized velocity = Va/D
a = characteristic length (thickness of specimen etc)
D = true diffusivity of H in lattice sites
τ = normalized time = Dt/a^2.

When the dislocation density increases with time, λ should be replaced by $\lambda(\tau)$ as discussed in section 4. It should be noted here that when the density of moving dislocations increases with time and a constant strain rate is used, the velocity of moving dislocation decreases with time.

Since the equations for moving traps contain the first derivative with respect to space, one boundary condition is necessary for each type of moving traps to obtain unique solutions from these equations. The following assumption was made here.

$$u^{m-}(x=1)=0 \qquad u^{m+}(x=0)=0 \qquad (29)$$

The permeation flux measured at the exit side, therefore, becomes

$$J)_{x=1} = - \frac{du_1}{dx})_{x=1} \quad + \quad v^+ u^{m+})_{x=1} \qquad (30).$$

The first term is the diffusional flux and the second term is the contribution from dislocation transport.

Various situations have been modeled, including the case where the boundary conditions simulate the hydrogen permeation experiment with the addition of plastic deformation.

The typical result of numerical calculations is shown in Figure 13. Note that in this case the arbitrarily chosen parameters were used to demonstrate the nature of the model. The dashed line describes the case of moving traps, half of which move toward the exit surface and half of which move toward the input surface, with the total trap site density remaining constant. Under these conditions, hydrogen transport by moving dislocations should increase the hydrogen permeation flux. The dotted line describes the case of a linearly increasing density of stationary traps which was discussed in section 4. The hydrogen permeation flux decreases in this case. In a real situation, dislocations move and also the dislocation density increases. A calculation for the combination of these effects, shown by the solid line, indicates that the flux decreases after some initial increase.

5.3. Estimation of the flux change by dislocation transport

The purpose of this section is to estimate the maximum contribution that dislocation transport may make to the flux change under the conditions of electrochemical hydrogen permeation experiment during plastic deformation. As discussed in the previous section, when both dislocation transport and the dynamic trapping effect contribute to the flux change, the

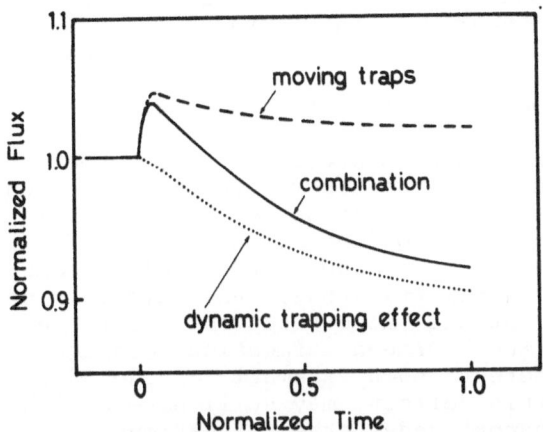

FIGURE 13. An example of the numerical calculation using the modified McNabb-Foster equations (25)-(28). The following parameters were used in these calculations ($\mu = \nu = 100$ for all case)

moving traps: $\lambda^+ = \lambda^- = 300$ $v^+ = v^- = 1$
dynamic trapping effect: $\lambda^S = 600 + 100\tau$ $v = 0$
combination: $\lambda^+ = \lambda^- = 300 + 50\tau$ $v^+ = v^- = 3/3 + 0.5\tau$

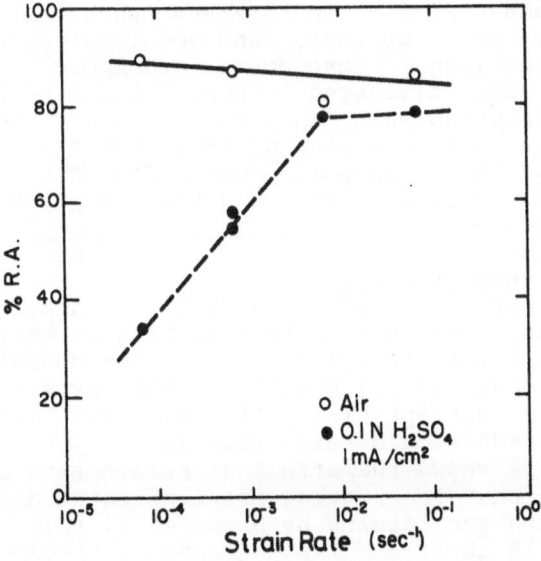

FIGURE 14. Effect of strain rate on reduction of area.

effect of dislocation transport becomes evident only at short times. Thus, the flux at short times was examined in detail. The conditions used in this calculation are those which simulate the experiment in Figure 8.

Using the parameters within the formalism of the McNabb-Foster equations estimated in section 3, the increase in flux expected from dislocation transport was estimated to be about 1% of the original flux at most. This change is well in the range of experimental scatter. Thus, under the experimental conditions used in this study, the contribution of dislocation transport has an undiscernable effect. This is because the lattice hydrogen concentration at the exit surface is controlled as zero under the electrochemical hydrogen permeation experiment, thus, the hydrogen concentration in moving dislocations also tends to be zero at this surface. Even though the hydrogen permeation technique is not accurate enough to detect such a small change due to dislocation transport, this effect may still have an important role on hydrogen transport under other situations or in other metals.

By analyzing the result of the numbers of calculations, the conditions which have a favourable contribution to dislocation transport may be summarized as follows.
(1) large mobile dislocation density
(2) high velocity (below critical velocity)
(3) high binding energy to dislocations
(4) low lattice hydrogen concentration
(5) slow kinetics
(6) strong anisotropy in dislocation motion

6. HYDROGEN EMBRITTLEMENT AND HYDROGEN TRANSPORT

The hydrogen embrittlement process has kinetic aspects so that the susceptibility to hydrogen embrittlement generally depends upon kinetic parameters such as strain rate, temperature, mode of load, and specimen geometry. In the previous sections, hydrogen transport during plastic deformation was discussed in detail and a general method to describe this phenomenon has been developed. In this section, the hydrogen embrittlement process itself will be studied to examine how the transport process discussed so far can be applied to explain the kinetic aspects of hydrogen embrittlement.

6.1. Experimental results

In order to study the kinetic aspects of hydrogen embrittlement, the slow straining tensile test was performed using the same material, Armco iron. The charging condition in this experiment is 0.1 N H_2SO_4 at the current density of 1 mA/cm^2. The susceptibility to hydrogen embrittlement was assessed by reduction of area (R.A.).

Figure 14 shows the effect of strain rate on the ductility of Armco iron. The strong strain rate dependence is in agreement with the results by Brown et al (2).

Figure 15 shows SEM fractographs of specimens fractured in a hydrogen environment at various strain rates. The specimen fractured at a slow strain rate exhibits a quasi-cleavage

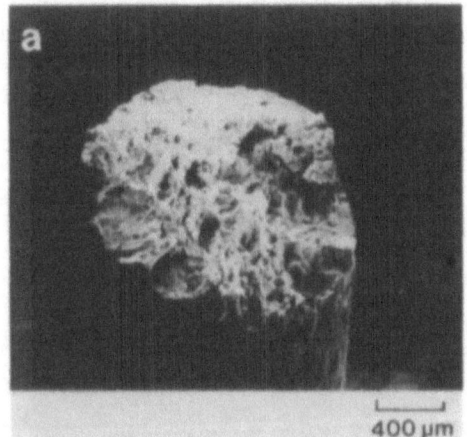

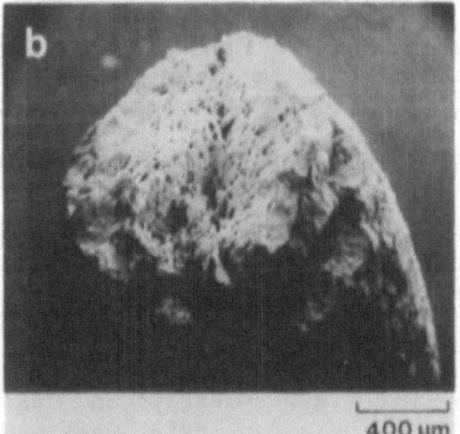

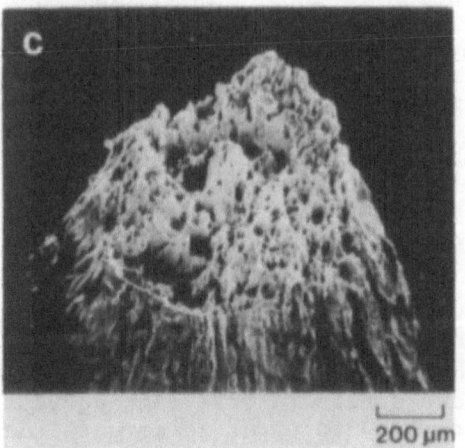

FIGURE 15. SEM fractographs of annealed Armco iron.

H charging: 0.1 N H_2SO_4 1 mA/cm^2.

(a)strain rate=6.67×10^{-5} (/sec)
(b)strain rate=6.67×10^{-4}
(c)strain rate=6.67×10^{-2}.

fracture surface, typical of hydrogen embrittlement, and much less evidence of plastic deformation. However, at a higher strain rate, 6.67×10^{-2}/sec, no effect of hydrogen was detected from the fractographic analysis. The fracture surface appears very similar for specimens deformed in air. Thus, from the fractographic analysis, it was also shown that at a higher strain rate, the hydrogen effect disappears, which is in agreement with the mechanical properties.

The commonly accepted argument about this strong strain rate dependence is that at high strain rates, the transport of hydrogen to critical locations is not sufficient and ductile failure occurs (24).

The model described in section 5 was applied to the above transport process during plastic deformation. As discussed in the previous sections, the dynamic trapping effect is dominant in the Fe-H system, therefore, this effect was examined in detail.

6.2. Theoretical explanation of strain rate dependence

The effect of hydrogen on the mechanical properties and the ductility is small up to the maximum load, while the most severe attack by hydrogen was observed after the maximum load. This fact leads to the view that the critical hydrogen concentration in the cracking process may be that at the maximum load. If the details of the cracking process were known as a function of hydrogen concentration, it may be possible to predict the cracking process from calculated values of the hydrogen concentration at the critical locations and this describes the whole cracking process for any given strain rate and boundary conditions. However, those details are not well known. Therefore, the critical hydrogen concentration defined above was used to examine the strong strain rate dependence of hydrogen embrittlement.

The model to describe the dynamic trapping effect and the estimated parameters described in section 3 were used to estimate this critical hydrogen concentration, $C_{critical}$. The situation used in this calculation is shown in Figure 16. Figure 16a shows a schematic of a strain-stress curve typical of Armco iron used in this study. Figure 16b shows the time dependent trap site density, N_{trap}. Luder's bands formation and propagation is a localized process, so that N_{trap} depends upon the location in the gauge section and the nature of Luder's bands propagation. In this calculation, an average was taken over the Luder's band deformation region. The sharp increase in the trap site density (Figure 16b) is, thus, the average behavior during the Luder's band region. $C_{critical}$ was calculated using the model in cylindrical coordinates. Figure 17 shows hydrogen concentration profile in a rod specimen at 25% plastic strain, which is the strain at the maximum load. Figure 17.a is for lattice hydrogen and Figure 17.b is for trapped hydrogen. R=0 is the center of the rod specimen and R=1 is the external surface. Four strain rates, ranging from 6.67×10^{-5} to 6.67×10^{-2} /sec, were used in order to compare the results here with the experimental data.

It is reasonable to assume that hydrogen associated with

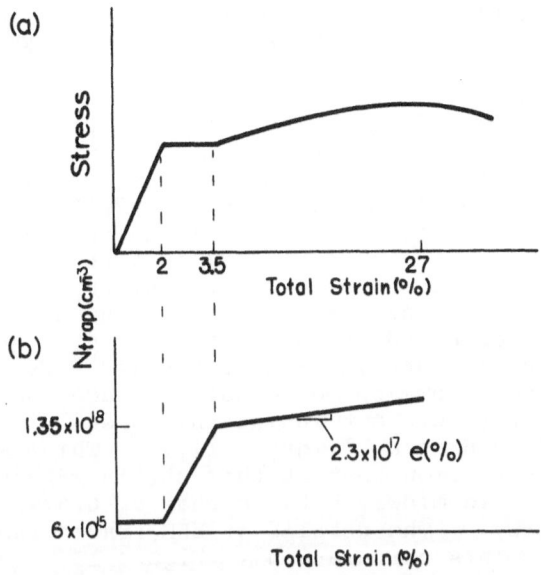

FIGURE 16. Parameters used to calculate the critical hydrogen concentration in rod specimen.

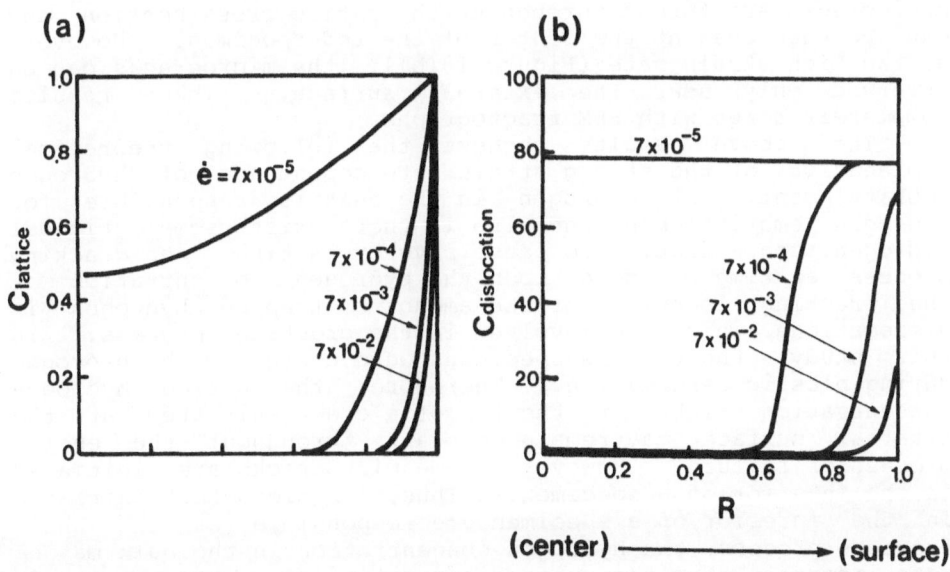

FIGURE 17. Hydrogen concentration profile in rod specimen for a given strain rate. (a) in lattice , (b) in dislocations.

dislocations but not in the lattice is responsible for hydrogen embrittlement because dislocation motion and distribution is essential for the cracking process. At the slowest strain rate, 6.67×10^{-5} /sec, it can be seen from the Figure 17 that dislocations at the maximum load are almost saturated by hydrogen throughout the entire cross section. This suggests that the supply of hydrogen from the environment through the external surface is large enough to fill the newly created dislocations with hydrogen. Thus, severe attack was observed. However, at the highest strain rate, 6.67×10^{-2} /sec, the time to achieve 25% strain is much less so that only the dislocations near the surface are filled with hydrogen. Since most of dislocations are free of hydrogen, this may explain the fact that no significant effect of hydrogen on R.A. and SEM fractographs was observed at this strain rate. An interesting point is that at the strain rate of 6.67×10^{-4} /sec, about 30% of the distance from the external surface has very high hydrogen concentration, but inside this area, dislocations are still free of hydrogen (Figure 17). This explains the experimental observation that at this strain rate the ductility measured by R.A. is midway between that at 6.67×10^{-5} /sec and complete ductility. The details of SEM fractographs show good agreement with this calculation. Figure 15.b shows an SEM photograph of the entire fracture surface. The outer area exhibits a brittle fracture, however, the inside area is still ductile.

The cross sectional views of the specimens fractured at the strain rates of 6.67×10^{-5} and 6.67×10^{-4} /sec are shown in Figure 18. Several transgranular microcracks can be seen in both specimens. At the low strain rate (Figure 18(a)), the microcracks are formed throughout the entire cross section and can be seen even at the center of the rod specimen. However, at the high strain rate (Figure 18(b)), the microcracks can be observed only near the external surface. These results completely agree with SEM fractographs.

The above results suggest the following theoretical explanation of the strong strain rate dependence of hydrogen embrittlement. If hydrogen in the bulk is responsible for hydrogen embrittlement and also if there exists some critical hydrogen concentration to cause severe cracking, the cracking process at time t depends upon the hydrogen concentration at the cracking location or the amount of trapped hydrogen in dislocations which are involved in the cracking process. In this study, the tensile specimens were charged with hydrogen during plastic deformation. Therefore, the lattice hydrogen concentration right at the tip of a crack initiated at the external surface may remain constant throughout the entire process of failure. However, some microcracks are initiated in the interior of a specimen. Thus, if microcrack initiated in the interior of a specimen are responsible for the whole fracture process, the hydrogen concentration in the bulk may be more important than the surface concentration. When the strain rate is high, a large number of dislocations are created in the specimen but these dislocations remain essentially free of hydrogen since the only supply of hydrogen is from the external

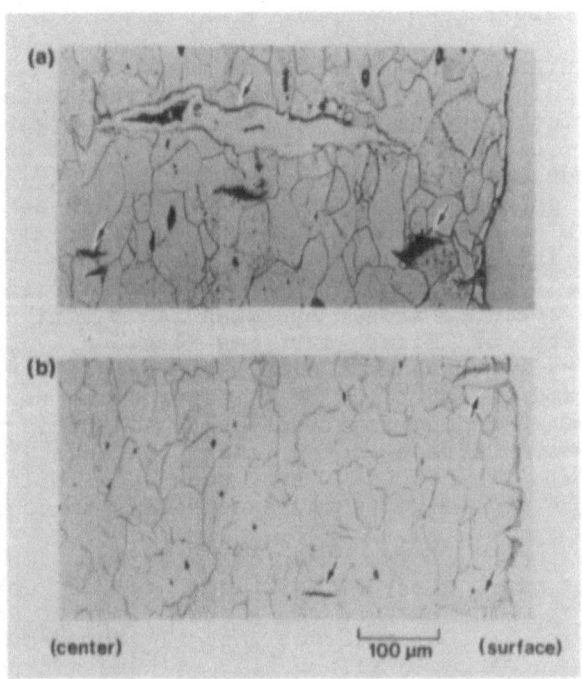

FIGURE 18. Cross sectional views of the rod specimens fractured at the strain rate of (a) 6.67×10^{-5}/sec, (b) 6.67×10^{-4}/sec.

surface. In section 2, it was shown that dislocations can accommodate many excess hydrogen atoms within the potential field around them. The only method to supply these excess hydrogen atoms is by the diffusion process. Therefore, when the multiplication rate of dislocations is high, the diffusion process can not provide enough supply of hydrogen to cause cracking within the timeframe of other failure mechanisms i.e. ductile failure. At slow strain rates, the diffusion process has enough time to fill these dislocations and can cause cracking.

Thus, the dynamic trapping effect can be used to explain the strong strain rate dependence of hydrogen embrittlement.

7. DISCUSSIONS AND CONCLUSIONS

At this point, there is no single mechanism which accounts for the whole spectrum of experimental observations as to hydrogen embrittlement of iron. Fracture due to hydrogen has a variety of modes depending upon metallurgical variables such as structure, chemical composition, and heat treatment. Therefore, it seems clear that the fundamental process of hydrogen embrittlement also depends upon these metallurgical variables. Since hydrogen in iron is very mobile compared to other solute atoms in iron, hydrogen is distributed rapidly over a variety of sites such as lattice, defects, and interfaces. Hydrogen at each site has a specific effect on the deformation and fracture at that site. Therefore, several effects may be operating simultaneously. At this point, it is not clear which effect is the most critical for hydrogen embrittlement. There is also some possibility that a combination of several effects is necessary to cause the final fracture.

Although the most critical effect on hydrogen embrittlement has not yet clarified, it is worthwhile to present the following picture of hydrogen embrittlement.

Nagumo et al (25) observed that microcracks are formed along very concentrated slip bands. These microcracks will then link together and result in fracture along the slip planes which are either {110} or {112}. The above experimental observation and the fact that quasi-cleavage fracture is typically along slip planes indicate that there is a close relationship between dislocation motion and embrittlement. Lynch(26) reported fractographic evidence of some crack tip plasticity even for the most brittle appearing intergranular cracking in steel.

Even though the exact mechanism for strain localization is not known, this effect creates a locally dense dislocation region. Recently, Kramer et al. (27) observed that the dislocation density in iron deformed in a hydrogen environment is higher than that in an inert environment. This experimental observation supports the creation of dense dislocation regions such as described above. This is also consistent with the report by Tabata et al. (28) which showed that hydrogen accelerates dislocation multiplication. Thus, the above argument leads to the view that dislocation motion and strain localization may be considered as precursors for microcrack formation at slip bands.

Since it is well documented in the literature that the motion and distribution of dislocations under the dynamic loading conditions is affected by the presence of hydrogen, it seems important to know the exact hydrogen concentration at these dislocation sites especially under the dynamic deformation conditions. This paper has, thus, summarized the several important features of hydrogen transport under these dynamic conditions.

First in section 2, it was shown that the hydrogen atoms are concentrated within a few Burger's distance from the dislocation core, as a result, that the core region accommodates most of the excess hydrogen atoms around a dislocation. It was also shown that the core is almost saturated by hydrogen under the typical experimental conditions.

In section 5.1, an atomistic model has been developed in order to describe the effect of dislocation motion on the local hydrogen atmosphere. It was concluded that under typical slow strain rate test conditions, most of the excess hydrogen atoms can be dragged by the moving dislocations. It was also found that even in the case of moving dislocations, most of the excess hydrogen atoms are still concentrated in the dislocation core region.

A macroscopic diffusion model, which includes several important interactions - the interactions of hydrogen with newly generated dislocations, the interactions of hydrogen with moving dislocations, and the effect of both of these on the diffusion process - has been developed in section 5.2. Although estimation of the parameters is not satisfactory in terms of an accurate determination of parameters, this model suggests the following important points to estimate the hydrogen concentration profile during plastic deformation.

First, it was shown that the dynamic trapping effect should be taken into account since dislocation generation is necessary to cause any plastic deformation. Dislocations generated in the bulk of the materials are initially free of hydrogen. Thus, there is no effect of hydrogen on these newly generated dislocations until hydrogen diffuses into these dislocations. This process necessarily creates the locally hydrogen depleted region in the surrounding lattice sites. Therefore, additional diffusion of hydrogen in the lattice sites is required in order to fill these continuouslly generated dislocations. Thus, this dynamic trapping effect is the first effect which should be taken into account in order to calculate the hydrogen concentration profile.

The effect of dislocation motion on the hydrogen concentration profile has also been modeled. The results indicate that the mechanism of dislocation transport by mobile dislocations owes its origin to the kinetics of hydrogen exchange between lattice sites and that in the moving dislocations. When the rate of redistribution is very high, this effect is very small. Thus, kinetic considerations must be carefully treated in order to reach a specific conclusion as to the significance of dislocation transport.

534

ACKNOWLEDGEMENTS

We wish to acknowledge the financial support for this particular experiment done at MIT provided by the Shell Companies Foundation.

We also wish to thank Dr. T. Murata and Dr. G.S. Frankel for valuable discussions.

REFERENCES

1. J.G. Morlet, H.H. Johnson, and A.R. Troiano, J. Iron and Steel Inst. 189 (1958) 37.
2. J.T. Brown and W.M. Baldwin, Trans. AIME, 200, (1954) 298.
3. G.M. Pressouyre, Hydrogen Problems in Steels, C.G. Interrante and G.M. Pressouyre eds., ASM, (1982) 18.
4. J.K. Tien, A.W. Thompson, I.M. Bernstein, and R.J. Richards, Metall. Trans. 7A, (1976) 821.
5. S.V. Nair, R.R. Jensen, and J.K. Tien, Metall. Trans. 14A, (1983) 385.
6. H.H. Johnson and J.P. Hirth, Metall. Trans. 7A, (1976)1543
7. R.A. Oriani, Fundamental Aspects of Stress Corrosion Cracking, R.W. Staehle ed.,(1967) 32.
8. J.P. Hirth and J. Lothe, Theory of Dislocation, 2nd ed., McGraw-Hill Book Co., New York, NY, (1976) 503.
9. J.P. Hirth, Metall. Trans. 11A, (1980) 861.
10. J.P. Hirth and B. Carnahan, Acta Met. 26, (1978) 1795.
11. T.N.L. Patterson, Maths. Comp. 22, (1968) 847.
12. R.B. Heady, Corrosion 34, (1978) 303.
13. A. McNabb and P.K. Foster, Trans. AIME, 227, (1963) 618.
14. M.A. Devanathan and Z. Stachurski, Proc. Roy. Soc., A270, (1962) 90.
15. A.J. Kumnick and H.H. Johnson, Acta Met. 28, (1980) 33.
16. M. Hashimoto and R.M. Latanision, submitted to Acta Met.
17. P. Bastien and P. Azou, Academie Des Sciences Comptes Rendus, 232, (1951) 1845.
18. R.C. Frank, Internal Stresses and Fatigue in Metals, G.M. Rassweiler and W.L. Grube eds., Elsevier Press, NY, (1959) 411.
19. M.R. Louthan, G.R. Caskey, J.A. Donovan, and D.E. Rawl, Mat. Sci. Eng. 10, (1972) 357.
20. J.A. Donovan, Metall. Trans. 7A, (1976) 1677.
21. L.M. Foster, T.H. Jack, and W.W. Hill, Metall. Trans. 1A, (1970) 3117.
22. G.S. Frankel and R.M. Latanision, International Congress on Metallic Corrosion, Toronto, 4, (1984) 466; Metall. Trans. in press.
23. H. Yoshinaga and S. Morozumi, Phil. Mag. 23, (1971) 1367.
24. H.K. Birnbaum, Atomistics of Fracture, R.M. Latanision and J. Pickens eds., (1981) 733.
25. M. Nagumo and K. Miyamoto, J. Jpn. Inst. Met. 45, (1981) 1309.
26. S.P. Lynch, Acta Met. 32, (1984) 79.
27. I.R. Kramer and J.P. Hirth, Scripta Met. 18, (1984) 539.
28. T. Tabata and H.K. Birnbaum, Scripta Met. 18, (1984) 231.

DISCUSSION

Comment by J. Chene:

The quantification of hydrogen trapping and transport mechanisms using permeation technique is strongly dependent on the possible evolution of hydrogen concentration on the surface. Can you assume that in your experiments the hydrogen evolution reaction on the surface is independent of the stress and/or strain state?

Reply:

In the case of predeformed specimens, the steady state hydrogen permeation is independent of plastic strain. This suggests that hydrogen evolution reaction is not affected by prestraining of the specimens. In the elastic region, the elastic interaction increases the H concentration at the surface which was taken into account in our model.

However, how hydrogen evolution reaction is affected by simultaneous plastic deformation is not well understood and more careful electrochemical experiments are needed to clarify it, although we believe that the dynamic trapping effect far exceeds the surface effect.

Comment by M. Daw:

How sensitive are your calculations to the way which you model the hydrogen-dislocation interaction? Because the majority of trapping occurs in the core, how the interaction is cut off there is important. Also, how important is the hydrogen-hydrogen interaction, which is neglected in your model? Perhaps because you fit to experiment, these effects somehow are included in the parameters obtained.

Reply:

In order to achieve more accurate calculations, in a microscopic scale, higher order approximations are necessary for the factors listed below.

(1) H-H interactions
(2) accurate activation energy
(3) not imaginary but the real interstitial structures
(4) exact core structure and potential profile
(5) not moving framework but laboratory framework
(6) higher order interaction such as stress relief by H.

This is especially important for strong core interactions such as those in the Fe-H system. However, better approximations are not available so only the simple discrete model described in the text was studied.

Comment by H. Birnbaum:

1) The strain rate effect you determined certainly will exist as the supply of hydrogen to the center of the specimen will be limited at high strain rates. However, this is not the entire strain rate effect since pre-charged specimens, with uniform hydrogen concentrations, also show an inverse strain rate effect on ductility.

2) Your calculation treated trapping at edge dislocations. In deformed iron at room temperature, the mobile dislocations have edge character; therefore, the majority of the dislocation structure has screw character. As a result, the dislocation trapping of H should be dominated by screw dislocations and the dislocation transport of H should be dominated by edge dislocations. How would this affect your theoretical treatment?

Reply:

1) We did pre-charge the specimen before deformation. Therefore the initial hydrogen distribution is uniform throughout the entire cross section of the specimen. However, as we deform the specimen, many dislocations are generated. Since the newly generated dislocations are initially free of hydrogen, those dislocations will take hydrogen from the surrounding lattice sites, and causing the dynamic trapping effect. When the initial lattice hydrogen concentration is not high enough to fill these dislocations, most dislocations will remain free of hydrogen unless the supply of hydrogen from the external surface can take place. Therefore the strain rate effect can be explained by the dynamic trapping effect even in the case of pre-charged specimen.

2) In the Fe-H system, the core interaction dominates the Cottrell type interaction. Therefore, it is not necessary to distinguish the screw dislocations from the edge dislocation if the core interactions of both types of dislocations are similar.
The kinetic parameters were experimentally determined; as a result, the macroscopic diffusion model is still valid, although rigorously speaking we should introduce two types of the core interactions to take into account the difference in the strength of core interactions.

Comment by H. Mughrabi:

Your stress-strain curves show a higher stress (which you did not mention) and a reduced ductility for specimens with hydrogen than in the case of specimens free of hydrogen. Now, the first detailed effects of hydrogen on the stress-strain behavior of iron known to me were the observations of Matsui, Kimura and others who showed that in iron crystals of very high purity the introduction of hydrogen reduced the flow stress and increased the ductility. Subsequent work by other Japanese workers on less pure iron showed the opposite effect and gave rise to a controversy. So, 1) what effect do impurities have on the stress-strain behavior of iron with and without hydrogen, and 2) can these effects be attributed to specific foreign atoms?

Reply:

Hydrogen reduces the yield strength and flow stress of high purity iron. This may be due to the change in the core structure of dislocations due to hydrogen, resulting, the reduction of Peierls potential. In less pure iron, some impurities such as C and N increase the yield strength. Hydrogen increases those impurity effects causing further strengthening, the mechanism for which is not well understood.

Comment by R. Kirchheim:

Besides core- and elastic-interaction one should also consider H-H interaction, because the local concentrations at the core of the dislocation become very high and the known H-H interaction energies are appreciably large (0.1 to 0.2 eV). This effect may be difficult to study in the Fe-H system due to the small solubility, but in systems like V-H, Nb-H and Pd-H it plays an important role.

Comment by J. F. Knott:

You have said nothing about the role of second-phase particles in providing traps for hydrogen, yet your material contains both inclusions and carbides. These crack or cavitate as plastic strain is increased and therefore might be expected to become more effective as traps with increase in strain. Do you have any comments?

Reply:

We have not taken into account the increase in the trapping capacity due to cavitation around second particles. However, trapping to the dislocations in iron seems to be much stronger than that to the voids.

Comment by H. J. Engell:

Is it actually correct in a strict sense that you need "enough hydrogen in dislocations to cause cracking"? Is it not also possible that the interstitial hydrogen causes crack formation, and its concentration (or activity) stays too low for cracking as long as the traps, i.e., the dislocation cores, are not yet saturated with hydrogen?

Reply:

It seems reasonable to say that hydrogen associated with dislocations but not in the lattice is responsible for hydrogen embrittlement because dislocation motion and redistribution is essential for the cracking process.
However, a more careful experiment should be done in order to determine the critical hydrogen concentration in dislocations for hydrogen embrittlement.

538

HYDROGEN INDUCED FRACTURE

J. P. Hirth
The Ohio State University
Department of Metallurgical Engineering
Columbus, Ohio 43201
U.S.A.

ABSTRACT
The phenomenology and mechanisms of hydrogen-induced fracture in metals and alloys are briefly reviewed. As a stimulus for the workshop sessions, key unresolved issues are identified for each form of fracture that is discussed.

1. INTRODUCTION

There has been considerable progress in the understanding of hydrogen embrittlement in recent years. The subject was addressed in the antecedent to the present conference and here we emphasize work subsequent to those proceedings (1). Aspects of hydrogen embrittlement are treated in reviews on effects in steels (2), enhanced dislocation mobility and ductile failure (3), and kinetics of embrittlement (4). With respect to kinetics, various processes (solution transport, adsorption, bulk diffusion, dislocation transport) have been identified as controlling the rate of embrittlement in different regimes of experimental variables (4,5). This work has implication for models of embrittlement but does not directly relate to atomistic mechanisms of crack initiation and propagation.

In the present work, we discuss research directly related to such atomistic mechanisms. An attempt is made to identify current issues that require further elucidation.

2. BRITTLE FRACTURE

2.1. Hydride formation and cracking

2.1.1. Phenomenology and mechanisms. Brittle fracture contributes to hydrogen embrittlement in metals that form hydrides, including the group V metals Nb, V, and Ta, alloys of Mg, Ti and Zr, and Pd as reviewed by Birnbaum (6). The fracture proceeds by the Westlake mechanism (7) of hydride formation in the stress concentrated region at a crack tip, cracking of the hydride and crack arrest at the hydride-matrix interface, and a cyclic repetition of the above events. Generally, what evidence is available suggests brittle cleavage behavior of single-phase hydrides below about room temperature (6). Measurements of the energy release rate for cleavage crack propagation on (100) planes in niobium hydride give values for the J integral of the order of twice the surface energy (6), similar to results for glass and brittle inorganic compounds. The implication is that plastic strain is limited at or below room temperature, with the factor of two being accounted for by surface step formation or very limited local plasticity. Bulk modulus measurements (8,9) indicate that hydrogen increases the binding energy relative to pure metals. In the ordered low-symmetry hydrides, dislocation mobility appears to be reduced both by blocking by twin boundaries and by a large fraction of the inherited dislocations being imperfect in the superlattice and hence pinned by antiphase boundaries (6).

The hydrides have an increased volume per metal atom. Hence the isoaxial tensile field at a loaded crack tip promotes hydride formation, as do shear stresses for low-symmetry hydrides (10). The lowering of the free-energy of formation of a hydride nucleus has been evaluated for both stress situations, including effects of local dislocation formation to relax the coherency strains (6,10,11,). The stress effects are such that hydrides can form at the crack tip under conditions where they would be thermodynamically unstable in the absence of stress.

When the hydrides form discontinuously near the crack tip, they can still crack and contribute to hole growth. In such a case the hydrides would reduce the critical stress intensity for ductile crack propagation in a manner analogous to inclusions or other second-phase particles.

2.1.2. Unresolved issues. Some details of the cracking process remain open for further work. While evidence is strong that hydrides form preferentially at crack tips, the effect on the critical stress intensity for crack propagation is complicated. The formation of a hydride creates a site vulnerable to brittle cracking. However, the transformation strain field relaxes the crack tip elastic field and produces transformation toughening (12,13). In addition, there remains the question of whether the microcrack in the brittle hydride nucleates at the tip of the microscopic crack or along the flank of the hydride, possible enhanced by dislocation pileup arrays in the matrix.

For the case of discontinuous hydrides, Puls and associates (14) have studied zirconium alloys. They find evidence for the propagation of a brittle microcrack in the matrix in a brittle manner, even though the matrix would not sustain a sharp crack under slow loading. This result indicates the possibility of the inertia of the brittle crack emerging from the hydride providing an added driving force for some brittle propagation into the matrix.

2.2. Single crystals

2.2.1. Phenomenology. As reviewed by Thomson and Lin (16), there is no question that brittle materials such as silicon, alkali halides, zinc (low temperatures) and tungsten (low temperatures) fail by cleavage. Yet transmission electron microscopy, etch pitting and decoration techniques reveal that some dislocation emission accompanies cleavage, the amount increasing with decreasing crack velocity (17,18). On the other hand, there are cases of hydrogen-induced flat fracture that appear cleavage-like at higher magnification but which seem to involve large local deformation and failure along a slip plane (see Sect. 3). Hence, there is a question whether hydrogen ever induces pure cleavage failure. The extensive work of Vehoff and coworkers, reviewed in (19) provides definitive evidence for such failure. Iron-silicon alloy single crystals undergo crack propagation on (100) planes in a purely ductile, wedge-opening manner, accommodated by the operation of two symmetric slip systems on planes inclined at 45 degrees to the crack propagation direction. If such crystals are exposed to hydrogen, the wedge angle decreases with increasing hydrogen chemical potential and the density of accommodation dislocations decreases. Hence an increasing portion of crack propagation occurs in a decohesion, crack opening manner. The above result is qualitatively consistent with theories for the competition between cleavage and blunting (e.g. ref. 20), provided that hydrogen has a greater influence on the former. The slow cleavage propagation would be consonant with a mechanism of overcoming the lattice trapping barrier (16) by a double-kink nucleation and growth mechanism (21).

With respect to the influence of hydrogen at the atomic level, some progress has been made. Rice and Thomson (20) considered the competitive processes of dislocation loop nucleation versus reversible extension of a crack by decohesion. One must recall that hydrogen could favor the decohesion-cleavage process either by weakening the decohering bonds or by making dislocation formation more difficult. In either case a detailed, first-principles theory for hydrogen interacting with the nonlinear region at a crack tip or dislocation core is not possible, as discussed by Messmer and Briant (22), because of the enormous computation time that would be required. Nevertheless, some guidance can be gained by considering either (22) exact calculations for hydrogen in small metal clusters (without relaxation of the cluster configuration) or larger scale atomic calculations using semiempirical potentials selected on the basis of cluster calculation results (23). The former approach (22) indicates a weak embrittling effect for hydrogen in nickel, the most-studied case, while the latter (23) indicates the possibility of hydrogen enhanced cleavage on (110) planes in nickel.

2.2.2. Unresolved issues. While the Vehoff results (19) clearly support the concept of hydrogen induced cleavage, consistent with a double-kink model, the details of the crack-tip process are not completely resolved. Some of the remaining questions are: does the hydrogen exert its effect at the very crack tip or within the bulk ahead of the crack, is the cracking continuous at the atomic scale or does it occur in discrete, rapid jumps, and, parallel with brittle cracking in general, to the extent that the J integral exceeds the appropriate surface energy, what are the dissipation processes or barriers that account for the difference? Also, pertinent to the cleavage-blistering competition, computer modelling would be valuable in providing insight at the atomic crack-tip level, but it would have to be three dimensional to resolve, for example, double kinks, and such 3-D calculations have not as yet been performed because of the great requirement for computation time. Until such calculations are performed, it seems unlikely that there will be any direct evidence for a lattice trapping barrier and a kink model for hydrogen induced cleavage. Pertinent to the atomic calculations, increased computer speed and memory should make possible further progress. Problems of interest would include: cluster calculations involving configurational relaxation and energy calculations; further development of semiempirical potentials to facilitate atomic calculations: and, ultimately, first principle calculations for nonlinear dislocation cores and crack tips.

2.3 Polycrystals

2.3.1. Phenomenology. Transgranular failure induced by hydrogen presumably occurs by hydrogen enhanced cleavage analogous to that described in the previous section. An example would be cracking of mild steel below the ductile-brittle transition temperature which is raised by hydrogen. We mention our own work (24) that shows a transition with increasing hydrogen fugacity from a mixed mode I-II ductile fracture to a mode I transgranular failure with minimal ductility in spheroidized carbon steels, thoriated nickel alloys, and quenched and tempered AISI 4340 steel (24). The other case of interest is intergranular fracture along current grain boundaries, or, for steels, along prior austenite grain boundaries.

Again, some cases of intergranular fracture occur with extensive local ductility (25) although the macroscopic crack appearance is flat, see Sect. 3. However, there is strong evidence as well for a decohesion-type hydrogen induced fracture at grain boundaries. Morgan and McMahon (26) discuss the extensive work of the McMahon group on the deleterious effect of hydrogen superposed on the similar effect of temper embrittling elements in quenched and tempered alloy steels. In particular, they present an extraction replica transmission electron micrograph of a boundary embrittled by tellurium, sulphur and hydrogen that shows no evidence for any local plastic deformation. Table 1 shows results for a quenched and tempered 1520 steel (27) where boron and hydrogen combine to give intergranular failure at zero macroscopic ductility in a plane-strain tensile test, even though without hydrogen, the material has extensive ductility. The fracture stress falls to less than one-half the yield stress in the most embrittled case.

Table 1. Plane-strain tensile properties for AISI 1520 steel, tempered at 400 C for 1 h, as a function of charging current density in poisoned 1 N sulfuric acid charged for 2 h.

Current Density A/m^2	Yield Stress (MPa)	Ultimate Tensile (MPa)	Uniform Strain	Fracture Strain
0	1040	1250	0.05	0.20
6	990	1140	0.05	0.05
20	----	431	0	0
50	----	425	0	0
170	----	420	0	0

Of course, analogous to the results for alkali halides discussed previously, limited dislocation injection would be expected to accompany cleavage-type failure. Also, for smooth or blunt-notched specimens, some plastic flow usually precedes brittle crack nucleation and the intersection of the crack-tip with preexisting dislocations or their attraction and popout at newly created crack surfaces could also produce ledges and pits on the crack surfaces that could be interpreted as suggesting dislocation motion.

The theoretical situation for intergranular separation is analogous to that discussed for single crystals with the added complications of the structure of the grain boundary and the presence of segregated solute or impurities there. There are useful first principle calculations of the energies of grain boundaries with segregants and for their rigid separation to form free surfaces, a process that gives an upper-bound estimate of the work of decohesion (28).

2.3.2. Unresolved issues. Further experimental work is needed to classify macroscopically appearing flat intergranular fractures as either cleavage accompanied (or preceded) by limited deformation or deformation enhanced cleavage-like separation in a local region that has deformed extensively. The same remarks about atomic mechanisms that were made for single crystals apply to the intergranular cracking case.

2.4. Mixed brittle-ductile fracture

2.4.1. Phenomenology. Qualitatively following an idea of Ashby's (29), we can depict the critical stress for fracture as shown in Fig. 1 as a function of temperature T or chemical potential of hydrogen μ_H. The stress can be the intensified stress near the tip of a precracked specimen. Below T_1 or above μ_1 the crack initiates and propagates essentially in a brittle manner as discussed above and crack nucleation is the critical event while above T_2 (or below μ_2) crack nucleation and growth or the propagation of a precrack occur in a ductile way. Between T_1 and T_2 or μ_1 and μ_2 plastic flow first takes place. This includes the precrack case where deformation

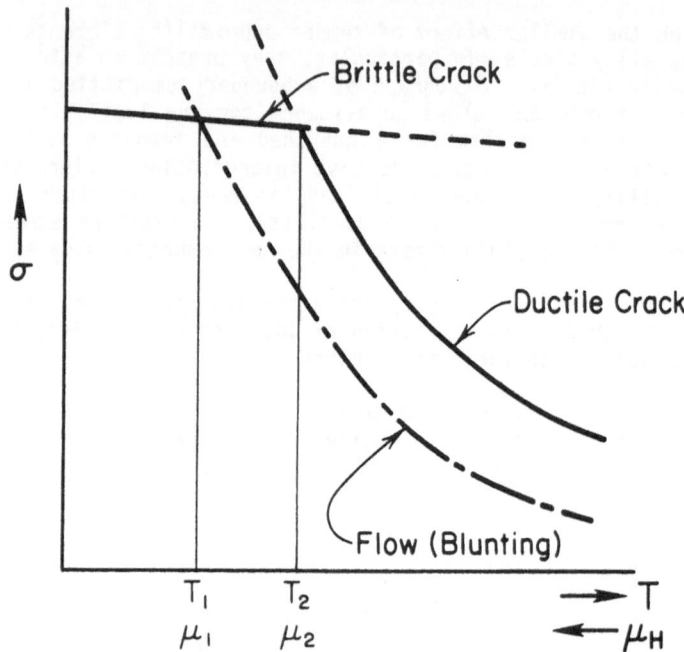

FIGURE 1. Representation of the critical local stress for cracking for ductile and brittle fracture as a function of temperature and hydrogen chemical potential.

occurs at the crack-tip (blunting) and there is stress elevation in the plastic region because of work hardening. In terms of either plastic relaxation in a continuum plasticity sense or dislocation motion in a crystal plasticity sense, the crack-tip is screened from the applied load and the tip stress intensity is less than the applied stress intensity. Despite the screening, the stress elevation is eventually sufficient to nucleate a microcrack either at the macroscopic crack tip or within the plastic zone ahead of the tip, usually by decohesion or cracking of a second phase particle. In the region between T_1 and T_2 or μ_1 and μ_2 the critical stress intensity for brittle fracture is less than that for ductile fracture and the crack then propagates in a brittle manner (30). In the crystal plasticity description, pertinent to the macroscopic tip propagation case, there has been intense activity in describing the screening fields of dislocations, for both the static (31,32,33) and dynamic (34) cases, as reviewed by Thomson and Lin (15). So far these have all been two-dimensional calculations (straight dislocations parallel to a straight crack tip). In the internal cracking case, statistics of particle sizes and interface strengths are important in establishing the critical crack nucleation event (35,36).

2.4.2. <u>Unresolved issues</u>. Three-dimensional screening calculation involving both non-straight (curved or kinked), crack tips, where a start has been made (21,37) and curved dislocations would be a desirable goal, probably beyond the capability of current numerical computers. Measurements of the cohesive energy of second-phase particle or inclusion interfaces, with and without impurities and hydrogen present, are needed. An improved theory for the nucleation of decohesion at the interface of such particles is needed. In general, these needs parallel those for the brittle fracture case near the ductile-brittle transition temperature in the absence of the hydrogen (35).

3. DUCTILE FRACTURE
3.1. Single crystals

3.1.1. Phenomenology. Below a critical hydrogen fugacity, hydrogen has no effect on the properties of single crystals (38,39), and iron, for example, necks to a double chisel point fracture. Above such a fugacity, for example for iron (40,41), hydrogen reduces the strain to fracture, and causes a flat fracture parallel to the primary (110) slip plane. The fracture is thus a mixed mode I-II fracture with the fracture plane parallel to the (110) plane with maximum resolved shear stress but also with a resolved normal stress corresponding to mode I loading. Such a failure is the crystal plasticity analog of the mixed mode shear fractures discussed later. The implication of the results is that intense local slip in a band weakens the material so that it can crack under the mode I load/component. Hydrogen is known to promote the planarity of (110) slip in iron, a factor that may contribute to the degradation process.

The details of the cracking process have not been completely elucidated. However, Wilsdorf and coworkers (42) observed crack propagation in-situ in transmission electron microscopy for gold and silver single crystals and found that the process involved intense deformation near the crack tip, void nucleation at cell boundary intersections, and connection of the void to the main crack by intense local slip. In situ observations for iron in the presence of hydrogen gave similar results (43), with crack propagation at reduced loads in the presence of hydrogen. In both cases, of course, compatibility constraints (e.g. satisfying the von Mises criterion) are relaxed for the thin specimens and bulk crack propagation may differ.

Pertinent to the hydrogen effects for iron (43), Birnbaum and coworkers find that dislocation mobility, as observed in situ, also is enhanced in the presence of high fugacity hydrogen in iron, nickel and other pure metals (3). This effect is suggested to directly connect to the fracture observations for iron (43). While it now seems incontrovertible that hydrogen enhances the mobility of screw dislocations and reduces the stress for equivalent dislocation motion at room temperature in high-purity iron (3, 44), hydrogen has the opposite effect in less pure iron (45). Since the flat fractures along slip plane traces are also observed in less pure iron (42), the influence of hydrogen on slip planarity or some other local phenomenon may be of importance in the fracture mechanism.

While only indirectly relating to fracture, several mechanisms for the influence of hydrogen on dislocation mobility have been suggested. Much evidence (46,47) favors an effect on double kink nucleation (and core structure) on screw dislocations (44,45), but a nonlinear elastic effect on dislocation intersection in the presence of hydrogen has also been proposed (3).

3.1.2. Unresolved issues. Some details of the process of hydrogen assisted fracture along slip plane traces after extensive local deformation are still not known. Does the mechanism directly observed for thin specimens in transmission electron microscopy apply for bulk specimens? Also, what is the relative importance of the influence of hydrogen on dislocation mobility, on slip planarity and on dislocation intersection in crack propagation in bulk specimens?

3.2. Polycrystals

3.2.1. Phenomenology. For polycrystals of pure metals or single-phase alloys, hydrogen-induced fracture analogous to that for single crystals occurs. The transgranular flat fracture facets in hydrogen affected fracture in such material appear to lie along slip plane traces, implying a local crack propagation much like that in a single crystal(49). For "intergranular" fracture in hydrogen-charged nickel, Birnbaum and coworkers (3) have found that the fracture actually occurs near but not at the boundary in a serrated pattern, with flat fracture facets parallel to (111) slip plane traces. They also find a hydrogen segregation field extending over many atomic distances near the boundary and postulate that the fracture occurs within one grain, influenced by the enhanced hydrogen concentration in the segregated region. As discussed in Sect. 2, even flatter intergranular fractures occurring at grain boundaries present some evidence for local slip (25).

3.2.2. <u>Unresolved issues.</u> The critical issues for polycrystals are the same as those discussed in Sect. 3.1 for ductile fracture of single crystals, involving the detailed mechanism of crack propagation, and in Sect. 2.3, involving the distinction between ductile and brittle decohesion events at crack tips.

3.3. Multiphase alloys

3.3.1. <u>Phenomenology.</u> In alloys containing second-phase particles, such as carbides in steels, or inclusions, such as manganese sulfides in steels, ductile failure involves void nucleation and growth at the particles in some manner. We restrict our discussion here to the example of steels. The voids nucleate by decohesion or cracking of the particles. In uniaxial tensile specimens, bluntly notched, or after necking of smooth bars, crack nucleation occurs in the specimen center where the isoaxial tensile stress is elevated, and growth in a mode I manner terminated by a mixed mode sliding off leads to the standard cup-cone fracture. The crack propagates by linkage of voids formed ahead of the crack with the main crack tip. Linkage occurs either by necking to a point of the remaining ligament or by some necking followed by mixed mode I-II shear cracking.

Under severe hydrogen charging conditions, damage in the form of voids, blisters or fissures can form even in the absence of stress (2). After charging with intermediate hydrogen fugacity, the mode of fracture and the initial stress-strain behavior are unchanged but failure occurs at a reduced strain, usually reflected in a decrease in percent reduction in area in tensile tests. There are trends not fully understood, from a mechanistic viewpoint, in hydrogen degradation as a function of hydrogen fugacity, microstructure, alloying elements, strain rate, impurity level, strength level and type of charging (50), but a discussion of these is beyond the scope of the present discussion. Some of the more detailed microstructural correlations have been performed on spheroidized steels, where hydrogen has been shown to influence void nucleation in some cases (51), void coalescence in others (52). The differences are possibly associated with amounts and distributions of phosphorus and sulfur at subboundaries or carbide interfaces. In the work where an effect or void condition was observed (51) the microstructure was such that many carbides were positioned on subboundaries, which would be consistent with the above possibility. Also it is interesting that tritium autoradiography of a spheroidized 1520 steel that is very low in P and S reveals no trapping of hydrogen at carbide interfaces until voids form there (53), while earlier work of a similar nature on steels with larger P and S contents did show hydrogen trapped at unvoided carbide interfaces (54).

Under plane strain loading, the mode of ductile failure is different, generally involving mixed mode I-II cracking with the cracks following traces of the characteristic slip line fields of plasticity theory. The failure mode is the same in the presence of hydrogen at intermediate fugacities, but the critical strain for crack initiation is lowered (24,55,56). Plane strain tensile tests and bend tests on AISI 1090 spheroidized steel show that the strain for surface roughening, the strain for the formation of shear bands as revealed by voids aligned along characteristic slip traces, the strain for the formation of surfaces microcracks, and the strain for the propagation of a crack to failure in the mixed mode I-II manner are all reduced about a factor of two in the presence of hydrogen (55). Similar effects are observed in compression bend tests (57).

The results for tests in the absence of hydrogen agree with predictions of continuum plasticity theory with respect to surface roughening shear localization, and differences in behavior in tension and compression (58). These findings are consistent with the shear bands following characteristic slip traces and with the seeming non-crystal plasticity nature of shear bond formation as indicated by transmission electron microscopy (59). However, since hydrogen does not have much effect on the stress-strain behavior up to the point of shear localization, the hydrogen effect is not directly consistent with continuum theory. Instead, the results suggest that hydrogen creates defects in the continuum sense (58). Continuum mechanics finite-element simulations show that the presence of voids enhances shear localization (60,61). Hence, if hydrogen were to enhance void nucleation, such a direct event would provide the continuum defect. However, the effect of hydrogen appears to be indirect since it influences the critical strain for surface roughening prior to void formation, since the alignment of voids along characteristic slip traces is consistent with void formation induced by incompatibility stresses caused by (not causing) shear localization, and since reduced strains for localization are also observed in compression.

The results are consistent with an enhanced effect of hydrogen on shear localization in the absence of void formation, perhaps because of enhanced slip planarity. Once voids form in the path of a shear band as a consequence of incompatibility they would further promote propagation of the shear band and both factors could ultimately contribute to failure. Relevant to the possible effect on shear localization, Bernstein and coworkers (62) have shown in transmission electron microscopy that the dislocation arrays around carbide particles in Armco iron and AISI 1520 steel change from multiple-slip, three-dimensional tangles to planar arrays containing dislocations from three slip systems in the presence of hydrogen.

3.3.2. Unresolved issues. Theoretical and experimental work is needed on the critical local stress-strain conditions, including inhomogeneous slip or dislocation pileups and the influence of particle size, for particle decohesion or cracking, for both clean interfaces and those with impurity adsorbates. This information is needed to better define the critical void formation events leading to ductile fracture. Such information would be important in understanding hydrogen effects as well as in understanding failure in the absence of hydrogen (35, 36). Further work is needed to define the trends of hydrogen in steels influencing void nucleation in some cases, void coalescence in others. Another issue is that of the mechanism by which hydrogen influences the onset of shear localization under plane-strain loading and whether this effect is independent of void formation.

4. SUMMARY

Since the benchmark of the previous NATO conference on the atomistics of crack propagation (1), the general phenomenology of hydrogen effects on fracture have been fairly well defined. Certainly, many of the seeming direct conflicts in hydrogen effects that were prevalent twenty years ago have been clarified. Nevertheless, many aspects of the influence of hydrogen on fracture, and particularly those pertinent to atomic-scale mechanisms at and near the crack tip, remain unresolved as indicated in the foregoing discussion.

Of the many mechanisms that have been suggested for hydrogen embrittlement, reviewed in refs. (2), (3) and (46), the results discussed here are consistent with decohesion at the crack tip for transgranular brittle fracture; decohesion at second-phase particles leading to ductile dimple rupture; deformation induced weakening in the crystal slip band for localized ductile flat fracture of single crystals and polycrystals along slip plane traces; and deformation-induced weakening in shear bands for mixed mode ductile fracture of engineering alloys. Slip softening (63) seems to be important in some cases of fracture along slip plane traces. Other models such as the pressure theory and that involving the decrease of surface energy in the presence of hydrogen may play some role in limited cases, for example in particle decohesion or void growth, but are not directly applicable to the work discussed here.

ACKNOWLEDGMENT

This work was sponsored by the National Science Foundation under Grant 8311620.

REFERENCES

1. Latanision, RM, and Pickens, JR: Atomistics of Fracture. New York: Plenum 1983.
2. Oriani, RA, Hirth, JP, and Smialowski, M: Hydrogen Degradation of Ferrous Alloys. Park Ridge, NJ: Noyes 1985.
3. Birnbaum, HK: IMD Lecture, March 1986; Metall. Trans. A, in press.
4. Wei. RP, and Gao, M: in ref. 2, p. 579.
5. Hwang, C., Bernstein, IM: Acta Metall., 34, 1001 (1986).
6. Birnbaum, HK: J. Less Common Met., 104, 31 (1984).
7. Westlake, DG: Trans. Am. Soc. Met., 62, 1000 (1969).
8. Amano, M, Mazzolai, FM, and Birnbaum, HK: Acta Metall., 31, 1549 (1983).
9. Springer, T: Topics Appl. Phys., 28, 75 (1978).
10. Puls, MP: Acta Metall. 32, 1259 (1984).
11. Birnbaum, HK, Grossbeck, ML, and Amano, M: J. Less Common Met., 49, 357 (1976).
12. Evans, AG, and Heuer, AH: J. Amer. Ceram. Soc., 63, 241 (1980).
13. Budiansky, B, Hutchinson, JW, and Lambropoulos, JC: Int. J. Solids Structure, 19, 337 (1983).
14. Puls, MP: this volume.
15. Thomson, R, and Lin, IH: in ref. 2, p. 454.
16. Thomson, R: Solid State Phys., in press.
17. Gilman, JJ, and Johnston, WG, Dislocations and Mechanical Properties of Crystals, eds. Fisher, JC, et al. New York: Wiley, 1957, p. 116.
18. Liu, JM, and Shen, BW: Metall. Trans. A, 15, 1247 (1984).
19. Vehoff, H, and Rothe, W: Acta Metall. 31, 1781 (1983).
20. Rice, JR, and Thomson, R: Philos. Mag., 29, 73, (1974).
21. Lin, IH, and Hirth, JP: J. Mater. Sci., 17, 447 (1982).
22. Messmer, RP, and Briant, CL: in ref. 2, p. 140.

23. Daw, MS, and Baskes, MI: Phys. Rev. B., 29, 6443 (1984).
24. Goldenberg, T, Lee, TD, and Hirth, JP: Metall. Trans. A, 9, 1663 (1978); 10, 199 (1979).
25. Lynch, SP: in ref. 1, p. 955.
26. Morgan, MJ, and McMahon, CJ, Jr.: in ref. 2, p. 608.
27. Chatterjee, A: Ph.D. Thesis. Columbus, OH, The Ohio State Univ., 1986.
28. Smith, JR: this volume.
29. Ashby, MF: personal communication, Aug. 1985.
30. Ritchie, RO, Knott, JF, and Rice, JR: J. Mech. Phys. Solids, 21, 395 (1973).
31. Thomson, R, and Sinclair, JE: Acta Metall., 30, 1325 (1982).
32. Lin, IH, Weertman, J, and Thomson, R: Acta Metall., 31, 473 (1983).
33. Hirth, JP, Hoagland, RG, and Popelar, CH: Acta Metall. 32, 371 (1984).
34. Lin, IH, and Thomson, R: research in progress.
35. Budiansky, B, Evans, AG, and Hutchinson, JW: Proc. DARDA Materials Research Council. Ann Arbor, MI: University of Michigan, 1983.
36. Williams, JC, and Hirth, JP: Rapid Solidification Processing, ed. Mehrabian, R, Gaithersburg, MD: National Bureau of Standards, 1982, p. 135.
37. Rice, JR: Int. J. Solids Structure, 21, 781 (1985).
38. Smialowski, M: in ref 2, p. 561.
39. Moriya, S., Matsui, H., and Kimura, H.: Matls. Sci. Eng., 40, 217 (1979).
40. Nakasoto, F, and Bernstein, IM: Metall. Trans. A, 9, 1317 (1978).
41. Takahashi, H, Takeyama, T, and Hara, T: Nihon-Kinz. Gakk., 43, 492 (1979).
42. Bauer, RW, Lyles, RL, Jr., and Wilsdorf, HGF: Z. Metallk., 63, 525 (1972).
43. Tabata, T, and Birnbaum, HK: Scripta Metall., 18, 23 (1984).
44. Matsui, H, Kimura, H, and Moriya, S: Mater. Sci. Eng., 40, 207 (1979).
45. Kimura, H: Trans. Jap. Inst. Metals, 26, 527 (1985).
46. Hirth, JP: Metall. Trans. A, 11, 861 (1980).
47. Seeger, A: Phys. Stat. Solidi, 55, 547 (1979).
48. Sato, A, and Meshii, M: Acta Metall., 21 753 (1973).
49. Boyer, HE, ed.: Metals Handbook, vol. 9. Metals Park, OH: Am. Soc. Metals, 1974, p. 64.
50. Thompson, AW, and Bernstein, IM: Adv. Corros. Sci. Technol., 7, 53, (1979).
51. Cialone, H, and Asaro, RJ: Metall. Trans. A, 10,367 (1979); 12, 1373 (1981).
52. Garber, RI, Bernstein, IM, and Thompson, AW: Metall. Trans. A, 12, 225 (1981).
53. Le, TD, and Bernstein, IM: Scripta Metall., in press.
54. Asaoka, T, Dagbert, C, Aucouturier, M, and Galland, J: Scripta Metall., 11, 467 (1977).
55. Onyewuenyi, O, and Hirth, JP: Metall. Trans. A, 14, 259 (1983); Chang, SC, and Hirth, JP: Metall. Trans. A, 16, 1425 (1985).
56. Kosko, TJ, and Thompson, AW: Scripta Metall., 16, 1367 (1982).
57. Rajan, V, and Hirth, JP: research in progress.
58. Hutchinson, JW, and Tvergaard, V: Int. J. Mech Sci., 22, 339 (1980).
59. Hatherly, M: Strength of Metals and Alloys, ed. Gifkins, RC, Oxford: Pergamon, 1983, p. 1181.
60. Gurson, AL: J. Eng. Mater. Technol., 99, 2 (1977).
61. Tvergaard, V: Int. J. Fracture, 17, 389 (1981).
62. Hwang, C, and Bernstein, IM: Acta Metall., 34, 1011 (1986).
63. Beachem, CD: Metall. Trans, 3, 437 (1972).

DISCUSSION

Comment by R. H. Jones:

Vanadium hydride was shown to be ductile at temperatures above ~250 K by Hishinuma. Is there any evidence in other hydriding systems where the ductility of the hydride may be a factor in hydrogen embrittlement?

Reply:

Birnbaum: Gahr and Birnbaum studied the fracture of single crystal $NbH_{0.75}$ and observed that it was very brittle with the K_{Ic} yielding a fracture surface energy equal to the Nb surface from the {100} or {110} cleavage surfaces. Russian workers and British scientists have studied Zr and Ti hydrides. While these have some ductility at T>300 K, they are generally rather brittle. As a general summary of common experiences, it may be said that most metal hydrides are very brittle.

Comment by J. R. Rice:

Can you discuss possible mechanisms by which H could promote planarity of slip?

Reply:

A plausible mechanism would be an influence on screw dislocation dissociation. Computer simulations indicate a three-fold dissociation in the absence of hydrogen, a configuration that promotes cross-slip. Possibly hydrogen interacts with screw dislocation cores (through the nonlinear dilatational field of the core) and changes the dissociation to a planar configuration. Atomic calculations would help greatly in this context.

Comment by J. R. Rice:

You showed examples of plastic shear localizations which seemed to have no reference to (average) crystal directions at a point along the shear band and, in polycrystals, to pass right through microstructural features like grain boundaries without change of orientation. Yet there are other examples in the literature on single crystal plasticity, involving shear localization, tensile necking patterns, crack tip response, etc., where the observed deformation patterns seem to be strongly controlled by the conventional crystallographic constraints on slip. What factors control whether the former or latter occurs? Are there ideas on how H might favor one over the other?

Reply:

The shear bands and microbands form at large strains where fine dislocation cells have formed and the angular misorientation across the cell walls has increased with strain. This structure is energetically not very stable and susceptible to shear localization. Factors which mitigate the tendency for cell formation, such as low stacking fault energy, would tend to stabilize the material against shear band formation. Hydrogen could help trigger shear band formation (in a defect sense) by promoting slip planarity (pileup stress concentration) or by enhancing void formation by particle decohesion.

Comment by S. M. Ohr:

Can you tell us why in some instances hydrogen will reduce the flow stress, say of iron, and in other instances it can increase the flow stress?

Reply:

A change of core dissociation of screw dislocations to a planar configuration under the influence of hydrogen would produce a softening effect, consistent with observations for very pure iron. For less pure iron, Birnbaum has shown that hydrogen is trapped on carbon. Since the hydrogen can move around the carbon to minimize the interaction energy with dislocations, it would then produce hardening.

Comment by D. Macdonald:

With respect to the nucleation and growth of hydrides by tensile stresses in front of the crack tip, what is the typical volume change for the reaction:
$$\text{Metal} + x\text{H (dissolved in metal)} \rightarrow MH_x \text{ (hydride)}$$

for hydride-forming metals? Is it large enough to be a viable contributor to embrittlement of the matrix in front of the crack?

Reply:

Howard Birnbaum tells me that the volume changes range from 12-24 percent. Indeed, as you note, the formed hydride shields the prior crack tip but enhances the tensile stress in the matrix in front of it. Rarely does this cause crack nucleation there, but it could contribute to some continued propagation into the matrix once the new microcrack passes through the hydride.

Comment by J. F. Knott:

You have said that the possible reason for the effect of H in promoting planar slip in iron is that it may in some way interact with the core of a (dissociated) screw dislocation. Since H interacts via s electrons which have spherical symmetry, how can it cause the (partial components of the) core to become planar?

Reply:

Computer calculations show that screw dislocation cores have a non-linear elastic dilatation equivalent to about an atomic volume per plane cut by the dislocation. Hydrogen has a large partial molar volume in iron and could interact in a nonlinear-elastic-size-effect way with screws. Also, there are some indirect indications of hydrogen redistribution under torsional loading that suggest the possibility of a tetragonal distortion field for a hydrogen solute atom, an effect that would lead to direct interactions with screws.

Comment by T. Watanabe:

Would you please tell us about the effect of texture on hydride formation (in polycrystals which have texture) and on H-induced fracture?

Reply:

Birnbaum: Very little work has been carried out on the effects of texture
on hydride embrittlement. The initiation of hydrides does depend on the
direction and sign of the stress relative to the crystallographic habit
planes of the hydride. The decrease of the hydride free energy depends on
the transformation strain tensor and the habit plane of the hydride. This
has been seen experimentally in the Zr-H and V-H systems.

Comment by R. Thomson:

Can you comment on what mechanism could lead to the quasi-cleavage or
shear induced flat fracture?

Reply:

This is an unresolved issue. There is debris in a slip band in the
form of dislocation dipoles, tangles and condensed vacancy arrays. These
could enhance void (microcrack) nucleation. Also, some secondary slip is
present and could lead to local cell formation and void nucleation at the
cell walls.

Comment by J. F. Knott:

I am intrigued by your comments on ductile fracture in notched bars.
Clearly, the critical strain for notch-surface initiation of voids in
compression can be less than that in tension, because rumpling is more
pronounced; but can they propagate to any marked extent below the notch?
In other words, to what extent are the fractures controlled simply by
shear stress/strain, without the need for tensile stress normal to the
shear plane/surface?

Reply:

The lower critical strains for compression that I referred to related to
the strain for rumpling. We did observe some voids and microcracks
following characteristic slip traces, but these did not propagate very far
because of the overall compressive loading; as, I think, you anticipated
in your question.

Comment by R. McMeeking:

In the case of hydride formers, both those with monolithic zones and
dispersed zones, are the zones larger than any plastic region that would
otherwise be present? If that is so, the problem can be treated as one of
elastic interaction. Are there cases where the plastic region is larger
than the hydride zone so that plasticity-hydride interactions are more
important?

Reply:

In most cases the hydride zone is larger. However, because of the
large volume increase on hydride formation, there are local plastic
(dislocation) regions around the hydride. This factor has been included
to some extent in the analysis of Puls referred to in my paper.

SURFACE HYDROGEN AND FRACTURE STRESS OF 4340 STEEL

D. R. BAER AND R. H. JONES
Battelle, Pacific Northwest Laboratories
Richland, WA 99352

INTRODUCTION

This paper summarizes a study examining interrelationships among surface hydrogen, sulfur surface contamination and gaseous hydrogen embrittlement of 4340 steel. Several measurements of the crack growth behavior of high strength steels[1-3] have suggested that crack growth is limited by the adsorption of hydrogen. Relatively little work has been done, however, that directly examines hydrogen adsorption on these materials.

This study combined the use of ion sputtering to control surface contamination, Auger Electron Spectroscopy (AES) to measure surface composition, Electron Stimulated Desorption (ESD) to monitor surface hydrogen, and an ultra-high vacuum tensile stage[4] to measure fracture strength for specimens in controlled environments with known surface conditions. A short paper focusing on the fracture aspects of the work has already been published[5].

EXPERIMENTAL

The 4340 steel used was obtained from R. P. Wei and had been used in a study of crack growth in hydrogen and water vapor[2]. The material had been heat treated to a yield strength of 1344 MPa[2]. Because the current work involved two related but different experiments, two types of specimens were prepared. Rectangular specimens approximately 1 cm^2 were prepared for mounting on a heating and cooling stage for studies of hydrogen adsorption and desorption. Flat dumb-bell shaped fracture specimens were prepared for mounting on the in situ tensile stage.

Specimens on both the hot/cold stage and the tensile stage were sputter cleaned, analyzed and exposed to hydrogen in a Perkin-Elmer Physical-Electronics 545 scanning Auger system. The chamber was also equipped with a Perkin-Elmer Physical-Electronics SIMS I ion detector used for ESD work as described by Joshi[6]. Auger data was generated using 3 or 5 kV electron beams with 0.1 μa current or less. The ESD measurements were made in the high sensitivity detection mode using a 1000 eV rastered electron beam. To minimize beam induced desorption, ESD measurements were taken intermittently.

Hydrogen and hydrogen sulfide were dosed onto the specimens through a calibrated leak and a capillary dosing tube creating effective pressures at the specimen surface in the range of 10^{-4} to 2×10^{-3} Pa. The hydrogen was Linde-Union Carbide research grade 99.9995% pure.

Specific coverages of sulfur were obtained on the specimens by dosing with H_2S. The coverages are determined relative to the saturated surface coverages. At saturation, the amplitude of the sulfur 150 eV peak amplitude divided by the amplitude of the iron 703 eV peak was 1.6 ±.05. For identical measurement conditions the saturation of H_2S on pure polycrystalline iron was 1.1 ±.07.

RESULTS AND DISCUSSION

I. Hydrogen Adsorption and Coverage as a Function of Temperature

The 'saturated' or steady state hydrogen signals observed during H_2 dosing of sputter cleaned 4340 steel were measured between -80°C and 400°C and compared to normalized surface coverages that would be expected on pure iron(7). The following two observations are of interest; first, for the gas pressures used in this work, desorption processes become important around 300°C and the hydrogen signal decreases with increasing temperature above 300°C, and second, although the rates at which equilibrium coverage is approached are very different, (at least 100x slower for 4340 than pure iron) the normalized hydrogen coverage on 4340 as a function of temperature is very similar to the equilibrium coverage calculated from adsorption parameters determined on iron(7).

II. Hydrogen Adsorption as a Function of Surface Sulfur

A series of H_2 adsorption measurements were made for surfaces precovered with various amounts of sulfur. For these measurements, specimens were heated to 800°C, sputter cleaned, covered with the desired amount of S, and then rapidly cooled to 30°C during hydrogen dosing. The H^+ ESD signal was then periodically measured until a relatively constant signal was observed. The specimen was then cooled to -30°C and a new saturation coverage measured. When all conditions were the same (dose rate, current, geometry, etc.) the -30°C measurements produced nearly identical H^+ ESD signals while those at 30°C varied considerably as shown in Figure 1. The saturation or steady state H^+ ESD signals for several measurements at +30°C and -30°C are shown in Figure 2.

Although the interpretation of ESD data can be complicated, it appears that low coverages of surface sulfur on 4340 steel at 30°C can enhance hydrogen adsorption while saturated sulfur coverage inhibits hydrogen adsorption. It should be noted that the high sulfur coverage inhibition of hydrogen adsorption generally agrees with single crystal work, while an enhanced adsorption at low sulfur coverages is not present in any single crystal study(8,9). However, low sulfur coverages can enhance H_2-D_2 equilibration on Pt(111) surfaces(10).

If it is assumed that this simple interpretation of the ESD data is correct and that the measurements on the thermally cycled specimens apply to tensile specimens in the hardened condition, the above results suggest

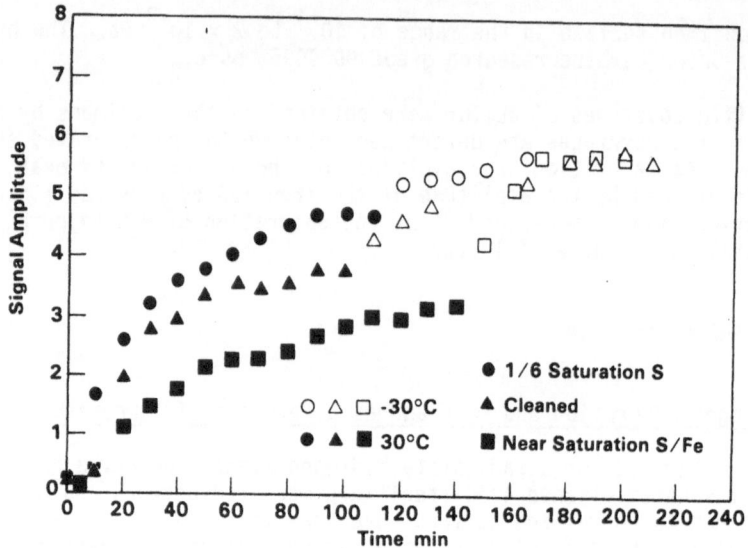

FIGURE 1. ESD Signals as a function of time from 4340 steel during H_2 adsorption with various amounts of sulfur preadsorbed on the steel surface.

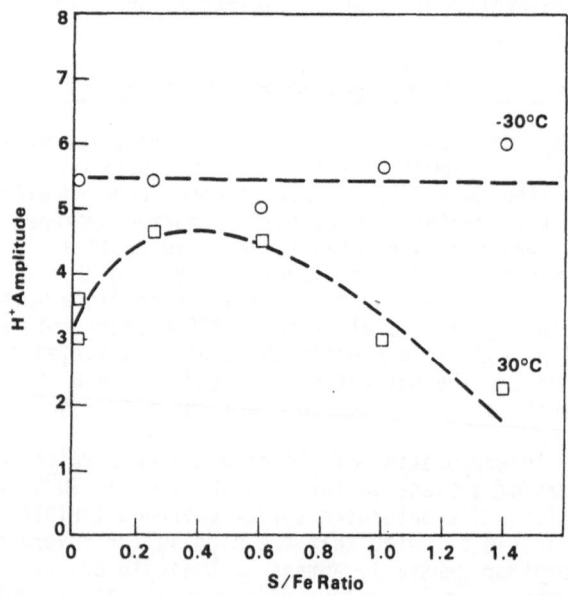

FIGURE 2. Steady state ESD H^+ signals at 30°C and -30°C as a function of sulfur coverage on the surface.

that small amounts of sulfur on the surface might enhance surface hydrogen and, thereby, hydrogen embrittlement. Larger amounts of sulfur might decrease the sensitivity to hydrogen.

III. Influence of Hydrogen and Sulfur on Fracture Stress

In situ tensile tests were performed using a continuously rising load of approximately 0.06 MPa/s. The duration of the tests varied from approximately 8 to 16 hours. The fracture stresses measured with hydrogen dosing varied with surface sulfur (Fig. 3) while the measurements without hydrogen present produced the values expected for the non-embrittled material. The first observation is that, for even the low gas pressures used in this work, hydrogen embrittlement could be measured. In addition, the variation of strength with sulfur coverage corresponds to the hydrogen surface coverage measurements in that the greatest decrease in strength occurred for a low value of sulfur coverage and the material was least affected by the hydrogen when the surface was saturated with sulfur. If the sulfur-hydrogen relationship of Figure 2 is assumed to apply for the fracture specimens, a linear relationship between hydrogen coverage and fracture strength is observed (Fig. 4). A linear dependence of the fracture stress on hydrogen concentration was predicted by Oriani and Josephic(11) and Gerberich, Livne and Chen(12). The surface coverage results are consistent with a linear bulk concentration dependence if it is assumed that the internal hydrogen is in equilibrium with the surface hydrogen.

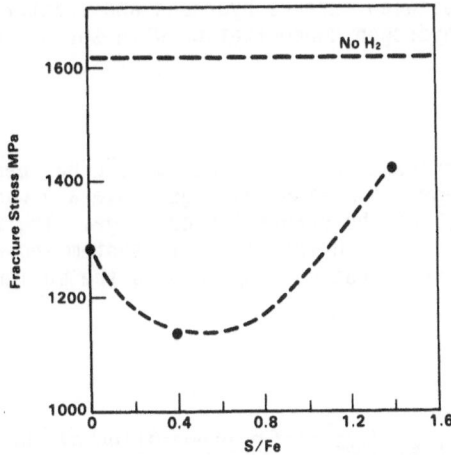

FIGURE 3. Fracture stress in H_2 as a function of sulfur on surface.

556

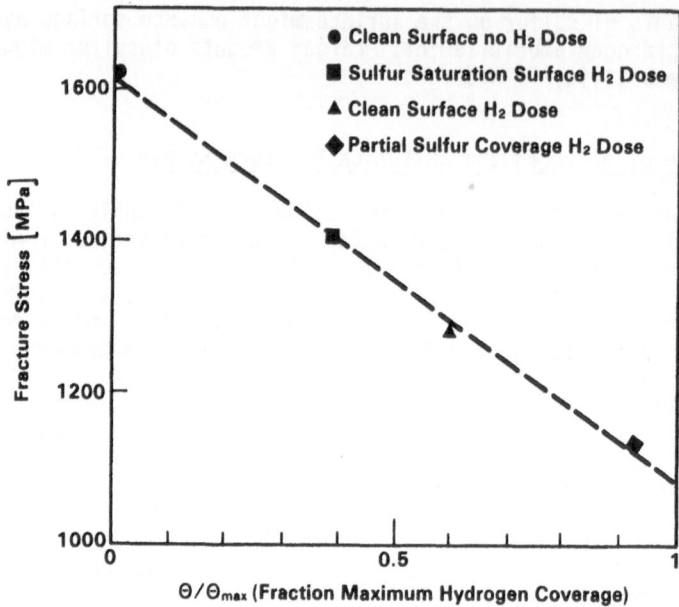

FIGURE 4. Fracture stress of 4340 steel as a function of hydrogen coverage. (Reproduced, with permission, from Ref. 5. Copyright 1986, Pergamon Press, Ltd.)

SEM micrographs show no evidence of surface initiated fracture due to the hydrogen, only microvoid coalescence. It is therefore concluded that the relationship between surface hydrogen and fracture stress resulted from a change in bulk hydrogen concentration with surface coverage.

CONCLUSION

Surface hydrogen coverage on 4340 steel at 30°C has been observed to vary with sulfur coverage. A maximum hydrogen coverage was observed at a sulfur coverage of 0.3 of the saturation coverage. The fracture stress of 4340 steel measured in the surface analysis system showed a maximum decrease in fracture stress that corresponded to the maximum hydrogen coverage.

ACKNOWLEDGMENTS

The assistance of J. L. Humason for preparation of the tensile test samples and R. P. Wei for supplying the material is gratefully acknowledged, as is the assistance of M. H. Engelhard and K. L. Steffens in data collection and M. T. Thomas for many helpful discussions. This work was supported by the National Science Foundation under grant DMR-8309465 with Battelle-Northwest Laboratory.

REFERENCES

1. Williams, D. P. and H. G. Nelson. Metall. Trans. 1, 63 (1970).
2. Simmons, G. W., P. S. Pao, and R. P.Wei. Metall. Trans. A. 9A, 1147 (1978).
3. Gangloff, R. P. and R. P. Wei. Metall. Trans. A. 8A, 365 (1977).
4. Baer, D. R., M. T. Thomas, and R. H. Jones. Metall. Trans. A. 15A, 853 (1984).
5. Jones, R. H. and D. R. Baer. Scripta Metallurgica 20 (1986) p. 929.
6. Joshi, A. and L. E. Davis. J. Vac. Sci. Technol. 14 (1977) p. 1310.
7. Shanabarger, M. R. Surface Science 150 (1985) p. 451.
8. Fisher, T. E. In Advanced Techniques for Characterizing Hydrogen in Metals. Ed. N. F. Fiore and B. J. Berkowitz. Warrendale, PA: The Metallurgical Society, 1982, p. 135.
9. Marcus, P. and J. Oudar. In Hydrogen Degradation of Ferrous Alloys. Ed. R. A. Oriani, J. P. Hirth and M. Smialowski. Park Ridge, NJ: Noyes Publications, 1985, p.36.
10. Pradier, C. M., Y. Bertheir and J. Oudar. Surface Science 130 (1983) 229-243.
11. Oriani, R. A. and P. H. Josephic. Acta Metall. 25 979 (1977).
12. Gerberich, W. W., T. Livne and X. Chen. In Modeling Environmental Effects on Crack Initiation and Propagation. Ed. R. H. Jones and W. W. Gerberich. Warrendale, PA: The Metallurgical Society, 1986.

DISCUSSION

Comment by P. Marcus:

1. As you pointed out, your observation that the H-coverage can be increased by adsorbed sulfur differs from previous studies in Fe, Ni, Pt, ... You attributed this apparent disagreement to the fact that you have studied a "metallurgical surface". What is your definition of such a surface?

2. Do you conclude from your correlation of the fracture stress with H-coverage on the surface that the H concentration in the bulk increases in the presence of S on the surface? Is that consistent with the fact that the external H_2 pressure was kept constant?

Reply:

1. The surface of the multicomponent and multiphase 4340 steel will have all alloy components at the surface. There is little or no possibility of having the types of ordered surfaces used in the single crystal work. For the single crystal work almost any surface impurity decreases hydrogen adsorption. Therefore, it seems that our cleanest surface is likely to adsorb hydrogen more slowly than a clean single crystal (as we observe). We are therefore looking at how sulfur influences an already slow adsorption rate. This may be rather different from what S could do on a crystal surface.

2. We are still analyzing how the surface concentration influences the fracture. One possibility is that the bulk hydrogen concentration that would be in equilibrium with the surface concentration varies directly with the surface hydrogen concentration. Some type of kinetic control due to a lower/higher driving force with lower/higher surface concentration may also be possible.

Comment by J. Lumsden:

Could the observed temperature effects be due to condensing of H_2O on surface at low temperature?

Reply:

We do not believe so. Our base vacuum condition during the heating stage part of the experiment was in the low 10^{-10} low range and we did not observe a buildup of oxygen during the H_2 dosing and in particular an increase upon cooling.

Comment by R. M. Latanision:

Just a comment on your observation that hydrogen coverage is associated with the presence of S on the surface of iron. It would be interesting to place on the surface a catalyst which is extreme in its behaviour from S, i.e., a dissociation catalyst such as Pt as opposed to the dissociation poison P.

Reply:

We hope to extend similar studies to other contaminants, but such work is not yet underway.

Comment by D. J. Duquette:

Some time ago, Thompson and Bernstein reported an effect of hydrogen on the size of dimples in steels. We have seen a similar effect in Ni. Did you attempt to correlate dimple size with either S or H coverage?

Reply:

Although we quickly looked at dimple size, we did not see any variation.

Comment by W. W. Gerberich:

Up to several years ago, the folklore in the materials community was that the residence time of hydrogen at the surface was greater allowing more to be absorbed, promoting embrittlement. I believe what you showed us that it was easier to adsorb hydrogen on free Fe than on S adsorbed Fe. Does this tell you anything about the absorption of H into Fe, and how does this impact on hydrogen embrittlement?

Reply:

Indeed, according to our results it is easier to adsorb hydrogen on a clean Fe surface than on Fe with adsorbed S. Adsorption at room temperature and below is only possible when atomic H is offered (generated by dissociation at a hot Pt-wire). The adsorbed S makes the dissociation of H_2 impossible, but obviously also the back reaction, the recombination of H-atoms will be hindered. So if atomic or ionic hydrogen is offered, e.g. in acid electrolytes, the hydrogen atoms can be adsorbed and absorbed by a S-covered iron surface, but they cannot recombine to evolve H_2. So high H-absorption is possible in the presence of adsorbed S and strong embrittlement, if atomic or ionic hydrogen is around.

Hydrogen Effects in Nickel Based Superalloy Single Crystals

J. Chene* and I.M. Bernstein**
*Laboratoire de Métallurgie Structurale
University of Paris-Sud Orsay, France 91405
**Department of Metallurgical Engineering and Materials Science
Carnegie Mellon University, Pittsburgh, PA 15213

INTRODUCTION

Hydrogen effects, particularly a degradation of mechanical properties, well known in ferrous alloys, are usually considered a consequence of associated large partial molar volume changes, a high lattice diffusivity and a direct effect on either the cohesive energy or dislocation behavior (1). The fact that similar changes are also observable in a wide number of crystal structures suggests not only constraints on operative mechanisms (2), but that it is of considerable interest to examine behavior in other materials.

We report here some aspects of hydrogen's interaction in oriented single crystals of a nickel-base superalloy, as they incorporate some quite general features which could help elucidate broader hydrogen effects, for example:

(a) The alloy contains a significant volume fractions of both weak (γ/γ' interfaces) and strong (grown-in porosity) hydrogen traps (3).

(b) The alloy exhibits reasonable solubility but low hydrogen diffusivities, quite opposite to the behavior in bcc metals.

(c) The alloy can exhibit both strain localization and a fracture mode change in the presence of hydrogen.

(d) The role of hydrogen on both ordered and disordered phases can be assessed.

EXPERIMENTAL PROCEDURE

2.1 Materials and Heat Treatment

The alloy currently being studied is designated CMSX-2 and contains as prominent solute additions, 8%Cr, 8%W, 6%Ta, 5.6%Al, 4.6%Co, 1.07%Ti, and 0.6%Mo. It was supplied as [001] oriented rods, solutionized for 3 hours at 1315°C and then usually heat treated as follows: 1050°C 16 h/air cool + 850°C/48/h/air cool.

2.2 Microstructure and Macrostructure

The macrostructure consist of a significant grown-in porosity, 0.1 to 0.3 volume percent in extent, with an average pore size 10-20 microns in diameter. A strong dendritic structure persisted with an accompanying solute variability, with the core enriched in W and Cr and depleted in Ta and Al (4).

The microstructure is dominated by cuboidal, ordered $L1_2 \gamma'$, with a volume fraction of 75-80% and an average size of 0.5 μm. The average γ' misfit at room temperature is ~ 0.4% although local differences are possible, particularly with the dendritic structure that is present (4).

Examples of the two main structural features, the pores and the γ' are shown in Figures (1) and (2). One goal of this study was to assess the relative importance of these suspected strong and weak traps on hydrogen-induced behavior.

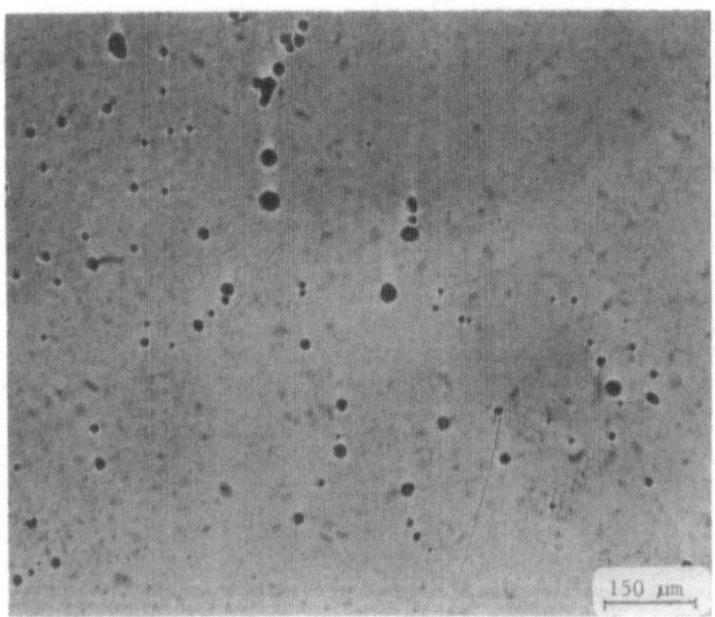

FIGURE 1. Typical As-Grown Porosity in CMSX-2

FIGURE 2. Typical Heat Treated Microstructure in this High γ' Volume Fraction Alloy (SEM).

562

2.3 Hydrogen Charging and Concentration

The primary method used to introduce hydrogen was by cathodic
potentiostatic charging in a mixture of two molten salts, 57% sodium
bisulfate and 43% potassium bisulfate at 150°C. Five hours of charging
introduced a 50 μm surface layer. Desorption at 150°C for 5 hours removed
about half the introduced hyrogen, suggesting the presence of strong
internal hydrogen traps (4).

RESULTS

3.1 Hydrogen Trapping

We have demonstrated and discussed elsewhere (4-6) that the grown-in
porosity is not only a strong trap, but it is one that can act as a center
for hydrogen-induced cracking under both tensile and fatigue loading.
Evidence for the role of pores as crack centers is given in Figures 3 and
4. Tensile tests on cathodically charged samples led to a profusion of
surface cracking, Figure 3. If the specimens were desorbed prior to
tensile testing the cracking was now much reduced and confined to the
vicinity of surface or subsurface voids. Similar results could be deduced
from fractographic analysis of charged and charged + desorbed specimens
pulled to failure (4).

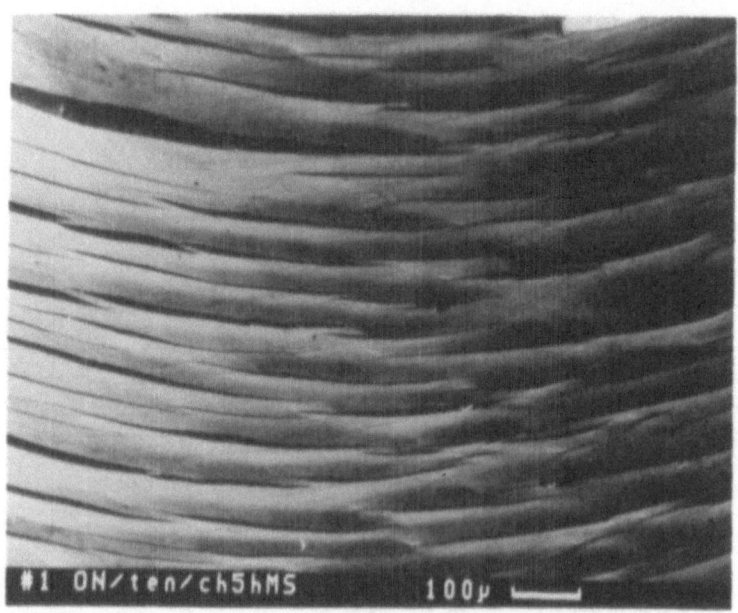

FIGURE 3. Hydrogen-Induced Surface Cracks in a Failed Tensile Specimen.

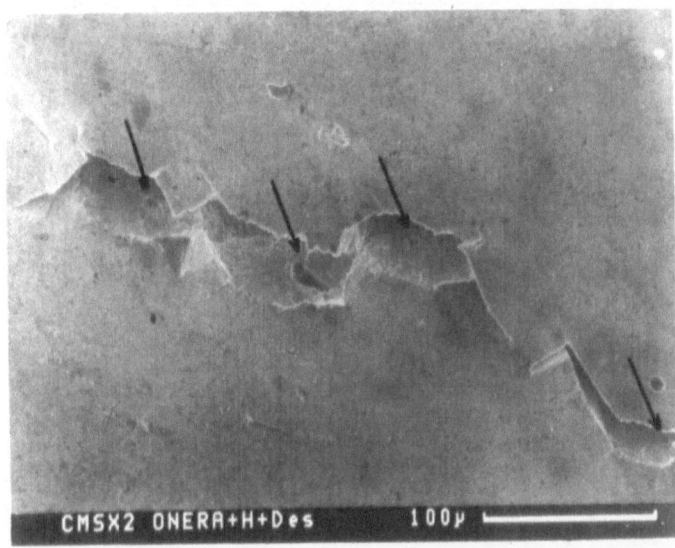

FIGURE 4. Void-Associated Surface Cracks in a Hydrogen Charged and then Desorbed Tensile Specimen.

These observations, along with hydrogen concentration measurements were used to estimate the effective hydrogen fugacity in the voids as well as the binding energy of hydrogen to the surface or in the volume of the pores (4).

For the former, the concentration of internal hydrogen in equilibrium with the surface hydrogen fugacity, obtained from desorption experiments, was about 150 ppm by weight (~1 atom per cent). This has been estimated to be equivalent to an effective surface hydrogen pressure of 14,000 atmospheres, a value suggestive of significant trapping (4). If the amount still trapped after desorption at 150°C (~300 ppm) is then presumed to be primarily associated with voids as diatomic hydrogen, this suggest the existence of pressures as high as 10,000 atmospheres, in reasonable agreement with the previously estimated value for the effective surface pressure.

Desorption at temperatures above 300°C was effective in removing all the hydrogen introduced, providing upper and lower limits for an estimate of the trapping enthalpy, H_b, in the voids. Using Fermi-Dirac statistics and assuming a low hydrogen equilibrium solubility, H_b was estimated to be in the range 0.7 to 0.8 eV (4).

We are currently studying the trapping characteristics of the γ/γ' interface which should be a weak trap considering the small misfit strains. However, there are indications that the fracture path is closely associated with γ/γ' interfaces in the presence of hydrogen (5), suggesting that a preferential distribution of hydrogen to such interfaces is possible. Work is continuing using tritium microradiography as a probe to examine the extent of interfacial trapping (7).

3.2 Hydrogen Effects on Deformation

The most striking effects of the presence of hydrogen on flow behavior were found in fully reversed room temperature fatigue loading at a constant plastic strain amplitude of 0.02% (8). Not only was the number of cycles to failure reduced after molten salt charging, but there was a

564

significant change in cyclic behavior (8). Load drops intermittently
occurred during both the tension and compression portion of the cycle
and were found to correlate with localized shear band formation. These
bands were present in both hydrogen-free and hydrogen-containing samples,
with a similar average amount of local strain accumulation. However,
with hydrogen the number of load drops were less frequent but considerably
larger, resulting in more highly localized and more intense shear bands.
This is indicative of hydrogen-induced strain localization observed in a
number of other materials (9). This difference in localization behavior
is illustrated in Figure 5; the larger step sizes in the charged sample
are consistent with the observed reduced ductility, although as will be
discussed failure does not occur along the octahedral slip planes.

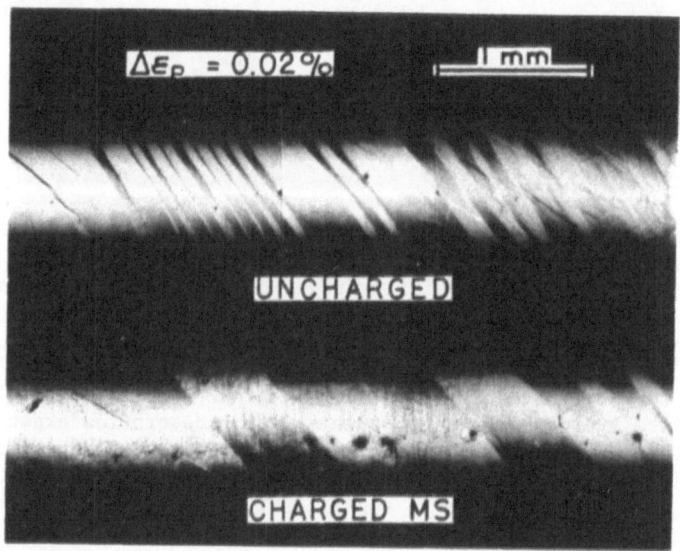

FIGURE 5. Localized Shear Bands in Hydrogen Charged and Uncharged Fatigue
Samples. (R = -1) T = 20°C.

3.3 Hydrogen Effects on Fracture

As previously mentioned, hydrogen introduced by molten salt charging
degrades the fatigue properties by reducing the number of cycles to
failure, often by as much as an order of magnitude. It also reduces the
tensile elongation. In both cases the magnitude of the reduction scales
with the relative volume of hydrogen-containing and hydrogen-free volume.
The low diffusivity of hydrogen at 150°C made it difficult to achieve a
uniform distribution except in very thin samples, so that specimens had
a surface rich hydrogen layer whose relative depth was found to scale
directly with the extent of embrittlement.

A most striking difference was a change in fracture characteristics
in the presence of hydrogen. Not only were the voids now the dominant
initiator, but cracking was directly associated with local shear bands, as
illustrated in Figure 6, showing microcrack formation in intense shear
bands in a failed tensile sample.

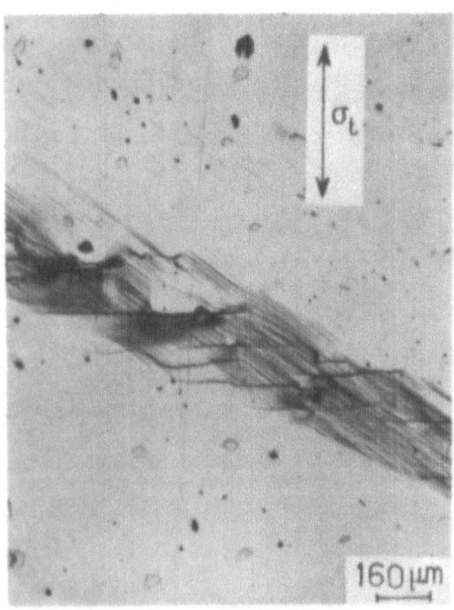

FIGURE 6. Cracks in Shear Bands in a Tensile Sample.

An even more dramatic difference, suggested by Figure 6, is that in contrast to uncharged fatigue and tensile specimens, fracture no longer occurred parallel to octahedral slip planes. Instead, the primary fracture plane was more normal to the [001] or [100] specimen axis, as illustrated for both tensile and fatigue in Figure 7. Two surface trace analysis of the accompanying surface cracks in the tensile sample showed conclusively that the fracture plane was parallel to {100}. In as much as the γ' faces are parallel to {100}, it is not yet certain if fracture is occurring on cube planes in the γ or by γ/γ' interfacial separation and thus the role of these interfaces as sites of hydrogen enrichment is still an important unanswered question.

CONCLUSIONS

Dissolved hydrogen is found to strongly affect the flow and fracture behavior of a single crystal nickel-base superalloy. Trapping enhanced strain localization and fracture mode changes in both tension and fatigue occur, demonstrating once again the pervasive effects possible due to hydrogen in non-ferrous alloys.

ACKNOWLEDGEMENTS

This work is being sponsored by NASA Lewis and we wish to especially thank Dr. R. Dreshfield who is acting as technical monitor. We also wish to acknowledge Dr. W. Kromp and Pr. F. Prinz for many useful discussions. One of the authors (J.C.) express his appreciation to the Centre National de la Recherche Scientifique and to the French Ministry of Defence (Contract DRET n° ERE 84/822) for financial support during his stay at Carnegie Mellon.

566

FIGURE 7. The Fracture Surface and Accompanying Surface Cracks in Hydrogen Charged Tensile and Fatigue Tests. The Crystallographic Nature of the Fracture is Evident in Both.

REFERENCES

1. Hirth, J.P.: Metall. Trans. A, 11(1980), 861.
2. Bernstein, I.M. and Thompson, A. W.: in Mechanisms of Environment Sen
   sitive Cracking of Materials, London, The Metals Society, 1977, p. 412.
3. Kedzierawski, P.: in Hydrogen Degradation of Ferrous Alloys, New
   Jersey, Noyes Publications, 1985, p. 271.
4. Baker, C.L., Chene, J., Bernstein, I.M., and Williams, J.C.: submitted
   to Metall. Trans.
5. Baker, C.L., Chene, J., Kromp, W. and Pinczolitis: in Structural
   Integrity and Durability of Reusable Space Propulsion Systems, NASA
   Conf. Publ. 2381, 1985, p. 171.
6. Chene, J., Baker, C.L., Bernstein, I.M., and Williams, J.C.: in
   Proceedings of Conference on High Temperature Alloys for Gas Turbines
   and Other Applications, Leige, Belgium, 1986 in press.
7. Lacombe, P., Aucouturier, M., and Chene, J.: in Hydrogen Embrittlment
   and Stress Corrosion Cracking, Metals Park, Ohio, ASM, 1984, p. 79.
8. Bernstein, I. M., Walston, S., Dollar, M., Domnanovich, A. and
   Kromp, W.: in Advanced Earth-to-Orbit Propulsion Technologies, NASA,
   1986, in press.
9. Hirth, J.P.: This conference proceedings.

DISCUSSION

Comment by D. J. Duquette:

Is there a correlation between your results where you have a brittle
surface layer and a ductile matrix with those of Sieradzki and Newman who
believe that a crack initiated in the brittle layers will propagate as a
brittle crack in the ductile matrix?

Reply:

The large number of cracks observed on the external brittle surface of
broken specimens and the fact that these cracks stop in the hydrogen
depleted zone by blunting or change in direction and proceed by shearing
along octahedral planes indicate that there is no direct correlation with
Sieradzki and Newman suggestions. It should be pointed out that the
condition may not be similar and the crack velocities are likely smaller
than the values they use in their calculations.

Comment by C. J. Altstetter:

On the precharged specimens, were the surface cracks due to the biaxial
tensile stresses which arise upon outgassing?

Reply:

Definitely not: no cracking is observed on the surface either after
cathodic charging or after additional desorption heat treatment at 150°C.
These cracks result from subsequent imposition of stress and/or plastic
strain.

HYDROGEN-INDUCED PHASE TRANSFORMATIONS IN THIN SPECIMENS OF AN AUSTENITIC STAINLESS STEEL

SEPPO TÄHTINEN, PERTTI NENONEN & HANNU HÄNNINEN
Technical Research Centre of Finland
Metals Laboratory, SF-02150 Espoo 15, Finland

1. INTRODUCTION

Cathodic hydrogen charging is a commonly applied experimental technique when hydrogen effects in austenitic stainless steels are studied. The interpretation of the experimental results is, however, complicated due to effects of changing stress state and hydrogen concentration in thin surface layers. It is generally accepted that dissolved hydrogen expands the austenite lattice, stabilizes hexagonal structure relative to austenite and results in martensitic transformations (1 - 3). In addition to martensitic phases also metastable hydrogen-rich phases have been reported to form in electrochemically hydrogen charged stainless steels. Crystallographically the hydrogen-induced martensitic structures seem to resemble those of known stress- or strain-induced martensites (4 - 7) although only a few detailed mechanistic or morphological studies are available (8).

The purpose of this study is to examine the hydrogen-induced changes in crystal structure of an austenitic stainless steel single crystal by transmission electron microscopy (TEM) in order to clarify the mechanisms and morphology of the hydrogen-induced phase transformations. A possible environment-sensitive cracking mechanism of austenitic stainless steels is proposed based on these observations.

2. HYDROGEN-INDUCED STRUCTURAL CHANGES

Electrochemical hydrogen charging (see experimental details Ref. 8) results in high density of near surface crystal defects which typically show an interstitial type of contrast behaviour. The size of these defects is below 10 nm and they have a clear triangular shape, see Fig. 1a. The formation of the defects is independent of surface orientation and therefore it is suggested that these crystal defects originate from local stress concentrations preceded by initial clustering or ordering of hydrogen atoms in austenite lattice.

The effect of charging time on phase transformations is dependent on the surface orientation of the specimen. In the (310) foils ε-martensite formed as thin plates associated with high dislocation density as shown in Fig. 1b. The transformation boundary was connected with a dislocation barrier indicating misfit between areas of ε-martensite and austenite. The ε-martensite plates seemed to grow from the surface into the specimen and most of them did not extend through the foil. Dislocation density and the amount of ε-martensite increased with increasing charging time without any sign of α'-martensite. Formation of dislocation pairs and stacking faults instead of ε-martensite in thin foil sections was a characteristic feature in the (211) foils, see Fig. 1c. In addition to prismatic dislocation pairs also glide dislocation pairs were found on the (111) and (111) planes clearly associated with α'-martensite in thicker foil sections as can be seen in Fig. 1d.

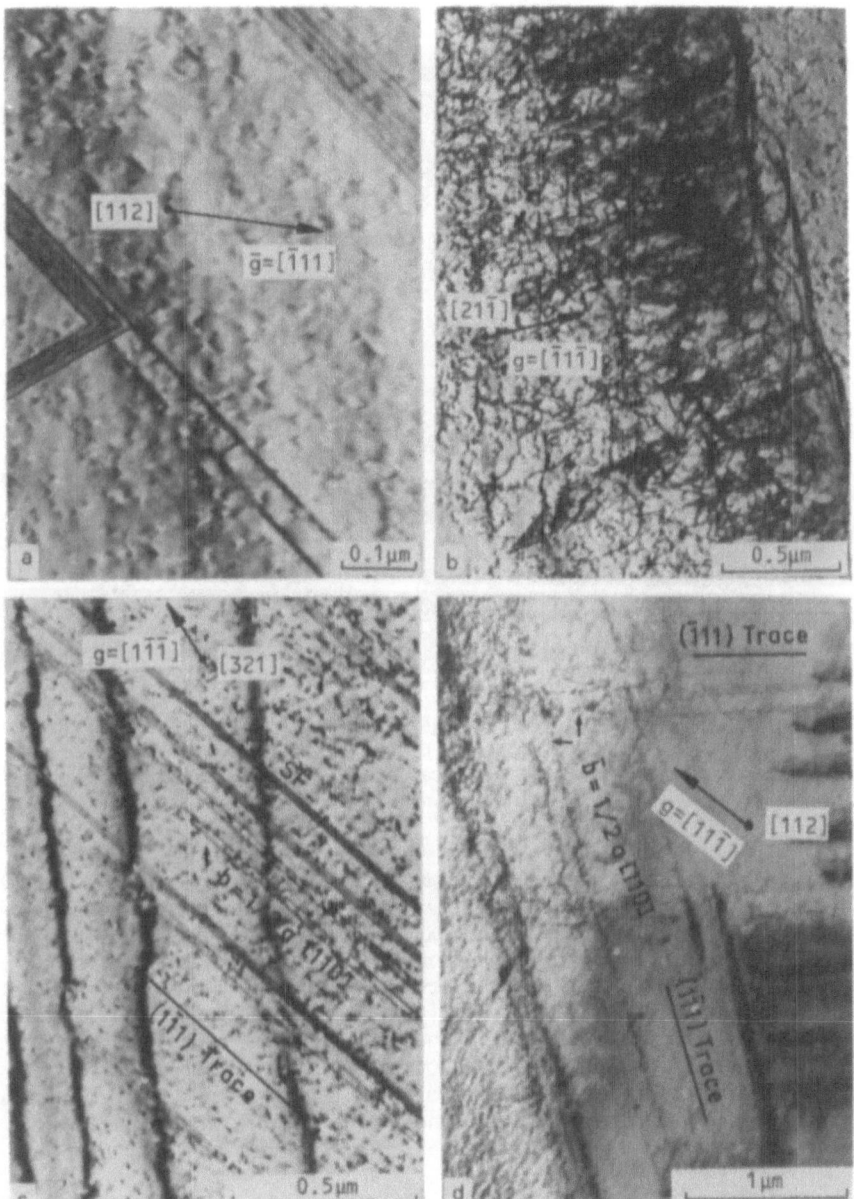

FIGURE 1. a) Hydrogen-induced near surface crystal defects in the (211) oriented foil after 60 sec hydrogen charging. b) Hydrogen-induced ε-martensite in the (310) oriented foil after 60 sec hydrogen charging. d) the (211) oriented foil after 90 sec hydrogen charging. c) A TEM micrograph showing hydrogen-induced prismatic dislocation pairs, stacking faults (SF) and near surface defects in thin area of the (211) oriented foil after 90 sec hydrogen charging. d) Hydrogen-induced α'-martensite and glide dislocation pairs in thick area of the (211) oriented foil after 90 sec hydrogen charging.

FIGURE 2. a) A dark field TEM micrograph showing hydrogen-induced α'-martensite structure in thick section of (211) oriented foil after 90 sec hydrogen charging. b) Hydrogen-induced α'- and ε-martensite in the same specimen.

The general appearance of the hydrogen-induced α'-martensite in the (211) foils can be seen in Fig. 2. The α'-martensite seemed to form in areas where the foil thickness was about 300 nm. The α'-martensite plates formed along two orientations deviating about 6 - 8 degrees from (111) and (111) planes which were almost perpendicular to the foil surface. The habit plane of the individual α'-plates was near {334} austenite planes. The α'-martensite plates were found to locate in the middle of the specimen volume where they formed a network structure without crossing each other, Fig. 2a. The α'-martensite formed mainly in areas of high dislocation density where also ε-martensite was formed. However, the independent contrast behaviour of ε-phase shows that it is mainly located in the surface layers, Fig. 2b.

3. DISCUSSION

Thin foil TEM specimens having a wedge shape offer a specimen geometry where the stress state changes with the hydrogen concentration gradient along the specimen thickness. Typically maximum thickness of stainless steel foils which can be observed by TEM is in the range of 0.3 to 0.4 μm. As the diffusivity of hydrogen is constant over the foil surface in the thinnest areas near the foil edge a constant hydrogen concentration through the specimen thickness is achieved and no concentration gradient assisted phase transformations are expected. As foil thickness increases also the hydrogen concentration gradient becomes steeper due to the high surface supersaturation and relatively slow diffusivity of hydrogen. In addition to hydrogen concentration gradient in the surface layer also a particular stress state is generated. Hydrogen expanded surface layer experiences compressive stress state balanced by a tensile stress state in the unexpanded interior of the specimen. This stress state is expected to bow existing dislocation lines to opposite directions in the surface and the interior of the foil, respectively, resulting in straight, parallel dislocation lines on opposite surfaces of the foil.

Depending on the foil orientation dislocation pairs, stacking faults or ε-phase were observed. When hydrogen concentration gradient and the stress state are large extensive ε-phase formation in the surface layers and α'-martensite formation in the interior of the foil is occuring. Extremely high tensile stress may be generated in the middle volume of the foil at a certain thickness where hydrogen concentration gradient reaches the maximum value on both surfaces. Thus, it seems to be the tensile stress state which plays a dominant role in the hydrogen-induced α'-martensite formation. Compressive stress state near the surface together with hydrogen favours the hexagonal phase formation.

These results can be transferred to crack tip conditions when the crack tip solution chemistry is condusive to hydrogen absorption on metal surfaces at the carck tip. A concept of crack tip mechanical and hydrogen-induced stresses is presented schematically in Fig. 3. Based on the fracture mechanics evaluations the local crack tip stress state reaches its maximum value ahead of the crack tip, Fig. 3a. Hydrogen solubility at the crack tip stress field is enhanced. Dissolved hydrogen expands the lattice and favours ε-phase formation. This expansion causes a compressive stress state at the crack tip and results in local relaxation of the stresses just at the crack tip. However, an additional accommodating tensile stress is produced ahead of ε-phase layer. When this tensile stress is superimposed into the original stress, a highly stressed region is obtained, Fig. 3b. Analogous to the TEM-observations this highly stressed region is expected to transform to α'-martensite. This phase transformation can produce still additional stresses and crack initiation ahead of the main crack tip resulting in linking to the main crack tip. Possible crack paths can be eg. ε- and α'-martensite interphases as well as cleavage in ε- and α'-martensite. These different crack paths should produce different crystallographies of the fracture surfaces either $\{111\}$, $\{334\}$, $\{225\}$ or $\{110\}$, respectively.

The crack advance distance of the individual discontinuous crack growth event for stress corrosion cracking of eg. stable AISI 310 stainless steel in boiling (155 $^\circ$C) $MgCl_2$ solution has been determined to be 0.5 μm every 15th second (9). Based on hydrogen diffusivity data at test temperatures the proposed mechanism can account for observed crack advance distances quite well. Since the maximum crack advance distance is about two times the crack opening displacement it will include both ε- and α'-phase layers and the resulting fine structure of the fracture surface will be in the very fine scale. This has made the unambiguous analysis of the fracture surface crystallography very difficult. Because the studied steel was relatively stable (M_s-temperature -226 $^\circ$C and stacking fault energy 70 mJ/m^2), the proposed crack growth model is expected to apply for a wide range of austenitic stainless steels.

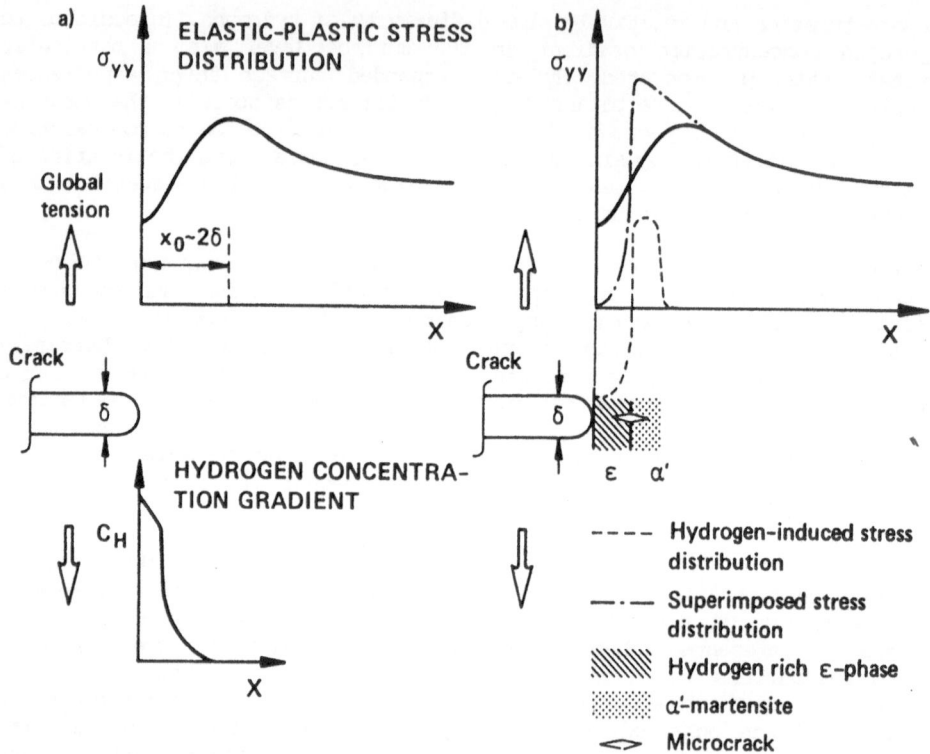

FIGURE 3. Schematic of hydrogen-induced fracture criteria at crack tip in microscopic terms when mechanical and hydrogen-induced stresses are super-imposed for austenitic stainless steels stress corrosion cracking.

4. ACKNOWLEDGEMENTS
  The financial support of Academy of Finland and Ministry of Trade and Industry of Finland is gratefully acknowledged.

5. REFERENCES

1. H. Mathias, Y. Katz and S. Nadiv, Metal-Hydrogen Systems, Ed. T. N. Veziroglu, p. 225, Pergamon Press, Oxford (1980).
2. S. Tähtinen, J. Kivilahti and H. Hänninen, Hydrogene et Materiaux, Ed. P. Azou, p. 185, Paris (1982).
3. N. Narita, C. J. Altstetter and H. K. Birnbaum, Met. Trans. 13A, 1355 (1982).
4. P. Pitcher, PhD Thesis, University of Wales (1979).
5. H. Hänninen, T. Hakkarainen and P. Nenonen, Hydrogen Effects in Metals, Eds. I. M. Bernstein and A. W. Thompson, p. 575, TMS-AIME, Warrendale, PA (1981).
6. E. Minkovitz and D. Eliezer, Scr. Metall. 16, 981 (1982).
7. M. Tanino, H. Komatsu and Funaki, J. Physique 43, C4-503 (1982).
8. S. Tähtinen, P. Nenonen and H. Hänninen, Scr. Metall. 20, 153 (1986).
9. D. V. Beggs, M. T. Hahn and E. N. Pugh, Hydrogen Embrittlement and Stress Corrosion Cracking, Eds. R. Gibala and R. Hehemann, p. 181, ASM, Metals Park, Ohio (1984).

DISCUSSION

Comment by R. C. Newman:

You state that the brittle fracture of an fcc phase is easier to understand if there is a phase transformation. In view of the brittle SCC failures in austenitic stainless steels at temperatures between 50 and 350°C, do you think these higher-temperature failures could involve phases such as ε?

Reply:

Basically yes, although the temperature dependence of hydrogen induced-ε-phase formation is not know, it seems to form quite readily under the influence of stress and hydrogen together. Eq. the SCC failure surfaces of austenitic stainless steels tested in boiling $MgCl_2$-solutions (155°C) show ζ-martensite.

Comment by C. J. Altstetter:

In the TEM specimens how can you tell which features formed during charging and which formed (disappeared) during subsequent gassing?

Reply:

Small near surface crystal defects can be traces of metastable phase reversions associated with outgassing. ε-jaζ-phase formation arises from relaxation of accommodation stresses between H-expanded surface layer and interior of the foil, and they are formed already during hydrogen charging.

A SYSTEM OF CRACKS IN BONDED HALF PLANES

S. ALTINTAS and M. B. CIVELEK
Bogazici University
Department of Mechanical Engineering
P.K. 2 Bebek
Istanbul, Turkey

1. INTRODUCTION

The structural strength of materials generally depends on the materials properties, the shape and size of defects as well as the orientation of such flaws in the medium. In the application of fracture mechanics other kinds of imperfections due to manufacturing are particularly important such as flat cavities which develop during casting, small cracks resulting from residual stresses in welded materials, and fatigue cracks under fluctuating external loads. Thus in a design dealing with such materials, it is necessary to have a good estimate of the stress state disturbed by the existence of these flaws.

In the last few decades a considerable progress has been made in the application of linear elastic fracture mechanics by employing plane elasticity theory. Many investigators solved infinite plane problems involving cracks with arbitrary orientations and locations. For example Isida (1), Datsyshin and Savruk (2) attacked the problem of arbitrary distributed cracks in an infinite plane. Their methods require that the cracks are sufficiently away from each other. A system of radial cracks in an infinite plane was analyzed by Aksoğan (3). A general formulation for a system of curvilinear cracks in bonded half planes was given by Ioakimidis and Theocaris (4) who used complex variable techniques. However, they obtained numerical results only for a straight or arc-shaped single crack. The problem of two bonded half planes one of which contains radial cracks have been recently studied by Aksoğan (5) and the single crack case was treated by Erdoğan and Aksoğan (6).

In this paper we consider the plane elastostatic problem of two bonded semi-infinite planes having different elastic constants and one of the semi-infinite planes contains arbitrarily distributed cracks. The method of analysis is based on the representation of a crack by a continuous distribution of dislocations. For this purpose, the stress solutions for a pair of edge dislocations are used as Green's functions which results in a set of singular integral equations. These integral equations are solved numerically to obtain stress intensity factors at the crack tip for various material combinations and crack distributions.

2. GENERAL FORMULATION OF THE PROBLEM

Consider two elastic half planes which are bonded along their interface. For the half planes, the elastic constants are denoted by μ_1, κ_1 and μ_2, κ_2, where μ is the shear modulus κ is $3-4\nu$ for plane strain and $3-\nu)/(1+\nu)$ for plane stress, ν being Poisson's ratio.

A line crack will be simulated by integrating the stress distributions of two edge dislocations with Burgers vectors $\mathbf{b}_x$ (parallel to x-axis) and $\mathbf{b}_y$ (parallel to y-axis) using the Airy stress functions given in (7). To simulate a line crack in s direction we use stress distributions of two dislocations with Burgers vector $\mathbf{b}_s$ and $\mathbf{b}_r$ as Green's functions.

The coordinates of a point on s-axis are $x = d+\alpha s$, $y = \beta s$, where

(α = cosθ, β = sinθ) and hence the normal and shear stress components σ_{rr} and σ_{sr} at this point in the {s,r} coordinate system can be expressed as

$$\sigma_{rr}(0,s) = \beta^2\sigma_{xx}(d + \alpha s, \beta s) + \alpha^2\sigma_{yy}(d + \alpha s, \beta s) - 2\alpha\beta\sigma_{xy}(d + \alpha s, \beta s)$$

$$\sigma_{sr}(0,s) = (\alpha^2 - \beta^2)\sigma_{xy}(d + \alpha s, \beta s) + \alpha\beta[\sigma_{yy}(d + \alpha s, \beta s) - \sigma_{xx}(d+\alpha s,\beta s)]$$

$$(1a,b)$$

where σ_{xx}, σ_{xy} and σ_{yy} are the stresses for $\vec{b}_s$ and $\vec{b}_r$. The expression for σ_{xx} is;

$$\sigma_{xx}(x,y) = (1/\pi)[(\beta D_r - \alpha D_s)G_{xx}(x,y - \beta t, d+\alpha t) - (\beta D_s + \alpha D_r)F_{xx}(\alpha, y - \beta t, d+\alpha t)]$$

where $G_{xx} = -\pi\sigma_{xx}^1(x,y,a_0)/D_x$, $\qquad$ $F_{xx} = -\pi\sigma_{xx}^2(x,y,a_0)/D_y$

$\qquad$ $D_r = 2\mu_1 b_r/(1 + \kappa_1)$, $\qquad\qquad$ $D_s = 2\mu_1 b_s/(1 + \kappa_1)$

with $\sigma_{xx}^1(x,y,a_0)$, $\sigma_{xx}^2(x,y,a_0)$ being stress distributions of dislocations with Burger's vectors b_x and b_y, respectively, located at $x = a_0$, $y = 0$. Expressions for σ_{xy} and σ_{yy} can be developed readily.

A line crack along the s-axis (Fig. 1) can be represented by continuous arrays of dislocations with density function

$$\dot{f}(s) = \frac{2\mu_1}{1 + \kappa_1} \frac{\partial}{\partial s} [v(0^+,s) - v(0^-,s)]$$

$$(2a,b)$$

$$g(s) = \frac{2\mu_1}{1 + \kappa_1} \frac{\partial}{\partial s} [u(0^+,s) - u(0^-,s)]$$

where u and v are the displacement components in the {s,r} coordinate system. As the crack extends from s = -a to s = a we obviously have

$$f(s) = g(s) = 0 \qquad \text{for} \qquad |s| > a \qquad\qquad (3)$$

To obtain the stresses due to distribution of edge dislocations we replace the terms D_s and D_r in equations (1a,b) by -f(t)dt and -g(t)dt, respectively and integrate from -a to a. The unknown density functions f(t) and g(t) will be determined from the conditions that stresses are prescribed on the crack surface which yield

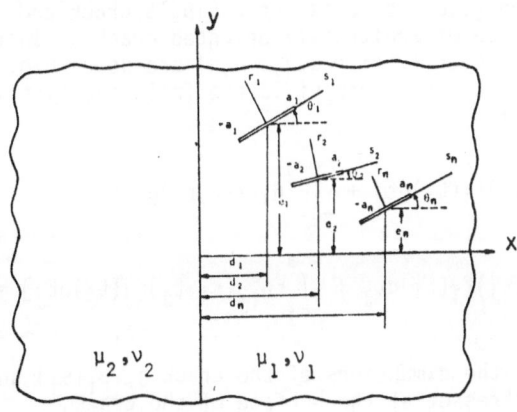

FIGURE 1. Arbitrary distribution of cracks in one of the half planes.

$$\sigma_{rr}(0,s) = \frac{1}{\pi} \int_{-a}^{a} k(s,t)f(t)dt + \frac{1}{\pi} \int_{-a}^{a} h(s,t)g(t)dt = p(s) \text{ for } -a<s<a$$

$$\sigma_{sr}(0,s) = \frac{1}{\pi} \int_{-a}^{a} m(s,t)f(t)dt + \frac{1}{\pi} \int_{-a}^{a} n(s,t)g(t)dt = q(s) \text{ for } -a<s<a \qquad (4a,b)$$

where $p(s)$ and $q(s)$ are normal and shear stresses at the crack surface and where the kernels are

$$k(s,t) = \beta^2(\alpha F_{xx} - \beta G_{xx}) + \alpha^2(\alpha F_{yy} - \beta G_{yy}) - 2\alpha\beta(\beta F_{xy} + \alpha G_{xy})$$

$$h(s,t) = \beta^2(\beta F_{xx} + \alpha G_{xx}) + \alpha^2(\beta F_{yy} + \alpha G_{yy}) - 2\alpha\beta(\beta F_{xy} + \alpha G_{xy})$$

$$m(s,t) = (\alpha^2 - \beta^2)(\alpha F_{xy} - \beta G_{xy}) + \alpha\beta[\alpha(F_{yy} - F_{xx}) - \beta(G_{yy} - G_{xx})]$$

$$n(s,t) = (\alpha^2 - \beta^2)(\beta F_{xy} + \alpha G_{xy}) + \alpha\beta[\beta(F_{yy} - F_{xx}) + \alpha(G_{yy} - G_{xx})]$$

where $F_{xx} = F_{xx}(d+\alpha s, \beta s-\beta t, d+\alpha t)$

$$G_{xx} = G_{xx}(d+\alpha s, \beta s-\beta t, d+\alpha t) \text{ etc.} \qquad (5a,b)$$

To provide the continuity of the material outside the crack we must impose two conditions,

$$\int_{-a}^{a} f(t)dt = 0, \quad \text{and} \quad \int_{-a}^{a} g(t)dt = 0 \qquad (6a,b)$$

It can easily be shown that the kernels $k(s,t)$, $n(s,t)$ have Cauchy singularities, namely

$$k(s,t) = \frac{1}{t-s} + k^*(s,t)$$

$$n(s,t) = \frac{1}{t-s} + n^*(s,t) \qquad (7a,b)$$

where $k^*(s,t)$ and $n^*(s,t)$ are bounded functions. Therefore, the equations (4a,b) form two singular equations of the first kind. These singular integral equations take simple forms when the crack is perpendicular or parallel to the interface, and the expressions obtained for these cases are in agreement with the results of Cook and Erdoğan(8) and Ashbaugh (9) and Erdoğan(10).

The formulation given above is for a single crack and can easily be extended for the case of arbitrarily oriented cracks. In this case the coordinate systems $\{s_i, r_i\}$, $(i = 1,2,...,n)$ are used (Fig. 1) and expressions of crack surface tractions yield the following system of singular integral equations:

$$\frac{1}{\pi} \sum_{j=1}^{n} [\int_{-a_j}^{a_j} k_{ij}(s_i,t_j)f_j(t_j)dt_j + \int_{-a_j}^{a_j} h_{ij}(s_i,t_j)g_j(t_j)dt_j] = p_i(s_i) \qquad (8a,b)$$

$$-a_i < s_i < a_i, (i = 1,..,n)$$

$$\frac{1}{\pi} \sum_{j=1}^{n} [\int_{-a_j}^{a_j} m_{ij}(s_i,t_j)f_j(t_j)dt_j + \int_{-a_j}^{a_j} n_{ij}(s_i,t_j)g_j(t_j)dt_j] = q_i(s_i)$$

$$-a_i < s_i < a_i, (i = 1,..,n)$$

where $(-a_j, a_j)$ are the dimensions of the crack j, $p_i(s_i)$ and $q_i(s_i)$ are the normal and shear stresses at the surface of the crack.

The systems of singular integral equations (4a,b) and (8a,b) require numerical solutions and the results are given next.

3. NUMERICAL RESULTS AND DISCUSSION

The systems of integral equations given above contain Cauchy singularities and their solution is possible through a numerical approach given by Erdoğan and Gupta (11). The equations are of general nature and have also been applied to the solution of the system of cracks in half-planes (12) and good agreement is found with the results of other investigators (13).

In this study, we first consider a single crack which is located perpendicular to the interface of two different materials. The materials differ in their elastic constants (μ_1, μ_2, ν_1 and ν_2). The dependence of the stress intensity factors at the two ends of the crack, placed in the material with μ_1, ν_1 as a function of $\Gamma = \mu_2/\mu_1$ is shown in Fig. 2a. The stress is applied parallel to the interface and we have taken $\nu_1 = \nu_2 = 0.3$. The stress intensity factors decrease monotonically as Γ increases. For $\Gamma = 1$, i.e., $\mu_1 = \mu_2$ we obtain the solution for the infinite plane. For materials with $\Gamma < 1$, i.e., $\mu_2 < \mu_1$, the stress intensity factor at the tip of the crack close to the interface is higher than the one away from the interface. This means that in material combinations, in which the crack is placed in the stronger one, stress intensity factor increases. For the case $\Gamma > 1$ the reverse trend is observed. The stress intensity factor at the tip of the crack, now placed in the weaker material decreases. The stress intensity factor at the tip close to the interface is smaller than the factor away from the interface. The effect of parallel cracks on the stress intensity factors as a function of Γ is shown in Fig. 2b. The general trend of Fig. 2a is followed with one important difference where the addition of a second crack decreases the stress intensity factors compared to the case of a single crack by about 15%.

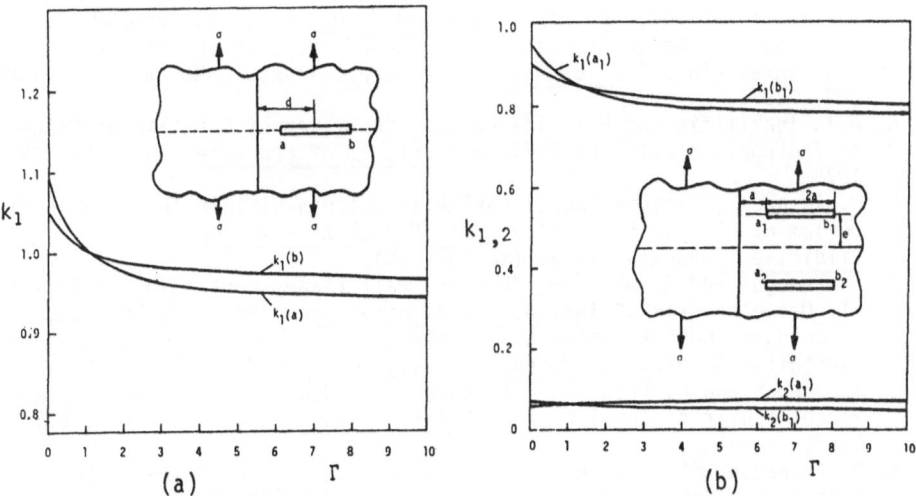

FIGURE 2. (a) Stress intensity factor as a function of Γ for a single crack (d/a = 2), (b) for two cracks (e/a = 1).

The effect of cracks distributed radially on the stress intensity factor as a function of Γ is shown in Fig. 3a. The loading condition is the same as in Fig. 2a. The general trend of decrease in k_1 values is again observed. For this case, $\theta = 30°$, k_1 values at the tip close to interface is smaller and decrease steeper than the values away from the interface. The effect of loading on the stress intensity factor is shown in Fig. 3b, where k_2 values are more significant than k_1 values.

578

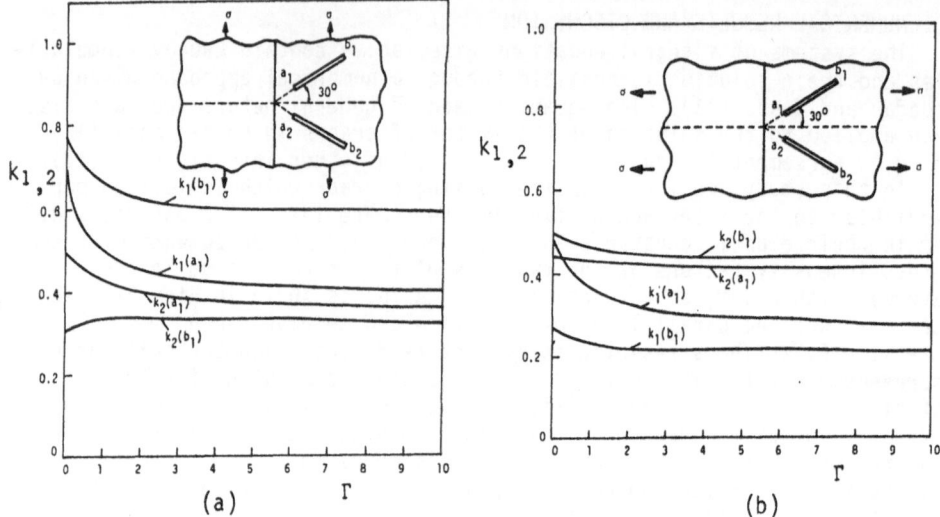

FIGURE 3. Stress intensity factors as a function of Γ for two symmetrical radial cracks. (a) Loading parallel, (b) perpendicular to the interface.

REFERENCES

1. M. Isida, "Analysis of Stress Intensity Factors for Plates Containing Random Array of Cracks," Bulletin of the JSME, Vol. 13, 635-642, 1970.
2. A.P. Datsyshin and M.P. Savruk, "A System of Arbitrarily Oriented Cracks in Elastic Solids," J. Appl. Math and Mech., Vol. 37, 326-332, 1973.
3. O. Aksoğan, "The Interaction of Colinear Arrays of Griffith Cracks on Two Radial Lines," J. Engng. for Industry, Trans. ASME, Vol. 98, 1086-1091, 1976.
4. N.I. Ioakimidis and P.S. Theocaris. "A System of Curvilinear Cracks in an Isotropic Elastic Half-Plane," Int. J. of Fracture, Vol. 15, 299-309, 1979.
5. O. Aksoğan, "Birinde Radyal Catlaklar Bulunan Bitisik Iki Yarim Düzlemin Düzlemsel Elastisite Problemi," 1. Ulusal Kirilma Konferansi Bildirileri, Ankara, 1981, (in Turkish).
6. F. Erdoğan and O. Aksoğan, "Bounded Half Planes Containing an Arbitrarily Oriented Crack," Int. J. Solids Structures, Vol. 10, 569-585, 1974.
7. J. Dundurs and G.P. Sendeckyj, "Behavior of an Edge Dislocation near a Bimetallic Interface," J. Apply Phys., Vol. 36, 3353-3354, 1965.
8. T.S. Cook and F. Erdoğan, "Stresses in Bonded Materials with a Crack Perpendicular to the Interface," Int. J. Engng. Sci., Vol. 10, 677-697, 1972.
9. N. Ashbaugh, "Stress Solution for a Crack at an Arbitrary Angle to an Interface," Int. J. of Fracture, Vol. 11, 205-219, 1975.
10. F. Erdoğan, Bonded Dissimilar Materials Containing Cracks Parallel to the Interface," Engng. Fracture Mechanics, Vol. 3, 231-240, 1971.
11. F. Erdoğan and G.D. Gupta, "On the Numerical Solution of Singular Integral Equations," Quart. Appl. Math., Vol. 29, 525-534, 1972.
12. M.B. Civelek and Ş. Altintas, "Birbirine Yapisik Iki Yaridülemde Olusmus Catlaklarin Uclarindaki Gerilme Siddeti Faktorleri," II. Ulusal Mekanik Kongresi, 203-219, Trabzon, 1981 (in Turkish).
13. F. Erdoğan and G.C. Sih, "On the Crack Extension in Plates Under Plane Loading and Transverse Shear," J. Basic Eng. Trans. ASME, Vol. 85, 519-527, 1963.

Comment by S. M. Ohr:

Have you applied your method to treat the case of a crack propagating along an interface between two elastic materials that differ in elastic modulus.

Reply:

No, we have not considered cracks along such interface.

INELASTIC NEUTRON SCATTERING AND RESISTIVITY OF HYDROGEN IN COLD-WORKED PALLADIUM

R. KIRCHHEIM, X.Y. HUANG, H.-D. CARSTANJEN[*] and J.J. RUSH[+]

Max-Planck-Institut für Metallforschung, Institut für
Werkstoffwissenschaften, D-7000 Stuttgart-1, FR-Germany

[*] Institut für angewandte und theoretische Physik, Universität
Stuttgart, D-7000-Stuttgart-80, FR-Germany

[+] National Bureau of Standards, Washington, DC 20234, USA

1. INTRODUCTION

The distribution of interstitial atoms around dislocations has to be described by Fermi-Dirac statistics, in order to take site saturation into
account. Calculations for hydrogen in the case of elastic interaction /1,2/
as well as experimental results /2,3/ have shown that local concentrations
near the dislocation core must be rather high. Thus H-H interaction should
play an important role which, however, complicates the theoretical discrip-
tion remarkably /4/.

In the present study measurements of the resistivity increment of hydrogen
in cold-worked palladium are presented which can be interpreted by a simple
model where part of the hydrogen atoms precipitate as a hydride in the de-
lated region of an edge dislocation and, therefore, do not contribute remarkably to resistivity. In order to distinguish between the precipitation
of a nucleus of a new phase at a dislocation according to Cahn's model /5/
and a concentrated segregation of hydrogen atoms without any appreciable
change of the positions of metal atoms, the local modes of hydrogen vibration were measured by inelastic neutron scattering. Preliminary results of
small angle neutron scattering are also presented. They were performed, in
order to determine the size of these regions of strong hydrogen segregation.

2. EXPERIMENTAL

For the resistivity measurements annealed Pd foils (1 hour at 1000 $^{\circ}$C and
10^{-5} mbar) were cold rolled and doped electrolytically with hydrogen. The
resistivity change due to hydrogen could be recorded in situ as described
elsewhere /3/.

For inelastic neutron scattering a heavily cold rolled Pd foil (about 99%
reduction in thickness) was cut into pieces of 2x4x0.01 cm. About 100 of
these sheets were clamped into an aluminum container which was evacuated
and backfilled with hydrogen. From the pressure drop within a calibrated
volume the concentration of dissolved hydrogen was determined to be H/Pd=
0.0085. The measurements were performed on the BT-4 triple axis spectrometer at the NBS reactor, using the (220) planes of Cu as monochromator and a
well-shielded, liquid nitrogen cooled Be filter as analyser. Over the range

of measurements the energy resolution was about 5 meV. After the scattering experiment 30% of the sheets were used for small angle neutron scattering and the remaining 70% were annealed for 1 h at 1000 °C and 10^{-6} mbar. The annealed material was treated in the same way as the cold-worked Pd before neutron scattering.

The experimental background was determined by measuring the scattering from the cold-worked or annealed Pd, respectively, before loading with hydrogen and this background was subtrated from each of the spectra.

Small angle neutron scattering was performed with deuterium instead of hydrogen, because the cross section for coherent neutron scattering is larger for D. Measurements were made in cold-worked V-D at the ILL in Grenoble, France and in cold-worked Pd-D at the NBS in Washington, U.S.A. A more detailed description of the experimental set up and the results will be given in Refs. 6 and 7.

3. RESULTS AND DISCUSSION

In Fig. 1 the resistivity increment of hydrogen ($\rho_H = \partial \rho / \partial c$) in different-ly cold-worked Pd foils is shown as a function of the hydrogen concentra-tion c. At very low concentrations most of the hydrogen is trapped by dis-locations and does not contribute to resistivity. This effect has been in-terpreted in Ref. 3 as the formation of a dense Cottrell cloud of all hy-drogen atoms at the dislocations, where similar to precipitates only small resisitivity changes are to be expected. With increasing content of hydro-gen a participation of H-atoms between the dense Cottrell cloud and sites

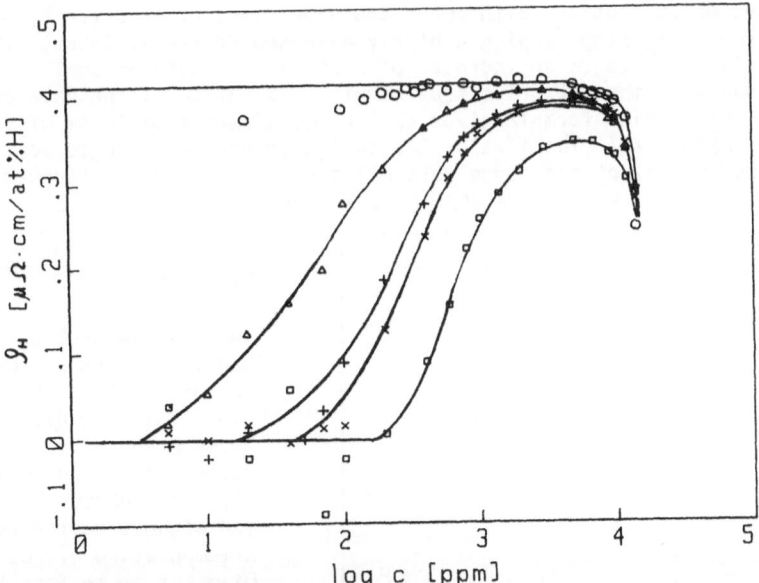

Fig. 1: Resistivity increment of hydogen (cf. Eq.2) versus hydrogen con-centration in annealed and cold-worked palladium at 293 K (o=an-nealed, Δ=15%, +=50%, x=90% and □=99% deformation).

582

far away from dislocations occurs, where only free H-atoms contribute to resistivity ρ as they do in well annealed Pd. Thus we have

$$(1) \qquad \rho = \rho_0 + c_f \rho_H^o$$

$$(2) \qquad \rho_H = \frac{\partial \rho}{\partial c} = \rho_H^o \frac{\partial c_f}{\partial c}$$

where ρ_0 is the resistivity of Pd, c_f is the concentration of free H-atoms and ρ_H^o their resistivity increment and ρ_H is the (differential) resistivity increment of all H-atoms.

At very high hydrogen concentrations additional hydrogen atoms go mostly into normal sites far away from dislocations but still some are used to increase the size of the dense Cottrell cloud which is ecspecially obvious for the sample with the highest dislocation density in Fig. 1. At concentrations above 10^4 ppm precipitation of the β-phase (may be at the Cottrell cloud) occurs.

From the difference between the total concentration and the concentration of free hydrogen as calculated from Eq. 1 the amount of trapped hydrogen can be evaluated. Using a dislocation density as determined in the same way as in Ref. 3 allows us to calculate the spatial extention of hydrogen trapping.

From these considerations a diameter of about 5 nm of the dense Cottrell cloud can be calculated for 1 at. % of hydrogen. This value of the diameter is in agreement with values obtained from solubility data /2/. Direct evidence of deuterium clusters with about the same diameter is given by small angle neutron scattering. Doping a highly deformed Pd-sample (about 99%) with deuterium results in an increase of neutron intensity at small angles in comparison with the deuterium free sample. Evaluation of the data according to the Guinier formula gives an average diameter of 10 nm of the scattering objects /7/. The idea of Carstanjen to use small angle scattering, in order to detect the dense Cottrell cloud, has also been successfully applied to the V-D system /6/. Fig. 2 shows the results in a Guinier

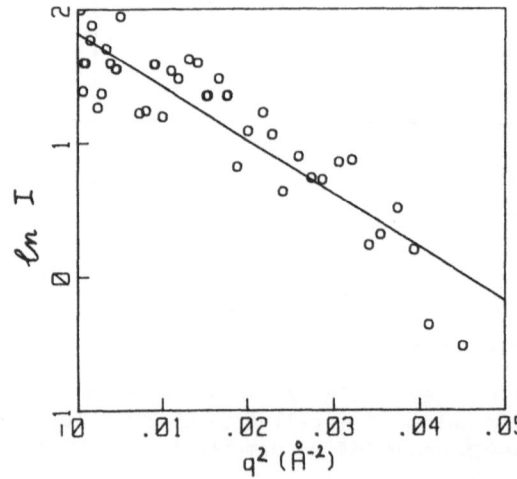

Fig. 2:

Small angle neutron intensity I versus momentum transfer q in a Guinier plot. I is the difference between vanadium 50% cold rolled with $2 \cdot 10^{-3}$ D/V and vanadium 50% cold rolled without deuterium at a temperature of 201 K. Diameter of the scattering objects is 3.0 nm.

plot. In the case of vanadium the deuterium cloud is only about 3.0 nm in diameter indicating a less pronounced trapping by dislocations in agreement with similar findings for the similar metal niobium /3/.

With the additional and direct evidence provided by small angle neutron scattering it is beyond doubt that an extended segregation of hydrogen at dislocations occurs. The driving force is the elastic interaction between hydrogen and the strain field of the dislocation which is enforced by an attractive H-H interaction. It seems to be a matter of definition whether we call these regions of high H-concentrations hydrides or dense Cottrell clouds. Nevertheless the H-H interaction may be strong enough to change the positions of the metal atoms slightly to form an ordered hydride. The latter case corresponds to two interstitial positions of H-atoms, one in the ordered hydride and one in the normal sites far away from dislocations. Whereas no ordering of the metal atoms in the neighborhood of the dislocation leaves the segregated H-atoms in a spectrum of different interstices (=disordered hydride) according to the rapidly changing strain field around dislocations. Inelastic neutron scattering was applied, in order to distinguish between the two different types of hydrides.

In well annealed α-Pd the three eigenstates of hydrogen vibration are degenerate due to the cubic symmetry of the octahedral site. Thus only one peak appears at 69 meV in the spectrum of inelastically scattered neutrons, as shown in Fig. 3. In the intermediate β-phase the hydrogen content is about 0.6 H/Pd at room temperature and the neutron peak shifts to 59 meV because of an overall lattice expansion and a corresponding softening of the H-Pd forces /8/. Therefore, in strongly deformed Pd with 0.008 H/Pd a main peak should appear at 69 meV (free hydrogen) with a smaller peak or a shoulder at 59 meV, if an ordered hydride similar to the β-phase is formed at the dislocations. However, Fig. 3 shows that the peak for the

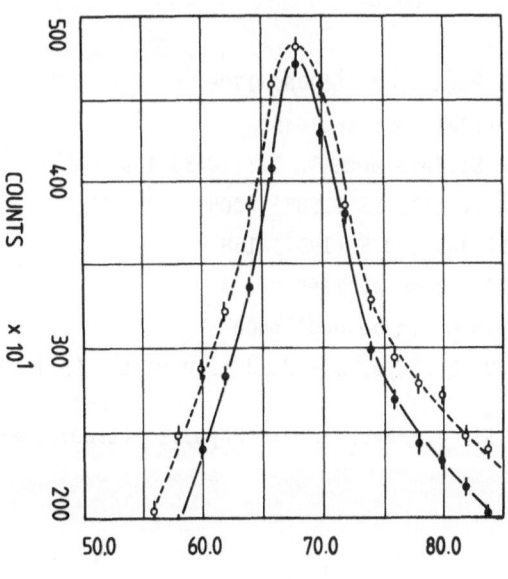

Fig. 3

Measured spectra for inelastically scattered neutrons in annealed Pd (o) and cold-worked Pd (o). H-concentration is 0.008 H/Pd in both cases, but the sample mass is 30% less for the annealed sample.

deformed Pd is broader with its maximum value at 69 meV. The intensity
at 69 meV is about the same for both the annealed and the deformed sample,
although the number of H-atoms was about 30% less in the annealed sample
due to its smaller weight. Neglecting neutron absorption in the thicker
sample of the deformed Pd, one can conclude that about 30% of the H-atoms
are trapped by the dislocations. This amount of hydrogen will come closer
to the value of 20% as obtained from resistivity measurements, if neutron
absorption by the thicker sample of the deformed Pd is taken into account.
The trapped atoms cause the broadening of the peak at 69 meV because they
occupy a spectrum of different sites. The trap sites are situated in the
rapidly changing strain field of the dislocation. Probably they are distorted octahedral sites where the degeneracy of the local modes of hydrogen
vibration no longer exists and two or three peaks per site occur. These
peaks of the distorted sites are distributed around the main peak for the
regular octahedron at 69 meV depending on the amount of distortion of the
site or the corresponding distance from the dislocation core. Thus the
behavior of hydrogen in trap sites of the dislocation is qualitatively
similar to amorphous metals /9/ although the broadening of the inelastic
peak is less pronounced for the dislocations.

CONCLUSIONS

Indirect evidence is provided by resistivity measurements in deformed
palladium that hydrogen forms a dense Cottrell cloud which grows with
increasing hydrogen concentration. By small angle neutron scattering the
existance and the diameter of these clouds could be determined directly
and by inelastic neutron scattering it could be shown that the formation
of the dense cloud was not accompanied by an ordering of the metal atoms
or a precipitation of an ordered hydride, respectively, but was rather an
extended segregation in the strain field of the dislocation.

REFERENCES

1. Hirth JP and Carnahan B, Acta metall. 26 (1978) 1795

2. Kirchheim R, Acta metall. 29 (1981) 835 and 845

3. Rodrigues JA and Kirchheim R, Scripta metall. 17 (1983) 159

4. Wolfer WG and Baskes MI, Acta metall. 33 (1985) 2005

5. Cahn JW, Acta metall. 4 (1956) 449 and 5 (1957) 168

6. Carstanjen H-D and Kirchheim R, to be published

7. Rush JJ, Glinka C and Kirchheim R, to be published

8. Rush JJ, Rowe JM and Richter D, Z. Phys. B - Condensed Matter 55 (1984)
   283

9. Rush JJ, Rowe JM and Maeland AJ, J. Phys.F:Metal Phys.,10 (1980) L283

DISCUSSION

Comment by P. Haasen:

Concerning the use of the Guinier plot in evaluating SANS of Cottrell clouds in Pd + D. Is this a valid approximation for cylindrical connected scatterers?

Reply:

It is the first and simplest approach we have used for the preliminary measurements. In the future, we will look more closely at the spatial distribution of the intensity at small angles in order to obtain information on the shape of the scatterers and to use better approximations for data evaluation.

FRACTURE INITIATION DUE TO HYDRIDES IN ZIRCALOY-2

M.P. PULS, B.W. LEITCH AND W.R. WALLACE
Atomic Energy of Canada Limited
Materials and Mechanics Branch
Whiteshell Nuclear Research Establishment
Pinawa, Manitoba, Canada ROE 1LO

1. INTRODUCTION

In hydride-forming metals, the presence of hydrides can sometimes lead to brittle fracture. Zirconium is a hydride-forming metal that forms the basis of a number of alloys used in CANDU[TM] nuclear reactors. Under certain circumstances, zirconium alloys are susceptible to a process of slow crack propagation called delayed hydride cracking (DHC). Extensive experimental investigations have shown that DHC involves the repeated preferential nucleation, growth and fracture of hydride platelets at the tip of a pre-existing crack (1,2). A concomitant theoretical description of the DHC process (3,4) has provided an explanation for many significant features of the observed cracking behaviour, such as the dependence of the crack velocity on temperature and Mode I stress intensity factor K_I. An important deficiency of the theoretical model, however, is that it cannot predict the critical stress intensity factor, K_{IH} below which DHC crack growth would stop. Experiments have shown that DHC crack velocity is approximately independent of K_I up to values approaching those of the fracture toughness of the bulk material, whilst at low K_I-values, a sharp drop-off occurs in crack velocity, suggesting a critical K_{IH} for DHC.

Fundamental to developing a theoretical understanding of K_{IH} is a criterion for the initiation of fracture at hydride platelets. Previous studies showed that the critical stress and strain to fracture hydrides could be quite different, depending on the uniaxial yield stress and on the stress state prevalent in the material tested. The conditions for fracture initiation at hydrides are, therefore, complex, requiring information on the internal stress and strains near the hydride at the time of fracture. A promising new technique to investigate this process involves the use of acoustic emission (AE) (5). The AE associated with hydride fracture can be correlated with the load-deflection curve and, when combined with finite element calculations, used to determine the stress and strain state at the onset of microcracking in the hydride particles.

The present work describes the results of such an approach to determine the effect of stress state on crack initiation in hydrides in Zircaloy-2. These results represent part of a larger study to investigate the effect of matrix strength, hydride size, hydrogen concentration and hydride/matrix constraint on hydride crack initiation (6).

2. EXPERIMENTAL

Plane-strain double notched or smooth tensile specimens were used to investigate the effect of stress state on hydride fracture. The tensile specimens were cut from flattened sections of pressure tubing with the tensile axis along the tranverse direction of the tube. The smooth specimens had a gauge length of 25.4 mm, a width of 9.53 mm and a thickness of 1.8 mm. The notched specimens had twice the thickness of the smooth specimens. The approximately semi-circular notches in the notched specimens were machined such that the width across the notches at the minimum cross-section was 1.8 mm. To increase the number of hydride platelets, hydrogen was added to concentration levels ranging from 0.18 to 0.90 at.%. A thermo-mechanical treatment was applied to create hydrides that had their plate normals predominantly oriented in the tensile axis direction.

Tensile testing, at an engineering strain rate of 1.6×10^{-4} s^{-1}, was carried out at room temperature in a screw-driven Instron machine. Acoustic emission was monitored using an Acoustic Emission Technology Corporation AET5000B. The overall gain was set at 93 dB, with the threshold for AE detection at 16 dB (for a 1 µV reference signal at the sensor).

3. EXPERIMENTAL RESULTS

Representative results of the tensile testing for smooth and notched specimens are given in Figures 1 and 2, respectively. Figure 1 shows that, for the smooth specimen, the AE rate generated in the early stages of deformation decreases to negligible values as the yield stress of the sample is reached. Beyond this point, the material work hardens very little and, therefore, the load increases by only a small amount, until the start of necking when the equivalent plastic strain has reached 4%. At this point, there is an increase in the AE event rate which, however, persists for only a short period of time. The remaining, extensive, deformation stage is characterized by a negligible accumulation of AE events up to the final rupture of the specimens. Using AE-event location techniques, it is possible to show (7) that the AE events generated up to the point where the yield stress is reached originate mostly from the grip area of the specimen, whilst the burst in AE at the onset of necking comes from the gauge section.

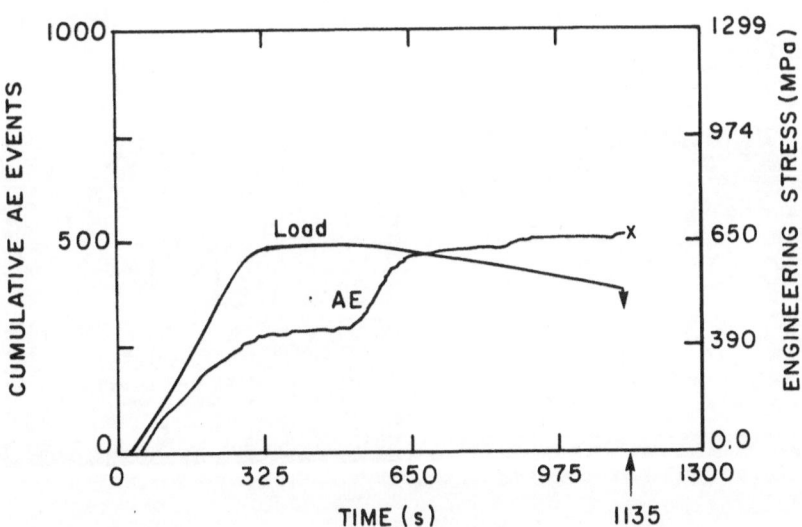

FIGURE 1. Cumulative AE events and engineering stress versus time for a smooth specimen.

In the notched specimens (Figure 2), the characteristics of AE-event accumulation in the early stages of deformation are similar to those of the smooth specimens. However, close to rupture, a dramatic and steady increase in AE occurs. This increase in AE commences around the point where the load-deflection curve becomes non-linear. As discussed later, at such a loading, a large part of the region between the notches has become plastic.

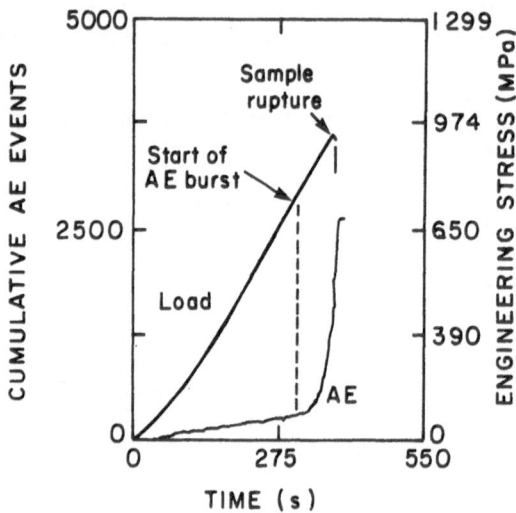

FIGURE 2. Cumulative AE events and net section engineering stress versus time for a notched specimen.

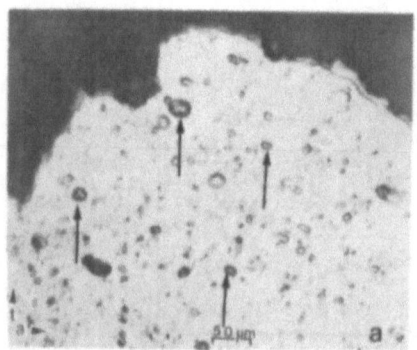
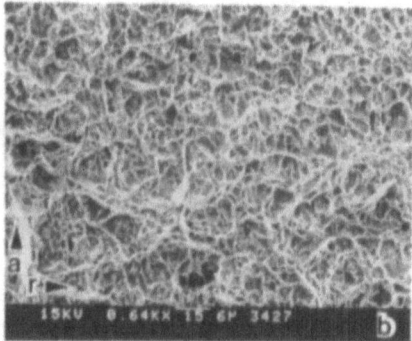

FIGURE 3. Smooth sample: a) Optical micrograph showing voids (arrows) below fracture surface (tensile axis is vertical). b) Corresponding fracture surface taken with the SEM.

The marked contrast in the AE characteristics between the smooth and notched specimens is also reflected in the different appearances of the specimens after fracture, as revealed by metallography and fractography. Figure 3a shows that numerous spherical voids near the fracture surface of a smooth specimen had nucleated and grown from fractured hydrides. The appearance of the corresponding fracture surface (Figure 3b) is consistent with this picture, showing a surface composed mainly of voids of different sizes. In contrast, the voids nucleated at hydrides in a notched specimen (Figure 4a) are elliptically shaped or crack-like with the major elliptical axes corresponding to the lengths of the hydrides in which they nucleated. The fracture surfaces (Figure 4b) consist of numerous irregularly shaped and brittle-appearing regions, the sizes of which correspond to the average dimensions of the hydrides in the alloy. These brittle regions are surrounded by small voids and void sheets. The origin of these smaller voids is not clear, but they may have nucleated at hydrides or second phase particles that are too small to be readily resolvable using standard metallographic techniques.

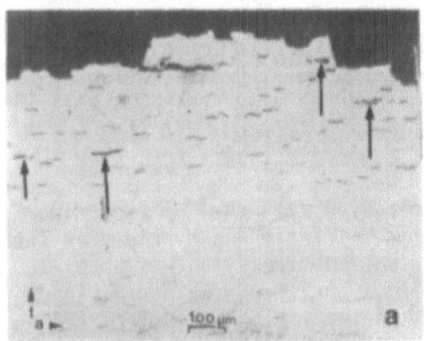 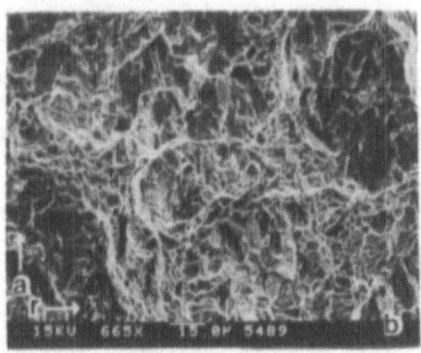

FIGURE 4. Notched sample: a) Optical micrograph showing fractured hydrides (arrows) below fracture surface (tensile axis is vertical). b) Corresponding fracture surface taken with the SEM.

4. FINITE ELEMENT CALCULATIONS

The stress and strain state at the onset of the AE burst, when the experimentally determined load-deflection curve starts to deviate from a straight line, was derived with the aid of finite element calculations. The calculations were carried out with the program MARC. To model the plastic response of the material, the program made use of the true stress-true plastic strain curve obtained from one of the smooth specimens. The yield stress obtained for this specimen was 645 MPa, which is close to the average value of 629 MPa obtained for all the specimens tested. The finite element results were used to generate a curve of the engineering net section stress versus load-point extension. The calculated engineering net section stress at the onset of non-linearity in this curve was found to be 760 MPa. This is close to the average, experimentally determined value of 800 MPa for specimens containing 0.18 at.% hydrogen. The plastic zone in the specimen, corresponding to this point of onset of non-linearity in the load-deflection curve, extends from the notch surface to cover about 75% of the distance between the notches. The equivalent plas-plastic strain ranges from zero at the boundary of the plastic zone to 0.8% within a small region at the root of the notch.

Experimentally it was found that rupture occurred 110 MPa higher (aver-age engineering net section stress) than the stress at which the AE burst begins. The finite element calculations showed that, at this loading, the equivalent plastic strain at the root notch is 1.7%. The plastic zone boundary covers the entire region between the notches and, near the centre line, has started to move towards the ends of the specimen.

5. CONCLUSIONS

The above results for notched and smooth specimens demonstrate that some plastic strain is always necessary before cracks are nucleated in hydrides. However, in a uniaxial state of stress (smooth specimens), much larger plastic strains are required to fracture hydrides than in a tri-axial state of stress. Moreover, in the uniaxial stress state, the matrix material surrounding the fractured hydrides can sustain much larger plas-tic deformation than it can in a triaxial state of stress. Thus hydride cracks formed in the smooth specimens have a chance to grow to become spherical voids before final rupture occurs, whilst those formed in the notched specimens remain crack-like.

Similar conclusions were deduced by Simpson (5) for Zr-2.5 wt% Nb and by Koss and co-workers (8,9) for titanium and Zircaloy-2. The latter work-ers, however, examined only cases where the hydride platelets have their plate normals perpendicular to the tensile axis direction. In this situation, the fracture surface always has a ductile appearance, even under multiaxial loading.

Comparison of the present results on Zircaloy-2 with the corresponding ones on Zr-2.5 wt% Nb (5,6) shows that crack initiation in hydrides in the latter alloy starts at an engineering net section stress that is about 100 MPa higher than in Zircaloy-2. However, since Zr-2.5 wt% Nb has a yield stress that is also about 100 MPa higher than the yield stress for Zircaloy-2, hydride fracture starts at an equivalent plastic strain, which is approximately the same in the two alloys. This shows the great import-ance that plastic strain plays in controlling the initiation of fracture in hydrides.

ACKNOWLEDGMENTS

The work reported here was partially funded by Canadian Utilities under the COG/CANDEV agreement.

REFERENCES

1. C.E. Coleman and J.F.R Ambler, Reviews of Coating and Corrosion 3, 105 (1979).
2. C.D. Cann and E.E. Sexton, Acta Metall. 28, 1215 (1980).
3. R. Dutton and M.P. Puls, "Effect of Hydrogen on Behaviour of Mate-rials", A.W. Thompson and I.M. Bernstein, eds., pp. 516, Metal Soci-ety, AIME, New York, 1976.
4. L.A. Simpson and M.P. Puls, Metall. Trans. 10A, 1093 (1980).
5. L.A. Simpson, Metall. Trans. 12A, 2113 (1981).
6. M.P. Puls, B.W. Leitch and W.R. Wallace, unpublished, WNRE, 1986.
7. M.P. Puls, unpublished, WNRE, 1986.
8. R.J. Bourcier and D.A. Koss, Acta Metall. 32, 2091 (1984).
9. Fan Yunchang and D.A. Koss, Metall. Trans. 16A, 675 (1985).

WORKSHOP SUMMARY: HYDROGEN EMBRITTLEMENT - SESSION A

PREPARED BY: I. M. Bernstein
DISCUSSION LEADERS: I. M. Bernstein, D. J. Duquette, J. P. Fidelle,
T. E. Fischer
RECORDERS: D. J. Alexander, D. R. Baer, J. Chene, M. Hashimoto

The first of the two workshops on hydrogen effects focused on examining four basic questions related to hydrogen. It was hoped that by concentrating our efforts and bringing to bear inputs and expertise from the diverse group of participants present at least a partial resolution to some of these fundamental issues could be achieved. The questions addressed included:

1. Can interaction (binding) energies of hydrogen to specific structural features (voids, interfaces, dislocations, etc.) be calculated by chemical and/or physical approaches? How valuable would such information be?

2. Hydrogen strongly affects the mechanical (and physical) behavior of a broad range of metals and alloys of different crystal structures, strength, toughness and hydrogen solubility and diffusivity. What differentiates hydrogen from other "embrittling" solutes which are usually more system specific?

3. Are hydrogen embrittling effects dominated, in general, by decohesion (cleavage) or plasticity controlled processes?

4. Most modeling approaches to hydrogen (or other environmental) embrittlement rely on coupling a kinetic analysis or permeability to the effect of local stress states on hydrogen distribution or redistribution. Can we do better?

Regarding the first of these, it was the consensus that such calculations were badly needed to understand how hydrogen behaves at or near the crack tip; in particular, to differentiate and contrast its influence on decohesion and slip. The relevant parameters needed are the equilibrium or kinetic changes in metal-metal bond energy, and dislocation formation energy or interfacial ledge energy due to the presence of hydrogen.

While current first principle calculations have so far only been able to rigorously treat perfect crystals with a surface as the only defect, recent progress, detailed at the conference, suggests that local point and line configurations should soon be treatable; in particular, the associated equilibrium positions and concomitant energies in the presence and absence of hydrogen. Examples of promising new approaches in this area include the embedded atom and tight bonding models, both already capable of treating the dynamics of crystals containing certain types of defects.

Such calculations must ultimately be able to predict a wide range of hydrogen-defect interactions, including edge and screw core effects, the role of differing hydrogen concentrations and dynamic and equilibrium consequences of dislocation and lattice transport, to be useful to the experimentalist. Support for such calculations could come from measurements of hydrogen-phonon interactions, of local strains around stationary and mobile hydrogen traps and of the fracture and plastic response of model systems with controlled interfaces, etc.

Discussions relative to the second question initially focused on the long-held phenomenological viewpoint that the uniqueness of hydrogen stems from its universally high lattice mobility. While this is clearly true in ferrous alloys, the observation of significant embrittlement in nickel and aluminum-based alloys, which exhibit very low lattice diffusivities near room temperature, demonstrates that this is not a necessary condition. However, the more general possibility of either a high permeability due to extensive solubility or alternate high-diffusivity processes such as grain boundary diffusion or, more likely, transport by mobile dislocations may indeed provide the necessary and sufficient conditions for the occurrence of specific embrittling phenomena.

An alternate atomistic viewpoint is that hydrogen is unique in that it can form bonds with only s orbitals. The nondirectional, delocalized nature of this bonding can allow hydrogen to reside at virtually any lattice position, including interstitial sites, grain boundaries, particles, interfaces, dislocations, etc. It was generally felt that such interactions with interfaces may prove to be most important, and therefore calculations of these interactions should be an early priority. One reason for this is that interfaces are sites for other impurities that can reduce the work of adhesion. Either these reductions can be additive to effects due to hydrogen, or more seriously, there can be synergistic interactions acting to multiply the embrittling tendencies. To differentiate between these possibilities would require carefully controlled experiments either where interfaces are absent or where they have a controlled structure and chemistry.

The third discussion area centered on the relative importance of cleavage or plasticity in hydrogen embrittlement. While there are at present no a priori approaches to predict among different materials whether crack growth will occur along potential or operative slip or cleavage planes, this distinction may not be critical. The reason for this is that on an atomic scale, calculations should be relatively insensitive to whether the bonds between some small aggregate of atoms are being broken by shear stresses or normal stresses. In this spirit, it is not surprising that hydrogen is predicted and often found to affect both decohesion and dislocation motion in the same system, although the evidence is often inferential, depending on changes in the value of terms like the measured fracture stress and yield stress and direct observations of dislocation motion in the electron microscope.

Nevertheless, there is likely an important discrimination between how hydrogen modifies the ease and extent of brittle and plastic processes. Except for the apparently special cases of intergranular or interfacial fracture in the presence of other embrittling species (e.g., temper embrittling solutes), or for special misorientations, there is overwhelming observational evidence that plasticity is a fundamental contributor to the hydrogen-induced embrittlement of virtually all susceptible metals and alloy systems. This fact must be taken into account in any attempt to calculate, model, or predict how hydrogen affects local and global mechanical properties and their associated dislocation behavior. Important problems to consider include the role of hydrogen in macroscopic and microscopic strain localization, the role of such localized plasticity in transgranular (and perhaps intergranular) fracture and the importance of hydrogen transport by mobile dislocations, not only as a kinetic enhancer but, more directly, as a way to modify dislocation character and response.

The final main discussion topic involved the best way(s) to model hydrogen embrittlement. Most previous efforts have focused on sustained load cracking of high-strength alloys, usually steels, in order to predict such parameters as the necessary local critical hydrogen concentration, the environmentally induced critical stress intensity and crack growth rate and the difference between internal and external hydrogen environments. There was general agreement that while hydrogen embrittlement kinetics, and to a lesser extent threshold stress intensities, often depend or scale with hydrogen diffusivity and/or permeability, no fundamental predictive relationship has yet been established. One major barrier is to be able to successfully rationalize why some alloy systems are more susceptible than others, yet have low diffusivities and permeabilities. A possible unifying approach is to extend current modeling efforts, which focus on lattice-driven hydrogen buildup in some part of the plastic or process zone, to consider dislocation transport of hydrogen as well as the detailed trapping behavior and fracture characteristics of environmentally sensitive microstructural heterogeneities. For example, it is now well established that strong traps can act either as failure embryos or as a means of redistributing an otherwise deleterious hydrogen concentration. The observation that specific traps may behave differently if the hydrogen is present as an external environment rather than as a dissolved species is another complication not readily explainable by current models.

Finally, no matter how successful any model is in characterizing the many parametric measures of hydrogen embrittlement, the critical and as yet unexplained question is precisely how some local hydrogen population in the lattice or at traps causes ductile or brittle crack initiation or growth. In other words, mechanistically how does hydrogen embrittle metals and alloys?

We have learned much about the detailed interactions of hydrogen with the lattice, with dislocations and with microstructure, but the answer to this most fundamental question is yet to be found. It was heartening to find that there is currently considerable optimism that current theoretical approaches, aided by new insights and larger, more robust computational facilities, will provide the answer to this and other important questions and very likely before the next Conference.

WORKSHOP SUMMARY: HYDROGEN EMBRITTLEMENT - SESSION B

PREPARED BY: H. K. Birnbaum
DISCUSSION LEADERS: F. R. N. Nabarro, M. Speidel, J. Knott, J. Rice
RECORDERS: S. Tahtinen, J. Hancock, M. Finnis, S. Thorpe

A. Importance of and evidence for dislocation transport in bcc and fcc systems.

The discussion suggested a general acceptance of dislocation transport of hydrogen in many systems over limited temperature ranges. However it was not generally accepted that this process plays an important role in the hydrogen embrittlement process as direct evidence for such a role is lacking. The principal issue seems to be the role of dislocation transport in increasing the flux of hydrogen and thereby resulting in a local supersaturation of hydrogen sufficient to drive the fracture process. Thus while individual dislocations may carry hydrogen atmospheres with them a net non-equilibrium flux of hydrogen requires a net flux of dislocations. This may occur in specialized cases, e.g., planar slip in stage I, but there was no agreement that it occurs in the general embrittlement situation. It was also emphasized that the process of dislocation transport has been studied in very few systems.

B. Role of hydrogen trapping at precipitates, grain boundaries, etc. in the fracture process.

There appears to be an increasing amount of evidence for hydrogen segregation and trapping at extended defects such as grain boundaries and precipitates. The hydrogen segregation appears to be closely coupled to the presence of other impurities at these sites. The role of trapping at coherent and incoherent precipitates appears to be important for the understanding of embrittlement of alloy systems.

C. Status of surface and bulk theories of hydrogen embrittlement.

This is difficult to assess due to a lack of information about the behavior of hydrogen either on the surface or in the interior of metals. There is some evidence for the adsorption theories of hydrogen embrittlement but these fail to answer the question of why the even stronger adsorption of gases such as O, N, S, etc. do not lead to embrittlement. One of the main issues appears to be the role of other surface impurities, such as S, on the role of H at the surface. It is not understood whether these adsorbed surface impurities affect the rates of hydrogen entry into the metal or the influence of hydrogen adsorbed on the surface on the fracture mechanism. This is an area where both critical experiments and calculations are badly needed.

Workshop Session 4:
Stress Corrosion and Corrosion Fatigue

Session Chairman: T. Magnin

FILM-INDUCED CLEAVAGE DURING STRESS-CORROSION CRACKING OF DUCTILE METALS AND ALLOYS

R.C. NEWMAN AND K. SIERADZKI [*]
Corrosion & Protection Center   [*]*Brookhaven National Laboratory*
U.M.I.S.T.   *Department of Applied Science*
Manchester M60 1QD   *Long Island, Upton, N.Y. 11973*
United Kingdom   *U.S.A.*

1. INTRODUCTION

There are several processes that can contribute to the growth of stress-corrosion cracks, including anodic dissolution, film growth and hydrogen-induced fracture. The text-book orthodoxy is a "spectrum" of behaviour ranging from obvious anodic control to obvious hydrogen embrittlement (1). Examples of extreme behaviour are the cracking of mild steel in nitrate solution (anodic control) and the cracking of high-strength, low-alloy steels in chloride solutions (a hydrogen effect).

Transgranular SCC has always occupied an uncomfortable position in the SCC "spectrum", as it invariably resembles cleavage (Figure 1) but is often placed with the anodically controlled systems because of its response to electrochemical potential (2,3). We hope to show that this situation can be resolved by treating all transgranular SCC as film-induced cleavage, and will present some recent evidence for this.

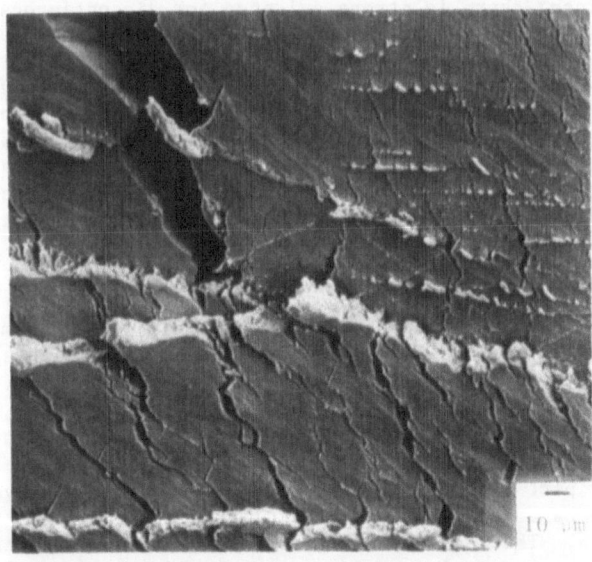

FIGURE 1. SCC fracture surface of a copper single crystal in sodium nitrite solution. Crack propagation left to right.

The occurrence of cleavage during transgranular SCC of brass, and the possible role of de-alloying, were first proposed by Edeleanu and Forty (4). Other important observations of brittle behaviour were made by Nielsen in austenitic stainless steels (5) and by Bakish in Cu-Au alloys (6). Edeleanu and Forty observed the side-faces of α-brass crystals during SCC in ammonia solution, and presented optical evidence for a stepwise growth in units of a few μm. Unfortunately the work attracted surprisingly little attention, and brass was gradually absorbed into the group of anodic dissolution/film rupture systems (7). The introduction of scanning electron microscopy did little to divert effort into possible cleavage mechanisms, probably because cleavage in an fcc alloy was too radical a departure from known mechanical behaviour. Another complicating factor was the obvious production of hydrogen during chloride SCC of austenitic steels (8), which prevented the realization that a general mechanism of transgranular SCC might be operating without assistance from hydrogen.

Increasing evidence for cleavage arose from a series of papers by E.N. Pugh and others (9-11). α-brass was again the best-behaved alloy, and by 1983 cleavage was well-established. The mechanism of cleavage initiation in fcc materials was not established, with hydrogen and de-alloying being invoked as possible critical processes. Later Pugh's group rejected hydrogen for brass or copper (12), while retaining it as a possibility for austenitic stainless steel. The main difficulty with non-hydrogen processes was the extremely high crack velocity thought to be necessary for cleavage in the fcc lattice; another complication was the observation of SCC in pure copper (13), later shown to be cleavage-like (14,15).

Our starting point for a new rationalization of transgranular SCC was to identify systems where hydrogen was clearly not involved. Copper, α-brass and copper-gold all fall into this category, and in all three (but particularly α-brass) there is strong evidence for discontinuous cracking. The next stage was to ask whether it was necessary to invoke a hydrogen effect in other, less noble alloys where hydrogen absorption at crack tips cannot be ruled out. We find that these systems (e.g. austenitic stainless steel, carbon steel in anhydrous ammonia) contain strong evidence for embrittlement processes other than hydrogen entry. Finally, a detailed study has been carried out on two systems (α-brass and Cu-Al in aqueous ammonia) to measure the relationship between the surface process responsible for the cleavage (de-alloying) and the average rate of crack propagation.

2. THEORY OF FILM-INDUCED CLEAVAGE

There are various ways in which thin films, a few tens of nm in thickness, can nucleate short-range cleavage. This is considered in detail elsewhere (16,17) and by Sieradzki in these proceedings (18). Briefly, we have shown that a ductile-to-brittle transition can be induced in an fcc metal by the presence of either brittle or ductile thin films around crack tips. Ductile films must be strongly bonded to the metal (preferably part of it, as in the case of a de-alloyed layer) and of different lattice parameter leading to local stresses and interfacial dislocation arrays. The requirements for brittle films, other than the strong bonding, are not so clear and require consideration of the mechanics of extremely fine-scale porous structures; present evidence suggests that de-alloyed layers may be in the brittle category even though they are composed of almost pure noble metal (Cu, Au, Ni for Cu-Zn, Au-Cu and Fe-Cr-Ni respectively) (19). Particularly important is the tensile stress induced in many de-alloyed layers by the lattice contraction during de-alloying, which leads to SCC at very low externally applied stress (17).

Once a fast-running crack has been nucleated by the thin film, it crosses the metal-film interface region and finds itself in an fcc lattice. We have shown (17), supported by molecular dynamic simulations, that such a crack can grow for a distance on the order of μm in an fcc lattice, provided that its initial velocity is on the order of 10% of the local dislocation velocity. The loss of kinetic energy preceding the crack arrest is due to emission of about 1 to 2 dislocations per lattice spacing as the crack grows.

The critical features of a film-induced cleavage process are :
(i) The presence of a surface film, formed by a solid-state reaction to give strong bonding to the metal (thus precipitated films are unsuitable) and of sufficient thickness (thus normal passive oxides of 2-5 nm thickness seem to be too thin).
(ii) Suitable mechanical properties in the film, possibly intrinsic brittleness arising from "nanoporosity" in de-alloyed layers.
(iii) Very favourable is an induced tensile stress in the film due to lattice contraction from de-alloying. The stress in the film influences the level and type of external loading required to cause SCC, e.g. films with induced compression are only effective under dynamic loading conditions, whereas some de-alloyed layers can cause SCC under zero applied load.

Alternative models of transgranular SCC lack the physical simplicity of film-induced cleavage, and have not so far led to any testable predictions. Adsorption-induced cleavage, while possibly a limiting case of film-induced cleavage when the adsorbate has extraordinary properties (hydrogen!), has little meaning when electrochemical measurements show a rapid film growth on new surfaces. Adsorption-induced plastic fracture (20) remains a possibility in some systems, and particularly in liquid metal embrittlement, but again relies on a bare metal surface which is demonstrably not present in many cases within milliseconds of exposing new metal at the crack tip. Neither can it account for the effect of potential. A new model based on surface diffusion has been produced by Galvele (21) but its value is unclear at present.

Our main experimental objective has been to establish the relationship between de-alloying and transgranular SCC in a system where there are no competing surface reactions and the rate of de-alloying can be measured electrochemically. The results conclusively support the film-induced cleavage concept and are shown below.

3. RELATIONSHIP BETWEEN DE-ALLOYING AND TRANSGRANULAR SCC

De-alloying occurs in solid solutions or intermetallic compounds where there is a large difference in reactivity between the components. Good examples are Cu-Al and Cu-Zn alloys, where copper is the more noble element and transgranular SCC is known to occur in ammoniacal environments.

Recent computer simulations and supporting experimental work (22,23) show that the only solid-state mass transport process required to sustain de-alloying is surface diffusion of the more noble element, as proposed earlier by Forty and Rowlands (24). Parting limits, or percolation thresholds, are observed in both the experiments and the computer simulations. The different behaviour of different alloys is related to the type of surface site (e.g. kink, step, etc.) from which the more reactive element can be dissolved.

An ideal environment in which to test for de-alloying and SCC was discovered by Bertocci et al. (25). They showed that the traditional corrosive solutions of cupric (Cu^{2+}) ions in ammonia could be replaced by a solution of cuprous (Cu^+) ions in ammonia without any essential change in the SCC behaviour of 70-30 brass. In the latter solution, copper is at equilibrium and it is easy to measure the rate of dezincification electrochemically.

Experimental Procedure

Single crystals of various Cu-Zn and Cu-Al alloys were made into tensile specimens for slow strain rate SCC testing. The SCC tests were done in a solution containing 15 molar ammonia and 0.05 molar cuprous ions, added as cuprous oxide. De-alloying was measured by scratching coated specimens of the alloys in the same solution, with the potential controlled at the equilibrium potential of a copper electrode, and recording the consequent anodic current transients. The depth of de-alloying was calculated using Faraday's laws, assuming that the de-alloyed layers were pure copper and that all the solute was replaced by an equivalent volume of porosity.

Results and Discussion

The results are shown in Figure 2 - the parting limits for the Cu-Al and Cu-Zn alloys were ~13 and ~17 at.% respectively, and these corresponded exactly to the onset of SCC. Absolutely no cracking was observed in any Cu-Al alloy containing $\leq$ 11% Al, or in any Cu-Zn alloy containing $\leq$17% Zn. The rates of crack growth, shown on the figure, were much higher than could be accounted for by successive growth and rupture of the de-alloyed layer without cleavage; typically the de-alloying comprises about 3% of the overall penetration during SCC, with cleavage accounting for the other 97%.

The demonstration of a correlation between de-alloying and transgranular SCC in two different systems will probably supersede the theory of intergranular-transgranular transitions based on stacking fault energy (26).

4. THE EFFECT OF POTENTIAL ON TRANSGRANULAR SCC

As already noted, transgranular SCC often follows closely the dependence of the anodic reaction rate on potential. This is easily understood, as the cleavage requires an anodic reaction to form the film, and the rate of this reaction depends on the potential. Models relating anodic reaction rates to SCC velocity generally contain unknowns or adjustable parameters which can mask the necessity for cleavage and lead to the conclusion that an anodic reaction is exclusively able to grow a cleavage-like crack. Alternatively, the experimental technique used to measure the bare-surface anodic reaction rates may be faulty as in the case of α-brass in sodium nitrite solution (2). As shown in Figure 3, the "intermediate strain rate technique" overestimates the current density flowing on strained surface by a factor of up to 50, possibly because the oxide film spalls from the surface during straining. The scratching technique, which also tends to overestimate the current density (but only slightly), gives results which are consistent with film-induced cleavage in that the charge flowing on a bare surface within a few seconds of its creation is a factor of 10 to 100 too low to account for the cracking without cleavage (27).

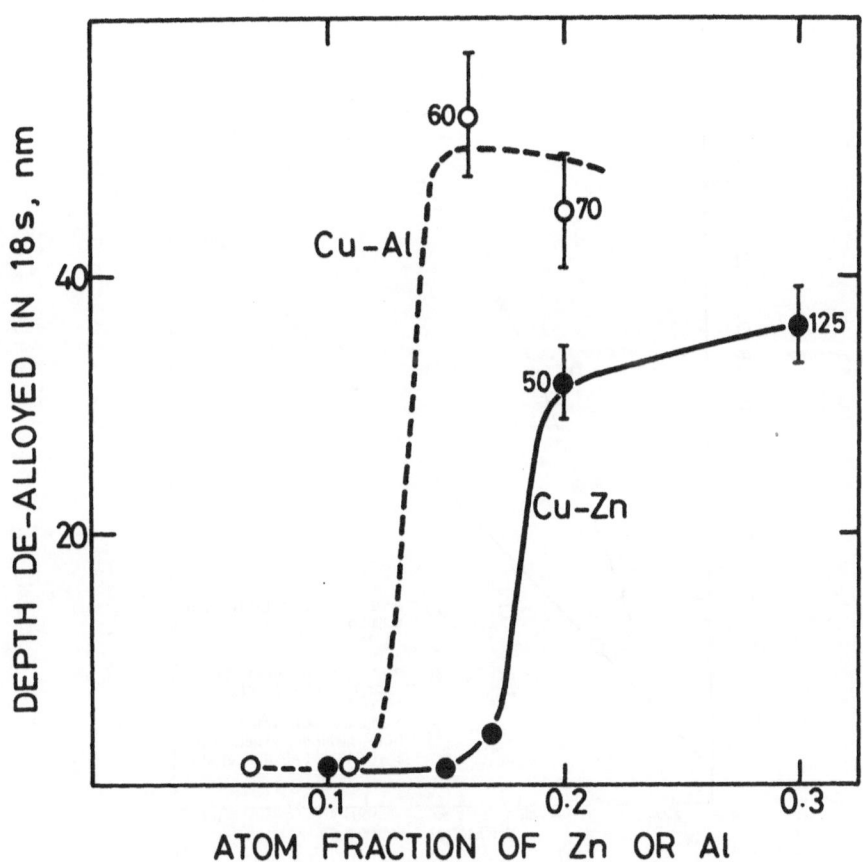

FIGURE 2. Dependence of the de-alloying rate on solute concentration for Cu-Zn and Cu-Al alloys at the equilibrium potential of copper in 15 M ammonia + 0.05 M cuprous ion solution at room temperature. The numbers next to the data points indicate the average SCC velocity in nm/s from slow strain rate tests of single crystals; no number indicates no SCC.

602

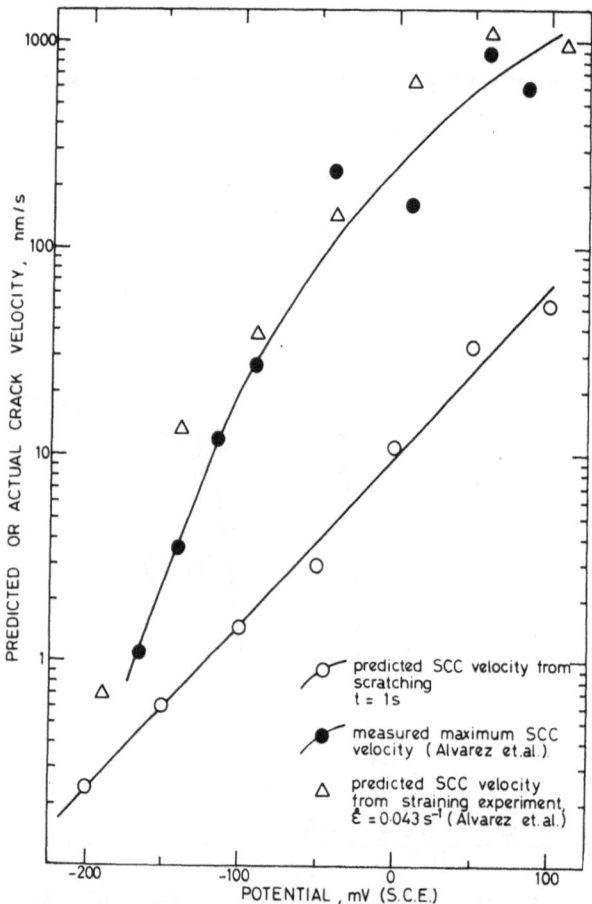

FIGURE 3. Comparison of predicted and actual SCC velocity for α-brass in
1 M NaNO₂ solution (27), showing that rapid straining tests agree
fortuitously (via Faraday's Laws) with the SCC velocities. The true
reaction rates on new surfaces are indicated by the scratching tests and
show current and charge densities too low by 1-2 orders of magnitude to
account for the SCC velocities. This is consistent with film-induced
cleavage.

These considerations are almost certainly applicable to other fcc systems such as copper, copper-gold and austenitic stainless steel, or to low-alloy steels in high-temperature water.

Mixed-mode Cracking

Intergranular cracking is often observed in systems where transgranular SCC is thought to be due to de-alloying (e.g. brass, copper-gold, stainless steel). We believe that this cracking does not involve brittleness in unattacked material; the de-alloyed layer grows more rapidly than within a grain, and is ruptured much more frequently by the strains at the crack tip.

5. RATIONALIZATION OF TRANSGRANULAR SCC MECHANISMS

Having shown conclusively that transgranular SCC is predominantly brittle (or at least mechanical) in copper alloys, we shall now present a more speculative rationalization of a wider range of transgranular SCC processes. This is intended as a stimulus to further work on the types of films that can nucleate cleavage, and is certainly less speculative than many other "unified" approaches to SCC.

We place the chloride SCC of austenitic stainless steels with the copper alloys as cleavage nucleated by de-alloying. The evidence for this is now being accumulated, and we summarize this below :

(i) The potential and pH within a crack (about -140 mV$_H$ and 1.5) are such that the metal surface must be in the active state. The oxide films found microscopically consist of precipitated material (mostly chromium hydroxide) which forms behind the crack tip.

(ii) Surface analysis shows that rapid de-alloying (nickel enrichment) occurs under these conditions. (These analyses are difficult because of precipitation of chromium hydroxide within the pores of the nickel sponge, and require comparison with binary Fe-Ni alloys [which also crack] to demonstrate the reality of the de-alloyed layer).

(iii) Electrochemical experiments done in simulated crack solutions show anodic reaction rates which are much too low to account for the cracking without cleavage.

(iv) Evidence from Pugh and others (10,11) clearly shows that the cracking is brittle.

One of the best features of this de-alloying model, as realized in a different context by Latanision and Staehle (29) is that it accounts neatly for the beneficial effect of raising the nickel content of the steel. Iron and nickel are not very different in reactivity, and the parting limit is around 75% iron rather than the very low values found in alloys such as Cu-Al; thus it is very difficult to crack steels with more than about 25% nickel. We have shown preliminarily that the addition of thiourea makes it more easy to de-alloy alloys with high nickel contents, such as Type 310, and it may be that the effect of thiourea or hydrogen sulphide is nothing to do with hydrogen [we expect this to be a rather controversial conclusion if it is confirmed!].

SCC of pure copper (13-15) occurs in sodium nitrite solution and seems to require dynamic straining even when a crack has already been initiated. This is to be expected from oxide-induced cleavage, where the induced stress from film growth is probably compressive. It is not clear why the nitrite ion is uniquely effective, but electrochemical measurements show that it does induce unusually rapid film growth - for example, the cuprous oxide film formed in a neutral sodium sulphate solution is 5-10 times thinner than that formed in nitrite over periods of a few tens of seconds at the same potential. More work is required here to establish the mechanism of oxide growth and the specific role of the nitrite ion.

Other examples of oxide-induced cleavage are rare because passive films formed at temperatures below 100°C are unsuitable for cleavage initiation. At higher temperatures the cracking of carbon or low-alloy steels is cleavage-like (29) and again dynamic straining is very effective though not as indispensable as in the case of copper. Again it is possible to obtain good agreement between anodic reaction rates on strained electrodes and the SCC velocity, but this must be treated with caution in view of the earlier experience with α-brass (2,27). Similar fractures can be obtained at low temperatures in phosphate solutions (30), and the films involved require characterization.

A very recent observation is of cleavage-SCC in pure copper crystals when strained at certain rates in cupric ammonia solution (31). This is slower than and distinct from the de-alloying-related cracking that occurs in alloys above the parting limit. It does not occur in the equilibrium cuprous ammonia solutions or under static loading. We believe that this cracking involves the production of a porous tunnelled metal layer at the crack tip by a critical rate of selective dissolution at emergent, moving dislocations.

Some unusual transgranular SCC phenomena occur in low-strength ferritic steels in exotic environments. Cracking in anhydrous ammonia (32) is strikingly cleavage-like but definitely does not involve hydrogen. Similar failures occur in anhydrous ammonia-methanol mixtures (33), and in both cases oxygen or an oxidizing potential is necessary for cracking to occur. Very similar fractures are found in CO/CO_2/water mixtures (34) and again the lack of strength dependence and the protective effect of cathodic polarization rule out a hydrogen effect. Finally, transgranular SCC has been obtained in practice in carbon disulphide solutions. We believe that all these can be unified as film-induced cleavage, where the films are thin nitride or carbide layers formed by oxidation of ammonia and reduction of carbon monoxide or carbon disulphide, respectively.

Finally we should briefly consider cleavage-like hydrogen fractures. In titanium alloys these can be associated with a definite hydride phase, but in other systems (aluminium alloys, ferritic steels under cathodic polarization, etc.) the cleavage must be nucleated by a near-surface effect. In most cases the quantity of hydrogen ahead of the crack, once the crack starts to run in a brittle fashion, probably has little influence on the deceleration and arrest process. In this respect the new understanding of film-induced cleavage in fcc alloys helps us to realize that "embrittlement" need only affect sufficient material to start a fast-running crack; subsequent propagation and arrest over distances on the order of μm is a property of the virgin metal or alloy lattice and does not require an embrittling agent ahead of the crack tip.

6. SUMMARY TABLE

As a basis for further research, we have constructed the following
table of suggested film-induced cleavage mechanisms.

Metal or Alloy	Environment	Initiating Layer
α-brass, Cu-Al	ammonia	de-alloyed layer (Cu)
Au-Cu	$FeCl_3$	de-alloyed layer (Au)
	acid sulphate	"
Fe-Cr-Ni	chloride	de-alloyed layer (Ni)
	hydroxide	de-alloyed layer or oxide
	H-T water	"
α-brass	nitrite etc	oxide
copper	nitrite	oxide
	ammonia (cupric)	porous dissolution zone
ferritic steel	H-T water	oxide
	phosphate	oxide (?)
	anh. ammonia	nitride
	$CO/CO_2/H_2O$	carbide
	CS_2/H_2O	"
Ti alloys	chloride etc.	hydride
Al alloys, steels	various	near-surface hydrogen

7. FUTURE DIRECTIONS

Outstanding problems in film-induced cleavage include the following :
(i) How does the crack reach near-sonic velocity in a film a few tens of
nm in thickness? Progress is expected here using de-alloyed layers as
model systems; there is already evidence that a de-alloyed layer formed
in solution without stress can subsequently initiate a small cleavage
crack when strained in air (35,36). A de-alloyed layer is a special case
of a porous solid (19), where the pore size approaches atomic dimensions.
(ii) For cases of oxide-induced cleavage, what are the special properties
of certain oxides that enable them to nucleate fast cracks?
(iii) The nitride-induced cleavage in ammonia cracking has not been
demonstrated using surface analysis; this work is in progress.

We do not feel that the continued propagation of the cleavage crack,
once nucleated in the film, is a major theoretical problem. Some
modification of the dynamic analysis presented earlier (17) should be
sufficient.

8. CONCLUSIONS

Transgranular SCC is a cleavage process. The step-wise brittle fracture
is triggered by a film which is usually a metal layer with nanometre
porosity but can also be an oxide or nitride in certain cases.

The relationship between de-alloying and SCC has been demonstrated
conclusively in Cu-Zn and Cu-Al alloys. This type of SCC depends on
electrochemical potential in the same general way as an anodic
dissolution process, but the anodic reaction cannot account exclusively
for the crack velocity.

9. ACKNOWLEDGEMENTS

Most of this work has been supported at Brookhaven National Laboratory
by the US Department of Energy, Division of Materials Sciences, Office of
Basic Energy Sciences, under contract number DE-AC02-76CH00016. The
contribution of R.C. Newman was made possible by a guest appointment on this
contract. Thanks are due to A.T. Cole (UMIST) and J.S. Kim (BNL) for their
contributions to the research.

606

REFERENCES

1. Parkins R.N.: Corrosion, ed. L.L. Shreir, p. 8:24, Newnes-Butterworths, London, 1976.
2. Alvarez, M.G., Manfredi, C., Giordano, M. and Galvele, J.R.: Corros. Sci. 24, 769 (1984).
3. Maier, I.A., Lopez Perez, E. and Galvele, J.R.: Corros. Sci. 22, 537 (1982).
4. Edeleanu, C. and Forty, A.J.: Phil. Mag. 46, 521 (1960).
5. Nielsen, N.A.: Fundamental Aspects of Stress-Corrosion Cracking, ed. R.W. Staehle, A.J. Forty and D. van Rooyen, p. 308, NACE, Houston, 1969.
6. Bakish, R.: J. Metals 9, 494 (1957).
7. Pugh, E.N.: The Theory of Stress-Corrosion Cracking in Alloys, ed. J.C. Scully, p. 418, NATO, Brussels, 1971.
8. Barnartt, S. and van Rooyen, D.: J. Electrochem. Soc. 108, 222 (1961).
9. Beavers, J.A. and Pugh, E.N.: Metall. Trans. A 11, 809 (1980).
10. Hahn, M.T. and Pugh, E.N.: Corrosion 36, 380 (1980).
11. Pugh, E.N.: Atomistics of Fracture, ed. R.M. Latanision and J.R. Pickens, p. 997, Plenum, New York, 1983.
12. Bertocci, U.: Embrittlement by the Localized Crack Environment, ed. R.P. Gangloff, p. 49, TMS-AIME, Warrendale, PA, 1984.
13. Pednekar, S.P., Agrawal, A.K., Chaung, H.E. and Staehle, R.W.: J. Electrochem. Soc. 126, 701 (1979).
14. Sieradzki, K., Sabatini, R.L. and Newman, R.C.: Metall. Trans. A 15, 1941 (1984).
15. Meletis, E.I. and Hochman, R.F.: Corros. Sci. 24, 843 (1984).
16. Newman, R.C. and Sieradzki, K.: Scripta Metall. 17, 621 (1983).
17. Sieradzki, K. and Newman, R.C.: Phil. Mag. A 51, 95 (1985).
18. Sieradzki, K.: These proceedings.
19. Li, R. and Sieradzki, K.: Phys. Rev. Letters 56, 2509 (1986).
20. Lynch, S.P.: Acta Metall. 29, 325 (1981).
21. Galvele, J.R.: J. Electrochem. Soc. 133, 953 (1986).
22. Sieradzki, K., Kim, J.S., Cole, A.T. and Newman, R.C.: The Relationship between De-alloying and Transgranular SCC of Cu-Zn and Cu-Al Alloys, J. Electrochem. Soc., submitted for publication.
23. Sieradzki, K. and Corderman, R.R.: unpublished results.
24. Forty, A.J. and Rowlands, G.: Phil. Mag. A 43, 171 (1981).
25. Bertocci, U., Thomas, F.I. and Pugh, E.N.: Corrosion 40, 439 (1984).
26. Pugh, E.N., Craig, J.V. and Montague, W.G.: Trans. ASM 61, 468 (1968).
27. Newman, R.C., Kim, J.S. and Sieradzki, K.: Stress-Corrosion Crack Growth by Film-Induced Cleavage : Some Kinetic Considerations, Modeling Environmental Effects on Crack Initiation and Propagation, ed. R.H. Jones, TMS-AIME, in press.
28. Latanision, R.M. and Staehle, R.W.: ref. 5, p. 214.
29. Congleton, J., Shoji, T. and Parkins, R.N.: Corros. Sci. 25, 633 (1985).
30. Parkins, R.N., Holroyd, N.J.H. and Fessler, R.R.: Corrosion 34, 253 (1978).
31. Kim, J.S. and Sieradzki, K.: unpublished results.
32. Deegan, D.C. and Wilde, B.E.: Stress Corrosion Cracking and Hydrogen Embrittlement of Iron Base Alloys, ed. R.W. Staehle, J. Hochmann, R.D. McCright and J.E. Slater, p. 663, NACE, Houston, 1977.
33. Omideyi, M.V. and Procter, R.P.M.: Proceedings of the 9th International Congress on Metallic Corrosion, Toronto, 1984, NRC, Ottawa, 1984.
34. Brown, A., Harrison, J. T. and Wilkins, R.: ref. 32, p. 686.
35. Cassagne, C.D., Flanagan, W.F., and Lichter, B.D.: Metall. Trans. A 17A, 703 (1985).
36. Cole, A.T.: unpublished results.

DISCUSSION

Comment by D. D. Macdonald:

How does stress enter into the percolation theory for stress corrosion cracking? Why should the porous structure you show fail by cleaving? I would have thought that the porous gold structure would have mechanical properties that could not support this type of fracture.

Reply:

The model of SCC by dealloying and cleavage was tested by assuming that dealloying at the crack tip was the same as that occurring on a scratched surface. Percolation theory relates to the static, pre-existing arrangement of atoms in the alloy, and stress is not relevant to this. Dealloying could theoretically be influenced by dynamic plastic strain, but our observations suggest that layers of a few hundred Å thickness are very plastically hard. In the earlier stages of layer growth, the population of monatomic steps could be enhanced by dislocation emergence or emission, leading to some enhancement of the dealloying rate, but this should be a small effect.

The porous structure which cleaves is the extremely fine-scale one near the interface in Lichter's long-immersion experiments. In SCC, the layer is only a few hundred Å in thickness and must have a pore size on the order of 10 Å; such a layer may well be truly brittle. Computer simulations are providing some answers here.

Comment by K. Nisancioglu:

Is a few monolayers of dealloying sufficient to induce cleavage? Or, do you require a thicker layer?

Reply:

Our experimental data indicate a requirement of at least 200 Å; this is consistent in general terms with Sieradzki's modelling.

Comment by B. D. Lichter:

1. Concerning the correlation between the percolation limit (the composition of less noble metal, below which dealloying is supposed not to occur) and the difference in reversible potential between noble and base metals, the Cu-Au system seems anomalous because $\Delta E°$ is large (~1 V or more) and yet the percolation limit is at least 0.75 Cu. Can your simulation account for this?

2. The actual composition for a "cluster" of atoms considered in your percolation-simulation can fluctuate from the mean. There can be diffusive jumps within the crystal which would permit this. How would the existence of such time-varying composition fluctuations be expected to affect your estimate of the percolation limit, and how can they be taken into account?

Reply:

1. The parting limit depends on potential. At $E°_{Au}$ it would be lower.

2. We believe that only surface diffusion is relevant at high rate characteristic of the early stages of de-alloying. Later on these processes might be relevant.

Comment by P. Neumann:

Dealloying in the very crack tip region (a few atomic distances in diameter) may be very effective to enhance decohesion. Is the maximum rate of dealloying imaginable at the very first atomic layer at the crack tip under stress (!) sufficient to enhance decohesion at the observed speed of ~100 nm/s (the dealloying rate of 40 nm/18s is an average taken over 18s).

Reply:

We plotted the dealloying rates by integrating over 18s, but the conclusions would be the same (within a factor of 2-3) even if we used the peak currents after 20 ms. The reason for this is that the transients are rather flat and do not show extremely high peak currents. The only solute atoms which may be removed at an extremely high rate are those in the very outermost atom layer.

The choice of 18s was not random, as this was our estimate of the interval between film fracture events at the crack tip.

Your suggested mode of crack advance is, of course, also inconsistent with the experimental observation of fast crack jumps.

Comment by P. Marcus:

ESCA measurements that we have done on a series of FeNi alloys (after exposure to acid solution at room temperature) show that there is clearly a Ni enrichment on the surface, but limited to the surface plane. Do you think that the results you presented are specific of Cl-containing solutions?

Reply:

Yes, but with some reservations -- to get bulk dealloying you need to have a low potential so that nickel does not dissolve too fast. To make a proper comparison, we should compare our results (15M LiCl, 140°, pH 1.5, E_{corr}) with the nearest equivalent situation without chloride. We have not made thorough measurements at room temperature in H_2SO_4 + NaCl solutions, but we believe there is dealloying in these. We think you should look at Fe-10% Ni in H_2SO_4 + NaCl.

We also see an effect of Cl^- in Fe-Cr-Ru alloys. In sulphuric acid Ru enriches in monolayer or extremely fine particulate form; in HCl the particles are much larger. This shows up as a degradation of the spontaneous passivation by Ru in the presence of chloride.

We believe that the "specific" effect of chloride on stainless steels is because it causes dealloying.

Comment by D. D. Vvedensky:

Is it clear that the percolation model is appropriate rather than a reaction-diffusion approach? What other models have you considered?

Reply:

We have considered divacancy diffusion, but it is much too slow to account for the observed rates of dealloying. Computer simulation using only surface diffusion reproduces all the gross observations including the very sharp parting limit and the existence of intermediate alloy compositions.

In some cases there may be a limited (local) dissolution/redeposition of the more noble metal, but this is clearly not generally necessary -- for example, in Au-Cu alloys at around 700 mV (SCE) where no gold oxidation is possible.

Comment by Engell:

1. Did you take into account the phase diagram of your alloy systems?
2. Did you account for atomic ordering in the alloys when calculating the concentration limits for dealloying?
3. Could you relate your i(t)-measurements with discontinuous COD?
4. How about the pronounced influence of chromium on TGSCC in Fe-Ni-Cr alloys?
5. Why in many systems, e.g., in stainless steels, is TGSCC so sharply related to the regime of electrochemical passivity?
6. Is not your approach also of importance for a deeper understanding of what is called the resistivity limit of noble metal alloys?

Reply:

1. All the alloys were single-phase.
2. Dealloying in ordered systems should show some subtle differences from disordered systems, but as yet we have not done the relevant experiments. Computer simulations are underway.
3. It should be possible to do this on small specimens, but we have not tried it.
4. Chromium has an influence, e.g. through acidification of the crack, but the dominant effect is that of Ni. Binary (martensitic) Fe-Ni alloys crack readily in hot chloride solutions, as do Fe-Cr-Ni alloys, and at constant Ni content in an austenitic steel the cracking is not very sensitive to Cr.
5. In the austenitic stainless steels, cracking does not require passivity, as shown many years ago -- cracking occurs very readily at pH 1-2, especially in dynamic loading. If the solution is initially neutral, there is a requirement to break down the passivity to establish the acid crack environment; thus there will always be a tendency for cracking in neutral solutions to correlate with incipient breakdown of passivity by pitting.
6. Percolation concepts are widely applicable to alloy properties, and this may be another example.

Comment by J. Lumsden:

How do you explain the increased resistance of ferritic ss to TGSCC, and the increased susceptibility to TGSCC when Mo is present in an 18/8 ss, in terms of the dealloying mechanism?

Reply:

Ferritic steels do crack if they contain some noble metal (Ni or Cu) that can form a sponge. We have not established the reasons for minor increases or decreases in TGSCC susceptibility, but this is part of our future program.

Comment by F. Nabarro:

Your lecture seems to involve porosity on an atomic scale growing to granularity on a micron scale at room temperature. What could be the mechanism of this?

Reply:

I don't think there is any problem in accounting for the initial sponge formation by surface diffusion of the noble metal, but we don't understand the coarsening process leading to the 0.1-1 µm pore size.

Comment by R. H. Jones:

Have the "cleavage-like" surfaces which show crack arrest marks been etched to determine their depth? Deep arrest marks may suggest a slip nature to these marks. Also, are the "cleavage-like" surfaces dealloyed? The electrochemical pulse associated with fracture would suggest that the surface should be dealloyed.

Reply:

The detailed structure of the crack arrest striations has not been established.
The issue of dealloying on fracture surfaces is a complex one. Our experiments on brass in nitrite are not strictly relevant as this process occurs at rather high potentials and involves an oxide film. Heldt and Hintz have shown that there is little dealloying near the crack tip for brass in ammonia, but this would be expected as the cleavage would involve a redeposition of sub-monolayer amounts of copper from copper ions in the crack; the crack tip dealloys normally owing to rupture of any such layer by plastic strain. Thus, Heldt and Hintz would have had to analyze exactly at the crack arrest positions to detect dealloying. Well behind the crack tip, the dealloying is obvious in surface analysis, as shown by several groups.

Comment by M. O. Speidel:

We observe fast (10^{-8}m/s) stress corrosion cracking of nuclear reactor pressure vessel steel in hot (288°C) water. If this was to be explained by oxide film rupture with subsequent cleavage--would the subsequent cleavage have to consider the effect of hydrogen on cleavage?

Reply:

The cracking in hot water plus oxygen may involve a large IR drop and hydrogen cannot be ruled out. However, if (as you have indicated) a very similar cracking occurs at high solution conductivities, then the crack potential approaches the free surface potential and it is likely that hydrogen is not involved. Hydrogen is certainly not necessary for the 3 μm jump of the crack once it is nucleated in the film, as similar processes occur in fcc alloys during SCC.

CORROSION FATIGUE OF METALS: A SURVEY OF RECENT ADVANCES AND ISSUES

RICHARD P. GANGLOFF[1] and DAVID J. DUQUETTE[2]

[1]Department of Materials Science
University of Virginia
Charlottesville, VA 22901
USA

[2]Materials Engineering Department
Rensselaer Polytechnic Institute
Troy, NY 12180-3590
USA

1. INTRODUCTION

Mechanisms of corrosion fatigue crack formation and advance, based on interactions between localized cyclic deformation and chemical reaction, were not emphasized during the 1981 NATO Research Workshop on The Chemistry and Physics of Fracture (1,2). The monotonic problems of stress corrosion cracking and hydrogen embrittlement were considered in detail, with the state of the art summarized by Professor Argon: "All this gives a rich picture and useful information for engineering application, but falls short of providing definitive understanding. Nevertheless, a picture emerges that indicates that the phenomenon is broad and encompasses several mechanisms." (3).

The unresolved issues cited in the 1981 Workshop Summary on Stress Corrosion Cracking were not specifically directed towards the corrosion fatigue problem, but are none-the-less relevant. Specific issues for future study, circa 1981, included (4):

Identification of the mechanism for transgranular SCC.

Definition of the chemistry and electrochemistry localized within the crack tip region.

Modeling of the atomistics of dissolution and passivation, with emphasis on the roles of stress and localized strain.

Increased emphasis on model system studies with well defined electrochemical conditions.

Increased emphasis on SCC initiation and the integration of controlled slow strain rate and precracked specimen experiments.

Consideration of the effects of stress state, local microstructure and misorientation at grain boundaries on environmental cracking.

Given the foundation established for environmental cracking at the first conference, it is now appropriate to consider the complex cyclic problem.

The purpose of this presentation is to survey recent approaches and progress on critical uncertainties in corrosion fatigue, some of which

were first cited in the proceedings of international conferences held in
the 1970s (5-7) and examined in more recent volumes (8-11). The above
issues pertaining to SCC are relevant to the corrosion fatigue problem,
with the cyclic character of the loading uniquely affecting the cracking
process. The current applicability of Argon's conclusion will be
demonstrated for corrosion fatigue, however, recent advances indicate that
basic understanding of damage is achievable. The end result in this
regard is the capability to reliably predict and control the performance
of new alloys, inhibited environments and components over long periods of
service and based on short term laboratory data.

The cumulative damage processes leading to environmentally assisted
fatigue failures can be subdivided in a number of ways. It is, perhaps,
most logical to divide the topic sequentially into four principle
categories: (1) pre-crack cyclic deformation, (2) microcrack initiation,
(3) small crack linkup with growth into the short crack regime and (4)
long crack growth. The specific demarcations between these regimes of
damage are the topic of debate (10). This survey is organized to consider
progress and uncertainties in each category. The advent of fracture
mechanics resulted in a great deal of work on the growth kinetics of long
cracks sized above 10 to 50 mm, category 4. The issues regarding the
interactions between cyclic deformation and chemical reactions,
environmentally assisted crack initiation mechanisms and the propagation
behavior of microstructurally small and physically short cracks are only
beginning to be addressed; recent findings are emphasized in this review.
Localized deformation, cracking and chemical damage processes are
emphasized.

In general the development of an integrated and basic view of the
corrosion fatigue problem is hindered by several factors recurrent
throughout this survey. Corrosion fatigue is influenced by a variety of
mechanical, chemical and microstructural variables which interact. It is
further necessary to investigate very slow rate deformation and cracking
phenomena in a finite and realistic time. Model system studies on
deformation and crack initiation have been conducted on relatively pure
materials, often monocrystalline, while crack propagation studies are
typically performed on complex structural materials including steels and
precipitation hardened aluminum or nickel based alloys. Corrosion fatigue
damage is highly localized at slip substructures or near the crack tip;
direct experimental observations are not available and behavioral
interpretations must be based on indirect or averaged measurements. As
the case for stress corrosion, corrosion fatigue is likely controlled by
both hydrogen based and dissolution/passivation mechanisms, the atomistics
of which are not known.

2. PRE-CRACK CYCLIC DEFORMATION AND FATIGUE CRACK INITIATION
2.1. Environmental Effects on Cyclic Deformation

When ductile metals are cyclically strained, the general deformation
associated with the first few (or few hundred) cycles changes continually
until a metastable dislocation substructure is established. For very pure
materials, the development of this structure requires measurable plastic
deformation, while for engineering alloys the fatigue related dislocation
substructure may be developed at cyclic stresses which are less than the
measured yield stress due to localized stress concentrations. In either
case, however, for annealed materials the development of the substructure
generally leads to cyclic hardening which saturates with time or number of
cycles. The level of the saturation strain (or stress if strain is
controlled) is a function of the applied stress (or strain) and correlates

with an increase in dislocation density with increasing numbers of cycles until both reach a constant value. For cold worked materials, cyclic softening is observed, but it also saturates after a given number of cycles. Since a saturation stress (or strain) is related to an applied strain (or stress), cyclic deformation is characterized by cyclic stress-strain curves which are similar to monotonic stress-strain data.

At, or just beyond, saturation the metastable dislocation structure breaks down locally and intense slip bands are formed in many metals and alloys. These features, which are called "persistent" slip bands (PSB), then cluster as either grooves or protrusions called intrusions or extrusions. While PSB's and intrusion/extrusion combinations are generally reported for pure metals such as copper or nickel, they have also been reported for solid solution alloys such as alpha brass or austenitic stainless steels and are particularly prominent in precipitation hardened aluminum alloys (12-15). There is at least one report of intrusions and extrusions in a low carbon steel.(16)

Fatigue cracks generally initiate at either PSB's or are associated with intrusion/extrusion bundles and grow into the metal or alloy along slip bands. An exception to this general observation is observed when high levels of plastic deformation are applied to polycrystals. In this case crack initiation occurs in grain boundaries, where deformation is intensified by geometric effects associated with the crystallographic discontinuities at the boundaries (17).

Cyclic deformation and crack initiation processes are profoundly affected by environments which actively corrode or passivate metals or alloys and recently have been shown to be affected by dissolved hydrogen, at least for the case of nickel (18,19). Gaseous environments, however, do not measurably affect pre-crack deformation processes. Some specific examples of the effects of active corrosion on fatigue deformation are discussed.

With reference to the effect of plasticity on pre-crack fatigue deformation, recent work utilizing monocrystalline and polycrystalline nickel shows that both the rate of cyclic hardening and the saturation stress level can be measurably affected by corrosive solutions (18,20). These experiments were performed using constant strain amplitudes and electrochemical control of potentials in deaerated 0.5 N sulfuric acid where bulk surface oxides are generally stable. Specifically, corrosion fatigue studies were performed at the open circuit corrosion potential (-280 mV, SCE), at a potential corresponding to active dissolution (+160 mV, SCE) and at a potential in the passive region (+650 mV, SCE) of the polarization curves. The cyclic stress-strain behavior of monocrystalline nickel as a function of potential is summarized in Figure 1, which clearly shows that the corrosivity of the environment affects the hardening behavior of the metal. In all cases either active corrosion or passivity preferentially hardens the nickel, at least at low applied cyclic strains. At higher cyclic strains active corrosion, at either the free corrosion potential or at the maximum corrosion rate, results in equivalent hardening. Passivity results in the greatest extent of hardening. A correlation between the extent of hardening and the specific condition of the surface remains to be quantified, but it is evident that a significant, readily measurable effect of environment exists. Laird and co-workers performed analogous experiments on copper, but did not observe similar behavior (21,22). It is possible that, due to experimental difficulties with the geometry and plasticity of the nickel samples, the cyclic stress/strain behavior shown in Figure 1 was not determined at saturation, whereas the response of the copper (Laird et al.) was

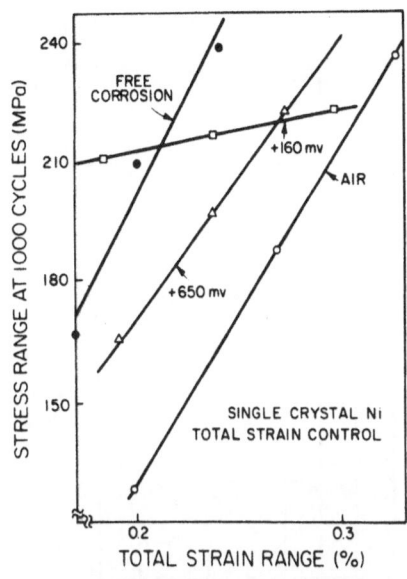

Figure 1. Cyclic stress-strain behavior of monocrystalline nickel in air and 0.5 N H₂SO₄ solution as a function of applied potential at low strains.

determined at the saturation stress. None-the-less, if the environmental effect on pre-crack deformation of nickel is observed for the pre-saturation regime, it indicates that the rate at which saturation is achieved is quite different for electrochemical control of the surface compared to straining in air.

In addition to affecting the cyclic stress-strain behavior of nickel, electrochemical control of the corrosivity of the environment results in significant effects on the morphology of surface deformation. Table 1 shows measured slip offset heights, based on the average of three monocrystalline fatigue specimens, for each of the environmental conditions examined.

TABLE 1. Environmental effects on surface slip morphology for pure Ni.

ENVIRONMENT	SLIP OFFSET HEIGHT (nm ± 10 nm)
Air	80
+650mV	60
Free Corrosion	30
+160mV	excessive corrosion

Figure 2 shows histograms of the PSB distributions produced at a constant strain amplitude. Experiments performed in air and at a passive potential resulted in similar, large slip offsets and a wide PSB distribution as shown in Figure 2a. In contrast experiments conducted at both the corrosion potential and at +160 mv (SCE) resulted in a reduction in the inter-PSB distance and a reduction in slip offset height, Figure 2b. These results are consistent since, for a constant shear strain amplitude, conservation of plasticity requires a larger number of small slip offsets

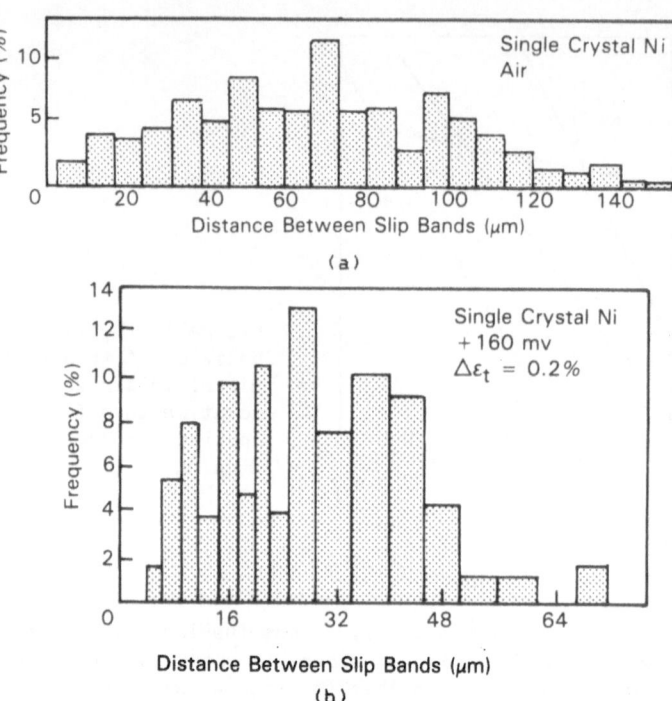

Figure 2. Histograms of distance between PSB clusters, from three replicate single crystal specimens fatigued for 1000 cycles under total strain control. Shear strain range = 0.12%, (a) air and (b) +160 MV (SCE) in 0.5 N H_2SO_4.

(active corrosion) as compared to smaller numbers of larger slip offsets (for air and passive surfaces).

Laird and co-workers reported a similar effect of environment on pre-crack deformation for monocrystalline copper (21,22). Under constant stress range conditions, where conservation of plastic strain is not a requirement, cyclic deformation can result in both larger slip offsets and a multiplication of PSB's (23). This is shown in Figure 3 which compares the surfaces of copper crystals, of the same orientation, cyclically deformed in air and under active corrosion conditions at a constant applied anodic current. This figure shows that the distribution of surface slip is altered by the corrosive environment, changing from dense clusters of more widely spaced PSB's for deformation in air, Figure 3a, to a more general distribution of finely spaced slip bands for the environmental exposure, Figure 3b. Figure 3c demonstrates that slip bands are preferentially corroded in contrast to the surrounding matrix. Confirmation of preferential corrosion and a demonstration of its severity are shown in Figure 4. In these micrographs the PSB cluster (extrusion) and a nascent, Stage I crystallographic crack are shown in Figure 4a for cyclic deformation in air. A corrosion groove associated with a Stage I crack under corrosive conditions is shown in Figure 4b.

2.2. Environmental Effects on Crack Initiation

When polycrystalline pure copper is tested under active corrosion conditions, the slip band distribution and height are affected, however, there is a transition in the crack path from trans- to intergranular

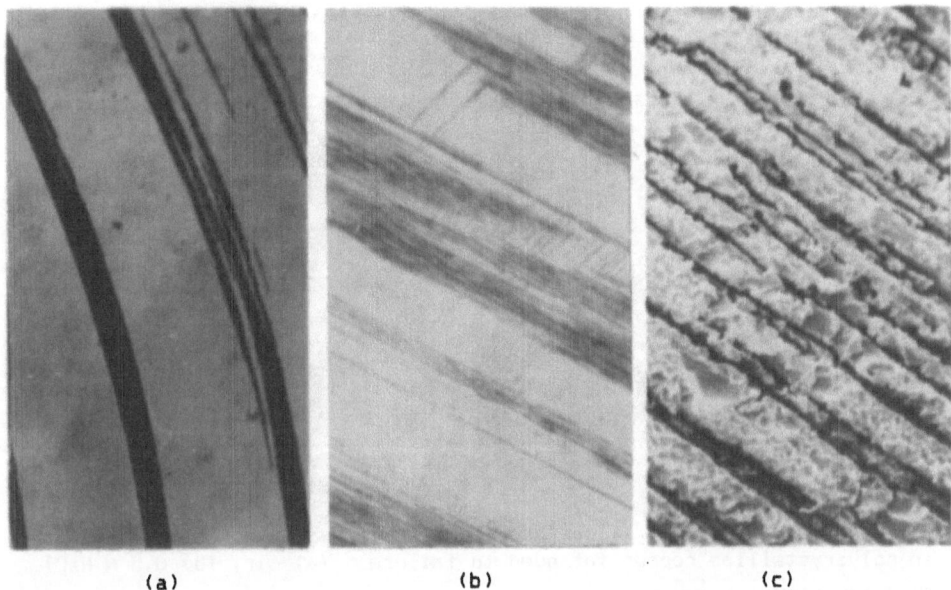

(a)　　　　　　　　　　(b)　　　　　　　　　　(c)

Figure 3. Copper single crystal, surface slip offset morphologies for (a) air, 105 cycles, (b) and (c) 0.5 N NaCl, i = 100 $\mu A/cm^=$, 105 and 106 cycles. After Laird et al. (21,22).

as demonstrated in Figure 5. Figure 5a shows the PSB distribution produced by cyclic deformation in air and also surface connected, Stage I, PSB cracks. This micrograph was obtained from plastic replicas stripped from fatigue surfaces; the ragged edges along PSB's are regions where the plastic intruded into microcracks. Figure 5b, the surface of a corrosion fatigue specimen, clearly shows slip band redistribution and multiplication, preferential corrosion attack and grain boundary crack initiation. Metallographic sections confirm that crack initiation is not a near-surface effect. Cross-sections through fatigue and corrosion fatigue cracks show the transcrystalline, crystallographic nature of Stage I cracks produced in air and the intergranular nature of corrosion fatigue

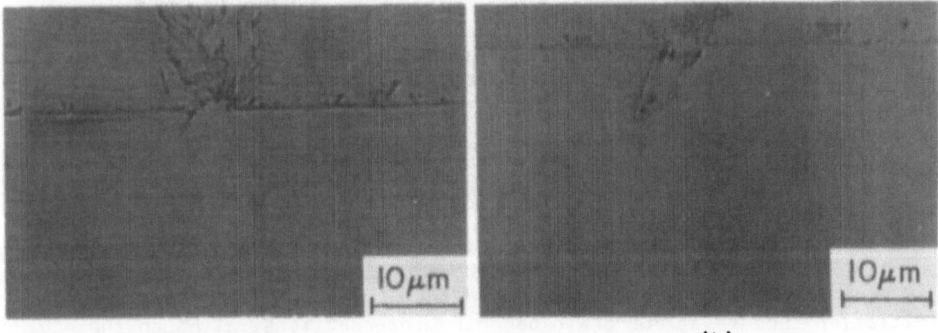

(a)　　　　　　　　　　　　(b)

Figure 4. Longitudinal metallographic sections of copper single crystals showing crack initiation and early propagation. (a) air, 105 cycles, (b) 0.5 N NaCl with applied anodic current of 100 $\mu A/cm^=$, 105 cycles (21,22).

618

cracks. Cross-section studies further confirm the large effect of
corrosion on PSB formation, as evidenced by extensive surface slip
associated with the metal surface.

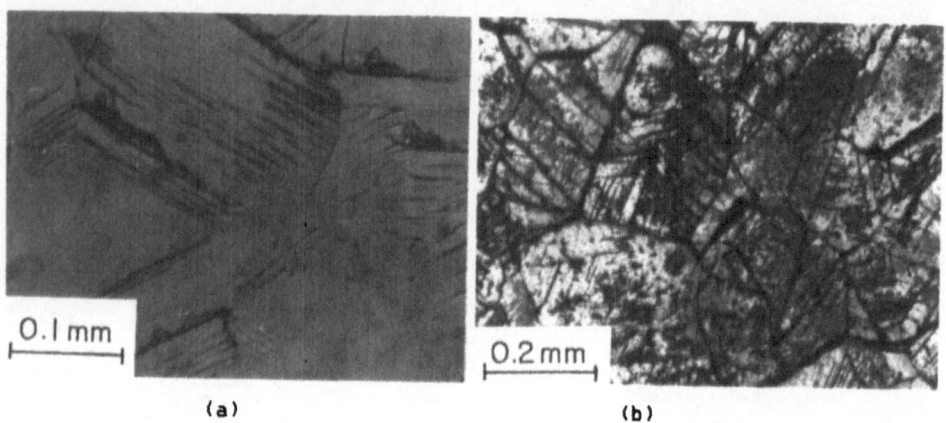

(a)   (b)

Figure 5.   Micrographs of surface replicas illustrating initiation sites
in polycrystalline copper fatigued to failure. (a) air, (b) 0.5 N NaCl
solution with i = 100 $\mu A/cm^2$.

       It is concluded that corrosion in conjunction with cyclic deformation
has a profound effect on pre-crack deformation processes as well as on
crack initiation and early propagation.

2.3. Electrochemistry of Straining Surfaces

       In addition to electrochemical effects on fatigue, there is also a
concomitant effect of fatigue deformation on the electrochemistry of metal
and alloy surfaces. The kinetics of electrochemical reactions are
sensitive to localized deformation processes, which are, in turn, alloy
composition and applied strain rate sensitive.
       In Fe-26Cr-1Mo alloys the deformation substructure generated by
fatigue is sensitive to cyclic strain rate (24-26). Superficial surface
damage, crack initiation mechanisms and fatigue lifetimes are markedly
different at low strain rates ($\dot{\epsilon} < 10^{-5}$ s^{-1}) compared to those which are
produced at high strain rates ($\dot{\epsilon} > 10^{-2}$ s^{-1}). Crack initiation life
decreases with decreasing applied strain rate as shown in Figure 6. At
low strain rates wavy slip PSB's are induced due to thermally activated
cross-slip processes and crack initiation occurs in the PSB's. At high
strain rates, cross slip is suppressed, plastic deformation is controlled
by screw dislocation mobility, glide occurs on different slip systems in
tension and compression, pencil glide is induced and strain localization
occurs at grain boundaries. Cracks subsequently initiate in grain
boundaries, but the initiation life is prolonged relative to transgranular
PSB cracking at low strain rates.
       When Fe-26Cr-1Mo alloys are exposed to chloride containing aqueous
solutions (3.5% NaCl, pH 6) the number of cycles to crack initiation and
the mode of cracking are altered. For slow cyclic strain rates, crack
initiation occurs at PSB-grain boundary intersections, although the
number of cycles to initiate cracks is not perceptibly affected compared
to fatigue in air, Figure 6. At higher strain rates the cracks remain
intergranular for neutral solutions although the number of cycles to crack
initiation is significantly reduced. Increasing the corrosivity of the

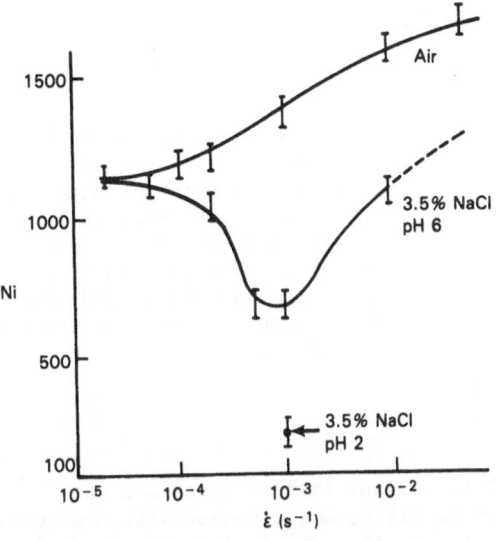

Figure 6. Influence of strain rate on the fatigue cycles for crack initiation, N_i, in Fe-26Cr-1Mo exposed to air and aqueous chloride at an imposed corrosion potential and strain ($\epsilon = 0.004$). After Magnin et al. (24-26).

environment by decreasing the pH, results in mixed intergranular and transgranular cracking and significantly lowers the number of loading cycles required for crack initiation.

By fixing the potential of the specimen surface at the free corrosion level, repassivation times after film rupture by localized deformation are measurable during cyclic loading (24-26). High strain rates result in high transient current densities (J) and low rates of repassivation, as shown in Figures 7 and 8. High strain rates also result in higher peak current densities compared to reaction at lower strain rates.

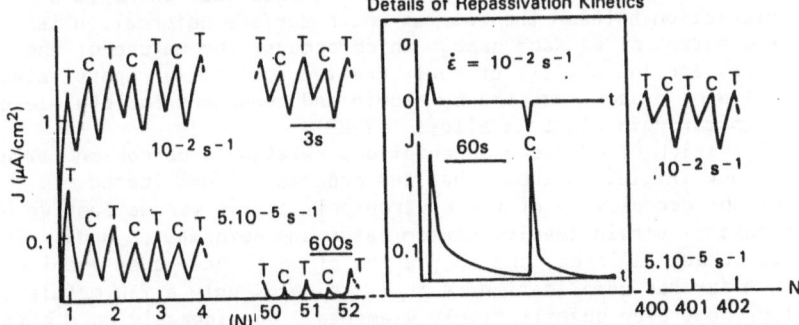

Figure 7. Influence of strain rate on the cyclic evolution of anodic current density at constant imposed electrode potential for different fractions of cyclic life and two strain rates at $\epsilon = 0.004$. (24-26).

Interestingly, repassivation is more rapid after compression than after tension. It is significant that the dissolution transients increase in the presaturation regime and decrease during saturation. Reaction increases again when cracks initiate.

The results of this study indicate that deformation modes, as controlled by strain rate, significantly affect corrosion fatigue

620

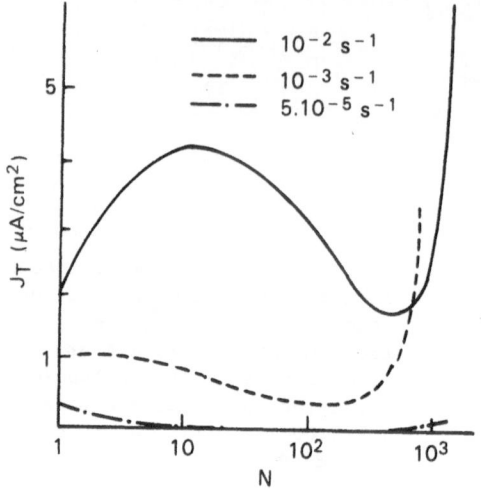

Figure 8. Evolution of the transient maximum current density, J_T, with the number of load cycles for three different strain rates at an imposed corrosion potential in 3.5% NaCl solution (pH 6) and strain (ϵ = 0.004). After Magnin et al. (24-26).

behavior. At low strain rates (PSB formation) fatigue lives are not appreciably affected because, for these alloys where repassivation occurs, slip steps are readily repassivated due to the long period of each cycle. Crack initiation occurs near grain boundaries but in PSB's. At high strain rates repassivation is not complete in each cycle and the number of cycles to initiate fatigue cracks is markedly reduced. Under these conditions, strain localization and, accordingly, crack initiation occur at grain boundaries. It is at intermediate strain rates, however, that the effects of environment are most pronounced. Here, strain localization in the form of persistent slip bands combines with an inadequate repassivation time between cycles to create large reductions in crack initiation (at grain boundaries) life. When repassivation is further reduced by reducing the pH of the environment, cracking becomes mixed transgranular and intergranular. These results show that there is a critical interaction between the rate at which surface deformation is produced, the extent of surface damage which occurs, the nature of the surface damage, and the ability of newly created surface to repassivate. Similar results have been reported by Magnin and coworkers for stainless steel and high strength aluminum alloys (27,28).

This discussion of critical observations related to corrosion fatigue damage and crack initiation shows that the process is complicated, depending on the corrosivity of the environment, active versus passive material behavior, strain levels, strain rates and deformation mode. The matrix of variables is large, and only a few of which have been examined in detail. A further complication is that, to date, only a few metals or alloy systems have been quantitatively examined. Considerably more effort must be expended before well understood models of corrosion fatigue pre-crack deformation and crack initiation can be developed.

3. SMALL CRACK LINKUP, SHORT CRACK GROWTH AND LONG CRACK PROPAGATION
3.1. Fracture Mechanics Similitude

The three categories of corrosion fatigue crack propagation are reasonably considered within the framework of linear elastic fracture mechanics (29-31). This approach relates corrosion fatigue crack growth rate (da/dN) to applied stress intensity range (ΔK), Figure 9, and with the environmental effect revealed by comparisons of data for embrittling

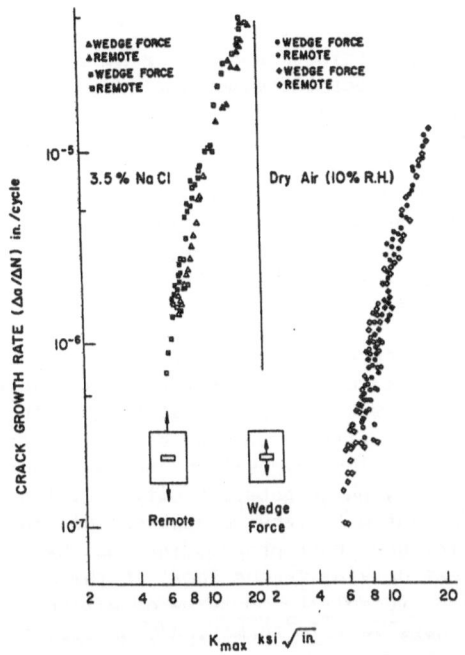

Figure 9. Fatigue crack growth rate versus applied stress intensity for K-increasing (remote) and K-decreasing specimen geometries of 7075-T6 aluminum alloys in either 3.5% NaCl or dry air. After Feeney et al. (31).

and benign gases or liquids. The da/dN-ΔK relationship, often sigmoidal in shape between a low stress intensity threshold and a high K cyclic toughness, is viewed as a material property, independent of crack size, specimen geometry and loading conditions. By the important concept of similitude, cracks grow at equal rates when subjected to equal applied stress intensities. As such, crack growth in a component can be predicted based on da/dN-ΔK data for the environment of interest and the appropriate stress intensity solution (29,30). An application of this method is described in Reference 32 and is summarized in Figure 10.

Similitude has been demonstrated for fatigue in moist air, but, to a lesser degree, for corrosion fatigue conditions. The classical experiment, first reported by Paris, is summarized in Figure 9. A unique

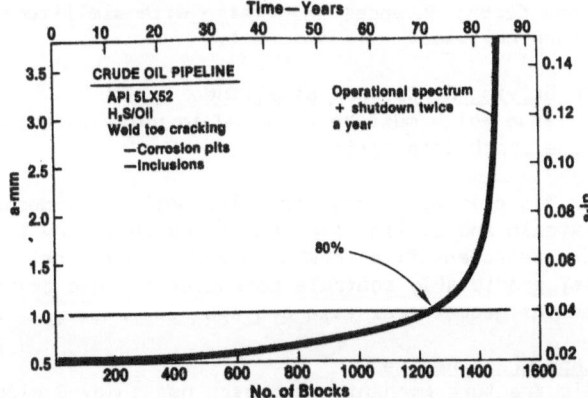

Figure 10. Fatigue crack length versus cumulative load cycles for a welded carbon steel pipeline carrying H₂S contaminated crude oil, predicted based on linear elastic fracture mechanics similitude (32).

crack growth law is observed for both constant remote loading, which produces increasing K with accelerating growth rates, and constant crack surface wedge loading, which produces decreasing K and decelerating crack growth. The remaining ligament tensile stress increases for each geometry; this parameter does not correlate with crack growth. While the results in Figure 9 support the similitude concept for fatigue in benign air and aqueous chloride, such data are otherwise limited for corrosion fatigue. Fracture mechanics has, in large part, been accepted without confirmation for environmental cracking and has become a standardized procedure for characterizing corrosion fatigue crack propagation (e.g. Reference 33). The centrally important issue is the extent to which fracture mechanics describes environmental effects for each region of fatigue crack growth.

Fracture mechanics similitude will correctly scale corrosion fatigue crack propagation kinetics in any of the categories of crack growth, when three conditions are met. (1) Applied stress intensity must accurately describe remote loading and crack geometry effects on the local crack tip stress, strain and strain rate field. (2) The composition and reactions of the environment local to the crack tip, while possibly different from bulk chemical conditions, must be constant with varying applied load and crack size. (3) Rates of corrosion fatigue crack propagation must be a function of the mechanical and chemical driving forces local to the crack tip. The chemical driving force may be formulated in terms of either hydrogen production/embrittlement or passive film formation/dissolution and rupture.

Investigations conducted over the past ten years indicate that stress intensity-based similitude is compromised by and must be modified to account for the effects of plasticity, crack motion, crack closure, crack deflection, crack size and occluded crack chemistry. The complexities associated with analysis of fatigue crack propagation in benign environments are well summarized in recent volumes emphasizing "crack closure" and "the small crack problem" (11,34,35). The controlling mass transport and chemical or electrochemical reactions within occluded, environmental fatigue cracks superimpose with mechanical problems to further obscure the relevant crack tip driving forces (9,36). These issues, while defined in scope, are currently unresolved and preclude the use of a single crack tip field parameter to scale the growth properties of corrosion fatigue micro- to long crack geometries. Critical uncertainties and recent advances associated with similitude and crack tip driving forces are reviewed.

3.2. Crack Tip Driving Forces and Similitude

The basic issue which must be resolved to unify each category of corrosion fatigue crack propagation is:

What crack tip driving force, including both local mechanical (stress, strain and strain rate) and crack chemistry (hydrogen production, transient film formation and dissolution) components, predictably controls corrosion fatigue crack growth independent of geometrical size and applied loading?

3.2.1. Long Crack Regime

The elastic fracture mechanics approach has enjoyed significant success in mechanistic and life prediction studies for long cracks sized above about 10 mm. None-the-less, two factors, extrinsic crack closure or more generally shielding, and time dependent localized crack environment chemistry, can greatly alter the true crack tip mechanical-chemical driving force from that predicted based on applied load and crack geometry through applied stress intensity.

    3.2.1.1. Crack Closure. Crack closure or shielding is the process wherein the range of crack tip stress and strain is reduced from the applied level because of extrinsic factors such as premature wake contact during unloading, crack tip processes, or crack surface forces from corrosion debris or fluids. Mechanisms of crack closure and shielding have been proposed based on the extensive research of Ritchie and coworkers (37), Figure 11. The importance of shielding to fatigue crack propagation, particularly in the near threshold regime, has been demonstrated repeatedly by studies for benign environments (10,11,34,35, 37). Mechanisms for most of the categories shown in Figure 11 have been advanced, at least qualitatively, and the similitude concept has been successfully modified based on measured effective stress intensity. The remaining challenge in this area centers on micromechanical modeling of the magnitude and loading history-geometry dependence of closure forces.

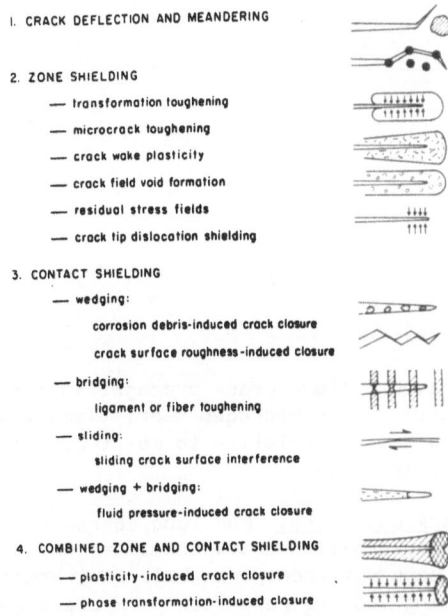

1. CRACK DEFLECTION AND MEANDERING

2. ZONE SHIELDING

    — transformation toughening

    — microcrack toughening

    — crack wake plasticity

    — crack field void formation

    — residual stress fields

    — crack tip dislocation shielding

3. CONTACT SHIELDING

    — wedging:

        corrosion debris-induced crack closure

        crack surface roughness-induced closure

    — bridging:

        ligament or fiber toughening

    — sliding:

        sliding crack surface interference

    — wedging + bridging:

        fluid pressure-induced crack closure

4. COMBINED ZONE AND CONTACT SHIELDING

    — plasticity-induced crack closure

    — phase transformation-induced closure

Figure 11. Mechanisms for fatigue crack tip shielding, emphasizing crack closure phenomena. After Ritchie et al. (37).

    In sharp contrast, chemically-based mechanisms for closure and shielding during the propagation of long corrosion fatigue cracks are undefined. Beneficial closure due to corrosion debris along crack surfaces has been understood for several years, particularly for steels in seawater (38). While laboratory cracks are easily arrested by calcium and magnesium containing phases, the magnitude of closure forces during complex loading, including random compressive cycles, is undefined (39). More generally, corrosion product wedging is not predictable quantitatively because of uncertainties regarding the compressive flow properties of the products and the unique chemistry within cracks. Ritchie provided experimental and analytical evidence for the importance of closure forces produced by viscous fluids within propagating fatigue cracks (40). The regimes of loading and geometry variables where this mechanism is important are predictable by analysis of the penetration and expulsion of fluid during a load cycle, combined with the hydrodynamic wedging action of the liquid in the crack.

Regarding long crack closure in corrosion fatigue, the critical issue is to identify important mechanisms based on environmental modification of the plasticity, surface roughness and crack tip process zone features summarized in Figure 11. For example, several studies have shown that near threshold fatigue in high strength steels exposed to gaseous hydrogen is slowed relative to moist air, while conventional hydrogen embrittlement is observed at higher stress intensities, Figure 12 (36,41). While several interpretations are possible as discussed in Section 4, hydrogen uptake may enhance either local plastic flow, mode II displacements or surface roughness to produce increased crack closure and reduced growth rates. In this case an effective stress intensity approach may provide for similitude in the engineering sense. Surely, environment could affect phase transformations, microcrack formation, voids and dislocation morphologies in the crack tip plastic zone, resulting in shielding. Future research is required to examine such phenomena.

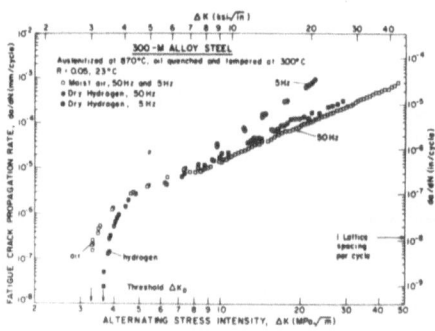

Figure 12. Gaseous hydrogen assisted fatigue crack propagation rates for ultra-high strength 300M alloy steel. Note hydrogen embrittlement at high ΔK contrasted with H_2 induced retardations relative to moist air in the near threshold regime. After Ritchie (36,41).

3.2.1.2. Localized Crack Chemistry. For long corrosion fatigue cracks, it is often argued that the localized crack chemistry which controls embrittlement, while perhaps different from bulk environment conditions, is constant with varying loading parameters and crack geometry once a steady state condition is achieved (29,30,42,43). In this case the chemical driving force is related to external environment activity and the stress intensity approach, modified for closure if necessary, provides an accurate localized driving force. While gaseous environments are likely constant with varying crack depth and treatable by K-based similitude, the situation is more complex for long cracks in electrolytes.

Discussion of crack chemistry effects in corrosion fatigue is complicated because of the variety of metal-environment systems and possible mechanisms of embrittlement. Two categories of behavior are important; electrochemical hydrogen production and embrittlement, or passive film formation, rupture and transient dissolution (9). For the former the corrosion fatigue component of crack growth rate is assumed to depend, often linearly, on the concentration of adsorbed and dissolved surface hydrogen. For the latter da/dN is modeled in terms of the amount

of Faradaic dissolution which occurs between film ruptures as controlled
by crack tip strain rate. Modeling of mass transport to the crack tip by
diffusion, convection and ion migration coupled with crack tip transient
reactions is required to develop the chemical driving force for either
embrittlement scenario.

A notable development since the Corsica meeting has been the
emphasis on modeling of crack solution chemistry, as discussed in detail
by Turnbull in this volume and by Pourbaix (44) at the previous NATO
workshop. The challenge from the corrosion fatigue perspective is to
relate crack chemistry predictions to brittle crack advance and to test
resulting models of the chemical-mechanical driving force. Crack
chemistry modeling has been limited to ferritic and martensitic steels in
room temperature chloride electrolytes which produce hydrogen.
Understanding of this system is relevant to stainless steels and aluminum
alloys in chlorides and perhaps to film rupture systems such as austenitic
stainless steels in high temperature dilute electrolytes (45).

For iron based alloys in chloride, crack chemistry was first
probed by Brown and coworkers (46). Figure 13 summarizes the specific
electrochemical reactions and the components of the transport-reaction
analysis (47-52). Within the crack, metal dissolution, cation hydrolysis

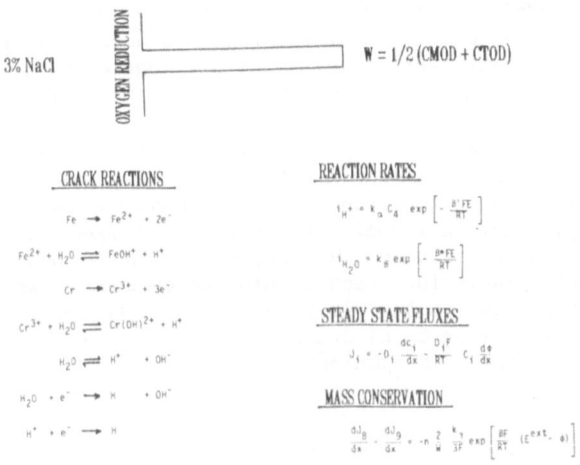

Figure 13. The electrochemical reactions and mass transport-reaction
equations which govern the chemistry within occluded cracks in active
steels exposed to aqueous chloride. After Turnbull (47-52).

and cathodic oxygen, hydrogen ion or water reduction reactions are
possible, with rates defined by polarization experiments. Crack opening
shape is given by fracture mechanics relations. Ionic concentrations,
particularly pH, and electrode potential gradients are predicted by
solutions to the coupled diffusion-convection and crack surface reaction
problem. Total rates of cathodic hydrogen production and anodic
dissolution at the crack tip are computed, and related to hydrogen uptake
and rates of brittle fracture. The issue here is to predict the extent to
which the concentration of adsorbed hydrogen varies with crack depth and
cyclic opening, factors which control the ease of mass transport and the
active surface available for chemical reaction.

The first step in understanding crack chemistry is, by necessity,

626

analysis of the static crack problem without the complicating effect of convective mixing for fatigue (47). For active steel in chloride, Turnbull demonstrated that the dominant source of embrittling hydrogen depends on applied electrode potential (53). Near the free corrosion potential, the crack tip acidifies due to iron or chromium dissolution and cation hydrolysis. The tip potential is 50 to 75 mV more active relative to external specimen surfaces. Since low pH and cathodic potential promote hydrogen ion reduction, the crack tip is the dominant source of adsorbed hydrogen, with the effect intensified by strain. In contrast for cathodic potentials below about -850 mV (SCE), the crack solution becomes alkaline and the tip potential is slightly more noble than external surface potentials. Water reduction is the dominant source of hydrogen, with this reaction increased in rate by up to 500-fold due to mechanical cleaning of the crack surface. As such, both the crack tip and the external surfaces are probable sources of embrittling hydrogen. Clearly, the chemical driving force depends on the source of embrittling hydrogen. For external surface control, time dependent crack growth will be controlled by the geometry of bulk hydrogen diffusion to the crack tip plastic zone.

A recent study by Gangloff and Turnbull coupled crack chemistry and static load brittle fracture for steel in NaCl at potentials near free corrosion where crack tip hydrogen dominates embrittlement (54). Detailed mass transport modeling, Figure 13, shows that the rate of hydrogen ion reduction and the concentration of adsorbed hydrogen (C_o) increases with decreasing static crack length for the single edge crack geometry at constant K. A relationship between the applied threshold stress intensity for embrittlement (K_{th}) and C_o was derived based on experimental data for gaseous hydrogen, where crack tip chemistry is easily related to bulk environment pressures. The model was confirmed by experimental measurements of K_{th} as a function of crack depth, as summarized in Figure 14. Critically, fracture is not described by a single material property, K_{th}, as required by similitude. Rather, K_{th} decreases with decreasing crack size, particularly for crack lengths below about 5 mm. The geometry dependence in this case is the result of the interaction between the crack tip stress field, defined by K, and crack chemistry defined by a complex function of crack depth and opening.

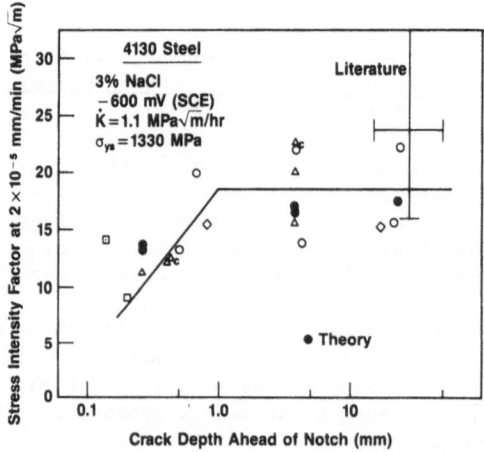

Figure 14. The effect of crack depth on K_{th} for steel in 3% NaCl, predicted based on transport-reaction modeling summarized in Figure 13 and compared to experimental measurements of static load, brittle crack growth (54).

Modeling of occluded chemistry for oscillating corrosion fatigue cracks is complicated by convective mixing contributions to mass transport, as reviewed by Turnbull herein and elsewhere (49). An initial approach to this problem was based on the concept of perfect mixing wherein fresh solution is drawn into the crack with perfect efficiency during the early stage of a single load cycle (50,55). Chemical reaction depletes or enriches species in the crack during the remainder of the cycle and a portion of the liquid is expelled efficiently upon unloading. The process repeats cyclically. This approach, which ignores diffusion to simplify the mathematical analysis, was developed for several assumed kinetics laws without specifying chemical or electrochemical reactions and subsequent embrittlement (55). More recently, Turnbull and coworkers analyzed the hydrodynamics of oxygen supply by convective mixing in conjunction with diffusion, and depletion by cathodic reduction within cracks for active steels in aerated aqueous chloride (48,49,51,52).

Both perfect mixing and advection modeling demonstrate that the concentration of reactants and products within an occluded fatigue crack depends on crack depth and opening, loading frequency, applied stress intensity range, mean stress and reaction rate parameters. Typical model predictions are shown in Figure 15 for the steel-aerated chloride system.

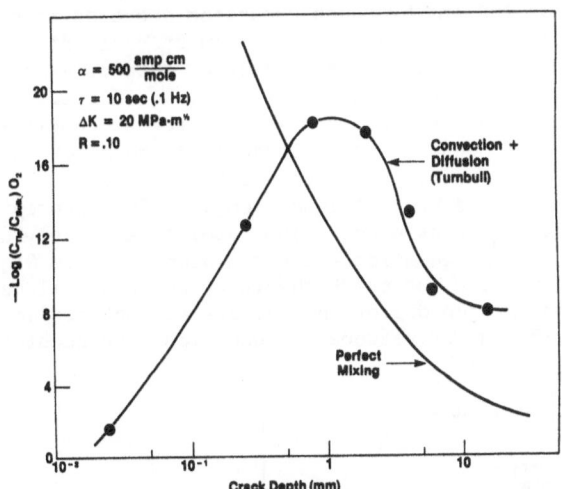

Figure 15. Predictions of the dissolved oxygen content near the corrosion fatigue crack tip at maximum load for steel in aerated chloride, based on perfect mixing and detailed hydrodynamic advection modeling (48,50-52,56).

The perfect mixing analysis predicts that the concentration of dissolved oxygen, present at the crack tip at maximum load in the cycle, decreases with decreasing crack depth. The detailed hydrodynamic model predicts a more complex behavior, but none-the-less, a strong crack geometry dependence of concentration. For deep cracks, convective mixing dominates and becomes more efficient with increasing crack depth to replenish cathodically consumed oxygen. For very shallow cracks, oxygen supply is dominated by diffusion which is enhanced for decreased crack depths. The net result of the diffusion and convection processes is the strong minimum in oxygen content. The perfect mixing analysis provides an upper bound on oxygen concentration for convection dominated deep cracks, but does not describe the diffusion regime. The characteristic crack length for the minimum in Figure 15 depends on oxygen reduction rate, loading frequency, stress intensity range and mean crack opening. While the modeling summarized in Figure 15 does not consider brittle cracking, it is clear that, if oxygen is involved in chemical corrosion fatigue damage, then the total driving force will not be defined solely by stress intensity. Simply stated, the crack chemistry depicted in Figure 15 is not constant with varying crack geometry, even within the long crack regime where fracture mechanics similitude is often invoked.

Two challenges are clear based on Figure 15. Firstly, transport and reaction modeling must be applied to a broader range of material and environment systems. Secondly, such models must be coupled with micromechanical descriptions of crack tip embrittlement to adequately predict corrosion fatigue.

For steel in aerated 3% NaCl, Gangloff integrated the reaction concepts in Figure 13 with a perfect mixing analysis of dissolved oxygen depletion, Figure 15, to develop a total driving force for corrosion fatigue by hydrogen embrittlement (56). It was argued that corrosion fatigue crack speed is proportional to the concentration of adsorbed hydrogen which depends linearly on solution pH. When the perfect mixing and reaction process favors O_2 depletion, for example at short crack depths, then pH is low and hydrogen reduction proceeds to promote brittle crack growth. Critically, the concentration of reactants or products, O_2 and by stoichiometry adsorbed H in this case, depends exponentially on the ratio of the area of crack surface participating in reaction to the volume of crack solution. For an idealized wedge crack at constant stress intensity, this ratio equals the reciprocal of the crack mouth opening displacement, a geometrical parameter which is proportional to stress intensity and the square root of crack length. This concept correlated the corrosion fatigue crack propagation rates of a variety of crack sizes as shown by the derived equation and the measurements in Figure 16. While indicating the direction for crack chemistry and embrittlement modeling, the perfect mixing oxygen depletion analysis was not confirmed by either experimentation (57) or by reasonable reduction rate constants (48,51,52).

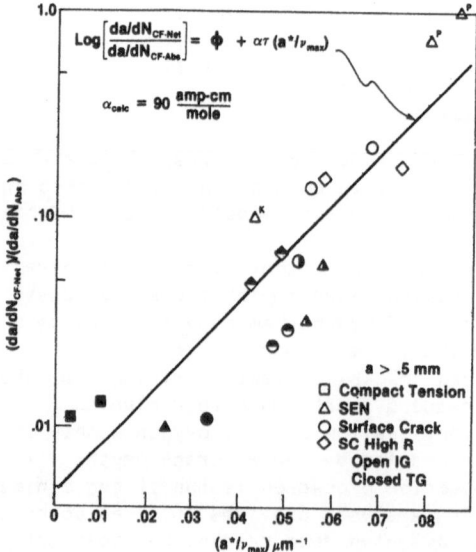

Figure 16. The influence of reciprocal crack mouth opening displacement on corrosion fatigue crack growth rate, normalized to account for ΔK differences, for high strength 4130 steel in aqueous NaCl (56).

These latter objections are clouded by the extent to which crack closure produces local turbulence and nearly perfect convective mixing.

If K-based similitude wholly describes corrosion fatigue, then the normalized crack growth rates in Figure 16 should equal 1.0 for all crack mouth openings, established by applied stress and crack depth. Such was not the case; the regimes where similitude breaks down are indicated. For the long crack case, crack mouth opening displacements are large and nearly constant with varying crack length. As such, crack chemistry changes are small and K similitude should approximately describe the driving force for corrosion fatigue of steels in seawater. In sharp contrast Figures 15 and 16 show that crack opening geometry, the ratio of crack surface area to occluded solution volume, and crack chemistry vary significantly with geometry and loading for the physically short crack regime, below about 5 mm. While the precise bounds of such behavior depend on chemical reaction kinetics and loading frequency, the potential for unpredictable short crack growth and small crack linkup is clear.

3.2.2. Short Crack Growth

The so-called "small crack problem" is based on the often demonstrated result that short and small fatigue cracks grow at unpredictably high rates and below threshold conditions inferred from long crack data with K-based similitude. This behavior has been researched extensively since the last NATO workshop and significant understanding has emerged, as summarized in the proceedings of a recent international conference (11). The importance of small crack propagation phenomena is graphically portrayed in Figure 10 where 80% of the cyclic life of a component is controlled by the growth of cracks smaller than 1 mm. The classes of crack size effects are summarized in Table 2 after Ritchie and Lankford (58). It has become accepted practice to refer to a chemically, physically or mechanically small crack, sized below 1 to 5 mm, as "short" if the crack intersects many grains along a wide crack perimeter. Microstructurally "small" cracks, totally encased in one or two grains and dimensionally limited in all directions, are discussed in Section 3.2.3.

TABLE 2. Classes of small fatigue cracks after Ritchie and Lankford (58)

Type of small crack	Dimension	Responsible mechanism	Potential solution
Mechanically small	$a \lesssim r_y$ [a]	Excessive (active) plasticity	Use of ΔJ, ΔS or crack-tip-opening displacement
Microstructurally small	$a \lesssim d_g$ [b]	Crack tip shielding, enhanced $\Delta \epsilon_p$, crack shape	Probabilistic approach
Physically small	$a \lesssim 1$ mm	Crack tip shielding (crack closure)	Use of ΔK_{eff}
Chemically small	Up to about 10 mm [c]	Local crack tip environment	

[a] r_y is the plastic zone size or plastic field of notch.
[b] d_g is the critical microstructural dimension, e.g. grain size, a is the crack depth and $2c$ the surface length.
[c] Critical size is a function of frequency and reaction kinetics.

Crack size effects originate with improper formulations of local crack tip mechanical and chemical driving forces. For short cracks, rapid rates are traceable to three basic causes, Table 2. For benign environments, large scale crack tip plasticity and underdeveloped crack wake closure (or shielding) are well established mechanisms which enhance growth rates. For corrosion fatigue, the occluded environments within short cracks differ from both the bulk chemistry and the conditions within long cracks and, for the cases examined to date, are unpredictably aggressive. These three mechanisms may interact to produce complex, short crack, corrosion fatigue behavior which is not simply predicted based on applied stress intensity.

Small crack-environment interactions relevant to corrosion fatigue were recently reviewed (36,59-61). Work exclusively emphasized steel and aluminum alloys in vacuum or moist air, and steels in aqueous electrolytes. Recent experimental data and causal mechanisms are surveyed.

Experiments which isolate fatigue crack size-environment interactions are complex, a factor which has limited progress to date. The first small crack studies replicated the surfaces of smooth specimens at selected percentages of total life to generate cyclic crack length data. While effective for very small crack sizes, this technique is tedious, requiring that loading and environmental exposure be interrupted, and as such has not been employed for monitoring corrosion fatigue cracking. Similarly, high resolution interrupted or insitu scanning electron microscopy of surface crack length and mouth opening is effective, but has not been applied to aggressive environments (62). Microscopic monitoring only yields surface crack length information. Electrical potential monitoring is very effective for short, through-thickness or surface cracks as small as 0.05 mm, provided that the initiation site is known (63). Crack area and depth information is obtainable and the method is applicable to aggressive gas and liquid environments. Crack closure forces can be inferred from potentials for vacuum and some liquids, however, interpretations are unclear and the method is questionable (64). Short crack closure forces have been accurately measured by sensitive displacement gauges (64), by microscopic methods (62) and by laser interferometry (65), however, such techniques have not been employed for complex and controlled environments. Potential drop and compliance techniques enable constant applied stress intensity experiments with short cracks to effectively isolate crack size effects. New experimental techniques are required to characterize short or small crack-environment interactions.

The rapid growth of short cracks in gases is understood based on crack closure concepts, however, the precise role of environment in establishing the magnitude and crack geometry dependence of closure forces is unclear. This problem was reviewed by Gangloff and Ritchie (36) and examined experimentally by Petit and coworkers (62). Results for high strength 4340 steel demonstrated that through-thickness edge cracks between 0.1 and 2 mm long grew up to five-fold faster compared to 20 to 50 mm long compact tension cracks, but only for hydrogen gas and at low stress ratios (36). Crack size had no effect on crack growth rate for helium or moist air environments, or for H_2 at high mean stress. Analysis indicated that hydrogen embrittlement occurred for all crack sizes, but significant closure developed for long cracks in hydrogen. The precise mechanism for H_2 enhanced closure is uncertain, however, the rougher intergranular surface and the possibility of enhanced plasticity may be factors. Recent data for lower strength 1020 steel show that long and short crack growth rates are correlated based on effective stress intensity for both moist air and vacuum (62). Closure levels were significant for air and the role of oxide induced shielding is likely.

Similar results on crack size effects in 7075 aluminum alloys have been reported for moist air, vacuum and nitrogen (62), however, the precise influence of the environment on crack closure forces has not been established.

Since gaseous corrosion fatigue crack growth rates are controlled by the slow step in a transport and reaction sequence, collisions between gas molecules and crack walls lower crack tip pressure and possibly decrease embrittlement. Impeded flow occurs when the molecular mean free path exceeds some fraction of the crack opening displacement; in this regime crack tip gas pressure will depend on crack depth, opening and on loading frequency. For cases where surface reactions are rapid, impeded transport will affect brittle crack growth rates, with the small crack less likely to have reduced environment pressures. This crack geometry sensitive process has not been considered systematically.

A significant data base demonstrates that short corrosion fatigue cracks (< 5 mm) in steels, exposed to aqueous chloride, grow unpredictably fast compared to long crack da/dN–ΔK data (36,59). The crack geometry effect is often not observed for a benign reference environment, suggesting an electrochemical origin. A survey of twelve investigations showed that the ratio of small to long crack growth rates is as high as 500 for high strength steels in 3% NaCl and decreases with decreasing yield strength to between 6 and 1.2 for HY130-type and low strength carbon steels, respectively (59). Limiting crack sizes for geometry independent, K-based growth vary between 1 and 10 mm, with precise values being difficult to define and dependent on mechanical variables such as loading frequency and chemical variables such as electrode potential. Typical data are shown in Figures 17 and 18 for limiting high and low strength steel (66,67). For the former, crack growth in vacuum and air is well defined by ΔK, independent of crack length between 0.1 and 50 mm because crack closure and plasticity effects are minimal. In contrast a wide

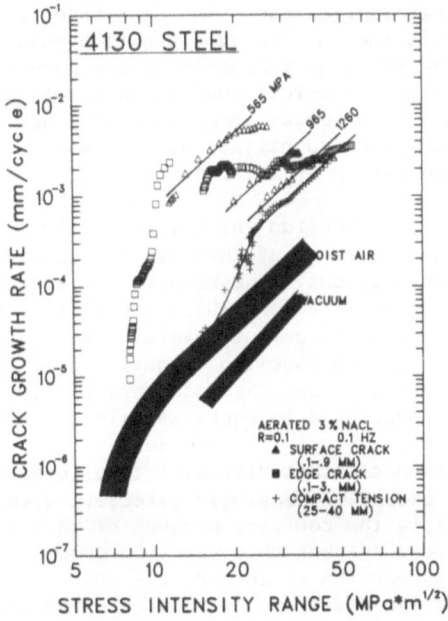

Figure 17. Crack growth rate-K results for variably sized corrosion fatigue cracks in high strength 4130 steel. Applied stress range is listed for surface cracks (66).

range of corrosion fatigue crack growth rates are observed for short edge
or surface elliptical and long compact tension cracks in NaCl. Stress
intensity does not correlate these data. Similar results were reported by
Saxena et al. for a high strength 18Mn-5Cr steel stressed in moist
hydrogen gas (68). For low strength API-2H steel, the crack size effect
is small and best revealed by constant stress intensity experiments. Note
that growth rate declines with increasing crack length, but only for
chloride, Figure 18.

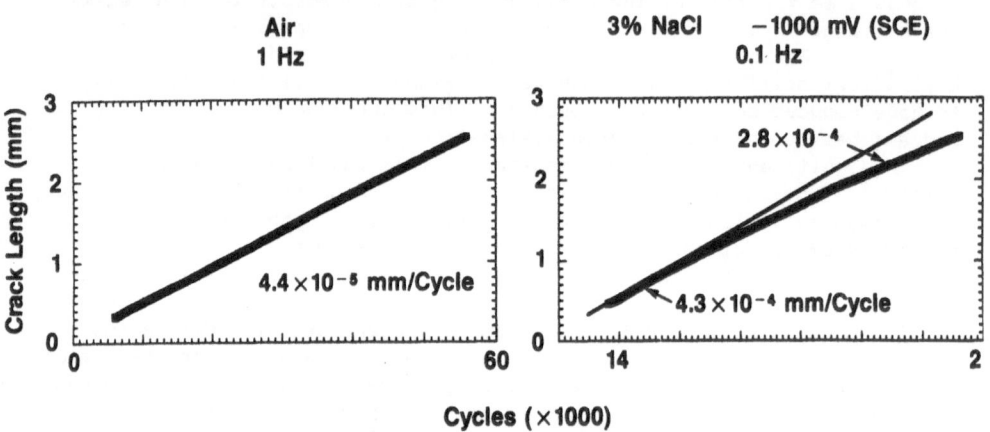

Figure 18. Cyclic short, through crack length data at constant applied
stress intensity (23 MPa·m$^{1/2}$, R=0.10) for low strength API-2H steel in
benign and corrosion fatigue environments (67).

     The importance of short crack phenomena in corrosion fatigue depends
on the strength level of the steel, on electrochemical and loading
variables and on the specific structural application. The magnitude of
the rates shown in Figure 17, the result that accelerated growth persists
to crack depths between 5 and 10 mm and the high incidence of breakdowns
in K-similitude, at least for steel in electrolytes, suggest that this
aspect of the driving force problem may be of equal importance to crack
closure. While corrective measures cannot be quantitatively stated,
directions have been indicated for formulation of the correct local
chemical and mechanical driving force.
     The origin of short crack effects in corrosion fatigue is surely
chemical and best understood within the framework of the mass transport-
reaction modeling discussed in previous sections. The results for active
steels in saline environments suggest several factors which are likely to
result in uniquely embrittling short crack environment chemistries.
     For static cracks, diffusion and reaction modeling (Figure 14 and
Reference 54) shows that geometry effects on local chemistry are subtle.
For steel in chloride, crack tip pH is predicted to decrease with
decreasing crack length, in all cases being equal or below bulk
environment levels. While intuition suggests that diffusion to eliminate
H$^+$ ion concentration gradients should become increasingly effective with
decreasing crack size, the model predicts the contrary because of the
dominant role of metal cations. For long cracks, this reaction product
accumulates and stifles the hydrolysis reaction which controls pH.
Clearly, detailed modeling is required for a variety of metal-environment
systems in order to generalize on crack chemistry, as suggested by
Turnbull and Newman (60).

Secondly, from the perfect mixing perspective for fatigue cracking, short cracks are characterized by extremely high crack surface area to occluded solution volume ratios. A high surface to volume ratio promotes rapid depletion of either embrittling or poisoning species, and the converse effect on the product(s) of electrochemical reaction. The precise role of this factor depends on material, environment and the embrittlement mechanism. For example, the diverse crack growth rate data in Figure 17 were correlated with the reciprocal mouth opening estimate of the area to volume ratio for constant stress intensity, Figure 16. Here, the depletion of dissolved oxygen was hypothesized but not proven (56,66). The surface area to volume ratio decreases dramatically with increasing short crack size at constant applied stress intensity, but is nearly constant with varying long crack geometry as shown in Figure 16. This dependence suggests that K-similitude will only approximately describe driving force for the longer crack, constant crack chemistry case.

The complete analysis of convective mixing, diffusion and electrochemical reaction highlights two additional characteristics of the short crack. Figure 15 indicates that convective mixing is not an important mode of mass transport for the short crack. Rather, crack chemistry is controlled by diffusion and ion migration. Two exceptions exist. The balance between diffusion and convection depends on crack opening shape, loading frequency and reaction kinetics. The bound between "short" and "long" crack chemistries and embrittlement varies and is blurred. As a generalization, Turnbull and Newman conclude that solution chemistry will be most sensitive to geometry for short crack depths below about 3 mm. Secondly, crack closure may cause local turbulence. In this case perfect mixing may better describe mass transport and the area to volume ratio concept will control local chemistry.

3.2.3   Small Crack Linkup and Growth

Studies of fatigue in benign environments clearly establish the unique growth kinetics of small (or micro-) cracks, of limited depth and surface length and encased in a few grains (58,61,69). The extensive work of Lankford and Davidson shows that such cracks grow at disturbingly high rates and below long crack threshold conditions because of high crack tip opening and associated high plastic strain. Crack wake closure and mode II displacements are initially small and develop with increasing crack size to retard subsequent growth rates. Small cracks also tend to arrest at grain boundaries. K-based similitude is of little use for such cases. These observations are generally relevant to a variety of materials, particularly aluminum alloys (11), however, analytical expressions for the governing crack tip driving force have not been developed.

Quantitative studies of environmental effects on the linkup and growth of small cracks have not been conducted, particularly for aqueous solutions. This deficiency is due to experimental difficulties with insitu monitoring of small cracks in aggressive environments and to the multiplicity of chemical (mass transport and reaction) and mechanical (closure, deflection and plasticity) processes which constitute the crack tip driving force. These complexities are illustrated by data on a 7075-T651 aluminum alloy where small cracks were produced in either purified nitrogen or moist air and monitored by interrupted replication or direct scanning electron microscopy (70). Equal, rapid small crack growth rates were observed for each environment, however, different deformation and fracture processes were operative. For air, cracking was slowed by Mode I blunting, but probably accelerated by adsorbed hydrogen embrittlement. In N_2, hydrogen embrittlement was not possible, but crack tips remained sharp

634

with mode II opening and growth by in plane crack tip shear.

Future studies of small crack-aqueous environment interactions must detail the local activity of the crack environment for either hydrogen production and uptake, or for anodic dissolution and passivation. The environmental effect on crack tip deformation and resulting in-plane and opening modes of crack extension must be assessed. As discussed in this volume by Sieradski and by Lynch at the previous NATO workshop (71), adsorbates such as hydrogen and films can affect crack tip deformation and fracture. Environment sensitive closure can also affect the crack tip driving force for this regime of corrosion fatigue (71). The methods of transport-reaction modeling and crack tip monitoring employed for short cracks and benign environments, respectively, should be useful.

For example, unique small crack-environment interactions are possible in metal-environment systems which are embrittled by the film rupture mechanism (72). This view, while not supported by specific crack growth data, is plausible because of the critical role of crack tip plasticity in both film rupture and small crack processes. Specifically, exceptionally high plastic strains and strain rates were measured at the tips of microstructurally small cracks in stainless steels and aluminum alloys (61,62,69,70,72). The slip dissolution/film rupture model equates corrosion fatigue crack growth rate to the transient amount of crack tip metal dissolved between film ruptures and predicts that da/dN is proportional to the frequency of ruptures and hence crack tip strain rate (72). The likelihood for enhanced strain and film rupture at the tips of microstructurally small cracks is high. Additionally, mass transport of reactants and products will probably be crack size and opening sensitive, with the resulting crack solution composition affecting the rate of near tip repassivation. These considerations are particularly relevant to stainless and carbon steels in high temperature water environments, as illustrated by the predictions of film rupture modeling in Figure 19 (72).

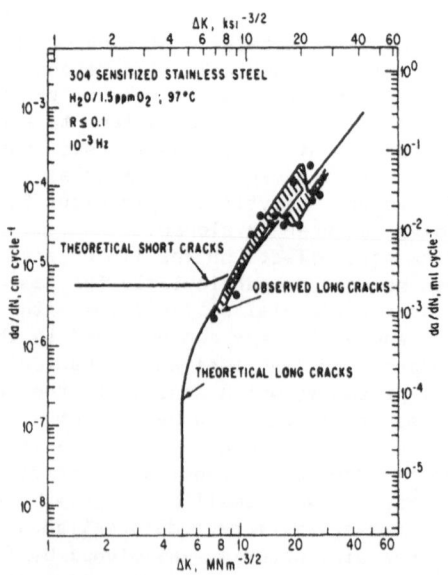

Figure 19. The predicted effect of crack size on corrosion fatigue crack propagation by film rupture for stainless steel in high temperature water. Data were only obtained for long cracks. After Ford and Hudak (72).

3.3. Additional Issues in Corrosion Fatigue Crack Propagation

While formulation of the correct chemical-mechanical crack tip
driving force, unifying the regimes of corrosion fatigue, is a critical
issue confronting the field, several additional recent advances and
uncertainties are important.

3.3.1. Crack Kinetics Predictions.

The fracture mechanics approach is well suited for reaction kinetics-
based, mechanistic studies of corrosion fatigue crack propagation (30,73).
Such research is aimed at the problem of predicting low frequency, long
term cracking behavior based on faster frequency, laboratory
experimentation and a fundamental model. The ferritic steel-aqueous
chloride system has been investigated in this regard, as represented by
the experimental work of Vosikovsky and Procter (74,75).

Wei and coworkers model the frequency effect in aqueous corrosion
fatigue with an approach derived from earlier work on steel and aluminum
alloys in gaseous, hydrogen producing environments (73,76). The corrosion
fatigue crack growth rate component is taken as proportional to the
fractional extent of surface electrochemical reaction during the rising
load portion of the fatigue cycle, a quantity which is related to the
concentration of adsorbed hydrogen produced by cathodic reaction within
the crack. Critically, the relevant crack tip reactions are unique to
surfaces cleaned by straining and depend exponentially on time, analogous
to gas phase kinetics. Wei argues that the reaction rate constant which
describes film formation and hydrogen production is determined by
measuring the current decay which results when the metal is fractured in
simulated crack solution electrolyte and when coupled to oxidized metal to
simulate the crack flanks at a mixed potential. Charge is calculated as a
function of time by integration of the current transient and one or more
rate constants are determined. The predictions of this approach are shown
in Figure 20 in conjunction with the measured frequency response for a
variety of steels in electrolytes. At intermediate frequencies, growth
rate increases with decreasing frequency, as predicted by the model and
incorporating two values of the reaction rate parameter which are typical
of electrochemical measurements. Secondly, a saturation crack growth rate
exists where further decreases in frequency have no degrading influence on

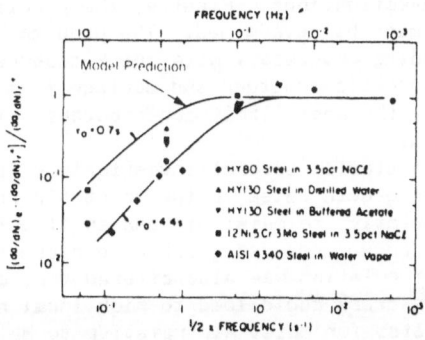

Figure 20. The predicted and measured effects of cyclic loading frequency
on corrosion fatigue crack growth rate for ferritic steels exposed to
aqueous chloride. After Wei et al. (76).

corrosion fatigue rates. Here, da/dN does not increase significantly because of the exponential decay of the rate of filming. That is, little additional hydrogen is produced for increasing loading times beyond the saturation level and corresponding to the long time "tail" of the reaction. The analysis involves one adjustable parameter which fixes the onset of the low frequency plateau.

The agreement between reaction rate theory and long crack fracture mechanics experiment, Figure 20, is encouraging. Additional work is, however, required. The current which is measured in the fracturing electrode simulation is a net current which reflects the difference between local anodic and cathodic rates. The relationship between this current and hydrogen production is not apparent. At the crack tip, hydrogen production is controlled by the rate of the cathodic reaction(s) and the composition of the crack solution. The fracturing electrode experiment must be conducted in simulated crack solution to include the effect of occluded mass transport on local reaction. To date, this complication has not been examined. The comparison in Figure 20 applies to high stress intensity corrosion fatigue. The reaction rate model does not implicitly define the effect of K on the kinetics of corrosion fatigue. None-the-less, this approach provides an important recent advance on the problem of long term corrosion fatigue prediction.

3.3.2. Near Threshold Corrosion Fatigue

Environmental effects on low growth rate, near threshold fatigue crack propagation have not been extensively examined despite the significant amount of work directed at benign environment threshold phenomena (34). The basic problem which hinders advances in this regime where, cycle-dependent corrosion fatigue occurs, is the excessive time required to measure low crack growth rates at the low cyclic frequencies which promote chemical effects and by standard fracture mechanics methods. The emphasis on physically short cracks has provided a solution to this deficiency. High resolution electrical potential monitoring enables characterization of crack growth rates at constant applied ΔK over increments of crack growth between 50 and 250 μm. With this method, data of the type shown in Figure 18 can be generated for rates below 10^{-6} mm/cycle at frequencies between 0.5 and 1 Hz, or alternately at rates on the order of 10^{-4} mm/cycle at frequencies as low as 10^{-3} Hz (59,77).

Two aspects of near threshold environmental effects should be examined in future investigations. Firstly, the precise mechanisms for environmental damage must be determined. The problem is complicated by the variety of contending processes; with crack closure, hydrogen embrittlement, film formation/rupture and surface film or hydrogen effects on cyclic deformation; the most likely contributors. This later factor warrants consideration.

For high strength steels, several investigators reported that gaseous hydrogen reduced crack growth rates in the threshold regime, but enhanced cracking at high ΔK relative to moist air and as illustrated in Figure 12 (36,41,78). While hydrogen enhanced crack wake closure was cited as important (41), the speculation was also offered that condensation of water vapor within the crack could lead to high local hydrogen activity and enhanced growth rates for moist air relative to H_2 (78). A recent study suggests a third and perhaps controlling mechanism (79). As summarized in Figure 21, crack growth in moist air, pure oxygen, moist oxygen and pure water vapor is equally rapid compared to vacuum for a moderate strength steel cycled at low stress intensity and 5 Hz. Gaseous hydrogen had an intermediate effect on da/dN, with growth rates increasing with decreasing frequency. Crack closure, inferred from compliance measurements, did not explain the trends in Figure 21. Rather, two mechanisms are suggested. Gaseous hydrogen embrittlement is clearly indicated. Critically, hydrogen from water vapor-steel oxidation reactions is not important because equal, fast crack growth rates are observed for the moist environments and pure oxygen. Several additional experiments confirmed that pure oxygen causes rapid crack growth. These data strongly suggest that surface films and perhaps the associated slip irreversibility play a key role in near threshold corrosion fatigue (80). Future research is required.

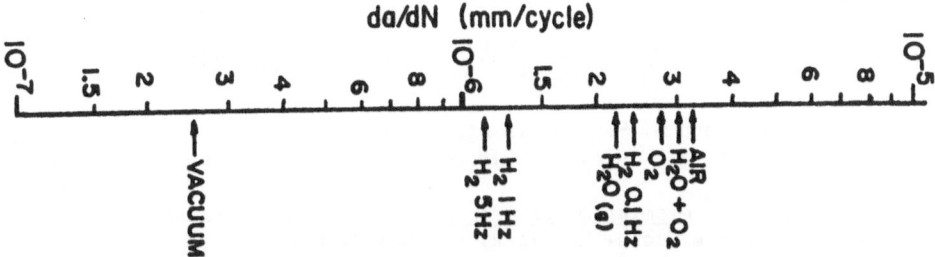

Figure 21. Corrosion fatigue crack growth rates in A537 steel exposed to several gaseous environments at constant applied stress intensity (8.2 MPa·m$^{1/2}$) at 5 Hz. Air: 100 kPa, O_2: 20 kPa, H_2: 50 kPa, H_2O: 80 Pa, Vacuum: 1 μPa. After Heubaum (79).

A second important issue is the integration of environmental effects on smooth specimen fatigue life and near threshold fatigue crack propagation. Uhlig and coworkers demonstrated that the cyclic life of the former type of specimen is significantly reduced by exposure to aqueous chloride, particularly at applied stresses below the inert environment endurance limit. (81-83). Electrochemical studies further showed that corrosion fatigue damage was produced only when a critical anodic current density was exceeded, due to either solution aeration or potentio-statically applied currents. In sharp contrast fracture mechanics studies of steels at similar strength levels show that the aqueous environment has little or no influence on fatigue crack propagation rates in the near threshold regime and for the mean stress and high frequency (30 Hz) conditions of the smooth bar specimens (33). Corrosion pitting, while a possible explanation for smooth specimen fatigue life reductions without environmental influences on crack growth, was not associated with crack initiation (83).

The likely rationale for environmental effects on S-N life is chemically stimulated, cyclic plastic deformation, as surveyed in Section 2. Precise mechanisms for environmental effects on pre-crack deformation and initiation are not available for complex alloys steels. Further, the reason for the lack of a damaging effect of environment on deformation and fracture at the crack tip in a fracture mechanics specimen is unclear. While localized chemistry may vary, modeling suggests that damaging anodic dissolution, implicated in the S-N studies, should occur at crack tips.

4. CLOSURE

Despite significant advances in corrosion fatigue over the past ten years, the most fundamental of problems remain unresolved. There is, however, hope in the direction which has been established by recent research. The stages of corrosion fatigue must be isolated. Chemical

638

effects on pre-crack cyclic deformation and crack initiation must be
determined for engineering alloys. Work on small or micro-crack linkup
must be extended to include the effects of aggressive environments.
Fracture mechanics concepts must be supplemented with mass transport and
reaction kinetics to quantitatively describe complex crack tip shielding
and crack geometry effects in corrosion fatigue. Modeling must integrate
the regimes of fatigue damage and provide methods for scaling short term
laboratory data to predict long term component performance.

REFERENCES

1. R.N. Parkins, in Atomistics of Fracture, Eds., R.M. Latanision and
    J.R. Pickens, Plenum Press, New York, p. 969 (1983).

2. R.O. Ritchie and S. Suresh, in Atomistics of Fracture, Eds., R.M.
    Latanision and J.R. Pickens, Plenum Press, New York, p. 835 (1983).

3. A.S. Argon, in Atomistics of Fracture, Eds., R.M. Latanision and J.R.
    Pickens, Plenum Press, New York, p. 1031 (1983).

4. E.N. Pugh, in Atomistics of Fracture, Eds., R.M. Latanision and J.R.
    Pickens, Plenum Press, New York, p. 1025 (1983).

5. Corrosion Fatigue, Chemistry, Mechanics and Microstructure, Eds.,
    A.J. McEvily and R.W. Staehle, NACE, Houston, TX (1971).

6. Corrosion Fatigue Technology, ASTM STP 642, Eds., H.L. Craig, Jr.,
    T.W. Crooker and D.W. Hoeppner, ASTM, Philadelphia, PA (1978).

7. Proceedings, Conference on the Influence of Environment on Fatigue,
    Inst. Mech. Engr., London (1978).

8. Corrosion Fatigue: Mechanics, Metallurgy, Electrochemistry and
    Engineering, ASTM STP 801, Eds., T.W. Crooker and B.N. Leis, ASTM,
    Philadelphia, PA (1984).

9. Embrittlement by the Localized Crack Environment, Ed., R.P. Gangloff,
    TMS-AIME, Warrendale, PA (1984).

10. Basic Questions in Fatigue, Vols. I and II, ASTM STP, Eds., R.P. Wei,
    J.T. Fong, R.J. Fields and R.P. Gangloff, ASTM, Philadelphia, PA, in
    press (1986).

11. Small Fatigue Cracks, Eds., R.O. Ritchie and J. Lankford, TMS-AIME,
    Warrendale, PA (1986).

12. P. J. E. Forsyth and C. A. Stubbington, J. Inst. Metals, Vol. 83,
    p. 395 (1954).

13. P. J. E. Forsyth and C. A. Stubbington, Acta Metall., Vol. 8, p. 811
    (1960).

14. D. J. Duquette and P. R. Swann, Acta Metall., Vol. 24, p. 241 (1976).

15. P. J. E. Forsyth, Proc. Roy. Soc., Vol. A242, p. 198 (1958).

16. D. J. Duquette, Ph.D. Dissertation, Massachusetts Institute of Technology (1968).

17. C. Laird, Ph.D. Thesis, University of Cambridge (1963).

18. C. Garcia, Ph.D. Thesis, Rensselaer Polytechnic Institute (1981).

19. H. Vehoff, in Basic Questions in Fatigue, Vols. I and II, ASTM STP, Eds., R.P. Wei, J.T. Fong, R.J. Fields and R.P. Gangloff, ASTM, Philadelphia, PA, in press (1986).

20. C. Garcia and D. J. Duquette, in Corrosion of Nickel Based Alloys, Ed., R.C. Scarberry, ASM, Metals Park, p. 127 (1985).

21. Yan Ben-da, Ph.D. Thesis, University of Pennsylvania (1984).

22. S. R. Ortner, Ph.D. Thesis, University of Pennsylvania (1985).

23. H. N. Hahn and D. J. Duquette, Acta Metall., Vol. 26, p. 279 (1978).

24. T. Magnin and L. Coudreuse, in Fatigue 84, Ed., C. J. Beevers, Vol. 3, EMAS, United Kingdom, p. 1447 (1984).

25. T. Magnin, L. Coudreuse and J. M. Lardon, in Basic Questions in Fatigue, Vols. I and II, ASTM STP, Eds., R.P. Wei, J.T. Fong, R.J. Fields and R.P. Gangloff, ASTM, Philadelphia, PA, in press (1986).

26. T. Magnin and L. Coudreuse, Matls. Sci. and Engr., Vol. 72, p. 125 (1985).

27. T. Magnin and J.M. Lardon, Matls. Sci. and Engr., Vol. 76, p. L7 (1985).

28. T. Magnin, C. Dubessy and P. Rieux, in Aluminum Alloys – Physical and Mechanical Properties, Eds., E.A. Starke, Jr. and T.H. Sanders, Jr., EMAS, United Kingdom, p. 1177 (1986).

29. A.J. McEvily and R.P.Wei, in Corrosion Fatigue: Chemistry, Mechanics and Microstructure, Eds., O. Deveraux, A.J. McEvily and R.W. Staehle, NACE, Houston, TX, p. 381 (1973).

30. R.P. Wei and G. Shim, in Corrosion Fatigue: Mechanics, Metallurgy, Electrochemistry and Engineering, ASTM STP 801, Eds., T.W. Crooker and B.N. Leis, ASTM, Philadelphia, PA, p. 5 (1983).

31. J.A. Feeney, J. C. McMillan and R.P. Wei, Met. Trans., Vol. 1, p. 1741 (1970).

32. O. Vosikovsky and R.J. Cooke, Int. J. Pres. Ves. & Piping, Vol. 5, p. 113 (1978).

33. C.E. Jaske, J.H. Payer and V.S. Balint, Corrosion Fatigue of Metals in Marine Environments, MCIC Report 81-42, Battelle, Columbus, OH (1981).

34. Fatigue Crack Growth Threshold Concepts, Eds., D.L. Davidson and S. Suresh, TMS-AIME, Warrendale, PA (1984).

35. Suresh and R.O. Ritchie, "The Propagation of Short Fatigue Cracks," Int. Metall. Rev., Vol. 29, p. 445 (1984).

36. R.P. Gangloff and R.O. Ritchie, in Fundamentals of Deformation and Fracture, Eds., B.A. Bilby, K.J. Miller and J.R. Willis, Cambridge University Press, Cambridge, UK, p. 529 (1985).

37. J.K. Shang, J.-L. Tzou and R.O. Ritchie, "Role of Crack Tip Shielding in the Initiation and Growth of Long and Small Fatigue Cracks in Duplex Microstructures", submitted to Metall. Trans. A. (1986).

38. W.H. Hartt and S.S. Rajpathak, "Formation of Calcareous Deposits Within Simulated Fatigue Cracks in Seawater", Paper No. 62 presented at Corrosion 83, Houston, TX, NACE (1983).

39. T.W. Thorpe, P.M. Scott, A. Rance and D. Silvester, Intl. J. Fatigue, Vol. 5, p. 123 (1983).

40. J.-L. Tzou, C.H. Hsueh, A.G. Evans and R.O. Ritchie, Acta Metall., Vol. 33, p. 117 (1985).

41. R.O. Ritchie, University of California-Berkeley, unpublished research (1983).

42. K. Tanaka and R.P. Wei, Engr. Fract. Mech., Vol. 21, p. 293 (1985).

43. R.P. Wei, G. Shim and K. Tanaka, in Embrittlement by the Localized Crack Environment, Ed., R.P. Gangloff, TMS-AIME, Warrendale, PA, p. 243 (1984).

44. M. Pourbaix, in Atomistics of Fracture, Eds., R.M. Latanision and J.R. Pickens, Plenum Press, New York, p. 835 (1983).

45. N.J.H. Holroyd, G.M. Scamans and R. Hermann, Embrittlement by the Localized Crack Environment, Ed., R.P. Gangloff, TMS-AIME, Warrendale, PA, p. 327 (1984).

46. B.F. Brown, in Stress Corrosion Cracking and Hydrogen Embrittlement of Iron Based Alloys, Eds., J. Hochmann, J. Slater, R.D. McCright and R.W. Staehle, NACE, Houston, TX, p. 747 (1976).

47. A. Turnbull and J.G.N. Thomas, J. Electrochem. Soc., Vol. 129 p. 1412 (1982).

48. A. Turnbull and D.H. Ferriss, in Proc. Conf. Corrosion Chemistry within Pits, Crevices and Cracks, Ed., A. Turnbull, NPL, Teddington, UK, in press (1986).

49. A. Turnbull, in Embrittlement by the Localized Crack Environment, Ed., R.P. Gangloff, TMS-AIME, Warrendale, PA, p. 3 (1984).

50. A. Turnbull, Br. Corros. J., Vol. 15, p. 162 (1980).

51. A. Turnbull, "A Theoretical Evaluation of the Influence of Crack Dimensions on Crack Electrochemistry", National Physical Laboratory, Report NPL DMA (D) 363, Teddington, UK (1983).

52. A. Turnbull, *Corr. Sci.*, Vol. 22, p. 877 (1982).

53. A. Turnbull, "Comparison of Hydrogen Charging of Structural Steel by Crack Tip Processes and by Bulk Reactions in the Cathodic Protection of Corrosion Fatigue Cracks", *Scripta Metall.*, in press (1986).

54. R.P. Gangloff and A. Turnbull, in *Modeling Environmental Effects on Crack Initiation and Propagation*, Eds., R.H. Jones and W.W. Gerberich, TMS-AIME, Warrendale, Pa., p. 55 (1986).

55. R.J. Taunt and W. Charnock, *Matls. Sci. and Engr.*, Vol. 35, p. 219 (1978).

56. R.P. Gangloff, in *Embrittlement by the Localized Crack Environment*, Ed., R.P. Gangloff, TMS-AIME, Warrendale, PA, p. 265 (1984).

57. R.P. Gangloff, in *Critical Issues in Reducing the Corrosion of Steel*, Eds., H. Leidheiser, Jr. and S. Haruyama, NSF/JSPS, Tokyo, Japan, p. 28 (1985).

58. R.O. Ritchie and J. Lankford, *Matls. Sci. and Engr.*, Vol. 84, in press (1986).

59. R.P. Gangloff and R.P. Wei, in *Small Fatigue Cracks*, Eds., R.O. Ritchie and J. Lankford, TMS-AIME, Warrendale, PA, p. 239 (1986).

60. A. Turnbull and R.C. Newman, in *Small Fatigue Cracks*, Eds., R.O. Ritchie and J. Lankford, TMS-AIME, Warrendale, PA, p. 269 (1986).

61. F.P. Ford and S.J. Hudak, Jr., in *Small Fatigue Cracks*, Eds., R.O. Ritchie and J. Lankford, TMS-AIME, Warrendale, PA, p. 289 (1986).

62. D.L. Davidson and J. Lankford, in *Small Fatigue Cracks*, Eds., R.O. Ritchie and J. Lankford, TMS-AIME, Warrendale, PA, p. 455 (1986).

63. R.P. Gangloff, in *Advances in Crack Length Measurement*, Ed., C.J. Beevers, EMAS, United Kingdom, p. 175 (1982).

64. A. Zeghloul and J. Petit, in *Small Fatigue Cracks*, Eds., R.O. Ritchie and J. Lankford, TMS-AIME, Warrendale, PA, p. 225 (1986).

65. J.J. Lee and W.N. Sharpe, Jr., in *Small Fatigue Cracks*, Eds., R.O. Ritchie and J. Lankford, TMS-AIME, Warrendale, PA, p. 323 (1986).

66. R.P. Gangloff, *Metall. Trans. A.*, Vol. 16A, p. 953 (1985).

67. R.P. Gangloff, C.N. Marzinsky and J.Y. Koo, unpublished research, Exxon Research and Engineering Company (1986).

68. A. Saxena, W.K. Wilson, L.D. Roth and P.K. Liaw, *Intl. J. Fract.*, Vol. 28, p. 69 (1985).

642

69. J. Lankford and D.L. Davidson, in Small Fatigue Cracks, Eds., R.O. Ritchie and J. Lankford, TMS-AIME, Warrendale, PA, p. 51 (1986).

70. J. Lankford, Fat. Engr. Matls. Struc., Vol. 6, p. 15 (1983).

71. S.P. Lynch, in Atomistics of Fracture, Eds., R.M. Latanision and J.R. Pickens, Plenum Press, New York, p. 955 (1983).

72. S.J. Hudak, D.L. Davidson and R.A. Page, in Embrittlement by the Localized Crack Environment, Ed., R.P. Gangloff, TMS-AIME, Warrendale, PA, p. 199 (1984).

73. R.P. Wei and G.W. Simmons, Int. J. Fract., Vol. 17, p. 235 (1981).

74. O. Vosikovsky, Trans. ASME, J. Engr. Matls. Tech., Vol. 97, p. 298 (1975).

75. B.R.W. Hinton and R.P.M. Procter, in Hydrogen Effects in Metals, Eds., I.M. Bernstein and A.W. Thompson, TMS-AIME, Warrendale, PA, p. 1005 (1981).

76. G. Shim, Y. Nakai and R.P. Wei, in Basic Questions in Fatigue, Vols. I and II, ASTM STP, Eds., R.P. Wei, J.T. Fong, R.J. Fields and R.P. Gangloff, ASTM, Philadelphia, PA, in press (1986).

77. R.P. Gangloff, unpublished research, Exxon Research and Engineering Co. (1985).

78. P.K. Liaw, S.J. Hudak, Jr. and J.K. Donald, in Fracture Mechanics: Fourteenth Symposium-Volume II: Testing and Applications, ASTM STP 791, Eds., J.C. Lewis and G. Sines, ASTM, Philadelphia, p. 370 (1983).

79. F.H. Heubaum, Ph.D. Dissertation, Northwestern University (1985).

80. R.D. Carter, E.W. Lee, E.A. Starke, Jr. and C.J. Beevers, Metall. Trans. A., Vol. 15A, p. 555 (1984).

81. D.J. Duquette and H.H. Uhlig, Trans. ASM, Vol. 61, p. 449 (1986).

82. D.J. Duquette and H.H. Uhlig, Trans. ASM, Vol. 62, p. 839 (1969).

83. H.H. Lee and H.H. Uhlig, Met. Trans., Vol. 3, p. 2949 (1972).

DISCUSSION

Comment by M. Speidel:

It is an old, almost classical observation that, as you go from fatigue to corrosion fatigue, the fatigue striations on fracture surfaces change from a "ductile" to a "brittle" appearance. Can you give a systematic explanation?

Reply:

The observation you describe has been reported primarily for engineering alloys, and in particular for aluminum alloys. We believe that corrosion fatigue of those alloys which show the transition from ductile to brittle striations are alloys where hydrogen is important to the fracture processes.

Comment by R. C. Newman:

Sieradzki has shown that oriented single crystals of copper undergo transgranular scc in ammonia solutions at very small anodic overpotentials and with dynamic straining. This fracture is associated with a "tunnelled" layer formed by selective dissolution at emergent dislocations. I wonder if similar processes could occur during corrosion fatigue? (But maybe you need a high exchange current density?)

Reply:

We have not considered the possibility of a "tunnelled" layer on corrosion fatigue processes. However, the experiments described by Sieradzki and yourself at this workshop should initiate a new and closer examination of fracture surfaces in corrosion fatigue.

Comment by R. H. Jones:

Is there evidence that the corrosion rates of heavily deformed metals are greater than annealed metals? If not, do the accelerated corrosion rates of PSB's result from the dynamic deformation process or the defect density of PSB's?

Reply:

The evidence related to enhanced corrosion due to deformation is mixed. In our case, however, pre-fatigue deformation, followed by corrosion, results only in general corrosion of a given metal or alloy. Dynamic straining and corrosion, on the other hand, results in preferential attack of slip bands. It can probably be concluded that the energy of the metal associated with emerging dislocations increases the active dissolution of the slip bands, and that continued deformation provides fresh, high energy metal to the corroding solution.

Comment by R. M. Latanision:

Concerning short cracks, my understanding is that, in the case of short cracks, the crack velocity increases as the stress intensity decreases. In effect the crack velocity passes through an effective minimum at about the threshold stress intensity.
Can you comment on this, and in particular, the implication of this observation on the significance of the threshhold stress intensity?

Reply:

Your understanding is correct, and the phenomenon occurs in the absence of environmental effects. It appears to be primarily associated with crack closure processes including roughness, debris, etc. The effect has not been widely studied under environmental conditions. However, it should be noted that, at constant ΔK, short cracks grow more rapidly than physically long cracks. This appears to be related only to the chemistry of the crack solution.

Comment by M. Bernstein:

Your observation that a corrosive environment enhances plasticity under constant plastic strain should consider that since the total strain is conserved, the slip height must decrease. Under such conditions, this observation could be related to activation of surface sources.

Reply:

I strongly agree that surface sources of PSB's are activated by the corrosive environment. Under stress controlled conditions, not only are more slip bands formed, but the slip step offsets are increased. This suggests that new PSB's are formed as well as an increase in the plasticity of the bands which are normally activated.

Comment by M. Bernstein:

Have you repeated the H_2 experiments in Ni bicrystals in an aqueous solution; particularly those that form passive films? It would be of interest to see if the crack morphology and path are similar in both cases.

Reply:

Regrettably no. I hope that the nickel bicrystal work which we initiated at the Max Planck Institut in Dusseldorf will be extended to corroding solutions as well as being continued in gaseous hydrogen.

Comment by B. D. Lichter:

On the question of the spacially localized corrosion of the PSB regions, we have observed a related phenomenon for α-brass single crystals in aqueous ammonia. In that case, prior deformation in air produces very little effect on subsequent immersion. However, if the crystals are strained while immersed, corrosion pits are observed to form along the slip bands. Under continuing straining in the media, the pits widen, producing {111} bounded slots. These can persist for some depth into the crystal, but eventually, a {110} stress-corrosion crack is nucleated.

Reply:

Thank you for your comment. It certainly agrees with our view that only mobile slip bands are preferentially attacked. This process may indeed be closely related to both SCC and CF and should be examined in more detail.

Comment by J. R. Rice:

In your discussion of a Ni bicrystal you showed that the PSB's taper to small or zero thickness as they reach the interface between crystals. Has there been a study of secondary slip near those PSB terminations to clarify plastic relaxation mechanisms at and near the interface?

Reply:

To my knowledge, there have been few studies of fatigue-generated slip with grain boundaries (in contrast to monotonic deformation). It would certainly appear to be important to initiate such a study, both from a theoretical as well as from an experimental point of view.

STRESS CORROSION CRACKING OF A 508 STEEL IN SATURATED MnS-SOLUTION

HANNU HÄNNINEN, HEIKKI ILLI & MARKKU KEMPPAINEN
Technical Research Centre of Finland
Metals Laboratory, SF-02150 Espoo 15, Finland

1. INTRODUCTION

Carbon and low alloy steels are susceptible to stress corrosion cracking (SCC) and corrosion fatigue (CF) in high temperature high pressure pure water. Mechanistically this susceptibility has recently been connected to MnS-inclusions of the steels, which can initiate cracking in SCC tests (1 - 4) and contribute to enhancement of crack growth both in CF (5 - 7) and in SCC tests (4). The role of MnS-inclusions in CF-crack growth has been schematically demonstrated in Fig. 1. Since in this model the dissolution of the uncovered MnS-inclusions and the change of local environment inside the crack is of primary importance, the tests in the environments believed to simulate the conditions near dissolving MnS-inclusions have been started. In the following the slow strain rate tests (SSRT) of an A 508 pressure vessel steel are presented in a range of corrosion potentials at 85° C in MnS-saturated pure water.

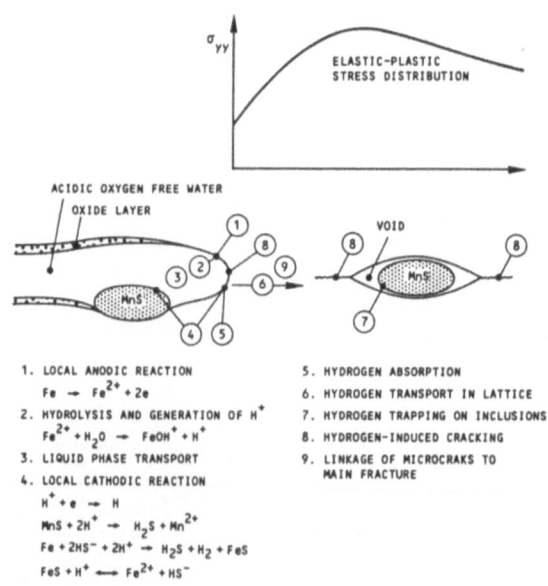

FIGURE 1. Schematic picture of the hydrogen-induced crack growth model proposed for corrosion fatigue crack growth in high temperature water (5).

2. EXPERIMENTAL

Notched round tensile specimens (gauge length 14 mm, gauge diameter 2.5 mm and V-shape 0.25 mm deep notch) of an ASTM A 508 Cl.3 steel, Table I, were tested by using an SSRT-technique (strain rate for gauge section 5.4 × 10^{-6} s^{-1}) in L- and T-orientations in MnS-saturated deionized pure water.

The SSRT-tests were performed without any O_2-control at controlled potentials in the potential range of -200...-1000 mV_{SHE}. Since the MnS-powder was on the bottom of the glass test cell it was also polarized during polarization experiments. The cracking susceptibility is evaluated by analysing the stress-strain curves, fractography and electrochemical polarization measurements with a SCE-reference electrode at 25° C.

TABLE 1. Chemical composition of A 508 Cl.3 forging used in the slow strain rate tests.

Element	C	S	P	Si	Mn	Ni	Cr	Mo	Cu	Co	V
wt-%	0.17	0.010	0.010	0.26	1.43	0.75	0.20	0.53	0.045	0.013	0.004

3. RESULTS

The cracking mode in the preliminary tests of smooth SSRT-specimens tested in MnS-saturated pure water at 85° C was completely ductile (1). In notched SSRT-specimens transgranular cracks initiating at the bottom of the notch are observed on a wide range of corrosion potentials, Fig. 2. The susceptibility to cracking is demonstrated by the strain to fracture (%) and the percentage of brittle fracture measured from SEM-fractographs as a function of corrosion potential. The effect of specimen orientation does not seem to be marked and the strain to fracture does not change very much in the various tests. The evaluation of the crack growth rates of different specimens resulted in the values of $10^{-5}...10^{-6}$ mms^{-1}.

The polarization curves of A 508 steel in the MnS-saturated pure water at 85° C are presented in Fig. 3a. The difference in corrosion potentials measured when polarized from cathodic or anodic directions is controlled by hydrogen absorption and sulfide (FeS_x) film and oxide (Fe_3O_4) film, respectively. In SSRT-tests slightly more positive corrosion potentials were measured (Fig. 2). Marked cracking susceptibility in the free corrosion potential SSRT-tests seems to be related to active-passive transition of the polarization curve in the studied environments. The pH-value on the steel surface as a function of potential at 85° C was measured by step-wise polarization from low potential value (-1450 mV_{SHE}). Stabilization of pH-value was weighted at each measuring potential about 2 min. The pH-value was varying between 5.5 and 6.0 in the range of interest. Fig. 3b shows the hydrogen concentration of iron as a function of electrode potential in acetate solution of pH 3.42 with different H_2S concentrations at 23° C (9). It clearly shows that cracking regime in our SSRT-tests is confined to the range where H-absorption is possible to take place. When the potential is lowered the hydrogen concentration saturates and thus the cracking susceptibility does not change markedly when electrode potential is lowered.

Fractography showed that cracks initiated on numerous locations on the bottom of the notch and grew transgranularly. The morphology of cracking is typical of hydrogen-induced cracking and similar to SCC and CF fractography obtained in pure high temperature high pressure water. In these tests only cracking initiated from the notches was observed and even at the most cathodic potentials no crack initiation inside the specimens at MnS-inclusions

648

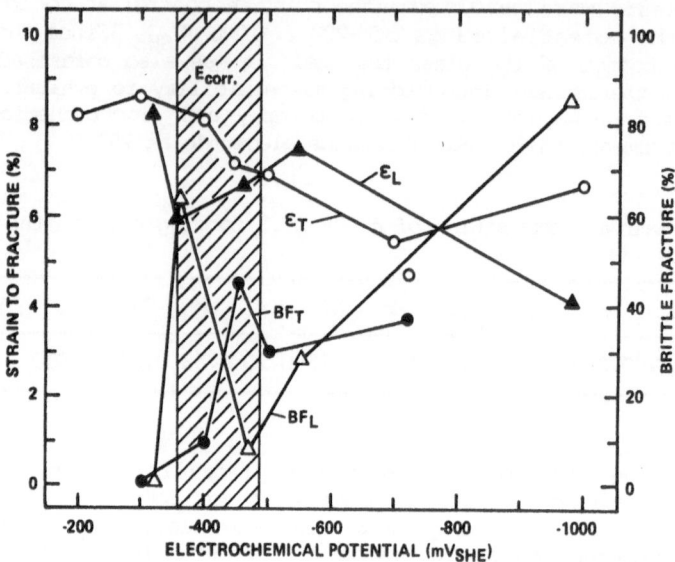

FIGURE 2. SSRT-test results of notched A 508 steel specimens (T- and L-orientations) obtained at different potentials in MnS-saturated pure water at 85° C. ε = strain to fracture and BF = brittle fracture ratio.

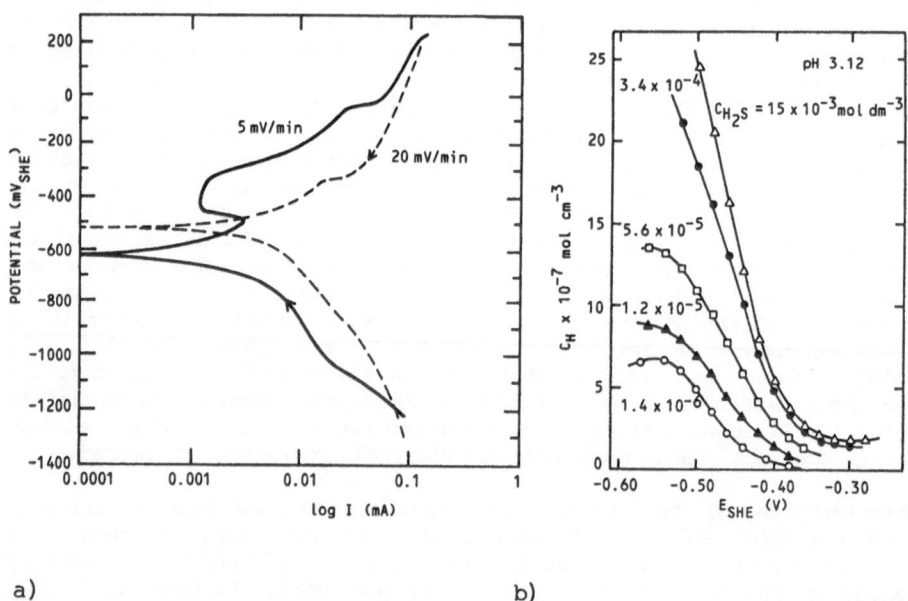

a)   b)

FIGURE 3. (a) Polarization curves of A 508 steel measured in MnS-saturated pure water at 85° C. (b) Hydrogen concentration (C_H) in iron as a function of the electrode potential in acetate solution of pH 3.42 with different H_2S concentrations at 23° C (Ref. 9).

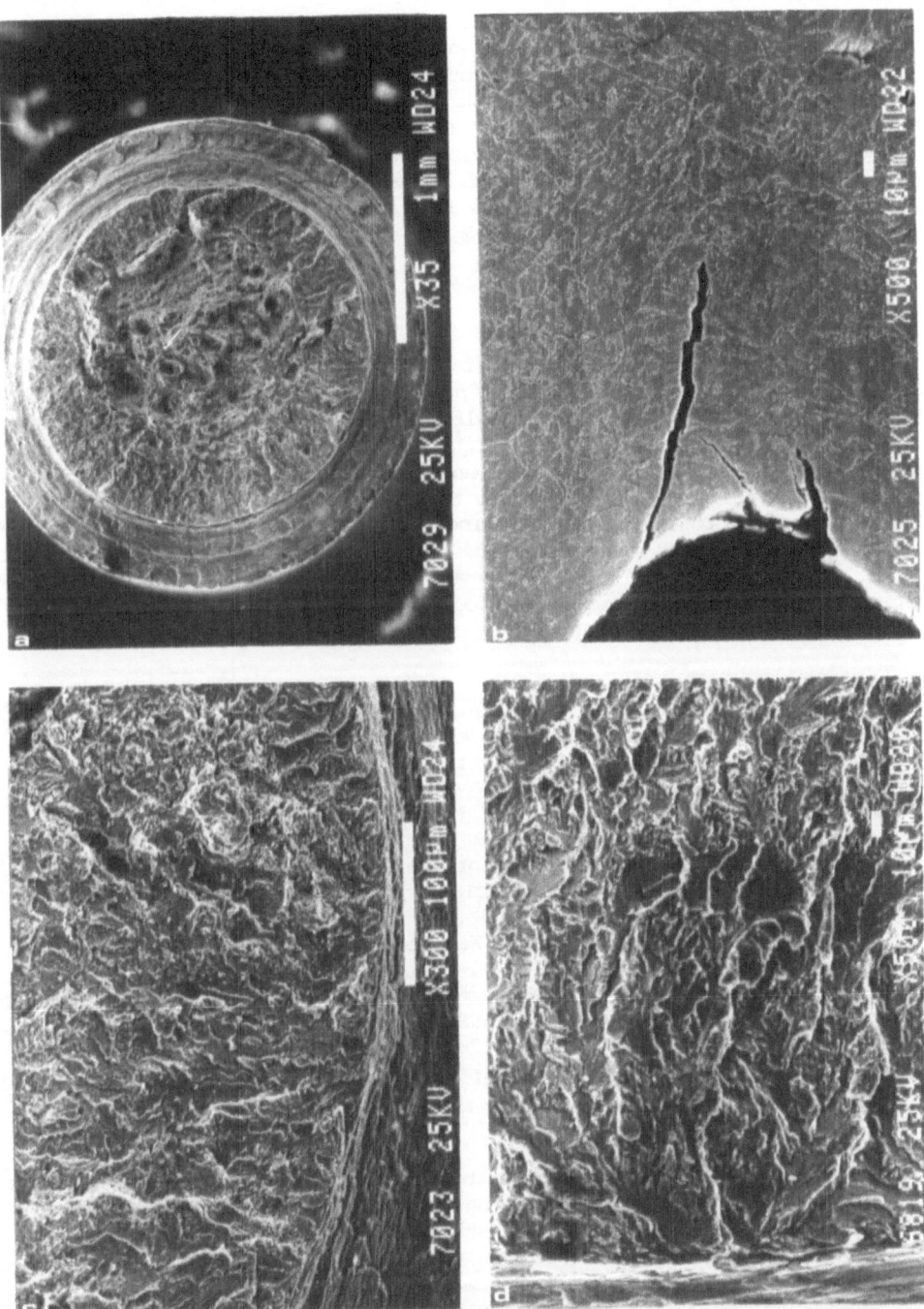

FIGURE 4. (a) Typical fracture surface of a SSRT-specimen (tested at -360 mV$_{SHE}$) showing the notch, brittle transgranular area and ductile final fracture. (b) A cross-section of a crack in terminated test at -570 mV$_{SHE}$. Fracture surface morphologies of specimens tested at -360 mV$_{SHE}$ (c) and -980 mV$_{SHE}$ (d).

was found. Cracking at corrosion potential (Fig. 4c) and at a cathodic potential -980 mV$_{SHE}$ (Fig. 4d) are very similar; the fracture surface at the corrosion potential has, however, corroded. The cross-section of a specimen tested at -570 mV$_{SHE}$ presented in Fig. 4b shows that crack growth is connected to the bainite packet structure of the steel so that the crack propagates through a packet on the same plane and the packet boundaries form a discontinuity to the crack growth. If the crack does jump through the individual packets at once or if it exhibits other type of discontinuous crack growth is not clear from the fractography. Fig. 4b shows also that the crack tip opening is about 2 μm.

4. DISCUSSION

In conditions supposed to prevail around dissolving MnS-inclusions in crevices and cracks cracking can be produced in notched specimens of A 508 steel at a wide range of potentials in SSRT-tests. Anodic polarization seems to prevent cracking. The critical potential about -300 mV$_{SHE}$ under which cracking takes place can be connected to the stability of H_2S/HS^- and absorption kinetics of hydrogen into the steel. Thus, cracking in this case is a type of hydrogen-induced cracking. Crack growth seems to occur at the tip of the main crack rather than at some distance ahead of the crack tip. Since the individual cracking events are probably very fast the real initiation sites of possible microcracks could not be revealed.

5. ACKNOWLEDGEMENTS

This work is part of the research program on Safety and Reliability of Nuclear Components and Materials performed at the Technical Research Centre of Finland and financed by the Ministry of Trade and Industry in Finland.

6. REFERENCES

1. K. Klemetti, & H. Hänninen, 2nd international symposium on environmental degradation of materials in nuclear power systems - water reactors, Monterey, CA, USA, 9.-12.9.1985.
2. M. E. Indig, J. E. Weber and D. Weinstein, Reviews in Coatings and Corrosion, Vol. 5, p. 173-225 (1982).
3. H. Choi, F. H. Beck, Z. Szklarska-Smialowska and D. D. Macdonald, Corrosion, Vol. 38, p. 136-144 (1982).
4. P. Hurst, ICCGR Meeting, Paris, 12-14 May 1986.
5. H. Hänninen, K. Törrönen, K. Kemppainen & S. Salonen, Corr. Sci., Vol. 23, p. 663-679(1983).
6. K. Törrönen, M. Kemppainen & H. Hänninen, EPRI NP-3483, Final Report, Palo Alto, CA, 1984.
7. F. P. Ford, P. Combrade, Proceedings of the second international atomic energy agency specialists' meeting on subcritical crack growth, Sendai, Japan, May 15-17, 1985, NUREG/CP-0067 MEA-2090 Vol.2, p. 231-268.
8. H. Hänninen, H. Illi, K. Törrönen, M. Vulli, ibid., p. 179-198.
9. H. Murayma, M. Sakashita & N. Sato, Proceedings JIMIS-2. Hydrogen in Metals, Ed. by T. Suzuki, The Japan Institute of Metals, Sendai, Japan, pp. 297-300.

DISCUSSION

Comment by A. Turnbull:

Do you have any evidence that the crack tip solution is saturated in MnS? Clearly the concentration will depend on the distribution of inclusions, dissolution rate and the rate of mass transport. Is it possible that by using a saturated solution that you are artificially changing the mechanism from dissolution to hydrogen? Is not the dissolution process also enhanced by H_2S?

Reply:

MnS-inclusion dissolution and FeS_x deposition on the corrosion fatigue fracture surfaces indicate that sulfide species are produced. Sulfides can enhance both the anodic dissolution process and the hydrogen absorption into the steel. The morphology of cracking and its potential dependence in saturated MnS-solutions support the hydrogen-induced cracking models.

Comment by R. C. Newman:

The SCC of pressure vessel steel in high-temperature water requires rather oxidizing conditions. How is this consistent with your measurements?

Reply:

These experiments simulate the conditions at the crack tip near dissolving MnS-inclusions. In stress corrosion the oxidizing conditions (>-300 mV_{SHE}) are needed to keep the macro-corrosion cell operating. Corrosion fatigue according to the suggested mechanism can also occur under reducing conditions.

Comment by W. Zheng:

Your results showed that the most severe HE occurred at around -500 mV (SCE) and cracking became less severe as the potential was moved to more cathodic value. It seems to me that this is not consistent with the general observation that HE severity increases as the potential is lowered. Could you give an explanation on it?

Reply:

Cracking susceptibility is marked at corrosion potential and increases only slightly at lower potentials. This seems to be related to the catalytic action of H_2S to enhance hydrogen absorption into the steel. The absorption seems to saturate somewhere around our corrosion potential and no marked increases are expected when the potential is lowered.

EXPERIMENTS ON BICRYSTALS CONCERNING THE INFLUENCE OF SEGREGATION AND SLIP ON STRESS CORROSION CRACKING

H. STENZEL, H. VEHOFF AND P. NEUMANN
Max-Planck-Institut für Eisenforschung
Max-Planck-Str. 1
D-4000
Dusseldorf, Federal Republic of Germany

1. INTRODUCTION

Reliable models which describe the mechanism of stress corrosion cracking (SCC) in even simple model systems are still lacking. The stationary crack geometry which develops with increasing crack length and time depends critically on localized changes in crack chemistry, local alloy composition, and on the interaction between localized slip and dissolution. Current models mainly describe the influence of crack geometry on changes in potential distribution and solution composition (1,2). Data on the influence of texture on the localized slip distribution and on the impurity segregation are difficult to obtain and hence, neglected in most models on SCC.

The purpose of the present study is to examine separately how grain boundary structure, segregation, and slip control SCC. The following questions should be answered:
1. Are clean grain boundaries sensitive to SCC and does SCC depend on grain boundary structure?
2. Does a simple relationship between grain boundary orientation, segregation, and SCC exist?
3. Do changes in the localized slip distribution alter the relationship between the localized strain rates, film formation, and dissolution rates?

To answer these questions, tests on bicrystals with defined tilt boundaries are done. The tilt angle is varied systematically, as well as, the alloy composition for special large angle boundaries, in order to separate the influence of segregation and slip on SCC. FeSi-bicrystals are used since these crystals can be cleaved intergranularly and transgranularly inside the Auger system. Therefore the impurity segregation can be examined after the test. Tests are done in carbonate solutions, since investigations on low strength steels have shown that SCC is little affected by the crack geometry in these environments (3).

2. RESULTS AND DISCUSSION

The experimental procedure, evaluation of the data, and derivation of the model which describes crack growth along a grain boundary are presented in a previous publication (4). This small note is concerned with the influence of segregation and slip on SCC.

Starting from a simple slip step dissolution mechanism, we assume that any slip event at the crack tip sequentially triggers a dissolution event. This yields:

$$a = (\frac{\Delta a_d + \Delta a_d}{\delta_s}) \delta \text{ with } \delta = \frac{\delta_s}{t_s},$$

where Δa_c and Δa_d are the corrosive and ductile crack growth increments which occur within the time interval t_s, between subsequent slip step events. δ is the crack tip opening rate expressed in single slip events, where δ_s is the plastic crack tip opening per slip event. With this simple assumption, δ_s and Δa_c can be directly evaluated from the crack tip opening angle:

$$ctg \ \frac{\alpha}{2} \ \overset{def}{=} \ \frac{\Delta a}{\delta_s}$$

A typical crack tip opening angle obtained by crack growth along a symmetric bicrystal boundary is shown in Figure 1a. The photograph demonstrates the drastic increase in the crack tip opening angle if the potential is switched from -310 mV$_{EH}$ to -110 mV$_{EH}$. Detailed measurements of the dependence of α on potential and opening rate δ are given in reference 4.

From the simple assumptions given above, it is evident that if, due to the unfavorable bicrystal orientations, slip is not completely confined to the crack tip, the crack growth rate can change drastically. FeSi preferentially dissolves along the grain boundary, therefore, any slip event slightly behind the active tip will only blunt the crack without triggering additional corrosive growth increments.

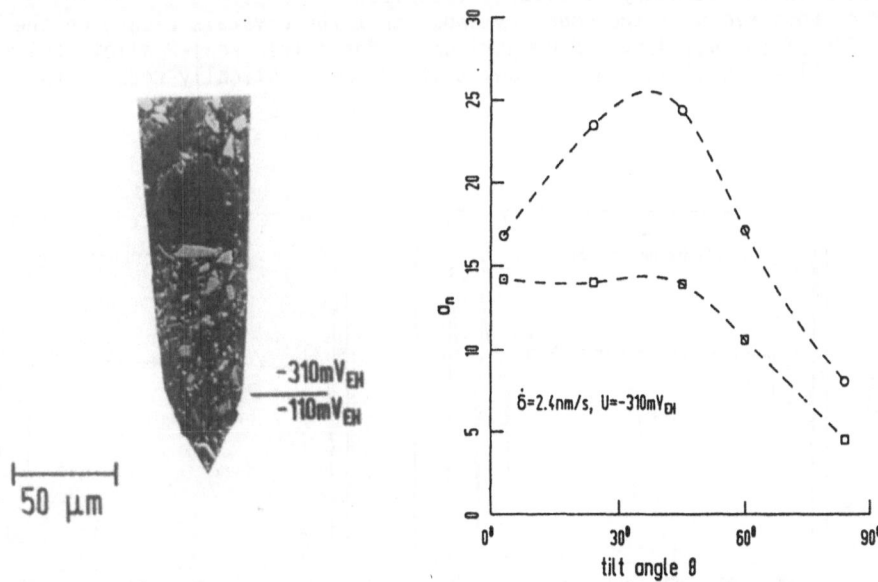

Figure 1. a) Section parallel to the side surface of a symmetric $\theta = 24°$ -<100>-tilt boundary; b) a_n as a function of the tilt angle θ for specimens notched at two different side surfaces.

To demonstrate this effect, the slip distribution at the crack tip must be varied under otherwise identical test conditions. Therefore crack growth tests are carried out with bicrystals of the same boundary orientation.

But the slip distribution at the tip is altered by systematic variation of the notch direction.

The dependence of a_0 on the orientation of symmetric <100>-tilt boundaries is given in Figure 1b for a fixed potential (-310 mV$_{EH}$) and opening rate (δ = 2.4 nm/s). For the chosen composition (30 ppm C, 30 ppm S, 2.7% Si, <50 ppm P, 0.15% Mn), Auger examinations of the grain boundaries show no C- and Si-enrichment compared to transgranular fracture surfaces. Only a small sulfur peak (approximately 5% of a monolayer) is observed. Therefore these results are first taken as evidence for the influence of grain boundary structure on SCC. But as demonstrated in the upper curve, for similar boundaries drastic changes in the growth rate are observed by simply altering the notch direction. For special boundaries changes up to a factor of three are obtained. The largest growth rates are found for orientations, in which only one slip system with a large Schmid factor operates in each grain. Hence, orientation effects can screen possible effects of the boundary orientation and of segregation on SCC. Therefore, all segregation studies are done on similar boundaries with the same notch direction.

Bicrystals with an asymmetric θ = 45° -<100>-tilt boundary were grown from three different FeSiP-melts (0.002%, 0.04%, 0.11% P, 20 ppm C, 10 ppm B, and 5 ppm S). Figures 2a and 2b show Auger diagrams of the intercrystalline fracture surfaces of the 0.11% P bicrystal. The enrichment of P and B on the boundary is clearly visible. The uneven distribution of P and B on both sides of the boundary shows that the crystals cleave on the (100) -side of the boundary. A comparison of Boron free FeSi-P alloys (5) with the alloy containing Boron shows thats Boron drastically reduces the P-segregation.

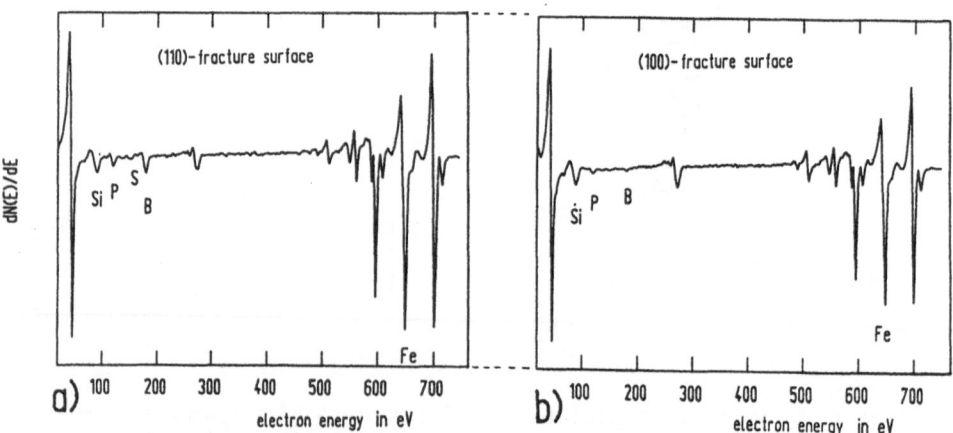

Figure 2. a,b: Auger spectra of a θ = 45° -<100>-tilt boundary in FeSi-0.11% P from opposite sides of the grain boundary at the same location.

In Figure 3 the crack growth rate is plotted vs. the crack tip opening rate for these three alloys. The plot demonstrates that even a small increase of the P coverage on the boundary (up to 10%) can alter the growth

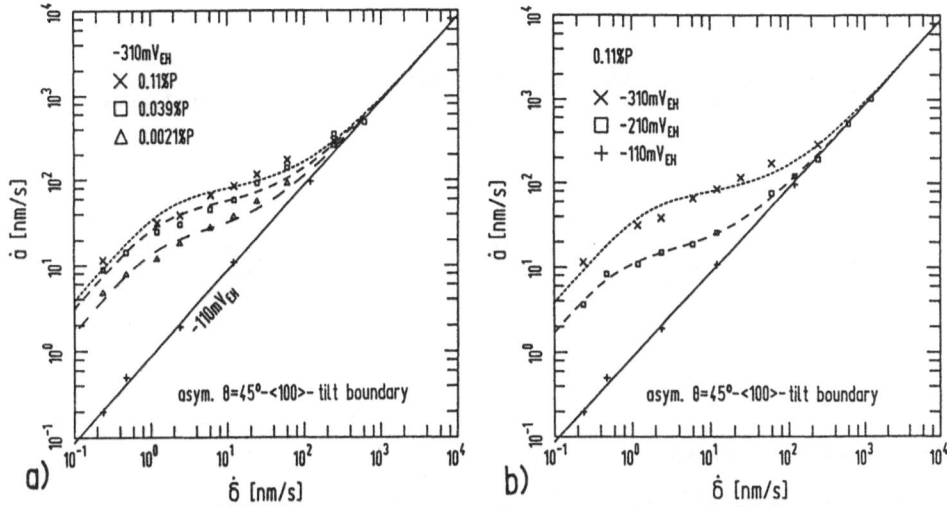

Figure 3. a) Stress corrosion crack growth rate, å, as a function of δ for three FeSi-P-alloys, b) The same plot of å vs δ for three different potentials.

rate by a factor of two. No intergranular corrosion is observed in these alloys. For the FeSi-0.11% P-alloy the influence of potential on the crack growth rate is shown in Figure 3b for the same boundary orientation. These results show that an increase of P coverage has the same effect on SCC as a potential shift to a more active region. A phosphorus coverage of 10% is approximately equivalent to a potential shift of 100 mV. Potentiodynamic polarization curves show a similar shift in the active-passive transition for a Fe-2% P-alloy compared to pure Fe or FeSi. Therefore electrochemical examinations of free surfaces with the appropriate boundary composition seem to be a good technique to quantify the influence of segregation on SCC.

SUMMARY AND CONCLUSIONS
    This investigation has shown the following:
-   For clean grain boundaries, the influence of grain boundary orien-
    tation on SCC is screened by the influence of orientation on slip.
-   For special boundaries the change of slip distributions at the tip
    by notching along different crystallographic direction alters the
    crack growth rate by a factor of three. If two systems with a large
    Schmid factor can operate in adjacent grains, the largest SCC sensi-
    tivity is observed.
-   Phosphorus segregation and potential changes influence similarly the
    strain rate dependence of SCC. It is concluded that a grain boun-
    dary with phosphorus behaves like a clean boundary at a more active
    potential.

    Based on these observations it can be concluded that texture and the purity of the alloy have only minor influence on improving SCC behavior of low strength steels in carbonate solutions. The best design against SCC seems to be controlled micro alloying of elements like Boron, which enhances the strength of the boundary and reduces segregation of detrimental elements on the boundary.

REFERENCES

1. Turnbull A and Thomas JG: J. Electr. Soc. 129, 1982.
2. Staehle RW: Stress Corrosion Cracking and Hydrogen Embrittlement of Iron Base Alloys, eds., R. W. Staehle, J. Hochmann, R. D. McCrigth and J. E. Slater, NACE-5, p. 180, 1973.
3. Parkins RN: Corrosion Sci. 20, p. 147, 1980.
4. Stenzel H, Vehoff H, and Neumann P: in: Modelling Environmental Effects on   Crack Initiation and Propagation, eds. R. H. Jones, and W. W. Gerberich, Fall Meeting of the TMS-AIME, Toronto, 1985.
5. Thompson RG, White CL, Wert JJ and Easton DS: Met. Trans. A 12, p. 1339, 1981.

DISCUSSION

Comment by R. C. Newman:

You alluded to experiments on transgranular SCC of austenitic stainless steel in magnesium chloride solution. Did your periodic unloading experiments indicate discontinuous cracking in this case? Hahn and Pugh showed evidence for discontinuous cracking based on a loss of correspondence between load pulse markings and the number of pulses, but I think this required quite high frequencies of load pulsing.

Reply:

Evaluation of our crack growth data on austenitic stainless steel single crystals using the slip step dissolution model yields in the limit of small crack growth rates, crack growth increments Δa_p of approximately 1 μm per slip step event. This value corresponds well to the findings of Hahn and Pugh for pulse times above 15 s. From our pulse tests we have found no clear correlation between markings and pulses. The fracture surface markings of the single crystals strained in magnesium chloride solutions are too ill defined on this scale. But again we have observed no evidence of instabilities, and for long times between pulses, markings of approximately 1 μm are expected due to slip step dissolution alone.

Comment by M. O. Speidel:

Was the intergranular SCC crack growth which you described fully continuous or could there possibly have been very small, regular mechanical brittle crack extensions (in addition to slip and dissolution)?

Reply:

In order to answer this question, we have carried out a series of tests with pulse loading. For different pulse widths T_1 (0.05s < T_1 < 0.4 s), the time between pulses was varied systematically between 0.1s and 100 s. In addition the mean stress and the pulse height was varied. The tests yielded the following results:
1. For a given positive mean stress and for large pulse heights transcrystalline cleavage in addition to intercrystalline cracking is observed. With increasing pulse height more and more transcrystalline cracks are

found. But for smaller pulse heights as well as in the continuous straining experiments no evidence for transcrystalline cleavage was observed.

2. For a given small pulse height and for a wide range of mean stresses,
the mean crack growth increment per pulse Δa_p is found to be independent of
the mean stress. Only for very low mean stresses (below $\sigma_y/2$) no crack
growth is obtained. From this result it can be concluded that the pulses
must be high enough to trigger single slip events.

3. For pulse times between 1 s and 100 s, the average increments per
pulse are found to be constant. But these increments Δa_p depend systematically on the applied potential.

4. From the average opening per pulse δ, the average pulse increment
Δa_p and the number of pulses per second N, a crack growth rate $\dot{a}$ vs δ
curve can be evaluated which corresponds to the curves obtained by
straining continuously.

5. The fracture surfaces show coarse slip markings with an average distance
of 1μm. But these markings correlate neither with the number of the
mechanical pulses nor with the number of the current density pulses. In
addition, the distance between these markings show no dependence on potential.

Thus it can be concluded that these rather detailed results can be
explained by the slip step dissolution model. We feel that there is no
need to postulate instable intercrystalline crack extensions.

Comment by J. P. Fidelle:

Did you -- or will you -- next investigate a large range of cathodic
potentials?, in aerated or desacerated solution? I am asking this because
(1) in aerated or desacerated solutions cracking morphology is likely to
change according to cathodic potentials and there might be one at which
would find a crack morphology similar to that found during the SCC tests.
(2) at low cathodic potentials in aerated solutions there is some cathodic
protection, which no longer exists when solutions are desacerated.

Reply:

Yes, we have investigated the influence of potential. For cathodic
potentials as well as for the corrosion potential we have observed enhanced
stable transcrystalline crack growth due to hydrogen (see Reference 4 of
this paper). For small anodic potentials, pitting together with hydrogen
assisted crack growth is found. By further increasing the potential,
pitting and intercrystalline fracture is observed. In the active- passive
transition range we observe only intercrystalline crack growth. In the
passive range we find only ductile crack growth.

Comment by R. H. Jones:

Have you considered the possible effect of transport within the electrolyte on your crack growth rate analysis.

Reply:

Yes, due to the low solubility of Fe^{2+} and due to buffering by carbonate
solution no change in the crack tip electrolyte is to be expected.
Measurements on simulated cracks in this solution show no potential drops.
In addition it was found that long fatigue cracks respond similarly to

external potential changes if compared to the relatively blunt cracks (crack tip opening angle $\geq 1°$), which we observe in the low strain rate tests. A simple calculation, assuming passive crack walls, gives only a potential drop of 5 mV.

Comment by X. G. Zhang:

You mentioned that anodic dissolution appear when slip occurs at the crack tip. How large is the current density and did you measure it?

Reply:

We were not able to measure the current density within the growing crack.

STRESS-CORROSION CRACKING OF Cu-25Au SINGLE CRYSTALS IN AQUEOUS
CHLORIDE MEDIA

T. B. CASSAGNE, W. F. FLANAGAN, AND B. D. LICHTER
Vanderbilt University
Department of Mechanical and Materials Engineering
Nashville, TN 37235

1. ABSTRACT

Transgranular stress-corrosion cracking (T-SCC) in disordered copper-25 atomic percent gold single crystals was studied in two chloride media: 2% (0.13M) $FeCl_3$ and oxygen-free 0.6M NaCl. Potentiostatic polarization, constant deflection tests, and high-resolution SEM examination were used to relate the electrochemical behavior with susceptibility to T-SCC in this system. Selective electro-dissolution of copper occurs over the investigated potential domain, and polarization is characterized by a region of potential in which the current is very low ($\sim$ 2-4 microamps/cm^2), increasing only very slowly with increasing potential up to a critical potential, E_c. Above E_c the current increases by a factor of 10^3 in a very narrow domain of rising potential ($\sim$ 50 mv). In both chloride media the critical potential was found to be $\sim$ +430 mv (sce). At and above E_c a porous gold-rich "sponge" is readily formed on the surface, and in aq. $FeCl_3$ metallographic evidence for sponge formation was found in the low-current region of potential, well below E_c. In both media, fracture surfaces display the characteristic cleavage-like features seen in other T-SCC systems. The lower limit of potential for susceptibility in aq. NaCl occurs at 400 mv (sce), slightly below E_c; whereas, for aq. $FeCl_3$ the limiting potential is 0 mv (sce), fully 430 mv below E_c. For both media, the upper limit of potential, if it exists, is greater than 600 mv (sce). These results are analyzed in order to distinguish between proposed mechanisms for T-SCC.

2. INTRODUCTION

While slip-dissolution models [1-3] may adequately describe inter-granular stress corrosion cracking (I-SCC), they fail to take into account several established features of transgranular stress corrosion cracking (T-SCC), in particular the decidedly non-Faradaic, discontinuous nature of crack advance, which appears to occur by mechanical cleavage. The subject has been reviewed by Bursle and Pugh [4], Pugh [5], Sieradzki and Newman [6], and by Meletis and Hochman [7]. A theoretical treatment has also been offered to account for these features [6,8]. Recently it was shown [9,10] that copper-gold alloys, tested in aqueous $FeCl_3$, NaCl, and acid-sulfate media, display many of the characteristics which support the mechanical cleavage mechanism of T-SCC. Past studies of copper-gold alloys were largely responsible for the "tunnel" (or "micropitting") model proposed by Pickering and Swann [11] and further developed by Swann and co-workers [12-14]. This model preserves many of the features of the slip-dissolution model but requires that spacially oriented selective dissolution produce a noble metal sponge which subsequently fails by ductile overload of the thin metal fibers comprising the sponge.

Recent studies of polycrystalline 15 and 25 at% Au alloys in aq. $FeCl_3$, aq. NaCl, and aq. acid-sulfate solutions [9] and of Cu_3Au single crystals

in aq. $FeCl_3$ [10] have shown that sponge formation on the fracture surface is not observed under high-resolution SEM (~ 200 Å), despite the
fact that massive sponge formation could be seen at low magnification in
the vicinity of occasional surface scratches.

The present work was undertaken to obtain further evidence with which
to distinguish between proposed mechanisms for T-SCC. In particular, Cu-
25 at% Au was studied in aq. $FeCl_3$ and aq. NaCl, and the susceptibility
to T-SCC in constant deflection tests (three-point bending) was determined
as a function of potential in both media. Fracture surfaces produced at
different potentials were examined using high-resolution SEM and electron
microprobe analysis techniques. To complement these studies, polarization
measurements were carried out in both media, and the role of sponge formation in the nucleation and growth of cracks was investigated.

3. EXPERIMENTAL PROCEDURES
3.1 Single Crystals
Experiments were performed on copper-25 atomic percent gold single
crystals (obtained from Prof. D. Pope, University of Pennsylvania). These
crystals had been produced in a vertical tube furnace using the Bridgman
technique by co-melting copper and gold of 99.99% purity in a high-purity
split graphite mold [15]. Thin crystals either 16mm x 3.0 mm x 1.5 mm or
23 mm x 1.5 mm x 0.6 mm were prepared with a slow-speed, high-density
diamond saw and electropolished in Morris solution (133 cm^3 glacial acetic
acid, 25 gm CrO_3, and 7 cm^3 H_2O). To insure that no cold work was present, an X-ray Laue diffraction pattern was obtained for each sample, and
further electropolishing was carried out until no asterism could be seen.
Finally, to eliminate compositional inhomogeneities, samples were annealed
in an argon atmosphere at 850C for 160 hours and brine-quenched to produce
disordered crystals. (Some short-range order, formed on quenching, is
most likely present).

3.2 Electrochemical Measurements
The solutions, aqueous 2% ferric chloride (0.13M, pH = 1.7) and 0.6M
NaCl (pH = 6.7), were prepared from reagent grade materials and triply-
distilled water. All potential measurements are reported on the saturated
KCl calomel scale (sce). Experiments with 0.6M NaCl were carried out
with oxygen-free solutions. For potentiostatic polarization measurements,
each current reading was performed after a time delay sufficient to reach
a quasi-steady state (5 minutes for $FeCl_3$ and 20 minutes for NaCl following
a 50 mv step).

3.3 Stress-Corrosion Tests
Samples were immersed while under constant deflection (three-point
bending) and potentiostatic control. These tests were performed with a
bending jig made of "Teflon" wih glass-rod end supports, spaced 15 mm
apart. A line-contact central loading wedge was driven by a hand-turned
screw giving a deflection of 1 mm per rotation. A deflection of 0.1 mm
was sufficient to attain a calculated stress in the tensile surface which
was comparable with the critical resolved shear stress for single crystals
[16]. Resulting fracture surfaces were observed without further preparation using an Hitachi X-650 Scanning Electron Microscope (SEM), with an
ultimate resolution of 60 Å.

4. RESULTS
4.1 Polarization Studies

4.1.1. <u>Aqueous Sodium Chloride.</u> (0.6M NaCl, O_2-free): Figure 1 presents results for anodic polarization from a rest potential of ~ -450 mv. This potential is ~ 730 mv more active than the reversible potential for copper with respect to Cu^+ and is due to the strong complexing effect of the chloride ion. This potential is also ~ 205 mv more noble than the reversible potential for hydrogen in this solution, precluding hydrogen-ion reduction on the alloy surface. On anodic polarization, a low-current region (~ 2-4 micro-amp/cm^2) is observed up to a "critical" potential, E_C = +430 mv, above which the current increases by several orders of magnitude within ~ 50 mv. Hashimoto et al. have reported similar results for this alloy in air-saturated NaCl [17,18]. Similar behavior has previously been reported for acid sulfate solutions [19,20], with E_C = +750 mv [20], a value considerably more noble than observed here in neutral chloride solutions. In all of these studies, a dealloyed layer consisting of a "sponge" of gold-rich filaments has been observed above E_C. Hashimoto et al. [17,18] observed the start of gold dissolution for this alloy in the range 600-800 mv. However, even at these high potentials and large currents (~ 0.5 A/cm^2) the dissolution is predominantly selective. Thus, both below and for a considerable potential range above E_C, corrosion occurs by exclusive selective dissolution of copper.

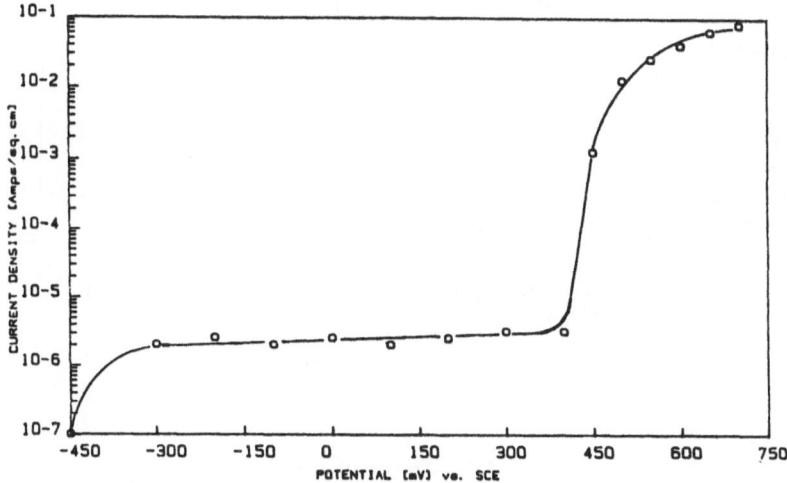

FIGURE 1. Quasi-steady-state potentiostatic anodic polarization curve for Cu-25 at% Au single crystal in O_2-free 0.6M NaCl.

4.1.2 Aqueous 2% Ferric Chloride (0.13M FeCl$_3$). Figure 2 presents results for both anodic and cathodic polarization of single crystal specimens in the range 0 to 800 mv. As is the case for NaCl, the anodic branch is due exclusively to the dissolution of copper, and the current rises abruptly as soon as the potential is raised above the corrosion potential of +460 mv.

A gold-rich sponge is produced on the surface during corrosion and the dissolution rate, i, was estimated from the measured film thickness, Δx formed in time t, using Faraday's Law: $i = (zF\rho\Delta X)/(Wt)$, assuming formation of Cu(I) ions (z = 1) and using the normal density and atomic mass (ρ,W) of metallic copper. Results are tabulated in Table I and show good agreement with the measured value at 600 mv.

Table I also shows that film formation occurs under cathodic polarization and is not confined to the potential region at or above the corrosion potential. Comparing the data of Fig. 1 and 2, it is seen that the anodic

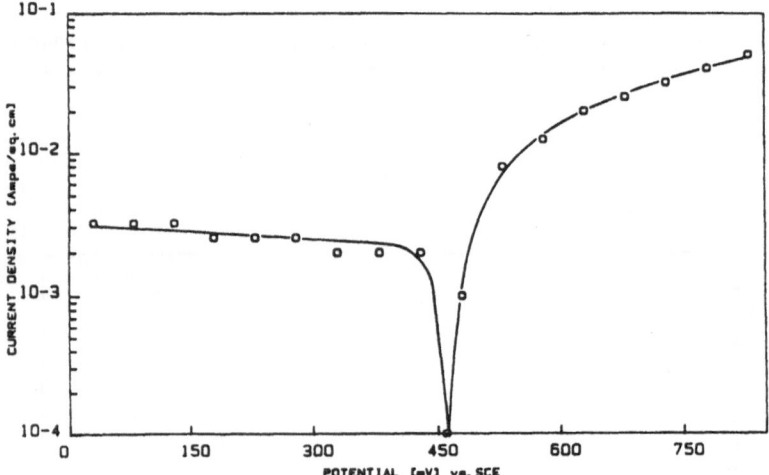

FIGURE 2. Quasi-steady-state potentiostatic polarization curve for Cu-25 at% Au single crystal in 2% (0.13M) FeCl$_3$.

TABLE I. Estimation of Electro-Dissolution Rates for Copper in 2% FeCl$_3$ From Measured Thickness of Dealloyed Layers	E (mv, sce)	t (days)	ΔX (μ)	$\Delta X/t$ (n/ms)	i (ma/cm^2)
	460	30	>750	0.30	>0.4
	460	10	30	0.025	0.05
	460	3	300	1.16	1.6
	600	0.08	80	11.6	16
	0	15	~ 4	0.0035	0.005

branch for $FeCl_3$ practically coincides with the data for NaCl above E_c. Thus, we may infer that a critical potential, close to that found for NaCl ($E_c \sim 430$ mv), also exists for reaction in aq. $FeCl_3$. The occurance of a rest potential slightly above this value is explained by Figure 3 which shows schematically the establishment of the corrosion potential as a mixed potential due to the interaction of the cathodic reaction ($Fe^{3+} + e^- = Fe^{2+}$) with the anodic reaction ($Cu = Cu^+ + e^-$). Because the anodic reaction increases abruptly above the critical potential, variations in the exchange current density or the reversible potential (fixed by the Fe^{3+}/Fe^{2+} ratio) of the cathodic reaction cause only relatively small variations of the corrosion potential but may produce large variations in the corrosion rate. These features are illustrated in Figure 3 and may explain the large variation ($\sim \times 10^3$) in the rate of sponge formation at the open-circuit potential as shown in Table I and reported in previous studies [9,10,15]. We have assumed that below E_c, the quasi-steady-state dissolution rate of copper in $FeCl_3$ is of the same order of magnitude as that in NaCl, since a value very nearly equal to this is also reported for acid sulphate [20]. Moreover, as shown in Table I the rate at 0 v was estimated from the measured thickness of the gold sponge layer formed on the alloy, and the calculated value (~ 5 $\mu amp/cm^2$) supports this assumption. Other published data for potentiodynamic studies in chloride media, reviewed by Kuhn [21], suggest that the rate may vary, and comparative studies of transient and quasi-steady-state electrodissolution are presently being conducted to resolve this question [22]. The significance of this issue to the mechanism of T-SCC is discussed below.

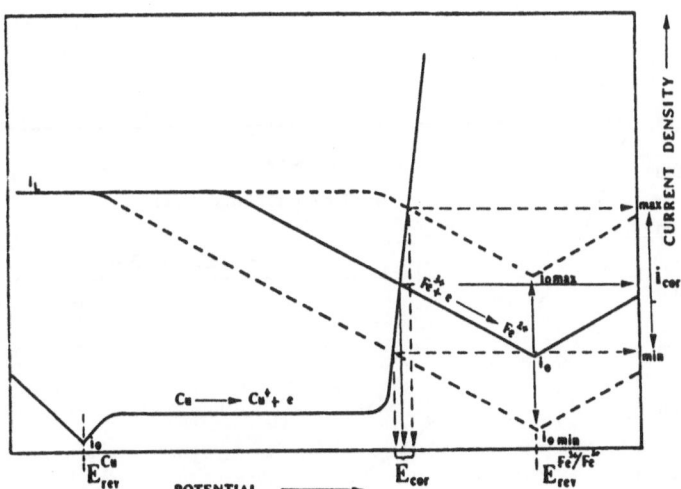

FIGURE 3. Schematic Evans diagram for Cu-25 at% Au in 2% $FeCl_3$. Diagram illustrates the large variation in corrosion rate due to possible variations in the exchange current for the Fe^{3+}/Fe^{2+} redox reaction. (A variation in E_{rev} for the redox reaction can produce a similar effect.)

4.2 Stress-Corrosion Experiments

For the constant deflection tests used in this study, failed samples appeared "brittle" over the entire fracture surface (Figure 4). Characteristic T-SCC fracture surfaces were produced in both media at potentials both below and above E_c. Except as noted below, failure occured at relatively low applied stress: the calculated initial stress in the tensile surface of the bent crystal was comparable to the critical resolved shear stress of ~ 50 MPa [16]. Figure 5 presents the time-to-failure results for tests carried out in both NaCl (O_2-free) and FeCl$_3$. These results suggest a distinct different in the potential-dependence of T-SCC for the two media: (i) for NaCl, cracking was not observed below 400 mv, a value only 30 mv more active than E_c; whereas, relatively rapid failure (1.5 hrs) occurred at (and above) 450 mv, a potential slightly more noble than E_c. At the potential of 400 mv the quasi-static anodic dissolution

FIGURE 4. Fracture surface for Cu-25 at% Au single crystal tested in 2% FeCl$_3$ under cathodic polarization (250 mv; t.t.f. ~ 5 hrs). Arrows indicate direction of crack growth.

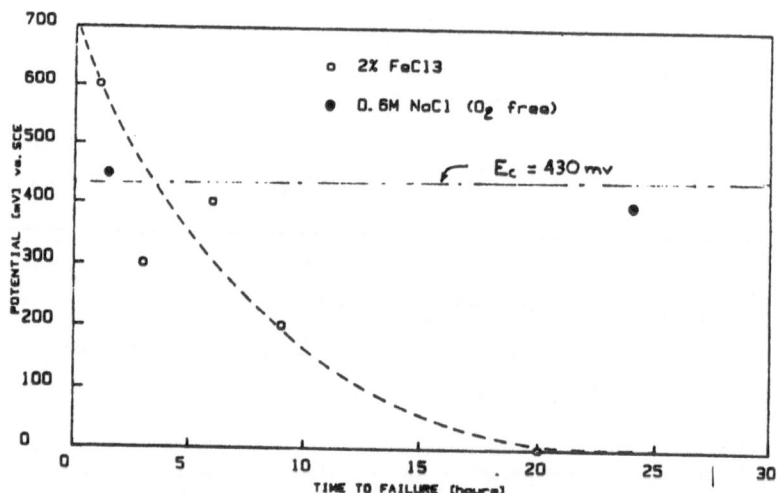

FIGURE 5. Effect of potential on time-to-failure of Cu-25 at% Au single crystals under constant deflection (three-point bending) in (i) O_2-free 0.6M NaCl (0.6 mm sample thickness) and (ii) 2% FeCl$_3$ (1.5 mm sample thickness).

rate lies in the low-current region (see Figure 1), and the measured value of ~ 4 micro-amp/cm^2 is only slightly greater than the rate at a potential of -300 mv (~ 2 microamp/cm^2); (ii) In sharp contrast to the above, in FeCl$_3$ the time-to-failure increases continuously as the potential decreases (becomes more active), with no apparent discontinuity occuring at E$_c$. The limit of susceptibility under constant deflection appears to be reached at a potential of 0 v. In repeated tests at this potential an increased deflection was sometimes required to promote failure which, moreover, did not always occur. That this was truly a limiting potential for T-SCC under the constant deflection was demonstrated by the following experiments: A sample under constant deflection was anodically polarized to 600 mv and maintained at this potential until multiple cracks were observed on the tensile surface. The potential was then lowered to 0 v and no further crack growth could be detected after 4 days.

Figure 6 is a SEM micrograph of fracture surfaces obtained in 0.6 M NaCl. No significant differences in morphology can be detected between fracture surfaces produced in NaCl or FeCl$_3$, and all fracture surfaces have substantially the same appearance above or below E$_c$. Cleavage-like facets separated by serrated steps are observed at all potentials, and undercutting of the steps is frequently seen. The orientation of the facets was determined as {110} from a single-surface trace analysis of the fracture surface produced in FeCl$_3$ [9]. (see also Figure 2 of Ref. 10). These features are strongly suggestive of cleavage events which are induced by the corrosive media and the applied stress. No evidence of massive dealloying (general presence of gold sponge) was detected in the SEM at a resolution of ~ 200 A. Microprobe analysis indicated the nominal composition of the fracture surface (effective sampling depth of 1 micrometer) to be equal to the composition of the uncorroded alloy (25 a/o gold). In other experiments it was possible to section through an arrested crack and to determine the composition of the alloy located in a region along the crack approximately 0.1 micrometer below the fracture surface. The nominal composition persists in this near-surface region for most of the length of the arrested crack.

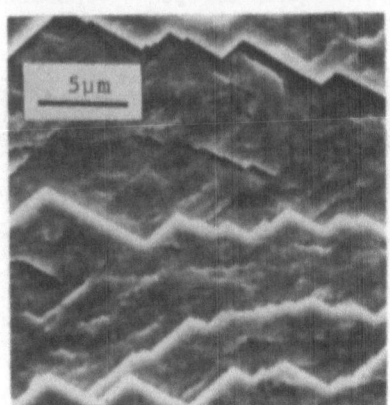 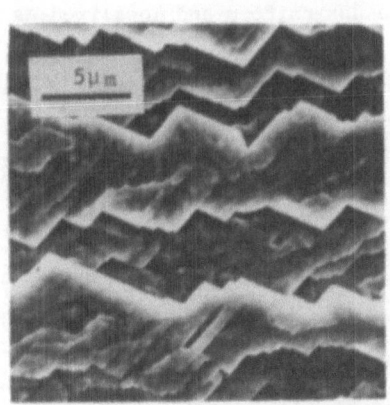

(a)   (b)

FIGURE 6. SEM micrographs of fracture surfaces produced on bending in 0.6M NaCl for crystals anodically polarized from the rest potential of -450 mv; viewed normal to the {110} fracture facets. (a) 400 mv (slightly below E$_c$); (b) 450 mv (slightly above E$_c$).

5. DISCUSSION

5.1. Selective Copper Electro-Dissolution

To a first approximation the electro-dissolution of copper in the two chloride media proceeds at the same rate over the investigated potential range. One might suspect that the large increase in rate occurring at and above E_c is due to gold dissolution; however, our previous observation [10] of a sample immersed in aq. $FeCl_3$ for 30 days at the corrosion potential (~ 460 mv) demonstrated exclusive copper dissolution. Further-more, Hashimoto et al. [17,18] found no appreciable gold dissolution below potentials of 600-800 mv for potentiodynamic polarization (0.67 v/min) of this alloy in air-saturated 0.6M NaCl, and our value of +430 mv for E_c agrees with their earlier finding.

5.2. Sponge Formation

Sponge formation appears to be characteristic of the dealloying process in aq. $FeCl_3$, both above and below E_c. Because the dissolution rates are comparable in both media, we may infer that sponge formation also occurs below E_c during electrodissolution of copper in aq. NaCl, although we have not directly observed this in the present study. Work is pre-sently in progress to determine the potential domain of sponge formation in aq. NaCl and to compare the transient dealloying rates in both media [22]. We can presently offer no explanation as to why the rate of selec-tive dissolution and consequent sponge formation increases discontinuously at a unique potential, characteristic of the media. The explanation given by Pickering [23] whereby sponge formation was predicted to occur only above E_c must be questioned in view of the present observation.

5.3. Role of Sponge Formation in T-SCC

Figure 7a, which shows the detailed nature of the gold sponge at high resolution, is the "fracture surface" of a sample which had experienced complete loss of copper after 30-day, stress-free immersion in $FeCl_3$ [10]. The sample failed in a brittle manner with only the smallest deflection. Our previous analysis [10] showed that although the sponge is "brittle" on a macro-scale, failure actually occurs by ductile shear of the thin fibres that comprise the sponge. As shown in Figure 7b, sponge formation can be non-uniform and localized as a consequence of plastic straining,

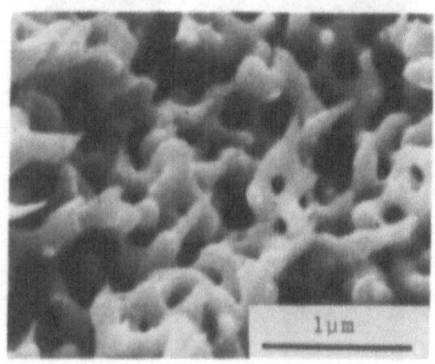 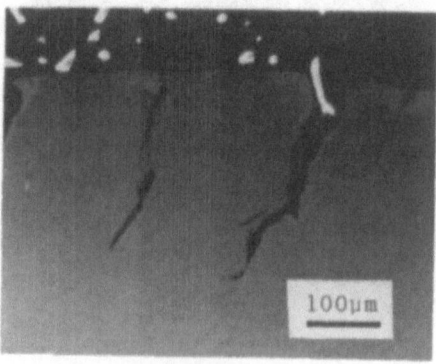

(a)   (b)

FIGURE 7. (a) High resolution SEM micrograph of gold sponge structure formed after 30-day immersion in aq. $FeCl_3$; (b) Non-uniform sponge formation on bent samples immersed in $FeCl_3$. Cracks appear to have nucleated in vicinity of gold sponge.

and stress-corrosion cracks appear to nucleate in these regions of accelerated sponge formation. This is understandable as a consequence of the observed "brittle" nature of the sponge. However, the greater part of the crack-propagation seems to occur <u>without</u> observable sponge formation occuring over the fracture surface (to within a resolution of ~ 20 nm). Thus, sponge formation may be an important source of surface-nucleated cracks, producing the necessary conditions for continued crack propagation, but an alternative mechanism may be needed to explain continuing crack growth. On the other hand, if sponge formation is also characteristic of the very early stages of reaction on a freshly exposed surface, then it is possible to imagine that a crack could propagate by the formation of a very thin coherent film (<20 nm) which, if it is of a brittle nature, could fail by local stresses induced by the coherency strain, the latter arising from the variation in composition and lattice parameter in the dealloyed layer. The crack growth rate would be determined by the dissolution rate, a requirement which is not supported by the present results (see calculations presented below). An additional difficulty of such a picture is that it cannot account for the morphological or crystallographic features of the fracture surface, nor can it account for crack-arrest distances, typically ~ 5 micrometers.

5.4. Role of Potential

The range of potential over which these observations have been made seems to preclude the involvement of hydrogen in the process for this system. Modeling calculations carried out by Bertocci [24] show that even at the crack tip the potential for the Cu/Cu^+ reaction remains well above the hydrogen evolution potential in spite of the complexing action of Cl^-. Thus, an explanation still needs to be sought based on the dealloying characteristics.

5.5. Mechanism

Slip-dissolution models require that the average rate of crack advance be of comparable magnitude to the rate of anodic dissolution at the crack tip. A mechanism which involves mechanical cleavage, on the other hand, may be conceived as involving a "film forming" anodic dissolution event, which induces the discontinuous propagation of a crack far into the normally ductile metal or alloy [6]. Recent evidence for just such a process has been presented by the authors [10]. Using data from Table I and Fig. 5, we may compare the average dealloying rate with the average crack-growth rate in $FeCl_3$ at two potentials. The results shown in Table II indicate that the crack growth rate far exceeds the dealloying rate at both potentials: by a factor of ~ 7000 at 0 v and by a factor of ~ 15 at 600 mv. This result provides additional support for the mechanical cleavage mechanism.

	E	$\Delta \dot{x}$ (nm/s)	v_c (nm/s)
TABLE II. Dealloying Rate, Δx, Compared with Avg. Crack-Growth Rate, v_c for Cu-25 Au in aq. $FeCl_3$ at Two Potentials	0	3×10^{-3}	20
	600	12	180

668

6. ACKNOWLEDGEMENTS

The authors appreciate helpful discussions with Drs. K. Sieradzki and
E. N. Pugh. This work was supported in part by the Joe Tatum Fund for
Corrosion Research at Vanderbilt University and also by a grant from the
National Science Foundation (Grant NO. NSF DMR 8420965).

REFERENCES

1. Champion, F. A., Symposium on Internal Stresses in Metals and Alloys,
    p. 468, Institute of Metals, London, 1948; H. L. Logan, J. Res. Nat.
    Bur. Stds., Vol. 48, p. 48 (1952).
2. Vermilyea, D. A., J. Electrochem. Soc., Vol. 119, p. 405 (1972).
3. Scully, J. C., Corrosion Science, Vol. 15, p. 207 (1975).
4. Bursle, A. J. and E. N. Pugh, Environment-Sensitive Fracture of
    Engineeirng Materials, (Ed.) Z. Foroulis, p. 18, The Metallurgical
    Society of AIME, 1979.
5. Pugh, E. N., Atomistics of Fracture, (Eds.) R. M. Latanision and J.
    R. Pickens, p. 997, Nato Conference Series VI, Vol. 5, Plenum Press,
    1983; Corrosion, Vol. 41, p. 517 (1985).
6. Sieradzki, K. and R. C. Newman, Philos. Mag., Vol. 51, p. 95, (1985).
7. Meletis, E. I. and R. F. Hochman, Corrosion Sci., Vol. 26, p. 63
    (1986).
8. Paskin, A., K. Sieradzki, P. K. Som, and G. J. Dienes, Acta Metall.,
    Vol. 30, p. 1781 (1982); Acta Metall., Vol. 31, p. 1253 (1983).
9. Lichter, B. D., T. B. Cassagne, W. F. Flanagan, and E. N. Pugh,
    Microstructural Science, Vol. 13, p. 361 (1985).
10. Cassagne, T. B., W. F. Flanagan, and B. D. Lichter, Metall. Trans.,
    Vol. 17A, p. 703 (1986).
11. Pickering, H. W. and P. R. Swann, Corrosion, Vol. 19, p. 373t (1963).
12. Swann, P. R., The Theory of Stress-Corrosion in Alloys", (Ed.) J. C.
    Scully, p. 113, NATO, Brussels, 1971.
13. Swann, P. R. and J. D. Embury, High Strength Materials, (Ed.) V. F.
    Zackay, p. 327, John Wiley, New York, 1965.
14. Silcock, J. M. and P. R. Swann, Environment-Sensitive Fracture of
    Engineering Materials, (Ed.) Z. Foroulis, p. 133, The Metallurgical
    Society of AIME, 1979.
15. Kuramoto, E., and D. P. Pope, Philos. Mag., Vol. 33, p. 675 (1976).
16. Bakish, R. and Robertson, W. D., Acta Metall., Vol. 4, p. 342 (1956).
17. Hashimoto, K., T. Goto, W. Suetaka, and S. Shimodaira, Trans. JIM,
    Vol. 6, p. 107 (1965).
18. Hashimoto, K. and T. Goto, Corros. Sci., Vol. 5, p. 597 (1965).
19. Gerischer, H. and H. Rickert, Z. Metallkunde, Vol. 46, p. 681 (1955).
20. Pickering, H. W. and P. J. Byrne, J. Electrochem. Soc., Vol. 118,
    p. 209 (1971).
21. Kuhn, A. T., Surface Technology, Vol. 13, p. 17 (1980).
22. Massinon, D., et al., Vanderbilt University, Nashville, TN, un-
    published research (1986).
23. Pickering, H. W., Corros. Sci., Vol. 23, p. 1107 (1983).
24. Bertocci, U., Embrittlement by the Localized Crack Environement,
    (Ed.) R. P. Gangloff, The Metallurgical Society of AIME (1984).

DISCUSSION

Comment by R. C. Newman:

1. The apparent change in the potential range for SCC in $FeCl_3$ solution (as compared with NaCl) is an IR drop effect due to the high cathodic current at the lower potentials in $FeCl_3$ solution.
Therefore, the observation of SCC and sponge formation at "low" potentials is an artifact and does not relate to the importance of the critical potential.
2. The deflection of the cracks propagated through the sponge (formed without stress) could be related to the requirement of a specific cleavage plane in the substrate but not in the coarse part of the sponge.
3. The observation of ligament necking in the coarse part of the sponge does not rule out brittle behavior in the fine-scale porous region near the interface.

Reply:

I agree with your comments (2) and (3). However, regarding comment (1) (sponge formation below E_c), I believe your claim must be considered speculative until a calculation of the IR drop is attempted. The definitive experiment is to see if sponge will form below E_c in NaCl. In fact, we have observed evidence of sponge at 400 mV in $NaCl$, which is slightly below E_c. Furthermore, since the sample is anodically polarized from -450 mV_c, the same reasoning concerning the IR drop would predict a <u>lower</u> potential at a propagating crack tip.

Comment by H. J. Engell:

What is the stress intensity for a metal sponge compound at the end of a crack in the sponge? What are the dynamic properties of a crack in the metal sponge? Are both appropriate to induce a brittle crack into the otherwise ductile metal?

Reply:

The question of the nature of the sponge near the alloy-sponge interface is still an open one. The pore size in this domain (1 to 10 nm) is obviously much smaller than that observed far away from the interface or in a "bulk sponge" produced by total dealloying of a sample. I would speculate that pore size would be of "atomic dimensions" at the interface, thus providing the necessary geometry for stress concentration. Theoretical analysis of crack dynamics for such a system is still incomplete. We can only present our observations of the crack propagating into the alloy.

ROLE OF SULFUR IN THE FORMATION AND THE BREAKDOWN OF PASSIVE FILMS.

P. MARCUS and J. OUDAR
Laboratoire de Physico-Chimie des Surfaces associé au C.N.R.S.
Ecole Nationale Supérieure de Chimie de Paris
11, rue Pierre et Marie Curie - 75005 PARIS - FRANCE

ABSTRACT

The role of sulfur in the passive film formation on nickel and nickel-base alloys has been extensively studied by using radiotracer and surface analysis techniques (AES and ESCA). The main conclusions are the following :
1. The effect of adsorbed sulfur is independent of its origin : preadsorption from gas or by surface segregation, surface enrichment by selective metal dissolution, adsorption from the electrolyte.
2. Sulfur increases the rate of metal dissolution in the active range of potentials.
3. Sulfur inhibits the formation of the passive film.
4. Sulfur induces the irreversible breakdown of the passive film to give localized or generalized corrosion.
The mechanisms of these effects are discussed.

INTRODUCTION

Extensive investigations of the effects of sulfur on the dissolution and passivation of nickel (1-3,7) and nickel alloys (4-6) have been conducted over the last decade. These studies were performed on well defined systems : high purity materials were used, single crystals were used in a first step, polycrystals in a second step to account for the role of the grain boundaries, sulfur was either preadsorbed in H_2S-H_2, dissolved in the bulk or added to the electrolyte. H_2S was labelled with radioactive sulfur (^{35}S) so that the sulfur concentrations could be quantitatively measured and the changes during the corrosion processes could be detected. The surface composition and structure were analysed by LEED, AES, ESCA and the radiochemical technique at the various stages of the electrochemical tests. This approach enabled us to elucidate the mechanisms of the action of sulfur on the dissolution and passivation of nickel. The main conclusions are summarized (8) and the detailed results can be found in the papers in the reference list.

1. ANODIC DISSOLUTION OF NICKEL : THE ROLE OF SULFUR

The main results are the following :
- In the active range of potentials sulfur adsorbed on nickel accelerates the dissolution (see Figures 1 and 2).
- The adsorbed sulfur remains on the surface, it is not consumed in the dissolution reaction, i.e. it acts as a catalyst (Figures 1 and 2). In this way a monoatomic layer of sulfur can have a promoting effect on the dissolution of macroscopic amounts of material.
- We suggested that this effect was due to i) a weakening of the metal-metal bonds (2), and ii) a possible effect of the nickel-sulfur dipoles which would assist the passage of the cations to the solution (2).

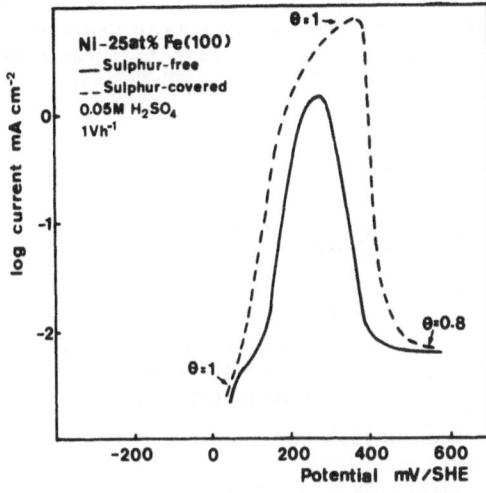

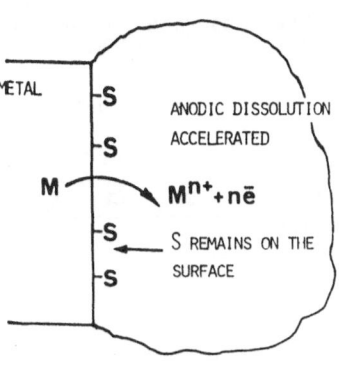

Figure 1 : Effects of adsorbed sulfur on the anodic polarisation curve of Ni-25%Fe alloy (from ref. 5).

Figure 2 : Schematic represent-ation of the accelerating effect of sulfur on the anodic dissol-ution.

2. PASSIVATION OF NICKEL : THE ROLE OF SULFUR

The main conclusions are the following :
- Sulfur blocks the passivation of nickel. The passive film cannot be formed on a surface covered by a complete monolayer of sulfur (see Fig.1).
- A desorption of > 20% of the monolayer is necessary to permit the formation of the passive film (see Fig.1).
- The mechanism of this effect is that sulfur blocks sites for adsorpt-ion of OH groups. Hydroxyl groups are still adsorbed, as observed by ESCA, but they are diluted by sulfur and therefore cannot recombine to yield the passive oxide. This is schematically shown in the following model :

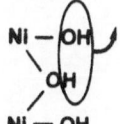

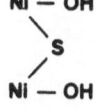

In this configuration (without sulfur) the formation of the passive film is favored.

In this configuration (with sul-fur) the formation of the passive film is precluded.

Figure 3 : The effect of sulfur on the passivation of nickel (from ref.7).

- After desorption of > 20% of the monolayer, the passive film can be formed but residual sulfur amounts of up to 80% of the monolayer can rem-ain at the metal/passive film interface. Sulfur at the interface can chan-ge the properties of the passive film : by changing e.g. the structure of the film, as shown by RHEED for nickel (2).

3. THE SULFUR SOURCES

Three sources of sulfur must be considered : sulfur can be adsorbed from a gas (this was the case for the experiments reported above), from a liquid, or it can originate from the bulk. From the bulk it can segregate to the surface by diffusion, at high temperature, or by anodic segregation. When adsorbed on the surface, sulfur has the same effects irrespective of its origin (2,5).

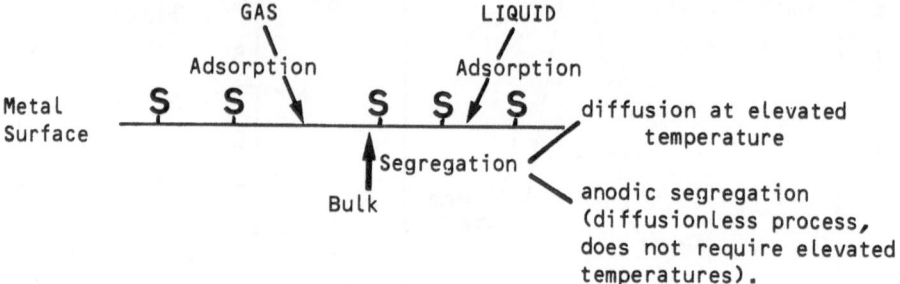

Figure 4 : The sulfur sources (S on the surface has the same effects irrespective of its origin).

4. ANODIC SEGREGATION OF SULFUR

The concept of anodic segregation of sulfur is the following :
- Bulk sulfur can accumulate on the surface during the dissolution of the metal (see Figure 5).

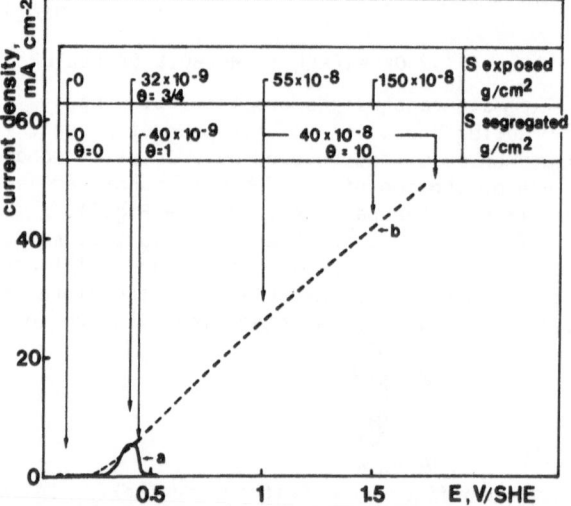

Figure 5 : The i-E curves of a) sulfur-free and b) sulfur-doped (S=50ppm) Ni(100). The table shows the measured values of the sulfur enrichment on the surface (S segregated) and the amount of sulfur contained in the dissolved part of the alloy (S exposed) (from ref. 2).

- In a first stage, an adsorption layer is built up, with all effects described before, and particularly the blocking of the passivation.
- In a later stage, a nickel sulfide is formed, which has no passivating property. This sulfide is formed <u>during</u> the corrosion process, and does not originate from sulfide inclusions. It reaches a stationary thickness (see Figure 6).

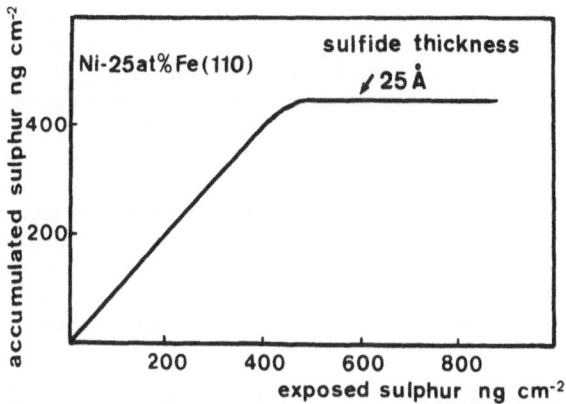

<u>Figure 6</u> : Surface enrichment of sulfur on Ni-25%Fe alloy (in ng cm^{-2}) vs the amount of bulk sulfur in the dissolved volume of material (from ref.4).

- The passive film is not stable on such surfaces.
- The surface enrichment of sulfur during the corrosion process is represented in Figure 7.

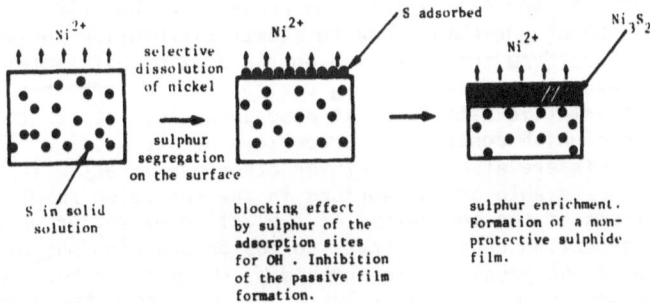

<u>Figure 7</u> : Anodic segregation of sulfur on nickel and consequences on the corrosion behaviour.

5. LOCAL EFFECTS

The results reported above were obtained on homogeneous surfaces but the same conclusions hold for local effects of sulfur on heterogeneous surfaces (2), e.g. at grain boundaries, with two additional factors to be considered : i) regions of structural defects dissolve more rapidly, which will cause a more rapid enrichment of sulfur by anodic segregation. The presence of sulfur will then locally promote the dissolution (resulting in an increased sulfur enrichment) and preclude the formation of the passive film, causing localized corrosion and ii) sulfur can be initially segregated in the grain boundaries, and this can considerably enhance the effects described above.

6. SULFUR-INDUCED BREAKDOWN OF THE PASSIVE FILM

In the passive state, cations slowly diffuse through the oxide film and go into the solution at the outer part of the film. Starting from a passive film formed on a bare metal surface, sulfur initially contained in the consumed metal is progressively enriched at the metal/oxide interface. For a critical sulfur concentration reached at this interface the breakdown of the oxide occurs. This critical concentration is of the order of one monolayer (9). After breakdown of the film the presence of sulfur on the metal surface definitively hinders the repassivation, in agreement with the preceding conclusions. Knowing the bulk sulfur content and the residual current in the passive state, the induction period before breakdown of the film can be predicted. This time to initiation is on the order of 100 hours for a bulk sulfur content of 50×10^{-6} (residual current 7 µA cm^{-2}) (9). The presence of segregated sulfur in structural defects, such as grain boundaries, shortens the incubation time for the appearance of localized attack (9).

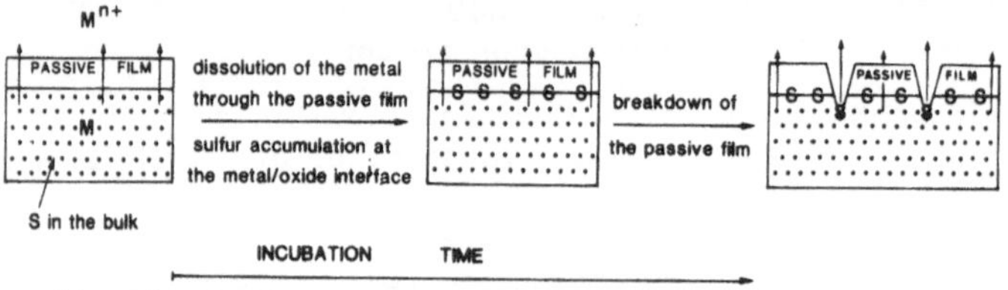

M^{n+}

PASSIVE FILM	dissolution of the metal	PASSIVE FILM		PASSIVE FILM
M	through the passive film		breakdown of	
	sulfur accumulation at		the passive film	
S in the bulk	the metal/oxide interface			

INCUBATION   TIME

Figure 8 : Mechanisms of the sulfur-induced breakdown of the passive films.

Finally it is concluded that the incubation time before the catastrophic corrosion starts, increases with (i) decreasing sulfur content (slower sulfur enrichment at the interface and lower probability of sulfur segregation in the defects) and (ii) increasing protective character of the film (slower rate of sulfur enrichment due to slower dissolution current).

Such observations draw the attention on the risk of extrapolating short time laboratory experiments to the long periods of time required in the practical use of metallic materials. It also shows the importance of knowing the mechanisms which control the properties of the passive film.

All these results are also relevant for other non metallic impurities as long as they are stable on the surface in the corrosion conditions. Direct implications of the mechanisms of the action of sulfur on the dissolution and passivation concern stress corrosion cracking phenomena. After local breakdown of the passive film induced by the stress, the repassivation can be hindered by adsorption of sulfur on the bare metal surface. Sulfur can come from the bulk metal by "anodic segregation", or from the electrolyte itself.

7. REFERENCES

(1) - P. Marcus, N. Barbouth and J. Oudar - C.R. Acad. Sci. Paris 280 (1975) 1183.

(2) - J. Oudar and P. Marcus - Appl. Surface Sci. 3 (1979) 48.

(3) - P. Marcus, J. Oudar and I. Olefjord - Mater. Sci. Eng. 42 (1980) 191.

(4) - P. Marcus and J. Oudar in Passivity of Metals and Semiconductors, Ed. M. Froment, Elsevier (1983) 119.

(5) - P. Marcus, A. Teissier and J. Oudar - Corrosion Sci. 24, (1984) 259.

(6) - P. Marcus, I. Olefjord and J. Oudar - Corrosion Sci. 24 (1984) 269.

(7) - P. Marcus and J. Oudar in Fundamental Aspects of Corrosion Protection by Surface Modification, eds. E. Mc Cafferty, C.R. Clayton and J. Oudar, The Electrochemical Society (1984) 173.

(8) - A similar presentation of sections I to V was given by P. Marcus at an EPRI Workshop on the role of sulfur species on the degradation of alloy 600 and similar alloys, Baltimore, 17-18 october 1985. Section VI is based on chapter IV of the thesis of H. Talah, presented at the University of Paris, 5 mai 1986.

(9) - H. Talah, Thesis, University of Paris (1986) and to be published.

DISCUSSION

Comment by R. C. Newman:

Russ Jones pointed out that the retention of sulphur on the walls of a crack leads to blunting and is not a severe IGSCC situation. However, in nickel alloys such as sensitized alloy 600, the wall has enough Cr to passivate and very severe SCC can occur in sulphur-containing environments. Many thousands of steam generator tubes cracked at Three Mile Island because of traces of thiosulphate ions. I wonder if segregated sulphur would have a strong effect on IGSCC of sensitized Alloy 600 in non-sulphur containing acids?

Reply:

The fact that S accelerates the dissolution may cause an acceleration of IGSCC if the walls of the crack are passivated and Cr may favor passivation of the walls of the cracks in Ni alloys in the presence of sulphur. Experiments on Ni-Fe-Cr alloys with sulphur are in progress.

Comment by D. D. Macdonald:

Exactly what form is the sulfur in on the nickel surfaces?

Reply:

Acceleration of the anodic dissolution and hindering of the passive film formation were observed with sulfur in the adsorbed state. Accumulation of bulk sulfur on the surface during dissolution of nickel, which we called "anodic segregation", leads to the formation of nickel sulfide. It is pointed out that this sulfide did not pre-exist, but was formed by surface reaction during the corrosion process.

Comment by J. P. Fidelle:

In one of your last slides, what was the driving force for sulphur diffusion?

Reply:

The sulfur enrichment at the metal/oxide film interface is a diffusionless process. It is caused by the selective dissolution of the metal. The electric field at anodic potentials is also in favor of this accumulation of S under the film.

Comment J. P. Fidelle:

Were you in a case where S did not prevent the formation of a passive film, but succeeded to damage it later? Or was the diffusionless accumulation of S at the metal/oxide interface due to residual corrosion?

Reply:

The passive film was formed on the bare surface. Then, the residual corrosion in the passive range is responsible for the progressive enrichment of S at the metal/oxide interface. Damaging of the film occurs when the concentration of S at the interface reaches a threshold of about one monolayer.

Comment by R. M. Latanision:

I wonder about the mechanism by which S will increase the oxidation rate of active nickel? Your view is that S reduces the strength of Ni-Ni bonds, therefore allowing an increased dissolution rate. In an electrochemical sense, one would expect that S should decrease the exchange current density for the cathodic partial process (hydrogen evolution) and this should decrease the corrosion rate at the mixed potential. Is it your view that, instead, S reduces the Tafel slope for the anodic or cathodic partial processes or effects the exchange current density for Ni dissolution?

Reply:

At anodic potentials, where there is no contribution of the cathodic reaction, we find that sulfur increases the dissolution current density, and does not cause striking changes of the Tafel slope. This supports the view that S weakens the Ni-Ni bond strength and lowers the energy barrier for the dissolution of Ni. At cathodic potentials, we find a structure dependent and S coverage-dependent effect on the hydrogen evolution reaction. On Ni (100), S reduces the cathodic current.

Comment by D. R. Baer:

We find that the supply of sulfur influences to some extent what is seen. In the active region, sulfur that comes to the surface during dissolution accumulates up to saturation coverage (as you observe); and at that point additional sulfur is dissolved so that only that saturation coverage remains. In this condition, excess bulk sulfur does not accumulate at the surface during dissolution. If Ni with a surface saturated in S is moved to a high enough potential in the passive region, some sulfur will be oxidized and the surface may passivate. To totally block passivation, some continued supply of S to the surface seems necessary in this case. These results do not contradict any you have presented, but the different type of experiments point out the different ways sulfur can act and how it can move around.

Reply:

We have shown indeed that in order to continuously block the passivation, S must be supplied to the surface because S present on the surface can be oxidized at high potentials. (We have shown this for Ni, Fe-Ni alloys and Pt). In the case of Ni and Fe-Ni alloys (and possibly other alloys), S in the bulk can act as a source of sulfur that compensate for the desorption (or oxidation) of S during dissolution. We have also shown that similar effects can be obtained with a H_2S-containing electrolyte.

Comment by R. H. Jones:

Your conclusion that S will accelerate SCC growth rates is not universally observed. In Ni with S segregated to the g.b.'s tested in 1 N H_2SO_4
at 900 mV (SCE), IGA and IGSCC only propagated about 0.1 mn. This was
interpreted as resulting from S on the IGA and IGSCC walls which kept them
in the active corrosion regime. SCC requires an active tip and passive
walls so the degradation effect of S turns out to be positive because the
S remains on the wall.

Reply:

My conclusion was that S does accelerate anodic dissolution and therefore may accelerate SCC growth rate. I agree that SCC requires an active
tip and passive walls. It is conceivable that this may be the case with
some Ni alloys. Experiments are underway to elucidate this.

BREAKDOWN AND FORMATION PROCESS OF TITANIUM OXIDE FILM DURING STRAINING IN AQUEOUS ELECTROLYTE

X. G. ZHANG and J. VEREECKEN
Dienst Metallurgie-Elektrochemie, Vrije Universiteit Brussel,
Brussels, Belgium

1. INTRODUCTION

Stress corrosion cracking (SCC) is a combined phenomenon due to the reaction at the interface of stressed material and electrolyte. Study of the behavior of the oxide film which separates the metal and the electrolyte is of great importance in understanding the mechanisms of SCC.

When mechanical stress is applied, the oxide film will be deformed together with the underlying material or it will crack depending on the plasticity of the oxide (1,2). However, an oxide film, even having very good plasticity, will crack when it is strained beyond certain extent. It is the case with SCC. Any oxide film at a crack tip will break due to the extremely high strain in the crack tip region. Then the protective property of an oxide film will mainly depend on the repairing process of the broken oxide film, i.e. on the repassivation kinetics of the metal-electrolyte system rather than on mechanical properties of the oxide film.

Beck (3,4) investigated intensively the passivation rate of bare titanium surface generated by fracturing the specimen in the electrolyte. He found that passivation of titanium bare surface in aqueous solutions is a very quick process and the oxide formation follows the high field mechanism. Schiebe (5) found similar results on bare titanium surface generated by mechanical planing the specimen in an electrolyte.

Although passivation of bare metal surface has been investigated intensively, little work has been done to study the oxide film breakdown and passivation process of a metal or alloy which undergoes a continuous damaging, for instance, during straining in an electrolyte.

This paper describes and compares the experiments made to study the electrochemical response of a titanium alloy when it is continuously strained in a neutral aqueous electrolyte and the experiments made to measure the passivation rate of bare surface generated by fracturing the specimen in the same electrochemical conditions.

2. EXPERIMENTS

Test specimens with a total length of 130 mm were machined from an 11 mm Ø Ti-6Al-4V alloy bar. The chemical composition of the alloy is given in Table I.

TABLE I. Chemical composition of the Ti-6Al-4V alloy in weight %

Ti	Al	V	Fe	C	O	N	H
bal.	6.5	4.4	0.21	0.01	0.13	0.01	0.002

The gauge part of the test specimen has a diameter of 3 mm and a length of 30 mm. Before testing, the specimens were polished with 1200 emery paper; and 3.5% NaCl was used as electrolyte. A titanium cylinder was used as an auxiliary electrode and a saturated calomel electrode (SCE) as reference. The tests were conducted at room temperature and in open air.

In the fracture experiment, the test specimen was broken with a tensile machine in the electrolyte. The transient current at a fixed potential on the fracture surface was digitally recorded with a storage oscilloscope. In the straining experiment, the test specimen was tensiled at a constant strain rate. The current response of the specimen during straining was plotted on an X-t recorder.

3. RESULTS AND DISCUSSION

Figure 1 shows the current density vs time curve from the fracture surface of Ti-6Al-4V in 3.5% NaCl aqueous solution at an imposed electrode potential of 0 v (SCE). The current density has been corrected for surface roughness of which the factor is about 3.4 and was determined with a double layer charging technique. The charging time constant for the system is an order of 10^{-5} s. The charging of the double layer of the fracture surface, therefore, has little influence on the measured current after it has reached its peak value, about 3×10^{-4} s.

It is noted that the high anodic current density starts to decrease within 1 ms after the fracture because of the passivation. After about 10 ms, log i starts to decay linearly with log t.

This linear relationship is characteristic of oxide film formation by the high field mechanism as described by Beck (4). In the first 10 ms the anodic current is due to the metal dissolution and the formation of the first monolayer of the oxide film (5). After the formation of this monolayer, the electrochemical dissolution rate is substantially reduced and almost all the current was consumed to growth of the oxide film. The linear part of the curve can be expressed as:

$$i = Kt^c \tag{I}$$

in which K and c are constants. For the curve in Figure 1 they have the value of 0.535 mA/s^c·cm^2 and 0.88 respectively.

When a specimen is continuously strained in 3.5% NaCl aqueous electrolyte at a fixed electrode potential, an anodic current can be measured. A typical current-time curve is shown in Figure 2. It is noted that when the specimen is strained in the elastic region there is no current. Only when the specimen undergoes plastic straining does anodic current occur. When the straining is stopped, the anodic current decreases towards zero. The characteristic of passivation of a material-electrolyte system does not depend on how the new surface is created. Hence the passivation rate on the new surface created during straining should follow the same equation as that on the fracture surface, i.e. $i = Kt^c$.

The total current measured at time t is the sum of the current distributed on the total new surface created in the whole time period from 0 to t:

$$I = \int_{0.01}^{t} idA = \int_{0.01}^{t} Kt'^c \alpha \pi DVdt' + I_0' \tag{II}$$

in which the lower integration limit is 0.01s because equation (I) is only valid after 0.01 s according to the result of the fracture experiment, I_0' is a constant depending on the initial condition, D is the gauge diameter,

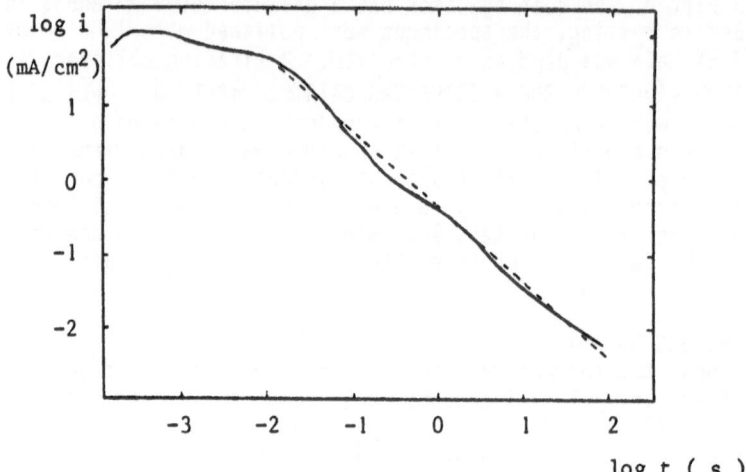

FIGURE 1. Current density-time curve measured on fracture surface of Ti-6Al-4V in 3.5% NaCl aqueous electrolyte at an electrode potential of $0 \ V_{sce}$

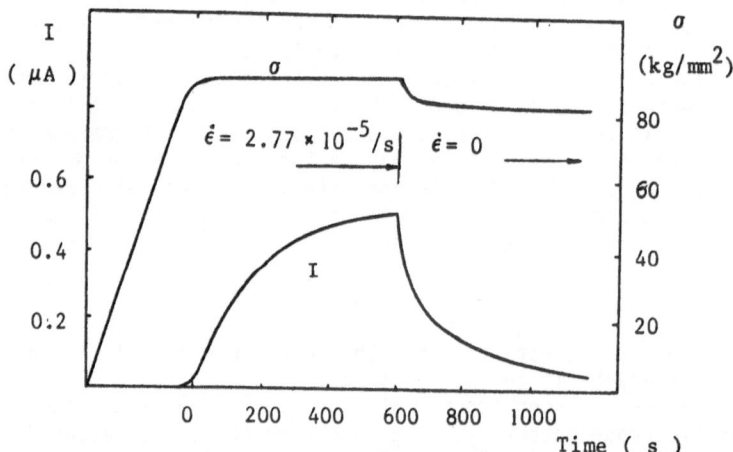

FIGURE 2. Current-time and stress-time curves measured on Ti-6Al-4V during straining in 3.5% NaCl solution at an electrode potential of $0 \ V_{sce}$

V is the crosshead speed of the tensile machine, and α is a constant. Integrating equation (II) gives:

$$I = K't^{1+c} + I_0 \qquad\qquad (III)$$

in which $K' = \dfrac{K}{1+c} \alpha \pi DV$ and $I_0 = I'_0 - \alpha \pi DV \dfrac{K}{1+c} 0.01^{1+c}$.

After the straining is stopped the current can be given by the following equation:

$$1 = \int_{t-t_c+t_0}^{t} Kt'^{c} \alpha \pi DV dt' \qquad\qquad (t > t_c)$$

then

$$I = K'[t^{1+c} - (t - t_c + t_0)^{1+c}] \qquad (IV)$$

in which t_c is the time at which straining is stopped, and t_0 is a constant. When $t = t_c$ equation (IV) is reduced to equation (III) and $K't_0^{1+c} = -I_0$.

The data measured in the experiment were fitted into equations (III) and (IV) taking $K = 0.535$ mA/s^c·cm^2, $c = -0.88$ from the fracture experiment, $D = 0.28$ cm and $V = 8.3 \times 10^{-5}$ cm/s.

Figure 3 shows the measured and calculated current-time curves. The good agreement between the calculated and experimental data indicates that the model reasonably describes the rate of passivation of the new surface created during straining as well as the intrinsic relationship between the straining experiment and the fracture experiment. Some misfitting between the calculated and the measured values in the beginning of the current appearance is due to the fact that before pure plastic straining is reached, there is always a transition region from elastic to plastic behavior, in which the damaged oxide film area is not proportional to the strain rate.

The difference between the current densities on the new surface created during straining and the fracture surface is given by α and $\alpha = 1.6$ for the case discussed above. This is probably caused by the surface roughness factor overestimation in the case of the fracture experiment or due to the real new surface area exceeding the apparent elongated surface area in the case of the straining experiment. The fact that the current density on the new surface created during straining is not less than that on the fracture surface indicates that the surface in both cases is the same in nature.

Passivation behavior at the crack tip plays a great role in SCC. Comparing the characteristics of the breakdown-repassivation process to the SCC susceptibility of different materials in different electrolytes would give significant information about the mechanism of SCC.

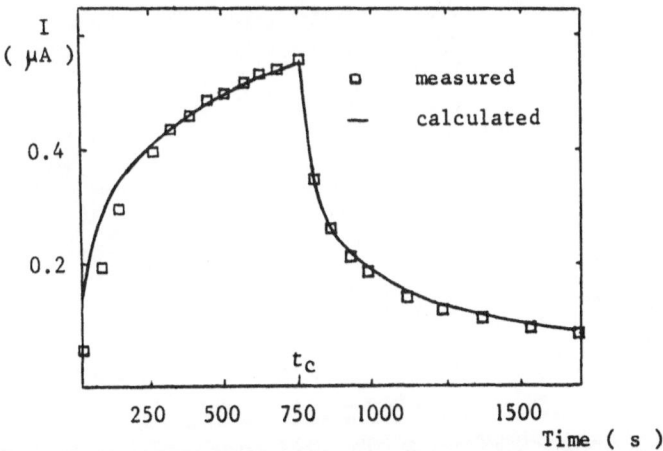

Figure 3. Measured and calculated current-time curves on strained specimen

682

4. CONCLUSIONS

1. The oxide film on the Ti-6Al-4V alloy only breaks when the material undergoes plastic deformation and the surface created during plastic deformation has the same bareness as the fracture surface.

2. The anodic current measured during straining originates from the passivation of the new surface. The passivation follows the high field mechanism of oxide film formation.

5. REFERENCES

1. S. F. Bubar and D. A. Vermilyea, J. Electrochem. Soc., 113, 892 (1966).

2. D. H. Bradhurst and J. S. Llewelyn Leach, J. Electrochem. Soc., 113, 1245 (1966).

3. T. R. Beck, Electrochem. Acta, 18, 815 (1973).

4. T. R. Beck, J. Electrochem. Soc., 129, 2500 (1982).

5. H.-J. Rätzer-Schiebe, Proc. of the 8th International Congress on Metallic Corrosion, 212 (1981).

INITIATION AND EVOLUTION OF MICROCRACKS DURING CORROSION FATIGUE OF BCC STAINLESS STEELS

T. MAGNIN
Ecole Nationale Superieure des Mines
Materials Department
St. Etienne
France

1. INTRODUCTION

The corrosion fatigue damage of metals and alloys is generally described in terms of surface dissolution-effects and/or hydrogen reduction reactions which are enhanced by the cyclic strain. In order to analyse the surface interactions between electrochemical damage and mechanical damage which control corrosion fatigue, it is necessary to compare quantitatively the evolution of the surface roughness during fatigue both in air and in the corrosive solution. Thus, the influence of the corrosive solution must be considered on both the nucleation of microcracks and the evolution of these microcracks into macrocracks leading to propagation and final failure.

In this paper, the influence of a 3.5 % NaCl solution on the initiation and the evolution of microcracks formed during cyclic plastic deformation of bcc Fe-26Cr-1Mo stainless steel is examined, paying a particular attention to the influence of the applied electrochemical potential E.

2. EXPERIMENTAL

Corrosion fatigue tests were carried out on a high purity ferritic stainless steel containing 25.5 to 26 wt% Cr, 0.94 to 1.15 wt% Mo and about 20 ppm C and 30 ppm N. This alloy has been previously studied in cyclic deformation in air, the evolution of the dislocation microstructures has been determined [1,2] and crack initiation has been shown to be intergranular [1]. A very sensitive testing equipment is used to simultaneously record the cyclic evolution of the mechanical and electrochemical parameters as it is described in detail elsewhere [3]. Symmetrical tension-compression tests were performed on smooth specimens under plastic strain control $\pm \Delta\varepsilon_p/2$ and at constant strain rate $\dot{\varepsilon}$, in dry air and in an aerated (or desaerated) 3.5 % NaCl solution at pH 6. During cycling at imposed potential, the current transients are recorded as a function

of the applied strain and the number of cycles. Figure 1 schematises the

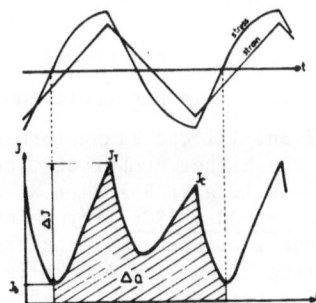

FIGURE 1. Evolution of the current transients during cycling.

evolution of the current density J ($J = I/1.5$ cm^2, where I is the recorded current) during one cycle. J_T, J_c, J_b and ΔQ are used to quantify the perturbation of the metal-solution interface.

To determine the surface roughness and to quantify the fatigue and corrosion fatigue damage, cellulose acetate replicas of the specimen surface are taken for different fractions of the fatigue life Ni corresponding to the 1 % rapid decrease of the saturation stress.

3. EXPERIMENTAL RESULTS

3.1.Electrochemical behaviour without cycling

The polarization curves of the Fe-26Cr-1Mo alloy in the aerated 3.5 % NaCl solution are described elsewhere [4]. They show that : (i) the free corrosion potential E_o is about - 150 mV/SCE. (ii) The pitting potential is about + 800 mV/SCE. Thus, the corrosion fatigue tests will be performed (i) in the passive region (at E = - 300 mV/SCE, E = E_o and E = + 500 mV/SCE and (ii) in the cathodic region at E $\leqslant$ - 500 mV/SCE.

3.2. Influence of E on the fatigue damage

The analysis of the replicas of the specimen surface during cycling in air and in the corrosive solution at imposed potential leads to classify the cracks in four types as a function of their length 1: (i) type I cracks for 1 $\leqslant$ 50 μm (typically the grain boundary size), (ii) type II cracks for 50 μm < 1 < 150 μm : they are due to the micropropagation in surface of type I cracks, (iii) type III cracks (150 μm < 1 < 500 μm) which result from the coalescence and the micropropagation in surface of type I and type II cracks (they are generally perpendicular to the stress axis), (iiii) type IV cracks (1 $\geqslant$ 500 μm) which lead to a macropropagation in volume, and the number of which is about 1 or 2 in our specimens).

Figure 2 shows the evolution of the number of the type I, II, and III cracks as a function of the number of cycles N, in air, at E = E_o and E = - 500 mV for $\Delta \varepsilon_p/2 = 4.10^{-3}$ and $\dot{\varepsilon} = 10^{-3}$ s^{-1}. It can be observed that :

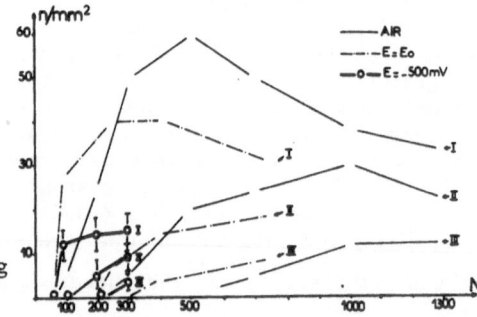

FIGURE 2. Evolution of the number of type I, II and III cracks during cycling at $\Delta \varepsilon_p/2 = 4.10^{-3}$ and $\dot{\varepsilon} = 10^{-3}$ s^{-1}

(i) at E = E_o, the type I cracks initiate for the same number of cycles (N ≃ 100) than in air but their number is much larger. (ii) at E = E_o, the type II and III cracks can form more rapidly in the corrosive solution because of the higher number of type I cracks, which explains the reduction of the fatigue life at E = E_o : Ni ≃ 700 at E = E_o and Ni ≃ 1400 in air. (iii) At E = - 500 mV/SCE (for which the hydrogen reduction is favoured [4]),the type I cracks also initiate at N = 100 but the transition from type I to type II and III cracks is considerably accelerated by the hydrogen reduction reaction. Ni is reduced by a factor 5 in comparison to air.

It has been shown elsewhere [4] that (1) Ni is reduced by a factor 9 in the deaerated solution for the same condition, which confirms the effect of the hydrogen reduction and (2) The reduction of the fatigue life at E = -500mV/SCE is much more pronounced at high than at low strain amplitude. (iiii) The number of cracks at N = Ni is quite similar at E = E_o and in air, but it is reduced at E = - 500 mV. Moreover, it must be noticed that many microcracks do not propagate in each case.

3.3. Cyclic evolution of the current transients

To correlate the mechanical and the electrochemical damages, the evolution of the saturation stress σ_s, the maximum microcracks length l and the electrochemical parameter J_T are simultaneously recorded on figure 3 at $\Delta\varepsilon_p/2 = 4.10^{-3}$, $\dot{\varepsilon} = 10^{-3} s^{-1}$ and E = E_o. In such conditions, the cyclic strain induces dissolution resulting from a depassivation-repassivation process [3] by breakdown of the passive film at slip band emergence and at grain boundaries. During the first cycles an increase of J_T is observed and related to a multiplication of the slip lines. At the beginning of the saturation regime J_T decreases which indicates a localization of the dissolution process. When type I cracks are formed (N > 100),

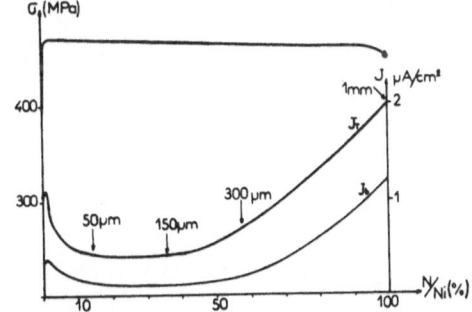

FIGURE 3. Correlation between microcracking and the evolution of J_T at $\Delta\varepsilon_p/2 = 4.10^{-3}$ and $\dot{\varepsilon} = 10^{-3} s^{-1}$, E = E_o

the decrease of J_T is slowing down. When type II and III cracks formation is starting, J_T increases till fracture (this results from a difficult repassivation at crack tips). Inversely, at E = - 500 mV, J_T decreases continuously till the macrocrack initiation, reflecting the hydrogen embrittlement due to the cyclic strain and the microcracks formation.

4. DISCUSSION AND CONCLUSIONS

The observed results are quite original and interesting conclusions can be drawn concerning the effect of an anodic dissolution and the hydrogen embrittlement on the corrosion fatigue damage of bcc stainless steels in a 3.5 % NaCl solution :
1) In the passive region (before pitting, i.e. at E = E_o and E_o<E<+500mV/SCE), the effect of the corrosive solution in the reduction of the fatigue lifetimes Ni is essentially associated with an acceleration of the micropropagation in surface of type I cracks (l ≃ 50 μm) which are more numerous than in air. This is due to a localized dissolution.
2) In the cathodic region (E < 500 mV/SCE), the marked reduction of the fatigue life at high strain amplitude and the decrease of this reduction at low strain amplitudes can be related to the mechanical formation of microcracks. The hydrogen embrittlement only occurs when type I cracks are mechanically formed. At high $\Delta\varepsilon_p/2$, type I cracks are quickly formed

and then the hydrogen reduction reactions can be very localized at the
crack tips. This induces a marked reduction of Ni which is related to the
localization of the electrochemical damage in the first type I cracks and
to their accelerated micropropagation. When $\Delta\varepsilon_p/2$ decreases, the formation
of type I cracks is delayed [4] and the effect of hydrogen is less and
less sensitive.

REFERENCES

1. Magnin T and Driver J.H : Mat. Sci. Eng., 39, 175, 1975.
2. Magnin T, Fourdeux A and Driver J.H : Phys. St. Sol., (a), 65, 301, 1981.
3. Magnin T and Coudreuse L : Mat. Sci. Eng., 72, 125, 1985.
4. Coudreuse L : Thesis, St-Etienne, France, 1986.

WORKSHOP SUMMARY: STRESS CORROSION AND CORROSION FATIGUE - SESSION A

PREPARED BY: H. E Haenninen
DISCUSSION LEADERS: B. D. Lichter, P. Neumann, D. Macdonald, H. J. Engell
RECORDERS: S. Altintas, R. M. McMeeking, A. T. Cole, J. Lumsden

Introduction

During the previous meeting (Atomistics of Fracture, eds. R. M.
Latanision and J. R. Pickens, 1981) Cu-Au was identified as a useful
system for mechanistic studies, since it undergoes both selective corrosion (of Cu) and transgranular SCC. Since new results on this system were
available, the groups discussed the propagation of cracks from the spongy
film into the underlying parent alloy associated with stress corrosion
cracking of Cu_3Au and other alloys as a main topic.

Brittle Film Induced Cleavage

This was identified as a new area requiring particular attention.
When the question of evidence for this phenomenon was considered, it was
felt to be poorly understood and a lot more evidence besides the Cu-Au
system are needed.

Synchronization of potential measurements and acoustic emission within
a few milliseconds seems to confirm that there are bursts of crack growth
taking place. It was agreed upon that there is enough evidence that the
crack does enter the ductile alloy. In addition, the rate at which cracking occurs over lengthier periods is faster than can be explained by
corrosion rates alone. It was noted that other systems show similar behavior in the sense of cracks propagating rapidly from brittle to normally
ductile regions without blunting. Such situations as carbide cracks running into the matrix in steel alloys and cracks propagating from ceramic
into niobium bonded to it were brought up. However, most cases involve
metals which can be brittle at high strain rate and due to hydrogen. It
was thought that no similar behavior had been observed in copper. In
contrast, cracks entering copper from a brittle phase usually blunt.

It was suggested that experiments should be sought in which cracks are
injected into the parent alloy by other means. This might involve a
brittle film on a flat surface, although it should be remembered that the
behavior in practice could depend on the geometry of the film. The nature
of the sponge close to the unattacked alloy is likely to be very different
from that observed in "bulk sponges" of the Cu-Au system. Direct observation of structure is needed in this domain. There is also a need for
calculations to show the specific mechanical behavior of these films,
including continuum calculations of local stresses, etc. Care would have
to be taken to ensure that the cracks in the film are made to run fast.
Additionally, experiments with the same alloys but different brittle materials would help to address the question of injecting fast cracks into the
relevant metal. Acoustic techniques are improving and so better measurements for the SCC problem will be available soon.

The deformation of the ligaments in the sponge was considered and it was observed that some are big enough to deform plastically, necking before breaking. Others are smaller than a whisker and there was some speculation about the presence of dislocations in them. It was thought that there might be some because of the history of deformation of the preligament material. In any case, even if the failure of the sponge is microscopically ductile, the macroscopic result will be a brittle type of crack propagation as the ligaments break or neck progressively. Thus, macroscopic plasticity can hardly be expected because there is no simple mechanism of, e.g., forming a shear band, which seems difficult since yielding in shear would shear off the ligaments.

The macroscopic elasticity of the sponge should be modelled. A difficulty is that the sponge properties vary with film depth and this should be taken into account. If macroscopic plasticity is absent, notions from shielding and antishielding can be used. For a film more compliant than the alloy and with the remote K level held fixed, the quasistatic K_{tip} will diminish as the crack propagates into the film. This could be a small effect and might be overcome by dynamic means. After the crack is arrested in the alloy, a film will grow on the surface of the new crack and this will reduce K_{tip} steadily. If it is K which controls propagation, this suggests that the crack should not propagate unless it follows the sponge-alloy interface. However, it is dangerous to speculate along these lines since the crack tip energy release rate will be affected by the reduction in elastic modulus, and the macroscopic fracture energy will vary spatially and temporally in the film. This indicates that improved modelling will help and augment the progress already made by Dr. Sieradzki and others.

The question of injecting the crack dynamically into the alloy was also considered. It was agreed that according to 2D elasticity, the crack has no inertia as long as there is no dispersion. This is in contrast to other defects such as dislocations. Dr. Sieradzki did demonstrate that his model, with pair potential effects, provides a crack with no inertia as well. For the kind of crack injection at issue here there are 2 models; one is the type used by Dr. Sieradzki which involves asymmetric emission and capture of dislocations, which manage to pull the crack along in something of a quasistatic manner; the other involves effects relevant at sonic crack propagation rates and is probably less relevant. With both these models, it is stopping the crack that becomes the issue. It was suggested that perhaps intact ligaments left behind by the propagating crack could be involved, tending to pin the crack closed on the flank and reducing the driving force.

Chemistry and Electrochemistry Within Cracks

It was generally agreed that the potential, pH and solution chemistry within cracks can be quite different from the bulk values; the local changes being mainly affected by local metallurgical inhomogeneities (MnS-inclusions, grain boundary segregants, etc.). It was felt that a far better understanding of crack tip chemistry exists than existed at the last of these meetings. The effective pH in steels containing Cr has been established as being in the acidic range, usually <1.5 in typical austenitic stainless steel SCC cracks. However, several areas require further investigation. Do current densities measured on flat surface relate directly to the actual crack current densities, how large a percentage of measured crack current doesn't directly contribute to crack growth (will it be dealloying or anodic dissolution) if any, etc.? Scratching tests simulate the freshly produced fracture surface adequately, although not completely; therefore, caution should be used in interpreting results.

Tests should be carried out more in environments which simulate the crack tip environment (e.g., MnS-saturated water for steels in pure water systems) and/or new techniques should be developed to assess both the thermodynamic and kinetic conditions existing at the crack tip. It will be especially essential to have the knowledge of chemical activity of the freshly created surfaces. Some extrapolation of surface interactions with gaseous environments are available already at present.

Miscellaneous Topics

1. The need to study the initiation stage of SCC was emphasized. As the initiation process is crucial in most industrial situations it is felt that much more work should be performed on elucidating this process. Is it obvious, for example, that initiation will always occur when slip occurs? If so, why doesn't every slip step stress corrode (or does it to some immeasurably small degree)? Also, it was noted that corrosion fatigue often initiates from corrosion pits. Therefore, in addition to the local changes inside the pits, the local stress intensity increases and possible local hydrogen absorption should be considered more carefully.

2. The often observed significantly higher velocity of the short surface crack geometry over the CT specimen configuration can be construed to be due either to an environmental crack size effect or to inelastic stress considerations in the determination of the stress intensity factor for the surface crack geometry or to a combination of both. The term "short crack" was thought, however, to be rather arbitrary, in electrochemical terms, since a significant potential drop can be observed in cracks of 100 μm. Is this a short crack in terms of LEFM? Can we differentiate the short crack problem between the chemical/electrochemical and LEFM effects? These are some of the current questions in this domain.

3. Evaluation of crack tip strain rates ($\dot{\varepsilon}_{ct}$) for both monotonic and cyclic loading could offer an analytical approach for relating SCC to CF phenomena. However, it was agreed that the present crack tip strain rate analyses need to be developed further, especially if applied to such technologically important areas such as short cracks and to the region of K_{th}, where a threshold $\dot{\varepsilon}_{ct}$ apparently exists. An evaluation of the mechanisms behind the critical crack tip strain rate values, which can trigger the environment sensitive cracking, was of primary interest.

4. Usually, experiments are conducted with Mode I conditions. Tests in Modes II and III should be more frequently pursued; in fact, a possible critical experiment to test the film-induced cleavage model is to observe the nature of fracture surfaces under such loading. Stress states are important in assessing the mechanical properties of thin films, alone or on substrates. This is important in both the slip dissolution and the

film-induced cleavage models as well as different hydrogen-induced cracking models.

5. In many environment-sensitive cracking systems, the influence of various parameters on cracking process is already understood to a reasonable extent. However, only recently have attempts been made to use our current understanding in predicting materials/component behavior exposed to aggressive environments. The main factors that need to be addressed are:

a. predictions of the K_{th}- and ΔK_o-values

b. why there is such a rapid increase in $(da/dN)_{\Delta K}$ - or $(da/dt)_K$- values in the near threshold (ΔK_o and K_{th}) region

c. prediction of the crack propagation rate in the "plateau" region

d. prediction of maximum possible environment-sensitive crack growth rates based on first principles.

If the mechanistically based quantitative prediction of the crack propagation rate is developed, then many diverse observations such as the critical potential for cracking, initiation sites, discontinuous/continuous mode of crack growth, etc., have to be explained.

WORKSHOP SUMMARY: STRESS CORROSION AND CORROSION FATIGUE - SESSION B

PREPARED BY: J. Lumsden
DISCUSSION LEADERS: D. Vvedensky, M. Speidel, A. W. Thompson, R. Thomson
RECORDERS: E. Kayali, C. Altstetter, D. Pettifor, J. De Hosson

There was much controversy concerning cracking mechanisms and concerning much of the evidence used to support particular mechanisms. However, there was general agreement that dissolution was a prerequisite to environmental cracking whether cracking was caused by either hydrogen or film rupture.

The correlation between passive film rupture and repassivation rate in SCC was considered to be conceptually acceptable, but the model is far from quantitative and predictive. Most of the discussion on mechanisms concerned the recently proposed "sponge model" for SCC. There was agreement that, at least in some cases, a dealloyed layer exists at the crack tip and that this layer has a sponge-like structure which is brittle on a macroscopic scale. However, in order to rationalize its brittle response, an understanding of the vacancy and dislocation content of the filaments interface with the environment is needed. A sponge structure is not observed on SCC fracture surfaces. Thus, this layer is thin and a nucleation site for a cleavage fracture. There was considerable skepticism about the possibility of a crack propagating from the sponge layer, transgranularly, with a sharp tip, through ductile material. There is no experimental evidence of brittle cracks propagating through ductile metals in what might be considered as model systems, such as a nitride layer on the surface of steel.

It was noted that the passive film rupture and dealloying models require different crack tip chemistries. In the former, the crack tip is in a passivating regime, whereas in the latter the crack tip is undergoing active dissolution. The possibility was also pointed out, particularly in the case of short cracks, that the crack tip chemistry could have cyclic behavior if discontinuous crack growth occurs. It was not decided whether such behavior would require refinement of either the slip dissolution or dealloying model.

The concensus concerning the effects of stress rate and loading mode on cracking was that data are too limited and incomplete to form definitive conclusions. Some general observations were that in extreme cases of crack branching it would seem that neither the stress intensity nor the loading mode is an important parameter. It was also noted that in some systems a true threshold stress intensity may not exist since K_{ISCC} seems to go lower and lower as investigators become more patient.

Summary Session

Session Chairman: R. H. Jones

CHEMISTRY AND PHYSICS OF FRACTURE
A CONFERENCE SUMMARY

Markus O. Speidel
Institute of Metallurgy
Swiss Federal Institute of Technology, ETH
Zurich, Switzerland

INTRODUCTION

The NATO Advanced Research Workshop on Chemistry and Physics of Fracture in Bad Reichenhall 1986 brought together outstanding scientists, specialists in widely different disciplines but with a common interest and acknowledged scientific achievements in the field of fracture. The contributions of these scientific luminaries to the conference were always very ably presented and often highly individualistic. Thus, it would be very difficult to summarize their individual contributions here. There is also no need to do this since their papers are contained in the present volume. Instead, it is attempted to summarize the topics of the objectives, the papers and the discussions in a generalized and thus simplified framework which should also serve to iden- tify directions for future fruitful research.

In the months preceding this Advanced Research Workshop on Chemistry and Physics of Fracture a letter was distributed to the participants concerning the "background and objectives" of the workshop. A few quotes from this letter appear to be helpful if one attempts to summarize the achievements of the workshop and to compare this with the stated objectives.

The letter began with a "recognition that engineering materials which are tough and ductile can be rendered susceptible to premature fracture by their reaction with the environment". It continued with the observation that "the mechanical properties of most engineering materials have been shown to be susceptible to hydrogen embrittlement or stress corrosion", and it concluded that the "ultimate goal" of the workshop was "increased understanding and control of technologically important fracture and embrittlement phenomena". This goal was to be achieved by, among other means, the identification of the "relevant microstructure, microchemistry, mechanics, surface chemistry and solution chemistry factors affecting the crack growth behavior of a given material". It was also recognized that "the interaction of the crack with complex microstructural features such as dislocations, dispersed phases, solute atoms, interfaces and free surfaces contribute to the collective na- ture of crack growth and can be a major factor controlling the crack propa- gation in a given material-environment combination". It was further recog- nized that "the kinetics of fracture processes can be studied by rate theory with the rate determining step describing the critical mechanism in crack propagation.

In the following a framework is described within which the above goals and ways to achieve them could possibly have been discussed. Secondly, an attempt is made to summarize some of the important topics discussed during the conference and some of the progress made. Thirdly, this is compared with the stated goals. Some of the issues which still remain to be solved as well as some of the areas which have emerged during the conference as fruitful in terms of future directions for research are commented upon.

THEORY OF FRACTURE AND SUBCRITICAL CRACK GROWTH

"A theory is a quantitative explanation of the observed facts, based on scientifically acceptable principles and methods". [1] Thus, if a theory is an explanation, it behooves us to start with that which is to be explained: the facts concerning fracture and subcritical crack growth. One of several possible ways to present such facts in a condensed manner is shown in figure 1. This is the crack velocity versus stress intensity curve, which, at stress intensities less than the fracture toughness K_{IC}, shows the familiar three stages [1]. The plateau in stage IV corresponds to the velocity of elastic waves which limits purely mechanical fast brittle cracks to several thousand meters per second in typical structural materials. It is our contention that diagrams of the type of figure 1 can conveniently and quantitatively represent most of the facts to be explained theoretically at a conference on the physics and chemistry of fracture. Such crack velocity versus stress intensity curves have been observed for all three important classes of materials: metallic, polymeric and ceramic.

Since chemistry, next to physics, is the main topic of the conference, environment induced subcritical crack growth is emphasized here and we see it sandwiched in between the two extremes of purely anodic dissolution without stress or strain (such as in pitting corrosion) and purely mechanical fracture without the assistance of any environment. Therefore, the stress intensity range between K_{IC} and K_{ISCC} in figure 1 deserves special attention.

It is of course possible to see the role of stress intensity on environment-induced subcritical crack growth mainly as a consequence of the strain and strain rate it imposes on the crack tip region, as has been pointed out in the NATO conference on Atomistics of Fracture 1981.

Figure 2 illustrates schematically the two important variations of the crack velocity versus stress intensity curve which need theoretical explanation for a given material. It appears that the workshop on the chemistry and physics of fracture could have attempted to answer two questions, both qualitatively and quantitatively, according to figure 2.

First, by which atomistic processes can changes in the solid or the presence of certain environments affect the threshold stress intensity for the onset of cracking - and how much?

Second, by which atomistic processes (rate processes, transport processes) can certain environments affect the subcritical crack growth rates - and how much?

For environment-induced subcritical crack growth, the quantitative answer could take the form [1]

$$\frac{\Delta a}{\Delta t} = f \ (K, \ R_{p0.2}, \ T, \ E, \ V, \ C_n, \ ...)$$

where $\frac{\Delta a}{\Delta t}$ is crack velocity, K is stress intensity, $R_{p0.2}$ is yield strength, T is temperature, E is electrode potential, V is viscosity of the environment, C_n is concentration of the n'th aggressive species, ... and so on for all parameters known to influence the crack growth rate.

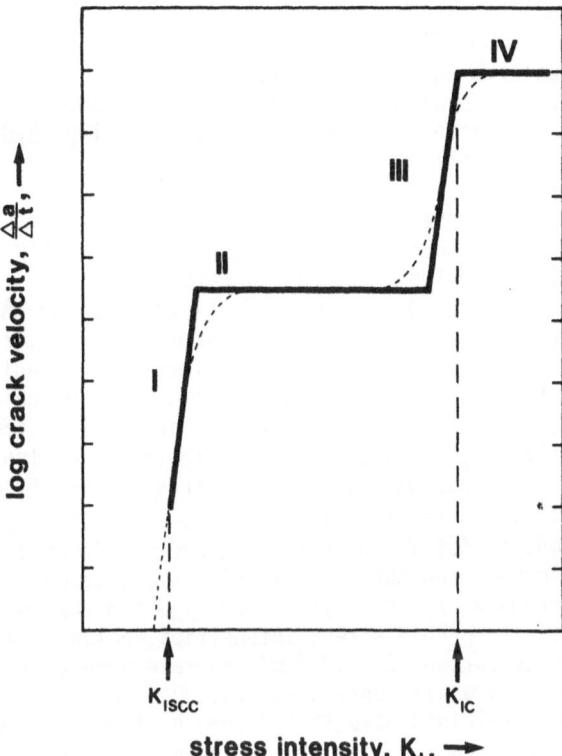

Fig. 1. Crack velocity versus stress intensity curve.

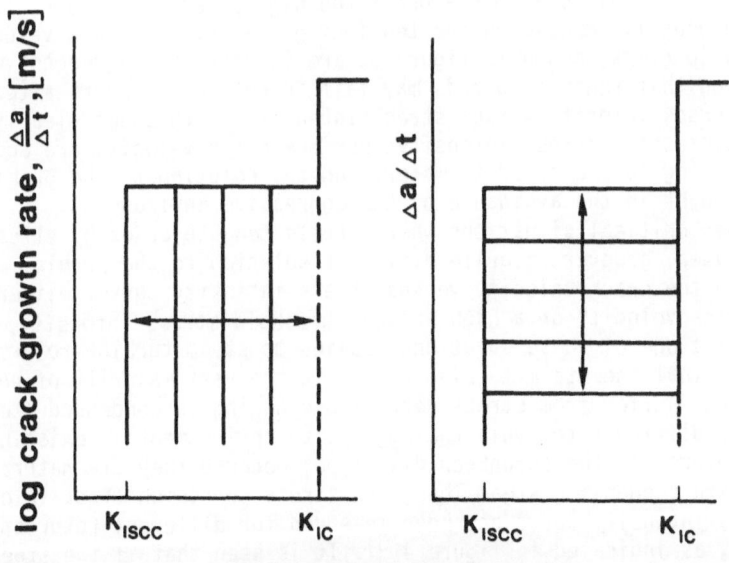

Fig. 2. Embrittlement may reduce the fracture toughness K_{IC}, environments may result in low threshold stress intensities K_{ISCC}. "Plateau" crack velocity studies could be related to chemical rate and transport theory.

EXPERIMENTAL DATA, ENGINEERING AND TECHNOLOGICAL IMPORTANCE

An "advanced research workshop on chemistry and physics of fracture" is not the proper place to present or ask for a lot of complex, service-related, experimental data on fracture, stress corrosion cracking, hydrogen embrittlement, liquid metal embrittlement and corrosion fatigue. This is because there is at present no generally acceptable theoretical treatment (in the above mentioned sense) of even the most simple and at the same time most important cracking phenomenon where chemistry and physics play equally important roles: the stress corrosion cracking of steels in water.

The need for a theoretical treatment of the physics and chemistry of stres corrosion cracking of steels in water is obvious for academic reasons alone. However, the need for such a theoretical treatment is strongly accentuated, to say the least, because of the importance stress corrosion cracking of steels in water has in engineering, technology and public safety considerations. A few examples may suffice to illustrate this and also to illustrate the usefulness of the crack-velocity versus stress intensity curves.

Example Nr. 1 is illustrated in figure 3. Steels of the type described ther are used in large quantities for prestressing and post-tensioning of concrete structures, rock-anchors and the like. Tensile yield strengths of around 1400 MN/m^2 are customary. It is obvious from figure 3 that such steels have low stress corrosion threshold stress intensities (below 20 $MN \cdot m^{-3/2}$) and high crack velocities (around 3×10^{-7} m/s in water) so that small precracks may lead to rapid failures in tensioning rods when water is present. Thus, about three hundred buildings have failed due to stress corrosion cracking of tensioning steels in water, many of them only hours after tensioning [2]. A fracture mechanics analysis indicates that small crack nuclei of only 0.2 mm depth (as often generated by corrosion) will exceed the threshold stress intensity of 20 $MN \cdot m^{-3/2}$ under the high stresses of about 1000 MN/m^2 which are usually imposed on the tensioning steels. The crack velocities of about 3×10^{-7} m/s, shown in figure 3, are in very good agreement with the observation that tensioning rods may fail in only a few hours after tensioning Thus the crack velocity versus stress intensity curve shows clearly that neither the threshold stress intensity nor the crack velocity are acceptable for tensioning rods exposed to water, and the solution to the problem is generally sought in the avoidance of the aggressive environment.

In other critical situations where steels tend to crack by stress corrosion in water, however, a quite different solution to the problem may be indicated by the crack velocity versus stress intensity curve: either a low enough crack velocity or a high enough threshold stress intensity.

The first one of these solutions applies to steam turbine rotors. It is well known that the steam turbine rotors in the vast majority of nuclear power plants suffer from stress corrosion cracking in condensed steam [3,4]. Here it is difficult to avoid the aggressive environment in existing steam turbine rotors (of the shrunk-on disc type) because they are naturally in contact with condensing steam. Thus, the stress corrosion crack velocity versus stress intensity curve has been measured for different rotor steels in hot water, as indicated in figure 4 [3]. It is seen that if the steel has a reasonably low yield strength, the crack velocity is reasonably slow, thus permitting an interim solution until the rotor is replaced. The interim solution consists in continued operation with slowly cracking rotors and setting periodic inspection intervals to make sure the crack size is always smaller by a certain safety margin than the critical crack size. Obviously

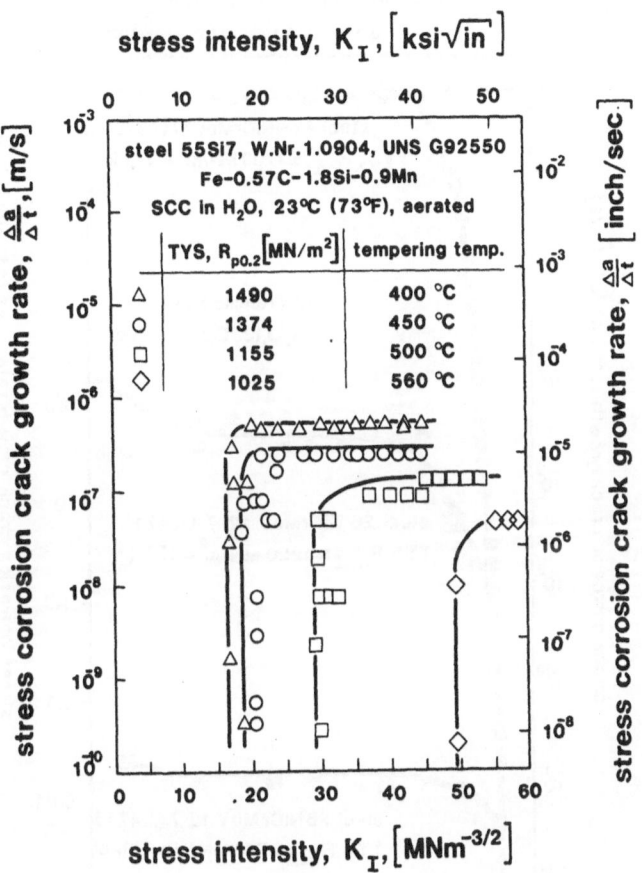

Fig. 3 High strength steels used for prestressing or posttensioning of concrete structures exhibit high stress corrosion crack velocities and low threshold stress intensities in water. Neither crack velocity nor threshold stress intensity are acceptable if the safety of the buildings and structures is to be guaranteed. The solution to the problem here is either a <u>significantly different environment</u> (good coverage of concrete, no oxygenated water) or a <u>significantly different steel</u>, for example a high strength steel which is not susceptible to stress corrosion cracking. At present, there exist not even the beginnings of a theory which could guide the development of such a steel.

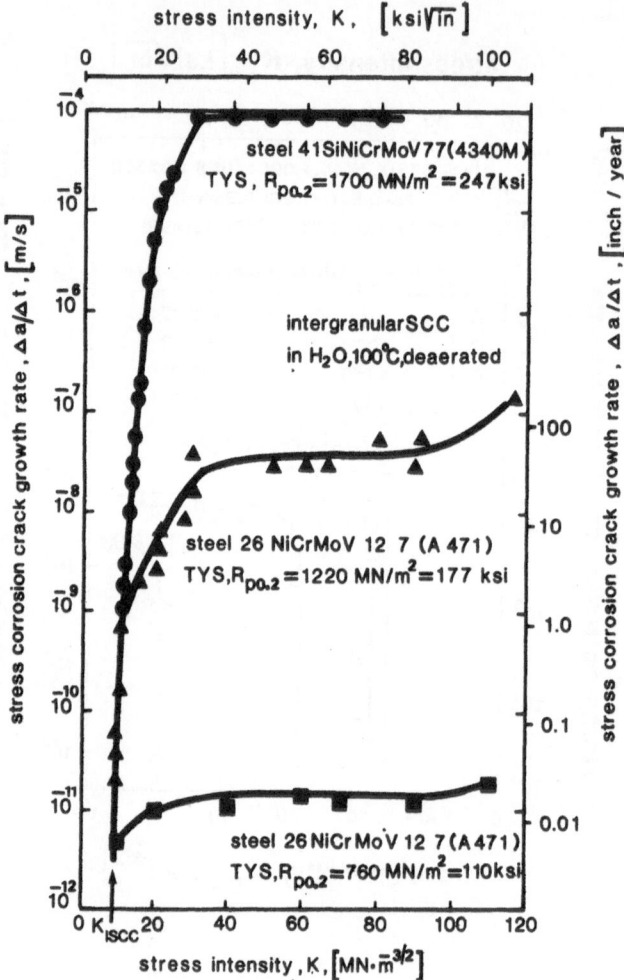

Fig. 4 In low alloy steels exposed to hot water, the yield strength has
a strong influence on the growth rates of stress corrosion cracks,
but the (very low) threshold stress intensity is not measurably
influenced. Steam turbine rotor steels with reasonably low yield
strengths may exhibit comparatively low stress corrosion crack
growth rates in water. This permits continued operation of
cracking steam turbine rotors within limited inspection intervals.
At present, there is no theory which could relate the "plateau"
stress corrosion crack growth rate to either yield strength or
tempering temperature.

this approach works only with a known and slow crack growth rate of, say, about 10^{-11} m/s, as shown in figure 4. Note also in figure 4 that the threshold stress intensity K_{ISCC} in the steels investigated is always near 10 MN·m$^{-3/2}$ and thus easily exceeded by small crack nuclei in highly stressed regions. Just simply lowering the design stress is therefore not a viable solution to the stress corrosion problem of steam turbine rotors.

Let us now turn to an important example where the crack growth rate is too high to be acceptable, but where the stress corrosion threshold stress intensity is larger than the stress intensities applied in service. This example is illustrated in figure 5. It shows two steels in high-temperature water, similar to nuclear reactor coolant. Note first the solid points corresponding to crack growth rate data for intergranular stress corrosion cracks in sensitized austenitic stainless steel. Welded austenitic pipes have experienced hundreds of stress corrosion service failures in high temperature water. Their crack growth rate of about 10^{-9} m/s is certainly near the upper limit of what can be accepted if temporary service with cracking components is to be permitted. The threshold stress intensity is very low, near 10 MN·m$^{-3/2}$ or possibly even lower.

Now compare the open circles with those solid points in figure 5. It is apparent that transgranular stress corrosion cracks in the ferritic reactor pressure vessel steel propagate almost a hundred times faster than the intergranular stress corrosion cracks in the sensitized austenitic pipe steel. With a growth rate of 5×10^{-8} m/s, a pressure vessel wall would be penetrated in less than two months. As this is much too short for an inspection period, care must be taken to exclude the actual occurrence of such stress corrosion cracking. As a matter of fact no stress corrosion service failures of nuclear reactor pressure vessels have become known so far and this may be attributed to the high standard of the water chemistry (keeping oxygen concentration and conductivity low) as well as to the low stress intensity levels resulting from careful stress relief annealing, low applied stress and careful nondestructive testing. The low stress level and the small acceptable flaw size combine to ensure a stress intensity which is smaller than the 20 to 28 MN·m$^{-3/2}$ threshold which we have observed, according to figure 5. There are, however certain nuclear reactor safety analyses which postulate the existence of a crack which penetrates already a certain fraction of the wall thickness [5]. The then resulting stress intensity K_I must always be smaller than the fracture toughness of the nuclear reactor pressure vessel steel so as to prevent purely mechanical failure, as indicated in figure 6. In this hypothetical case, however, the stress corrosion cracking threshold stress intensity of 20 to 28 MN·m$^{-3/2}$ would certainly be exceeded as also indicated in figure 6. Thus, without realizing it, this safety analysis presupposes the absence of stress corrosion cracking of reactor pressure vessel steel in high temperature water. A fan-shaped transgranular stress corrosion crack in a typical reactor pressure vessel steel is illustrated in figure 7. It developed in water at 288°C with a slightly higher oxygen concentration and a slightly higher conductivity than normal nuclear reactor coolant [6]. Because of the safety implications, it appears highly desirable to substantiate by detailed theoretical analysis the major influential variables on both, crack growth rate and threshold stress intensity of this type of stress corrosion cracking of steel in water. At this point in time we are not even in a position to say whether the perfect safety record of the nuclear pressure vessels is due to good water chemistry or due to low applied threshold stress intensities. And we have no idea how wide our safety margin really is.

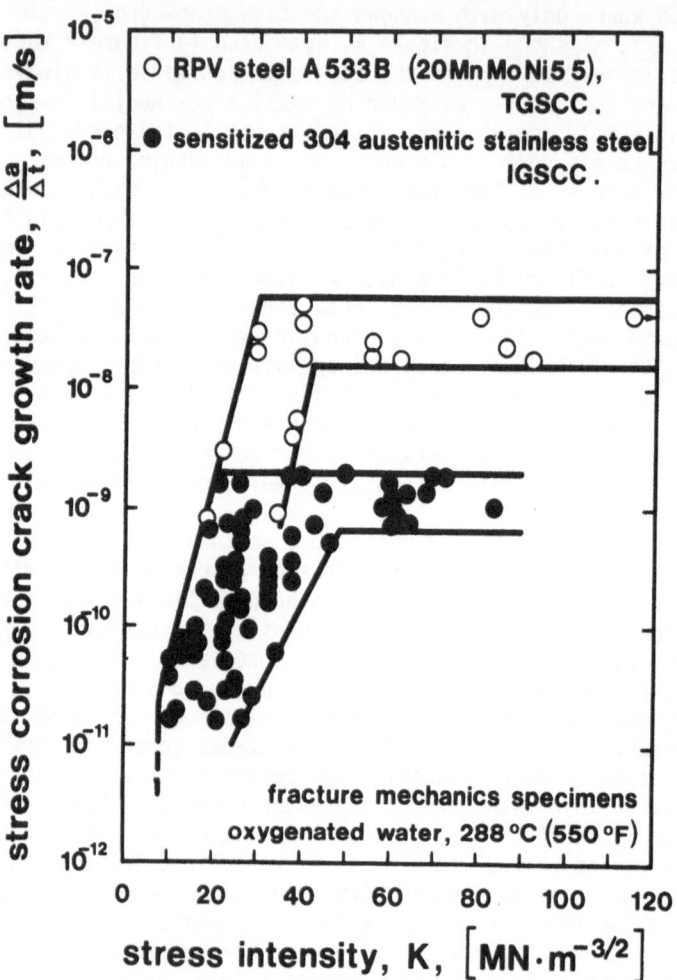

Fig. 5 Reactor pressure vessel steel exhibits almost a hundred times
faster transgranular stress corrosion crack growth rates than
sensitized austenitic stainless steel in high temperature water.

Let us finish this series of examples of important stress corrosion problems with cold worked, high strength, austenitic nonmagnetic steels for ge-
nerator-rotor retaining rings. Their stress corrosion behaviour in water is
illustrated in figure 8. Steel B was used world wide in over 10'000 generator rotors and it has resulted in the stress corrosion service failure of
many of them. In a few instances the consequential damage was in the hundreds of millions of dollars. The crack growth versus stress intensity curve
of steel B in figure 8 indicates that both the threshold stress intensity
and the plateau crack growth rate are not acceptable for continued service
if such a steel is in contact with water.

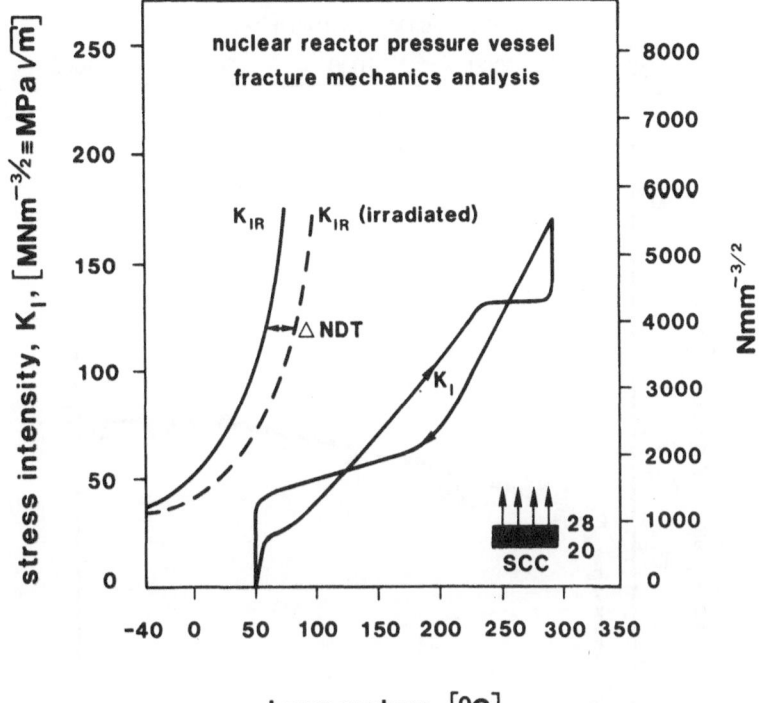

Fig. 6 Stress corrosion cracking (SCC) could occur at lower stress
intensities than certain safety analyses permit, provided the
water chemistry at the crack tip is condusive to it.

Fig. 7 Transgranular **stress corrosion** crack in reactor pressure vessel
steel exposed to 288°C water. The fan-shaped stress corrosion crack
developed from a fatigue precrack.

704

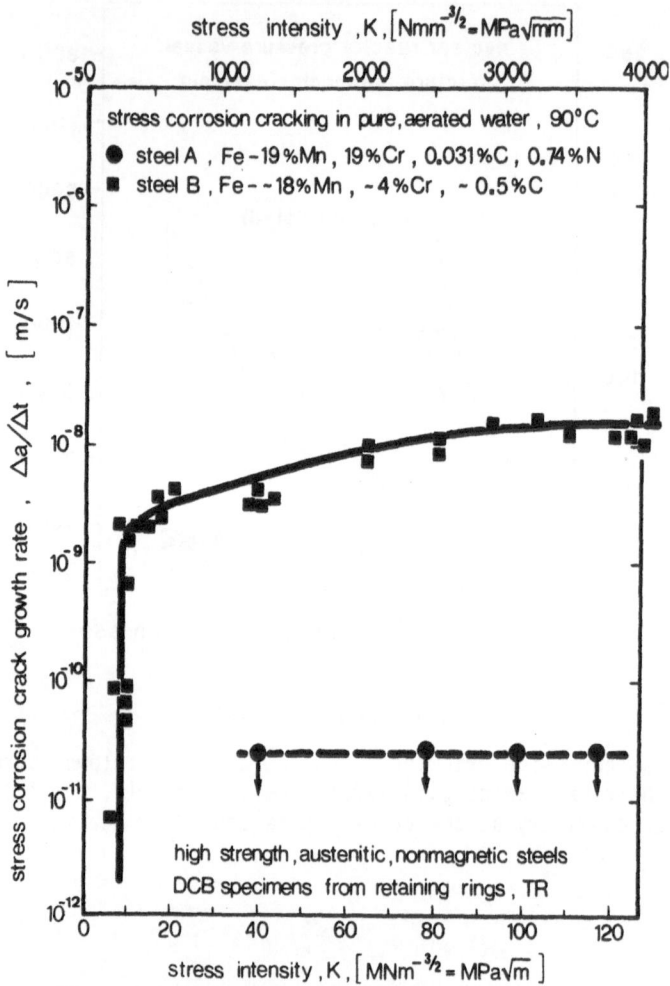

Fig. 8 Two cold worked manganese-austenitic steels. B is highly
susceptible to stress corrosion cracking in water while
A is fully resistant.

Thus, if water (or aqueous solutions) cannot be excluded with certainty,
steel B is not fit for purpose. Fortunately, continued steel development has
resulted in the high-nitrogen steel A which is entirely resistant to stress
corrosion cracking in water, as shown in figure 8. This highly important
steel development was done strictly by trial and error and without any detailed theoretical analysis of the atomistic processes near the crack tip.
If only we could derive a theoretical understanding of the reasons for the
difference of the two curves in figure 8, we would be closer, by a giant
step, to the "ultimate goal" of the workshop which was "increased understanding and control of technologically important fracture and embrittlement
phenomena".

PROGRESS IN UNDERSTANDING FRACTURE

The majority of the presentations and discussions at the conference were focussed on the problem of "dry" fracture. In terms of figure 1 this means an effort to understand why fracture ensues when the fracture toughness K_{IC} is reached, what mechanisms occur at the crack tip, what is the absolute value of K_{IC} and how can it be influenced. Such influences are possible through external physical parameters (e.g. temperature, loading rate) or through internal changes of composition or microstructure ("embrittlement", "toughening") or through the chemistry of the environment and its reaction with the crack tip surface ("environment-induced fracture", stress corrosion cracking, "external" hydrogen embrittlement). In all three cases of influencing the threshold stress intensity for fracture, it is necessary to consider separately three different fracture paths: fracture by void coalescence, fracture by cleavage, and fracture by grain boundary separation. For none of these nine cases were quantitative theories presented at the conference, but in several papers progress towards such quantitative theories or calculations was shown and discussed.

1. Fracture without chemical assistance - void coalescence

Models designed to permit an estimation of the fracture toughness, K_{IC}, were reviewed by Thompson [7]. In recent models the toughness is equated to the product of yield strength, local strain and the critical distance between the void nuclei:

$$J_{IC} \propto \sigma_y \, \varepsilon_f \cdot L$$

Problems with K_{IC} estimates based on this equation include the difficulty to calculate the appropriate local strain necessary to nucleate the voids, and also the strain to make them grow to coalescence. One then has still to measure the critical distance L to estimate the fracture toughness.

A more detailed consideration has recently led Uggowitzer [8] to the following expression for the fracture toughness of materials fracturing by void coalescence:

$$K_{IC} = \frac{R_p}{1-2\nu} \left[s\Pi \, (1+n) \left(\phi^* \, \frac{E}{R_p} \right)^{1+n} \right]^{1/2}$$

where R_p is the yield strength, ν is Poissons constant, n is the strain hardening exponent, E is the modulus of elasticity, ϕ^* is the critical strain (to be determined by a plane strain tensile test) and s is the distance between void nuclei. This equation permits a far reaching understanding of the effects of solid solution hardening, work hardening and precipitation hardening on the fracture toughness of ductile metallic materials [8]. The remaining theoretical challenge, however is still the calculation of the two components of the local strain for void nucleation and void growth.

2. Fracture without chemical assistance - cleavage

The competition between brittle cleavage crack extension and the emission of dislocations was the dominant theme in many presentations and discussions at the conference. Argon [9], discussing the effect of temperature on the brittle to ductile transition compared the energy barrier G_{IC} for cleavage with the energy barrier ΔG for dislocation nucleation and noted a difference between theory and observed facts: Pb, Au, Cu, Ag, Al, Ni could spontaneously

nucleate dislocations from a crack tip and are, as expected, noncleaveable.
On the other hand, Fe, W, LiF, MgO, Si, Zn, Mo, V, have high enough energy
barriers for dislocation nucleation and thus, theoretically, there should be
cleavage without any ductility in these solids, unless dislocations are already introduced. The facts are different. Dislocation emissions from crack
tips have been directly observed as well as brittle to ductile transitions
in nearly perfect crystals, well below the melting temperature.

Haasen [10] presented a well researched example to this: the fracture and
deformation behaviour of pure silicon. Si is brittle below 800°C and fails
by cleavage with a fracture toughness of about 0.9 MN$\cdot$m$^{-3/2}$. At higher temperatures, precracked Si crystals just shear off in a ductile fashion. It
was observed that dislocation generation at the crack tip is necessary but
not sufficient for ductility. To really make the materials ductile one has
to have a fast enough dislocation velocity.

Similar data to Haasen's fracture toughness of 0.9 MN$\cdot$m$^{-3/2}$ for Si could
be derived from other presentations at the conference. Values of 2γ and work
of fracture considered in Knott's [11] contribution lay in the range of
2 - 14 J/m^2. Using the simple relation EG = K^2, we could estimate a lower
limit for the fracture toughness of steels at about K$_{IC}$ = 0.6 MN$\cdot$m$^{-3/2}$ if
cleavage occurs from sharp precracks without plasticity. Similar values of
fracture toughness may be expected for grain boundary separation based on a
surface energy slightly smaller than 1 J/m^2 calculated by DeHosson and Vitek
[12] using computer modeling with pair potentials.

For better results on intrinsic ductile-brittle transitions, better potential calculations are needed. The tight binding bond model presented by
Finnis [13] and the embedded atom method presented by Daw [14] appear to be
progress in this direction.

It is thought that brittle to ductile transitions occur when subcritically
blunted cracks generate additional dislocations of fast enough velocity.
Knott [11] notes that the competition between cleavage and the emission of
dislocations does not occur exactly at the tip of a fatigue precrack, even
in brittle high strength steels, since about 10^4 dislocations are already
generated there just to give the necessary crack tip opening displacement.
Cleavage cracks in steels nucleate in oxide or silicate particles by the
pileup of dislocations. Thus, K$_{IC}$ is much higher than expected solely from 2γ.

Whitesides [15] emphasized that brittle fracture due to bond separation
should be considered separately in polymers, ceramics and metals, because
their bond strengths are a hundred times different and thus, these materials
cannot be in the same category (the diamant bond strength being $\approx$ 10 eV and
the Van der Waals bond strength in polymers being $\approx$ 0.1 eV).

3. Fracture without chemical assistance - grain boundary separation

Neither theoretical nor experimental values of fracture toughness for
brittle grain boundary separation without embrittling effects have been presented at the conference. To do this systematically remains a theoretical
challenge. The data obtained by DeHosson [12] for surface energies of grain
boundaries, however, indicate they are in the same range as for cleavage
fracture.

4. Internal changes of composition - void coalescence

Thompson [7], Hancock [16], McMeeking [17] and Hirth [18], discussing void nucleation, growth and coalescence models for ductile crack growth note that hydrogen may

- lower the void nucleation strain
- lower the void growth strain   and
- alter the number of effective void nucleating particles

It remains a challenge to put forward quantitative theories how these phenomena finally affect the fracture toughness and the threshold stress intensity for subcritical crack growth of hydrogen containing materials which fail by void coalescence.

5. Internal changes of composition - cleavage

The theoretical cleavage stress intensities are already so low (less than 1 $MN \cdot m^{-3/2}$, according to point 2 above) that perhaps this is a case where hydrogen, in the words of Fidelle [19] "has nothing to decrease". Rather, it appears that the interaction of hydrogen with dislocations and thus plasticity, as discussed by Hirth [18] and Hashimoto [20] influences the competition between cleavage and plasticity in favor of the former. If hydrides form in an otherwise noncleavable material, then cleavage in the hydride and the resulting stress intensity for crack propagation merit further theoretical efforts with the aim of a quantitative prediction of the threshold stress intensity.

6. Internal changes of composition - grain boundary separation

The separation of precontaminated grain boundaries has been discussed at the workshop as yet another problem of competition between crack growth and plasticity which results in blunting, shielding and brittle to ductile transitions. Rice [21] and Grabke [22] discussed the question whether something which segregates to the grain boundaries will always also embrittle them, or whether "anomalous segregators" could indeed toughen the grain boundaries. While it is quite clear that traces of boron can effectively prevent premature intergranular separation of grain boundaries in steels at elevated temperatures [23], there was no hard evidence presented that boron is an effective grain boundary toughener for steels at low temperatures. Further theoretical investigations to elucidate the role of possible grain boundary tougheners appear to be highly desirable. Two questions need to be answered here: How can future grain boundary tougheners be theoretically predicted and how much will they increase the fracture toughness.

The grain boundary embrittling effect of hydrogen and phosphorous was discussed intensively at the Bad Reichenhall NATO workshop. Computer simulation of crack tip processes using molecular dynamics with pair potentials, as discussed at the Corsica NATO workshop cannot give a realistic picture of the effect of hydrogen or phosphorous on metal-metal binding. Finnis [13] sees a role for the tight-binding bond model to compute the competition between bond-breaking and dislocation emission at the crack tip, once fast computation is possible which will also allow to introduce the third dimension. The role of hydrogen might be to weaken the metal-metal bond from each of the neighbours of the hydrogen atom. Quantitative data for this and calculations for grain boundaries have not been reported and thus, K_{IC} cannot be estimated. Daw [14] presented the embedded atom method which is no more complicated than the pair potential method but contains more "physics" and also appears to be a progress over what was known in the Corsica-NATO workshop. It appears possible to calculate the effect of hydrogen on the surface energy and on cleavage and on grain boundary separation. This has not yet been translated into K_{IC} data. In summary, it appears to be a remote but very worthwhile goal for those concerned with molecular dynamics to calculate quantitatively the fracture toughness of materials which cleave or exhibit brittle grain boundary separation. Further calculations should then include the effect of boron and phosphorous and the prediction of other unusual segregators. Finally, extensions beyond closed packed metallic materials will have to be developed, particularly for materials with band gaps.

Knott [11] presented an interesting case of grain boundary separation at "constant chemical potential" whereby low alloy steels can exhibit slow subcritical intergranular crack growth at temperatures around 500°C. The crack growth rate is controlled by the diffusion of sulphur from within the grain to a position of high triaxial stresses just ahead of the crack tip. It is possible to calculate an activation energy for this type of cracking, by measuring crack growth rates at constant stress intensity at different temperatures, and the value obtained is reasonably close to that for the diffusion of sulphur in iron. This work constitutes one of the very few cases presented at the workshop where chemical rate theory has been used to elucidate the rate determining step and thus provide a better understanding of crack growth rates - as outlined schematically in figure 2.

7. External chemical influences on fracture - void coalescence

It is well known that hydrogen from gaseous or aqueous environments may reduce the ductility of materials which still fracture by void coalescence. The NATO workshop has not produced new quantitative theories of this phenomenon beyond what has been discussed in relation to internal hydrogen embrittlement. One important aspect, however, has been discussed in detail, namely hydrogen transport, particularly dislocation transport [20]. This may eventually help to understand the strain rate dependence and the temperature dependence of hydrogen embrittlement. A quantitative theoretical treatment, however, is still lacking.

8. External chemical influences on fracture - cleavage

Sieradski and Newman [25] presented a modernized version of the Edeleanu and Forty model for transgranular cleavage-like stress corrosion cracking. In their work [24], [25] they develop a quantitative theory which appears to be capable to eventually predict K_{ISCC} values. Naturally, their basic assumption that cleavage cracks can run even a short length in otherwise ductile materials has raised an intensive discussion at the workshop. Nevertheless, there appears to be a wide range of applicability of these ideas to many transgranular stress corrosion cracking situations [25]. Further quantitative developments appear to be very desirable.

This theory is based on the idea that cleavage cracks originate in brittle corrosion-product films and penetrate as cleavage cracks for a short distance at high velocity into the underlying matrix even if the latter is normally ductile. This mechanism could perhaps explain the type of cracking shown in figures 5, 6 and 7 if the crack-starting film is assumed to be oxide. In this way, the electrochemical influences on the crack velocity such as oxygen content, electrode potential, conductivity and temperature [6], [26] could be understood as controlling the formation of the oxide film while the striated transgranular fracture surface could be understood as resulting from repeated short cleavage bursts.

Hirth [18] and Neumann [27] discussed the effect of hydrogen on the growth of cracks in iron and nickel base materials where the hydrogen induced cracks follow low-indexed crystallographic (cleavage- and slip-) planes. In some cases [28] it is now possible to describe the pressure and temperature dependence of the crack growth rate. It appears that a full theoretical description of crack velocity as a function of stress intensity and the other environmental parameters should now be within reach.

9. External chemical influences on fracture - grain boundary separation

Under this heading fall the well known cases of intergranular stress corrosion cracking, hydrogen embrittlement and liquid metal embrittlement. The first two could be considered as concurrent segregation of damaging species to the grain boundaries where diffusion could be the rate-limiting process with a consequent effect of loss of cohesion.

At the workshop, Watanabe [29] put new emphasis on the effect different grain boundaries could have on environment-induced intergranular cracking. While no quantitative prediction of this effect was given, it is at least known in some detail how the slip distribution at the grain boundary can affect the stress corrosion crack growth rate [30].

The effect of crack-tip electrochemistry on stress corrosion cracking has been discussed with particular emphasis on the kinetics and thermodynamics of the reactions itself, Turnbull [31], MacDonald [32], Pourbaix [33]. What remains to be done is to show in a quantitative way how our much improved understanding of the crack-tip electrochemistry can help to predict the crack velocity as a function of stress intensity and the external electrochemical and fluid flow parameters. Of particular importance, both theoretical and practical, is a future quantitative theory of the effect of potential on the growth rates and thresholds of stress corrosion cracking. The most important point to be elucidated is the critical potential below which SCC appears to cease.

CONCLUSION

The most positive accomplishment of the conference was the exchange of divergent views between conferees of different disciplines, as in the preceding NATO workshop in Corsica. The most important future work - as seen by this particular and perhaps prejudiced author - is not only to continue with the highly interesting and sophisticated theoretical mechanistic studies of fracture but in each case to make an effort to extend those to a stage where they can contribute to a quantitative understanding and theoretical prediction of crack velocity versus stress intensity curves.

REFERENCES

1. Speidel,M.O.:Current Understanding of Stress Corrosion Crack Growth in
   Aluminum Alloys, in "The Theory of Stress Corrosion Cracking in Alloys",
   NATO ScientificAffairs Division, Brussels, 1971, pp289-344
2. Speidel, M.O.: Spannungsrisskorrosion hochfester Stähle für das Bauwesen,
   Material+Technik, 1986/3, pp 97-101
3. Speidel, M.O. and Bertilsson, J.E.: Stress Corrosion Cracking of Steam
   Turbine Rotors, in "Corrosion in Power Generating Equipment", Markus O.
   Speidel and Andrejs Atrens, ed., Plenum Press, New York, 1984, pp 331-357
4. Jaffee, R.J.: "Materials and Electricity", Met.Trans. 17A(1986)
   pp 755-775
5. German nuclear standard KTA 3201.2
6. Speidel, M.O.: Stress Corrosion Cracking of Nuclear Pressure Vessel Steel,
   in Proceedings of the Second International Atomic Energy Agency Specialists
   Meeting on Subcritical Crack Growth, May 15-17, 1985, Sendai, Japan,
   NUREG/CP-0067 MEA-2090, 1986, Vol. 1, pp 465-476
7. Thompson, A.W.: in this volume
8. Uggowitzer, P.J.: "Mikrostruktur und Bruch metallischer Werkstoffe",
   Proceedings DVM Arbeitskreis Bruchvorgänge, February 18-19, 1986, Aachen,
   Germany
9. Argon, A.S.: in this volume
10. Haasen, P.: in this volume
11. Knott, J.F.: in this volume
12. DeHosson, J.Th.M. and Vitek, V.: in this volume
13. Finnis, M.W.: in this volume
14. Daw, M.: in this volume
15. Whitesides, G.: in this volume
16. Hancock, J.: in this volume
17. McMeeking, R.M.: in this volume
18. Hirth, J.: in this volume
19. Fidelle, J.P.: in this volume
20. Hashimoto, M.: in this volume
21. Rice, J.: in this volume
22. Grabke, H.J.: in this volume
23. Ernst, P., Uggowitzer, P., Speidel, M.O.:"Improved boron-containing 12%Cr
    steels with high creep rupture strength for steam turbine, gas turbine
    and other applications", in "High Temperature Alloys for Gas turbines and
    other Applications 1986", D.Reidel Publishing Co. Dordrecht,1986,
    pp 1357-1369, See also J.Mat.Sci.Letters 5(1986), pp 835-839
24. Sieradski, K., and R.C. Newman: "Brittle behaviour of ductile metals
    during stress-corrosion cracking". Phil.Mag.A (1985), Vol.51, pp 95-132
25. Newman, R.C. and Sieradski. K.: in this volume
26. Speidel, M.O.: Stress Corrosion Cracking and corrosion Fatigue of Steels
    in Hot Water, Proc. Intern.Conf. on Fatigue, Corrosion Cracking, Fracture
    Mechanics, Failure Analysis, Dec.2-6,1985, Salt Lake City, Utah, USA,
    American Society for Metals, 1986, pp 55-60
27) Neumann, P. : in this volume
28) Vehoff, H., and Klameth, H.K.: "Hydrogen embrittlement and trapping at
    crack tips in Ni-single crystals", Acta Met. 33(1985), pp 955-962

29) Watanabe, T.: in this volume
30) Stenzel, H., Vehoff, H., Neumann, P.: in this volume
31) Turnbull,A.: in this volume
32) MacDonald,D.: in this volume
33) Pourbaix, M.: in this volume

List of Participants

Alexander, D.   University of Cambridge, Department of
Metallurgy and Materials Science, Pembroke
Street, Cambridge CB2 3QZ, United Kingdom

Altintas, S.   Bogazici University, Department of Mechanical
Engineering, P.K. 2 Bebek, Istanbul, Turkey

Altstetter, C.J.   University of Illinois-Urbana Champaign, 1304
W. Green Street, Urbana, Illinois 61801,
U.S.A.

Argon, A.S.   Massachusetts Institute of Technology,
Department of Mechanical Engineering, Room
1-306, Cambridge, MA 02139, U.S.A.

Baer, D.R.   Pacific Northwest Laboratory, Materials
Science and Technology Dept., P.O. Box 999,
Richland, WA 99352, U.S.A.

Bernstein, I.M.   Carnegie-Mellon University, Department of
Metallurgical Engineering and Materials
Science, Pittsburgh, PA 15213, U.S.A.

Birnbaum, H.K.   University of Illinois, Department of
Metallurgy, 1304 W Green Street, Urbana, IL
61801, U.S.A.

Bompard, H.   Laboratoire des Materiaux, Ecole Centrale des
Arts et Manufactures, Grande Voie des Vignes,
92290 Chatenay Malabry, France

Brown. L.M.   University of Cambridge, Cavendish Laboratory,
Madingley Road, Cambridge CB3 OHE, United
Kingdom

Chene, J.   Universite de Paris - Sud, Laboratoire de
Metallurgie Structurale, Batiment 413, 91405
Orsay Cedex, France

Cole, A.T.   University of Manchester Institute of Science
and Technology, Corrosion Centre, 88 Sackville
Street, Manchester, M60 1QD, United Kingdom

714

Daw, M.   Sandia National Laboratory, Department 8341,
                                 Livermore, CA   94550, U.S.A.

De Hosson, J. Th. M.   University of Groningen, Applied Physics
                                 Department, Nijenborgh 18, 9747 AG Groningen,
                                 The Netherlands

Duquette, D.J.   Rensselaer Polytechnic Institute, Materials
                                 Engineering Department, Troy, NY   12181,
                                 U.S.A.

Eberhart, M.E.   Los Alamos National Laboratory, Department
                                 MSG 730, Los Alamos, NM   87545, U.S.A.

Engell, H.-J.   Max-Planck-Institut für Eisenforschung GMBH,
                                 D-4000 Düsseldorf 1, Federal Republic of
                                 Germany

Ferreira, M.G.S.   Instituto Superior Tecnico, Lisboa 1096 Codex,
                                 Portgual

Fidelle, J.P.   Commissariat a L'Energie Atomique, Service
                                 Metallurgie, 91680- Bruyeres-le-Chatel, France

Finnis, M.W.   Theoretical Physics Division, AERE Harwell,
                                 Oxfordshire, OX11 ORA, United Kingdom

Fischer, T.E.   Exxon Research and Engineering Company,
                                 Clinton Township, Route 22 East, Annandale, NJ
                                 08801, U.S.A.

Gerberich, W.W.   University of Minnesota, Department of
                                 Chemical Engineering and Materials Science,
                                 Minneapolis, MN   55455, U.S.A.

Grabke, H.-J.   Max-Planck-Institut für Eisenforschung GMBH,
                                 D-4000 Düsseldorf 1, Federal Republic of
                                 Germany

Haasen, P.   Institut für Metallphysik der Universität
                                 Göttingen, 3400 Göttingen, Hospitalstrasse
                                 3/5, Federal Republic of Germany

Hancock, J.W.   University of Glasgow, James Watt Engineering
                                 Laboratories, Glasgow   G12 80Q, Scotland,
                                 United Kingdom

Hänninen, H.   Technical Research Centre of Finland, Metals
                                 Laboratory, SF-02150 Espoo 15, Finland

Harris, T.M.   Massachusetts Institute of Technology,
                                 Department of Materials Science and
                                 Engineering, Room 4-217, Cambridge, MA   02139,
                                 U.S.A.

Hashimoto, M.   Nippon Steel Corporation, R&D Laboratories-1,
1618 Ida, Nakahara-ku, Kawasaki, 211, Japan

Hirth, J.P.   The Ohio State University, Department of
Metallurgical Engineering, 116 W 19th Avenue,
Columbus, Ohio 43201, U.S.A.

Jones, R.H.   Pacific Northwest Laboratories, Materials
Science and Technology Department, P.O. Box
999, Richland, WA 99352, U.S.A.

Kayali, E.S.   Department of Metallurgical Engineering,
Istanbul Technical University, Maslak,
Istanbul, Turkey

Kirchheim, R.   Max-Planck-Institut für Metallforchung,
Werkstoffwissenschaften Institut, Seestrasse
92, 7000 Stuttgart 1, Federal Republic of
Germany

Knott, J.F.   University of Cambridge, Metallurgy and
Materials Science, Pembroke Street, Cambridge
CB2 30Z, United Kingdom

Latanision, R.M.   Massachusetts Institute of Technology,
Department of Materials Science and
Engineering, Room 8-202, Cambridge, MA 02139,
U.S.A.

Lichter, B.   Vanderbilt University, Box 1612 Station B,
Nashville, Tennessee 37235, U.S.A.

Losch, W.   COPPE, Universidade Federal, Programa
Engenharia Metallurgica de Materials, C.P.
1191-ZC-00, Rio de Janeiro, Brazil

Lumsden, J.   Rockwell International, 1049 Camino Dos Rios,
Thousand Oaks, CA 91360, U.S.A.

MacDonald, D.D.   SRI International, Chemistry Department, 333
Ravenswood Avenue, Menlo Park, CA 94025,
U.S.A.

Marcus, P.   Ecole Nationale Superiure de Chemie de Paris,
Laboratoire de Physico-Chimie des Surfaces, 11
Rue Pierre et Marie Curie, 75005, Paris,
France

McMeeking, R.M.   University of California-Santa Barbara,
Materials Program, College of Engineering,
Santa Barbara, CA 93106, U.S.A.

Mughrabi, H.
Universitat Erlangen-Nürnberg, Institut für Werkstoffwissenschaften, Martenstrasse 5, D-8520 Erlangen, Federal Republic of Germany

Nabarro, F.R.N.
Council on Scientific and Industrial Research, National Institute for Materials Research, P.O. Box 395, Pretoria 001, South Africa

Neumann, P.
Max-Planck-Institut für Eisenforschung GMBH, D-4000 Düsseldorf 1, Federal Republic of Germany

Newman, R.C.
Corrosion & Protection Center, U.M.I.S.T., Manchester M60 1QD, United Kingdom

Nisancioglu, K.
Norwegian Institute of Technology, Department of Chemistry, Laboratory G, N. 7034 Trondheim-NTH, Norway

Ohr, S.M.
Department of Materials Science and Engineering, State University of New York at Stony Brook, Stony Brook, NY 11794, U.S.A.

Oudar, J.
Ecole Nationale Superieure de Chemie de Paris, Laboratoire de Physico-Chimie des Surfaces, 11 Rue Pierre et Marie Curie, 75005 Paris, France

Pettifor, D.G.
Imperial College of Science and Technology, Department of Mathematics, Queen's Gate, London SW7 2BZ, United Kingdom

Pourbaix, M.
Centre Belge d'Etude de la Corrosion, Avenue Paul Heger, Grille 2, 1050 Brussels, Belguim

Reuschle, T.
Institut de Physique du Globe, CNRS, Universite Louis Pasteur, 5, rue Rene Descartes, 67084 Strasbourg, France

Rice, J.R.
Harvard University, Division of Applied Sciences, Cambridge, MA 02138, U.S.A.

Sieradzki, K.
Brookhaven National Laboratory, Department of Applied Science, Long Island, Upton, NY 11973, U.S.A.

Smith, J.R.
General Motors Research Laboratory, Physics Department, Warren, MI 48089, U.S.A.

Speidel, M.O.
ETH-Zentrum, CH-8092 Zurich, Switzerland

Strehlau, J.
Institut für Geophysik, Universität Kiel, 2300 Kiel, Federal Republic of Germany

Tähtinen, S.   Technical Research Centre of Finland, Metals
                                Laboratory, SF-02150 Espoo 15, Finland

Thompson, A.W.   Carnegie-Mellon University, Department of
                                Metallurgical Engineering and Materials
                                Science, Pittsburgh, PA 15213, U.S.A.

Thomson, R.   Center for Materials Science, National Bureau
                                of Standards, Gaithersburg, MD 20899, U.S.A.

Thorpe, S.J.   NATO Science Fellow, 40 Crang Avenue, Toronto,
                                Ontario, M6E 2Z9, Canada

Turnbull, A.   National Physical Laboratory, Teddington,
                                Middlesex, TW11 0LW, United Kingdom

Vvedensky, D.D.   The Blackett Laboratory, Imperial College,
                                London SW7 2BZ, United Kingdom

Watanabe, T.   Tohoku University, Department of Materials
                                Science, Faculty of Engineering, Sendai, 980,
                                Japan

Whitesides, G.   Harvard University, Chemistry Department,
                                Cambridge, MA 02138, U.S.A.

Zhang, X.G.   Department of Metallurgy and Electrochemistry,
                                The Free University of Brussels, Pleinlaan 2,
                                1050 Brussels, Belgium

Zheng, W.   Corrosion and Protection Centre, UMIST, P.O.
                                Box 88, Manchester M60 IQD, United Kingdom

Subject Index